CALCULUS AND LINEAR ALGEBRA

CALCULUS AND LINEAR ALGEBRA

VOLUME I

Vectors in the Plane and One-Variable Calculus

Wilfred Kaplan

Donald J. Lewis

Department of Mathematics
University of Michigan

Scholarly Publishing Office, University of Michigan University Library   Ann Arbor

Published by The Scholarly Publishing Office
The University of Michigan University Library
Copyright © 1970, Wilfred Kaplan and Donald J. Lewis
Preface Copyright © 2007, Wilfred Kaplan and Donald J. Lewis

1st paperback edition, 2007 The Scholarly Publishing Office
First published by Wiley & Sons, Inc.

Scholarly Publishing Office
300 Hatcher North
Ann Arbor, MI 48109
lib.spo@umich.edu

ISBN: 978-1-4255-8913-4 (softcover)

The University of Michigan University Library, through its Scholarly Publishing Office, is committed to providing academic publishing services that are responsive to the needs of both producers and users, foster a sustainable economic model, and support robust author rights.

The Scholarly Publishing Office seeks to disseminate high-quality scholarly content through both print and electronic publishing. As part of this mission, the Library is pleased to re-issue Calculus and Linear Algebra by Wilfred Kaplan and Donald Lewis. This text was an invaluable part of the education of a past generation of math students, and through this republication it will become a vital part of teaching a new generation.

Paul N. Courant (signature)

Paul N. Courant
University Librarian and Dean of Libraries
Harold T. Shapiro Collegiate Professor of Public Policy
Professor of Economics and of Information
The University of Michigan

Wilfred Kaplan
Emeritus Professor of Mathematics
University of Michigan

Wilfred Kaplan, born in 1915, is a native of Boston.
He graduated from Harvard College in 1936, then
remained at Harvard for a Ph.D. in mathematics in
1939. From 1940 until his retirement in 1986, he
taught mathematics at the University of Michigan.
His research has dealt with differential equations,
complex function theory, and applied mathematics.
He is the author of many influential textbooks, including Advanced Calculus (1952 to 2003), Ordinary
Differential Equations (1958), Operational Methods
for Linear Systems (1962), Advanced Mathematics
for Engineers (1981), and Maxima and Minima with

*Photo courtesy of the
Mathematics Department,
University of Michigan*

Applications (1999). Over a long period of time, Pro-
fessor Kaplan has been active in the affairs of the American Association of University Professors (AAUP), a national organization dedicated to defending faculty
rights, primarily academic freedom and tenure. An accomplished musician, he has
enjoyed playing piano or violin in chamber music groups, and has done much to
promote the arts, especially music and drama, in the Ann Arbor area.

Donald J. Lewis
Emeritus Professor of Mathematics
University of Michigan

Donald J. Lewis was born in Adrian, Minnesota in
1926. He majored in mathematics at the College
of St. Thomas (St. Paul, Minnesota), then earned a
Ph.D. in mathematics from the University of Michigan in 1950. He taught at Ohio State University
and the University of Notre Dame, and held several
research appointments before joining the faculty of
the University of Michigan in 1961. He served as
Chair of the Mathematics Department for ten years,
followed by four years as Director of the Division
of Mathematical Sciences of the National Science
Foundation in Washington, D.C. Professor Lewis has
done extensive research in algebraic number theory
and diophantine equations, and has directed 27 Ph.D.

*Photo courtesy of the Bentley
Historical Library, University
of Michigan (Box 6, T. H.
Hildebrandt Papers)*

theses at Michigan. He has played an active role in mathematical educational
innovations and is author of the textbook Introduction to Algebra (1965). Outside
of mathematics, he has a particular interest in gardening and has enjoyed maintaining an elaborate garden at his home in Ann Arbor.

Biographical notes written by Peter Duren, Professor of Mathematics, University of Michigan

REISSUANCE OF CALCULUS TEXT BY KAPLAN AND LEWIS

By Wilfred Kaplan

The power of the Internet in meeting scholarly demands is illustrated by
the Google Book Project, which aims to make books of all kinds available
online. A particular need arose at the University of Michigan, when Associate Professor Stephen DeBacker proposed to use a long out-of-print
work by his colleagues Wilfred Kaplan and Donald J. Lewis (now both
emeritus professors) for a course he was to teach. The work he selected
was Calculus and Linear Algebra, originally published by John Wiley &
Sons; volume I appeared in 1970 and volume II in 1971. The first volume
covers vectors in the plane and one-variable calculus; the second treats
vector space, many-variable calculus and differential equations. The two
volumes provide material for a freshman-sophomore course in calculus in
which linear algebra is gradually introduced and blended with the calculus. The work introduces many novel ideas and proofs.

In order to make the work available for instruction, assistance was needed.
The Scholarly Publishing Office of the University of Michigan Library
has agreed to partner with the University's Mathematics Department to
reissue it as originally printed. Thus two volumes will be published, of
the quality needed for classroom use, but with costs held down as far as
possible. The authors have agreed to this and are receiving no royalties
for the publication. Both proclaim their happiness that the work, a labor
of love created 40 years ago, should again be used by students.

As a byproduct of the production of the books, an online version of the
texts is being created, to be listed in Google Books. Thus the work
becomes freely available to the world in its entirety.

PREFACE

This book is the first of a two-volume text on calculus and linear algebra that is intended to provide enough material for a freshman-sophomore course.

Our principal objective is the integration of linear algebra and calculus. Although these subjects can be treated independently, each gains in depth and significance by being related to the other. In the more advanced aspects of calculus (functions of several variables, differential equations), linear algebra is especially valuable; through it the theory is greatly simplified. For elementary calculus, linear algebra is less important. However, it is of value particularly for the study of curves in the plane. Here vector algebra simplifies the theory, reveals the geometric meaning of the formulas, and relates the theory to physical concepts such as velocity and acceleration. For the study of linear algebra (vector spaces and matrices), the calculus provides an inexhaustible well of significant examples to illustrate and to clarify the theory.

In this first volume, linear algebra appears in two aspects: (1) vectors in the plane (Chapter 1), and (2) linear independence and basis for sets of functions (introduced in Section 2-9). The first topic is given much more weight, and applications occur throughout the volume. The second is treated lightly, but sufficiently often to provide familiarity with the concepts, and confidence in working with them. The early encounter with the idea of linear independence will make a later in depth study of vector spaces easier.

Where possible, the text stresses the geometric aspects of the theory, both in the calculus and in the algebra. Indeed, it is gradually revealed that linear algebra is an essential tool for developing geometry and its relation to the calculus. Geometry is frequently used to motivate a proof and to emphasize the qualitative aspect of a theorem. At the same time, the computational aspect of both the calculus and the linear algebra are fully developed, and the student is motivated toward the use of computers.

We believe that to be effective, both mathematicians and the users of mathematics must have a qualitative feeling for the theory as well as a facility in the procedures that give quantitative results. This principle has motivated our approach throughout this book.

The mathematical development includes a rigorous and essentially self-contained treatment of the subject. However, in general, the difficult ideas are first presented intuitively, then are formulated precisely and are illustrated. Finally they are established completely. The difficult proofs are in separate sections, marked by a double dagger (‡), and can easily be omitted. In addition, sections of average difficulty that can be omitted without affecting continuity are marked by a single dagger (†). The double dagger is also used occasionally to indicate an especially difficult problem.

PLAN OF THE TEXT

Chapter 0, an introductory chapter, is intended for reference and review. Part of it (or all of it) can be studied in depth in accordance with the background of the students.

Chapter 1 introduces vectors in the plane; the presentation relies heavily on plane geometry.

Chapter 2 reviews and develops the idea of a function and presents the concept of limit as the first stage in developing the calculus. The least upper-bound axiom is introduced at the end and, in a † section, it is used to prove the main theorems.

Chapter 3 is a systematic development of differential calculus, with some applications to geometry and to the sciences. The derivatives of sin x, cos x, ln x, and e^x are given with intuitive justification and are used often; rigorous proofs are deferred until Chapter 5. Accordingly, they are available for early use by students in engineering and physics. A student who has completed this chapter has a solid grounding in differential calculus. Vectors appear at several points, especially for curves in parametric form.

Chapter 4 is a similarly complete treatment of integral calculus. The introductory sections explain the ideas of definite and indefinite integrals. Then the main techniques for finding indefinite integrals are developed. Finally, the third and longest part is devoted to the definite integral, with some applications, especially to area and arc length. The definition of the integral is based on upper and lower estimates and leads to a simple proof of the main theorems for integrals of continuous functions. The Riemann integral is also defined and (in a ‡ section) is shown for continuous functions to be equivalent to the definite integral. Throughout there is emphasis on computational procedures and computers.

Chapter 5 is a brief, rigorous treatment of the trigonometric, logarithmic, exponential, and related functions. This chapter can be omitted without affecting continuity, since all the main results are given elsewhere in the text.

Chapter 6 provides further applications of differential calculus—tests for maxima and minima, graphs of plane curves in rectangular and polar coordinates, Newton's method, Taylor's formula, and indeterminate forms. Much of this chapter can be studied immediately following Chapter 3, if so desired, since integration appears only occasionally. In particular, Sections 6-1 to 6-5 make no reference to integration.

Chapter 7 presents applications of the definite integral to area in rectangular and polar coordinates, volumes and surface area of solids of revolution, moments of mass distributions, and centroids. Line integrals are introduced at several points. The role of integration in the physical sciences is well illustrated. There are discussions of improper integrals and the trapezoidal and Simpson's rules. Six sections are devoted to differential equations; they are included here: (1) because their development is a natural extension of earlier theory and (2) to make them available at an early stage for students of engineering and physics. The material covered is adequate for most of the problems that these students will encounter in the early years of their studies. Much of the material in this chapter is not essential for later chapters. The Instructor should choose the topics to be covered in accordance with his and his students' interest and the time available.

Chapter 8 is concerned with infinite sequences and series, convergence tests, rearrangement and product theorems, power series, Taylor's formula and series, and Fourier series. Some reference is made to complex series. To a considerable extent, this chapter is independent of the others and can be studied earlier or later.

In Volume II (to be published) the chapter headings are as follows.

Chapter 9. Vector Spaces
Chapter 10. Matrices and Determinants
Chapter 11. Euclidean Geometry
Chapter 12. Differential Calculus of Functions of Several Variables
Chapter 13. Integral Calculus of Functions of Several Variables
Chapter 14. Differential Equations

Numerous problems sets are provided throughout. Answers to selected problems appear at the end of this volume. The problems for which answers are provided are labeled by a boldface number or letter.

We thank the publisher for fine cooperation and, especially, John B. Hoey for his help and encouragement. We express our appreciation to Mrs. Helen M. Ferguson for her excellent work in typing the manuscript.

Wilfred Kaplan
Donald J. Lewis

Ann Arbor, 1969

CONTENTS

CHAPTER 2
LIMITS

CHAPTER 3
DIFFERENTIAL CALCULUS

CHAPTER 4
INTEGRAL CALCULUS

CHAPTER 5
THE ELEMENTARY TRANSCENDENTAL FUNCTIONS

CHAPTER 6
APPLICATIONS OF DIFFERENTIAL CALCULUS

CHAPTER 7
APPLICATIONS OF THE INTEGRAL CALCULUS

CHAPTER 8
INFINITE SERIES

INTRODUCTION
REVIEW OF ALGEBRA, GEOMETRY, AND TRIGONOMETRY

In this chapter we present a summary of the topics of algebra, geometry, and trigonometry that are essential for the rest of the book. At the end of the chapter, references are given in which the topics are covered in full. We also provide a number of exercises by which the reader can refresh his knowledge and can test his background for what is to follow.

0-1 THE REAL NUMBERS

The real numbers arise naturally in measuring distances or more generally in giving the relative positions of points on a line. We choose a reference point O on the line, choose a unit of distance, and then indicate the position of each point on the line by giving its distance from O, in terms of the chosen unit; to distinguish the points on one side of O from those on the other side, we assign a plus sign to the former and a minus sign to the latter. The plus sign is, however, usually not written. The result is the familiar *number axis* of Figure 0-1. Every real number is to be represented by one point on the axis,

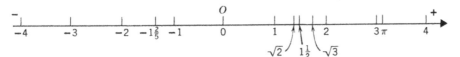

Figure 0-1 Number axis.

and every point on the axis represents one number. The point O itself represents the number 0 (zero), the points at one unit from O correspond to the numbers $+1$ and -1, and the points two units from O correspond to the numbers 2, -2, and so on.

The numbers that correspond to a whole number of units on the plus side are the *positive integers:* 1, 2, 3, ..., 10, ..., 3567, ...; those that correspond to a whole number of units on the negative side are the *negative integers:* $-1, -2, -3, \ldots, -10, \ldots, -50, \ldots$. The numbers 0, 1, $-1, 2, -2$, 3, $-3, \ldots$ are called the *integers.* Since we can divide any line segment into a given number of equal parts, we have fractional numbers: $3\frac{1}{2}, 5\frac{1}{4}, -\frac{2}{3}, \ldots$. As in arithmetic, each fraction can be written in the form m/n, where m and n are integers and n is positive. We call these numbers *rational numbers.* A real number that is not rational is said to be *irrational.*

Two expressions that represent numbers are said to be equal if they correspond to the same point on the number axis. Thus, $\frac{1}{2}$ and $\frac{2}{4}$ are equal because both correspond to the midpoint of the segment from 0 to 1. In general, rational numbers m/n and p/q are equal precisely when $mq = np$; for every rational number m/n, we can find an equal rational number p/q for which p and q have no common divisor. The equals sign $(=)$ obeys these rules:

$$a = a; \qquad \text{if } a = b, \text{ then } b = a; \qquad \text{if } a = b \text{ and } b = c, \text{ then } a = c.$$

We write $a \neq b$ to indicate that a is not equal to b.

By finding the length of a hypotenuse of a right triangle, one of whose sides has unit length, we are led to the numbers $\sqrt{2}$, $\sqrt{3}$, ..., as suggested in Figure 0-2. It can be shown that $\sqrt{2}$ is not a rational number, and is therefore

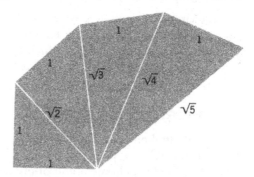

Figure 0-2   Irrational numbers.

irrational. There are many more irrational numbers (for example, $\sqrt{3}$, $\sqrt{5}$, $\sqrt{6}$), each giving the length of a line segment, and hence represented by a point on our number axis. Also π, the circumference of a circle of diameter 1, can be shown to be irrational.

The collection of all these numbers—the integers, the rational numbers, the irrational numbers—forms the class of *all real numbers*. We can assign to each real number a decimal representation, for instance,

$$137.56214\ldots, \qquad -33.33333\ldots, \qquad 3.141592\ldots.$$

Usually the representation is unending, so that we speak of an infinite decimal. A terminating decimal is a rational number; for example, $1.79 = \frac{179}{100}$. However, not every rational number equals a terminating decimal.

Throughout this book the word "number" will normally mean "real number."

The positive integers and zero play a special role in counting, that is, in finding out how many objects there are in a given collection. When there are none, we say that there are zero objects or that the collection is *empty*. When, for some positive integer n, we can count off the objects by using the integers $1, 2, 3, \ldots, n$, we say that there are n objects and that the collection is *finite*. (An empty collection is also called finite.) When we cannot count the objects in this way, we say that the collection is *infinite*. For example, the collection of all integers is infinite, as is the set of even integers, and as is the set of real numbers between 0 and 1.

When we have numbers, we expect to add, multiply, and possibly, to subtract and divide them. A collection of numbers is said to form a *number system* provided that addition and multiplication are defined within the system, and provided that the following properties hold true for arbitrary numbers a, b, c in the system.

1. $a + b = b + a$.   (commutative law for addition)
2. $a + (b + c) = (a + b) + c$.   (associative law of addition)
3. 0 is in the system and $a + 0 = a$.
4. To each a in the system, there is a unique solution of the equation $a + x = 0$. We denote this solution by $-a$.
5. $ab = ba$.   (commutative law for multiplication)
6. $a(bc) = (ab)c$.   (associative law for multiplication)
7. 1 is in the system and $a \cdot 1 = a$.
8. For each $a \neq 0$ in the system, there is a unique solution of the equation $ax = 1$. This solution is denoted by a^{-1}.
9. $a(b + c) = ab + ac$.   (distributive law)

From the rules we can show that we can subtract any two numbers and can divide any two numbers, except for division by zero. Furthermore, $ab = 0$ if, and only if, $a = 0$ or $b = 0$. Also $(-a)(-b) = ab$ and $(-a)b = -(ab)$.

Since we know that the collection of all real numbers obeys the laws $1 \ldots 9$, we can state that the real numbers form a number system. (There are other number systems, for example, that formed of all rational numbers and that formed of all complex numbers; see Section 0-17.)

0-2 INEQUALITIES

The numbers on the plus side of O in Figure 0-1 are called positive numbers, the ones on the minus side are called negative numbers. A number that is nonnegative is thus either positive or 0; a number that is nonpositive is either negative or 0.

For two numbers a, b, we write $a < b$ (a is less than b) or $b > a$ (b is larger than a) when $b - a$ is positive. If a and b are both positive, $a < b$ means that b corresponds to a larger distance from O than does a. If a is negative and b is positive, then $b - a$ is positive; thus every negative number is less than every positive number. If a and b are both negative, then $a < b$ when b is closer to O than a. In all cases, $a < b$ when a lies to the left of b on the number axis in Figure 0-1. This is illustrated in Figure 0-3.

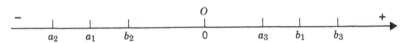

Figure 0-3 $a_1 < b_1$, $a_2 < b_2$, $a_3 < b_3$.

The signs $<$ and $>$ are called *inequality signs*, and they satisfy the following rules.

10. If $a \neq b$, then either $a < b$ or $a > b$.

11. If $a < b$ and $b < c$, then $a < c$.

12. If $a < b$ and c is a real number, then $a + c < b + c$.

13. If $a < b$ and $c > 0$, then $ac < bc$.

14. If $a < b$ and $c < 0$, then $ac > bc$.

15. For no a is $a < a$.

16. If $0 < a < b$, then $\dfrac{1}{a} > \dfrac{1}{b} > 0$.

17. If $a \neq 0$, then $a^2 > 0$.

If a and b are positive, then $0 < a$ and $0 < b$ so that, by rule 12, $b < a + b$ and hence, by rule 11, $0 < a + b$; similarly, by rules 13 and 11, $0 < ab$. Thus the sum and product of positive numbers are positive. Since $a + (-a) = 0$, it is impossible for both a and $-a$ to be positive.

We can combine the equality and inequality signs: The symbol "$a \leq b$" means "either $a = b$ or a is less than b" or, equivalently, "a is not greater than b." We also write, for example, $a < b \leq c$ to indicate that b is larger than a but not larger than c. An expression such as $a < b$ is called an *inequality*; an expression such as $a < b < c$ is called a *double inequality*.

0-3 ABSOLUTE VALUE

The rules for inequality permit us to define the absolute value of a real number a, which we denote by $|a|$. We define

18. $|a| = \begin{cases} a & \text{if } a \geq 0, \\ -a & \text{if } a < 0. \end{cases}$

Thus $|2| = 2$, $|-1.3| = 1.3$. In general, $|a|$ is the larger of a, $-a$. From our definition it follows that

19. $|a| \geq 0$, and $|a|$ is equal to 0 only if $a = 0$.

The absolute value obeys these additional rules:

20. $|a| = |-a|$.

21. $|ab| = |a||b|$.

22. $|a + b| \leq |a| + |b|$.

Consequently, we observe that

$$|a| = |(a + b) - b| \leq |a + b| + |-b| = |a + b| + |b|$$

whence

$$|a| - |b| \leq |a + b|$$

But, similarly, we obtain

$$|b| - |a| \leq |a + b|$$

and thus we have the rule:

23. $||a| - |b|| \leq |a + b|$.

The real numbers possess one other essential property that cannot be derived from properties 1 to 9; this property is the least upper bound principle. We do not need this principle immediately and, therefore, we postpone its discussion until Chapter 2, where it is needed. By using this principle, we can show that each real number is representable as an infinite decimal. For the present, we can assume that the real numbers and the infinite decimals are the same number system.

PROBLEMS

1. (a)[1] Find an integer x so that $10\sqrt{2} < x < 10\sqrt{3}$.
   (b) Find an integer x so that $-5\sqrt{2} < x < -3\sqrt{3}$.
   (c) Find a rational number x so that $\sqrt{2} < x < \sqrt{3}$.
   (d) Find a rational number x so that $\pi < x < \pi + 0.01$.
2. Determine whether $x < y$, $x = y$, or $x > y$ is correct for each of the following cases.
   (a) $x = -3, y = -2$ (b) $x = 1, y = -2$
   (c) $x = \sqrt{5} - \sqrt{3}, y = \sqrt{7} - \sqrt{2}$
   (d) $x = \dfrac{1}{\sqrt{3} - \sqrt{11}}, y = \dfrac{1}{\sqrt{3} - \sqrt{13}}$
3. Evaluate: (a) $|-3.5|$, (b) $|0.2|$, (c) $||x||$, (d) $|-|x||$, (e) $|x - y| - |y - x|$.
4. Show that $|a - b|$ can be interpreted as the distance between a and b on the number axis.
5. Find x in each of these cases: (a) $|x| = 0$, (b) $|x| = 2$, (c) $|x - 1| = 2$, (d) $|x + 1| = 1$.
6. The symbol $\sqrt{x}$ denotes 0 if $x = 0$ and the positive square root of x, if $x > 0$. Justify the following rules for all real x and y.
   (a) $\sqrt{x^2} = |x|$ (b) $\sqrt{x^4} = x^2$
   (c) $(x|x|)^2 = x^4$ (d) $\sqrt{x^2 - 2xy + y^2} = |x - y|$
7. Show that the rules 20 and 21 are valid for all real numbers a and b.
8. (a) If $a < b$, must $a^2 < b^2$?   (b) If $a < b$, must $a^3 < b^3$?
   (c) If $|a| < |b|$, must $a^2 < b^2$?   (d) If $a \neq b$, must $|a| \neq |b|$?
   (e) If $|a| \neq |b|$, must $a \neq b$?   (f) If $a < b$, must $1/a > 1/b$?
9. Prove that, if x and y are rational, then so are xy and $x + y$.

0-4 SETS

In mathematics the words collection, class, and set are synonymous. However, the word set is most commonly used.

A set of numbers thus means a collection of real numbers—for example, the numbers 1, 2, 3, 4; or all positive integers; or all negative numbers.

We can specify a set of numbers by giving a property common to the numbers in the set and only to those numbers; for example, the set of all numbers which are positive even integers, or the set of all numbers x for which $1 < x < 2$. A finite set can simply be listed; for example, the set 5, 7, 11.

A shorthand generally used to describe sets is suggested by the following examples:

[1] Problems numbered in boldface have answers provided at the end of the volume.

$\{x \mid x > 1\}$ means the set of all numbers x which are
                                            greater than 1.

$\{x \mid x^2 + 2x - 1 = 0\}$ means the set of all numbers x for which
                                            $x^2 + 2x - 1 = 0$, hence, the set that consists
                                            of both roots of this quadratic equation.

A set, once specified, can be denoted by a single letter. Thus we often write R for the set of all real numbers.

By the *union* of two sets A and B we mean the set C which consists of the objects in A and the objects in B. Thus, x is in C exactly when x is in A, or x is in B (and perhaps in both). For example, R is the union of the rational numbers and the irrational numbers. The union of A and B is usually denoted by $A \cup B$ (see Figure 0-4).

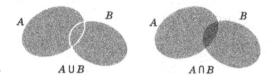

Figure 0-4   Union and intersection. $A \cup B$ $A \cap B$

By the *intersection* of two sets A, B we mean the set C which consists of the objects in both A and B. Thus, if A is the set of even integers (all integers divisible by 2) and B is the set of all positive integers, then the intersection of A and B consists of the positive even integers. The intersection of A and B is usually denoted by $A \cap B$ or AB (see Figure 0-4).

A set with no objects is called an *empty set*. Since, any two empty sets are indistinguishable, we speak of the empty set; it is usually denoted by $\varnothing$.

The objects in a set are called *elements* of the set. If every element in set A is also an element of set B, then A is said to be included in B, or we say that A is a *subset* of B. The symbol $A \subset B$ is used to indicate this relationship.[1] If $A \subset B$ and $B \subset A$, then $A = B$.

Intervals. By an interval we mean a set that consists of all real numbers which lie between two given numbers, and which perhaps includes one or both of the given numbers. Thus an interval is described by a double inequality (Section 0-2). For example, $0 \leq x \leq 1$ describes the set consisting of all real numbers x which are between 0 and 1, including the end values 0 and 1; the set can also be denoted by $\{x \mid 0 \leq x \leq 1\}$. Similarly, $-3 < x < -1$ (or $\{x \mid -3 < x < -1\}$) describes the set of all real numbers between -3 and -1, excluding -3 and -1.

Let a and b denote given real numbers, with $a < b$. Then the interval $a \leq x \leq b$ is called a *closed interval* and is denoted by $[a, b]$. The interval $a < x < b$ is called an *open interval* and is denoted by (a, b). The intervals $a \leq x < b$ and $a < x \leq b$ are called *half-open intervals* and are denoted by $[a, b)$ and $(a, b]$, respectively. In each of these four cases, a and b are called the *end points* of the interval; each x for which $a < x < b$ is called an *interior point* of the interval.

---

[1] The notations for sets such as $\{ \mid \}$, $\cup$, $\cap$, will be used sparingly. However, the concept of set will occur often in this book.

For many purposes we need intervals not of finite length. For example, the inequality $x \geq 0$ describes an *infinite interval*, consisting of all real numbers x which are positive or zero. The general cases are as follows

$$a \leq x \qquad a < x \qquad x \leq b \qquad x < b \qquad \text{all real numbers } x.$$

The last of these intervals, the whole real number system R, is often described symbolically by the double inequality

$$-\infty < x < \infty$$

Similarly, the interval $a \leq x$ can be described by the double inequality $a \leq x < \infty$. We shall not use the terms open or closed for the infinite intervals, but the term end point or interior point can be used where appropriate. (The concept of ∞, or *infinity*, will be considered in Chapter 2.)

All types of intervals are illustrated in Figure 0-5, in which the end points included are marked by a tiny $\times$.

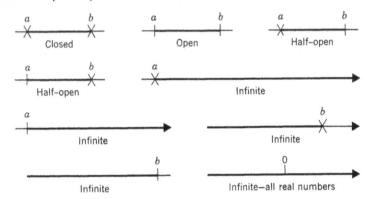

Figure 0-5   Types of intervals.

PROBLEMS

1. Let N be the set of all positive integers. Show that each of the following sets is finite and list its elements:
   (a) $\{x \mid x \text{ is in } N \text{ and } x < 5\}$,
   (b) $\{x \mid x \text{ is in } N \text{ and } 11 < x^2 < 134\}$,
   (c) $\{x \mid x \text{ is in } N \text{ and } x^2 + x - 5 < 0\}$,
   (d) $\{x \mid |x| < \sqrt{2} \text{ and } x \text{ or } -x \text{ is in } N\}$.
2. Determine whether 3 belongs to each of the following sets:
   (a) $\{x \mid x > -2\} \cup \{x \mid x < 0\}$.
   (b) $\{x \mid x^2 < 5\} \cap \{x \mid x^2 - 1 \text{ is an even integer}\}$.
   (c) The empty set.
3. Describe all the subsets of each of the sets:
   (a) The set consisting of 0 and 1.
   (b) The set consisting of a pen, a pencil, and an eraser.
   (c) The set consisting of all pairs (x, y), where $x = 0$ or 1 and $y = 0$ or 1.
4. For each of the following inequalities describe the set of real numbers x for which the inequality is valid.
   (a) $x^2 < 4$
   (b) $x(x - 1) > 0$

(c) $x(x - 1) \leq 0$ (d) $(x - 1)(x - 2) < 0$

(e) $x^2 + x + 1 > 0$ (f) $\dfrac{x}{x - 1} \geq 0$

(g) $\dfrac{x}{x + 1} - \dfrac{2}{x + 1} \geq 0$ (h) $\dfrac{1}{x} < -1$

5. Classify each of the following intervals in terms of the types shown in Figure 0-5.
   (a) $-1 \leq x \leq 1$ (b) $-2 < x$ (c) $3 < x < 100$
   (d) $x \geq 0$ (e) $x < 0$

6. Find the intersection of each of the following pairs of intervals and classify:
   (a) $[-1, 1]$ and $[0, 2]$ (b) $[2, 5)$ and $(0, 4]$
   (c) $0 < x$ and $x < 1$ (d) $-5 \leq x \leq -3$ and $-4 < x$
   (e) $[-7, -2]$ and $[-1, 0]$

7. (a) Show that the intersection of two closed intervals is either a closed interval, a point, or the empty set.
   (b) Show that the intersection of two open intervals is either an open interval or the empty set.
   (c) What can be said of the intersection of an open interval and a closed interval?

8. Describe the set of real numbers x for which each of the following equalities or inequalities holds true:
   (a) $|x - 1| > 0$ (b) $|x - 3| = |x + 2|$

   (c) $0 < |x - 2| < 1$ (d) $\left| \dfrac{x}{x + 1} \right| = 2$

   (e) $|(x - 1)(x - 2)| = 2$

0-5   PLANE AND SOLID GEOMETRY

We shall assume familiarity with the important axioms and theorems of plane geometry (see Reference No. 5 at the end of the chapter). The following theorems will be used frequently.

PYTHAGOREAN THEOREM. *The square on the hypotenuse of a right triangle is equal to the sum of the squares on the two legs.*

TRIANGLE INEQUALITY. *The length of each side of a triangle is less than the sum of the lengths of the other two sides.*

We shall also assume familiarity with a *Cartesian coordinate system* in the plane, as in Figure 0-6. The units of distance on the x-axis and y-axis are equal.

Solid Geometry. The theorems of solid geometry will be used only rarely in this book. However it is important to be familiar with the simplest properties of lines, planes, and spheres in space: in particular, to know that three mutually perpendicular lines can be constructed through a given point, to be used as the axes of an xyz-coordinate system in space (see Figure 0-7). In addition, we shall occasionally use formulas for surface area S and volume V (see Figure 0-8).

For a sphere of radius r, $V = \frac{4}{3}\pi r^3$, $S = 4\pi r^2$.

For a prism, cylinder, or parallelepiped, $V = $ base times altitude.

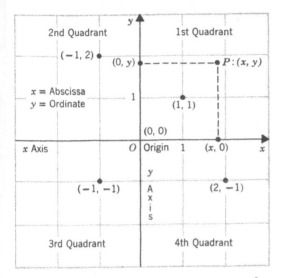

Figure 0-6 Cartesian coordinate system in the plane.

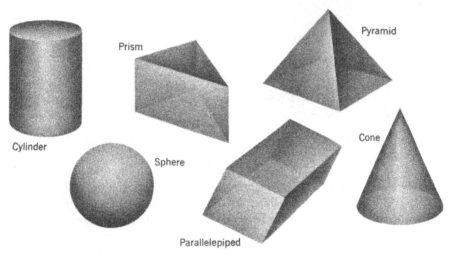

Figure 0-7 Cartesian coordinates in space.

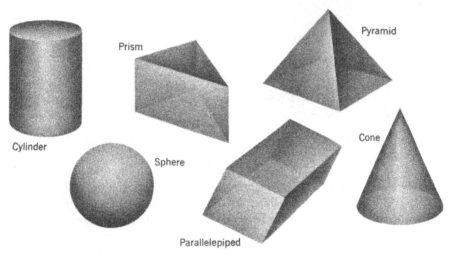

Figure 0-8 Important solids.

For a right circular cylinder of radius r and altitude h, $S = 2\pi r h$ (area of curved surface).

For a cone or pyramid, $V = \frac{1}{3}$ base times altitude.

For a right circular cone, the lateral area is $S = \frac{1}{2}$ slant height times circumference of base.

0-6 ANALYTIC GEOMETRY

For plane analytic geometry we use a Cartesian coordinate system, as in Figure 0-6. All statements of plane geometry can be translated into statements about sets of points given by their coordinates. Thus, by the Pythagorean theorem, the distance between two points (x_1, y_1), (x_2, y_2) is found to be

$$d = \sqrt{(x_2 - x_1)^2 + (y_2 - y_1)^2}$$

An equation in x and y has an associated graph (or locus) in the plane: the graph consists of those points, and only those points, whose coordinates satisfy the equation. Several graphs with associated equations are shown in Figure 0-9; they include three straight lines and a circle.

We can also graph an inequality in x and y. The shaded portion in Figure 0-10 is the graph of the inequality $x^2 + y^2 < 1$. The points concerned are all the ones inside the circle $x^2 + y^2 = 1$.

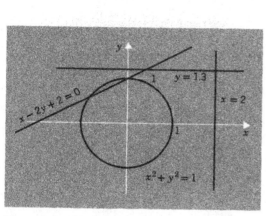

Figure 0-9   Graphs of equations.

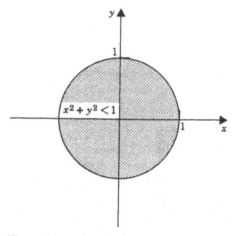

Figure 0-10   Graph of inequality.

We can also consider the points that satisfy two equations or two inequalities. This is a case of intersection of two sets (Section 0-4). Figure 0-11 illustrates the intersection of the graphs of $4x + 7y = 2$, $2x - 3y = 5$; just one point $(\frac{41}{26}, -\frac{8}{13})$ satisfies both equations. Figure 0-12 illustrates the inter-

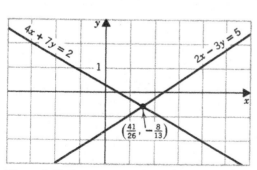

Figure 0-11   Intersection of two loci.

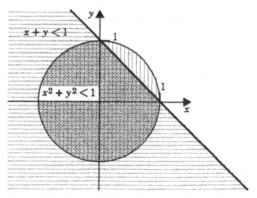

Figure 0-12   Points satisfying two inequalities.

section of the graphs of two inequalities; the shaded portion (inside the circle and below the line) satisfies both.

The union of two sets also appears in graphing. For example, the graph of the equation $(x + y - 1)(x + y + 1) = 0$ is the set of all points (x, y) for which either $x + y - 1 = 0$ or $x + y + 1 = 0$; hence, it is the union of two graphs, as in Figure 0-13.

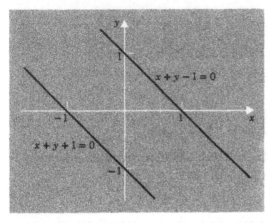

Figure 0-13   Graph of
$(x + y - 1)(x + y + 1) = 0$.

The *x-intercepts* of a graph are the *x*-coordinates of the points of intersection of the graph with the *x* axis. The *y-intercepts* are defined similarly. For example, the line $2x - 3y = 5$ in Figure 0-11 has *x*-intercept $\frac{5}{2}$ and *y*-intercept $-\frac{5}{3}$.

Notice that different equations may have the same graph. Thus

$$(x^2 + y^2 + 1)(x + y - 2) = 0$$
$$3x + 3y - 6 = 0$$
$$x^2 + 2xy + y^2 - 4x - 4y + 4 = 0$$
$$x + y - 2 = 0$$

all have the same graph. The first equation is obtained from the last by multiplying both sides by $x^2 + y^2 + 1$, which is 0 for no point (x, y) in the plane; the second is obtained from the last by multiplying both sides by 3; and the third is obtained from the last by squaring both sides. Equations that have the same graph are considered to be *equivalent*.

We can generalize this discussion to solid analytic geometry with the aid of the *xyz*-coordinate system of Figure 0-7. The distance d between points (x_1, y_1, z_1) and (x_2, y_2, z_2) is found to be

$$d = \sqrt{(x_2 - x_1)^2 + (y_2 - y_1)^2 + (z_2 - z_1)^2}$$

Accordingly, the equation $x^2 + y^2 + z^2 = a^2$ has as graph a sphere whose center is the origin $(0, 0, 0)$ and whose radius is a (see Figure 0-14).

0-7 LINEAR EQUATIONS IN x AND y

By a linear equation in x and y is meant an equation of the form:

$$Ax + By + C = 0, \qquad A \text{ and } B \text{ not both } 0, \tag{0-70}$$

A, B, and C being fixed real numbers.

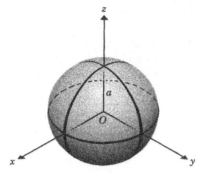

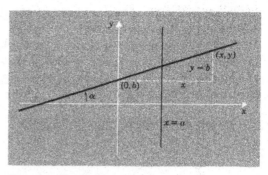

Figure 0-14   Sphere $x^2 + y^2 + z^2 = a^2$.   **Figure 0-15**   Graph of linear equation.

THEOREM. *The graph of a linear equation is always a straight line, and every straight line is the graph of a linear equation.*

This is proved, for example, by noticing first that a straight line perpendicular to the x axis at $(a, 0)$ has the equation

$$x = a \tag{0-71}$$

See Figure 0-15. A line not perpendicular to the x-axis must cross the y-axis at a point $(0, b)$; if (x, y) is another point on the line, then by similar triangles the ratio $(y - b)/x$ has the same value for all choices of (x, y). This ratio m is called the *slope* of the line and the equation $(y - b)/x = m$, or

$$y = mx + b \quad \text{(slope-intercept)} \tag{0-72}$$

is then an equation whose graph is the line in question. Both equations (0-71) and (0-72) can be written in an equivalent form (0-70), and every equation (0-70) can be written in an equivalent form (0-71) or in an equivalent form (0-72). If $B \neq 0$, $Ax + By + C = 0$ is equivalent to

$$y = -\frac{A}{B}x - \frac{C}{B}$$

which is the same as (0-72) if we write

$$m = -\frac{A}{B}, \qquad b = -\frac{C}{B} \tag{0-73}$$

This gives the slope m and y-intercept b for the equation (0-70). If $B = 0$, then $A \neq 0$, and $Ax + C = 0$ is equivalent to $x = -C/A$, of form (0-71).

For a line parallel to the x-axis, $m = 0$ and the equation becomes

$$y = b.$$

A line of the form (0-71) is perpendicular to the x-axis; in this case, we say that the slope is infinite. By similar triangles, we verify that the slope can be found for a line not perpendicular to the x-axis by choosing two distinct points (x_1, y_1), (x_2, y_2) on the line; then

$$m = \frac{y_2 - y_1}{x_2 - x_1} \tag{0-74}$$

By the same reasoning, an equation for the line is

$$y - y_1 = m(x - x_1) \qquad \text{(point-slope)} \tag{0-75}$$

or

$$y - y_1 = \frac{y_2 - y_1}{x_2 - x_1}(x - x_1) \qquad \text{(2-point)} \tag{0-76}$$

If the line has nonzero x-intercept a and nonzero y-intercept b, then the points $(a, 0)$ and $(0, b)$ can be used as (x_1, y_1), (x_2, y_2) and (0-76) becomes

$$\frac{x}{a} + \frac{y}{b} = 1 \qquad \text{(intercept)} \tag{0-77}$$

The different forms of the equation are labeled slope-intercept, and so on, as indicated next to the equations.

The slope m can be interpreted as

$$m = \tan \alpha,$$

where α is the angle of inclination of the line, shown in Figure 0-15. If lines L_1, L_2 have slopes m_1, m_2 and angles of inclination α_1, α_2, respectively, then L_1 is parallel to L_2 precisely when $\alpha_1 = \alpha_2$ and $m_1 = m_2$; L_1 is perpendicular to L_2 precisely when α_1, α_2 differ by $90°$ and, hence,

$$m_2 = \tan \alpha_2 = \tan(\alpha_1 \pm 90°) = -\cot \alpha_1 = -\frac{1}{\tan \alpha_1} = -\frac{1}{m_1}$$

that is, L_1 is perpendicular to L_2 precisely when $m_1 m_2 = -1$. (This reasoning fails for lines parallel to the axes, for which one slope is 0, and the other is infinite.)

It follows that a line parallel to the line $Ax + By + C = 0$ and passing through (x_0, y_0) is given by

$$A(x - x_0) + B(y - y_0) = 0$$

while a line perpendicular to the line $Ax + By + C = 0$ and passing through (x_0, y_0) is given by

$$-B(x - x_0) + A(y - y_0) = 0$$

PROBLEMS

1. In the xy-plane, graph the points $(3, 0)$, $(0, -2)$, $(0, 0)$, $(-1, -2)$, $(4, -1)$.
2. Show that the triangle with vertexes $(2, 2)$, $(5, 7)$, $(10, 4)$ is a right triangle.
3. Show that the triangle with vertexes $(1, 1)$, $(3, 5)$, $(10, -1)$ is isosceles.
4. Show that the points $(1, 2)$, $(2, 4)$, $(4, 8)$ lie on a straight line.
5. Graph: (a) $(x^2 + y^2)(x - y) = 0$, (b) $(x + y)(x - y) = 0$, (c) $(x + y)^2 + (x - y)^2 = 0$, (d) $x - y > 0$, (e) $x^2 - y^2 \leq 0$.

6. For each of the following linear equations, graph, find the slope and the intercepts:
    (a) $2x - 3y = 6$ (b) $x - 2 = 0$ (c) $y + 3 = 0$

    (d) $y - 5 = -3(x + 2)$ (e) $\dfrac{x}{-1} + \dfrac{y}{-3} = 1$

7. Find the equation of a straight line L satisfying the given conditions:
    (a) $(4, 2)$ is on L, slope $m = 5$.
    (b) $(2, 0)$ and $(0, 3)$ are on L.
    (c) $(5, 1)$ and $(7, 2)$ are on L.
    (d) L has slope 5 and y-intercept -1.
    (e) $(1, 3)$ is on L, L is parallel to the line $x - 5y = 0$.
    (f) $(1, 3)$ is on L, L is perpendicular to the line $2x + 3y = 1$.
    (g) $(2, 2)$ is on L, L has angle of inclination $\pi/4$.
    (h) $(3, 2)$ is on L, L is parallel to the x-axis.
    (i) $(4, 1)$ is on L, L is perpendicular to the x-axis.
8. Graph the linear inequalities:
    (a) $3x - 5y + 7 < 0$ (b) $2x + y - 2 \leq 0$
9. (a) Show that if k, h are real numbers, not both 0, then

$$k(2x - 7y + 5) + h(x + 3y - 15) = 0$$

   is the equation of a line containing the point A of intersection of the lines
   $2x - 7y + 5 = 0$, $x + 3y - 15 = 0$.
   (b) Find a line containing A and parallel to the line $2x - y = 0$.
   (c) Find a line containing A and the point $(1, 1)$.
   (d) Find a line through A which has 2 as its x-intercept.

0-8 SIMULTANEOUS LINEAR EQUATIONS

For two unknowns, two simultaneous linear equations have the form
$$\begin{aligned} a_1x + b_1y &= k_1 \\ a_2x + b_2y &= k_2 \end{aligned} \tag{0-80}$$
We here assume that at least one of a_1, b_1 is not 0 and at least one of a_2, b_2 is not 0, so that each equation represents a line.

"In general," the equations represent two straight lines intersecting at a single point (x, y), which is the solution of the pair of equations. In that case, the point (x, y) can be found by elimination, as in the following example.

EXAMPLE 1

$$\begin{aligned} 2x - 3y &= 5 \\ 4x + 7y &= 2 \end{aligned}$$

By multiplying the first equation by 2 and by subtracting from the second, we obtain
$$13y = -8$$
so that $y = -\tfrac{8}{13}$; this value can be substituted in the first equation to give $x = \tfrac{41}{26}$. The solution can be checked graphically, as shown in Figure 0-11.

The procedure fails when the two lines (0-80) are parallel or coincident or, equivalently, when the coefficients are proportional: $a_1/b_1 = a_2/b_2$ or

$$a_1b_2 - a_2b_1 = 0$$

In this case, we may have no solution (the case of distinct parallel lines), or we may have infinitely many solutions (the case of two coincident lines). The following two examples illustrate these alternatives.

EXAMPLE 2. Let $2x - 3y = 5$, $2x - 3y = 7$. The lines are distinct and parallel, and there is no solution.

EXAMPLE 3. Let $2x - 3y = 5$, $4x - 6y = 10$. The lines are coincident. Both equations are satisfied by every point on the line $2x - 3y = 5$.

Homogeneous Equations in x and y. If $k_1 = 0$ and $k_2 = 0$ in (0-80), the equations are called *homogeneous*. Thus, homogeneous equations have the form:

$$\begin{aligned} a_1x + b_1y &= 0 \\ a_2x + b_2y &= 0 \end{aligned} \tag{0-81}$$

If $a_1b_2 - a_2b_1 \neq 0$, then the lines cannot be parallel or coincident and, therefore, meet in just one point. That point must be the origin $(0, 0)$; $x = 0$, $y = 0$ is called the *trivial solution* of (0-81). If $a_1b_2 - a_2b_1 = 0$, the lines have the same slope and both pass through $(0, 0)$; hence, the lines coincide and there are infinitely many solutions, given by all points on one line.

Three Equations in Three Unknowns. Here we have equations of the form:

$$\begin{aligned} a_1x + b_1y + c_1z &= k_1 \\ a_2x + b_2y + c_2z &= k_2 \\ a_3x + b_3y + c_3z &= k_3 \end{aligned} \tag{0-82}$$

It is shown in solid analytic geometry that "in general" these equations represent three planes in space, and that three planes usually meet at a single point (as at the apex of a triangular pyramid).

We shall discuss the exceptional cases in the next section and here we give an example of the typical case of a unique solution.

EXAMPLE 4

$$\begin{aligned} 2x + y + z &= 4 \\ 3x - y - 2z &= 1 \\ x + 2y + 3z &= 5 \end{aligned}$$

We use an elimination procedure. We eliminate y by adding the first two equations to obtain

$$5x - z = 5$$

and by adding twice the second equation to the third to obtain

$$7x - z = 7$$

By subtracting the last two equations, we get $2x = 2$, so that $x = 1$. The previous equations then give $z = 0$ and finally $y = 2$. Thus our unique solution is $x = 1$, $y = 2$, $z = 0$.

Homogeneous Equations. When $k_1 = 0$, $k_2 = 0$, $k_3 = 0$ in (0-82), the equations are called homogeneous. In this case, we always have the trivial solution $x = 0$, $y = 0$, $z = 0$. If this is not the only solution, then one has infinitely many solutions. We give an example of the latter case:

EXAMPLE 5

$$3x + 5y + z = 0$$
$$2x - y + 2z = 0$$
$$4x + 11y = 0$$

The elimination of z from the first two equations gives $4x + 11y = 0$, which is the same as the third equation. Hence, we can obtain all solutions by choosing x arbitrarily and by then taking $y = -4x/11$ and

$$z = -3x - 5y = -3x + (20x/11) = -13x/11$$

The solutions $(x, y, z) = (x, -4x/11, -13x/11)$ can be shown to fill a straight line in space, passing through the origin $(0, 0, 0)$.

0-9 DETERMINANTS

In studying the two simultaneous linear equations (0-80), we found the expression $a_1b_2 - a_2b_1$, formed from the coefficients of the unknowns, to be important. This expression is a second-order determinant that is usually denoted by

$$\begin{vmatrix} a_1 & b_1 \\ a_2 & b_2 \end{vmatrix}$$

By definition

$$\begin{vmatrix} a_1 & b_1 \\ a_2 & b_2 \end{vmatrix} = a_1b_2 - a_2b_1 \tag{0-90}$$

When numerical values are given to a_1, b_1, a_2, b_2, the determinant gets a numerical value, for example,

$$\begin{vmatrix} 3 & 5 \\ 7 & -2 \end{vmatrix} = -6 - 35 = -41$$

We rewrite the equations (0-80):

$$a_1x + b_1y = k_1$$
$$a_2x + b_2y = k_2 \tag{0-91}$$

The usual elimination procedure leads to a solution that can be expressed in terms of determinants:

$$x = \frac{D_1}{D}, \qquad y = \frac{D_2}{D} \tag{0-92}$$

where

$$D = \begin{vmatrix} a_1 & b_1 \\ a_2 & b_2 \end{vmatrix}, \qquad D_1 = \begin{vmatrix} k_1 & b_1 \\ k_2 & b_2 \end{vmatrix}, \qquad D_2 = \begin{vmatrix} a_1 & k_1 \\ a_2 & k_2 \end{vmatrix} \tag{0-93}$$

The equations (0-92) and (0-93) are known as *Cramer's rule* for the solution of (0-91). When $D = 0$, (0-92) become meaningless; this is the exceptional case $(a_1 b_2 - a_2 b_1 = 0)$ of parallel or coincident lines that is discussed in Section 0-8.

The results for 2 equations in 2 unknowns can be generalized to n equations in n unknowns, although the rules become more complicated to state, even with the help of determinants. Here, we give only a few of the most important rules.

An nth order determinant is written

$$
\begin{vmatrix}
a_{11} & a_{12} & a_{13} & \cdots & a_{1n} \\
a_{21} & a_{22} & a_{23} & \cdots & a_{2n} \\
\vdots & & & & \\
a_{n1} & a_{n2} & a_{n3} & \cdots & a_{nn}
\end{vmatrix}
$$

Its value (to be explained) depends on all n^2 quantities $a_{11}, \ldots, a_{nn}$.

Expansion by Minors of the First Column. This is a procedure that reduces a determinant to an expression which involves determinants of order one less. We illustrate it for $n = 3$:

$$
\begin{vmatrix}
a_{11} & a_{12} & a_{13} \\
a_{21} & a_{22} & a_{23} \\
a_{31} & a_{32} & a_{33}
\end{vmatrix}
= a_{11}
\begin{vmatrix}
a_{22} & a_{23} \\
a_{32} & a_{33}
\end{vmatrix}
- a_{21}
\begin{vmatrix}
a_{12} & a_{13} \\
a_{32} & a_{33}
\end{vmatrix}
+ a_{31}
\begin{vmatrix}
a_{12} & a_{13} \\
a_{22} & a_{23}
\end{vmatrix}
$$

Thus, each entry in the first column is multiplied by a second-order determinant, and alternately by 1 and -1. The second-order determinant in each case is the *minor* of the entry in the first column; the minor is obtained from the original determinant by striking out the row and column that contain the entry selected. If we expand the second-order determinants, we obtain finally

$$
\begin{vmatrix}
a_{11} & a_{12} & a_{13} \\
a_{21} & a_{22} & a_{23} \\
a_{31} & a_{32} & a_{33}
\end{vmatrix}
$$

$$
= a_{11}(a_{22}a_{33} - a_{32}a_{23}) - a_{21}(a_{12}a_{33} - a_{32}a_{13}) + a_{31}(a_{12}a_{23} - a_{22}a_{13})
$$
$$
= a_{11}a_{22}a_{33} + a_{12}a_{23}a_{31} + a_{13}a_{21}a_{32} - a_{13}a_{22}a_{31} - a_{12}a_{21}a_{33} - a_{11}a_{23}a_{32}
$$

We can use this expansion to evaluate a determinant with numerical entries. For example,

$$
\begin{vmatrix}
1 & 2 & 0 \\
3 & -1 & 5 \\
2 & 1 & 2
\end{vmatrix}
= 1(-2 - 5) - 3(4 - 0) + 2(10 + 0) = 1
$$

In the general case, we denote by M_{ij} the *minor* of a_{ij}, that is, the determinant of order $n - 1$ obtained by deleting the row and column containing a_{ij}. Then

$$
\begin{vmatrix}
a_{11} & a_{12} & \cdots & a_{1n} \\
a_{21} & & \cdots & a_{2n} \\
a_{31} & & \cdots & a_{3n} \\
\vdots & & & \\
a_{n1} & & \cdots & a_{nn}
\end{vmatrix}
$$

$$
= a_{11}M_{11} - a_{21}M_{21} + a_{31}M_{31} - \cdots + (-1)^n a_{n1}M_{n1} \quad (0\text{-}94)
$$

A determinant of order 1 is $|a_{11}| = a_{11}$. The expansion of a determinant of order 2 is then consistent with the general rule (0-94):

$$
\begin{vmatrix}
a_{11} & a_{12} \\
a_{21} & a_{22}
\end{vmatrix}
= a_{11}M_{11} - a_{21}M_{21} = a_{11}a_{22} - a_{21}a_{12}
$$

We can verify that a determinant of order n has $n! = n(n - 1)(n - 2) \cdots 1$ terms; for $n = 10$, this means 3,628,800 terms! Hence, the determinant is a very concise way of writing this particular expression (a polynomial in n^2 variables—see Section 0-12).

For n equations in n unknowns $x_1, \ldots, x_n$:

$$
\begin{aligned}
a_{11}x_1 + a_{12}x_2 + \cdots + a_{1n}x_n &= k_1 \\
a_{21}x_1 + a_{22}x_2 + \cdots + a_{2n}x_n &= k_2 \\
&\ \vdots \\
a_{n1}x_1 + a_{n2}x_2 + \cdots + a_{nn}x_n &= k_n
\end{aligned}
\quad (0\text{-}95)
$$

we form the determinant of coefficients D. This is precisely the determinant in (0-94). We also form the determinants $D_1, \ldots, D_n$ by replacing the 1st, 2nd, ..., and nth column, respectively, of D by $k_1, k_2, \ldots, k_n$. Thus

$$
D_2 =
\begin{vmatrix}
a_{11} & k_1 & a_{13} & \cdots & a_{1n} \\
a_{21} & k_2 & a_{23} & \cdots & a_{2n} \\
\vdots & \vdots & \vdots & & \vdots \\
a_{n1} & k_n & a_{n3} & \cdots & a_{nn}
\end{vmatrix}
$$

Then, if $x_1, \ldots, x_n$ satisfy (0-95), elimination leads to the equations

$$
Dx_1 = D_1, \qquad Dx_2 = D_2, \qquad \ldots, \qquad Dx_n = D_n \quad (0\text{-}96)
$$

If $D \neq 0$, the unique solution is then given by Cramer's rule:

$$
x_1 = \frac{D_1}{D}, \qquad x_2 = \frac{D_2}{D}, \qquad \ldots, \qquad x_n = \frac{D_n}{D} \quad (0\text{-}97)
$$

If $D = 0$, there is no solution unless $D_1 = 0, D_2 = 0, \ldots, D_n = 0$ (but even then there may be no solution). If all these determinants are 0 and there is a solution, then there are infinitely many solutions.

When $k_1 = 0, \ldots, k_n = 0$, the equations are *homogeneous*. In this case, if $D \neq 0$, then there is only the trivial solution $x_1 = 0, x_2 = 0, \ldots, x_n = 0$. If $D = 0$, then there are infinitely many solutions.

Additional properties of determinants and their applications are given in Chapter 10 (see also Section 1-12).

PROBLEMS

1. Find all solutions and check graphically:

    (a) $2x + y = 3$ (b) $2x - y = 6$ (c) $x - y = 0$
         $x - 2y = 7$ $4x - 2y = 5$ $2x - 2y = 0$

    (d) $x + 2y = 0$ (e) $4x + 2y = 1$
         $5x - 3y = 0$ $2x + y = 2$

2. Find all solutions:

    (a) $x + y - z = 1$ (b) $3x - y - z = 2$
         $2x + 3y - 2z = 3$ $4x + y + 2z = 7$
         $x + y + z = 3$ $x + 2y + 3z = 4$

    (c) $x + 2y - z = 0$ (d) $s + t = 2$
         $2x - y + z = 0$ $t - u = 3$
         $3x + y = 0$ $u - z = 2$
                                  $s - z = 7$

3. In each case evaluate the determinant:

(a) $\begin{vmatrix} 2 & 1 \\ 0 & 3 \end{vmatrix}$
    (b) $\begin{vmatrix} 5 & 1 \\ 1 & 6 \end{vmatrix}$
    (c) $\begin{vmatrix} 7 & 4 \\ 14 & 8 \end{vmatrix}$

(d) $\begin{vmatrix} 2 & 0 & 1 \\ 1 & 0 & 3 \\ -1 & 5 & 2 \end{vmatrix}$
    (e) $\begin{vmatrix} 1 & 1 & 1 \\ 2 & 2 & 2 \\ 3 & -4 & 5 \end{vmatrix}$
    (f) $\begin{vmatrix} 1 & 0 & 0 & 2 \\ 2 & 0 & 3 & 0 \\ 0 & 1 & 0 & 0 \\ 5 & -1 & 0 & 2 \end{vmatrix}$

4. Use determinants to determine whether there is a unique solution:

    (a) The equations of Problem 1(a).   (b) The equations of Problem 1(b).
    (c) The equations of Problem 1(c).   (d) The equations of Problem 1(d).
    (e) The equations of Problem 1(e).   (f) The equations of Problem 2(a).
    (g) The equations of Problem 2(b).   (h) The equations of Problem 2(c).
    (i) The equations of Problem 2(d).

0-10 FUNCTIONS

A central idea in mathematics is that of a function.

Definition. A *function* is an assignment, to each element of a given set X, of an element on another set Y. The set X is called the *domain* of the function. The elements of Y that are assigned to at least one element of X form a set called the *range* of the function.

The definition is indicated schematically in Figure 0-16. Each element in X is joined to the corresponding element in Y by a curved arrow.

A function is usually denoted by a single letter: f, g, F, G, ϕ, ... If f is a function with domain X, we say that "f is defined in X" (or "on X"). For each x in X we denote by $f(x)$ the value y in Y assigned to x, that is, $y = f(x)$.

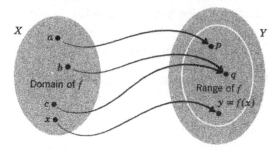

Figure 0-16   Function f.

EXAMPLE 1. To each real number x, we assign its square: x^2. If we call the function f, then $f(x) = x^2$. We say here that "the function f is defined by the equation or formula $y = x^2$." The domain of the function consists of all real numbers (described symbolically by $-\infty < x < \infty$). For this function f, we have $f(0) = 0$, $f(1) = 1$, $f(2) = 4$, $f(-2) = 4$, $f(3) = 9$, $f(a) = a^2$ and $f(x) = x^2$.

EXAMPLE 2. To each of five men named Brown, Jones, Smith, Holland, and King, we assign the first letter of the surname. If we call the function g, then $g(\text{Brown}) = B$, $g(\text{Jones}) = J$, $g(\text{Smith}) = S$, $g(\text{Holland}) = H$, $g(\text{King}) = K$. This example illustrates the great generality in the concept of a function. The objects in the domain and range can be of any sort whatsoever.

When a function is specified by a formula, we often omit specification of the domain in cases in which the context indicates this set. For example, in the context of real numbers the function $y = x^2$ would be understood to have as domain the set of all real numbers; the function $y = \sqrt{x}$ (positive square root) is understood to have as domain the set of all nonnegative real numbers. When we wish to use a domain smaller than the obvious one, we must specify it. For example, $y = x^2$, $0 \le x \le 1$ specifies a function whose domain is the interval $[0, 1]$.

It is always best to refer to a function by a single letter, such as f. However, it is general practice to say, for example, "the function $y = f(x)$," "the function $f(x)$," "the function $y(x)$," "the function $\sin x$." Occasionally, we shall also use these less precise phrases.

Two functions f, g are said to be *equal*, $f = g$, if they have the same domain and range and $f(x) = g(x)$ for every x in the domain. Thus, the formulas $y = x^2 + 2x + 1$ and $y = (x + 1)^2$ describe equal functions. Notice, however, that the second function could equally well be given by the formula $x = (t + 1)^2$. We are free to use whatever notation we wish for a typical element of the domain (often called "independent variable") and for the corresponding element ("dependent variable") in the range. The function has a meaning independent of the notation chosen. In the example, g assigns to each real number the square of the number increased by one. No matter whether we write $y = (x + 1)^2$, $x = (t + 1)^2$, or $z = (u + 1)^2$, we are still considering the same function.

A function is often called a *mapping*. More specifically, a function with domain X and range contained in Y is said to be a *mapping of X into Y*; we also say that f maps X into Y. The terminology is suggested by the fact that constructing a map, as in geography, entails assigning a point on the map to each point of the region that is being described. When the range of f is all of Y, we speak of a mapping of X *onto* Y.

Let f be a mapping of X into Y. Then f is said to be a *one-to-one* mapping (or function) if no element of Y is assigned to more than one element of X. Thus, the elements of X are paired off with the elements of the range of f, as suggested in Figure 0-17. The function suggested in Figure 0-18 is not one-to-one, nor is the function of Example 1. However, the function of Example 2 is one-to-one. A one-to-one mapping is also called a *one-to-one correspondence*.

A one-to-one mapping f always has an *inverse mapping* (or *inverse function*). The inverse mapping, denoted by f^{-1}, assigns to each y in the range of f

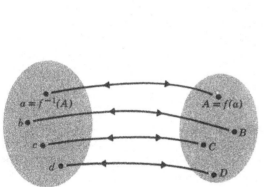

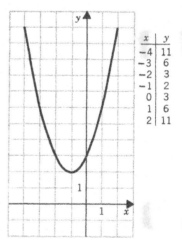

x	y
-4	11
-3	6
-2	3
-1	2
0	3
1	6
2	11

Figure 0-17 One-to-one mapping and inverse mapping.

Figure 0-18 The function $y = x^2 + 2x + 3$.

the unique element x for which $f(x) = y$. This is illustrated in Figure 0-17. For the function of Example 2, $f^{-1}(B) = $ Brown, $f^{-1}(J) = $ Jones, and so on.

The function $y = x^2$, $-\infty < x < \infty$, is not one-to-one. However, if we restrict the domain to the interval $x \geq 0$, we obtain a one-to-one mapping: $y = x^2$, $x \geq 0$; the inverse mapping is given by $x = \sqrt{y}$, $y \geq 0$.

For each set X the *identity mapping* is the one that assigns each element of X to itself. We generally denote this mapping by I, so that $I(x) = x$ for all x in X (notice that I has X as domain and range). The mapping I is clearly one-to-one, and its inverse is again I, that is, $I = I^{-1}$.

0-11 REAL FUNCTIONS OF A REAL VARIABLE

A large part of the calculus is concerned with functions whose domain and range are sets of real numbers. Each such function is called a "real function of a real variable." The word "real" can be omitted when the context makes the meaning clear, and we often write simply "a function of one variable" or, by using the notation described above, "a function $y = f(x)$." Usually such a function is given by a formula such as $y = x^2 + 2x + 3$. However, as we shall learn, the function can be described in other ways.

The *graph* of a real function consists of all points (x, y) in the xy-plane for which x is in the domain of the function and y is the value assigned to x. Since each x has only one y assigned, each vertical line in the plane meets the graph in, at most, one point.

The function given by the formula $y = x^2 + 2x + 3$ can be represented graphically, as shown in Figure 0-18. This figure also shows a table of values of the function. Each pair x, y in the table corresponds to a point on the graph.

The *zeros* of a function f are those values x for which $f(x) = 0$.

Among the real functions of a real variable there are important classes, which we describe below.

Linear functions. $y = mx + b$, where m and b are fixed real numbers. The graph of each function is a straight line of slope m (Section 0-7).

Polynomial functions. $y = a_n x^n + \cdots + a_1 x + a_0$, where $a_n, \ldots, a_0$ are fixed real numbers, called the *coefficients*, with $a_n \neq 0$. The polynomial is said to have *degree* n. Here n may be 0 or any positive integer. In writing a polynomial, we generally suppress terms with 0 coefficient. The function $y = 0$ is also considered to be a polynomial, called the *zero* polynomial; it is convenient to assign to it the degree -1. For $n \geq 1$, a polynomial f has at most n (real) zeros (see Section 0-18). If $x_1, \ldots, x_m$ are the zeros, then one has a factorization:

$$f(x) = (x - x_1)^{k_1} \cdots (x - x_m)^{k_m} g(x)$$

Here, $k_1, \ldots, k_m$ are positive integers, called the *multiplicities* of the zeros, with $k_1 + \cdots + k_m = q \leq n$; $g(x)$ is a polynomial of *even* degree $n - q$ and can be factored into quadratic factors that correspond to the complex zeros of f (Section 0-18). The following is an example:

$$f(x) = (x - 1)(x - 2)^3(x^2 + x + 1)(x^2 + 2x + 2)^2$$

Here the real roots would be listed as 1, 2, 2, 2. Determining the zeros of f is equivalent to solving the algebraic equation of degree n:

$$a_n x^n + \cdots + a_1 x + a_0 = 0$$

For $n = 2$, this is the problem of solving the quadratic equation $ax^2 + bx + c = 0$. The solutions are given by the *quadratic formula*:

$$\frac{-b \pm \sqrt{b^2 - 4ac}}{2a} \tag{0-110}$$

(When $b^2 - 4ac < 0$, we obtain complex roots; see Section 0-17.)

Rational functions. $y = p(x)/q(x)$, where $p(x)$ and $q(x)$ are polynomials and q is not the zero polynomial. Here the domain is the set of all real numbers except the (real) zeros of $q(x)$. An example is

$$y = \frac{x^2 + 3}{x^2 - 2x - 3}, \qquad x \neq 3, x \neq -1$$

Trigonometric functions. $y = \sin x$, $y = \cos x$, $y = \tan x, \ldots$ (see Section 0-15).
Exponential functions. $y = 2^x$, $y = 3^x, \ldots$ (see Section 0-19).
Power functions. $y = x^3$, $y = x^{3/2}$, $y = x^{-2}, \ldots$ (see Section 0-19).
Logarithmic functions. $y = \log_{10} x$, $y = \log_e x, \ldots$ (see Section 0-19).

These classes in no way exhaust the class of all real functions.

0-12 REAL FUNCTIONS OF SEVERAL REAL VARIABLES

The equation $z = x^2 + y^2$ assigns a real number z to each pair (x, y) of real numbers. We call this a real function of two real variables. In general, a real function of two real variables assigns a real number to each (x, y) of a certain set (domain) of such pairs; we can consider the domain as a set in the xy-plane of analytic geometry. Similarly, a real function of k variables assigns a real number z to each member of a set of ordered k-tuples $(x_1, \ldots, x_k)$. Thus, $z = x_1 x_2 x_3 - x_4 x_5$ defines a function of five variables.

The discussion of the preceding section can be extended to functions of several variables. Below we mention only the analogues of three of the classes.

Linear functions. $z = a_1 x_1 + \cdots + a_k x_k + b$.
Polynomial functions. $z = $ a sum of terms of form $a x_1^{n_1} x_2^{n_2} \ldots x_k^{n_k}$, where $n_1, \ldots, n_k$ are nonnegative integers and a is a real number. For example, $z = x_1^2 + x_2^2 - x_1 x_2 x_3$ defines a polynomial in x_1, x_2, x_3.
Rational functions. $z = $ a ratio of two polynomials.

Remark. It is general practice to regard polynomials and rational functions simply as algebraic expressions and to speak of the polynomial $x^2 + 2x + 3$, for example.

PROBLEMS

1. Each of the following functions assigns the value y to x as in the table given. Find the domain and range of the function.

(a)

x	0	1	2	3	4
y	2	3	5	4	7

(b)

x	a	b	c	d	e
y	b	a	d	c	b

(c)

x	5	6	7	8	9
y	2	2	2	2	2

(d)

x	2	4	6	8
y	3	4	5	no value

2. Let X be the set of all right triangles in the plane. For x in X, let $f(x)$ be the area of x. Evaluate $f(x)$ in the following cases.
   (a) x has legs 2, 3.   (b) x has hypotenuse 2 and is isosceles.
   (c) x is inscribed in a circle of radius 5 and one angle is $30°$.

3. For each of the following real functions of a real variable given by a formula, state the domain and range.

(a) $y = -2 - x^2$ (b) $y = x^3 - x$

(c) $y = \sqrt{x^2 - 1}$ (d) $y = \dfrac{1}{x - 1}$

(e) $y = x^{-3}$ (f) $y = 2 \sin x - 2$

4. Let f be the function given by the equation $y = x^2 - x + 2$. Evaluate:
   (a) $f(0)$ (b) $f(1)$ (c) $f(-2)$ (d) $f(b)$

5. Let g be the function that is defined by the equation $x = 16t^2 + 30t$. Evaluate:
   (a) $g(0)$ (b) $g(1)$ (c) $g(a)$ (d) $g(x)$

6. Let F be the function that is defined by the equation $y = 1/(1 + \sin x)$, where x is in radians. Evaluate:
   (a) $F(\pi/2)$ $F(\pi)$ (c) $F(0)$ (d) $F(1)$

7. State which of the functions of Problem 1 are one-to-one mappings, and for each such function give a table for the inverse function.

8. State which of the following functions are one-to-one and, for each, give a formula for the inverse function:
   (a) $y = 2x$ (b) $y = 5x + 7$ (c) $y = x^3$

   (d) $y = x^4$ (e) $y = \dfrac{1}{x}$ (f) $y = \dfrac{x}{x - 1}$

9. *Absolute-value function.* To each x, we assign the absolute value of x, $|x|$, thereby defining a function: $y = |x|$.
   (a) What are the domain and range?
   (b) Evaluate $|-3|$, $|\sqrt[3]{8}|$, $|-7|$.
   (c) Graph the function.

10. For each of the following geometric quantities give an appropriate formula, and interpret the formula as a function by giving the domain and range in each case:
    (a) The area of a square of given side.
    (b) The volume of a sphere of given radius.
    (c) The surface area of a sphere of given radius.
    (d) The altitude of an equilateral triangle of given side.
    (e) The hypotenuse of an isosceles right triangle of given legs.

11. Find all (real) zeros of each of the polynomials and give the multiplicity of each zero:
    (a) $x - 2$ (b) $x^2 - 2$ (c) $x^2 - 2x + 1$
    (d) $x^2 - 3x - 4$ (e) $(x - 1)^3(x + 2)^5(x^2 + 1)^2$ (f) $x^5 + 1$

12. Interpret as a function of several variables and state which is a polynomial:
    (a) The area of a rectangle of given sides.
    (b) The hypotenuse of a right triangle of given legs.
    (c) The altitude on side c of an isosceles triangle of sides a, a, c.
    (d) The speed after t seconds of a particle starting from rest and moving on a straight line with acceleration a (in feet per second per second).
    (e) The product of the roots plus twice the sum of the roots of the quadratic equation $ax^2 + bx + c = 0$.

0-13 GRAPH OF A SECOND DEGREE POLYNOMIAL

We consider a function given by

$$y = ax^2 + bx + c, \qquad a \neq 0 \tag{0-130}$$

so that the function is a second degree polynomial in x and is thus defined for all x. We shall see that the graph fits the locus definition of a *parabola*, and hence we at once refer to it as a parabola.

The x-intercepts of the graph are obtained by solving the quadratic equation $ax^2 + bx + c = 0$; thus, the intercepts are given by (0-110) above. When $b^2 - 4ac < 0$, there are no (real) intercepts; when $b^2 - 4ac = 0$, there is one intercept $-b/(2a)$; when $b^2 - 4ac > 0$, there are two intercepts. In the last two cases, we can write our function as

$$y = a(x - x_1)(x - x_2) \tag{0-131}$$

where x_1, x_2 are the intercepts (0-110) ($x_1 = x_2$ if $b^2 - 4ac = 0$).

By completing the square, we can write (0-130) as follows:

$$y = a\left(x + \frac{b}{2a}\right)^2 + c - \frac{b^2}{4a} \tag{0-132}$$

From this equation we observe that, if $a > 0$, then y has its *smallest* value for $x = -b/(2a)$ and y becomes larger and larger (as large as desired) as x moves away from this value. If $a < 0$, then y has its *largest* value for $x = -b/(2a)$ and y becomes smaller and smaller (that is, minus a large positive number) as x moves away from this value. Hence, we have typical graphs, as in Figure 0-19. The lowest or highest point on the graph is $(-b/(2a), c - [b^2/(4a)])$, the *vertex* of the parabola.

Locus Definition of Parabola. A parabola is defined as the locus of a point moving so that its distance from a fixed point (*focus*) equals its distance

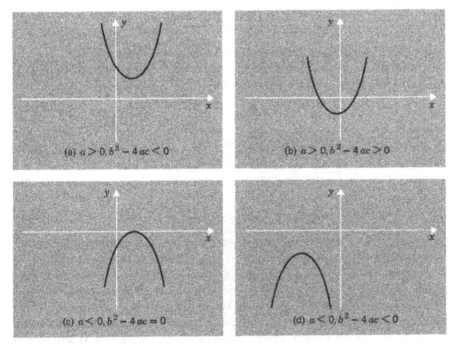

(a) $a > 0, b^2 - 4ac < 0$

(b) $a > 0, b^2 - 4ac > 0$

(c) $a < 0, b^2 - 4ac = 0$

(d) $a < 0, b^2 - 4ac < 0$

Figure 0-19 Parabolas $y = ax^2 + bx + c$.

from a fixed line (*directrix*) not containing the focus. Let a parabola be given and let axes be chosen so that the directrix is parallel to the *x*-axis. We can then choose h, k and p so that the focus is at $(h, k + p)$ and the directrix is

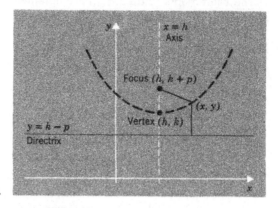

Figure 0-20 Parabola as a locus.

the line $y = k - p$ (see Figure 0-20). The point (x, y) then satisfies the locus condition precisely when

$$\sqrt{(x - h)^2 + (y - k - p)^2} = |y - k + p|$$

If we square and simplify, we obtain the equation

$$4p(y - k) = (x - h)^2 \tag{0-133}$$

This can be written in the form (0-130) or in the form

$$y = \frac{(x - h)^2}{4p} + k \tag{0-133'}$$

A comparison with (0-132) shows that the vertex is at (h, k).

Conversely, as (0-132) shows, every equation (0-130) can be written in the form (0-133') with

$$h = -\frac{b}{2a}, \qquad k = c - \frac{b^2}{4a}, \qquad p = \frac{1}{4a}$$

Accordingly, parabolas are precisely the curves representable by equations of form (0-130), for appropriate choice of coordinate axes.

The parabola equation (0-130) can be considered as a special case of the equation

$$Ax^2 + Cy^2 + Dx + Ey + F = 0 \tag{0-134}$$

where $A \neq 0$, $C = 0$, $E \neq 0$. More generally, we can consider all equations (0-134) in which one, but not both, of A, C is zero. When $A \neq 0$, $C = 0$, $E \neq 0$, we are led to (0-130) again; when $A = 0$, $C \neq 0$, $D \neq 0$, we get a similar equation with the roles of the *x*- and *y*-axes interchanged. Thus our equation (0-134) describes all parabolas opening upward, downward, to the left, or to the right. However, the equation also includes certain degenerate cases: when $A \neq 0$, $C = 0$, and $E = 0$, and when $A = 0$, $C \neq 0$, and $D = 0$; they are the equations

$$Ax^2 + Dx + F = 0, \qquad (A \neq 0)$$
$$Cy^2 + Ey + F = 0, \qquad (C \neq 0)$$

They represent two lines parallel to one of the axes, or two coincident lines, or no locus at all (two imaginary lines); these loci are considered to be *degenerate parabolas*.

0-14 CIRCLE, ELLIPSE, HYPERBOLA

A *circle* is the locus of a point whose distance from a given point (the *center*) is a given positive number a (the *radius*). If the center is at the origin $(0, 0)$, a point (x, y) satisfies the locus condition precisely when $\sqrt{x^2 + y^2} = a$, or when

$$x^2 + y^2 = a^2 \qquad \text{(0-140)}$$

Accordingly, (0-140) is the equation of a circle with center $(0, 0)$ and radius a (Figure 0-21).

If the center is at the point (h, k), the equation is

$$(x - h)^2 + (y - k)^2 = a^2 \qquad \text{(0-141)}$$

On expanding, this equation takes the form:

$$x^2 + y^2 - 2hx - 2ky + h^2 + k^2 - a^2 = 0$$

Hence, every circle has an equation of form

$$Ax^2 + Ay^2 + Dx + Ey + F = 0 \qquad \text{(0-142)}$$

where $A \neq 0$. Equation (0-142) is termed the *general form of the equation of the circle*. From an equation in general form we can recover the center and the radius by completing the square:

$$A\left(x^2 + \frac{D}{A}x + \frac{D^2}{4A^2}\right) + A\left(y^2 + \frac{E}{A}y + \frac{E^2}{4A^2}\right) + F - \frac{D^2 + E^2}{4A} = 0$$
$$\left(x + \frac{D}{2A}\right)^2 + \left(y + \frac{E}{2A}\right)^2 = \frac{D^2 + E^2 - 4AF}{4A^2}$$

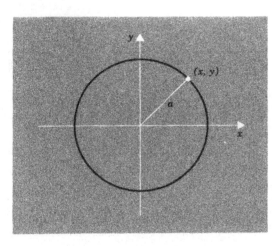

Figure 0-21   Circle.

Thus the center is at

$$\left(-\frac{D}{2A}, -\frac{E}{2A}\right)$$

and the radius is

$$a = \frac{\sqrt{D^2 + E^2 - 4AF}}{2|A|}$$

However, the quantity under the square root sign may be 0; we then refer to the locus as a *point-circle*—its graph is just the center. The quantity under the square root may also be negative; in this case the graph is the empty set.

Accordingly, the general equation (0-142), with $A \neq 0$, represents a true circle only when $D^2 + E^2 - 4AF > 0$.

The Ellipse. An ellipse can be defined as the locus of a point moving so that the sum of its distances from two points (the *foci*) is a given positive number $2a$ (greater than the distance between the foci). If we choose axes so that the foci are at $F_1:(c, 0)$, $F_2:(-c, 0)$, where $c > 0$, we obtain the equation

$$\frac{x^2}{a^2} + \frac{y^2}{b^2} = 1 \tag{0-143}$$

where $b = \sqrt{a^2 - c^2} > 0$. For each point (x, y) on the graph, the points $(x, -y), (-x, y), (-x, -y)$ are also on the graph (Figure 0-22). Hence the ellipse is symmetrical with respect to both axes. A typical ellipse is shown in Figure 0-22, which also gives other information. The circle can be regarded as

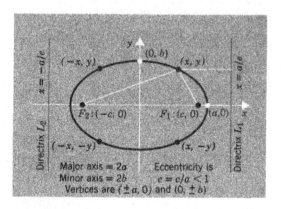

Figure 0-22   Ellipse $(x^2/a^2) + (y^2/b^2) = 1$.

a limiting case of an ellipse, in which the foci come together. Then $c = 0$, $b = a$, and the equation becomes $x^2 + y^2 = a^2$; the eccentricity e for the circle is 0.

If the coordinate axes are chosen more generally, but so that the foci lie on a line parallel to the x-axis, then the equation takes the form

$$\frac{(x - h)^2}{a^2} + \frac{(y - k)^2}{b^2} = 1 \tag{0-144}$$

analogous to (0-141); we say that the ellipse has *center* at (h, k). If we expand (0-144), we obtain an equation of form

$$Ax^2 + Cy^2 + Dx + Ey + F = 0 \qquad (0\text{-}145)$$

with $A > 0$ and $C > 0$. By completing the square as for (0-142), we can reduce (0-145) to the form

$$\frac{(x - h)^2}{a^2} + \frac{(y - k)^2}{b^2} = q$$

where $q = 1$, 0, or -1. When $q = 1$, we have an ellipse; when $q = 0$, a point-ellipse (h, k); when $q = -1$, an imaginary locus.

The Hyperbola. A hyperbola can be defined as the locus of a point which moves so that the difference of its distances from two points (the *foci*) equals a given positive number $2a$ (less than the distance between the foci). If we choose axes so that the foci are $F_1:(c, 0)$ and $F_2:(-c, 0)$, then the equation of the locus is found to be

$$\frac{x^2}{a^2} - \frac{y^2}{b^2} = 1 \qquad (0\text{-}146)$$

where $b = \sqrt{c^2 - a^2} > 0$. Again, there is symmetry with respect to the x-axis and the y-axis. However, x must be $\geq a$ or $\leq -a$, as we can observe from the expression for y:

$$y = \pm b \sqrt{\frac{x^2}{a^2} - 1} \qquad (0\text{-}147)$$

The same equation shows that if $|x|$ is very large, then approximately

$$y = \pm \frac{bx}{a} \qquad (0\text{-}148)$$

This equation describes two straight lines, called the *asymptotes* of the hyperbola (Figure 0-23). As $|x|$ becomes large, the hyperbola approaches its asymptotes.

If the axes are chosen only to make the foci lie on a line parallel to the x-axis, the equation takes the form

$$\frac{(x - h)^2}{a^2} - \frac{(y - k)^2}{b^2} = 1 \qquad (0\text{-}149)$$

with center (h, k). Expanding out, we get an equation of form (0-145) where $AC < 0$. From an equation (0-145) with $AC < 0$, we can complete the square to obtain an equation

$$\frac{(x - h)^2}{a^2} - \frac{(y - k)^2}{b^2} = q$$

with $q = 1$, 0, or -1. When $q = 1$, we again have the hyperbola (0-149); when $q = -1$, we have an equation

$$\frac{(y - k)^2}{b^2} - \frac{(x - h)^2}{a^2} = 1$$

for a hyperbola with the foci on a line parallel to the y-axis; when $q = 0$, we have an equation

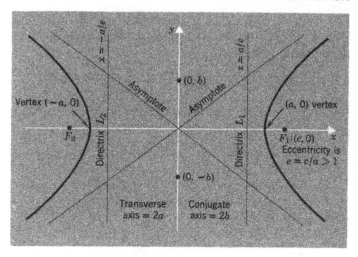

Figure 0-23
Hyperbola.

$$\frac{(x-h)^2}{a^2} - \frac{(y-k)^2}{b^2} = 0$$

which can be written

$$\left(\frac{x-h}{a} + \frac{y-k}{b}\right)\left(\frac{x-h}{a} - \frac{y-k}{b}\right) = 0$$

The locus consists of two straight lines intersecting at (h, k).

A Unified Definition of Ellipse, Parabola and Hyperbola. Let a point move so that the ratio of its distance from a fixed point (focus) to its distance from a fixed line (directrix), not containing the focus, is a fixed positive number e (eccentricity). Then the locus is:

An ellipse is $e < 1$.
A parabola if $e = 1$.
A hyperbola if $e > 1$.

For the parabola, this is as indicated in Section 0-13 above. For the ellipse and hyperbola, an explanation is needed since, in the original definition, we had two foci. In fact, there are also two directrices, and we can obtain the ellipse or hyperbola by using directrix L_1 and focus F_1, or directrix L_2 and focus F_2 (see Figures 0-22 and 0-23). In both figures, the directrices are the lines $x = \pm a/e$ (see Problem 11 below).

From the general locus definition it follows that two ellipses with the same eccentricity differ only in scale and position in the plane; that is, they are similar figures. A similar remark applies to two parabolas ($e = 1$) and to two hyperbolas with the same eccentricity.

We can also obtain all three curves as plane sections of a cone or *conic sections;* the circle is also included in this definition, as are some of the degenerate cases (point-circle, intersecting lines, etc.). (See Reference 6 at the end of this chapter.)

PROBLEMS

1. Graph, and find the focus, vertex, and directrix:

   (a) $y = x^2 + 2x + 5$ (b) $y = -x^2 - 4x + 7$

   (c) $y - 3 = 2(x - 1)^2$ (d) $y^2 - 2x + 3 = 0$

   (e) $y^2 - 2y - 3 = 0$ (f) $x^2 = 0$

2. Find the center and radius of each of the following circles and graph:

   (a) $x^2 + y^2 = 7$ (b) $2x^2 + 2y^2 = 5$

   (c) $x^2 + y^2 - 5x + 6y = 0$ (d) $3x^2 + 3y^2 + 4x + 6y + 52 = 0$

3. Show that a circle $x^2 + y^2 + ax + by + c = 0$ is tangent to the x-axis if, and only if, $4c = a^2$.

4. Find the equation of a straight line of slope 2 tangent to the circle $x^2 + y^2 = 180$.

5. Find the equation of the circle through the points $(1, 1)$, $(1, -2)$, and $(2, 3)$.

6. Find the points of intersection of the circles $x^2 + y^2 - 2x = 0$, $x^2 + y^2 - 3y = 0$.

7. Let two circles $x^2 + y^2 + ax + by + c = 0$, $x^2 + y^2 + Ax + By + C = 0$ be given and form the equation:

   $$(°) \quad x^2 + y^2 + ax + by + c - (x^2 + y^2 + Ax + By + C) = 0$$

   (a) Show that if the circles meet in a single point, then $(°)$ represents a common tangent.

   (b) Show that, if the circles do not intersect and are not concentric, then $(°)$ represents a line perpendicular to the line that joins the centers of the circles.

8. Find the vertexes, center, eccentricity, foci, and directrices, and graph:

   (a) $3x^2 + 4y^2 = 12$ (b) $9x^2 + 5y^2 = 45$

   (c) $5x^2 - 4y^2 = 20$ (d) $9x^2 - 16y^2 = 12$

9. Find the equation of an ellipse that satisfies the stated conditions and graph:

   (a) Focal distance 4, major axis 6.

   (b) Major axis 4, minor axis 2.

   (c) Eccentricity $\frac{1}{2}$, major axis 4.

10. Reduce to a standard form for an ellipse or hyperbola and graph:

    (a) $2x^2 + 4y^2 - 3x - 4y - 2 = 0$ (b) $3x^2 - 6y^2 + 6x + 6y - 2 = 0$

    (c) $4x^2 - y^2 - 16x + 2y + 15 = 0$ (d) $x^2 + 2y^2 - 2x + 12y + 19 = 0$

11. Let a point (x, y) move so that the ratio of its distance from $(c, 0)$ to the line $x = a/e$ is e, where a, c, e are given positive numbers, $e \neq 1$, and $c = ae$. Show that the locus is an ellipse or hyperbola with center at $(0, 0)$ according as $e < 1$ or $e > 1$. Show further that every ellipse and hyperbola with center $(0, 0)$ and foci on the x-axis is obtained in this way. Show that the same curve is obtained if $(c, 0)$ is replaced by $(-c, 0)$ and the line $x = a/e$ by the line $x = -a/e$.

12. Show that a straight line in the xy-plane meets a nondegenerate conic section in at most two points. Describe conditions under which it can meet the conic section in just one point.

0-15 TRIGONOMETRY

Angles can be measured in degrees or in radians. One radian is the angle subtended at the center of a circle by an arc equal to the radius; hence, in a circle of radius 1, the arc has length 1. For a central angle of α radians the arc is $r\alpha$ (see Figure 0-24). An angle of d degrees has radian measure α, where

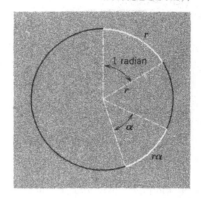

Figure 0-24   Radian measure.

$$\frac{d}{180} = \frac{\alpha}{\pi} \tag{0-150}$$

Because of its importance in calculus, radian measure is preferred in this book.

Let θ be a given real number that is considered as the radian measure of an angle. The trigonometric functions of θ are then defined by the following procedure. In the xy-plane, we construct an angle whose radian measure is θ, formed by the positive x-axis and a ray L that starts at the origin O, as in Figure 0-25. If $P:(x, y)$ is a point other than O on the ray, then we define

$$\sin \theta = \frac{y}{r}, \qquad \cos \theta = \frac{x}{r}, \qquad \tan \theta = \frac{y}{x}$$

$$\csc \theta = \frac{r}{y}, \qquad \sec \theta = \frac{r}{x}, \qquad \cot \theta = \frac{x}{y} \tag{0-151}$$

where $r = \sqrt{x^2 + y^2} > 0$. By similar triangles, these values do not depend on the choice of P on L. In particular, we can (if we like) always choose P to have distance 1 from O.

When θ is positive, the ray L can be thought of as being obtained by

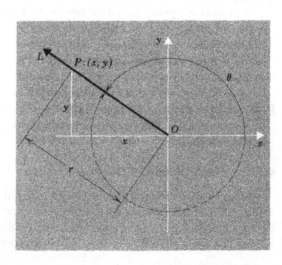

Figure 0-25   Trigonometric functions.

rotating a ray from the OX direction through θ radians in the counterclockwise direction, as suggested in Figure 0-26. Equivalently, we can locate the point $P:(x, y)$ on L by moving a distance $r\theta$ in the counterclockwise direction around the circle with center O and radius r. When θ is negative, we move in the clockwise direction a distance $r|\theta|$. Accordingly, we can end at the same

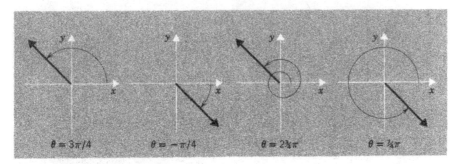

Figure 0-26 Angles in trigonometry.

point $P:(x, y)$ for many different values of θ. Since one revolution corresponds to 2π radians, if θ is one value leading to P, then

$$\theta \pm 2\pi, \qquad \theta \pm 4\pi, \qquad \theta \pm 6\pi, \qquad \theta \pm 8\pi, \ldots$$

are the other values. All these numbers lead to the same x, y and r, and thus to the same values of the trigonometric functions. Therefore, all six trigonometric functions (0-151) have the same value at θ and $\theta + 2\pi$:

$$\sin(\theta + 2\pi) = \sin \theta, \qquad \cos(\theta + 2\pi) = \cos \theta, \ldots$$

We have now given a meaning to $\sin \theta$, $\cos \theta$, $\tan \theta$, $\csc \theta$, $\sec \theta$, $\cot \theta$ for real θ, with the exception that the definitions (0-151) for $\tan \theta$, $\csc \theta$, $\sec \theta$, $\cot \theta$ break down when the denominator is 0. Thus, $\tan \pi/2$, $\cot 0$, $\csc 0$, $\sec \pi/2$ are undefined.

The essential features of the six functions are shown in the graphs of Figure 0-27. Notice that the sign of each function depends only on the quadrant in which P lies (Figure 0-25).

Identities. From the definition, we deduce a long series of identities, such as $\sin^2\theta + \cos^2\theta = 1$, $\tan \theta = \sin \theta/\cos \theta$. The more important ones are listed in Appendix V at the end of this volume.

0-16 POLAR COORDINATES

For each point $P:(x, y)$ in the xy-plane, we have an angle $\theta = \sphericalangle XOP$, determined as in Section 0-15, and a distance r from the origin. The distance r is nonnegative and is uniquely determined; however, θ is not unique but is determined up to addition or subtraction of multiples of 2π. One calls (r, θ) *polar coordinates* of (x, y). By (0-151),

$$x = r \cos \theta, \qquad y = r \sin \theta \tag{0-160}$$

so that the Cartesian coordinates (x, y) can be found from the polar coordi-

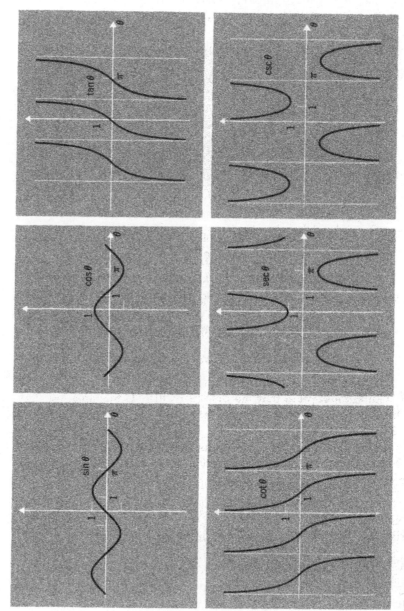

Figure 0-27 The trigonometric functions.

nates. If the Cartesian coordinates are known, r can be found, and (0-160) then gives $\cos\theta$ and $\sin\theta$ from which (by tables) θ can be found. The origin is an exception; here, $x = 0$, $y = 0$, $r = 0$, and θ can be given any value.

For a point P with polar coordinates (r, θ), we call θ the *polar angle* (or *argument*, or *amplitude*) of P, r the *polar distance* (or *modulus*) of P. The directed line OX, from which angles are measured, is called the *polar axis*; the origin O is the *pole*. In general, to introduce polar coordinates in a plane, we need only select a point O as a pole, a directed line through O as the polar axis, a positive direction for angles, and a unit of distance.

0-17 COMPLEX NUMBERS

The study of quadratic and higher degree equations leads us naturally to introduce "imaginary" numbers, such as $i = \sqrt{-1}$, and the more general complex numbers $a + bi$, where a and b are real. These numbers can be added and multiplied:

$$(a + bi) + (c + di) = (a + c) + (b + d)i$$
$$(a + bi) \cdot (c + di) = (ac - bd) + (ad + bc)i$$

These operations obey the same algebraic rules as the ones for real numbers, that is, the rules 1 to 9 of Section 0-1. Therefore, the complex numbers form a number system. The "zero," denoted by 0, is the number $0 + 0i$; the "one," denoted by 1, is the number $1 + 0i$. In general, we can identify the number $a + 0i$ with the real number a; that is, we can consider the complex numbers as a number system that *includes* all real numbers.

The following are examples of algebraic operations with complex numbers:

$$(1 + 2i)(3 - 5i) = 3 - 10i^2 + 6i - 5i = 3 - 10(-1) + i = 13 + i$$

$$\frac{1 + 4i}{2 + 3i} = \frac{1 + 4i}{2 + 3i} \cdot \frac{2 - 3i}{2 - 3i} = \frac{2 - 12i^2 + 5i}{4 - 9i^2} = \frac{14 + 5i}{13}$$

$$z^2 + 2z + 2 = 0 \quad \text{if} \quad z = \frac{-2 \pm \sqrt{-4}}{2} = \frac{-2 \pm 2\sqrt{-1}}{2} = -1 \pm i$$

We do not have an inequality ($<$ or $>$) for complex numbers. Hence, in the context of complex numbers, a "positive" number can mean only a positive real number.

Complex numbers can be represented graphically by the points in the xy-plane, the number $z = x + yi$ being represented by the point (x, y). The real numbers thus appear as the points on the x-axis, as indicated in Section 0-1; the numbers on the y-axis have form yi and are called *pure imaginary* (for $y \neq 0$). The number $x - yi$, symmetric to $z = x + yi$ in the x-axis, is called the *conjugate* of $z = x + yi$ and is denoted by $\bar{z}$ (see Figure 0-28). The distance from the origin to (x, y), $\sqrt{x^2 + y^2}$, is called the *absolute value* or *modulus* of $z = x + yi$, and is denoted by $|x + yi|$ or $|z|$. This is a natural generalization of the absolute value of real numbers, since

$$|x + 0i| = \sqrt{x^2} = |x|$$

The addition of complex numbers is in accordance with the parallelogram law of physics (Figure 0-29).

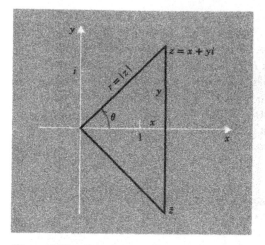

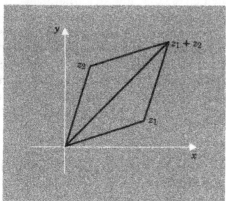

Figure 0-28 Complex numbers.

Figure 0-29 Addition of complex numbers.

We can associate with each $z = x + yi$ the polar coordinates (r, θ) of the point (x, y). Then

$$z = x + yi = r \cos \theta + ir \sin \theta = r(\cos \theta + i \sin \theta)$$

Since r is the distance from the origin to (x, y), $r = |z|$. The angle θ (in radians) is called the *argument* (or *amplitude*) of z, and we write

$$\theta = \arg z$$

For given z, θ is not uniquely determined, as is to be expected for a polar coordinate angle.

By the definition of multiplication, we verify the identity:

$$(\cos \alpha + i \sin \alpha)(\cos \beta + i \sin \beta) = \cos(\alpha + \beta) + i \sin(\alpha + \beta) \quad (0\text{-}170)$$

By the repeated application of this rule, we deduce *De Moivre's formula*:

$$(\cos \theta + i \sin \theta)^n = \cos n\theta + i \sin n\theta \quad (0\text{-}171)$$

If $z = x + yi$, we write $\bar{z} = x - iy$ as above, and

$$x = \mathrm{Re}(z), \qquad y = \mathrm{Im}(z)$$

We call x the *real part* of z, and y (not iy) the *imaginary part* of z. These quantities obey the following rules.

$$z + \bar{z} = 2 \, \mathrm{Re}(z) \qquad z - \bar{z} = 2i \, \mathrm{Im}(z)$$
$$\mathrm{Re}(z_1 + z_2) = \mathrm{Re}(z_1) + \mathrm{Re}(z_2)$$
$$\mathrm{Im}(z_1 + z_2) = \mathrm{Im}(z_1) + \mathrm{Im}(z_2)$$
$$\overline{(z_1 + z_2)} = \bar{z}_1 + \bar{z}_2 \qquad (0\text{-}172)$$
$$\overline{(z_1 z_2)} = \bar{z}_1 \cdot \bar{z}_2$$
$$z \bar{z} = |z|^2$$
$$\overline{(\bar{z})} = z$$

We notice also that, from the definition of complex numbers, $z_1 = z_2$ precisely when $\mathrm{Re}(z_1) = \mathrm{Re}(z_2)$ and $\mathrm{Im}(z_1) = \mathrm{Im}(z_2)$.

PROBLEMS

1. Convert from degrees to radians: $90°$, $360°$, $-180°$, $11°$, $(180/\pi)°$.
2. Convert from radians to degrees: $\pi/4$, $3\pi/4$, π, 3π, $-\pi$, 1, 1.7, -7.3.
3. What is the area of a circular sector of central angle α (radians) in a circle of radius r?
4. A regular polygon of n sides is inscribed in a circle of radius r.
    (a) Find its area.   (b) Find its perimeter.
5. Evaluate:
    (a) $\sin(\pi/4)$ (b) $\cos(\pi/3)$ (c) $\sin \pi$ (d) $\cos(-\pi/6)$
    (e) $\tan(7\pi/4)$ (f) $\csc(3\pi/2)$ (g) $\cot(-7\pi/4)$
6. Prove the identities:
    (a) $\sin(x + y) \cdot \sin(x - y) = \sin^2 x - \sin^2 y$ (b) $\sin 3\theta = 3 \sin \theta - 4 \sin^3 \theta$
    (c) $\cos^4 \theta = (\frac{1}{8})(3 + 4 \cos 2\theta + \cos 4\theta)$ (d) $\cot \theta = \csc 2\theta + \cot 2\theta$
7. Solve for θ: (a) $2 \sin^2 \theta + \sin \theta - 1 = 0$, (b) $\tan \theta + \cos \theta = 2$.
8. Graph the points with the given polar coordinates:

    (a) $(3, 0)$ (b) $\left(2, \dfrac{\pi}{4}\right)$ (c) $(1, 5\pi)$

    (d) $\left(2, \dfrac{-\pi}{3}\right)$ (e) $(1, -2.2)$ (f) $(1, 1)$

9. Find a set of polar coordinates for each of the following points with given Cartesian coordinates (x, y):
    (a) $(2, 2)$ (b) $(-1, 0)$ (c) $(0, -2)$ (d) $(3, -2)$
10. Find the Cartesian coordinates of the points with given polar coordinates:

    (a) $\left(1, \dfrac{\pi}{6}\right)$ (b) $(3, \pi)$ (c) $(5, 2\pi)$ (d) $\left(2, \dfrac{13\pi}{4}\right)$

11. Evaluate:
    (a) $(3 + 5i) + (2 + 7i)$ (b) $(1 - i) - (-3i)$ (c) $(1 + i)(1 - i)$

    (d) $(3 - i)^3$ (e) $\dfrac{2 + i}{1 - i}$ (f) $\dfrac{(2 + i)^3}{(1 + 2i)}$

    (g) $\left(\cos \dfrac{\pi}{4} + i \sin \dfrac{\pi}{4}\right)^8$ (h) $\cos 1 + i \sin 1$

12. Prove that $|z_1 - z_2|$ equals the distance between the points z_1, z_2:
    (a) From the geometric meaning of the addition of complex numbers (Figure 0-29).
    (b) By expressing $|z_1 - z_2|$ in terms of x_1, y_1, x_2, y_2.
13. Let $z = r(\cos \theta + i \sin \theta) \neq 0$. Show that

    (a) $\dfrac{1}{z} = \dfrac{\bar{z}}{|z|^2}$ (b) $\dfrac{1}{z} = \dfrac{1}{r}(\cos \theta - i \sin \theta)$

14. Prove the rule (0-170).
15. (a) Set $n = 2$ in (0-171) and take real and imaginary parts on both sides to prove that

$$\cos 2\theta = \cos^2\theta - \sin^2\theta, \qquad \sin 2\theta = 2 \sin \theta \cos \theta$$

   (b) Set $n = 3$ in (0-171) and take real and imaginary parts on both sides to prove that

$$\cos 3\theta = \cos^3 \theta - 3 \cos \theta \sin^2 \theta$$
$$\sin 3\theta = 3 \cos^2 \theta \sin \theta - \sin^3 \theta$$

16. (a) Prove that $z_1 \cdot z_2 = 0$ implies $z_1 = 0$ or $z_2 = 0$.
    (b) Show that, although for real numbers $x^2 + y^2 = 0$ implies $x = 0$ and $y = 0$, for complex numbers $z_1{}^2 + z_2{}^2 = 0$ does not imply $z_1 = 0$ or $z_2 = 0$.

0-18 ALGEBRAIC EQUATIONS

By an algebraic equation (or polynomial equation), we mean an equation of the form

$$a_n z^n + a_{n-1} z^{n-1} + \cdots + a_1 z + a_0 = 0 \qquad (0\text{-}180)$$

where n is a nonnegative integer, with $a_n \neq 0$, $a_0, a_1, \ldots, a_n$ are fixed complex numbers, and z is the variable or indeterminate. The equation has degree n.

The left side of Equation (0-180) is a *complex polynomial of degree n* in z. To solve Equation (0-180) is to find all zeros of this polynomial. As for real polynomials, if z_0 is a zero, then $z - z_0$ is a factor of the polynomial.

FUNDAMENTAL THEOREM OF ALGEBRA. *Every algebraic equation of degree $n \geq 1$ has at least one root.*

For a proof see page 155 of Reference 3 at the end of this chapter. From the fundamental theorem, we conclude that we can extract successive factors $z - z_1$, $z - z_2$, $\ldots$, of the left side of (0-180) until the equation is reduced to the form

$$a_n(z - z_1)(z - z_2) \cdots (z - z_n) = 0$$

This shows that there are *always exactly n complex roots of an equation of degree n*. However, the z_i need not be distinct, and hence, some of the roots may be repeated. If a root is repeated k times, we say the root has *multiplicity k*. Thus we have the rule: an equation of degree n has exactly n roots, where each root is counted as many times as its multiplicity.

When the coefficients $a_0, \ldots, a_n$ are real numbers, the roots found will include the real roots (as discussed in Section 0-11) but will usually also include some complex roots. However, when we have real coefficients, the complex roots always come in conjugate pairs; that is, if $a + bi$ is a root of multiplicity k (with $b \neq 0$), then $a - bi$ is also a root of multiplicity k.

Finding the roots of an algebraic equation is generally a time-consuming computational problem (made much easier by use of a digital computer). For an equation with real coefficients, we can seek real roots by graphing $y = a_n x^n + \cdots + a_0$ and by trying to find all x-intercepts. The calculus is of much help in this regard (see Section 6-9). Since the complex roots come in pairs, such an equation of odd degree always has at least one real root. Descartes' rule of signs and other similar rules also are helpful (see Chapter VI of Reference 3). For an equation whose coefficients are integers, it is valuable to know that every integer root is an exact divisor of the coefficient a_0, and more generally every rational root can be expressed as p/q, where p is a divisor of a_0 and q a divisor of a_n. For further information on general rules and on numerical procedures, see Reference 3 at the end of the chapter.

PROBLEMS

1. For each of the following equations two roots are given. Find the remaining roots:
    (a) $z^4 - 3z^3 + 3z^2 - 3z + 2 = 0$, $z = 1$, $z = 2$

(b) $z^4 - z^3 - z + 1 = 0$, $z = 1$, $z = 1$

(c) $z^5 + 3z^4 - 7z^3 + 13z^2 - 8z + 10 = 0$, $z = 1 + i$, $z = i$

2. Construct an algebraic equation whose roots are the given numbers:

   (a) $0, 0, 1, 1 \pm i$,   (b) $2 \pm 3i, 2 \pm 3i$.

3. Prove that if $a \neq 0$, the sum of the roots of the equation $az^2 + bz + c = 0$ is $-b/a$, and the product of the roots is c/a.

4. Prove that if $a_n \neq 0$, the sum of the roots of $a_n z^n + a_{n-1} z^{n-1} + \cdots + a_1 z + a_0 = 0$ is $-(a_{n-1}/a_n)$. (*Hint.* Write the equation in factored form.)

5. A real polynomial f is said to be *irreducible* (relative to real polynomials) if it cannot be factored as a product of two real polynomials of lower degree. Show that f is irreducible if, and only if, f is linear or $f(x) = ax^2 + bx + c$ with $b^2 - 4ac < 0$.

0-19 EXPONENTS AND LOGARITHMS

Let x be a real number. If n is a positive integer, x^n is defined as $x \cdot x \cdots x$ (n times), and x^{-n} is defined as $1/x^n$, provided that $x \neq 0$. We also define x^0 to be 1, for $x \neq 0$. For $x \neq 0$, $y \neq 0$, we have then the rules

$$x^{m+n} = x^m \cdot x^n, \qquad x^{m-n} = \frac{x^m}{x^n}$$

$$(x^m)^n = x^{mn}, \qquad x^n y^n = (xy)^n \tag{0-190}$$

for all integers m, n.

Fractional powers of x can be defined for positive x (and in certain cases for x zero or negative). For example, every positive number x has a *positive* square root that is denoted by $\sqrt{x}$ or $x^{1/2}$. A proof of this familiar fact can be given by geometry or by showing that one of the standard procedures for calculating $\sqrt{x}$ as a decimal determines a definite number whose square is x. This question and related ones about cube roots, fifth roots, and x^{α} for general real α are settled much more easily with the aid of the calculus (Chapters 3 and 5). Here we simply state some of the important results.

For each positive number x and each positive integer n, there is a unique positive real number y for which $x = y^n$; y is called the positive nth root of x and is denoted by $\sqrt[n]{x}$ or $x^{1/n}$. For every integer m, positive or negative, we then define

$$x^{m/n} = (x^{1/n})^m$$

and can verify that also

$$x^{m/n} = (x^m)^{1/n}$$

and that $x^{m/n} = x^{p/q}$, if p and q are integers such that $m/n = p/q$.

Furthermore, if α is any real number, we can define x to the power α or x^{α} (x always positive). For example, we can calculate $x^{\sqrt{2}}$ for any given x, to any desired accuracy, by replacing $\sqrt{2}$ by enough digits of its decimal expansion. For instance, $5^{\sqrt{2}}$ equals $5^{1.41}$ approximately and, with better accuracy, $5^{\sqrt{2}}$ equals $5^{1.414}$, and so on. Since $1.41 = \frac{141}{100}$, a rational number, $5^{1.41}$ is a case of $5^{m/n}$, as discussed above; the same applies to $5^{1.414}$. Thus, $x^{\sqrt{2}}$ is found by approximation by $x^{m/n}$ for certain rational numbers m/n.

The result is a definition of x^{α} for every positive x and every real number α. Furthermore $x^{\alpha} > 0$ always, and we have the rules that parallel (0-190):

$$x^{\alpha+\beta} = x^{\alpha}x^{\beta}, \qquad x^{\alpha-\beta} = \frac{x^{\alpha}}{x^{\beta}}$$

$$(x^{\alpha})^{\beta} = x^{\alpha\beta}, \qquad (xy)^{\alpha} = x^{\alpha}y^{\alpha} \tag{0-191}$$

For fixed α, $y = x^{\alpha}$ then defines a function of x (the α-power function). This is graphed for several cases in Figure 0-30.

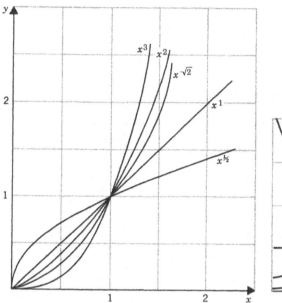

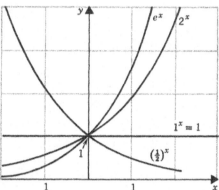

Figure 0-30   The function x^{α}.   **Figure 0-31**   The function a^{x}.

We can also fix a positive number a, and can consider the function defined by

$$y = a^{x}$$

for all real x. This function is called the *exponential function of base a*. Its graph is shown, for several values of a, in Figure 0-31. For the calculus, one case is of crucial importance: the one when the base is

$$e = 2.71828\ 18285\ldots$$

In the calculus e^{x} is referred to as *the* exponential function.

Logarithms. We have observed that for *fixed a*, a^{x} is always a positive number. As Figure 0-31 suggests, except for the case $a = 1$, we can choose x uniquely to make $a^{x} = y$ for any given positive y. This value of x is called the logarithm of y to the base a, and we write

$$x = \log_{a} y$$

Thus, by definition,

$$a^{\log_{a}y} = y \qquad \text{and} \qquad \log_{a} a^{x} = x, \qquad a > 0,\ a \neq 1 \tag{0-192}$$

The equation $x = \log_{a} y$ defines x as a function of y, $y > 0$.

We now interchange x and y and write our *logarithmic function* as

$$y = \log_a x, \qquad x > 0$$

Here a is always a fixed positive number, not 1.

From the properties of the exponential function, we deduce properties of the logarithmic function:

$$\log_a(x_1 x_2) = \log_a x_1 + \log_a x_2 \qquad \log_a \frac{x_1}{x_2} = \log_a x_1 - \log_a x_2$$

$$\log_a x^c = c \log_a x \qquad \frac{\log_a x}{\log_b x} = \log_a b \tag{0-193}$$

Here, x_1, x_2, x are positive and a and b are positive and different from 1.

When a is chosen as 10, we obtain the function $\log_{10} x$, which is the *common logarithm* of x, used in numerical work. When $a = e$, we obtain the *logarithmic function* of the calculus, which is denoted simply by $\log x$ in older literature and, at present, generally by $\ln x$; this function is also called the *natural logarithm* or *Napierian logarithm* of x (after the inventor, John Napier, 1550–1617).

PROBLEMS

1. Simplify:

(a) $\left(\dfrac{x^3 y^2}{x^{-1} y^{-5}} \right)^2$ (b) $(x^{-2})^{-2}$

(c) $\sqrt[3]{x^6 y^{27}}$ (d) $\dfrac{x^{3/2} + 2x^{1/2} + x^{-1/2}}{x + 1}$

2. Simplify:
   (a) $\log_{10} 10^{1.72}$ (b) $10^{\log_{10} \pi}$ (c) $e^{\log_e x}$
   (d) $\ln x + \ln 2x - 3 \ln x^2$

3. Under what conditions can the definition of $x^{m/n}$ be extended to x zero or negative?

0-20 INDUCTION

To justify many formulas and theorems, we need the procedure of *mathematical induction*, which we describe here. The validity of the process is generally accepted as an axiom fundamental to all mathematics.

Principle of induction. *The truth of a proposition relating to all positive integers n (1, 2, 3, 4, . . .) can be concluded from*

(a) *truth of the proposition for $n = 1$;*
(b) *a demonstration that truth of the proposition for an arbitrary n implies its truth for $n + 1$.*

This principle can be justified intuitively by reasoning as follows. By (a) we know the proposition is true for $n = 1$. By (b) it must thus be true for $n = 2$ also. By (b) again it must be true for $n = 3$, and so on. The crucial words are the last three: "and so on." We must be sure we really get all cases in this way;

that is, that going from 1 to $1 + 1 = 2$, from 2 to $2 + 1 = 3$, and so on, describes a definite process that sweeps out all positive integers. The principle of induction is a precise way of stating that every positive integer is a finite sum of 1's.

EXAMPLE 1. *Sum of an arithmetic progression.* The proposition to be proved is

$$a + (a + d) + (a + 2d) + \cdots + [a + (n - 1)d] = \frac{n[2a + (n - 1)d]}{2}$$

for $n = 1, 2, \ldots$.

(a) For $n = 1$ the assertion is $a = a$, which is true.
(b) For $n + 1$ the assertion is

$$a + (a + d) + \cdots + [a + (n - 1)d] + [a + nd] = \frac{(n + 1)(2a + nd)}{2}$$

If the assertion is true for n, then the left side is equal to

$$\frac{n[2a + (n - 1)d]}{2} + a + nd$$

which simplifies to

$$\frac{(n + 1)(2a + nd)}{2}$$

which is the same as the right side.

Hence the proposition is true for all positive integers n.

EXAMPLE 2. Prove that every complex polynomial of degree $n \geq 1$ can be factored in the form

$$a_n(z - z_1) \cdots (z - z_n)$$

(see Section 0-18 above).

(a) For $n = 1$, the assertion is $a_1 z + a_0 = a_1(z - z_1)$ for appropriate z_1, provided that $a_1 \neq 0$. This is true, since

$$a_1 z + a_0 = a_1\left(z + \frac{a_0}{a_1}\right) = a_1(z - z_1), \quad \text{with } z_1 = -\frac{a_0}{a_1}$$

(b) Let the proposition be assumed true for a particular n and let $a_{n+1}z^{n+1} + \cdots + a_0$ be a polynomial of degree $n + 1$ $(a_{n+1} \neq 0)$. By the fundamental theorem of algebra, this polynomial has a zero, which we call z_{n+1}. By the rule on factoring, the polynomial has $z - z_{n+1}$ as a factor. Therefore

$$a_{n+1}z^{n+1} + \cdots + a_0 = (a_{n+1}z^n + b_n z^n + \cdots + b_0)(z - z_{n+1})$$

where the degree and the first term of the second factor are determined by the way polynomials are multiplied. By the induction hypothesis, the first factor can be replaced by $a_{n+1}(z - z_1)(z - z_2)(\cdots)(z - z_n)$. Thus

$$a_{n+1}z^{n+1} + \cdots a_0 = a_{n+1}(z - z_1) \cdots (z - z_n)(z - z_{n+1})$$

Therefore, the proposition is proved true for degree $n + 1$. Hence, by induction, the proposition is true for all n.

Inductive Definition. Many of the algebraic expressions for general cases depend on an arbitrary integer n. We tend to jump to the general case without a detailed explanation. A more careful procedure is to use inductive (or recursive) definition: the expression is defined for $n = 1$; then a procedure is given for deriving the expression for $n + 1$ from the expression for n.

EXAMPLE 3. *Definition of* x^n ($n = 1, 2, 3, \ldots$). We write in Section 0-19, above, that $x^n = x \cdot x \cdots x$ (n times). A clearer procedure is as follows:

(a) **Definition.** $x^1 = x$.
(b) **Definition.** $x^{n+1} = x^n \cdot x$ for $n \geq 1$.

EXAMPLE 4. *Definition of* $n!$ ($n = 0, 1, 2, 3, \ldots$).

(a) **Definition.** $0! = 1$, $1! = 1$.
(b) **Definition.** $(n + 1)! = (n + 1)n!$ for $n \geq 1$.

We can describe this process in general terms and prove, by the principle of induction, that it does define our expression for all positive integers n (see Reference 7, pp. 28, 42–45).

EXAMPLE 5. Another example is the "sum of n terms" as in Examples 1 and 2. In general, we consider

$$f(1) + f(2) + \cdots + f(n)$$

where $f(n)$ is a given function whose domain is the set of all positive integers. The sum itself is a new function, say $g(n)$. Then we define

$$g(1) = f(1)$$
$$g(n + 1) = g(n) + f(n + 1)$$

This defines $g(n) = f(1) + f(2) + \cdots + f(n)$ for all positive integers n. It is common practice to write $\sum_{k=1}^{n} f(k)$ for this sum.

0-21   THE BINOMIAL THEOREM, PERMUTATIONS AND COMBINATIONS

The binomial theorem is one more proposition that can be proved by induction. The proposition is

$$(a + b)^n = a^n + \binom{n}{1} a^{n-1}b + \binom{n}{2} + \cdots + \binom{n}{k} a^{n-k}b^k$$

$$+ \cdots + \binom{n}{n-1} ab^{n-1} + b^n, \qquad n = 1, 2, 3, \ldots$$

or with the Σ notation mentioned above

$$(a + b)^n = \sum_{k=0}^{n} \binom{n}{k} a^{n-k}b^k$$

where

$$\binom{n}{k} = \frac{n!}{k!(n-k)!}, \qquad k = 0, 1, 2, \ldots, n$$

We call the numbers $\binom{n}{k}$ the *binomial coefficients*.

Examples of binomial expansions are:

$$(a + b)^1 = a + b$$
$$(a + b)^2 = a^2 + 2ab + b^2$$
$$(a + b)^3 = a^3 + 3a^2b + 3ab^2 + b^3$$
$$(a + b)^4 = a^4 + 4a^3b + 6a^2b^2 + 4ab^2 + b^3$$
$$(a + b)^5 = a^5 + 5a^4b + 10a^3b^2 + 10a^2b^3 + 5ab^4 + b^5$$

Relation to Permutations and Combinations. The number of *permutations* (different orders of arrangement) of n different objects is $n!$. Thus, 3 objects a, b, c have the $3! = 6$ permutations

$$abc, \qquad acb, \qquad bca, \qquad bac, \qquad cab, \qquad cba$$

More generally, the number of permutations of all groups of k different objects taken from n different objects $(n \geq k)$ is

$$n(n - 1)(n - 2) \cdots (n - k + 1) = \frac{n!}{(n - k)!}$$

Thus, the permutations of two letters from the 3 letters a, b, c are

$$ab, \qquad ac, \qquad ba, \qquad bc, \qquad ca, \qquad cb$$

and $3!/1! = 6$. We can justify the general rule by reasoning that there are n choices for the first position; for each such choice there are $n - 1$ choices for the second position, and so on (by induction); thus, in all $n(n - 1)(n - 2) \cdots (n - k + 1)$ permutations are possible.

If we disregard the order in each permutation, then we are counting *combinations* of k distinct objects taken from n. Since each combination gives rise to $k!$ permutations, there are

$$\frac{n(n - 1) \cdots (n - k + 1)}{k!} = \frac{n!}{k!(n - k)!} = \binom{n}{k}$$

combinations in all. This can be justified also by reasoning that the coefficient of x^k in the expansion of

$$(1 + x)^n = (1 + x)(1 + x) \cdots (1 + x) \qquad (n \text{ factors})$$

is simply the number of ways of choosing k positions out of n; but, by the binomial theorem,

$$(1 + x)^n = \sum_{k=0}^{n} \binom{n}{k} x^k$$

Hence $\binom{n}{k}$ is the number of combinations of n objects taken k at a time.

PROBLEMS

1. Prove De Moivre's formula (Section 0-17) by induction.
2. Prove the rule for the sum of a geometric progression:

$$a + ar + \cdots + ar^{n-1} = \frac{a(1 - r^n)}{1 - r}, \qquad r \neq 1, n = 1, 2, \ldots$$

3. Prove that

$$1^2 + 2^2 + \cdots + n^2 = \frac{n(n + 1)(2n + 1)}{6}$$

4. Prove that

$$1^3 + 2^3 + \cdots + n^3 = \frac{n^2(n + 1)^2}{4}$$

5. Prove that

$$(\tfrac{1}{2}) + \cos \theta + \cos 2\theta + \cdots + \cos n\theta = \frac{\sin(n + \tfrac{1}{2})\theta}{2 \sin \tfrac{1}{2}\theta}$$

for $\theta \neq 0, \pm 2\pi, \pm 4\pi, \ldots$.

6. Prove that $a^{n+1} - b^{n+1} = (a - b)(a^n + a^{n-1}b + \cdots + ab^{n-1} + b^n), n = 1, 2, 3, \ldots$.
7. Prove that

(a) $\dbinom{n}{k} = \dbinom{n}{n - k}$ (b) $\displaystyle\sum_{k=0}^{n} (-1)^k \dbinom{n}{k} = 0,$ (c) $\dbinom{n}{k} + \dbinom{n}{k - 1} = \dbinom{n + 1}{k}.$

8. How many different bridge hands are there?

REFERENCES

1. Allendoerfer, C. B., and Oakley, C. O., *Principles of Mathematics*, 2nd ed. New York: McGraw-Hill, 1963.
2. Brumfiel, C. R., Eicholz, R. E., Fleenor, C. R., and Shanks, M. E., *Pre-Calculus Mathematics*. Reading, Mass.: Addison-Wesley, 1965.
3. Dickson, L. E., *First Course in the Theory of Equations*. New York: Wiley, 1922.
4. Dolciani, M. P., Beckenbach, E. F., Donnelly, A. J., Jurgenson, R. L., and Wooton, W. *Modern Introductory Analysis*. Boston: Houghton-Mifflin, 1967.
5. Keedy, M. L., Jameson, R. E., Smith, Stanley A., Mould, Eugene H., *Exploring Geometry*. New York: Holt, Rinehart and Winston, 1967.
6. Lehmann, C. H., *Analytic Geometry*. New York: Wiley, 1942.
7. McCoy, N. H., *Introduction to Modern Algebra*. Boston: Allyn and Bacon, 1965.
8. Vance, E. P., *Trigonometry*. Reading Mass.: Addison-Wesley, 1954.
9. Wilder, R. L., *Introduction to the Foundations of Mathematics*, 2nd ed. New York: Wiley, 1965.

1
TWO-DIMENSIONAL
VECTOR GEOMETRY

1-1 INTRODUCTION

Geometry was developed more than 2000 years ago by the Greeks. They developed a systematic way of analyzing the properties of points, lines, triangles, circles, and other configurations. Their work was beautifully synthesized in Euclid's *Elements*, which has formed the basis of plane and solid geometry even to the present day. In recent times, other sets of axioms and postulates have been introduced, and the logical structure has been improved; but the essential subject matter remains the same.

By contrast, algebra as a logical structure has been developed only in modern times, mainly during the last 100 years. The formulation is surprisingly simple: a very few axioms suffice to organize all the procedures of ordinary algebra.

Furthermore, it has been discovered that essentially all of geometry can be reformulated in algebraic language. Instead of combining points and lines in the usual geometric fashion, we perform algebraic operations on certain objects called *vectors*. The vectors obey certain algebraic laws similar to the ones that govern numbers; for example, if **a** and **b** are vectors, then **a** + **b** = **b** + **a**. In fact, the study of vectors can be developed systematically from a few axioms such as the ones for numbers. The theorems of geometry become theorems of the algebra of vectors, with the emphasis on equations, identities, and inequalities, rather than on geometric concepts such as congruence, similarity, and the intersection of lines.

Here we develop the algebra of vectors for plane geometry. The discussion is informal, and is intended to convey an intuitive feeling for the subject. However, at several points, we also indicate how the theory can be built up in strictly logical fashion.

Later, we shall learn that this vector algebra, or *linear algebra* (the name indicating its geometric origin), is a valuable tool for the calculus. It is in fact the foundation for a large part of modern mathematics.

1-2 DIRECTED LINE SEGMENTS AND VECTORS

We consider a fixed plane, in which all geometric objects will lie. We also assume a unit of distance has been chosen.

By a *directed line segment* in the plane, we mean a line segment with a direction chosen on it. We denote by AB the segment that joins point A to point B, with direction from A to B; we denote by BA the segment with direction from B to A. In a figure, we can show the direction by an arrowhead, as in Figure 1-1. We call A the *initial point* and B the *terminal point* of the directed line segment AB.

Occasionally, we also write AB for an undirected line segment, as in geometry. The context will always make the meaning clear.

The velocity of a ship crossing a lake can be indicated by a directed line segment PQ, as shown in Figure 1-2. Here the *length* of PQ is chosen to equal

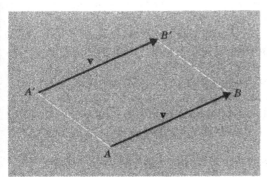

Figure 1-1 Directed line segments and vectors.

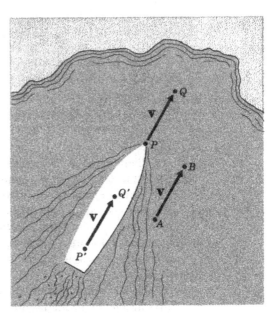

Figure 1-2 Velocity vector of a ship.

the speed of the ship, in appropriate units (for instance, a length of 1 inch corresponding to 10 mph), and the *direction* of PQ indicates the direction of the ship's motion.

Now, we could equally well represent the motion of the ship by another directed line segment $P'Q'$, as in Figure 1-2. Here the length of PQ equals that of $P'Q'$, and both point in the same direction. In fact, the desired information is conveyed by any directed line segment AB, as in Figure 1-2, where AB has the same length as PQ, is parallel to PQ, and has the same direction. In other words, we need to know only the length and the direction, but not the actual position of the directed line segment. We simply say that the velocity of the ship is a *vector*.

In general, a vector in the plane is a combination of a *length* (nonnegative real number) and a *direction*. We denote vectors by boldface letters: **v**, **w**, **r**, **a**, ... (In writing, it is common to denote a vector by $\vec{v}$ or $\underline{v}$.) Each vector **v** can be represented by a directed line segment AB having the given length and direction, but **v** is equally well represented by any other line segment that has the same length and direction. When $\mathbf{v}$ is represented by AB, we write

$$\mathbf{v} = \overrightarrow{AB}$$

Thus, $\overrightarrow{AB}$ denotes the vector determined by the directed line segment AB. For the case of Figure 1-2, we would write

$$\mathbf{v} = \overrightarrow{AB} = \overrightarrow{PQ} = \overrightarrow{P'Q'}$$

as suggested in the figure.

We can think of a vector as a directed line segment that is free to be moved about in the plane, as long as the length and direction are unaffected. A physical example is the needle of a compass being moved about; the positions change, but the *vector* represented by the needle is always the same.

Notice in particular that, given a vector $\mathbf{v}$, represented, for instance, by AB, and given an initial point A', we can always represent $\mathbf{v}$ uniquely by a directed line segment $A'B'$. For, as in Figure 1-1, we simply choose B' so that AB and $A'B'$ are the opposite sides (similarly directed) of a parallelogram; the point B' is uniquely determined by this condition. By the same reasoning, $\mathbf{v}$ can be represented uniquely by a directed line segment with a given terminal point.

In this chapter we often choose a fixed origin O and then represent each vector $\mathbf{u}$ as $\overrightarrow{OP}$; that is, $\mathbf{u}$ is represented by the directed line segment from O to P. Thus, to each vector $\mathbf{u}$ we assign a unique point P, and each point P is assigned to a unique vector $\mathbf{u}$. In the language of Section 0-10, we obtain a *one-to-one correspondence between vectors and points in the plane*. This is illustrated in Figure 1-3, in which $\mathbf{u}$ corresponds to P, $\mathbf{v}$ to Q, and $\mathbf{w}$ to S. If we choose a new origin, we obtain a different correspondence between vectors and points.

There are many examples of vectors in physics: velocity, acceleration, force, and momentum. For example, a force has magnitude and direction and, therefore, can be represented by a directed line segment. Another illustration is the concept of a *displacement:* If we make a trip and at its end are 50 miles north of our starting point, we say that we have been displaced "50 miles to the north." Since only distance and direction are being considered, we again have a vector $\mathbf{u}$. We can represent $\mathbf{u}$ by the directed line segment AB from the initial point A to the destination B.

Terminology for Vectors. We denote the length of the vector $\mathbf{v}$ (in the given units) by $|\mathbf{v}|$. The use of the absolute value symbol is justified by the fact that $|\mathbf{v}|$ obeys rules like the ones for absolute values (and, in fact, $|\mathbf{v}|$ is a generalization of absolute value). We also call $|\mathbf{v}|$ the *magnitude* of $\mathbf{v}$, or the *norm* of $\mathbf{v}$.

If $|\mathbf{v}| = 1$, so that $\mathbf{v}$ has length 1, $\mathbf{v}$ is called a *unit vector*.

For convenience, we introduce a zero vector $\mathbf{0}$. This corresponds to a directed line segment from a point to itself, that is,

$$\mathbf{0} = \overrightarrow{PP} = \overrightarrow{QQ} = \overrightarrow{SS} = \ldots$$

We assign the length 0 to this vector and allow it to have every direction. We then have the rule:

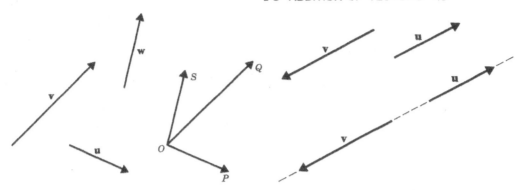

Figure 1-3 Correspondence between vectors and points.

Figure 1-4 Parallel or collinear vectors.

For every vector v, $|v| \geq 0$; and $|v| = 0$, if and only if, $v = 0$.

In general, two vectors **u**, **v** are *equal* precisely when they have the same length and direction; that is, **u** = **v** precisely when **u** and **v** are representable by the same directed line segments.

Two vectors **u**, **v** are said to be *parallel* or *collinear* if they can be represented by directed line segments on the same line (Figure 1-4). Thus, **u**, **v** are parallel when they have the same or opposite direction. In particular, **0**, **v** are parallel for every **v**.

In vector theory, numbers are usually referred to as *scalars*. Thus, 3, 7.8, and $- \sqrt{2}$ are scalars.

1-3 ADDITION OF VECTORS

Given vectors **u** and **v**, we select an initial point A; next we choose B so that $u = \overrightarrow{AB}$; then we choose C so that $v = \overrightarrow{BC}$ (Figure 1-5). The vector $w = \overrightarrow{AC}$ is then termed the *sum* of **u** and **v**, and we write $w = u + v$. Thus, by our definition,

$$\overrightarrow{AB} + \overrightarrow{BC} = \overrightarrow{AC}$$

If we change the initial point to A' and represent **u** by $A'B'$, **v** by $B'C'$, then the sum **w** is represented by $A'C'$. As suggested in Figure 1-5, AC and $A'C'$ have the same length and direction, so that the same vector **w** is obtained regardless of how we choose the initial point.

The addition of vectors is in accordance with the *parallelogram law* for

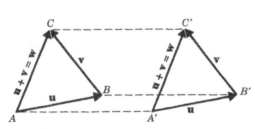

Figure 1-5 Addition of vectors.

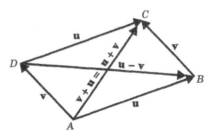

Figure 1-6 Addition and subtraction of vectors.

combining forces, as suggested in Figure 1-6. The definition of addition is also consistent with the way we combine displacements: a displacement that takes us from A to B which is followed by one that takes us from B to C is equivalent to a single displacement that takes us from A to C.

The addition of vectors satisfies the following laws:

$$\mathbf{u} + \mathbf{v} = \mathbf{v} + \mathbf{u} \qquad \text{(commutative law)} \qquad (1\text{-}30)$$
$$(\mathbf{u} + \mathbf{v}) + \mathbf{w} = \mathbf{u} + (\mathbf{v} + \mathbf{w}) \qquad \text{(associative law)} \qquad (1\text{-}31)$$
$$\mathbf{u} + \mathbf{0} = \mathbf{u} = \mathbf{0} + \mathbf{u} \qquad\qquad\qquad\qquad\qquad (1\text{-}32)$$
$$|\mathbf{u} + \mathbf{v}| \le |\mathbf{u}| + |\mathbf{v}| \qquad \text{(triangle inequality)} \qquad (1\text{-}33)$$

These rules are familiar rules of the ordinary algebra of numbers (scalars). We cannot assume that they are true for vectors, since vectors are not numbers; they are new objects.

The proof of rule (1-30) is suggested in Figure 1-6; $\mathbf{u}$ and $\mathbf{v}$ are each represented by a pair of opposite sides of the parallelogram, properly directed, and each way of adding them gives the diagonal vector $\overrightarrow{AC}$. (When $\mathbf{u}$, $\mathbf{v}$ are collinear, the parallelogram collapses and a new proof is needed; see Problem 7 below.)

For (1-31), we write $\mathbf{u} = \overrightarrow{AB}$, $\mathbf{v} = \overrightarrow{BC}$, $\mathbf{w} = \overrightarrow{CD}$. Then $(\mathbf{u} + \mathbf{v}) + \mathbf{w} = (\overrightarrow{AB} + \overrightarrow{BC}) + \overrightarrow{CD} = \overrightarrow{AC} + \overrightarrow{CD} = \overrightarrow{AD}$ and, similarly, $\mathbf{u} + (\mathbf{v} + \mathbf{w}) = \overrightarrow{AD}$ (see Figure 1-7).

For (1-32), we write $\mathbf{u} = \overrightarrow{AB}$, $\mathbf{0} = \overrightarrow{BB}$, so that $\mathbf{u} + \mathbf{0} = \overrightarrow{AB} + \overrightarrow{BB} = \overrightarrow{AB} = \mathbf{u}$. By (1-30), $\mathbf{0} + \mathbf{u} = \mathbf{u} + \mathbf{0}$.

The rule (1-33) simply expresses the fact that each side of a triangle is less than the sum of the other two sides (Figure 1-6). The triangle collapses when the vectors are collinear, and a separate proof is needed for that case (Problems 6 and 7 below). The equals sign holds true only when $\mathbf{u}$ and $\mathbf{v}$ have the same direction (in particular, when one is $\mathbf{0}$).

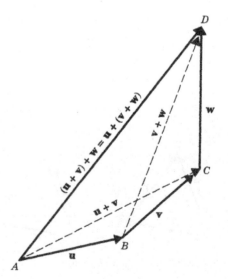

Figure 1-7   Proof of associative law.

1-4 SUBTRACTION OF VECTORS

To subtract **u** from **w** is to find a vector **v** which, when added to **u**, gives **w**. If we set $\mathbf{u} = \vec{AB}$, $\mathbf{w} = \vec{AC}$, then $\mathbf{v} = \vec{BC}$ is such a vector and is, clearly, the only one (Figure 1-5). We write $\mathbf{v} = \mathbf{w} - \mathbf{u}$. Thus, by our result,

$$\vec{AC} - \vec{AB} = \vec{BC} \tag{1-40}$$

In particular, if $\mathbf{u} = \vec{AB}$, then

$$\mathbf{0} - \mathbf{u} = \vec{AA} - \vec{AB} = \vec{BA}$$

The vector $\vec{BA}$ is obtained from $\mathbf{u} = \vec{AB}$ by reversing its direction (Figure 1-8). We denote this vector by $-\mathbf{u}$. Thus

$$\mathbf{0} - \mathbf{u} = -\mathbf{u} \tag{1-41}$$

Equation (1-40) can now be written

$$\vec{AC} - \vec{AB} = \vec{BA} + \vec{AC} = -\vec{AB} + \vec{AC} = \vec{AC} + [-\vec{AB}]$$

that is,

$$\mathbf{w} - \mathbf{u} = \mathbf{w} + (-\mathbf{u}) \tag{1-42}$$

In the parallelogram of Figure 1-6, notice that one diagonal is $\mathbf{u} + \mathbf{v}$, and that the other (directed as shown) is $\mathbf{u} - \mathbf{v}$.

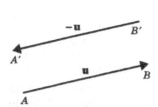

Figure 1-8   The vector $-\mathbf{u}$.

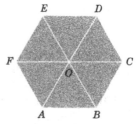

Figure 1-9

PROBLEMS

1. A car travels 100 miles east, then 200 miles north, then 100 miles east. Determine graphically the total displacement of the car.
2. Let A, B, P, Q be four distinct points on a line. List all possible orders in which the points can lie if AB and PQ have the same direction.
3. In Figure 1-9, $ABCDEF$ is a regular hexagon with center O. Group the following in sets of equal vectors:
   (a) $\vec{AB}$, $\vec{DE}$, $\vec{AE}$, $\vec{OE}$, $\vec{OF}$, $\vec{AC}$, $\vec{BC}$, $\vec{AF}$, $\vec{CD}$, $\vec{AD}$, $\vec{FD}$, $\vec{BO}$, $\vec{OC}$
   (b) $\vec{AB} + \vec{OE}$, $\vec{AF} + \vec{BC}$, $\vec{AO} + \vec{CD}$, $\vec{ED} + \vec{AF}$
   (c) $\vec{AD} - \vec{AF}$, $\vec{FE} - \vec{BA}$, $\vec{AO} - \vec{OF}$, $\vec{FD} - \vec{DB}$
4. Draw 4 nonzero vectors **u**, **v**, **w**, **r**, so that $\mathbf{u} + \mathbf{v} + \mathbf{w} + \mathbf{r} = \mathbf{0}$. What general rule does this suggest?

5. (a) Show that, if A, B, C are vertexes of a triangle, then $|\overrightarrow{AC}| < |\overrightarrow{AB}| + |\overrightarrow{BC}|$.

   (b) Show that, if $\mathbf{u}$, $\mathbf{v}$ are noncollinear vectors, then $|\mathbf{u} + \mathbf{v}| < |\mathbf{u}| + |\mathbf{v}|$. [See part (a).]

6. (a) Show that if $\mathbf{u} = \mathbf{0}$, then $|\mathbf{u} + \mathbf{v}| = |\mathbf{u}| + |\mathbf{v}|$.

   (b) Show that if $\mathbf{u} = \mathbf{0}$, then $\mathbf{u} + \mathbf{v} = \mathbf{v} + \mathbf{u}$.

7. Let $\mathbf{u}$, $\mathbf{v}$ be nonzero collinear vectors.

   (a) Prove that, if $\mathbf{u}$, $\mathbf{v}$ have the same direction, then $\mathbf{u} + \mathbf{v}$ has the same direction as both, and has length $|\mathbf{u}| + |\mathbf{v}|$; if $\mathbf{u}$, $\mathbf{v}$ have opposite directions, then $\mathbf{u} + \mathbf{v} = \mathbf{0}$ when $|\mathbf{u}| = |\mathbf{v}|$, and otherwise $\mathbf{u} + \mathbf{v}$ has the direction of the longer vector and has length $|\,|\mathbf{u}| - |\mathbf{v}|\,|$.

   (b) From the results of part (a), show that $\mathbf{u} + \mathbf{v} = \mathbf{v} + \mathbf{u}$.

   (c) From the results of part (a), show that $|\mathbf{u} + \mathbf{v}| = |\mathbf{u}| + |\mathbf{v}|$ when $\mathbf{u}$, $\mathbf{v}$ have the same direction, and that

$$|\mathbf{u} + \mathbf{v}| = |\,|\mathbf{u}| - |\mathbf{v}|\,| < |\mathbf{u}| + |\mathbf{v}|$$

when $\mathbf{u}$, $\mathbf{v}$ have opposite directions.

8. (a) Show that for all $\mathbf{u}$, $\mathbf{v}$ in the plane

$$|\mathbf{u} + \mathbf{v}| \geq |\,|\mathbf{u}| - |\mathbf{v}|\,|$$

   (b) When does equality occur in the rule of part (a)?

1-5 MULTIPLICATION OF VECTORS BY SCALARS

We write

$$\mathbf{u} + \mathbf{u} = 2\mathbf{u}, \qquad \mathbf{u} + \mathbf{u} + \mathbf{u} = 3\mathbf{u}$$

and so on. We notice that $2\mathbf{u}$ is then a vector twice as long as $\mathbf{u}$ and has the same direction as $\mathbf{u}$, and that $3\mathbf{u}$ is a vector 3 times as long as $\mathbf{u}$ and with the same direction (Figure 1-10). This suggests defining $k\mathbf{u}$ for each *positive* number (scalar) k as a vector that has the length $k|\mathbf{u}|$ and the same direction as $\mathbf{u}$. For positive k, we can interpret $(-k)\mathbf{u}$ as $k(-\mathbf{u})$, and hence we obtain a vector k times as long as $\mathbf{u}$ and having the *opposite* direction. We interpret $0\mathbf{u}$ to be $\mathbf{0}$. We are thus led to a general definition.

Definition of scalar times vector. If a is a real number (scalar) and $\mathbf{u}$ is a vector, then $a\mathbf{u}$ is $\mathbf{0}$ for $a = 0$ or $\mathbf{u} = \mathbf{0}$ and otherwise $a\mathbf{u}$ is a vector whose magnitude is $|a|\,|\mathbf{u}|$, having the same direction as $\mathbf{u}$ or the opposite direction, according to whether $a > 0$ or $a < 0$.

From this definition it follows that $\mathbf{u}$ and $a\mathbf{u}$ can always be represented by directed line segments on the same line; that is, $\mathbf{u}$ and $a\mathbf{u}$ are always collinear vectors. In general, if $\mathbf{u}$, $\mathbf{v}$ are collinear and $\mathbf{u} \neq \mathbf{0}$, then $\mathbf{u}$, $\mathbf{v}$ can be represented by directed line segments on the same line, so that $\mathbf{v} = a\mathbf{u}$ for some a. Here a is positive or negative according to whether these segments have the same or the opposite direction (with $a = 0$ for $\mathbf{v} = \mathbf{0}$); since $|\mathbf{v}| = |a|\,|\mathbf{u}|$, we have $|a| = |\mathbf{v}|/|\mathbf{u}|$.

The multiplication of vectors by scalars obeys these rules:

$$1\mathbf{u} = \mathbf{u}. \tag{1-50}$$

$$0\mathbf{u} = \mathbf{0} \qquad \text{and} \qquad a\mathbf{0} = \mathbf{0}. \tag{1-51}$$

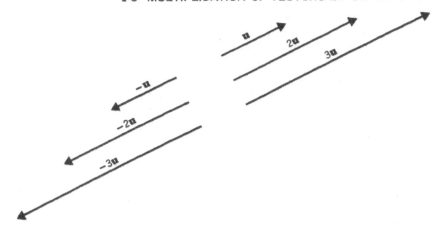

Figure 1-10 Scalar times vector.

If $a\mathbf{u} = \mathbf{0}$,   then $a = 0$ or $\mathbf{u} = \mathbf{0}$.   (1-52)

$(ab)\mathbf{u} = a(b\mathbf{u})$.   (1-53)

$(-a)\mathbf{u} = -(a\mathbf{u}) = a(-\mathbf{u})$.   (1-54)

$|a\mathbf{u}| = |a|\,|\mathbf{u}|$.   (1-55)

$(a + b)\mathbf{u} = a\mathbf{u} + b\mathbf{u}$.   (1-56)

$a(\mathbf{u} + \mathbf{v}) = a\mathbf{u} + a\mathbf{v}$.   (1-57)

$(a - b)\mathbf{u} = a\mathbf{u} - b\mathbf{u}$.   (1-58)

$a(\mathbf{u} - \mathbf{v}) = a\mathbf{u} - a\mathbf{v}$.   (1-59)

The rules (1-50)–(1-55) follow quite directly from the definition (see Problem 9 below). To prove rule (1-56) for $\mathbf{u} \neq \mathbf{0}$, we can relate multiplication of $\mathbf{u}$ by scalars to a number axis, as suggested in Figure 1-11. We represent $\mathbf{u}$ by a directed line segment on a line L, and introduce a scale along L so that the

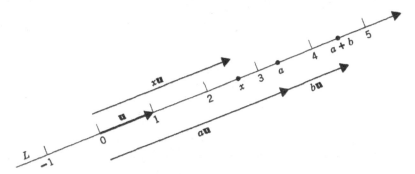

Figure 1-11 Scalar times vector and number axis.

initial point of the segment is at 0, and the terminal point at 1; thus, $|\mathbf{u}|$ is taken as the unit of length. Every point on L now has a coordinate x, so that L becomes an x-axis; we can refer to the points on L by their coordinates: the point 0, the point 3, and so on. The directed line segment from 0 to x then represents the vector $x\mathbf{u}$. The vector $a\mathbf{u} + b\mathbf{u}$ can now be obtained by first representing $a\mathbf{u}$ as the directed line segment from 0 to a and then moving b units from a, in the positive or negative direction along L, according as b is positive or negative; thus $au + bu$ is represented by the directed line segment from 0 to $a + b$ and, therefore, $au + bu = (a + b)u$. (In particular, $bu + au = (b + a)u = (a + b)u = au + bu$; this gives another proof of the commutative law of addition for collinear vectors; see Problem 7 above.) We can say that, in terms of the number axis of Figure 1-11, the addition of scalar multiples of u corresponds to the ordinary addition of numbers.

If $u = 0$, the number axis cannot be used and, in this case, (1-56) follows from (1-51).

To prove (1-57), we assume first that $a > 0$ and that u, v are noncollinear, and represent u as $\overrightarrow{AB}$, and v as $\overrightarrow{BC}$, as in Figure 1-12. We then extend the

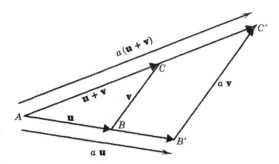

Figure 1-12   Proof of
$a(u + v) = au + av$.

sides of the triangle ABC to B', C', as indicated, to obtain a triangle $AB'C'$ similar to ABC, with sides a times as long. By geometry, $B'C'$ is parallel to BC, and hence $\overrightarrow{B'C'} = av$. Accordingly

$$\overrightarrow{AB'} + \overrightarrow{B'C'} = au + av = \overrightarrow{AC'} = a(u + v)$$

If a is negative, there is a similar proof in which the sides of triangle ABC are extended in the opposite direction; if $a = 0$, the rule follows from (1-51). If u, v are collinear and $u \neq 0$, then v can be represented by a directed line segment from 0 to x on a number axis, as in Figure 1-11 above. Thus, $v = xu$ and

$$a(u + v) = a(u + xu) = a(1 + x)u = (a + ax)u = au + axu$$
$$= au + a(xu) = au + av$$

by rules (1-50), (1-53), and (1-56). Finally, if $u = 0$, then (1-57) follows from (1-32).

The rules (1-56) and (1-57) are called *distributive laws*.

Rule (1-58) follows from (1-56), and rule (1-59) from (1-57) (see Problem 9 below).

Remark. Occasionally it is convenient to use the notation of division of vector u by scalar a to denote multiplication of u by $1/a$; for example, $u/2 = (\tfrac{1}{2})u$.

PROBLEMS

1. Let triangle ABC be given, let $\overrightarrow{AB} = u$ and $\overrightarrow{AC} = v$, and let points P, Q be chosen so that $\overrightarrow{AP} = 3u$, $\overrightarrow{AQ} = 2v$, as in Figure 1-13. Express the following vectors in terms

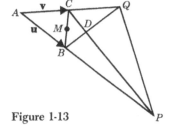

Figure 1-13

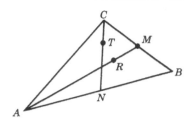

Figure 1-14

of **u** and **v**: (a) $\overrightarrow{BC}$, (b) $\overrightarrow{PB}$, (c) $\overrightarrow{PQ}$, (d) $\overrightarrow{PC}$, (e) $\overrightarrow{BQ}$, (f) $\overrightarrow{AM}$, where M is the midpoint of $\overrightarrow{BC}$, (g) $\overrightarrow{BD} + \overrightarrow{DC} + \overrightarrow{CQ}$.

2. Let ABC be a triangle, let M be the midpoint of side BC, and N the midpoint of side AB (Figure 1-14). Let R be chosen on AM and T on CN so that $|\overrightarrow{AR}|/|\overrightarrow{AM}| = \frac{3}{4}$, and $|\overrightarrow{CT}|/|\overrightarrow{CN}| = \frac{1}{3}$. Let $\mathbf{u} = \overrightarrow{AB}$, $\mathbf{v} = \overrightarrow{AC}$. Express the following vectors in terms of **u** and **v**: (a) $\overrightarrow{BA}$, (b) $\overrightarrow{BC}$, (c) $\overrightarrow{AN}$, (d) $\overrightarrow{AM}$, (e) $\overrightarrow{AR}$, (f) $\overrightarrow{BR}$, (g) $\overrightarrow{CN}$, (h) $\overrightarrow{CT}$, (i) $\overrightarrow{TM}$.

3. Let O be a given point. Show that, if M is the midpoint of the line segment AB, then

$$\overrightarrow{OM} = \frac{1}{2}(\overrightarrow{OA} + \overrightarrow{OB})$$

4. Let C be a point on the straight line through O and B. Let $\mathbf{u} = \overrightarrow{OB}$. Find $\overrightarrow{OC}$ in terms of **u** if
   (a) C lies between O and B, and $|\overrightarrow{OC}|/|\overrightarrow{CB}| = m/n$ ($m > 0$ and $n > 0$).
   (b) C lies on OB extended past B, and $|\overrightarrow{OC}|/|\overrightarrow{CB}| = m/n$ ($0 < n < m$).

5. Let O be a given point. Show that if C divides the line segment AB in the ratio $m:n$, then

$$\overrightarrow{OC} = \frac{1}{m + n}(n\,\overrightarrow{OA} + m\,\overrightarrow{OB})$$

6. (a) Show that R is on the line determined by the points P and Q if, and only if, $\overrightarrow{PR}$ is a scalar multiple of $\overrightarrow{PQ}$.
   (b) Show that R is on the line segment determined by the points P and Q if, and only if, $\overrightarrow{PR} = k\,\overrightarrow{PQ}$ with $0 \le k \le 1$.
   (c) Show that R is inside or on a side of parallelogram $ABCD$ if, and only if, $\overrightarrow{AR} = a\mathbf{u} + b\mathbf{v}$, with $0 \le a \le 1$, $0 \le b \le 1$, where $\mathbf{u} = \overrightarrow{AB}$, $\mathbf{v} = \overrightarrow{AD}$.

7. Let $\mathbf{u}, \mathbf{v}, \mathbf{w}, \mathbf{z}$ be unit vectors, no two being equal, and let $\mathbf{u} + \mathbf{v} + \mathbf{w} + \mathbf{z} = \mathbf{0}$. Show that the vectors can be paired so that each pair consists of a vector and its negative. (*Hint*. Show first that the vectors can be represented as the sides of a quadrilateral, properly directed.)

8. Wandering William reached a crossroad A in a plain. One sign pointed to a road leading straight to two towns B and D, at distances of 1 mile and 5 miles, respectively. Another sign pointed to a road leading straight to towns C and E, at distances of 1 mile and 7 miles, respectively (Figure 1-15). His goal was the tower at Q on the road from D to E. Should he go via D or via E? He sighted the tower in the distance and noticed that it was lined up with a flagpole at M, halfway between B and C. He calculated rapidly with vectors and concluded that Q was $\frac{5}{12}$ of the way from D to E, so that his best route was via D. Show that his conclusion was correct.

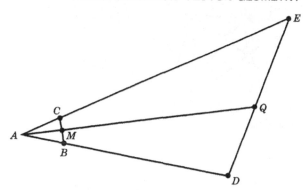

Figure 1-15

9. Justify each of the following rules:
   (a) (1-50)   (b) (1-51)   (c) (1-52)   (d) (1-53)   (e) (1-54)   (f) (1-55)
   (g) (1-58) [*Hint*. Use (1-54) to write the right-hand side as $au + (-b)u$ and apply (1-56)]   (h) (1-59) [*Hint*. Use (1-54) and (1-57)]

† 1-6 GEOMETRIC APPLICATIONS

The concept of vector enables us to give rather elementary proofs for certain geometric theorems. We give two illustrations.

EXAMPLE 1. Prove that, if the midpoints of two sides of a triangle are joined by a line segment, then that segment is parallel to the third side and half as long.

PROOF. Let the triangle be ABC, as in Figure 1-16. Let M be the midpoint of AB, N the midpoint of BC, and draw MN. Then

$$\overrightarrow{MB} = \frac{1}{2}\,\overrightarrow{AB}, \qquad \overrightarrow{BN} = \frac{1}{2}\,\overrightarrow{BC}$$

by the definition of scalar multiplication. Hence

$$\overrightarrow{MN} = \overrightarrow{MB} + \overrightarrow{BN} = \frac{1}{2}\,\overrightarrow{AB} + \frac{1}{2}\,\overrightarrow{BC} = \frac{1}{2}(\overrightarrow{AB} + \overrightarrow{BC}) = \frac{1}{2}\,\overrightarrow{AC}$$

Therefore, MN must be parallel to AC and half as long.

EXAMPLE 2. Prove that the diagonals of a parallelogram bisect each other.

PROOF. Let the parallelogram be $ABCD$ (Figure 1-17). Let the midpoint of diagonal AC be M, the midpoint of BD be N. Then

$$\overrightarrow{AM} = \frac{1}{2}\,\overrightarrow{AC}$$

$$\overrightarrow{AN} = \overrightarrow{AB} + \frac{1}{2}\,\overrightarrow{BD} = \overrightarrow{AB} + \frac{1}{2}(\overrightarrow{BA} + \overrightarrow{AD})$$

$$= \left(\overrightarrow{AB} - \frac{1}{2}\,\overrightarrow{AB}\right) + \frac{1}{2}\,\overrightarrow{AD}$$

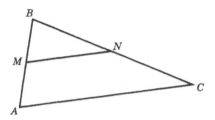

Figure 1-16 Example 1.

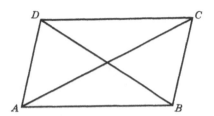

Figure 1-17 Example 2.

$$= \frac{1}{2}(\overrightarrow{AB} + \overrightarrow{AD}) = \frac{1}{2}(\overrightarrow{AB} + \overrightarrow{BC}) = \frac{1}{2}\overrightarrow{AC}$$

Thus $\overrightarrow{AM} = \overrightarrow{AN}$, so that M and N coincide and the diagonals do bisect each other.

PROBLEMS

1. Side AC of triangle ABC is extended past C its own length to D, and side BC is extended its own length past C to E (Figure 1-18). Show that ED and AB are equal and parallel.

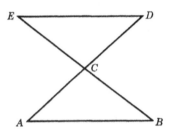

Figure 1-18

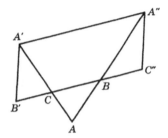

Figure 1-19

2. Let $ABCD$ be a quadrilateral, let P, Q, R, S be the midpoints of the sides AB, BC, CD, and DA. Prove that $PQRS$ is a parallelogram.
3. Prove that the medians of a triangle meet at a point two thirds of the way from each vertex to the opposite side.
4. Let $ABCDEF$ be a hexagon. Let sides AB, DE be equal and parallel, and let sides BC and EF be equal and parallel. Show that sides CD and FA are equal and parallel.
5. The sides of triangle ABC are extended as in Figure 1-19 so that $\overrightarrow{AA'} = 3\overrightarrow{AC}$, $\overrightarrow{B'C} = \overrightarrow{CB}$, $\overrightarrow{AA''} = 3\overrightarrow{AB}$, and $\overrightarrow{CB} = \overrightarrow{BC''}$. Show that $A'B'C''A''$ is a parallelogram.
6. Given a triangle ABC as in Figure 1-20, let points P on AC, Q on CB, R on AB, M on PQ, N on QR be such that

$$|\overrightarrow{AP}| = 2|\overrightarrow{PC}|, \quad |\overrightarrow{CQ}| = 2|\overrightarrow{QB}|, \quad |\overrightarrow{BR}| = 2|\overrightarrow{RA}|, \quad |\overrightarrow{PM}| = 2|\overrightarrow{MQ}|, \quad |\overrightarrow{QN}| = 2|\overrightarrow{NR}|$$

Prove that MN is parallel to AC and $|\overrightarrow{AC}| = 3|\overrightarrow{MN}|$.
7. In Figure 1-21, let OAB be a triangle, let $\overrightarrow{OC} = (\frac{2}{3})\overrightarrow{OB}$, $\overrightarrow{AD} = (\frac{2}{3})\overrightarrow{AB}$, and let E be the midpoint of OA. Prove that the line segments OD, AC, BE meet at a point P, which is three fourths of the way from O to D, three fourths of the way from A to C, and halfway between B and E.

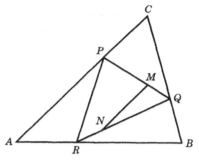

Figure 1-20

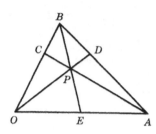

Figure 1-21

8. In Figure 1-22, *ABCD* is a parallelogram, *M* is the midpoint of *DC*. Show that *AM* meets *DB* at a point *E*, two thirds of the way from *A* to *M* and one third of the way from *D* to *B*.

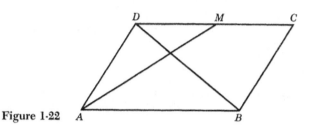

Figure 1-22

1-7 LINEAR INDEPENDENCE, BASIS

Two vectors in the plane are said to be *linearly dependent* if one of the two is a scalar multiple of the other. Thus **u**, **v** are linearly dependent if **u** = *a***v** or **v** = *a***u** for some scalar *a*. Hence, **u**, **v** are linearly dependent precisely when **u**, **v** are collinear, as in Figure 1-23. In particular, **0**, **v** are linearly dependent for every **v**, since we can write **0** = 0**v**.

The two vectors **u**, **v** are said to be *linearly independent* when they are not linearly dependent—hence, when they are *not* collinear (Figure 1-24). There are many pairs of linearly independent vectors; in fact, every pair of (directed) sides of a triangle provides such a pair. In particular, for each nonzero vector **u**, we can find many vectors **v**, such that **u**, **v** are linearly independent.

Criterion for Linear Independence. Two vectors **u**, **v** are linearly independent precisely when the following condition holds true:

$$\text{If } a\mathbf{u} + b\mathbf{v} = \mathbf{0}, \qquad \text{then} \qquad a = 0 \qquad \text{and} \qquad b = 0 \qquad (1\text{-}70)$$

Figure 1-23 Linearly dependent vectors.

Figure 1-24 Linearly independent vectors.

The criterion is essentially a rephrasing of the definition. If **u, v** are linearly independent, then (1-70) must hold true since, if not, we could find a, b *not both* 0 so that $a\mathbf{u} + b\mathbf{v} = \mathbf{0}$ and, therefore, either

$$\mathbf{u} = -\frac{b}{a}\mathbf{v} \qquad \text{or} \qquad \mathbf{v} = -\frac{a}{b}\mathbf{u}$$

in contradiction to the linear independence of **u, v**. Conversely, if (1-70) holds true, then **u, v** must be linearly independent since, otherwise, either $\mathbf{v} = k\mathbf{u}$ or $\mathbf{u} = k\mathbf{v}$; that is

$$k\mathbf{u} + (-1)\mathbf{v} = \mathbf{0} \qquad \text{or} \qquad (-1)\mathbf{u} + k\mathbf{v} = \mathbf{0}$$

and (1-70) would be violated.

We can restate the criterion in several ways.

The vectors **u, v** are linearly independent precisely when the only scalars a, b, for which $a\mathbf{u} + b\mathbf{v} = \mathbf{0}$ holds true, are $a = 0$, $b = 0$.

The vectors **u, v** are linearly dependent precisely when scalars a, b can be found, not both 0, for which $a\mathbf{u} + b\mathbf{v} = \mathbf{0}$.

Rule for Comparing Coefficients. Let **u, v** be linearly independent vectors. Then

$$a\mathbf{u} + b\mathbf{v} = \alpha\mathbf{u} + \beta\mathbf{v} \qquad \text{implies} \qquad a = \alpha, b = \beta \qquad (1\text{-}71)$$

For the first equation in (1-71) can be written

$$(a - \alpha)\mathbf{u} + (b - \beta)\mathbf{v} = \mathbf{0}$$

By (1-70) this implies $a - \alpha = 0$, $b - \beta = 0$; that is, $a = \alpha$, $b = \beta$.

An expression $a\mathbf{u} + b\mathbf{v}$ is called a *linear combination* of **u, v**. We say that **u, v** form a *basis* for vectors in the plane if every vector **w** in the plane can be expressed uniquely as a linear combination of **u, v**. Two linearly dependent vectors **u, v** can never form a basis. For then, **u, v** are collinear and, hence, so are all the linear combinations of **u** and **v**; thus, no vector **w** not collinear with **u** and **v** can be expressed as a linear combination of **u** and **v**. However, any two linearly independent vectors form a basis.

BASIS THEOREM. *If* **u, v** *are linearly independent vectors in the plane, then* **u, v** *form a basis for vectors in the plane.*

PROOF. Represent **u** as $\overrightarrow{AB}$, **v** as $\overrightarrow{AC}$ (see Figure 1-25). Since **u, v** are linearly independent, $\overrightarrow{AB}$, $\overrightarrow{AC}$ are not collinear. Represent **w** as $\overrightarrow{AP}$. Through P draw a line parallel to AB, intersecting the line through A, C at M. We can write $\overrightarrow{AM} = y\mathbf{v}$. Similarly, the line through P parallel to AC meets the line through A, B at a point N, and we can write $\overrightarrow{AN} = x\mathbf{u}$. Finally, as Figure 1-25 shows, $\mathbf{w} = \overrightarrow{AN} + \overrightarrow{AM} = x\mathbf{u} + y\mathbf{v}$. Since **u, v** are linearly independent, (1-71) holds true; therefore, the representation of **w** as a linear combination of **u, v** is unique. Our theorem is proved.

The significance of a basis is shown in Figure 1-26. Here our vectors are represented as $\overrightarrow{OP}$, where O is a fixed origin and P varies. Thus, we have a one-to-one correspondence between vectors and points, as in Section 1-2. As

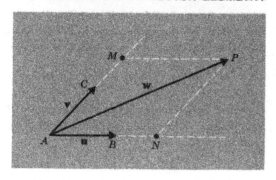

Figure 1-25 $w = xu + yv.$

P moves along the line on which $b = 1$, we obtain the linear combinations $au + v$ for different values of a; as P moves on the line for which $a = 2$, we obtain the linear combinations $2u + bv$. Thus, in effect we have an "oblique" coordinate system in the plane.

EXAMPLE 1. Let u, v be linearly independent vectors, hence a basis. Show that $w = 3u + v$, $z = 2u - v$ also form a basis.

Solution. We must show that w, z are linearly independent, and apply the criterion (1-70). If $aw + bz = 0$, then

$$a(3u + v) + b(2u - v) = 0$$

Hence

$$(3a + 2b)u + (a - b)v = 0$$

But u, v are linearly independent. Therefore [by (1-70) again]

$$3a + 2b = 0, \qquad a - b = 0$$

These are two simultaneous linear equations for a, b. By solving, we find that $a = 0$, $b = 0$. Therefore, w, z are linearly independent.

EXAMPLE 2. Let u, v be linearly independent vectors. For what values of k are $w = 2u + v$, $z = ku - 5v$ linearly dependent?

Solution. We want to be able to find a, b not both 0 so that

$$a(2u + v) + b(ku - 5v) = 0$$

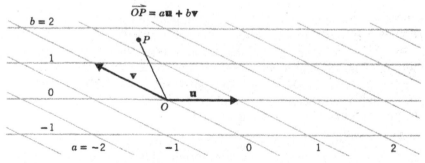

Figure 1-26   Basis in the plane.

As for Example 1, the coefficients of **u**, **v** must be 0:

$$2a + kb = 0, \qquad a - 5b = 0$$

Hence, $a = 5b$ and the first equation gives $10b + kb = 0$ or $(10 + k)b = 0$. Therefore, either $b = 0$ (so that $a = 5b = 0$), or $k = -10$. Hence, $k = -10$ is the only possibility. Indeed, for $k = -10$, the equations for a, b are both equivalent to $a - 5b = 0$ and are satisfied, for example, by $a = 5$, $b = 1$. Thus, the vectors are linearly dependent for $k = -10$. In fact, for this value of k,

$$\mathbf{z} = -10\mathbf{u} - 5\mathbf{v} = -5(2\mathbf{u} + \mathbf{v}) = -5\mathbf{w}$$

Generalizations. We can extend the concept of linear independence to 3 or more vectors in the plane. Guided by (1-70), we say that **u**, **v**, **w** are linearly independent if

$$a\mathbf{u} + b\mathbf{v} + c\mathbf{w} = \mathbf{0} \qquad \text{implies} \qquad a = 0, b = 0, c = 0$$

Otherwise, we term **u**, **v**, **w** linearly dependent. However, we obtain nothing interesting in this way. *Every set of three vectors* **u**, **v**, **w** *in the plane is linearly dependent.* For either **u**, **v** are linearly independent, or they are linearly dependent. In the former case, the Basis Theorem applies, and $\mathbf{w} = a\mathbf{u} + b\mathbf{v}$, or

$$a\mathbf{u} + b\mathbf{v} + (-1)\mathbf{w} = \mathbf{0}$$

This is an equation of the form $a\mathbf{u} + b\mathbf{v} + c\mathbf{w} = \mathbf{0}$ with not all coefficients a, b, c equal to 0. Thus, **u**, **v**, **w** are linearly dependent. In the latter case, we can choose a, b not both 0, so that $a\mathbf{u} + b\mathbf{v} = \mathbf{0}$. Then

$$a\mathbf{u} + b\mathbf{v} + 0\mathbf{w} = \mathbf{0}$$

so that again **u**, **v**, **w** are linearly dependent.

The theory of vectors extends to 3-dimensional space. In space we find that we can find 3 linearly independent vectors **u**, **v**, **w** and that each such set forms a basis; that is, every vector **z** in space can be expressed uniquely as a linear combination of **u**, **v**, **w**:

$$\mathbf{z} = a\mathbf{u} + b\mathbf{v} + c\mathbf{w}$$

In space, every set of 4 vectors **p**, **q**, **r**, **s** is linearly dependent; that is, there is a relation

$$a\mathbf{p} + b\mathbf{q} + c\mathbf{r} + d\mathbf{s} = \mathbf{0}$$

with not all coefficients 0.

These results suggest that the *dimension* of a space equals the number of vectors in a basis. We can even consider vectors in one-dimensional space, that is, vectors on a line. Here every single nonzero vector **u** is considered to be "linearly independent," and each such **u** forms a basis: every vector **v** on the line can be expressed as $a\mathbf{u}$.

1-8 VECTORS AS NUMBER PAIRS

Now let a basis **u**, **v** be chosen and be kept fixed in the following discussion. Then for each vector **w** there is a unique pair of numbers a, b for which **w** =

$a\mathbf{u} + b\mathbf{v}$. Conversely, each pair of numbers a, b determines a vector $a\mathbf{u} + b\mathbf{v}$ in the plane. Thus, we specify all vectors in the plane by giving all *ordered pairs* (a, b) of numbers; each pair determines a vector, and every vector has a unique ordered pair. The order is essential since, in general, $a\mathbf{u} + b\mathbf{v}$ is different from $b\mathbf{u} + a\mathbf{v}$ (when are they equal?).

We write $\mathbf{w} \leftrightarrow (a, b)$ to indicate that $\mathbf{w}$ corresponds to the pair a, b. We call a, b the *components* of $\mathbf{w}$ with respect to the basis $\mathbf{u}$, $\mathbf{v}$. If we choose another basis, then we obtain a new assignment of components. For example, in Figure 1-27,

$$\mathbf{w} = -\sqrt{2}\mathbf{u} - \sqrt{2}\mathbf{v} \leftrightarrow (-\sqrt{2}, -\sqrt{2})$$

But, if we use the new basis $\mathbf{u}^*$, $\mathbf{v}^*$, then

$$\mathbf{w} = -2\mathbf{u}^* + 0\mathbf{v}^* \leftrightarrow (-2, 0)$$

Notice that, if we interchange the order of vectors in the basis ($\mathbf{v}$, $\mathbf{u}$ instead

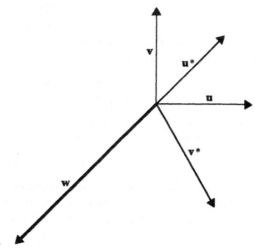

Figure 1-27   Different bases.

of $\mathbf{u}$, $\mathbf{v}$), then the ordered number pair associated with $\mathbf{w}$ is also interchanged.

Thus, the representation of vectors by number pairs has meaning only if we fix a basis, in a definite order.

With respect to our fixed basis $\mathbf{u}$, $\mathbf{v}$, let

$$\mathbf{w} = a\mathbf{u} + b\mathbf{v} \leftrightarrow (a, b)$$
$$\mathbf{z} = x\mathbf{u} + y\mathbf{v} \leftrightarrow (x, y)$$

Then

$$\mathbf{w} + \mathbf{z} = (a + x)\mathbf{u} + (b + y)\mathbf{v} \leftrightarrow (a + x, b + y)$$
$$c\mathbf{w} = c(a\mathbf{u} + b\mathbf{v}) = ca\mathbf{u} + cb\mathbf{v} \leftrightarrow (ca, cb)$$

Thus, the ordered number pair associated with $\mathbf{w} + \mathbf{z}$ is found by adding the respective components of $\mathbf{w}$ and $\mathbf{z}$, and the pair associated with $c\mathbf{w}$ is obtained by multiplying each component of $\mathbf{w}$ by c.

We therefore define addition and scalar multiplication of ordered number pairs as follows:

$$(a,\ b) + (x,\ y) = (a + x,\ b + y) \qquad \text{(1-80)}$$
$$c(a,\ b) = (ca,\ cb) \qquad \text{(1-81)}$$

It then follows that our vector operations can be carried out completely in terms of the ordered pairs. In a sense, we have a new language for the algebra of vectors.

EXAMPLE. In terms of a fixed basis **u**, **v**, let **w** $\leftrightarrow (3, 5)$, **z** $\leftrightarrow (2, -4)$. Then

$$\mathbf{w} + \mathbf{z} \leftrightarrow (5,\ 1), \qquad \mathbf{w} - \mathbf{z} \leftrightarrow (1,\ 9), \qquad 2\mathbf{w} \leftrightarrow (6,\ 10)$$
$$3\mathbf{w} + 2\mathbf{z} \leftrightarrow (9,\ 15) + (4,\ -8) = (13,\ 7)$$

PROBLEMS

Note: Throughout these problems, **u**, **v** form a pair of linearly independent vectors in the plane; hence, they form a basis for all vectors in the plane.

1. Trace Figure 1-28 on a sheet of paper and construct graphically:
   (a) $2\mathbf{u} - 3\mathbf{v}$ (b) $-\mathbf{u} + \mathbf{v}$ (c) $0\mathbf{u} + 2\mathbf{v}$
   Also represent each of the following vectors in terms of the basis:
   (d) $\overrightarrow{OC}$ (e) $\overrightarrow{AC}$ (f) $\overrightarrow{CB}$ (g) $2\overrightarrow{AB}$
2. Show that each of the following pairs of vectors is linearly independent, and illustrate this with a graph.
   (a) $\mathbf{u},\ \mathbf{u} + \mathbf{v}$ (b) $\mathbf{u} + \mathbf{v},\ \mathbf{u} - \mathbf{v}$
   (c) $2\mathbf{u} - 3\mathbf{v},\ 5\mathbf{u} - 7\mathbf{v}$ (d) $-\mathbf{u},\ -4\mathbf{v}$

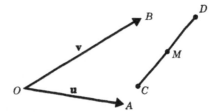

Figure 1-28

3. For each of the following pairs of vectors, determine whether it is a linearly independent set, and illustrate this with a graph.
   (a) $2\mathbf{u} + \mathbf{v},\ \mathbf{u} + 2\mathbf{v}$ (b) $\mathbf{u} + 3\mathbf{v},\ 3\mathbf{u} + 6\mathbf{v}$
   (c) $9\mathbf{u} + 6\mathbf{v},\ 6\mathbf{u} + 4\mathbf{v}$ (d) $3\mathbf{v},\ 5\mathbf{v}$
4. In each of the following cases, determine k (if possible) so that the given pair of vectors is linearly dependent.
   (a) $\mathbf{u},\ 3\mathbf{u} + k\mathbf{v}$ (b) $\mathbf{u} - 2\mathbf{v},\ 3\mathbf{u} + k\mathbf{v}$
   (c) $(k + 1)\mathbf{u} + \mathbf{v},\ 4\mathbf{u} + (k + 1)\mathbf{v}$ (d) $\mathbf{u},\ k\mathbf{u} + \mathbf{v}$
5. Let vectors be represented by number pairs, in terms of the given basis **u**, **v**, so that $\mathbf{u} \leftrightarrow (1, 0)$, $\mathbf{v} \leftrightarrow (0, 1)$. Let $\mathbf{w} \leftrightarrow (2, 0)$, $\mathbf{r} \leftrightarrow (3, 4)$, $\mathbf{s} \leftrightarrow (1, -4)$.
   (a) Show that **w**, **r** are linearly independent.
   (b) Find a, b so that $\mathbf{s} = a\mathbf{w} + b\mathbf{r}$.
   (c) In Figure 1-28 let M be the midpoint of CD and let $\overrightarrow{OC} \leftrightarrow (a_1, b_1)$, $\overrightarrow{OD} \leftrightarrow (a_2, b_2)$, $\overrightarrow{OM} \leftrightarrow (a, b)$. Show that $a = \frac{1}{2}(a_1 + b_1)$, $b = \frac{1}{2}(a_2 + b_2)$.
6. (a) Show graphically that $\mathbf{w} = \mathbf{u} - \mathbf{v}$, $\mathbf{z} = \mathbf{u} + \mathbf{v}$ are linearly independent and, thus, form a basis.

(b) Express the vectors $r = 2u + v$ and $s = 3u - v$ in terms of z and w.

7. (a) Prove that, if a, b, c, d are numbers such that $ad - bc \neq 0$, then $au + bv$, $cu + dv$ are linearly independent and, hence, form a basis.

(b) Prove that, if $au + bv$, $cu + dv$ are linearly dependent, then $ad - bc = 0$ [see part (a)].

(c) Prove that, if $ad - bc = 0$, then $au + bv$, $cu + dv$ are linearly dependent.

8. Let w be a nonzero vector and let (a, b) be an ordered number pair. Show that a basis u, v can be chosen for which $w \leftrightarrow (a, b)$ in this basis, for each of the following cases:

(a) $a \neq 0$, $b \neq 0$ (*Hint*. Choose u so that u, w are linearly independent, then show that v can be chosen so that $w = au + bv$.)

(b) $a \neq 0$, $b = 0$

1-9 ANGLE BETWEEN VECTORS, ORTHOGONAL BASIS

Let u, v be two nonzero vectors in the plane. The *angle between* u *and* v is defined to be the angle BOA, where $u = \overrightarrow{OA}$, $v = \overrightarrow{OB}$; the angle will usually be measured in radians, and the value will always be chosen between 0 and π. The value will be denoted by $\sphericalangle(u, v)$. It does not depend on the reference point O (Figure 1-29). For if we choose another reference point O' and construct A', B', so that $\overrightarrow{O'A'} = u$, $\overrightarrow{O'B'} = v$, then the two angles have their sides parallel and similarly directed; hence, they are equal. We do not define the angle between u and v when either vector is 0.

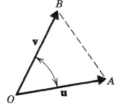 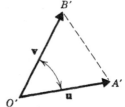

Figure 1-29   Angle between vectors.

In the triangle OAB of Figure 1-29, the sides are $|u|$, $|v|$ and $|u - v|$; let $\varphi = \sphericalangle(u, v)$. The law of cosines gives

$$|u - v|^2 = |u|^2 + |v|^2 - 2|u||v| \cos \varphi \tag{1-90}$$

Thus

$$\cos \varphi = \frac{|u|^2 + |v|^2 - |u - v|^2}{2|u||v|} \tag{1-91}$$

When u, v are linearly dependent, the triangle collapses, but (1-90) and (1-91) still hold true, as we easily verify (see Problem 7 below).

When $\varphi = \pi/2$, (1-90) reduces to the Pythagorean theorem:

$$|u - v|^2 = |u|^2 + |v|^2 \tag{1-92}$$

Conversely, when (1-92) holds true, (1-91) gives $\cos \varphi = 0$ and, since $0 \leq \varphi \leq \pi$, we have $\varphi = \pi/2$.

When $\varphi = \pi/2$, the directed line segments representing u are always perpendicular to the ones that represent v. We therefore say that two nonzero vectors with $\pi/2$ as angle are *perpendicular* or *orthogonal*. Clearly, two nonzero orthogonal vectors in the plane are linearly independent and, therefore, always constitute a basis. Such a basis we call an *orthogonal basis*. When given a nonzero vector **u**, we can always find a nonzero vector **v** orthogonal to **u**, so that **u**, **v** form an orthogonal basis; any two such vectors **v** are linearly dependent.

By a *unit vector* we mean a vector of magnitude 1. An *orthonormal basis* for vectors in the plane is a basis of orthogonal unit vectors. If **u** is a unit vector, then there are exactly two vectors **v** and $-$**v** that with **u** form an orthonormal basis (Figure 1-30). Now let a Cartesian coordinate system be chosen in the plane (Section 0-5), with coordinates (x, y) and corresponding axes OX and OY. Let A be the point $(1, 0)$ and B the point $(0, 1)$. Then the vectors **i** $= \overrightarrow{OA}$ and **j** $= \overrightarrow{OB}$ are orthogonal unit vectors, as in Figure 1-31; hence,

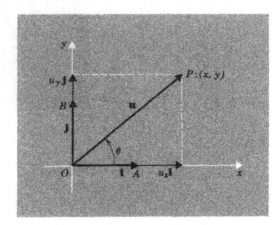

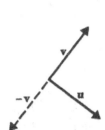

Figure 1-30   Orthonormal basis.   **Figure 1-31**   The vectors **i** and **j**.

they form an orthonormal basis. An arbitrary vector **u** can be expressed as a linear combination of **i** and **j**. We write

$$\mathbf{u} = u_x \mathbf{i} + u_y \mathbf{j} \tag{1-93}$$

so that u_x, u_y are the components of u with respect to the basis **i**, **j**, and we can write **u** $\leftrightarrow (u_x, u_y)$ as in Section 1-8. We also call u_x the *x-component* of **u**, u_y the *y-component* of **u**. If we represent **u** as $\overrightarrow{OP}$, as in Figure 1-31, then (u_x, u_y) are simply the Cartesian coordinates of P. By the Pythagorean theorem

$$|\overrightarrow{OP}|^2 = |\mathbf{u}|^2 = u_x{}^2 + u_y{}^2 \tag{1-94}$$

If we introduce a signed angle $\theta = \sphericalangle XOP$, as in trigonometry (Figure 1-31), then (provided that $|\mathbf{u}| \neq 0$),

$$u_x = |\mathbf{u}| \cos \theta, \qquad u_y = |\mathbf{u}| \sin \theta \tag{1-95}$$

Equations (1-94) and (1-95) permit us to determine the magnitude and the direction of **u** ($|\mathbf{u}|$ and θ) from its x- and y-components u_x, u_y, or to determine the components u_x, u_y from the magnitude and the direction.

The notation **i**, **j** is standard for an orthonormal basis, obtained as in Figure

1-31, from a Cartesian coordinate system, and will be used frequently. In terms of a Cartesian coordinate system (Figure 1-32), let points $P_1:(x_1, y_1)$, $P_2:(x_2, y_2)$ be given. Then

$$\overrightarrow{OP_1} = x_1\, \mathbf{i} + y_1\, \mathbf{j}$$
$$\overrightarrow{OP_2} = x_2\, \mathbf{i} + y_2\, \mathbf{j}$$

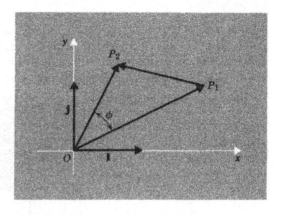

Figure 1-32 Vector $\overrightarrow{P_1P_2}$.

so that

$$\overrightarrow{P_1P_2} = \overrightarrow{OP_2} - \overrightarrow{OP_1} = (x_2 - x_1)\mathbf{i} + (y_2 - y_1)\mathbf{j} \qquad (1\text{-}96)$$

The vector represented by the directed segment P_1P_2 has components that are obtained by subtracting the x- and y-coordinates of P_1 from those of P_2.
Accordingly, by our rule (1-94), which is now applied to $\mathbf{u} = \overrightarrow{P_1P_2}$,

$$|\overrightarrow{P_1P_2}|^2 = (x_2 - x_1)^2 + (y_2 - y_1)^2 \qquad (1\text{-}97)$$

or

$$|\overrightarrow{P_1P_2}| = \sqrt{(x_2 - x_1)^2 + (y_2 - y_1)^2} \qquad (1\text{-}98)$$

This is the *distance formula* of analytic geometry. It expresses the distance between two points in terms of their Cartesian coordinates.

We remark that the choice of origin and coordinate axes is at our disposal. If we are given an orthogonal basis $\mathbf{u}$, $\mathbf{v}$, we obtain an orthonormal basis $\mathbf{i}$, $\mathbf{j}$ by multiplying by scalars (see Figure 1-33):

$$\mathbf{i} = \frac{1}{|\mathbf{u}|}\mathbf{u}, \qquad \mathbf{j} = \frac{1}{|\mathbf{v}|}\mathbf{v}$$

We can now select an origin and can choose coordinate axes so that $\mathbf{i}$ is a unit

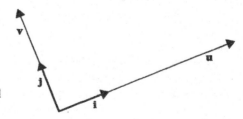

Figure 1-33 Formation of orthonormal basis.

vector directed along the positive x-axis, and $\mathbf{j}$ is a unit vector directed along the positive y-axis. Thus, each orthogonal basis and choice of origin give a corresponding set of coordinate axes.

PROBLEMS

1. Verify: (a) $\sphericalangle(\mathbf{u}, \mathbf{v}) + \sphericalangle(\mathbf{u}, -\mathbf{v}) = \pi$;   (b) $\sphericalangle(\mathbf{u}, \mathbf{v}) = \sphericalangle(-\mathbf{u}, -\mathbf{v})$;   (c) $\sphericalangle(\mathbf{u}, \mathbf{v}) = \sphericalangle(a\mathbf{u}, a\mathbf{v})$ for $a \neq 0$.

2. Let $\mathbf{u} = 3\mathbf{i} - \mathbf{j}, \mathbf{v} = 2\mathbf{i} + \mathbf{j}$. Evaluate and check graphically: (a) $\mathbf{u} + \mathbf{v}$,   (b) $3\mathbf{u} - \mathbf{v}$, (c) $|\mathbf{u}|, |\mathbf{v}|$,   (d) $\sphericalangle(\mathbf{u}, \mathbf{v})$.

3. In terms of Cartesian coordinates, let O be the origin $(0, 0)$, let A, B, C be the points $(1, 1), (-1, 1), (2, -1)$, respectively. Evaluate and check graphically: (a) $|\overrightarrow{OA}|$; (b) $|\overrightarrow{AC}|$;   (c) $\cos \sphericalangle ACB$;   (d) $|2\,\overrightarrow{OA} + \overrightarrow{AB}|$.

4. Find the unit vectors which make an angle of $2\pi/3$ with $\mathbf{u} = \mathbf{i} - \mathbf{j}$.

5. Show that the triangle with vertexes $(-1, -1), (6, -2), (7, 5)$ is an isosceles triangle.

6. Let $\mathbf{u} = u_x\mathbf{i} + u_y\mathbf{j}$. Prove that (a) $|\mathbf{u}| \leq |u_x| + |u_y|$. (b) $|\mathbf{u}| \geq (\sqrt{2}/2)(|u_x| + |u_y|)$. [*Hint.* Square both sides and use (1-94).]

7. Show that Equation (1-90) is valid for $\mathbf{u}, \mathbf{v}$ nonzero but linearly dependent. [*Hint.* We can write $\mathbf{v} = k\mathbf{u}$. Substitute $k\mathbf{u}$ for $\mathbf{v}$ in both sides of (1-90); show that $\cos \varphi = 1$ for $k > 0$, $\cos \varphi = -1$ for $k < 0$, and that both sides are equal.]

1-10 INNER PRODUCT

The quantity $|\mathbf{u}||\mathbf{v}| \cos \varphi$, which appears in Equation (1-90) above, is very important for vector theory and, therefore, receives a special name: the *inner product of* $\mathbf{u}$ *and* $\mathbf{v}$. Generally, the inner product is denoted by $\mathbf{u} \cdot \mathbf{v}$ or by $(\mathbf{u}, \mathbf{v})$, and because of the first notation, it is also called the dot product. We now give a general definition:

$$\mathbf{u} \cdot \mathbf{v} = \begin{cases} 0, & \text{if } \mathbf{u} \text{ or } \mathbf{v} \text{ is } \mathbf{0} \\ |\mathbf{u}||\mathbf{v}| \cos \varphi, & \text{if } \mathbf{u} \neq \mathbf{0}, \mathbf{v} \neq \mathbf{0} \end{cases} \tag{1-100}$$

where $\varphi = \sphericalangle(\mathbf{u}, \mathbf{v})$.

From this definition we conclude that if $\mathbf{u} \cdot \mathbf{v} = 0$, then $\mathbf{u} = \mathbf{0}, \mathbf{v} = \mathbf{0}$, or $\mathbf{u}, \mathbf{v}$ are perpendicular. For if $\mathbf{u} \cdot \mathbf{v} = 0$, and $\mathbf{u} \neq \mathbf{0}, \mathbf{v} \neq \mathbf{0}$, then $\cos \varphi = 0$, so that $\varphi = \pi/2$. Conversely, if $\mathbf{u}$ and $\mathbf{v}$ are perpendicular, then $\varphi = \pi/2$ and $\mathbf{u} \cdot \mathbf{v} = 0$. If $\mathbf{u}$ or $\mathbf{v}$ is $\mathbf{0}$, we also agree to call the vectors perpendicular. We thus obtain the rule:

$$\mathbf{u} \cdot \mathbf{v} = 0 \quad \textit{if, and only if,} \ \mathbf{u}, \mathbf{v} \ \textit{are perpendicular vectors} \tag{1-101}$$

If $\mathbf{v}$ is a unit vector, so that $|\mathbf{v}| = 1$, $\mathbf{u} \cdot \mathbf{v} = |\mathbf{u}| \cos \varphi$. If φ is an acute angle (Figure 1-34a), $|\mathbf{u}| \cos \varphi$ is simply the length of the projection of $\mathbf{u}$ on a line on which $\mathbf{v}$ lies. If φ is an obtuse angle (Figure 1-34b), $|\mathbf{u}| \cos \varphi$ is minus the length of the projection. We call this signed projection the *component of* $\mathbf{u}$ *in the direction of* $\mathbf{v}$, and abbreviate it by $\text{comp}_v\mathbf{u}$. Thus

$$\mathbf{u} \cdot \mathbf{v} = \text{comp}_v\mathbf{u} \quad \text{when} \quad |\mathbf{v}| = 1 \tag{1-102}$$

Rule (1-102) continues to hold true when $\mathbf{u} = \mathbf{0}$, both sides reducing to 0. If $\mathbf{v}$ is taken as $\mathbf{i}$ or $\mathbf{j}$, we obtain

$$\mathbf{u} \cdot \mathbf{i} = |\mathbf{u}| \cos \theta = u_x = \text{comp}_i \, \mathbf{u}$$

$$\mathbf{u} \cdot \mathbf{j} = |\mathbf{u}| \cos \left(\frac{\pi}{2} - \theta \right) = u_y = \text{comp}_j \, \mathbf{u} \tag{1-103}$$

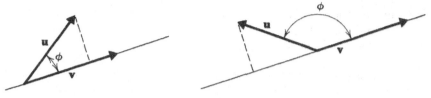

Figure 1-34   Inner product.

Thus, the x- and y-components of $\mathbf{u}$ are the components of $\mathbf{u}$ in the directions $\mathbf{i}$ and $\mathbf{j}$, respectively.

If $\mathbf{v}$ is not necessarily a unit vector, but is not $\mathbf{0}$, then (1-100) shows that

$$\mathbf{u} \cdot \mathbf{v} = |\mathbf{v}| \, \text{comp}_v \, \mathbf{u} \tag{1-104}$$

Thus, *the inner product is the length of* $\mathbf{v}$ *times the signed projection of* $\mathbf{u}$ *in the direction of* $\mathbf{v}$. This expression appears in mechanics as a measure of work done by a force. If (Figure 1-35) an object moves from A to B under the influence of a constant force $\mathbf{F}$, then the "work done by $\mathbf{F}$" is the product of

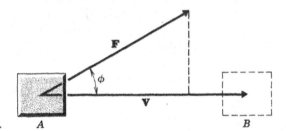

Figure 1-35   Work. A

the component of $\mathbf{F}$ in the direction of motion by the distance moved; that is,

$$\text{work} = (\text{comp}_v \, \mathbf{F}) \, |\mathbf{v}| = |\mathbf{F}| \cos \varphi \, |\mathbf{v}| = \mathbf{F} \cdot \mathbf{v} \tag{1-105}$$

1-11 PROPERTIES OF THE INNER PRODUCT

Let an orthonormal basis $\mathbf{i}$, $\mathbf{j}$ be chosen and let $\mathbf{u} = u_x \mathbf{i} + u_y \mathbf{j}$, $\mathbf{v} = v_x \mathbf{i} + v_y \mathbf{j}$. Then we have the basic rule:

$$\mathbf{u} \cdot \mathbf{v} = u_x v_x + u_y v_y \tag{1-110}$$

To prove this rule, we first assume that $\mathbf{u}$, $\mathbf{v}$ are nonzero vectors, and we represent $\mathbf{u}$ as $\overrightarrow{OP_1}$, $\mathbf{v}$ as $\overrightarrow{OP_2}$, as in Figure 1-32. Hence, $x_1 = u_x$, $y_1 = u_y$, $x_2 = v_x$, $y_2 = v_y$. By the law of cosines, if $\varphi = \sphericalangle(\mathbf{u}, \mathbf{v})$,

$$|\overrightarrow{P_1 P_2}|^2 = |\mathbf{u}|^2 + |\mathbf{v}|^2 - 2|\mathbf{u}| \, |\mathbf{v}| \cos \varphi = |\mathbf{u}|^2 + |\mathbf{v}|^2 - 2\mathbf{u} \cdot \mathbf{v}$$

By (1-97) and (1-94), we can also write

$|\overrightarrow{P_1P_2}|^2$

$$= (x_2 - x_1)^2 + (y_2 - y_1)^2 = x_2{}^2 - 2x_1x_2 + x_1{}^2 + y_2{}^2 - 2y_1y_2 + y_1{}^2$$
$$= (x_1{}^2 + y_1{}^2) + (x_2{}^2 + y_2{}^2) - 2(x_1x_2 + y_1y_2)$$
$$= (u_x{}^2 + u_y{}^2) + (v_x{}^2 + v_y{}^2) - 2(u_xv_x + u_yv_y)$$
$$= |\mathbf{u}|^2 + |\mathbf{v}|^2 - 2(u_xv_x + u_yv_y)$$

If we equate the two expressions for $|\overrightarrow{P_1P_2}|^2$, we obtain

$$-2\mathbf{u} \cdot \mathbf{v} = -2(u_xv_x + u_yv_y)$$

and the rule (1-110) follows. If either $\mathbf{u}$ or $\mathbf{v}$ is $\mathbf{0}$, then both sides of (1-110) are 0, so that the rule generally holds true.

From (1-110) and (1-94) we conclude that $\mathbf{u} \cdot \mathbf{u} = u_x{}^2 + u_y{}^2 = |\mathbf{u}|^2$. Hence

$$\mathbf{u} \cdot \mathbf{u} = |\mathbf{u}|^2 \quad \text{or} \quad |\mathbf{u}| = \sqrt{\mathbf{u} \cdot \mathbf{u}} \qquad (1\text{-}111)$$

This rule also follows from the definition (1-100).

EXAMPLE 1. Show that the triangle of vertexes A: $(3, 5)$, B: $(4, 8)$, C: $(6, 6)$ is isosceles (Figure 1-36).

First solution. Let $\mathbf{u} = \overrightarrow{AC}$, $\mathbf{v} = \overrightarrow{AB}$, $\mathbf{w} = \overrightarrow{BC}$. Then $\mathbf{u} = 3\mathbf{i} + \mathbf{j}$, $\mathbf{v} = \mathbf{i} + 3\mathbf{j}$, $\mathbf{w} = 2\mathbf{i} - 2\mathbf{j}$. Hence, by (1-111) and (1-110),

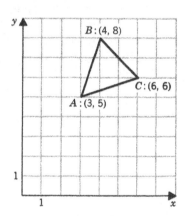

Figure 1-36   Example 1.

$$|\mathbf{u}|^2 = \mathbf{u} \cdot \mathbf{u} = 9 + 1 = 10, \qquad |\mathbf{v}|^2 = \mathbf{v} \cdot \mathbf{v} = 10, \qquad |\mathbf{w}|^2 = \mathbf{w} \cdot \mathbf{w} = 8$$

Therefore $|\mathbf{u}| = |\mathbf{v}|$, so that AB and AC are equal and the triangle is isosceles.

Second solution. Let $\beta = \sphericalangle ABC = \sphericalangle(\mathbf{w}, -\mathbf{v})$. Let $\gamma = \sphericalangle ACB = \sphericalangle(\mathbf{u}, \mathbf{w})$. Then

$$\mathbf{w} \cdot (-\mathbf{v}) = |\mathbf{w}||\mathbf{v}| \cos \beta, \qquad \mathbf{w} \cdot \mathbf{u} = |\mathbf{w}||\mathbf{u}| \cos \gamma$$

Therefore

$$\cos \beta = \frac{\mathbf{w} \cdot (-\mathbf{v})}{|\mathbf{w}||\mathbf{v}|} = \frac{(2\mathbf{i} - 2\mathbf{j}) \cdot (-\mathbf{i} - 3\mathbf{j})}{\sqrt{8}\sqrt{10}} = \frac{-2 + 6}{\sqrt{80}} = \frac{1}{\sqrt{5}}$$

$$\cos \gamma = \frac{\mathbf{w} \cdot \mathbf{u}}{|\mathbf{w}||\mathbf{u}|} = \frac{(2\mathbf{i} - 2\mathbf{j}) \cdot (3\mathbf{i} + \mathbf{j})}{\sqrt{8}\sqrt{10}} = \frac{6 - 2}{\sqrt{80}} = \frac{1}{\sqrt{5}}$$

Thus, $\beta = \gamma$ and the triangle is isosceles.

EXAMPLE 2. Let $\mathbf{u} = \mathbf{i} + \sqrt{3}\mathbf{j}$, $\mathbf{v} = 2\mathbf{i} + 2\mathbf{j}$. Find $\text{comp}_v \, \mathbf{u}$.

Solution. We have

$$\text{comp}_v \, \mathbf{u} = |\mathbf{u}| \cos \varphi = \frac{|\mathbf{u}||\mathbf{v}| \cos \varphi}{|\mathbf{v}|} = \frac{\mathbf{u} \cdot \mathbf{v}}{|\mathbf{v}|} = \frac{2 + 2\sqrt{3}}{\sqrt{8}} = 1.9$$

EXAMPLE 3. Show that the triangle with vertexes A: $(2, 1)$, B: $(6, 3)$, C: $(4, 7)$ is a right triangle.

Solution. $\overrightarrow{AB} = 4\mathbf{i} + 2\mathbf{j}$, $\overrightarrow{BC} = -2\mathbf{i} + 4\mathbf{j}$, $\overrightarrow{AC} = 2\mathbf{i} + 6\mathbf{j}$. Thus, $\overrightarrow{AB} \cdot \overrightarrow{BC} = 4(-2) + 2(4) = 0$, and $\overrightarrow{AB}$ is perpendicular to $\overrightarrow{BC}$.

The inner product obeys several other rules:

$$\mathbf{u} \cdot \mathbf{v} = \mathbf{v} \cdot \mathbf{u} \qquad \qquad (1\text{-}112)$$
$$\mathbf{u} \cdot \mathbf{u} \geq 0 \qquad \qquad (1\text{-}113)$$
$$\mathbf{u} \cdot \mathbf{u} = 0 \qquad \text{if and only if} \qquad \mathbf{u} = \mathbf{0} \qquad \qquad (1\text{-}114)$$
$$\mathbf{u} \cdot (\mathbf{v} + \mathbf{w}) = \mathbf{u} \cdot \mathbf{v} + \mathbf{u} \cdot \mathbf{w} \qquad \qquad (1\text{-}115)$$
$$(a\mathbf{u}) \cdot \mathbf{v} = \mathbf{u} \cdot (a\mathbf{v}) = a(\mathbf{u} \cdot \mathbf{v}) \qquad \qquad (1\text{-}116)$$

The proofs are left as exercises (Problem 4 below). These rules show that the inner product resembles ordinary multiplication. However, there are differences; for example, there is no associative law $(\mathbf{u} \cdot \mathbf{v}) \cdot \mathbf{w} = \mathbf{u} \cdot (\mathbf{v} \cdot \mathbf{w})$—in fact, here, neither side makes sense!

From the definition of the inner product we have, for an orthonormal basis $\mathbf{i}, \mathbf{j}$, the rules

$$\mathbf{i} \cdot \mathbf{i} = \mathbf{j} \cdot \mathbf{j} = 1, \qquad \mathbf{i} \cdot \mathbf{j} = \mathbf{j} \cdot \mathbf{i} = 0 \qquad \qquad (1\text{-}117)$$

(Problem 4 below). We can now compute inner products as follows:

$$\begin{aligned}(a\mathbf{i} + b\mathbf{j}) \cdot (c\mathbf{i} + d\mathbf{j}) &= (a\mathbf{i}) \cdot (c\mathbf{i}) + (b\mathbf{j}) \cdot (c\mathbf{i}) + (a\mathbf{i}) \cdot (d\mathbf{j}) + (b\mathbf{j}) \cdot (d\mathbf{j}) \\ &= (ac)(\mathbf{i} \cdot \mathbf{i}) + (bc)(\mathbf{j} \cdot \mathbf{i}) + (ad)(\mathbf{i} \cdot \mathbf{j}) + (bd)(\mathbf{j} \cdot \mathbf{j}) \\ &= ac + bd.\end{aligned}$$

Here we have used the algebraic rules and (1-117). If we use the ordered pair notation, our result is

$$(a, b) \cdot (c, d) = ac + bd \qquad \qquad (1\text{-}118)$$

This is really the same as (1-110). In both cases, we are referring to an orthonormal basis.

PROBLEMS

1. Let $\mathbf{u} = \mathbf{i} + \mathbf{j}$, $\mathbf{v} = 2\mathbf{i} + \mathbf{j}$. Evaluate, checking graphically:

(a) u_x, u_y, v_x, v_y;   (b) $|\mathbf{u}|, |\mathbf{v}|$;   (c) $\mathbf{u} \cdot \mathbf{v}$;

(d) $\cos \varphi$, $\quad \varphi = \sphericalangle(\mathbf{u}, \mathbf{v})$;   (e) $\text{comp}_v \, \mathbf{u}$, $\text{comp}_u \, \mathbf{v}$.

2. Show that the four points A: $(1, 2)$, B: $(2, 3)$, C: $(1, 4)$, D: $(0, 3)$ are vertexes of a square.

3. Show that the triangle with vertexes $(-1, -1)$, $(6, -2)$, $(7, 5)$ is an isosceles right triangle.

4. Prove: (a) (1-112), (b) (1-113), (c) (1-114), (d) (1-115) [*Hint.* Use (1-110)], (e) (1-116), (f) (1-117).

5. An object weighing 5 pounds slides 6 feet down a $60°$ incline. Find the work done by the force of gravity.

6. In raising an object in a vertical plane, we exert a force $\mathbf{F}$ to overcome gravity. Let the plane be the xy-plane, with the y-axis pointing upward, and represent the force $\mathbf{F}$ by $w\mathbf{j}$, where w is the weight of the object.

   (a) Find the work done by force $\mathbf{F}$ in going from P: $(a, 2a)$ to Q: $(3a, 5a)$ along the segment PQ.

   (b) Show that the work done in going from A to B and then from B to C along the segments AB, BC is the same as that done in going directly from A to C along the segment from A to C.

   (c) Extend the result of (b) to a general broken line. (This illustrates a general rule: the work done by a constant force is "independent of the path.")

7. Prove that the diagonals of a rhombus are perpendicular. (*Hint.* Let $\mathbf{u}$, $\mathbf{v}$ be vectors forming the sides, so that $|\mathbf{u}| = |\mathbf{v}|$.)

8. Prove that, if the diagonals of a parallelogram are perpendicular, then the parallelogram is a rhombus.

9. Let AB be a diameter of a circle of center O. Let P be any point of the circumference other than A or B. Prove that $\sphericalangle APB = \pi/2$. (*Hint.* Use $\overrightarrow{AO}$ and $\overrightarrow{OP}$ as a basis.)

10. Prove that for a parallelogram the sum of the squares of the diagonals is equal to the sum of the squares of the sides.

11. Let $\mathbf{u}$, $\mathbf{v}$ be a basis (not necessarily orthogonal) and let $\mathbf{w} = a\mathbf{u} + b\mathbf{v}$, $\mathbf{z} = c\mathbf{u} + d\mathbf{v}$, so that $\mathbf{w} \leftrightarrow (a, b)$, $\mathbf{z} \leftrightarrow (c, d)$. Let $\alpha = \mathbf{u} \cdot \mathbf{u}$, $\beta = \mathbf{u} \cdot \mathbf{v}$, and $\gamma = \mathbf{v} \cdot \mathbf{v}$.

    (a) Show that $|\mathbf{w}|^2 = \alpha a^2 + 2\beta ab + \gamma b^2$.

    (b) Show that $\mathbf{w} \cdot \mathbf{z} = \alpha ac + \beta(ad + bc) + \gamma bd$.

    (c) Show that, for all x, y, $\alpha x^2 + 2\beta xy + \gamma y^2 \geq 0$, with equality if, and only if, $x = 0$, $y = 0$.

1-12 LEFT TURN OPERATION, DIRECTED ANGLE OF TWO VECTORS, AREA FORMULA

Let $\mathbf{u} = a\mathbf{i} + b\mathbf{j}$ be a nonzero vector. We denote by $\mathbf{u}^{\dashv}$ the vector obtained from $\mathbf{u}$ by rotating it through an angle of $\pi/2$ in the counterclockwise direction (Figure 1-37). As in the figure, $a = |\mathbf{u}| \cos \theta$, $b = |\mathbf{u}| \sin \theta$, so that

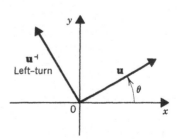

Figure 1-37 Left turn.

$$\mathbf{u} = |\mathbf{u}| \cos\theta\,\mathbf{i} + |\mathbf{u}| \sin\theta\,\mathbf{j}$$

$$\mathbf{u}^\dashv = |\mathbf{u}| \cos\left(\theta + \frac{\pi}{2}\right)\mathbf{i} + |\mathbf{u}| \sin\left(\theta + \frac{\pi}{2}\right)\mathbf{j}$$

$$= -|\mathbf{u}| \sin\theta\,\mathbf{i} + |\mathbf{u}| \cos\theta\,\mathbf{j} = -b\mathbf{i} + a\mathbf{j}$$

Accordingly, we have the general rule:

$$\text{If } \mathbf{u} = a\mathbf{i} + b\mathbf{j}, \qquad \text{then} \qquad \mathbf{u}^\dashv = -b\mathbf{i} + a\mathbf{j} \qquad (1\text{-}120)$$

We also write $\mathbf{0}^\dashv = \mathbf{0}$. Then (1-120) is valid for all vectors. Since $\mathbf{u}^\dashv$ is obtained from $\mathbf{u}$ by "making a left turn," we read $\mathbf{u}^\dashv$ as "$\mathbf{u}$ left turn."

Notice that we can write symbolically

$$\mathbf{u}^\dashv = \begin{vmatrix} a & b \\ \mathbf{i} & \mathbf{j} \end{vmatrix} \qquad (1\text{-}21)$$

For, if we expand the determinant as usual, we obtain $a\mathbf{j} - b\mathbf{i} = -b\mathbf{i} + a\mathbf{j}$. Also, if $\mathbf{v} = c\mathbf{i} + d\mathbf{j}$, then

$$\mathbf{u}^\dashv \cdot \mathbf{v} = \begin{vmatrix} a & b \\ c & d \end{vmatrix} = \begin{vmatrix} u_x & u_y \\ v_x & v_y \end{vmatrix} \qquad (1\text{-}122)$$

For

$$\mathbf{u}^\dashv \cdot \mathbf{v} = (-b\mathbf{i} + a\mathbf{j}) \cdot (c\mathbf{i} + d\mathbf{j}) = -bc + ad = ad - bc.$$

EXAMPLES

$$(2\mathbf{i} - 3\mathbf{j})^\dashv = 3\mathbf{i} + 2\mathbf{j}, \qquad (\mathbf{i} + \mathbf{j})^\dashv = -\mathbf{i} + \mathbf{j}$$

Now let $\mathbf{u}$ be a nonzero vector. Then $\mathbf{u}$, $\mathbf{u}^\dashv$ form an orthogonal basis. We divide these vectors by $|\mathbf{u}|$ and obtain the orthonormal basis $\mathbf{u}/|\mathbf{u}|$, $\mathbf{u}^\dashv/|\mathbf{u}|$. An arbitrary vector $\mathbf{v}$ can be expressed in terms of this basis. For $\mathbf{v} \neq \mathbf{0}$, we can write

$$\mathbf{v} = |\mathbf{v}| \cos\psi \frac{\mathbf{u}}{|\mathbf{u}|} + |\mathbf{v}| \sin\psi \frac{\mathbf{u}^\dashv}{|\mathbf{u}|} \qquad (1\text{-}123)$$

Here the angle ψ is the trigonometric angle from the direction of $\mathbf{u}$ to that of $\mathbf{v}$, as in Figure 1-38; thus, ψ can be positive or negative, greater than π, and so on. If, for example, $3\pi/4$ is one value for ψ, then other values are $2\pi + (3\pi/4)$, $-5\pi/4$, and so on. We call ψ the *directed angle* from $\mathbf{u}$ to $\mathbf{v}$, in contrast with the undirected angle $\varphi = \measuredangle(\mathbf{u}, \mathbf{v})$.

From (1-123)

$$\mathbf{u} \cdot \mathbf{v} = |\mathbf{v}| \cos\psi \frac{\mathbf{u} \cdot \mathbf{u}}{|\mathbf{u}|} + |\mathbf{v}| \sin\psi \frac{\mathbf{u} \cdot \mathbf{u}^\dashv}{|\mathbf{u}|}$$

But the second term is 0, since $\mathbf{u}$, $\mathbf{u}^\dashv$ are orthogonal. If we replace $\mathbf{u} \cdot \mathbf{u}$ by $|\mathbf{u}|^2$, we conclude that

$$\cos\psi = \frac{\mathbf{u} \cdot \mathbf{v}}{|\mathbf{u}||\mathbf{v}|} \qquad (1\text{-}124)$$

This relation shows that $\cos\psi = \cos\varphi$. The conclusion is not surprising, since one choice of ψ must be either φ or $-\varphi$, and both of these angles have the same cosine.

From (1-123) we obtain, in the same way,

$$\mathbf{u}^{\dashv} \cdot \mathbf{v} = |\mathbf{v}| \sin \psi \frac{\mathbf{u}^{\dashv} \cdot \mathbf{u}^{\dashv}}{|\mathbf{u}|} = |\mathbf{u}||\mathbf{v}| \sin \psi$$

since $\mathbf{u}^{\dashv} \cdot \mathbf{u}^{\dashv} = |\mathbf{u}^{\dashv}|^2 = |\mathbf{u}|^2$. Hence

$$\sin \psi = \frac{\mathbf{u}^{\dashv} \cdot \mathbf{v}}{|\mathbf{u}||\mathbf{v}|} \tag{1-125}$$

From (1-124) and (1-125), we obtain both $\cos \psi$ and $\sin \psi$ and, therefore, can find ψ (up to multiples of 2π).

Area formula. From (1-125) and (1-122), we have

$$|\mathbf{u}||\mathbf{v}| \sin \psi = \mathbf{u}^{\dashv} \cdot \mathbf{v} = \begin{vmatrix} u_x & u_y \\ v_x & v_y \end{vmatrix} \tag{1-126}$$

Here, the left side is, except for a possible minus sign, the area of the parallelogram whose sides, properly directed, are $\mathbf{u}$ and $\mathbf{v}$ (see Figure 1-38). We thus obtain the useful rule: *the determinant of second order*

$$\begin{vmatrix} u_x & u_y \\ v_x & v_y \end{vmatrix}$$

equals $\pm A$, *where A is the area of the parallelogram whose sides are* $\mathbf{u}$ *and* $\mathbf{v}$. The + sign is used when ψ, the directed angle form $\mathbf{u}$ to $\mathbf{v}$, is in the first and second quadrants; the − sign is used otherwise.

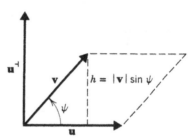

Figure 1-38 Directed angle from u to v.

When $\psi = 0$ or π, $\sin \psi = 0$, the parallelogram collapses to one of zero area, and $\mathbf{u}$, $\mathbf{v}$ are linearly dependent. If $\mathbf{u}$ or $\mathbf{v}$ is $\mathbf{0}$, ψ is not defined, but the determinant is clearly 0 (as is the area). Thus, in general, *the determinant is zero precisely when* $\mathbf{u}$, $\mathbf{v}$ *are linearly dependent*.

Sometimes we have occasion to introduce two different orthonormal bases: $\mathbf{i}, \mathbf{j}$ and $\mathbf{i}', \mathbf{j}'$. It then may happen that the two sets are *similarly oriented*, as in Figure 1-39, or *oppositely oriented*, as in Figure 1-40. Each orthonormal basis has an associated positive direction for angles (the direction in which θ increases in Figure 1-31). For similarly oriented bases, the direction is the same for both; otherwise the directions are opposite, as suggested in Figure 1-40. If we let $\mathbf{u} = \mathbf{i}'$, $\mathbf{v} = \mathbf{j}'$ and u_x, u_y, v_x, v_y be the components of $\mathbf{u}$, $\mathbf{v}$ with respect to the basis $\mathbf{i}, \mathbf{j}$, then the bases are similarly oriented or not according as the determinant (1-126) is positive or negative. Thus, when they are similarly oriented, the angle ψ from $\mathbf{u}$ to $\mathbf{v}$ is $+\pi/2$; in the other case, it is $-\pi/2$.

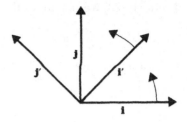

Figure 1-39 Similarly oriented orthonormal bases.

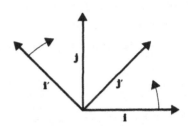

Figure 1-40 Oppositely oriented orthonormal bases.

PROBLEMS

Throughout assume a fixed orthonormal basis **i**, **j**, with associated positive direction for angles.

1. For each pair of vectors given, find $\varphi = \measuredangle(\mathbf{u}, \mathbf{v})$ and ψ, the angle from **u** to **v**:
   - (a) $\mathbf{u} = \mathbf{i}, \mathbf{v} = \mathbf{j}$
   - (b) $\mathbf{u} = \mathbf{i}, \mathbf{v} = -\mathbf{j}$
   - (c) $\mathbf{u} = \mathbf{i} - \mathbf{j}, \mathbf{v} = -\mathbf{i}$
   - (d) $\mathbf{u} = \sqrt{3}\mathbf{i} - \mathbf{j}, \mathbf{v} = -\mathbf{i} + \mathbf{j}$

2. Find the area of each of the figures:
   - (a) The parallelogram $ABCD$ with vertexes A: (0, 0), B: (2, 3), C: (5, 7), D: (3, 4).
   - (b) The parallelogram $ABCD$ with vertexes A: (5, 2), B: (7, -1), C: (2, -4), D: (0, -1).
   - (c) The triangle ABC with vertexes A: (2, 1), B: (5, 3), C: (0, -4).
   - (d) The quadrilateral $ABCD$ with vertexes A: (2, 3), B: (4, 0), C: (5, 7), D: (1, 8).

3. (a) Show that P: (x, y) is on the line through P_1: (x_1, y_1), P_2: (x_2, y_2) if, and only if,

$$\begin{vmatrix} x - x_1 & y - y_1 \\ x_2 - x_1 & y_2 - y_1 \end{vmatrix} = 0$$

   (b) Where is P if the determinant is greater than 0? Less than 0?

4. In each case, graph the vectors $\mathbf{i}'$, $\mathbf{j}'$, verify that they form an orthonormal basis, and determine whether it is oriented similarly or oppositely to **i**, **j**:

   - (a) $\mathbf{i}' = \mathbf{j}, \mathbf{j}' = -\mathbf{i}$
   - (b) $\mathbf{i}' = \dfrac{\mathbf{i} + \mathbf{j}}{\sqrt{2}}, \mathbf{j}' = \dfrac{-\mathbf{i} + \mathbf{j}}{\sqrt{2}}$
   - (c) $\mathbf{i}' = \dfrac{3\mathbf{i} - 4\mathbf{j}}{5}, \mathbf{j}' = \dfrac{4\mathbf{i} + 3\mathbf{j}}{5}$
   - (d) $\mathbf{i}' = \dfrac{4\mathbf{i} + 3\mathbf{j}}{5}, \mathbf{j}' = \dfrac{3\mathbf{i} - 4\mathbf{j}}{5}$

5. In each case show that **u** is a unit vector, and find **v** so that **u**, **v** form an orthonormal basis oriented similarly to **i**, **j**:

   - (a) $\mathbf{u} = \dfrac{5\mathbf{i} - 12\mathbf{j}}{13}$
   - (b) $\mathbf{u} = \dfrac{\sqrt{3}\mathbf{i} + \mathbf{j}}{2}$

(c) $u = \dfrac{-i - \sqrt{3}j}{2}$ (d) $u = \dfrac{-7i + j}{5\sqrt{2}}$

6. Prove the rules: (a) $(u + v)^{\dashv} = u^{\dashv} + v^{\dashv}$;   (b) $(au)^{\dashv} = au^{\dashv}$;   (c) $(u^{\dashv})^{\dashv} = -u$; (d) $i^{\dashv} = j$, $j^{\dashv} = -i$; (e) the angle from u to v equals the angle from $u^{\dashv}$ to $v^{\dashv}$; (f) $|u - v| = |u^{\dashv} - v^{\dashv}|$;   (g) if α is the angle from u to v and β is the angle from $u^{\dashv}$ to v, then $\sin \alpha = \cos \beta$.

7. Equation (1-123) expresses v in the form $au + bu^{\dashv}$ and, hence, expresses v as the sum of a vector collinear with u and a vector orthogonal to u. Show with the aid of dot products that this expression is unique; that is, if $au + bu^{\dashv} = cu + du^{\dashv}$, then $a = c$ and $b = d$.

8. Prove that given a line L and a point P not on L, there is a unique point P_0 on L for which PP_0 is perpendicular to L. (*Hint.* Choose distinct points Q_1, Q_2 on L, let $u = \overrightarrow{Q_1Q_2}$, $v = \overrightarrow{Q_1P}$, write $v = au + bu^{\dashv}$, as in Problem 7, and choose P_0 on L so that $\overrightarrow{Q_1P_0} = au$.)

1-13 PHYSICAL APPLICATIONS, STATICS

In physics it is shown that velocity, acceleration, and force can be represented by vectors. In each case, the direction of the vector is that of the physical quantity, and the length of the vector equals the magnitude of the physical quantity, in appropriate units. For velocities we have the rule that for a motion that is compounded of several motions, the velocity vector is the vector sum of the velocity vectors for the several motions. The meaning of this rule will be made clear by an example.

EXAMPLE 1. A man jumps from a moving car in such a way that, if the car were standing still, his velocity would have magnitude 10 (in miles per hour) and would make an angle of 60° with the forward direction. If the car is moving at 20 mph, with what velocity does he leave the car?

Solution. Let u be the velocity vector of the car, v the velocity vector the man would have if the car were standing still. Then the actual velocity of the man is $u + v$. We choose an orthonormal basis i, j (Figure 1-41), so that $u = 20i$, $v = 5i - 5\sqrt{3}j$, and $u + v = 25i - 5\sqrt{3}j$, the desired velocity vector.

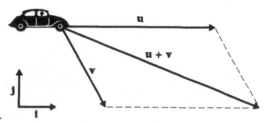

Figure 1-41   Combining velocities.

For forces acting on a particle, the particle is in equilibrium (and, therefore, has no acceleration) precisely when the vector sum (*resultant*) of the forces is 0. A similar rule applies to a rigid body that is subject to forces whose lines of action pass through a single point.

EXAMPLE 2. Four members of a bridge truss act on a pin with forces as shown in Figure 1-42; F_1 and F_2 are horizontal forces, F_3 is vertical, and F_4 makes an angle of 30° with the horizontal. Find $|F_1|$, $|F_4|$ and determine the sense of F_1 and F_4 if $|F_2| = 3900$, $|F_3| = 1500$ (all in pounds).

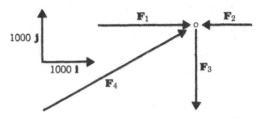

Figure 1-42   Forces in equilibrium.

Solution. In terms of an orthonormal basis (Figure 1-42),

$$F_1 = ai, \qquad F_2 = -3900i, \qquad F_3 = -1500j, \qquad F_4 = ci + dj$$

Thus, $a - 3900 + c = 0$, $-1500 + d = 0$. In addition, $d/c = \tan 30° = 1/\sqrt{3}$. Consequently, $d = 1500$, $c = \sqrt{3}\, d = 1500\sqrt{3}$, $a = 3900 - c = 1302$. Hence, $|F_1| = 1302$, $|F_4| = \sqrt{c^2 + d^2} = 3000$, F_1 has sense opposite to F_2, and F_4 acts upward.

PROBLEMS

1. A man walks at a speed of 4 mph in the aisle of a moving train, which is moving along a straight track at 40 mph. Find his velocity vector relative to the ground if (a) he is walking to the front of the car, and (b) he is walking to the rear of the car.

2. A small plane can fly 150 mph in still air. The wind is blowing 25 mph from the southwest. What course should the pilot set and how long will it take him to fly 200 miles (a) due north, (b) due east?

3. Two forces of magnitude 8 lb and 10 lb act on a particle at an angle of $\pi/4$. Find the direction and magnitude of the resultant.

4. Two forces of 10 lb each act on a particle at an angle of $\pi/2$. A third force of 15 lb also acts on the particle, in the same plane as the first two. What direction should the third force have so that the resultant of all three forces has the greatest magnitude?

5. Three forces act on a particle in a plane and are in equilibrium. If the magnitudes are 50 lb, 100 lb, and 200 lb, respectively, find the angles between the forces.

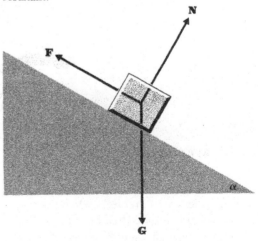

Figure 1-43

6. For a body resting on an inclined plane (Figure 1-43), three forces act: gravity **G**, a reaction force **N** perpendicular to the plane, a friction force **F** directed upward along the plane. The coefficient of friction μ is defined as the ratio of $|\mathbf{F}|$ to $|\mathbf{N}|$ when the angle of inclination α is such that the body is about to slip. Show that $\mu = \tan \alpha$ for this α.

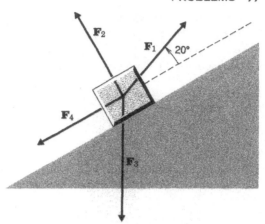

7. A weight of 250 lb. rests on a plane inclined 30° to the horizontal. It is acted on by a

Figure 1-44

force $\mathbf{F}_1$ of magnitude 200 (in pounds) directed upward along a line making a 20° angle with the plane, by the gravitational force $\mathbf{F}_3$ acting downward, by a reaction force $\mathbf{F}_2$ acting perpendicular to the plane, and by a force $\mathbf{F}_4$ acting downward along the plane (Figure 1-44). Find $\mathbf{F}_2$ and $\mathbf{F}_4$. (*Hint.* Use an orthonormal basis **i**, **j** with **i** along the plane.)

8. A span AB is free to turn in a vertical plane through AC, and is fastened by a cable BC to a point C, vertically above A (Figure 1-45). Suppose that $\measuredangle BAC = 45°$, $AB = 20'$, and $AC = 12'$. Determine the forces that act at B if a 1000 lb weight is hung from B and if the system is at rest.

9. *Center of mass.* Let $P_1, \ldots, P_k$ be k distinct points in the plane, and suppose that positive masses $m_1, \ldots, m_k$, respectively, are located at these points (Figure 1-46). Let A be a chosen reference point. The *center of mass* C is then defined by the equation

$$(m_1 + \cdots + m_k)\overrightarrow{AC}$$
$$= m_1\overrightarrow{AP_1} + \cdots + m_k\overrightarrow{AP_k}$$

(a) Show that C is independent of the choice of A; that is, if $(m_1 + \cdots + m_k)\overrightarrow{A'C'} = m_1\overrightarrow{A'P_1} + \cdots + m_k\overrightarrow{A'P_k}$, then $\overrightarrow{AC} = \overrightarrow{AC'}$, so that $C = C'$.

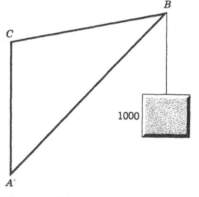

Figure 1-45

(b) Let $k = 3$ and let C_{12} be the center of mass of m_1 at P_1, m_2 at P_2. Show that C is the center of mass of mass $m_1 + m_2$ at C_{12} and mass m_3 at P_3.

(c) Let $k = 3$, P_1: $(1, 2)$, P_2: $(3, 5)$, P_3: $(2, 7)$, C: $(2, 6)$, $m_1 + m_2 + m_3 = 7$ units. Find m_1, m_2, m_3.

0. In mechanics it is shown that, if two objects collide, their total *momentum* is unaffected by the collision. For the case of two particles, the total momentum is $m_1\mathbf{v}_1 + m_2\mathbf{v}_2$, where m_1, m_2 are the masses and $\mathbf{v}_1$, $\mathbf{v}_2$ are the velocity vectors. Let a particle of mass 1000 lb collide with one of mass 10 lb. Before the collision, let both particles

move along a line in opposite directions at speeds 5 ft/sec and 3 ft/sec respectively; after the collision let the first particle have speed 4 ft/sec in the same direction as before. Find the speed and the direction of the second particle after collision. (The example might be illustrated by a fully laden freight car bumping an empty one.)

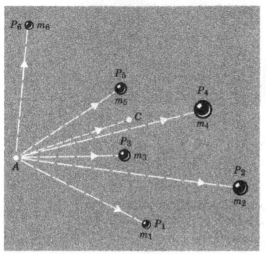

Figure 1-46   Center of mass.

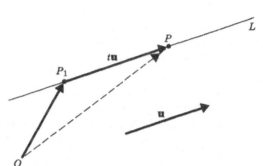

Figure 1-47   Vector equation of line.

1-14 EQUATION OF STRAIGHT LINE

Let a straight line L be given and let P_1 be a given point on L. Let O be a fixed origin. Let P be an arbitrary point on L (Figure 1-47). For all choices of P, the vectors $\overrightarrow{P_1P}$ are linearly dependent, so that we can write $\overrightarrow{P_1P} = t\mathbf{u}$, where t is a scalar and $\mathbf{u}$ is one nonzero vector representable by a directed line segment on L. Since $\overrightarrow{OP} = \overrightarrow{OP_1} + \overrightarrow{P_1P}$, we have

$$\overrightarrow{OP} = \overrightarrow{OP_1} + t\mathbf{u} \tag{1-140}$$

Here O, P_1 and $\mathbf{u} \neq \mathbf{0}$ are given and t is allowed to vary. As t varies over all real numbers, P takes on all positions on L, each position having a unique t-value. We call (1-140) a *vector equation* of line L. Other vector equations for the same line are obtained by changing the point P_1 and by replacing $\mathbf{u}$ by a vector $k\mathbf{u}$ for some scalar k ($k \neq 0$); we can also change the origin O.

EXAMPLE. Let Cartesian coordinates be chosen and let O be the origin, P_1 the point $(2, 1)$, and $\mathbf{u}$ the vector from $(2, 1)$ to $(5, 2)$. The positions of P for different values of t are shown in Figure 1-48, and are also tabulated in the accompanying table.

In Equation (1-140) we can interpret t as *time*, measured in a given unit. The equation then describes the motion of a point P along the line. At time $t = 0$, P is at P_1; at time $t = 1$, P has been displaced from P_1 by $\mathbf{u}$, and so on. Thus, $\mathbf{u}$ is the displacement per unit time, and we call $\mathbf{u}$ the *velocity vector*

t	P
0	$(2, 1)\ P_1$
1	$(5, 2)$
2	$(8, 3)$
3	$(11, 4)$
-1	$(-1, 0)$
-2	$(-4, -1)$

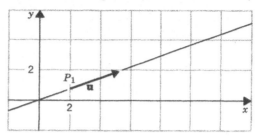

Figure 1-48   Graph of vector equation of line.

of the motion. The *speed* is the distance moved per unit time; accordingly, the speed is $|\mathbf{u}|$. We summarize: Equation (1-140) describes the motion of a point P along a straight line at speed $|\mathbf{u}|$, with velocity vector $\mathbf{u}$. Since $\mathbf{u}$ is a given vector, our speed is fixed. Motion with changing speed (and changing direction) is studied in the calculus.

Let us now consider two lines, with corresponding equations:

$$\overrightarrow{OP} = \overrightarrow{OP_1} + t\mathbf{u}, \qquad \overrightarrow{OQ} = \overrightarrow{OQ_1} + s\mathbf{v} \tag{1-141}$$

For the second line we denote the variable point by Q, the fixed point by Q_1, the "time" by s, and the given vector by $\mathbf{v}$.

There are then three possibilities (proofs are left to Problem 10 below):

(a) The two lines may coincide. In this case $\mathbf{u}, \mathbf{v}$ must be linearly dependent and $\overrightarrow{P_1Q_1}, \mathbf{u}$ must also be linearly dependent. Conversely, if both pairs—$\mathbf{u}, \mathbf{v}$ and $\overrightarrow{P_1Q_1}, \mathbf{u}$—are linearly dependent, the lines coincide.

(b) The two lines may be parallel and noncoincident. In this case $\mathbf{u}, \mathbf{v}$ must again be linearly dependent, but $\overrightarrow{P_1Q_1}, \mathbf{u}$ must be linearly independent. Conversely, if $\mathbf{u}, \mathbf{v}$ are linearly dependent, but $\overrightarrow{P_1Q_1}, \mathbf{u}$ are linearly independent, the lines are parallel.

(c) The lines intersect at a single point. In this case, $\mathbf{u}, \mathbf{v}$ are linearly independent and, conversely, if $\mathbf{u}, \mathbf{v}$ are linearly independent, the lines intersect at a single point. In this case, we define the angle between the lines to be $\varphi = \sphericalangle(\mathbf{u}, \mathbf{v})$ or its supplement $\pi - \varphi$ (see Figure 1-49). We also use the same definition for cases (a) and (b), so that the angle between parallel or coincident lines is taken to be 0 or π.

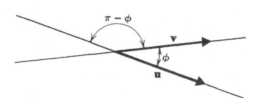

Figure 1-49   Angle between lines.

1-15 PARAMETRIC EQUATIONS FOR A LINE

Let a Cartesian coordinate system be chosen with origin O and let a straight line L have vector equation (1-140). Let P_1 be the point (x_1, y_1), $\mathbf{u}$ the vector $a\mathbf{i} + b\mathbf{j}$, let $P: (x, y)$ be a general point on L. Then $\overrightarrow{OP_1} = x_1\mathbf{i} + y_1\mathbf{j}$, $\overrightarrow{OP} = x\mathbf{i} + y\mathbf{j}$, and (1-140) becomes

$$x\mathbf{i} + y\mathbf{j} = x_1\mathbf{i} + y_1\mathbf{j} + t(a\mathbf{i} + b\mathbf{j}) = (x_1 + at)\mathbf{i} + (y_1 + bt)\mathbf{j}$$

Hence, P is on L precisely when

$$x = x_1 + at, \qquad y = y_1 + bt \tag{1-150}$$

Equations (1-150) are called *parametric equations* of L. In (1-150), t is called the *parameter,* and t is allowed to take on all real values.

If we write $\mathbf{u} = |\mathbf{u}| \cos \theta\, \mathbf{i} + |\mathbf{u}| \sin \theta\, \mathbf{j}$, so that θ gives the direction of $\mathbf{u}$, then $\tan \theta$ is called the *slope* of the line and is usually denoted by m. Since $a = |\mathbf{u}| \cos \theta$, $b = |\mathbf{u}| \sin \theta$, we have

$$m = \tan \theta = \frac{b}{a} \tag{1-151}$$

When $\theta = \pi/2$, for example, $\tan \theta$ is undefined; we say that the line L has *infinite slope.*

A nonzero vector $\mathbf{v}$ perpendicular to $\mathbf{u}$ is called a *normal vector* of the line L. Therefore, $\mathbf{u}^{\dashv} = -b\mathbf{i} + a\mathbf{j}$ is a normal vector, as is each vector $k\mathbf{u}^{\dashv}$ with $k \neq 0$.

In Equations (1-150), as t varies, x and y vary and locate the moving point P. For the example in Section 1-14 (Figure 1-48) $\mathbf{u} = 3\mathbf{i} + \mathbf{j}$ and the parametric equations are

$$x = 2 + 3t, \qquad y = 1 + t$$

The x and y values for different t are given in the table in Figure 1-48. The slope is $b/a = 1/3$.

As in the preceding section, we can change the initial point P_1 and can replace $\mathbf{u}$ by a vector $k\mathbf{u}$ ($k \neq 0$). Thus we obtain other sets of parametric equations for the same line. We can also express the conditions for coincidence, being parallel and intersecting in terms of the parametric equations (See Problem 11 below).

The angle between two lines

$$\begin{aligned} x = x_1 + a_1 t, \qquad y = y_1 + b_1 t \\ x = x_2 + a_2 t, \qquad y = y_2 + b_2 t \end{aligned} \tag{1-152}$$

can also be found from the parametric equations. The vectors associated with the lines (1-152) are

$$\mathbf{u}_1 = a_1\mathbf{i} + b_1\mathbf{j} \qquad \mathbf{u}_2 = a_2\mathbf{i} + b_2\mathbf{j}$$

Accordingly, if $\varphi = \sphericalangle(\mathbf{u}_1, \mathbf{u}_2)$

$$\cos \varphi = \frac{\mathbf{u}_1 \cdot \mathbf{u}_2}{|\mathbf{u}_1||\mathbf{u}_2|} = \frac{a_1 a_2 + b_1 b_2}{\sqrt{a_1^2 + b_1^2}\sqrt{a_2^2 + b_2^2}}$$

and φ, or $\pi - \varphi$, is the angle sought. In particular, the lines are perpendicular when $\varphi = \pi/2$ or $\mathbf{u}_1 \cdot \mathbf{u}_2 = 0$ or

$$a_1 a_2 + b_1 b_2 = 0 \tag{1-153}$$

1-16 LINEAR EQUATION FOR A STRAIGHT LINE

A straight line is commonly represented by a linear equation in x and y: that is, by an equation

$$Ax + By + C = 0$$

with A, B not both 0 (Section 0-7). We can go from the parametric equations (1-150) to a linear equation by eliminating t. We find

$$-b(x - x_1) + a(y - y_1) = 0 \quad \text{or} \quad -bx + ay + (bx_1 - ay_1) = 0 \tag{1-160}$$

We can also write this equation in the point-slope form:

$$y - y_1 = \frac{b}{a}(x - x_1) = m(x - x_1) \tag{1-161}$$

We can also obtain the linear equation by the following reasoning: the point P: (x, y) is on L precisely when $\overrightarrow{P_1P}$ is orthogonal to the normal vector $\mathbf{u}^{\perp}$. Hence, P is on L precisely when

$$[(x - x_1)\mathbf{i} + (y - y_1)\mathbf{j}] \cdot [-b\mathbf{i} + a\mathbf{j}] = 0$$

or

$$-b(x - x_1) + a(y - y_1) = 0$$

We again obtain the first of the equations (1-160).

We can also go from a linear equation to parametric equations. For example, given the equation $2x - y + 3 = 0$, we set $x = t$ and find that $y = 2t + 3$; thus

$$x = t, \quad y = 2t + 3$$

are parametric equations for the line. Here the slope is $2/1 = 2$. For the general equation $Ax + By + C = 0$ we find, in the same way, that the slope is

$$m = -\frac{A}{B} \tag{1-162}$$

PROBLEMS

1. Graph Equation (1-140) for the case in which $|\overrightarrow{OP_1}| = 2$ and $\mathbf{u}$ is a unit vector perpendicular to $\overrightarrow{OP_1}$.
2. Write parametric equations and graph for each of the following lines L:
   (a) $(3, 2)$ on L, $\mathbf{u} = 5\mathbf{i} - \mathbf{j}$ along L.
   (b) $(2, 1)$ and $(3, -2)$ on L.
   (c) (x, y) is $(3, 0)$ for $t = 0$; (x, y) is $(2, 2)$ for $t = 1$.
   (d) $(x, y) = (0, 0)$ for $t = -1$; $(x, y) = (1, 3)$ for $t = 2$.
3. (a) Graph the velocity vector and find the speed for the equations $x = 3 + 2t$, $y = 5 - t$.
   (b) A point P moves on the line through $(2, 3)$ and $(5, 7)$ at speed 10 units per unit of time. At $t = 0$, P is at $(2, 3)$ moving toward $(5, 7)$. Find parametric equations for the motion.

4. Let P_1, P_2 be distinct points on line L. Show that the vector equation of L can be written

$$\overrightarrow{OP} = (1 - t)\overrightarrow{OP_1} + t\overrightarrow{OP_2}$$

Show that, for $0 \leq t \leq 1$, P is on the segment $P_1 P_2$ and divides it in the ratio $t : (1 - t)$.

5. Find a general linear equation for the line $x = 4 + 3t$, $y = -1 - t$.

6. Find parametric equations for the line $x + 3y - 5 = 0$.

7. Find a line L through the origin satisfying the given condition:
   (a) L is parallel to the line $2x - y + 14 = 0$.
   (b) L is perpendicular to the line $2x - y + 14 = 0$.

8. A point P moves according to the equations: $x = 3 + 5t$, $y = 3 - 2t$. A second point Q moves according to the equations $x = 3 + t$, $y = 3 + t$.
   (a) At what time and where do P and Q coincide?
   (b) When are P and Q two units apart?

9. Find the point of intersection of the line $x = 2 - 3t$, $y = 1 + 4t$, and the line $x = 5 + 2s$, $y = -2 - 3s$.

10. Prove the correctness of the conditions given in Section 1-14 for the lines (1-141)
    (a) to coincide   (b) to be parallel   (c) to intersect at one point

11. Let two lines be given by parametric equations (1-152) and let

$$D = \begin{vmatrix} a_1 & b_1 \\ x_2 - x_1 & y_2 - y_1 \end{vmatrix}, \quad E = \begin{vmatrix} a_1 & b_1 \\ a_2 & b_2 \end{vmatrix}$$

Show that the lines are:
    (a) Coincident precisely when $D = 0$ and $E = 0$.
    (b) Parallel and noncoincident precisely when $D \neq 0$ and $E = 0$.
    (c) Intersecting precisely when $E \neq 0$.
    (*Hint.* As in Section 1-12, $a\mathbf{i} + b\mathbf{j}$ and $c\mathbf{i} + d\mathbf{j}$ are linearly dependent precisely when $ad - bc = 0$.)

12. Find the equation of a straight line from the given information:
    (a) $(1, 3)$ on L, $\mathbf{v} = 2\mathbf{i} - \mathbf{j}$ normal to L
    (b) $(5, 2)$ on L, $\mathbf{u} = \mathbf{i} + 3\mathbf{j}$ along L
    (c) $(0, 0)$ on L, $\mathbf{v} = \mathbf{i}$ normal to L
    (d) $(1, 0)$ and $(1, 2)$ on L

13. For each of the following lines, graph and find the slope and a normal vector:
    (a) $2x - y - 3 = 0$ (b) $y - 2 = 4(x - 7)$

    (c) $y - 5 = 0$ (d) $\dfrac{x}{7} + \dfrac{y}{2} = 1$

14. Let line L have unit normal vector $\mathbf{v}$ and let N be the foot of the perpendicular to L from O, as in Figure 1-50 (if L passes through O, N is taken to be O). Let $\overrightarrow{ON} = p\mathbf{v}$, so that $|\overrightarrow{ON}| = |p|$.
    (a) Show that L has equation $\overrightarrow{OP} \cdot \mathbf{v} = p$. This is called the *normal form* of the equation of L.
    (b) Show that in rectangular coordinates the equation becomes $x \cos \omega + y \sin \omega = p$, where $\mathbf{v} = \cos \omega \mathbf{i} + \sin \omega \mathbf{j}$.
    (c) Show that the linear equation $Ax + By + C = 0$ can be written in normal form as in (b), where

$$\cos \omega = \frac{A}{\sqrt{A^2 + B^2}}$$

$$\sin \omega = \frac{B}{\sqrt{A^2 + B^2}}$$

[*Hint.* Write the equation first as $(x\mathbf{i} + y\mathbf{j}) \cdot (A\mathbf{i} + B\mathbf{j}) = -C$, then multiply both sides by $1/|A\mathbf{i} + B\mathbf{j}|$.]

(d) Show that, if $p_1 \neq p_2$, the equations

$$x \cos \omega + y \sin \omega = p_1,$$
$$x \cos \omega + y \sin \omega = p_2$$

represent parallel lines at distance $|p_1 - p_2|$ apart.

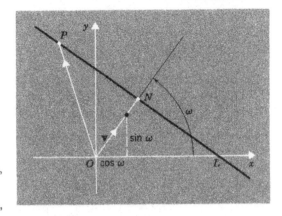

Figure 1-50

(e) Find the distance between the parallel lines $3x + y - 5 = 0, 3x + y + 7 = 0$. [*Hint.* Apply the results (c) and (d).]

(f) Show that the distance from point $P_1: (x_1, y_1)$ to the line $L: Ax + By + C = 0$ is

$$d = \frac{|Ax_1 + By_1 + C|}{\sqrt{A^2 + B^2}}$$

[*Hint.* Construct a line through P_1 parallel to L and proceed as in (e).]

2
LIMITS

2-1 CONCEPT OF A FUNCTION, TERMINOLOGY, COMPOSITION

The concept of a function is defined and is illustrated in Section 0-10. We assume familiarity with this section and with the important terms used: *domain, range, one-to-one, inverse function, identity function, equality* of functions, and *mappings into* or *onto* a set.

Generally, functions are denoted by letters such as f, g, F, G, φ, ... For the function f, we denote the value assigned to x by $f(x)$; in addition, we frequently write $y = f(x)$ for this value, and also refer to the function whose equation is $y = f(x)$ or, more simply, to the function $y = f(x)$.

Occasionally, we use arrows to indicate the assignment of values by a function. Thus, instead of $f(2) = 7$, we can write $2 \xrightarrow{f} 7$.

In this chapter we emphasize real functions of a real variable, that is, functions whose domain and range are sets of real numbers. We call them, briefly, "functions of one variable." The equation $y = x^2$ defines such a function f, whose domain is the set of all real numbers and whose range is the set of all nonnegative real numbers. We can describe these sets by the respective inequalities $-\infty < x < \infty$, $y \geq 0$. By a *zero* of a function f, we mean a value x for which $f(x) = 0$.

Later, we shall also consider functions that are not real functions of a real variable. For example, we shall consider a function assigning a vector to each real number of an interval; we call this a *vector function*. In addition, we shall meet functions of several real variables. For example, $z = \sqrt{x^2 + y^2}$ is such a function. It can also be interpreted as $|x\mathbf{i} + y\mathbf{j}|$, the norm of the vector $\mathbf{r} = x\mathbf{i} + y\mathbf{j}$, and, therefore, as a function whose domain is the set of vectors in the plane, and whose range is a set of real numbers. In general, a function whose range is a set of real numbers is called a *real-valued function*.

Occasionally, we do not specify the domain of a particular function, with the understanding that the domain is the set of all values for which the function is defined by a given formula; for example, $y = 1/x$ has domain all real numbers except 0.

The equations $y = x^2$, $y = t^2$, and $u = v^2$ all serve as formulas for functions. However, all three functions are the same function f, which assigns to each real number the square of that number. Thus we have a free choice of notation for the typical values in the domain and the range of a function. In each particular problem, it is important to be consistent and to indicate each change of notation.

For a function given by a formula, we occasionally may not use the whole domain for which the formula is valid, but may restrict ourselves to a smaller domain. For example, we may consider the function $y = \sin x$, $0 \leq x \leq \pi$.

Composition of Functions. A familiar trick begins: "Choose a number from 1 to 10, then double it, then square what you have, then . . . ." We leave the analysis of the trick to the exercises (Problem 7 below) and observe here that we are dealing with several functions in a special way. The first function is given by $u = 2x$, x being the chosen number. The second is given by $y = u^2$. After the second step, the value of y is expressible in terms of x:

$$y = u^2 = (2x)^2$$

Our example is a special case of the following general situation. Let $u = f(x)$ and $y = g(u)$ be two functions. If for each x, $u = f(x)$ is in the domain of g, then each x determines a u which determines a y:

$$x \xrightarrow{\ f\ } u \xrightarrow{\ g\ } y$$

and $y = g[f(x)]$. Thus we have assigned a y to each x by the equation $y = g[f(x)]$. The new function is called the *composite* of g and f and is denoted by $g \circ f$. Thus

$$x \xrightarrow{\ g \circ f\ } g[f(x)]$$

EXAMPLE 1. $u = f(x) = \sin x$, $y = g(u) = u^2$. Then, $g \circ f$ is given by the equation $y = (\sin x)^2$. Thus we are simply substituting one function in another. We can best picture what is happening not by the usual graphs but by a typical function diagram (Figure 2-1).

In general, the composition of two functions g, f is the formation of a new function $g \circ f$ whose value at each x is

$$g[f(x)]$$

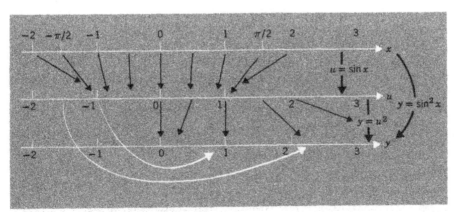

Figure 2-1 Composition of $y = u^2$ and $u = \sin x$.

that is, it is the value of the function g at the point $f(x)$. We point out that g could also be given by an equation $y = g(x)$. The notation we choose is a matter of convenience. Thus the composite of x^2 and $\sin x$ is $[\sin x]^2$. However, confusion is avoided in each case when we write, as above, $y = g(u)$ and $u = f(x)$ (of course, other letters can be used).

It is important to notice that *the order of operations is significant* here; that is, $g \circ f$ gives the function $[\sin x]^2$, whereas $f \circ g$ would be the function obtained from

$$y = \sin u, \qquad u = x^2$$

that is, $y = \sin x^2$.

EXAMPLE 2. $y = \sqrt{1 + \cos x}$. This is the composition $g \circ f$, where

$$y = g(u) = \sqrt{u}, \qquad u = f(x) = 1 + \cos x$$

EXAMPLE 3. $y = \log_{10}(1 - x)$. This is the composition $g \circ f$, where

$$y = g(u) = \log_{10} u, \qquad u = f(x) = 1 - x$$

Domain of Composite Function. In our examples we have ignored the question of where the composite function is defined. For Examples 1 and 2 the function f is defined for all real x and the range of f lies in the domain of g, so that $g \circ f$ has the same domain as f. However, for Example 3, the range of $f(x) = 1 - x$ is the set of all real numbers, whereas the domain of $g(u) = \log_{10} u$ is the set of *positive* real numbers. To form the composite function, we must first restrict the domain of f to those x for which $1 - x$ is positive, that is, to all $x < 1$. Thus we can write

$$g[f(x)] = \log_{10}(1 - x), \qquad x < 1$$

By restricting the domain of f, we have really changed the function and should change the symbol used. However, it is general practice to use the same symbol, since the context makes the meaning clear.

PROBLEMS

1. Express each of the following functions as a composite function $g \circ f$, and also give the domain of f, g, and $g \circ f$.
   (a) $y = \sin(x^2 + 1)$ (b) $y = (\log_{10} x)^2$
   (c) $y = (x + 2)^{10}$ (d) $y = 1/(1 + x^3)$
2. Let $y = f(x) = x^3$, $y = g(x) = \sin x$, $y = h(x) = 2^x$. Form each of the composite functions:
   (a) $f \circ g$ (b) $g \circ h$ and $h \circ g$ (c) $f \circ h$
   (d) $f \circ f$ (e) $f \circ (g \circ h)$ and $(f \circ g) \circ h$
3. Find two functions f, g (not the same function) for which $f \circ g = g \circ f$.
4. State the domain for each of the functions:
   (a) $y = \log_{10} \log_{10} x$ (b) $y = \log_{10} \log_{10} \log_{10} x$ (c) $y = \sqrt{1 - \sqrt{x - 1}}$
   (d) $y = 1/(1 - \sin x)$ (e) $y = \sqrt{-1 - 3x - x^2}$
5. Let $f(x) = x^x$ and let $g = f \circ f$. Which is a correct expression for $g(x)$?
   (a) x^{x^x} (b) $x^{[x^{(x^x)}]}$ (c) $x^{[x^{x+1}]}$ (d) $(x^x)^{(x^x)}$

6. Show that, if f maps X into Y, g maps Y into Z, and h maps Z into W, then $h \circ (g \circ f)$ and $(h \circ g) \circ f$ are the same function.

7. John was asked to pick a number from 1 to 10, then to double it, then to square his result, then to divide his result by the number he started with, then to add 4, then to divide by 4, and then state the number he obtained. He stated: "6," and was told that he had started with 5. Let x be the number he started with, u the number obtained by doubling, and so on. Show that the final number is $x + 1$.

8. (a) Give an example to show that the range of $g \circ f$ need not be the same as the range of g.

   (b) Show that, if f and g are one-to-one and $g \circ f$ is defined, then $g \circ f$ is one-to-one.

   (c) Show that for all f

   $$I \circ f = f \quad \text{and} \quad f \circ I = f$$

   (*Remark. I* stands for a different *identity function* in each equation.)

2-2 QUALITATIVE ANALYSIS OF FUNCTIONS OF ONE VARIABLE

By the *graph* of a function f of one variable, we mean the set of all points (x, y) in the xy-plane for which $y = f(x)$. To illustrate the concepts developed in this chapter, we first consider them qualitatively, as suggested by a graph.

Figure 2-2 shows the graph of a function f defined for $1 \leq x \leq 4$. We observe that the graph is smooth and has no break; we say that f is *continuous* in the interval. At $x = 3$, $y = -1$, and $f(x) \geq -1$ for all x sufficiently close to 3, and we say that f has a *local minimum*, equal to -1, at $x = 3$. In fact, $f(x) \geq -1$ for $1 \leq x \leq 4$ and $f(x) = -1$ only at the point $x = 3$ of this interval; we say that, in this interval, f has its *absolute minimum*, equal to -1, at $x = 3$. We notice that at $(3, -1)$ the *tangent line* to the graph is horizontal; for the present we can think of a tangent line as one that simply "hugs" the graph as closely as possible. In the given interval $[1, 4]$, f has its largest value, equal to 1, only at $x = 1$, and we say that f has its *absolute maximum*, equal to 1, at $x = 1$; however, at this point, the tangent line is not horizontal (this is

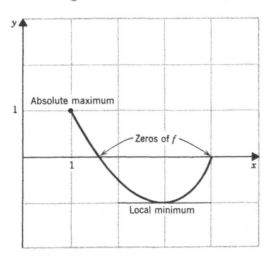

Figure 2-2 Function defined on $[1, 4]$.

related to the fact that the maximum occurs at an end point of the interval). The function has *zeros* at $x = 1.6$ and $x = 4$; that is, 1.6 and 4 are zeros of f.

We now examine the function f given graphically in Figure 2-3. Its domain is considered to be all real numbers. Between $x = 3$ and $x = 6$, $f(x)$ rises as x increases. We say that f is *monotone strictly increasing* in the interval [3, 6]. At the point (9, 6), we cannot draw a tangent, since the curve changes direction suddenly; there is a *corner* at this point, and f has a local maximum at $x = 9$. At the point (12, 3) a tangent line would have to be vertical, and the curve doubles back along such a line. There is a *cusp* at this point, and f has a local minimum at $x = 12$. At both the corner and the cusp, there is no break in the curve and, therefore, f is still continuous at these points.

As x increases toward 15, $f(x)$ seems headed for the value 7; we say that f has *limit* 7 as x approaches 15 from the left. Similarly, f has limit 10, as x approaches 15 from the right. Since the two limits disagree, we say that f has a *jump discontinuity* at $x = 15$. As x approaches 20 from the left, f shoots off to "infinity" in the negative direction. We say that f has *limit* $-\infty$ as x approaches 20 from the left; similarly, f has *limit* ∞ or $+\infty$ as x approaches 20 from the right. The line $x = 20$ serves as a *vertical asymptote* of the graph. Because of the break in the graph at $x = 20$, f is said to be discontinuous at $x = 20$.

As x increases indefinitely beyond 20, $f(x)$ appears to approach the value 7; we say that f has limit 7 as x approaches $+\infty$, and that there is a *horizontal asymptote* $y = 7$. At $x = -4$, the graph illustrates how a function can oscillate rapidly, while remaining continuous; the graph has no break. As x approaches -8 from the right, the oscillations become more and more rapid and maintain their size, so that f has no definite limit. This illustrates an *oscillatory discontinuity*.

2-3 OPERATIONS ON FUNCTIONS OF ONE VARIABLE

Let f and g be functions defined in the same interval. Then we denote by $f + g$ the function h, defined in this interval, for which $h(x) = f(x) + g(x)$ for each x; that is, $f + g$ is the function whose value is $f(x) + g(x)$ for each x. For example, if

$$f(x) = x^2 - 2, \qquad g(x) = x^3 + 1, \qquad 0 \le x \le 1$$

then $f + g = h$ is defined by

$$h(x) = x^3 + x^2 - 1, \qquad 0 \le x \le 1$$

We denote by $f \cdot g$ or fg the function F such that $F(x) = f(x)g(x)$ in the given interval. Thus for f and g, as above,

$$F(x) = (x^2 - 2)(x^3 + 1), \qquad 0 \le x \le 1$$

Similarly, we can form $f - g$ and $f \div g$, provided that, in the last case, g has no zero so that there is no division by zero. Also, if c is a fixed real number, we can form cf, the function whose value at each x is $cf(x)$.

Let us start with the constant functions (the functions defined by $y = c$, where c is fixed) and the function $y = x$, all defined for $-\infty < x < \infty$. By

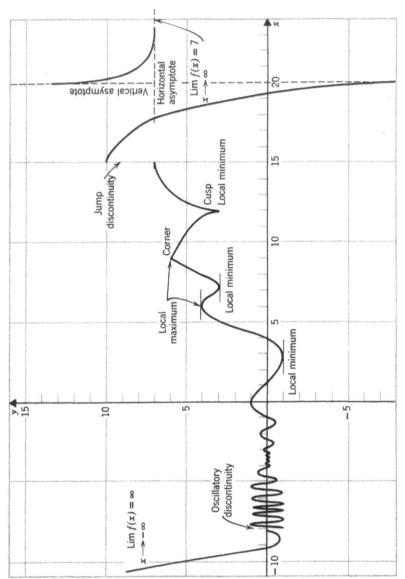

Figure 2-3 Features of functions of one variable.

successive addition and multiplications, we can build from these functions the functions

$$f(x) = x + 1, \quad g(x) = x + 2, \quad h(x) = x^2, \quad F(x) = 3x^2 + 2x + 5, \ldots$$

and so on. In this way, we obtain all *polynomial functions*: that is, all functions of form

$$p(x) = a_0x^n + a_1x^{n-1} + \cdots + a_{n-1}x + a_n, \qquad -\infty < x < \infty$$

If we also allow the operation of division, then we obtain all ratios of polynomials, that is, all rational functions, such as

$$G(x) = \frac{x^3 + 5}{x^2 - 1}$$

The domain of this last function must exclude $x = 1$ and $x = -1$, the zeros of the denominator.

Other classes can be built up in the same way by starting with other functions. By starting with the constant functions and the trigonometric functions $\sin x$ and $\cos x$, we obtain the functions:

$$2 \sin x + 3 \cos x, \quad 2 \cos^2 x - 1 = \cos 2x, \quad 2 \sin x \cos x = \sin 2x.$$

We can combine the previous operations with composition (Section 2-1). From the functions x^2 and $\sin x$, successive operations of addition or composition yield the functions:

$$x^2 + \sin x^2, \quad \sin(\sin x + x^2) \quad (\sin^2 x + x^2)^2$$

Graphing of Functions Constructed from Others. If $h = f + g$, then the graph of h can be obtained from the graphs of f and g by "addition of ordinates," as suggested in Figure 2-4. For each x, we simply add to $g(x)$ the value $f(x)$ by copying the distance $f(x)$ above (or below, when $f(x)$ is negative) the point $[x, g(x)]$. Similarly, the graph of $h = f - g$ is obtained by subtracting ordinates (Figure 2-5). The construction of the graph of $h = f \cdot g$ from the graphs of f and g is illustrated in Figure 2-6. Here, we are forced to measure

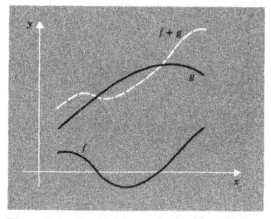

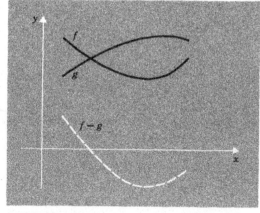

Figure 2-4 Graph of $f + g$ obtained by addition of ordinates.

Figure 2-5 Graph of $f - g$.

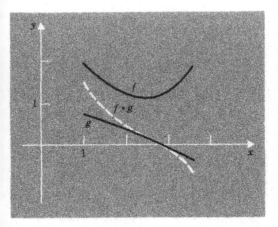

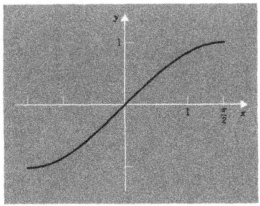

Figure 2-6 Graph of $f \cdot g$.

Figure 2-7 Monotone strictly increasing function: $y = \sin x$, $-\pi/2 \leq x \leq \pi/2$.

$f(x)$ and $g(x)$ for each x, and to multiply to obtain $h(x)$. However, a rough sketch of the graph of h is usually easy to construct. A similar statement applies to the graph of $h = f/g$.

2-4 INVERSE FUNCTIONS

A function f, defined in a given interval, is said to be *monotone strictly increasing* if

$$x_1 < x_2 \qquad \text{implies} \qquad f(x_1) < f(x_2)$$

for all x_1, x_2 in the interval (Figure 2-7). If $x_1 < x_2$ implies only $f(x_1) \leq f(x_2)$, f is said to be *monotone nondecreasing*. Similarly, f is *monotone strictly decreasing* (or *monotone nonincreasing*) if $x_1 < x_2$ implies $f(x_1) > f(x_2)$ [or $f(x_1) \geq f(x_2)$].

A monotone strictly increasing function f is necessarily one-to-one. For the unequal values x_1, x_2 correspond to the unequal values $f(x_1)$, $f(x_2)$. Therefore, a monotone strictly increasing f always has an *inverse function*, which we denote by f^{-1}, and f^{-1} is also monotone strictly increasing. If $y = f(x)$, then $x = f^{-1}(y)$ (Figure 2-8). Thus, to graph the inverse function in the usual way, we must reflect the graph of f in the line $y = x$; this has the effect of interchanging axes. Notice that the *domain* of f^{-1} is the *range* of f. Also $f[f^{-1}(b)] = b$, $f^{-1}[f(c)] = c$ for each b or c for which the appropriate function is defined. In general, f is the inverse of f^{-1}.

There is a similar discussion of monotone strictly decreasing functions.

We can now add to our list of operations: formation of the inverse function (when it exists). For example, $y = x^2$ is monotone strictly increasing for $x \geq 0$; hence, it has an inverse: $x = \sqrt{y}$, $y \geq 0$. (Notice that $\sqrt{y}$ is the *positive* square root, for $y > 0$.) The inverse is simply a new function, and we can represent it equally well by $y = \sqrt{x}$, $\sqrt{x} \geq 0$. In the same way, from $y = x^n$ in general ($n = 1, 2, 3, \ldots, x \geq 0$), we obtain the new functions $y = x^{1/n}$, $x \geq 0$.

From $y = \sin x$, $-\pi/2 \leq x \leq \pi/2$, we obtain the inverse $x = \text{Sin}^{-1} y$, $-1 \leq y \leq 1$, as in Figures 2-7 and 2-8. We call this function the *principal*

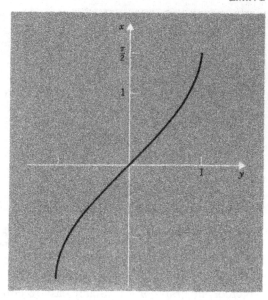

Figure 2-8 Inverse of function
of Figure 2-7: $x = \text{Sin}^{-1} y$,
$-1 \le y \le 1$.

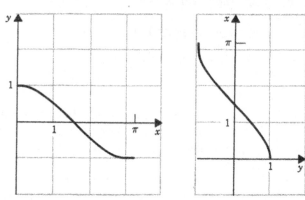

Figure 2-9 $y = \cos x$ and $x = \text{Cos}^{-1} y$.

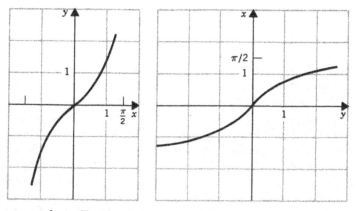

Figure 2-10 $y = \tan x$ and $x = \text{Tan}^{-1} y$.

value of the *inverse sine*. The symbol $\sin^{-1} y$ (small s) denotes an "angle whose sine is y," chosen in some other way. The functions $x = \text{Cos}^{-1} y$ and $x = \text{Tan}^{-1} y$ are defined similarly (see Figures 2-9 and 2-10).

Finally, we observe that, by definition, the inverse of the function $y = a^x$, $-\infty < x < \infty$, is the function $x = \log_a y$, $y > 0$ (Figure 2-11). Here a is a fixed positive number different from 1.

We have here assumed it known that each of the functions $\sin x$, $\cos x$, $\tan x$, a^x is monotone strictly increasing or decreasing in the appropriate interval. This will be proved in later chapters.

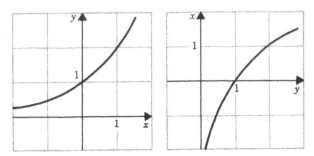

Figure 2-11 a^x and its inverse $x = \log_a y$.

PROBLEMS

1. For each of the following functions state (i) the intervals in which the function is monotone strictly increasing, (ii) the intervals in which the function is monotone strictly decreasing, (iii) the local maxima or minima, if any, and (iv) the absolute maximum and minimum.

   (a) The function of Figure 2-12(a).   (b) The function of Figure 2-12(b).
   (c) The function of Figure 2-12(c).   (d) The function of Figure 2-12(d).

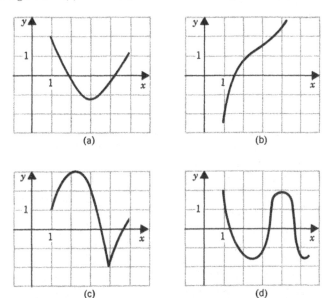

Figure 2-12

2. $(a) \ldots (e)$ Draw the graph of a continuous function $y = f(x)$ with the properties indicated in Table 2-1.

Table 2-1

Function	a	b	c	d	e
Domain	$0 \leq x \leq 5$	$0 \leq x \leq 5$	$0 \leq x \leq 5$	$(-\infty, \infty)$	$(-\infty, \infty)$
Local maximum	$x = 1, y = 2$	$x = 0, y = 2$	$x = 0, y = 2$ $x = 3, y = 1$ $x = 5, y = 2$	None	None
Local minimum	$x = 0, y = -1;$ $x = 5, y = 1$	$x = 5, y = 0$	$x = 2, y = 0$ $x = 4, y = 0$	$x = 1, y = 0$	None
Absolute maximum	$y = 2$ at $x = 1$	$y = 2$ at $x = 0$	$y = 2$ at $x = 0, 5$	None	None
Absolute minimum	$y = -1$ at $x = 0$	$y = 0$ at $x = 5$	$y = 0$ at $x = 2, 4$	$y = 0$	None

3. Draw an accurate graph of the function $y = x^2$ in the interval $0 \leq x \leq 3$, with the same scale on both axes. Then graph the tangent lines at $x = 0, x = 1, x = 2$, $x = 3$ and determine their slope by measurement. Compare your results with the ones given by the formula of calculus: slope of the tangent line at (x, y) is $2x$.

4. Justify by geometry the statement: the slope of the tangent at a point (x, y) on the circle $x^2 + y^2 = 1$ is $-x/y$. Verify graphically at the points $(\frac{3}{5}, \frac{4}{5}), (-\frac{1}{2}, \sqrt{3}/2)$.

5. Draw an accurate graph of the function $y = \sin x$ in the interval $-\pi \leq x \leq \pi$, with the same scale on both axes. Determine graphically the slopes of the tangent lines for $x = -\pi, -2\pi/3, -\pi/2, -\pi/3, 0, \pi/3, \pi/2, 2\pi/3, \pi$. Compare your results with the ones given by the rule of calculus: slope at (x, y) is $\cos x$.

6. Graph the function: $y = |x|$ for $-1 \leq x \leq 1$, and explain why the graph has no tangent at $(0, 0)$.

7. Graph the function $y = \sqrt{|x|}$ for $-1 \leq x \leq 1$. Does the graph have a tangent at $(0, 0)$? If so, what is its slope?

8. Let $y = f(x) = x^2, y = g(x) = 1$. Show how each of the following functions can be constructed from f and g by repeated addition, subtraction, and multiplication:
   (a) $y = x^2 - 2$ (b) $y = x^4 - 2x^2 + 3$ (c) $y = 5x^6 - 2x^4$

9. We define the *zero function* as the constant function always equal to 0. We denote the function by 0 (even though, to be precise, we should distinguish between the number 0 and the zero function). We write $-f$ for $(-1)f$ (constant function -1 times f). Justify the statements, for given functions f, g, h defined on an interval:
   (a) $f + 0 = f$ (b) $f + (-f) = 0$ (c) $0f = 0$
   (d) If $f + g = 0$, then $g = -f$ (e) If $f + g = h$, then $g = h + (-f)$

10. In the notation of Problem 9, does $f \cdot g = 0$ imply that either $f = 0$ or $g = 0$?

11. From the graphs of Figure 2-13 obtain the graphs of the following:
   (a) $f + g$ (b) $2f + g$ (c) $f + h$ (d) $g - f$ (e) $2f - h$
   (f) $f \cdot g$ (g) f^2 (h) $g \cdot h$ (i) f/g (j) g/h
   (k) $F \circ f$ (l) $F \circ g$ (m) $F \circ (F \circ g)$ (n) f^{-1} (o) F^{-1}

12. For each of the following functions obtain an expression for the inverse function, and graph function and inverse:

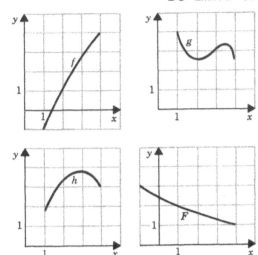

Figure 2-13

(a) $y = 2x$, all x (b) $y = x^4$, $0 \leq x$

(c) $y = x^2 + 2x - 1$, $x \geq -1$ (d) $y = 2^x$,   all x

13. Justify the statement: If $y = f(x)$, $a \leq x \leq b$, is monotone strictly increasing (or decreasing), then the inverse function is also strictly increasing (or decreasing).

14. Let the function f be defined as follows:

$$f(x) = x \text{ for } 0 \leq x < 1, \quad f(x) = 1 + x \text{ for } 1 \leq x \leq 2$$
$$f(x) = -x + 4 \text{ for } 2 < x \leq 3$$

Graph and show that the function is discontinuous, is not monotone strictly increasing or decreasing, but does provide a one-to-one correspondence between the interval $0 \leq x \leq 3$ and the interval $0 \leq y \leq 3$. Give formulas for the inverse function.

15. Evaluate: (a) $\text{Sin}^{-1} 1$, (b) $\text{Sin}^{-1}(-\frac{1}{2})$, (c) $\text{Cos}^{-1} 0$, (d) $\text{Cos}^{-1} 1$.

16. The function $x = \text{Cot}^{-1} y$ is defined as the inverse of $y = \cot x$, $0 < x < \pi$. Graph function and inverse.

17. A real function f is said to be periodic and to have *period* $c > 0$ if f is defined for all x and $f(x + c) = f(x)$ for all x. State which of the following functions is periodic and give a period for each that is.

(a) $\sin x$ (b) $x + \sin x$ (c) $2 \cos x - \sin x$ (d) $x^2 \cos 5x$

(e) $\sin(x/3)$ (f) $\cos(x/5)$ (g) $\sin(x + \sin x)$ (h) $2^{\sin x}$

18. A real function f is said to be *bounded* if $|f(x)| \leq K$, where K is a constant, for all x in the domain of f. Determine which of the following functions is bounded and give a value of K for each function that is.

(a) x^2 (b) $\sin x$ (c) $\dfrac{1}{1 + x^2}$ (d) $\sin(x + 2^x)$

2-5 LIMITS

The concept of limit is suggested in our qualitative discussion in Section 2-1. This concept is basic for the calculus. We give an illustration in an everyday setting and then formulate a precise definition.

EXAMPLE 1. Let us imagine a traveler on a train due in New York at 5:17 P.M. He must be there on time, so he continually looks at his watch and checks with the timetable. He notes: at 5:01 P.M., the train is 10 miles out; at 5:10 P.M., the train is 4 miles out; at 5:13 P.M., the train is 2 miles out; at 5:15 P.M., the train is 1 mile out, and all is well—at 5:17 P.M. exactly the train pulls up at its platform in Grand Central Terminal. The traveler has been observing a limit process, which we suggest graphically in Figure 2-14. In mathematical

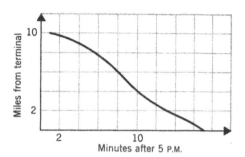

Figure 2-14   Train approaching its destination.

language, we say that the distance from the terminal approached the limit 0 as the time approached 5:17 P.M. In order for this to happen, the distance had to become very small as the time neared 5:17 P.M.

Definition. Let a real function f be given, for instance, by an equation $y = f(x)$, where the function is defined for $a < x \le b$, and perhaps also at $x = a$. Then we say that f has *limit c* as x approaches a if the values of $y = f(x)$ can be made as close to c as desired by restricting x to a sufficiently small interval $a < x < a + q$; thus, for all x of the interval $a < x < a + q$, the values of y lie within a prescribed small interval $c - p < y < c + p$, as suggested in Figure 2-15. We write in symbols: $f(x) \to c$ as $x \to a$, or

$$\lim_{x \to a} f(x) = c$$

There is a similar definition for $f(x) \to d$ as $x \to b$, or

$$\lim_{x \to b} f(x) = d$$

when $y = f(x)$ is defined in an interval $a \le x < b$, and perhaps also at $x = b$. This is also suggested in Figure 2-15. A function may also be defined on both "sides" of the value of x in question—that is, the function is defined for all x of an interval $a < x < b$, except perhaps x_0, where $a < x_0 < b$. We then write $f(x) \to c$ as $x \to x_0$ or

$$\lim_{x \to x_0} f(x) = c$$

if the values of $y = f(x)$ can be made as close to c as desired by restricting x to a sufficiently small interval $x_0 - q < x < x_0 + q$, except for the value x_0 (that is, $x_0 - q < x < x_0$ or $x_0 < x < x_0 + q$). This is illustrated in Figure 2-16.

Remark. To distinguish the limit at an end point (a or b) from that at an interior point, we normally write

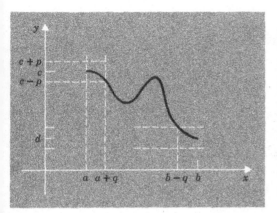

Figure 2-15 Limits as $x \to a$ or $x \to b$.

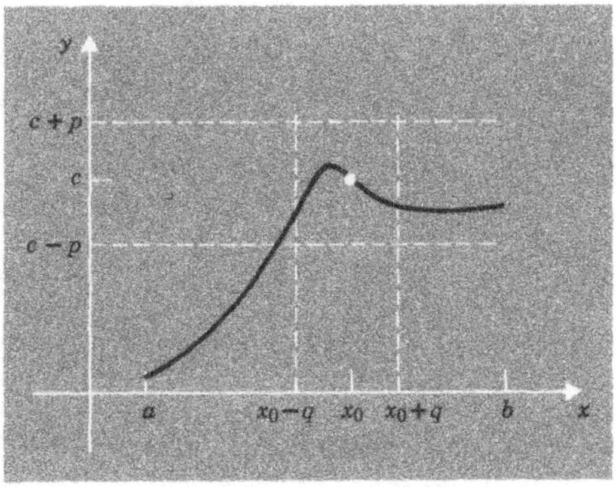

Figure 2-16 Limit at an interior point.

$$\lim_{x \to a+} f(x), \qquad \lim_{x \to b-} f(x)$$

for the limits at the end points, and call them limits to the right and left, respectively. The $a+$ is meant to suggest that, in the limit process, x takes on values greater than a; similarly, the symbol $b-$ indicates that x is to take on values less than b. These symbols can also be used at interior points. For example, if $a < x_0 < b$, then

$$\lim_{x \to x_0+} f(x) = c$$

means that the values of $f(x)$ can be made as close to c as desired by making x sufficiently close to x_0 and greater than x_0. Figure 2-17 shows a function f with jump discontinuity at $x = 2$ for which

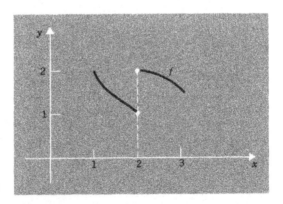

Figure 2-17

$$\lim_{x \to 2-} f(x) = 1, \qquad \lim_{x \to 2+} f(x) = 2$$

In all cases, *in the limit process we ignore the value of the function at the chosen value of x; it may or may not be defined.*

In Example 1 we let y be the number of miles out, and t the time in minutes after 5:00 P.M., so that we can write $y = f(t)$. Then

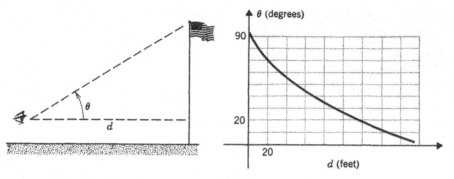

Figure 2-18 Angle θ as function of distance d.

$$\lim_{t \to 17-} f(t) = 0$$

In this example, also $f(17) = 0$.

EXAMPLE 2. Consider a person who observes the angle of elevation of the top of a flagpole that he is approaching (Figure 2-18). As his distance d (in feet) approaches 0, the angle θ approaches 90°, but never quite reaches that value, since d cannot quite become 0. We say that the limit of θ is 90 (degrees) as d approaches 0 or, with $\theta = f(d)$,

$$\lim_{d \to 0+} f(d) = 90$$

Here $f(0)$ is not defined, since the person cannot stand right at the flagpole.

EXAMPLE 3. We consider a table of sines, with the angles in radians. The first 11 entries in such a table are given in the accompanying Table 2-2. If we round off the values of sin x to the nearest hundredth, we find that in every

Table 2-2   Limit of (sin x)/x as x → 0

x	$\sin x$	$\dfrac{\sin x}{x}$
0.00	0.00000000	—
0.01	0.00999983	0.999983
0.02	0.0199987	0.99993
0.03	0.0299955	0.99985
0.04	0.0399893	0.99973
0.05	0.0499792	0.99958
0.06	0.0599640	0.99940
0.07	0.0699428	0.99917
0.08	0.0799147	0.99893
0.09	0.0898785	0.99865
0.10	0.0998333	0.99833

case the value is the same as that for x: $\sin(0.1) = 0.01$, $\sin(0.02) = 0.02$ to the nearest hundredth. Of course this "coincidence" works only for certain angles, namely those very close to 0 (radians). For example, sin $1 = 0.8415$, sin π = sin 3.1416 = 0. In fact, the accuracy of the statement sin $x = x$ improves the closer x is to 0. In order to show this, we have tabulated in the third column the values of (sin x)/x. It is clear that this ratio approaches 1 as x approaches 0; at $x = 0$, the ratio is 0/0 and, therefore, is undefined. Clearly, we again have a limit process: *the limit of* (sin x)/x, *as x approaches* 0, *is* 1. The truth of this statement, which we have only empirically deduced, is established as an important theorem in the calculus (Chapters 3 and 5).

We can also consider the function (sin x)/x for negative values of x. Since sin($-x$) = $-$sin x, the graph is symmetrical in the y-axis (Figure 2-19). Thus the values of (sin x)/x must approach 1 as x approaches 0 through positive or negative values, and we can write

$$\lim_{x \to 0} \frac{\sin x}{x} = 1$$

We notice that $f(0) = 0/0$ is not defined, as indicated by the "hole" in the graph.

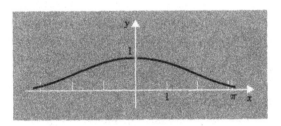

Figure 2-19 The function (sin x)/x.

EXAMPLE 4. We consider the measurement of the speed of a car traveling on a straight highway and suppose that the car is accelerating. The usual definition: "speed equals distance divided by time" is unsatisfactory. For the distance traveled in 10 seconds is more than double the distance traveled in the first five seconds; we get different values for the speed according to the time interval chosen. Yet the speedometer shows a speed at each instant. How can we attach a meaning to this "instantaneous speed?" We let x be the distance of the car from a fixed reference point, in miles, and let t be the time in hours after some chosen reference time, so that $x = f(t)$ (Figure 2-20). To find the speed at the instant t_0, we first consider a time t after t_0 and compute the distance traveled: $d = f(t) - f(t_0)$, as in Figure 2-20. Then we divide by the time to obtain what we can call the average speed in the interval from t_0 to t:

$$\text{average speed} = \frac{f(t) - f(t_0)}{t - t_0}$$

Now we must take the limit of this value as t approaches t_0:

$$\text{instantaneous speed} = \lim_{t \to t_0 +} \frac{f(t) - f(t_0)}{t - t_0}$$

We could also consider a time interval just before t_0, for instance, from t to t_0, where $t < t_0$. The distance traveled is $f(t_0) - f(t)$ and the average speed is

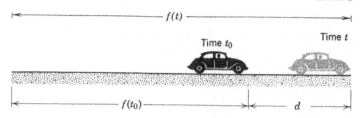

Figure 2-20 Speed of a car.

$$\text{average speed} = \frac{f(t_0) - f(t)}{t_0 - t} = \frac{f(t) - f(t_0)}{t - t_0}$$

Thus because of two changes in sign we have the same expression as above. Hence we can write, including both cases:

$$\text{instantaneous speed at time } t_0 = \lim_{t \to t_0} \frac{f(t) - f(t_0)}{t - t_0}$$

Remarks on Limits. From the definition of limit, it follows that *there can be only one limit value of a function;* that is, there can be at most one value y_0 so that $f(x)$ can be made as close as desired to y_0 by making x sufficiently close to x_0 (see Problem 8 below). Thus a limit, when it exists, is *unique,* and we can refer to *the* limit. This remark applies to all limit definitions in the calculus.

If in the interval considered (excepting always x_0 itself) *we know that* $f(x) < K$, *where K is constant, and* $\lim\limits_{x \to x_0} f(x)$ *exists, then*

$$\lim_{x \to x_0} f(x) \leq K$$

(see Problem 9 below). Equality must be allowed, as illustrated by $(\sin x)/x$ at $x = 0$, with $K = 1$. The conclusion is unaffected if we assume $f(x) \leq K$. There is a similar statement for $f(x) > K$ and for the case $f(x) \geq K$.

More generally, if $f(x) < g(x)$ [or $f(x) \leq g(x)$] in the interval (except, always, at x_0), and both limits exist, then

$$\lim_{x \to x_0} f(x) \leq \lim_{x \to x_0} g(x)$$

There is a similar statement for $f(x) > h(x)$ [or $f(x) \geq h(x)$] (see Problem 10 below).

If $h(x) \leq f(x) \leq g(x)$ in the interval $(x \neq x_0)$ and g and h have the same limit:

$$\lim_{x \to x_0} h(x) = \lim_{x \to x_0} g(x) = c$$

then f also has limit c:

$$\lim_{x \to x_0} f(x) = c$$

As suggested in Figure 2-21, f is "trapped" between g and h, and is forced to the same limit. To achieve $c - p < f(x) < c + p$, we first choose $q_1 > 0$ so

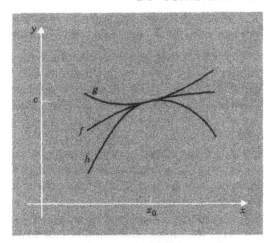

Figure 2-21 Function f trapped between functions g and h.

that $c - p < g(x) < c + p$ for $x_0 - q_1 < x < x_0 + q_1$ $(x \neq x_0)$ and then choose $q_2 > 0$ so that $c - p < h(x) < c + p$ for $x_0 - q_2 < x < x_0 + q_2$ $(x \neq x_0)$. Then let q be the smaller of q_1, q_2. It then follows that for $x_0 - q < x < x_0 + q$ $(x \neq x_0)$,

$$c - p < h(x) \leq g(x) < c + p$$

and, therefore, since $h(x) \leq f(x) \leq g(x)$.

$$c - p < f(x) < c + p$$

2-6 CONTINUITY

The limit concept now permits us to define precisely what we mean by continuity. Figure 2-22 shows a discontinuous function g and a continuous one f. In Figure 2-22(a) at a there is a jump discontinuity. Clearly there can be no limit at a; that is,

$$\lim_{x \to a} g(x)$$

does not exist. For there is no value l for which $g(x)$ remains very close to l

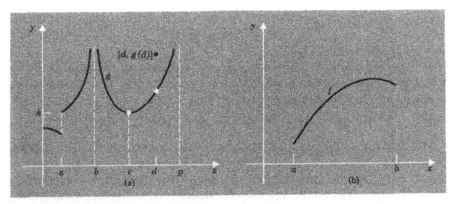

Figure 2-22 Discontinuous and continuous functions.

for x sufficiently close to a. It is also clear that there is no limit at b or p [at p, it is $\lim\limits_{x \to p-} g(x)$ which does not exist]. We might say that "the limit is infinite" at these points, but we still say the limit does not exist. At c the tiny circle on the graph is meant to suggest that there is no point on the graph for $x = c$; that is, $g(c)$ is undefined. However, it appears that we could fill in the tiny gap in the graph by a point (c, k) and make it smooth. In particular, it appears that

$$\lim_{x \to c} g(x) = k$$

that is, $g(x)$ has a limit, as x approaches c, and the limit is precisely the value needed to yield a continuous graph. At d the value assigned is not the one expected; here we would have to say that g has a limit, as x approaches d, but the limit is not equal to $g(d)$.

To summarize: it appears from all our examples that discontinuities arise whenever the limit of the function fails to exist and whenever the limit exists but does not equal the value of the function. This leads us to our goal:

Definition of continuity. Let a function $y = f(x)$ be defined for $a \le x \le b$ and let $a < x_0 < b$. Then the function f is said to be *continuous at* x_0 if $\lim\limits_{x \to x_0} f(x)$ exists and equals $f(x_0)$ or, more concisely, if

$$\lim_{x \to x_0} f(x) = f(x_0) \tag{2-60}$$

If f is defined for $a \le x \le b$, then f is continuous at a if $\lim\limits_{x \to a+} f(x) = f(a)$ and f is continuous at b if $\lim\limits_{x \to b-} f(x) = f(b)$.

Finally, we can say what we mean by a function being continuous in an interval—there must simply be no points of discontinuity in the interval:

Definition of continuity in an interval. The function $y = f(x)$, defined in the closed interval $a \le x \le b$, is said to be continuous in that interval if f is continuous at every point of the interval; that is, if for all x_0 satisfying $a < x_0 < b$ $\lim\limits_{x \to x_0} f(x)$ exists and equals $f(x_0)$, $\lim\limits_{x \to a+} f(x)$ exists and equals $f(a)$, and $\lim\limits_{x \to b-} f(x)$ exists and equals $f(b)$.

We can also extend the previous definition to open intervals, half-open intervals, and infinite intervals. For example, a function f defined in the interval $0 < x < \infty$ is continuous in that interval if (2-60) above holds true at every x_0 for which $0 < x_0 < \infty$.

It is important always to be clear about the interval in which the function is considered given—that is, the domain of the function. For example, the function $h(x)$ defined as follows for $-\infty < x < \infty$:

$$h(x) = 0 \text{ for } x < 0, \qquad h(x) = 1 \text{ for } x \ge 0$$

is discontinuous at $x = 0$ and, therefore, is not continuous in the interval given. However, we can consider the same function in the interval $0 \le x < \infty$. Then $h(x) = 1$ for every x in the new domain, and the function is now continuous in the interval considered. Actually, by changing the domain, we have changed the function, and we should use a different notation. However, in such cases, usually the context makes the meaning clear, and we frequently write, for example: the function tan x is not continuous in the interval $-\infty < x < \infty$, but is continuous in the interval $-\pi/4 \leq x \leq \pi/4$.

It follows from the definition of continuity that, if a function is continuous in a given interval, then it is also continuous in each interval contained in the given interval.

PROBLEMS

1. Discuss informally the concept of limit involved in each of the following situations:
   (a) A runner in a race.
   (b) The density of the atmosphere as a function of altitude above sea level.
   (c) The elongation of a wire that supports a weight as a function of the weight supported.
   (d) The temperature of a given volume of gas as a function of the pressure.
   (e) The acceleration of a car moving on a straight highway.
   (f) The slope of the tangent to the graph of a function $y = f(x)$ at a point (consider chords through the point and a nearby point).

2. For each of the following use Table 2-2.
   (a) Tabulate the function $y = \dfrac{x - \sin x}{x}$ and empirically determine $\lim\limits_{x \to 0+} \dfrac{x - \sin x}{x}$.

   (b) Proceed as in (a) for $\lim\limits_{x \to 0+} \dfrac{x - \sin x}{x^2}$.

   (c) Proceed as in (a) for $\lim\limits_{x \to 0+} \dfrac{x - \sin x}{x^3}$.

   (d) Proceed as in (a) for $\lim\limits_{x \to 0+} \dfrac{\sin^2 x}{x^2}$.

3. For each of the following functions, graph and state whether the function is continuous in the interval given and, if not, state the points of discontinuity.
   (a) Interval $0 \leq x \leq 2$; $f(x) = 1$ for $0 \leq x \leq 1$, $f(x) = 2 - x$ for $1 < x \leq 2$.
   (b) Interval $0 \leq x < \infty$; $f(0) = 0$, $f(x) = 1/x$ for $x > 0$.
   (c) Interval $0 \leq x \leq 3$; $f(x) = 1/(x - 3)$ for $0 \leq x \leq 2$, $f(x) = -1/(x - 1)$ for $2 < x \leq 3$.
   (d) Interval $-1 \leq x \leq 1$; $f(x) = 2^{1/x}$ for $x \neq 0$, $f(0) = 1$.

4. (a) Show that the definition of limit can be restated as follows: Let f be defined for $a < x < b$ except perhaps at x_0, where $a < x_0 < b$. Then f has limit c as x approaches x_0 if, for every number $\epsilon > 0$, a number $\delta > 0$ can be found so that for every x in (a, b) for which $0 < |x - x_0| < \delta$ we have $|f(x) - c| < \epsilon$.
   (b) Show that the definition of continuity at x_0, as in part (a), can be restated: f is continuous at x_0 if $f(x_0)$ is defined and, for every number $\epsilon > 0$, a number $\delta > 0$ can be found so that for every x in (a, b) for which $|x - x_0| < \delta$ we have $|f(x) - f(x_0)| < \epsilon$.

5. Prove that each of the following functions is continuous for every x:
   (a) $y = x$, $-\infty < x < \infty$ (b) $y = 2x + 1$, $-\infty < x < \infty$

6. We illustrate a proof that $y = f(x) = x^2$ is continuous at $x = 2$. We wish to show that, if x is sufficiently close to 2, the values of $f(x)$ are as close to $f(2) = 4$ as desired. We can indicate a value of x close to 2 as $2 + h$, where h will be chosen close to 0.

The corresponding value of y is $(2 + h)^2 = 4 + 4h + h^2$. This differs from 4 by $4h + h^2$. We wish to show that $4h + h^2$ can be made as small as we wish by choosing h sufficiently small. For example, we must show that, for every given positive number ϵ, we can make $|4h + h^2| < \epsilon$ or $|h(4 + h)| < \epsilon$ by choosing $|h| < \delta$, for δ sufficiently small and positive. We choose δ as the smaller of the two numbers $1, \epsilon/5$. Then we reason: $|h| < \delta$ implies $|h| < 1$, so that $|4 + h| < 5$, and $|h| < \delta$ also implies $|h| < \epsilon/5$, so that

$$|4h + h^2| = |h(4 + h)| = |h||4 + h| < (\epsilon/5)5 = \epsilon$$

as asserted (see Figure 2-23).

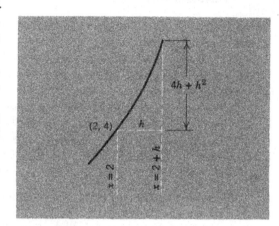

Figure 2-23

(a) Carry out a similar reasoning to show that f is continuous at $x = 3$.

(b) Do the same for $x = -2$.

(c) Do the same for an arbitrary x_0.

7. Let $f(x) = (x^2 - 1)/(x - 1)$ wherever the right-hand side has meaning.

(a) Show that f is defined for all x except 1.

(b) Show that for $x \neq 1$, $f(x) = x + 1$. Why isn't this true for $x = 1$?

(c) Show that $\lim_{x \to 1} f(x) = 2$.

(d) Is f continuous at $x = 1$?

8. Prove *the uniqueness of limits*. That is, let $\lim_{x \to x_0} f(x) = c$ and let k be a number different from c. Prove it cannot happen that $\lim_{x \to x_0} f(x) = k$. [*Hint*. Let $2p = |k - c|$, so that $p > 0$. Choose q so that $c - p < f(x) < c + p$ for $x_0 - q < x < x_0 + q$ $(x \neq x_0)$, and show that for x in this interval $y = f(x)$ is *excluded* from the interval $k - p < y < k + p$.]

9. Prove that, if f is defined in the interval $[a, b]$ (except perhaps at x_0) and $f(x) \leq K$ for all x (except perhaps x_0), and $\lim_{x \to x_0} f(x)$ exists, then $\lim_{x \to x_0} f(x) \leq K$. [*Hint*. Let c be the limit, and suppose that $c > K$. Let $p = c - K$, so that $p > 0$. Choose q as in the definition of limit and show that $x_0 - q < x < x_0 + q$ implies $f(x) > K$, so that there is a contradiction.]

10. Prove that if f and g are defined in the interval $[a, b]$ (except perhaps at x_0) and $f(x) \leq g(x)$ for $x \neq x_0$, then $\lim_{x \to x_0} f(x) \leq \lim_{x \to x_0} g(x)$, provided that both limits exist.

11. A rowboat is crossing a river, at an angle of $\pi/4$ with the banks. Let the banks be represented by the lines $y = 0$, $y = b$ in the xy-plane, and let the path of the boat be the line $y = x$, $0 \leq x \leq b$.

(a) Let u be the y-coordinate of the point on land nearest to the boat, so that u becomes a function of x. Graph u and show that it is discontinuous; discuss the discontinuity.

(b) Let v be the shortest distance from the boat to land. Graph v as a function of x and show that the graph has a corner.

2-7 THEOREMS ON LIMITS AND CONTINUITY

Previously, we discussed continuity in intuitive fashion and, then, gave a precise definition in terms of limits. For a function given by a formula, our definition is awkward to apply (see Problem 6 above). It is most desirable that we easily establish continuity (or discontinuity) for these functions. To that end, there is a group of theorems on limits and continuity which we shall now discuss. Formal proofs are given later in this chapter (Section 2-14).

THEOREM A. *Let f and g be functions defined in the interval* $a \leq x \leq b$, *except perhaps at* x_0, *where* $a < x_0 < b$. *Let f and g have limits as x approaches* x_0:

$$\lim_{x \to x_0} f(x) = c, \qquad \lim_{x \to x_0} g(x) = k$$

Then $f + g$, $f - g$, fg, f/g *all have limits at* x_0:

$$\lim_{x \to x_0} [f(x) + g(x)] = c + k = \lim_{x \to x_0} f(x) + \lim_{x \to x_0} g(x) \qquad (2\text{-}70)$$

$$\lim_{x \to x_0} [f(x) - g(x)] = c - k = \lim_{x \to x_0} f(x) - \lim_{x \to x_0} g(x) \qquad (2\text{-}71)$$

$$\lim_{x \to x_0} f \cdot g = c \cdot k = \left[\lim_{x \to x_0} f(x) \right] \cdot \left[\lim_{x \to x_0} g(x) \right] \qquad (2\text{-}72)$$

$$\lim_{x \to x_0} \frac{f(x)}{g(x)} = \frac{c}{k} = \frac{\left[\lim_{x \to x_0} f(x) \right]}{\left[\lim_{x \to x_0} g(x) \right]}, \qquad \text{provided that } k \neq 0 \quad (2\text{-}73)$$

The different assertions (2-70), . . . , (2-73) are often stated in words:

"limit of the sum of two functions = sum of the limits of the two functions"

and so on or, more briefly, "limit of sum = sum of limits," and so on.

EXAMPLE 1. Let $f(x) = (\sin x)/x = g(x)$. Then, as in Example 3 in Section 2-5,

$$\lim_{x \to 0} f(x) = \lim_{x \to 0} \frac{\sin x}{x} = 1$$

Hence by (2-72)

$$\lim_{x \to 0} \frac{\sin^2 x}{x^2} = \lim_{x \to 0} f(x) \cdot f(x) = \lim_{x \to 0} f(x) \cdot \lim_{x \to 0} f(x)$$

$$= \lim_{x \to 0} \frac{\sin x}{x} \cdot \lim_{x \to 0} \frac{\sin x}{x} = 1 \cdot 1 = 1$$

EXAMPLE 2. Let $f(x) = (1/x) \sin x$, let $g(x) = 3$ (constant function). Then the constant function g has as limit the number 3 at every x and

$$\lim_{x \to 0} \frac{3 \sin x}{x} = \lim_{x \to 0} 3 \cdot \lim_{x \to 0} \frac{\sin x}{x} = 3 \cdot 1 = 1$$

$$\lim_{x \to 0} \left(3 - \frac{\sin x}{x} \right) = \lim_{x \to 0} 3 - \lim_{x \to 0} \frac{\sin x}{x} = 3 - 1 = 2$$

EXAMPLE 3. Let $f(x) = x$ for all real x. Then, as an immediate consequence of the definition,

$$\lim_{x \to x_0} f(x) = x_0$$

for every x_0. Therefore

$$\lim_{x \to x_0} (3 + 2x) = \lim_{x \to x_0} 3 + \lim_{x \to x_0} 2x = \lim_{x \to x_0} 3 + \lim_{x \to x_0} 2 \lim_{x \to x_0} x = 3 + 2x_0$$

$$\lim_{x \to x_0} x^2 = \lim_{x \to x_0} x \cdot \lim_{x \to x_0} x = x_0 \cdot x_0 = x_0^2$$

$$\lim_{x \to x_0} (5 + 3x + 6x^2)$$

$$= \lim_{x \to x_0} 5 + \left(\lim_{x \to x_0} 3 \right) \left(\lim_{x \to x_0} x \right) + \left(\lim_{x \to x_0} 6 \right) \left(\lim_{x \to x_0} x^2 \right) = 5 + 3x_0 + 6x_0^2$$

EXAMPLE 4. Let $f(x) = 1$ (constant function), $g(x) = x$ for all x. Then

$$\lim_{x \to x_0} \frac{1}{x} = \lim_{x \to x_0} \frac{f}{g} = \frac{\lim_{x \to x_0} f}{\lim_{x \to x_0} g} = \frac{1}{x_0}$$

provided that $x_0 \neq 0$.

Remark. When $\lim g(x) = 0$, $\lim [f(x)/g(x)]$ may or may not exist:

$$\lim_{x \to 0} \frac{x}{x} = 1, \qquad \lim_{x \to 0} \frac{\sin x}{x} = 1$$

while

$$\lim_{x \to 0} \frac{1}{x}, \qquad \lim_{x \to 0} \frac{|x|}{x}$$

do not exist. However, when $\lim g(x) = 0$, $\lim [f(x)/g(x)]$ can never equal $(\lim f)/(\lim g)$, since the last expression is meaningless.

Justification of Theorem A. As stated above, a formal proof is given in Section 2-14. Here we give an intuitive basis for the theorem. For (2-70), we know that f has limit c, and that g has limit k as x approaches x_0. Thus for x sufficiently close to (but not at) x_0, both $f(x)$ and $g(x)$ are close to their respective limits. For example, for x sufficiently close to x_0, $f(x) < c + 0.1$, $g(x) < k + 0.1$, so that $f(x) + g(x) < c + k + 0.2$ (the "error" in the sum is at most the sum of the errors). Similarly, for x sufficiently close to x_0, $f(x) > c - 0.1$, $g(x) > k - 0.1$, and $f(x) + g(x) > c + k - 0.2$. Thus, for x sufficiently close to x_0

$$c + k - 0.2 < f(x) + g(x) < c + k + 0.2$$

By replacing 0.1 by 0.01, we can ensure, by appropriate shrinking of the interval about x_0, that

$$c + k - 0.02 < f(x) + g(x) < c + k + 0.02$$

Clearly, we can make $f(x) + g(x)$ as close as desired to $c + k$ by suitably restricting the interval about x_0 in which x is allowed to lie (with $x \neq x_0$).

The rule (2-71) is proved in the same way. The rule (2-72) requires a multiplication of inequalities. Let us take a particular case, $\lim f(x) = 2$,

$\lim g(x) = 3$. We wish to show that $f(x) \cdot g(x)$ is as close as we wish to 6 for x sufficiently close to (but not at) x_0. Now, as previously, $f(x) > 2 - 0.1$, $g(x) > 3 - 0.1$ for x sufficiently close to x_0, so that (by the rules for inequalities, see Section 0-2)

$$f(x)g(x) > (2 - 0.1)(3 - 0.1) = 6 - 0.5 + 0.01 = 5.51$$

For x even closer to x_0, $f(x) > 2 - 0.01$, $g(x) > 3 - 0.01$, so that

$$f(x)g(x) > (2 - 0.01)(3 - 0.01) = 6 - 0.05 + 0.0001 = 5.9501$$

Similarly, for x sufficiently close to x_0, we have also $f(x) < 2 + 0.01$, $g(x) < 3 + 0.01$, so that [$f(x)$ and $g(x)$ being necessarily positive]

$$f(x)g(x) < 6 + 0.05 + 0.0001 = 6.0501$$

or, as a consequence of the last two conclusions,

$$5.9501 < f(x)g(x) < 6.0501$$

for x sufficiently close to x_0. It is clear that we can make $f(x)g(x)$ as close to 6 as we wish by keeping x sufficiently close to x_0.

The rule (2-73) is proved similarly by operations on inequalities.

We can describe all the rules as simply expressing the fact that in adding, in subtracting, and in multiplying and dividing, a slight error in the numbers involved has only a slight effect on the result, except for the division by numbers very close to zero. If we divide 1.004 by 0.002 and make an error of 0.001 in numerator and denominator, so that we divide 1.003 by 0.001, we get 1,003 as a result instead of 502—a tremendous error!

Theorem A has been stated only for limits at an interior point of an interval. It can be formulated in similar fashion for limits at an end point. Thus, for example,

$$\lim_{x \to a+} [f(x) + g(x)] = \lim_{x \to a+} f(x) + \lim_{x \to a+} g(x)$$

provided that both limits on the right exist.

We now turn to the parallel theorem on continuity.

THEOREM B. *Let $f(x)$ and $g(x)$ both be defined in the interval $a \le x \le b$. If both f and g are continuous at the point x_0 of the interval, then so are $f + g, f - g, f \cdot g, f/g$—provided, for f/g, that $g(x_0) \ne 0$. If both f and g are continuous in the whole interval, then so are $f + g, f - g, f \cdot g, f/g$— provided, for f/g, that $g(x) \ne 0$ in the interval.*

This theorem is an immediate consequence of Theorem A. For example, if f and g are continuous at x_0, and x_0 is an interior point, then

$$\lim_{x \to x_0} f(x) = f(x_0), \qquad \lim_{x \to x_0} g(x) = g(x_0)$$

so that, by (2-70),

$$\lim_{x \to x_0} [f(x) + g(x)] = f(x_0) + g(x_0)$$

Hence $f + g$ is continuous at x_0. Similar reasoning applies to the end points *a* and *b*. The statements for $f - g$, $f \cdot g$, f/g are proved in the same way. For f/g, the condition that *g* not be 0 at the point or points considered is essential; in fact, f/g becomes undefined wherever $g(x) = 0$.

The statements about continuity throughout the interval follow from the first part of Theorem B. For a function is continuous throughout an interval precisely when it is continuous at every point of the interval. We could also use open intervals, half-open intervals or infinite intervals here.

THEOREM C. *Let $y = f(x)$ be defined for $a \leq x \leq b$ except perhaps at x_0, where $a \leq x_0 \leq b$. Let the range of f be contained in the interval $c \leq y \leq d$ and let $u = g(y)$ be defined in this interval, so that $g \circ f$ is defined. Let $F = g \circ f$. Let g be continuous at y_0, where $c \leq y_0 \leq d$ and let $\lim_{x \to x_0} f(x) = y_0$. Then*

$$\lim_{x \to x_0} F(x) = \lim_{x \to x_0} g[f(x)] = g(y_0) \tag{2-74}$$

or equivalently

$$\lim_{x \to x_0} g[f(x)] = g\left[\lim_{x \to x_0} f(x)\right] \tag{2-74'}$$

If f is continuous at x_0, then so is $F = g \circ f$. If f is continuous for $a \leq x \leq b$, and g is continuous for $c \leq y \leq d$, then $F = g \circ f$ is continuous for $a \leq x \leq b$.

To justify (2-74'), we remark that, with $g(y_0) = u_0$, we can ensure that the values $u = g(y)$ lie as close to u_0 as desired by restricting *y* to a sufficiently small interval about y_0, as suggested in Figure 2-24. This follows from the

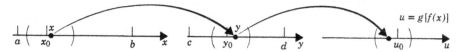

Figure 2-24   Limit of composite function.

continuity of *g* at y_0. Since *f* has limit y_0 as *x* approaches x_0, by making *x* sufficiently close to x_0, we can ensure that the values of $y = f(x)$ lie in the chosen interval about y_0, so that the values of $F(x) = g[f(x)]$ lie in the chosen interval about u_0. Thus (2-74') follows. (We have treated x_0 and y_0 as interior points of their respective intervals. The same reasoning applies when either or both are end points.)

If *f* is continuous at x_0, then $y_0 = f(x_0)$ and

$$\lim_{x \to x_0} F(x) = g[f(x_0)] = F(x_0)$$

so that *F* is continuous at x_0. If *f* is continuous at every point of the interval $a \leq x \leq b$, it follows that *F* is also, as asserted.

Remark. This theorem can also be extended to other types of intervals, for example open intervals $a < x < b$ and $c < y < d$.

THEOREM D. *Let $y = f(x)$ be defined and continuous in the interval $a \leq x \leq b$ and let x_1, x_2 be two points of this interval. Let $f(x_1) = y_1$, $f(x_2) = y_2$, $y_1 \neq y_2$. Then for every number y between y_1 and y_2 there is a number x between x_1 and x_2 for which $f(x) = y$.*

Theorem D is known as the *Intermediate Value Theorem*. Its significance is suggested in Figure 2-25. A continuous function cannot "skip any values";

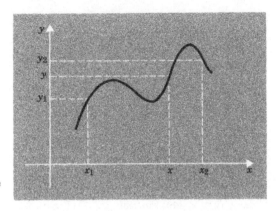

Figure 2-25 Intermediate Value Theorem.

as x varies from x_1 to x_2, $f(x)$ must pass through every value y between y_1 and y_2 at least once. The proof (given in Section 2-14) consists essentially in showing that, if one value y_0 were skipped, then we would have a value x, between x_1 and x_2, at which there is a transition between the values less than y_0 and the values greater than y_0; x would be a discontinuity of f, contrary to assumption.

EXAMPLE 5. It is easily shown by Theorem B that every polynomial $f(x)$ is continuous for all x. The polynomial $f(x) = x^3 + x - 3$ has the value -1 for $x = 1$ and the value 7 for $x = 2$. Thus, by Theorem D, $f(x) = 0$ for some x between 1 and 2; that is, the equation $x^3 + x - 3 = 0$ has a root between 1 and 2.

Continuity of the Inverse Function. We have seen in Section 2-4 that, if f is monotone strictly increasing or monotone strictly decreasing in an interval, then f has an inverse. We now ask whether continuity of f implies that the inverse is continuous.

THEOREM E. *Let $y = f(x)$ be defined and continuous in the interval $a \le x \le b$ and let f be monotone strictly increasing. Then the range of f is the interval $c \le y \le d$ where $c = f(a)$, $d = f(b)$, and the function $x = f^{-1}(y)$ is continuous and monotone strictly increasing in the interval $c \le y \le d$.*

PROOF. The fact that the range of f is the interval $c \le y \le d$ follows from Theorem D and the fact that f is monotone strictly increasing (Problem 11 below). The inverse function f^{-1} is thus defined for $c \le y \le d$. It is monotone strictly increasing. For if $y_1 < y_2$ and $f^{-1}(y_1) = x_1$, $f^{-1}(y_2) = x_2$, then $f(x_1) = y_1$, $f(x_2) = y_2$. If $x_1 \ge x_2$, then we would have $y_1 \ge y_2$, since f is monotone strictly increasing. Hence, $x_1 < x_2$. Accordingly, $f^{-1}(y)$ is monotone strictly increasing. Given a value y_0, $c < y_0 < d$, and an interval $(x_0 - p, x_0 + p)$ containing x_0, we can choose an interval $(y_0 - q, y_0 + q)$ containing y_0 so that, for y in this interval, $f^{-1}(y)$ lies in the interval $(x_0 - p, x_0 + p)$; we simply choose q as the smaller of the numbers $f(x_0) - f(x_0 - p)$, $f(x_0 + p) - f(x_0)$; in Figure 2-26, the first is the smaller, so that $y_0 - q = f(x_0 - p)$,

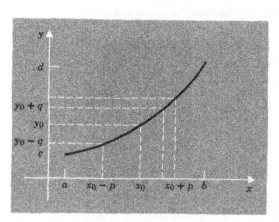

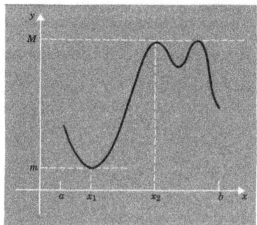

Figure 2-26 Continuity of inverse function.   **Figure 2-27** Absolute minimum and maximum.

$y_0 + q < f(x_0 + p)$. In any case $y_0 - q < y < y_0 + q$ implies $x_0 - p < f^{-1}(y) < x_0 + p$, as needed. Accordingly, $f^{-1}(y)$ has limit $f^{-1}(y_0) = x_0$ as $y \to y_0$, and $f^{-1}(y)$ is continuous at y_0. Similarly, we prove that $f^{-1}(y)$ is continuous at c and d (Problem 12 below).

Remark. The theorem can be extended to half-open intervals, open intervals, and infinite intervals. The analogous theorem also holds true for monotone strictly decreasing functions.

THEOREM F. *Let* $y = f(x)$ *be defined and continuous in the interval* $a \leq x \leq b$. *Then there are two numbers* m *and* M *so that* $f(x_1) = m$, $f(x_2) = M$ *for some* x_1, x_2 *in the interval* $a \leq x \leq b$, *and so that*

$$m \leq f(x) \leq M \qquad for \qquad a \leq x \leq b$$

The meaning of Theorem F is suggested in Figure 2-27. The value m is called the *absolute minimum* of f, and M is called the *absolute maximum* of f (Section 2-2). By Theorem D we know that f takes on every value between m and M. Accordingly, by Theorem F, *the range of a function defined and continuous on a closed interval is a closed interval* $m \leq y \leq M$. The conclusion of Theorem F is false if the domain of definition of f is an open interval or half-open interval (Problem 15 below). The proof of the theorem is given in Section 2-14.

2-8 CONTINUITY OF POLYNOMIALS AND OTHER COMMON FUNCTIONS

Theorems B, C, and E now permit us to establish the continuity of many functions. For example, the function $y = f(x) = x$ is defined and continuous for all real x, as follows at once from the definition (Problem 5 following Section 2-6). Hence, so also is $f \cdot f$, that is, the function $y = x^2$ (all x). In the same way, we show by induction that $y = x^3, y = x^4, \ldots$ are all continuous for all x. Since $y = f(x) = k$ (constant function) is also continuous for all x, so are the functions kx, kx^2, kx^3, . . . (multiplication of continuous functions). Then, by addition, the functions

$$1 + 2x, \qquad 1 + 2x + x^2, \qquad 3 + 5x + 6x^2 + x^3$$

and so on, are continuous for all x. Accordingly, *all polynomials in x are continuous for all x.* By division we now obtain the rational functions:

$$\frac{1 + 2x}{1 + 3x + 5x^2}, \qquad \frac{x^2 + 2}{x^3 + 5x + 1}, \ldots$$

In general, $f(x) = $ polynomial/polynomial. Here we must be careful about division by 0. We conclude from Theorem B: *every rational function is continuous in every interval in which the denominator has no zeros.* Thus

$$f(x) = \frac{1 + 2x}{1 + 2x + x^2} = \frac{1 + 2x}{(1 + x)^2}$$

is continuous in every interval not containing $x = -1$, the only zero of the denominator.

We shall prove in Chapter 5 that the functions $\sin x$ and $\cos x$ are continuous for all x. Thus, by Theorem C, so are the functions $\sin 2x$, $\sin 3x$, . . . , $\cos 2x$, $\cos 3x$, . . . and, as above, so are the "trigonometric polynomials":

$$a_0 + a_1 \cos x + b_1 \sin x + \cdots + a_n \cos nx + b_n \sin nx$$

The function $\tan x = (\sin x)/(\cos x)$ is continuous, by Theorem B, except where $\cos x = 0$, that is, with the exception of $x = \pm \pi/2$, $\pm 3\pi/2$, ...

The function $y = \sin x$, $-\pi/2 \le x \le \pi/2$ is monotone strictly increasing and has an inverse function $x = \operatorname{Sin}^{-1} y$, $-1 \le y \le 1$ (Figure 2-8 in Section 2-4 above) and by Theorem E, the continuity of the inverse function follows from that of the function $y = \sin x$. In the same way, we verify that the functions $\operatorname{Tan}^{-1} x$ and $\operatorname{Cos}^{-1} x$ (Section 2-4) are continuous; for $\operatorname{Tan}^{-1} x$, we need the analogue of Theorem E for an infinite interval.

By combining Theorems B, C, and E, we can now show that many other functions are continuous. For example, the function

$$y = \sin(x^2 + 2x + 3) + 3 \cos\left(\frac{x}{x^2 + 1}\right)$$

is continuous. In Chapter 5, we shall also show that the exponential function $y = a^x$ (a being a given positive number, $a \neq 1$) is continuous for all x. It follows from Theorem E that the inverse function $x = \log_a y$ is continuous for $y > 0$; that is, $y = \log_a x$ is continuous for $x > 0$. Accordingly, by Theorems B and C, functions such as

$$y = 2^{3x + x^2}, \qquad y = \log_{10}(1 + x^2)$$

are continuous for all x, whereas the function

$$y = \log_{10} x(1 - x)$$

is defined and continuous only for $0 < x < 1$.

The function $y = x^2$ is monotone strictly increasing for $x \ge 0$. Hence by Theorem E the inverse function $x = \sqrt{y}$ is continuous in the interval $0 \le$

$y < \infty$ or, equivalently, the function $y = \sqrt{x}$ is continuous for $x \geq 0$. Similarly, $y = \sqrt[3]{x} = x^{1/3}$, $y = x^{2/3}, \ldots$, $y = x^{-1/2}, \ldots$ and, in general, all powers of x are continuous for $x > 0$. It follows that $y = \sqrt{x + 1}$ is continuous for $x \geq -1$, $y = \sqrt{1 + \sqrt[3]{x}}$ is continuous for $x \geq 0$, $y = x + \sqrt{x - x^2}$ is continuous for $0 \leq x \leq 1$, and so on.

For all those functions for which continuity has been established one can evaluate limits by substituting the proper value:

$$\lim_{x \to x_0} f(x) = f(x_0)$$

provided x_0 is a point of continuity of f. For example, $\lim_{x \to 0} (x^2 - 2) = -2$, $\lim_{x \to \pi} \cos x = \cos \pi = -1$.

PROBLEMS

1. Evaluate each of the following limits.

   (a) $\lim_{x \to 0} \dfrac{\sin^3 x}{x^3}$

   (b) $\lim_{x \to 0} \dfrac{x}{\sin x}$

   (c) $\lim_{x \to 0} \left[2 + 3\dfrac{\sin x}{x} + 5\dfrac{\sin^2 x}{x^2} \right]$

   (d) $\lim_{x \to 1} (3 + x + 5x^2)$

   (e) $\lim_{x \to 2} \dfrac{2}{x^2 + 1}$

2. Let $\lim_{x \to x_0} f(x) = c$, $\lim_{x \to x_0} g(x) = k$. Evaluate each of the limits.

   (a) $\lim_{x \to x_0} [2f(x) + 4g(x)]$

   (b) $\lim_{x \to x_0} [2f(x) - 3g(x)]$

   (c) $\lim_{x \to x_0} \dfrac{f(x)}{1 + [g(x)]^2}$

   (d) $\lim_{x \to x_0} \{[f(x) - c][g(x) - k]\}$

   (e) $\lim_{x \to x_0} 2^{f(x)}$

3. Show, on the basis of Theorems B and C and the results derived in the text, that each of the following functions is continuous.

   (a) $y = x + \sin x$, all x

   (b) $y = \dfrac{x}{2 - \sin^2 x}$, all x

   (c) $y = x \tan x$, $-\dfrac{\pi}{2} < x < \dfrac{\pi}{2}$

   (d) $y = \dfrac{x \log_{10} x}{1 + 2^x}$, $x > 0$

   (e) $y = \dfrac{x^2}{2^x - 1}$, $x > 0$

   (f) $y = \dfrac{x}{x^3 + x + 1}$, $x \geq 0$

   (g) $y = \sqrt{1 - \sqrt{x}}$, $0 \leq x \leq 1$

   (h) $y = \sqrt{x - \sqrt{x}}$, $x \geq 1$

   (i) $y = \sin \dfrac{x}{x - 1}$, $0 \leq x < 1$

   (j) $y = \log_{10} \sin x$, $0 < x < \pi$

4. For each of the following functions, state whether the function is continuous in the interval stated and give a reason.

   (a) $y = \dfrac{1}{x^2 - 4}$, $0 \leq x \leq 1$

(b) $y = \dfrac{x}{x^2 - 2x - 3}$, $-2 \le x \le 2$

(c) $y = \sin\dfrac{1}{x}$, $0 < x \le 1$, $y = 0$ for $x = 0$

(d) $y = |x|$, $-\infty < x < \infty$

(e) $y = |x^2 + x - 3|$, $-\infty < x < \infty$

(f) $y = [x]$, the integer part of x, $0 \le x \le 5$; ($[1.3] = 1$, $[2.76] = 2$, etc.)

(g) $y = [x]$, $0 \le x < 1$

(h) $y = [x]$, $0 < x \le 1$

(i) $y = (x - 1)[x]$, $0 < x < 2$

5. Prove in detail the statements in Theorem C that concern the continuity of $g \circ f$.

6. (a) Let $y = f(x)$ be defined and continuous for $a \le x \le b$; let $y = g(x)$ be defined and continuous for $b \le x \le c$; let $f(b) = g(b)$. Justify the statement that the function $F(x)$, with domain $a \le x \le c$, so that $F(x) = f(x)$ for $a \le x \le b$, $F(x) = g(x)$ for $b < x \le c$ is continuous.

   (b) Show, on the basis of (a), that the function $y = F(x)$ so that $y = 2 + x$ for $0 \le x \le 1$, $y = x^2 + 5x - 3$ for $1 < x \le 2$ is continuous for $0 \le x \le 2$ and graph the function.

7. (a) Let $y = f(x)$ be defined and continuous in the half-open interval $a < x \le b$ and let $\lim\limits_{x \to a+} f(x)$ exist, $\lim\limits_{x \to a+} f(x) = c$. Show that the function F defined by

$$y = f(x) \text{ for } a < x \le b, \qquad y = c \text{ for } x = a$$

is continuous for $a \le x \le b$.

   (b) Show on the basis of (a) that the function

$$y = \dfrac{\sin x}{x}, \; 0 < x \le \pi, \qquad y = 1 \text{ for } x = 0$$

is continuous.

8. Let $f(x)$ and $g(x)$ be defined in the interval $a \le x \le b$.

   (a) If f is continuous and g is not continuous, can $f + g$ be continuous? Can $f - g$ be continuous? Can $f \cdot g$ be continuous?

   (b) If f and g are both not continuous, can $f + g$ be continuous? Can $f - g$ be continuous? Can $f \cdot g$ be continuous? Can f/g be continuous?

9. Let us assume that the continuity of the functions a^x (a fixed and greater than 1) and $\log_a x$ have been established.

   (a) Show that for any real number k and for $x > 0$,

$$x^k = 10^{k \, \log_{10} x}$$

   (Hint. Use the rule $\log a^b = b \log a$ and the definition of the logarithm.)

   (b) Show on the basis of Theorems B and C that, for fixed k, the function $y = x^k$ is continuous for $x > 0$. [Hint. Use the result of part (a).]

   (c) Let f and g be continuous in the interval $[a, b]$, let $f(x) > 0$ for x in $[a, b]$, and let $F(x) = f(x)^{g(x)}$. Show that

$$F(x) = 10^{g(x) \log_{10} f(x)}$$

and that F is continuous in $[a, b]$.

10. (a) The function $y = \text{Cos}^{-1} x$ is defined for $-1 \leq x \leq 1$ in Section 2-4 above. Show on the basis of the continuity of $y = \cos x$ and Theorem E that $\text{Cos}^{-1} x$ is continuous.

(b) Show that the function

$$y = \text{Sin}^{-1}\left(\frac{1}{1 + x^2}\right)$$

is continuous for all x and graph.

(c) Show that the function $y = \text{Sin}^{-1}(\sin x)$ is continuous for all x and graph.

11. Let $y = f(x)$ be monotone strictly increasing and continuous for $a \leq x \leq b$ and let $c = f(a), d = f(b)$. Show that the range of f is the interval $c \leq y \leq d$. (*Hint.* Use Theorem D.)

12. Extend the proof of Theorem E to the end point $y_0 = c$.

13. Show on the basis of Theorem D that each of the following equations has a solution in the interval given.

(a) $x^3 + x = 3, 0 \leq x \leq 2$ (b) $2^x + x = 2, 0 \leq x \leq 1$

(c) $\tan x - x = 0, \dfrac{\pi}{2} < x < \dfrac{3\pi}{2}$

14. Find the absolute minimum and absolute maximum.

(a) $y = x^2, 1 \leq x \leq 2$ (b) $y = \sin x, 0 \leq x \leq \pi$

(c) $y = \dfrac{1}{x}, 1 \leq x \leq 2$ (d) $y = 2x^2 - x - 3, -2 \leq x \leq 2$

15. Show that the function has no absolute maximum.

(a) $y = 1 - \dfrac{1}{x}, 1 \leq x < \infty$ (b) $y = \dfrac{1}{1 - x^2}, -1 < x < 1$

(c) $y = x, 0 < x < 1$

16. Show that the function has an absolute maximum even though it is defined in an interval which is not closed.

(a) $y = \sin x, 0 < x < 2\pi$ (b) $y = \dfrac{1 - x^2}{1 + x^4}, -\infty < x < \infty$

17. Assume we have proved that $\lim\limits_{h \to 0} (\sin h)/h = 1$.

(a) Prove that $\lim\limits_{h \to 0} \sin h = 0$, so that $\sin x$ is continuous at $x = 0$. [*Hint.* $\sin h = h\{(\sin h)/h\}$.]

(b) Prove that $\sin x$ is continuous for all x. [*Hint.* By a trigonometric identity (Appendix, Table V), $\sin(x + h) - \sin x = 2 \sin(h/2) \cos[(2x + h)/2]$. Hence $|\sin(x + h) - \sin x| \leq 2 |\sin(h/2)|$. Now use the result of part (*a*).]

(c) Prove that $\cos x$ is continuous for all x. {*Hint.* $\cos x = \sin[(\pi/2) - x]$. Use the result of part (*b*).}

2-9   VECTOR SPACES OF FUNCTIONS

We learned in Section 2-3 that real functions defined on a given interval can be *added* and can be *multiplied by constants;* that is, given f and g, we can form $f + g$ and cf. These operations obey the same rules as the ones for vectors in the plane:

$$f + g = g + f, \ f + (g + h) = (f + g) + h, \ f + 0 = f, \ 1f = f,$$
$$c_1(c_2 f) = (c_1 c_2)f, \ c(f + g) = cf + cg, \ (c_1 + c_2)f = c_1 f + c_2 f$$

Here equality of two functions means that they have the same value for each x of the interval; 0 denotes the function equal to 0 for all x. Furthermore, subtraction is always possible: given f and g, there is a unique h [denoted by $g - f$ and equal to $g + (-1)f$] for which $f + h = g$. Verification of the rules is mechanical. For example, $f + g = g + f$, since $f(x) + g(x) = g(x) + f(x)$ for all x by the commutative law of addition of numbers.

Because of the analogy with vectors, we say that the set of all functions that are defined on a given interval form a *vector space*.

Instead of considering *all* functions on a given interval, we can consider a selected set of functions—for example, all polynomials or all continuous functions. To each of these sets of functions all the preceding remarks apply. The sum of two polynomials is again a polynomial, a scalar times a polynomial is again a polynomial, and the commutative law, etc., hold true for polynomials simply because they hold true for functions in general. For continuous functions on an interval, the assertion follows from Theorem B.

Definition. A *vector space* of functions is a nonempty set V of functions having the same domain such that, if f and g are in V and c is a real number (scalar), then $f + g$ and cf are in V.

Thus the polynomials (domain $-\infty < x < \infty$) form a vector space of functions, as do the continuous functions on a given interval. Another example is the set of polynomials of degree at most N, where N is a fixed positive integer. For the sum of two such polynomials also has degree at most N, as does a constant times such a polynomial. Notice that the polynomials of *fixed* degree, for instance, degree 2, do *not* form a vector space. For the sum of two such polynomials may have degree less than 2:

$$(x^2 + 1) + (-x^2 + x) = x + 1.$$

The concepts of linear independence and basis (Section 1-7) can be applied to a vector space of functions, V. The functions $f_1, \ldots, f_k$ in V are considered to be *linearly independent* if

$$c_1 f_1 + \cdots + c_k f_k = 0 \qquad \text{implies} \qquad c_1 = 0, \ldots, c_k = 0$$

If $f_1, \ldots, f_k$ are not linearly independent, they are said to be *linearly dependent*.

EXAMPLE 1. Let V be the vector space of all polynomials. Then $1, x, \ldots, x^{k-1}$ are linearly independent. If not all of $c_1, \ldots, c_k$ are 0, then $c_1 1 + c_2 x + \cdots + c_k x^{k-1}$ is a polynomial of degree at least 0 and at most $k - 1$. Hence it has at most $k - 1$ zeros and, therefore, this polynomial cannot equal the function 0, which is 0 for every value of x.

EXAMPLE 2. In the same vector space, the polynomials

$$f_1(x) = x^2 - 2, \qquad f_2(x) = x^2 + x, \qquad f_3(x) = 3x^2 + x - 4$$

are linearly dependent. For

$$2(x^2 - 2) + 1(x^2 + x) + (-1)(3x^2 + x - 4) = 0$$

as we can easily verify.

We note that, if $f_1, \ldots, f_k$ are linearly independent, then none of the functions can be the function 0. Also (for $k \geq 2$) $f_1, \ldots f_k$ are linearly dependent precisely when one of these functions can be expressed as a linear combination of the others: for example, $f_k = a_1 f_1 + \cdots + a_{k-1} f_{k-1}$. If $f_1, \ldots, f_k$ are linearly independent, then each set of l functions ($l < k$) chosen from $f_1, \ldots, f_k$ is also linearly independent. The proofs of these remarks are left as exercises [Problem 8(a) to (d) below].

Notice that many common identities express the linear dependence of sets of functions; for example, the identities

$$\sin^2 x + \cos^2 x - 1 = 0, \qquad \cos 2x - 2\cos^2 x + 1 = 0$$

show the linear dependence of the sets

$$\{\sin^2 x, \cos^2 x, 1\} \qquad \text{and} \qquad \{\cos 2x, \cos^2 x, 1\},$$

respectively.

A set $f_1, \ldots, f_k$ is said to form a *basis* for vector space V, if every f in V is expressible uniquely as a linear combination of $f_1, \ldots, f_k$. For example, $1, \ldots, x^{k-1}$ form a basis for the vector space of all polynomials of degree at most $k - 1$. The functions $f_1, \ldots, f_k$ of a basis must be linearly independent [Problem 8 (e) below].

Not every vector space V has a basis, in the sense defined. For example, the vector space V of *all* polynomials does not have such a basis. We must allow for an infinite basis to cover these cases; for the vector space of all polynomials, the infinite set $1, x, \ldots, x^k, \ldots$ serves as a basis. For a definition of infinite basis, see Chapter 9.

THEOREM G. *Let V be a vector space of functions having basis $f_1, \ldots, f_k$ and let $g_1, \ldots, g_k$ be linearly independent members of V. Then $g_1, \ldots, g_k$ also form a basis.*

PROOF. For simplicity we give a proof here for the case $k = 2$. A general proof is given in Chapter 9. Since f_1, f_2 are a basis, we can write

$$g_1 = af_1 + bf_2, \qquad g_2 = cf_1 + df_2$$

Here we can eliminate f_2 and f_1 as usual to obtain

$$dg_1 - bg_2 = (ad - bc)f_1, \qquad -cg_1 + ag_2 = (ad - bc)f_2 \quad (2\text{-}90)$$

If $ad - bc = 0$, then we would have $dg_1 - bg_2 = 0$, $-cg_1 + ag_2 = 0$. Since g_1, g_2 are linearly independent, we would then have $d = 0, b = 0$, $c = 0, a = 0$ and, therefore, $g_1 = 0, g_2 = 0$, which is absurd. Consequently, $ad - bc \neq 0$. Hence, we can use Equations (2-90) to express f_1, f_2 as linear combinations of g_1, g_2. But every f in V is expressible as a linear combination of f_1, f_2 and, thus, by the previous sentence, as a linear combination of g_1, g_2: $f = c_1 g_1 + c_2 g_2$. Since g_1, g_2 are linearly independent, c_1, c_2 are unique [Problem 8 (f) below]. Hence, g_1, g_2 form a basis.

COROLLARY. *If vector space V has one basis of k functions, then every basis for V has exactly k functions.*

Thus, let $f_1, \ldots, f_k$ be a basis, let $g_1, \ldots, g_h$ be a basis, and let $h > k$. Then $g_1, \ldots, g_h$ are linearly independent and, hence, so are $g_1, \ldots, g_k$. Consequently, by Theorem G, $g_1, \ldots, g_k$ are a basis. Therefore, $g_{k+1}, \ldots, g_h$ are expressible as linear combinations of $g_1, \ldots, g_k$; this contradicts the linear independence of $g_1, \ldots, g_h$. Hence, $h \le k$ and similarly $k \le h$, so that $k = h$.

The corollary suggests the following definition:

Definition. A vector space V of functions has *dimension k* if V has a basis consisting of k functions.

For example, the polynomials of degree at most N form a vector space of dimension $N + 1$; for a basis consists of the $N + 1$ functions $1, x, \ldots, x^N$.

We observe that there is one very simple vector space of functions—that consisting of the single function 0. We assign the dimension 0 to this vector space (there is no basis in the usual sense). When V has functions other than 0 but has no (finite) basis, we say that V has dimension ∞.

PROBLEMS

1. Show that each of the following sets of functions is a vector space:
   (a) All polynomials $a_0 + a_1 x^2 + \cdots + a_k x^{2k}$ containing no term of odd degree.
   (b) All functions continuous on $[0, 1]$ and having a zero at 1.
   (c) All functions defined on $[0, 1]$ with limit 0 as $x \to 0+$.
   (d) All functions defined on $[0, 1]$ and having a limit as $x \to 0+$.
   (e) All linear combinations of $\cos x$, $\cos 2x$, $\cos 3x$ (domain $-\infty < x < \infty$).

2. Show that each of the following sets of functions is not a vector space:
   (a) All functions f defined and nonnegative: $f(x) \ge 0$ for all x, on a given interval.
   (b) All functions defined and not continuous on a given interval.
   (c) All functions continuous on $[0, 1]$ and having value 1 at $x = 1$.
   (d) All functions defined and having a finite number of zeros on $[0, 1]$.

3. For the vector space of Problem 1 (e), show that $\cos x$, $\cos 2x$, $\cos 3x$ are linearly independent and, hence, form a basis. [*Hint.* Let $c_1 \cos x + c_2 \cos 2x + c_3 \cos 3x = 0$ and show, by choosing three different values of x, that $c_1 = 0, c_2 = 0, c_3 = 0$.]

4. Test for linear independence in the vector space V of all polynomials:
   (a) $1 + x, 1 + 2x, 1 + 3x$ (b) $x^2 - 1, 2x^2 - 4, x^2 + 1$
   (c) $x^2 + 1, x^2 - x, x^2 + x$ (d) $x, x + x^2, x + x^2 + x^3, x^4$
   (e) $x^3 - 1, 2x^3 - 2, x^4$ (f) $x^4 - 2x^2, x^4 + 2x^2, -x^4 - 2x^2$

5. Can we find 5 linearly independent polynomials of degree 3?

6. Let V be the set of all rational functions

$$\frac{ax + b}{(x - 1)(x - 2)}, \qquad x \ne 1, x \ne 2$$

   (a) Show that V is a vector space of dimension 2.
   (b) Show that $g_1(x) = 1/(x - 1)$, $g_2(x) = 1/(x - 2)$ are in V, are linearly independent, and form a basis of V.

7. Let V be the set of all rational functions

$$\frac{ax^2 + bx + c}{x(x^2 - 4)}, \qquad x \neq 0, x \neq 2, x \neq -2$$

   (a) Show that V is a vector space of dimension 3.
   (b) Show that $g_1(x) = 1/x$, $g_2(x) = 1/(x + 2)$, $g_3(x) = 1/(x - 2)$ are in V, are linearly independent, and form a basis of V.

8. Let $f_1, \ldots, f_k$ be members of a vector space V of functions. Prove the following:
   (a) If $f_1 = 0$, then $f_1, \ldots, f_k$ are linearly dependent.
   (b) If f_1 is expressible as a linear combination of $f_2, \ldots f_k$, then $f_1, \ldots, f_k$ are linearly dependent.
   (c) If $k \geq 2$ and $f_1, \ldots, f_k$ are linearly dependent, then one of the functions is expressible as a linear combination of the others.
   (d) If $f_1, \ldots, f_k$ are linearly independent and $h < k$, then $f_1, \ldots, f_h$ are linearly independent.
   (e) If $f_1, \ldots, f_k$ are a basis for V, then $f_1, \ldots, f_k$ are linearly independent.
   (f) If $f_1, \ldots, f_k$ are linearly independent and $c_1 f_k + \cdots + c_k f_k = c_1' f_1 + \cdots + c_k' f_k$, then $c_1 = c_1', \ldots, c_k = c_k'$.

2-10 LIMITS AS x APPROACHES $+\infty$ OR $-\infty$

Let a function $y = f(x)$ be defined in an interval $a \leq x < \infty$. Thus the value of y is given for every x, no matter how large, as long as $x \geq a$. It may happen that the values of $y = f(x)$ approach a definite value c as x gets larger and larger. More precisely it may happen that, for each prescribed interval $c - p < y < c + p$ including c, we can find a value $x_0 \geq a$ so that

$$c - p < f(x) < c + p \qquad \text{for} \qquad x > x_0$$

This is illustrated in Figure 2-28; the graph of $y = f(x)$ remains in the white region for $x > x_0$. Under these conditions we say that f has *limit c as $x \to \infty$* (or as $x \to +\infty$) and write

$$\lim_{x \to \infty} f(x) = c$$

We also say that the line $y = c$ is a *horizontal asymptote* of the graph of f.

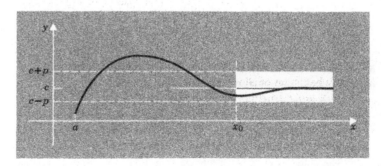

Figure 2-28 Limit as $x \to +\infty$.

EXAMPLE 1. $y = f(x) = 1/x^2$ has limit 0 as $x \to \infty$. For, given $p > 0$, we have

$$-p < \frac{1}{x^2} < p$$

for $x > x_0 = \dfrac{1}{\sqrt{p}}$. Then $x^2 > 1/p$, so that $1/x^2 < p$; and $x^2 > 0$, so that $1/x^2 > 0 > -p$.

By similar reasoning we can show that

$$\lim_{x \to \infty} x^{-n} = 0$$

for every positive n (not necessarily an integer) (see Problem 3 below).

We have a similar definition for

$$\lim_{x \to -\infty} f(x) = c$$

In this case f must be defined in an interval $-\infty < x \le b$ for some b, and the values of f must lie in a prescribed interval $(c - p,\ c + p)$, provided that $x < x_0$, where x_0 is suitably chosen. We can verify that

$$\lim_{x \to -\infty} x^{-n} = 0$$

provided that n is a positive integer (or more generally provided that n is of form p/q, where p and q are positive integers and q is odd) (see Problem 3 below).

Theorem A remains valid for limits as x approaches ∞ or $-\infty$; see Section 2-14. Essentially the same proof is needed as for the case $x \to x_0$. The only difference is that we must replace the condition "x is close to x_0" by one for "x is close to ∞" (or to $-\infty$). To say that x is close to ∞ is to say that x is sufficiently large: $x > x_0$; to say that x is close to $-\infty$ is to say that x is (algebraically) sufficiently small: $x < x_0$.

Theorem C also remains valid for x approaching ∞ or $-\infty$, and again for the same reason as for the case $x \to x_0$.

EXAMPLE 2. Let $y = f(x) = x^2/(2 + x^2)$. Then we can write (for $x > 0$)

$$f(x) = \frac{1}{\dfrac{2}{x^2} + 1}$$

We now apply Theorem A for x approaching ∞, observing that the limit of a constant function equals the constant value:

$$\lim_{x \to \infty} \frac{1}{\dfrac{2}{x^2} + 1} = \frac{\lim 1}{\lim \dfrac{2}{x^2} + \lim 1} = \frac{\lim 1}{2 \lim \dfrac{1}{x^2} + \lim 1} = \frac{1}{2 \cdot 0 + 1} = 1$$

Hence

$$\lim_{x \to \infty} f(x) = 1$$

2-11 INFINITE LIMITS OF A FUNCTION

The function $y = f(x) = 1/x$ has the property that, as x approaches 0 through positive values, y becomes larger and larger, eventually exceeding any prescribed positive number:

$$f(1) = 1 \qquad f(0.1) = 10, \qquad f(0.01) = 100, \qquad f(0.000001) = 1{,}000{,}000$$

We cannot say that y is approaching any real number. However, following custom, we use the symbolism of limits and write: $y \to \infty$ (or $y \to +\infty$) as $x \to 0+$, or

$$\lim_{x \to 0+} \frac{1}{x} = \infty$$

Despite our use of the limit symbol, we still say that, in this case, "$f(x)$ has no limit," or we may say that "$f(x)$ has no finite limit" or that "$f(x)$ has limit ∞."

Definition. Let $y = f(x)$ be defined for $a \le x \le b$, except perhaps at one point x_0 $(x_0 \ne b)$. Then we write

$$\lim_{x \to x_0+} f(x) = \infty$$

if, for every number K, we can find an interval $x_0 < x < x_0 + q$ in which $f(x) > K$. Similarly, we write

$$\lim_{x \to x_0+} f(x) = -\infty$$

if, for every number L, we can find an interval $x_0 < x < x_0 + q$ in which $f(x) < L$.

Notice that $\lim f(x) = -\infty$ holds true precisely when $\lim[-f(x)] = \infty$.

Analogous definitions are given for $\lim f(x) = \infty$ as $x \to x_0-$, $x \to x_0$, $x \to \infty$ and $x \to -\infty$, and for $\lim f(x) = -\infty$ in the various cases. Many of these cases are illustrated in Figure 2-29. We observe that $\lim f = \infty$ as $x \to x_0 +$ (or $x \to x_0 -$ or $x \to x_0$) corresponds to a vertical asymptote,

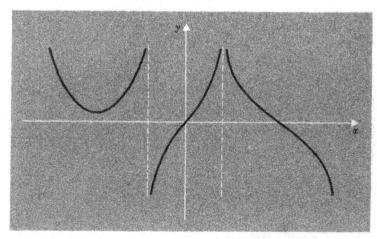

Figure 2-29 Infinite limits.

while $\lim f = c$ as $x \to \infty$, or $x \to -\infty$, corresponds to a horizontal asymptote.

EXAMPLE 1. $y = f(x) = \dfrac{1}{(x-1)^2}$. Here $\lim\limits_{x \to 1} f(x) = \infty$, since we can make $f(x) > K$, for $K > 0$, by making

$$(x-1)^2 < \frac{1}{K} \qquad \text{or} \qquad |x-1| < \frac{1}{\sqrt{K}}$$

that is, $x = 1$ being excluded, for

$$1 < x < 1 + \frac{1}{\sqrt{K}} \qquad \text{or for} \qquad 1 - \frac{1}{\sqrt{K}} < x < 1$$

We need not consider the case of $K \leq 0$ since, if we can make $f(x) > 1$, for example, then we have also made $f(x)$ greater than every negative number and 0. In general, we must only show that we can satisfy $f(x) > K$ for all arbitrarily large K.

Does the analogue of Theorem A hold true for infinite limits? Thus, if

$$\lim f = \infty, \qquad \lim g = \infty$$

for instance, as $x \to x_0+$, is

$$\lim(f + g) = \lim f + \lim g = \infty + \infty?$$

In fact, what should we mean by $\infty + \infty$? In this case, the answer is simple: $\infty + \infty = \infty$; that is, if $\lim f = \infty$ and $\lim g = \infty$, then $\lim(f + g) = \infty$. For, given $K > 0$, we can choose $q_1 > 0$ so that $f(x) > \frac{1}{2}K$ for $x_0 < x < x_0 + q_1$, and can choose $q_2 > 0$ so that $g(x) > \frac{1}{2}K$ for $x_0 < x < x_0 + q_2$, therefore,

$$f(x) + g(x) > \frac{1}{2}K + \frac{1}{2}K = K$$

for $x_0 < x < x_0 + q$ where q is the smaller of q_1, q_2. A similar reasoning holds true for limits as $x \to x_0-$, $x \to x_0$, $x \to \infty$, $x \to -\infty$.

However, if $\lim f = \infty$ and $\lim g = \infty$, what is $\lim(f - g)$? Here, no general rule can be given, as we show by an example:

EXAMPLE 2. $f(x) = 1/x$, $g(x) = (1/x) - k$, where k is a constant. Here $f(x) \to \infty$ as $x \to 0+$, $g(x) \to \infty$ as $x \to 0+$, but

$$f(x) - g(x) = \frac{1}{x} - \left(\frac{1}{x} - k\right) = k$$

so that $f(x) - g(x) \to k$ as $x \to 0+$. Thus the limit is different for different choices of k:

$$\lim_{x \to 0+} \left[\frac{1}{x} - \left(\frac{1}{x} - 1\right)\right] = 1, \qquad \lim_{x \to 0+} \left[\frac{1}{x} - \left(\frac{1}{x} - 2\right)\right] = 2, \dots$$

Examples can be given in which $f \to \infty$, $g \to \infty$, and $f - g \to \infty$, or $f - g \to -\infty$, or $f - g$ has no limit at all (see Problem 9 below).

Thus we must be very careful in combining functions with infinite limits. The principal facts are summarized in Table 2-3. The first entry is the rule for $\lim(f + g)$ when $\lim f = \infty$, $\lim g = \infty$, as considered above; $F(x)$ here stands for $f(x) + g(x)$, and we seek $\lim F(x)$, that is, $\lim(f + g)$. The fact that $\lim(f + g) = \infty$ is given in the fourth column, and the last column gives a "shorthand" version of the result, which is easier to remember: $\infty + \infty = \infty$.

The table indicates by a question mark in column five the cases in which there is no rule; then the corresponding "algebraic operation" with ∞ is undefined. The table also includes cases in which f or g has a finite limit, including 0. In particular, there are cases in which 0 is the limit of a denominator.

Table 2-3   Infinite Limits and Related Ones

No.	$\lim f(x)$	$\lim g(x)$	function $F(x)$	$\lim F(x)$	Shorthand
1	∞	∞	$f + g$	∞	$\infty + \infty = \infty$
2	∞	∞	$f - g$	?	$\infty - \infty$ undef.
3	∞	k	$f + g$	∞	$\infty + k = \infty$
4	$-\infty$	k	$f + g$	$-\infty$	$-\infty + k = -\infty$
5	∞	∞	fg	∞	$\infty \cdot \infty = \infty$
6	∞	$-\infty$	fg	$-\infty$	$\infty \cdot (-\infty) = -\infty$
7	∞	$k > 0$	fg	∞	$\infty \cdot k = \infty, k > 0$
8	∞	$k < 0$	fg	$-\infty$	$\infty \cdot k = -\infty, k < 0$
9	∞	0	fg	?	$\infty \cdot 0$ undef.
10	k	∞	$f \div g$	0	$k \div \infty = 0$
11	∞	∞	$f \div g$	?	$\infty \div \infty$ undef.
12	∞	$-\infty$	$f \div g$	?	$\infty \div (-\infty)$ undef.
13	$k > 0$	$0+$	$f \div g$	∞	$k \div (0+) = \infty$, if $k > 0$
14	∞	$0+$	$f \div g$	∞	$\infty \div (0+) = \infty$
15	$k > 0$	$0-$	$f \div g$	$-\infty$	$k \div (0-) = -\infty$, if $k > 0$
16	∞	$0-$	$f \div g$	$-\infty$	$\infty \div (0-) = -\infty$
17	0	0	$f \div g$	?	$0 \div 0$ undef.

Note: In Nos. 13 and 14 $\lim g(x) \doteq 0+$ means that g has limit 0 and $g > 0$ for x sufficiently close to the value it is approaching; for Nos. 15 and 16 the same condition holds true, with the exception that $g(x) < 0$.

The cases for which there is no rule are called "indeterminate forms." The differential calculus has as one of its goals the development of procedures for finding the limits in these cases, especially for the indeterminate form $0/0$, No. 17 in the table (see Chapters 3 and 6).

For brevity, the table does not include cases which, by a simple rewriting, can be reduced to those of the table. Thus, if $\lim f = -\infty$ and $\lim g = \infty$, then $\lim(fg)$ can be reduced to No. 6 by writing

$$\lim[f(x)g(x)] = \lim[\{-f(x)\}\{-g(x)\}]$$

and hence to $\infty \cdot (-\infty) = -\infty$.

The proofs of the rules and illustrations are given in the exercises.

PROBLEMS

1. Prove:

(a) $\lim\limits_{x \to \infty} \dfrac{1}{x} = 0$

(b) $\lim\limits_{x \to -\infty} \dfrac{1}{x} = 0$

(c) $\lim\limits_{x \to \infty} \dfrac{1}{x^3} = 0$

(d) $\lim\limits_{x \to -\infty} \dfrac{1}{x^3} = 0$

(e) $\lim\limits_{x \to \infty} \dfrac{1}{\sqrt{x}} = 0$

(f) $\lim\limits_{x \to -\infty} \dfrac{1}{\sqrt[3]{x}} = 0$

2. Evaluate, with the aid of the results of Problem 1 and theorems on limits:

(a) $\lim\limits_{x \to \infty} \left(\dfrac{1}{x} + \dfrac{1}{x^2}\right)$

(b) $\lim\limits_{x \to \infty} \left(\dfrac{2}{x} - \dfrac{3}{x^2}\right)$

(c) $\lim\limits_{x \to \infty} \left(2 - \dfrac{1}{x^3}\right)$

(d) $\lim\limits_{x \to -\infty} \left(\dfrac{1}{x^2} + \dfrac{2}{x^3}\right)^4$

(e) $\lim\limits_{x \to \infty} \dfrac{2x + 1}{x^2}$

(f) $\lim\limits_{x \to \infty} \dfrac{x^2 - 5x + 1}{x^3}$

(g) $\lim\limits_{x \to \infty} \dfrac{2x^2 + 3x + 1}{x^2}$

(h) $\lim\limits_{x \to \infty} \dfrac{5x^3 - 1}{x^3}$

(i) $\lim\limits_{x \to \infty} \dfrac{x^2 - x + 1}{2x^2 - 1}$

(j) $\lim\limits_{x \to \infty} \dfrac{3x^3 - 4x + 1}{4x^3 + 3x^2 - 2}$

(k) $\lim\limits_{x \to \infty} \dfrac{x}{\sqrt{x^2 + 1}}$

(l) $\lim\limits_{x \to -\infty} \dfrac{x}{\sqrt{x^2 + 1}}$

(m) $\lim\limits_{x \to \infty} \dfrac{x\sqrt{x} + \sqrt[3]{x} + 1}{\sqrt{x^3 - 1} + x}$

(n) $\lim\limits_{x \to \infty} \dfrac{x^{-1/2} + x^{-3/2}}{(2x + 1)^{-1/2} + x^{-1}}$

‡3. Let n be a positive integer. Prove that (a) $\lim\limits_{x \to \infty} x^{-n} = 0$, (b) $\lim\limits_{x \to \infty} x^{-1/n} = 0$,

(c) $\lim\limits_{x \to \infty} x^{-\alpha} = 0$ for every positive α. (*Hint.* Choose the integer n so that $0 < n < \alpha$
or $0 < 1/n < \alpha$. Then either $0 < x^{-\alpha} < x^{-n}$ or $0 < x^{-\alpha} < x^{-1/n}$ for $x > 0$.)

(d) $\lim\limits_{x \to -\infty} x^{-\alpha} = 0$ for α a positive integer or α a rational number p/q where p and q are positive and q is odd.

‡4. Prove that, if $a > 1$, then $\lim\limits_{x \to \infty} a^{-x} = 0$. [*Hint.* Show that $0 < a^{-x} < p$ if $x > [\log_{10}(1/p)][\log_{10} a]^{-1}$.]

5. Evaluate, with the aid of the results of Problems 3 and 4:

(a) $\lim\limits_{x \to \infty} \dfrac{2^x}{3 \cdot 2^x + 1}$

(b) $\lim\limits_{x \to \infty} \dfrac{3 + 2^{-x} + 5^{-x}}{2 + 4^{-x}}$

(c) $\lim\limits_{x \to \infty} \dfrac{2x5^x}{(3x + 1)5^x + 1}$

(d) $\lim\limits_{x \to \infty} \dfrac{x^\pi}{2x^\pi + \pi}$

6. Evaluate where possible:

(a) $\lim\limits_{x \to \infty} \dfrac{\sin x}{x}$

(b) $\lim\limits_{x \to \infty} \dfrac{\sin^2 x}{x}$

(c) $\lim\limits_{x \to \infty} \dfrac{x \sin x}{x^2 + 1}$

(d) $\lim\limits_{x \to \infty} \sin x$

(e) $\lim\limits_{x \to \infty} \dfrac{x \cos x}{2 + x}$

(f) $\lim\limits_{x \to \infty} \dfrac{2x}{x^2 + \cos x}$

7. Evaluate with the aid of Table 2-3:

(a) $\lim_{x \to 2} \dfrac{x^2}{(x-2)^2}$

(b) $\lim_{x \to 0} \dfrac{\cos x}{x \sin x}$

(c) $\lim_{x \to 2} \dfrac{\cos x}{\sin^2 \pi x}$

(d) $\lim_{x \to 0+} \csc x$

(e) $\lim_{x \to 1+} \dfrac{\log_{10}(1+x)}{x-1}$

(f) $\lim_{x \to 1-} \dfrac{e^{-x}}{x-1}$

(g) $\lim_{x \to \infty} \dfrac{x^3}{2x^2+1}$

(h) $\lim_{x \to \infty} \dfrac{x^5}{(x-1)^4}$

(i) $\lim_{x \to \infty} \dfrac{x^2 + \sin x}{x+1}$

(j) $\lim_{x \to \infty} \dfrac{x^3+1}{x^2+x^{1/2}}$

(k) $\lim_{x \to \infty} \sin\left(\dfrac{\pi x^2 + 1}{2x^2 + x - 1}\right)$

8. Let $p(x) = a_0 x^n + a_1 x^{n-1} + \ldots$ and $q(x) = b_0 x^m + b_1 x^{m-1} + \ldots$ be polynomials of nonnegative degrees n, m, respectively. Show that

$$\lim_{x \to \infty} \frac{p(x)}{q(x)} = \begin{cases} \pm\infty & \text{if} \quad n > m \\ a_0/b_0 & \text{if} \quad n = m \\ 0 & \text{if} \quad n < m \end{cases}$$

with $+$ if $a_0 b_0 > 0$ and $-$ if $a_0 b_0 < 0$.

9. With regard to entry No. 2 in Table 2-3, choose functions f and g so that f has limit ∞, and g has limit ∞, as x approaches 0, and a prescribed one of the following holds true:

(a) $\lim_{x \to 0} [f(x) - g(x)] = \infty$

(b) $\lim_{x \to 0} [f(x) - g(x)] = -\infty$

(c) $f(x) - g(x)$ has no limit, finite or infinite, at $x = 0$.

10. With regard to the entry No. 17 in Table 2-3, choose functions f and g so that both have limit 0 as x approaches 1 and a prescribed one of the following holds true:

(a) $\lim_{x \to 1} \dfrac{f(x)}{g(x)} = 2$

(b) $\lim_{x \to 1} \dfrac{f(x)}{g(x)} = 0$

(c) $\lim_{x \to 1} \dfrac{f(x)}{g(x)} = \infty$

11. Justify the following entries in Table 2-3:

(a) No. 3 (b) No. 5 (c) No. 7 (d) No. 8 (e) No. 13

12. Prove that, if f has limit ∞ as $x \to a$ (or $x \to \infty$, etc.) then $g = 1/f$ has limit 0. Does the converse hold true?

13. Prove that, if f has limit c as $x \to 0$, then $g(x) = f(1/x)$ has limit c as $x \to \infty$. Does the converse hold true?

14. Prove that, if g has limit 0 and f/g has limit c as $x \to a$, then f has limit 0 as $x \to a$.

15. Which of the following sets of functions, defined for $0 < x < \infty$, forms a vector space?

(a) All functions having a limit as $x \to \infty$.

(b) All functions having a limit as $x \to 0+$.

(c) All functions having limit 0 as $x \to \infty$.

(d) All functions having limit 0 as $x \to 0+$.

(e) All functions having limit ∞ as $x \to \infty$.

(f) All functions having limit $-\infty$ as $x \to 0+$.

2-12 LIMITS OF INFINITE SEQUENCES

By a *finite sequence of numbers,* we mean a listing in a definite order of n numbers, where n is a positive integer. We can write the list as

$$x_1, x_2, \ldots, x_n$$

where x_1 is the first, x_2 is the second, and so on. Or we may give the numbers explicitly:

$$3, 5, 2, 1.9, 7.8.$$

Since the order is crucial, we must always think of an index value attached to each number, indicating whether the number is first, second, third, and so on. We can thus describe a finite sequence of numbers as an assignment of a real number to each of the integers $1, 2, \ldots, n$, for some positive integer n; that is *a finite sequence of numbers is a real function whose domain is the set of integers $1, \ldots, n$.* This is illustrated graphically in Figure 2-30.

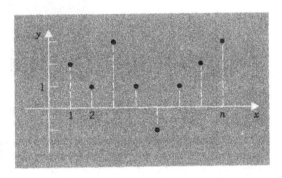

Figure 2-30 Finite sequence as a function.

Occasionally, it may be more convenient to allow the index to start at 0, or at some other integer value, even negative. Thus the rule

$$f(k) = \frac{1}{3 + k}, \qquad k = -2, -1, 0, 1, 2$$

defines a finite sequence.

EXAMPLE 1. *The binomial coefficients.* For each positive integer n, let $\binom{n}{k}$ be the coefficient of $x^k y^{n-k}$, in the binomial expansion of $(x + y)^n$ (see Section 0-22). Then the numbers $\binom{n}{0}, \binom{n}{1}, \ldots, \binom{n}{n}$ form a finite sequence. We give some sample cases:

$$
\begin{array}{ll}
n = 1: & 1, 1 \\
n = 2: & 1, 2, 1 \\
n = 3: & 1, 3, 3, 1 \\
n = 4: & 1, 4, 6, 4, 1
\end{array}
$$

By an *infinite sequence* we mean an assignment of a real number x_k to each positive integer k; thus we obtain a never-ending list:

$$x_1, x_2, x_3, \ldots, x_k, \ldots \qquad (2\text{-}120)$$

As above, we can interpret the list as a function: *an infinite sequence is a function whose domain is the set of all positive integers $1, 2, 3, \ldots$* This is illustrated in Figure 2-31. Occasionally, as previously, we may start the index at 0, or -1, or some other integer.

EXAMPLE 2. $n!$ (*factorial*). The factorial is a sequence, assigning to each integer n the value $n! = 1 \cdot 2 \cdot 3 \cdots n$. Thus the sequence is

$$1,\ 2,\ 3,\ 6,\ 24,\ 120,\ 720, \ldots$$

As customary, we also define $0!$ to be 1 and start the list with $0!$.

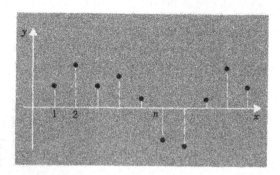

Figure 2-31 Infinite sequence as a function.

In the calculus, most sequences of interest are infinite and, hence, the word "sequence" by itself is normally understood to mean "infinite sequence." To indicate the sequence (2-120) as a whole, we denote it by $\{x_n\}$.

For some sequences, the value may approach a definite number as the index increases. More precisely, $\{x_n\}$ may have a limit as $n \to \infty$:

Definition. The infinite sequence $\{x_n\}$ has limit c as $n \to \infty$,

$$\lim_{x \to \infty} x_n = c$$

if, for every $p > 0$, we can find a positive integer N so that $x - p < x_n < c + p$ for $n > N$.

This definition parallels the one for

$$\lim_{x \to \infty} f(x) = c \tag{2-121}$$

and the two types of limit are closely related. In the one case, the function is defined only at integer values; in the other, the function is defined say for $a \le x < \infty$. Otherwise, the two definitions are the same. In fact, whenever (2-121) holds true, we can obtain a sequence with limit c by considering f only for integer values. For example

$$\lim_{x \to \infty} \frac{x^2}{x^2 + 2} = 1$$

as shown in Section 2-10 above. Hence

$$\lim_{n \to \infty} \frac{n^2}{n^2 + 2} = 1$$

that is, the sequence

$$\frac{1}{3},\ \frac{4}{6},\ \frac{9}{11}, \ldots,\ \frac{n^2}{n^2 + 2}, \ldots$$

has limit 1.

A sequence with limit c is said to be *convergent*, and *to converge to* c. A sequence with no limit is said to be *divergent*.

Our Theorem A on limits can be rephrased in terms of sequences: *if*

$$\lim_{n \to \infty} x_n = c, \qquad \lim_{n \to \infty} y_n = k$$

then

$$\lim_{n \to \infty} (x_n \pm y_n) = c \pm k$$

$$\lim_{n \to \infty} (x_n \cdot y_n) = c \cdot k$$

$$\lim_{n \to \infty} \frac{x_n}{y_n} = \frac{c}{k}, \qquad provided \qquad k \neq 0$$

The formal proof (Section 2-14) of this statement is exactly the same as for the original Theorem A.

Theorem C also has an analogue: *If, for n sufficiently large, all values x_n lie in the interval $a \leq x \leq b$, in which the function $g(x)$ is defined, if $\lim x_n = x_0$ and g is continuous at x_0, then*

$$\lim_{n \to \infty} g(x_n) = g(x_0) = g\left(\lim_{n \to \infty} x_n\right)$$

The justification (Section 2-11) and formal proof (Section 2-14) are the same as for the original theorem.

Definition. A sequence $\{x_n\}$ is said to be *bounded above* by B if $x_n \leq B$ for some fixed B, for $n = 1, 2, \ldots$, and to be *bounded below* by A if $x_n \geq A$ for some fixed A, for $n = 1, 2, \ldots$.

Definition. A sequence $\{x_n\}$ is said to be *monotone strictly increasing* if

$$x_n < x_{n+1} \qquad for \qquad n = 1, 2, \ldots,$$

and to be *monotone nondecreasing* if

$$x_n \leq x_{n+1} \qquad for \qquad n = 1, 2, \ldots$$

The sequence $\{x_n\}$ is said to be *monotone strictly decreasing* if

$$x_n > x_{n+1} \qquad for \qquad n = 1, 2, \ldots$$

and to be *monotone nonincreasing* if

$$x_n \geq x_{n+1} \qquad for \qquad n = 1, 2, \ldots$$

EXAMPLE 3. The sequence $\{n/(n + 1)\}$ is bounded above by 1 and is monotone strictly increasing (Problem 3 below).

EXAMPLE 4. The sequence $\{n!\}$ is monotone strictly increasing (for $n \geq 1$) but is not bounded above (Problem 3 below).

The following theorem on monotone sequences has many applications in the calculus:

THEOREM H. *Let $\{x_n\}$ be an infinite sequence which is bounded above by B and monotone nondecreasing. Then $\{x_n\}$ is convergent:*

$$\lim_{n \to \infty} x_n = c \leq B$$

Similarly, if $\{x_n\}$ is bounded below and monotone nonincreasing then $\{x_n\}$ is convergent.

The first case is illustrated in Figure 2-32. A proof is given in the next section. Here we give an informal discussion. We have

$$x_1 \leq x_2 \leq x_3 \leq x_4 \leq \cdots \leq x_n \leq x_{n+1} \leq \cdots$$

and all numbers are less than or equal to B. Hence, as n increases, x_n moves to

Figure 2-32 Monotone nondecreasing sequence bounded above.

the right or stands still. Therefore, it is plausible that the numbers x_n must "pile up" on some one number—either on B itself, or on some number less than B; the number on which they pile up is the desired limit.

In fact, we can obtain the decimal representation of the limit. We indicate this by a numerical example. Let us take $B = 5$. We "observe" the numbers $x_1, x_2, \ldots$ and ask whether they ever reach or exceed 4, or 3, etc.; we here test each integer between x_1 and 5. Suppose we find that the largest integer reached or exceeded is 3. Then we ask which is the largest of the numbers $3.0, 3.1, \ldots, 3.9$ reached or exceeded. Suppose we find that the answer is 3.4. Then we ask which is the largest of the numbers $3.40, 3.41, \ldots, 3.49$ reached or exceeded, and so on. In this way, we find a decimal, for example, $3.4275842002 18\ldots$. This number x_0 must be the limit of the sequence. For by the process we have used, the members of the sequence differ from x_0 by less than 1, less than 0.1, less than 0.01, $\ldots$ and so on, provided that, in each case, we take n sufficiently large.

EXAMPLE 5. $x_n = (\frac{1}{2}) + (\frac{1}{4}) + \cdots + (1/2^n)$, so that the sequence is

$$\frac{1}{2}, \frac{3}{4}, \frac{7}{8}, \ldots$$

We can show (Problem 4 below) that

$$x_n = 1 - \frac{1}{2^n}$$

and, therefore, that $x_n < 1$ for all n; thus the sequence is bounded above. It is clearly monotone strictly increasing:

$$x_n = \frac{1}{2} + \frac{1}{4} + \cdots + \frac{1}{2^n} < \frac{1}{2} + \frac{1}{4} + \cdots + \frac{1}{2^n} + \frac{1}{2^{n+1}} = x_{n+1}$$

Hence, by Theorem H, x_n is convergent. In fact, we can clearly make $1/2^n$ less than any prescribed quantity by choosing n sufficiently large; consequently, x_n can be made as close to 1 as desired:

$$\lim_{n \to \infty} x_n = \lim_{n \to \infty} \left(\frac{1}{2} + \frac{1}{4} + \cdots + \frac{1}{2^n} \right) = 1$$

EXAMPLE 6

$$x_n = \left(1 + \frac{1}{n} \right)^n$$

so that the sequence is

$$2, \quad \left(\frac{3}{2}\right)^2 = 2\frac{1}{4}, \quad \left(\frac{4}{3}\right)^3 = 2\frac{10}{27}, \quad \left(\frac{5}{4}\right)^4 = 2\frac{113}{256}, \ldots$$

We shall show in Chapter 5 that the sequence is monotone strictly increasing and bounded above by 3. *The limit is the number e:*

$$e = \lim_{n \to \infty} \left(1 + \frac{1}{n}\right)^n = 2.7182818285\ldots$$

There is an analogue of Theorem H for functions defined on an interval:

THEOREM H'. *Let $y = f(x)$ be defined for $a \leq x < \infty$ and let f be monotone nondecreasing:*

$$f(x') \leq f(x'') \qquad \text{for} \qquad x' < x''$$

If f is bounded above: $f(x) \leq B$ for all x, then f has a limit as $x \to \infty$:

$$\lim_{x \to \infty} f(x) = c \leq B$$

The proof is left to Problem 6 below. Theorem H' can be restated for monotone nonincreasing functions and for limits as $x \to x_0+$, $x \to x_0-$, or $x \to -\infty$.

2-13 THE LEAST UPPER BOUND AXIOM

In order to prove Theorem H—that every bounded monotone sequence converges—we are forced to look very carefully at the real number system. We can do this by representing real numbers as decimals, as we did in the reasoning of the preceding section. An alternative is to formulate a general principle that gives us all the results obtainable by decimals but that is easier to formulate and to use. One such principle is the *least upper bound axiom*, formulated below.

Definition. A set E of numbers is *bounded above* if there is a number B for which $x \leq B$ for every x in the set E. The number B is called *an upper bound for E*.

We observe that, if B is an upper bound for E, and $B < B'$, then B' is also an upper bound for E.

Least Upper Bound Axiom. *If a nonempty set E of real numbers is bounded above, then E has a least upper bound $\bar{B}$; that is, there is a number $\bar{B}$ so that*

(1) $\bar{B}$ *is an upper bound for E.*
(2) *If B is an upper bound for E, then $B \geq \bar{B}$.*

EXAMPLE 1. Let E be the interval: $0 < x < 1$. Then $\bar{B} = 1$ is the least upper bound of E.

EXAMPLE 2. Let E be the set of all numbers x for which $x^2 < 2$. Then E is bounded above, since $B = 2$ is an upper bound (if x were greater than 2 and in E, then x^2 would exceed 4, which is impossible for x in E). In fact, 1.5 is an upper bound, as is 1.45, ... We know what the least upper bound is; namely, $\bar{B} = \sqrt{2}$. It is because we know that $\sqrt{2}$ (and square roots of positive numbers generally) exist as real numbers, that we obtain our least upper bound.

Remark. If we mark the rational numbers $0, \pm 1, \pm 2, \ldots, \pm\frac{1}{2}, \pm\frac{3}{2}, \ldots,$ $\pm\frac{1}{3}, \pm\frac{2}{3}, \ldots,$ on a number axis, they appear to gradually fill up the axis. However, there are "holes" left, corresponding to the irrational numbers $\sqrt{2}$, $\sqrt{3}$, π, e, ... The least upper bound axiom assures us that, after the irrational numbers have been filled in, there are no holes left. We say that the real numbers form a "continuum,"—an array with no gap or break.

Throughout the rest of this book we shall use the least upper bound axiom as a known property of real numbers. This axiom, along with the algebraic properties noted in Sections 0-1 and 0-2, will suffice for all the theory.

A set E is said to be *bounded below* if there is a number A for which $x \geq A$ for every x in E. We can then state:

GREATEST LOWER BOUND THEOREM. *A nonempty set E of real numbers which is bounded below has a greatest lower bound $\bar{A}$; that is, there is a number $\bar{A}$ so that*

(1) $\bar{A}$ *is a lower bound for E.*
(2) $\bar{A} \geq A$ *for every lower bound A of E.*

This theorem could have been used instead of the least upper bound axiom; each is a logical consequence of the other.

To prove the theorem, we let E_1 be the set of *all* lower bounds of E. Then E_1 is not empty by assumption (E has a lower bound). Also E_1 is bounded above, since each element x of E is an upper bound for E_1 (see Figure 2-33). Hence,

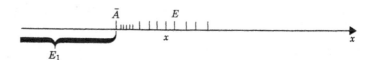

Figure 2-33   Greatest lower bound.

by the least upper bound axiom, E_1 has a least upper bound which we call $\bar{A}$. We assert that $\bar{A}$ is a lower bound for E. Indeed, if we had $x_1 < \bar{A}$ for some x_1 in E, then $\bar{A}$ could not be the least upper bound of E_1, since x_1 is also an upper bound of E_1. Hence, $\bar{A}$ is a lower bound for E. Also $\bar{A} \geq A$ for every A in E_1; that is, $\bar{A} \geq A$ for every lower bound of E. Thus, $\bar{A}$ is the greatest lower bound of E.

Uniqueness of Least Upper Bound and Greatest Lower Bound. Since the least upper bound is the smallest number with a certain property, clearly, there can be only one least upper bound. For a similar reason, there can be only one greatest lower bound.

Notation. As customary, we write glb E and lub E for the greatest lower bound and least upper bound of E. Other notations used are inf E and sup E:

$$\text{inf } E = \text{glb } E, \qquad \text{sup } E = \text{lub } E$$

PROOF OF THEOREM ON MONOTONE SEQUENCES. We prove Theorem H of Section 2-12 for the case of a monotone nondecreasing sequence $\{x_n\}$, bounded above, for instance, by B. The set E we take to be the set of all numbers that occur in the sequence. Then E is also bounded above by B and, by the least upper bound axiom, E has a least upper bound $\bar{B}$, at most, equal to B. Hence $x_n \le \bar{B}$ for every n. If x' is a number less than $\bar{B}$, then $x_n > x'$ for some n, say n'; otherwise $\bar{B}$ would not be the least upper bound. Since the sequence is nondecreasing, we have $x_n > x'$ for $n > n'$, that is

$$x' < x_n \le \bar{B} \qquad \text{for} \qquad n > n'$$

Since x' was an arbitrary number less than $\bar{B}$, we conclude that

$$\lim_{n \to \infty} x_n = \bar{B} = c \le B$$

PROBLEMS

1. List as finite sequences:
   (a) The terms of an arithmetic progression of 5 terms, the first being 6 and the difference being 7.
   (b) The numbers x between 0 and 2π for which $\sin^2 2x = 1$.
   (c) The coefficients of a general polynomial of degree n.
   (d) The number of distinct prime factors of the integers $1, 2, \ldots, 20$.
   (e) The positive square root of the integers $1, 2, \ldots, n$.
2. Determine whether the infinite sequence $\{x_n\}$ is convergent and find the limit when it is:

   (a) $x_n = (-1)^n$ (b) $x_n = \dfrac{1}{n}$ (c) $x_n = \dfrac{1}{2^n}$ (d) $x^n = n^2$

   (e) $x_n = \dfrac{n}{n+1}$ (f) $x_n = \dfrac{n}{n^2+1}$ (g) $x_n = n!$

3. Which of the sequences of Problem 2 are:
   (a) Monotone strictly increasing?   (b) Monotone strictly decreasing?
   (c) Bounded above?   (d) Bounded below?
4. Show that, for the sequence of Example 5 in Section 2-12, x_n is the sum of a geometric progression and, therefore, $x_n = 1 - 2^{-n}$.
5. Show that the set is bounded above and find the least upper bound:
   (a) The interval $-1 < x < 1$.   (b) the interval $-1 \le x \le 1$.
   (c) All rational numbers whose square is less than 3.
   (d) The numbers $0.3, 0.33, 0.333, 0.3333, \ldots$.
   (e) The numbers $0.9, 0.99, 0.999, 0.9999, \ldots$.
   (f) The range of the function $y = 1/(x^2 + 1)$, $-\infty < x < \infty$.
   (g) The range of the function $y = x^2/(x^2 + 1)$, $-\infty < x < \infty$.
6. Prove Theorem H'. (*Hint.* Imitate the proof of Theorem H in Section 2-13.)
7. Prove that, on the basis of the least upper bound axiom, if a and b are positive real numbers, then there is a positive integer n so that $na > b$. (This property of the real number system is called Archimedean order. To prove it, assume that $na \le b$ for every n and apply the axiom.)
8. (a) Prove that, for every positive real number x, there is a largest integer n so that $n \le x$. (This is the integral part of x, $[x]$; see Problem 4 following Section 2-8.)

(b) Prove that, for every positive real number x and each positive integer h, there are integers $n, k_1, k_2, \ldots, k_h$, with $0 \leq n, 0 \leq k_1 \leq 9, 0 \leq k_2 \leq 9, \ldots, 0 \leq k_h \leq 9$, so that

$$n + \frac{k_1}{10} + \frac{k_2}{100} + \cdots + \frac{k_h}{10^h} \leq x < n + \frac{k_1}{10} + \cdots + \frac{k_h + 1}{10^h}$$

(This is the main step in representing x as a decimal.)

9. (a) Show that the function $y = 1/x$ is monotone strictly decreasing for $0 < x < \infty$.

(b) Show that $\lim_{x \to \infty} (1/x)$ exists.

(c) Show that the limit in (b) is 0. [*Hint.* Show that $1/(x + 1)$ must have the same limit c as $1/x$, but $1/(x + 1) = (1/x)[1 + (1/x)]^{-1}$, so that $c = c/(1 + c)$.]

10. (a) Use the known properties of exponential functions to show that $y = 2^{-x}$ is monotone strictly decreasing for $-\infty < x < \infty$, and that $2^{-x} > 0$.

(b) Show that $\lim_{x \to \infty} 2^{-x}$ exists.

(c) Show that $\lim_{x \to \infty} 2^{-x}$ must be 0. [*Hint.* Let the limit be c. Show that $2 \cdot 2^{-x}$ also has limit c but should have limit $2c$.]

11. Consider all infinite sequences as indexed starting with 1. Show that each of the following is a vector space of functions:

(a) All infinite sequences.   (b) All convergent sequences.

(c) All sequences having limit 0.   (d) All sequences which are bounded above and below.

‡2-14 PROOFS OF THEOREMS ON LIMITS AND CONTINUITY

For convenience, we state the definition of limit here in terms of δ and ϵ (delta and epsilon). In our previous discussion we have often had to specify "in a sufficiently small interval about x_0" or "sufficiently close to y_0." These conditions are expressed by inequalities such as $|x - x_0| < \delta, |y - y_0| < \epsilon$, where δ and ϵ are positive numbers. The inequality $|x - x_0| < \delta$ makes x within distance δ of x_0, hence, is equivalent to

$$x_0 - \delta < x < x_0 + \delta$$

To state that x is within distance δ of x_0 but $x \neq x_0$, we use the double inequality

$$0 < |x - x_0| < \delta$$

Definition. Let $y = f(x)$ be defined in the interval $[a, b]$ except perhaps at the point x_0 of this interval. Then we write

$$\lim_{x \to x_0} f(x) = c$$

if, for every $\epsilon > 0$, we can find a $\delta > 0$ so that for every x in $[a, b]$ for which $0 < |x - x_0| < \delta$ we have

$$|f(x) - c| < \epsilon$$

When $x_0 = a$ or b, we also write, respectively,

$$\lim_{x \to a+} f(x) = c \qquad \text{or} \qquad \lim_{x \to b-} f(x) = c$$

The new definition of limit is equivalent to the previous one; the p of the old definition has been replaced by ϵ, the q by δ. We stress that the choice of δ depends on the ϵ.

The definition of continuity can be reformulated: f is continuous at x_0 if, for each $\epsilon > 0$, there is a $\delta > 0$ so that $|f(x) - f(x_0)| < \epsilon$ for all x in $[a, b]$ for which $|x - x_0| < \delta$. For the new condition is equivalent to the old one: $\lim f(x) = f(x_0)$ as $x \to x_0$.

As before, the type of interval in which f is defined can be varied. We shall give the proofs only for the case of a closed interval.

PROOF OF THEOREM C. As we shall learn, this theorem is the fundamental one, so that we prove it first. We are given f as above, with $\lim_{x \to x_0} f(x) = y_0$. We are also given a function $g(y)$ which is defined in the interval $c \le y \le d$, in which y_0 lies, and g is continuous at y_0. Finally, the range of f lies in the interval $c \le y \le d$, so that we can form the composite function

$$F(x) = g[f(x)], \qquad a \le x \le b$$

We then want to show that

$$\lim_{x \to x_0} F(x) = g(y_0) \tag{2-140}$$

or equivalent that $\lim g[f(x)] = g[\lim f(x)]$.

Since g is continuous at y_0, given $\epsilon > 0$, we can choose δ_1 so that $|g(y) - g(y_0)| < \epsilon$ for $|y - y_0| < \delta_1$. Since $\lim_{x \to x_0} f(x) = y_0$, we can then choose δ_2 so that $|f(x) - y_0| < \delta_1$ for $0 < |x - x_0| < \delta_2$. Hence for $0 < |x - x_0| < \delta_2$, we have

$$|g[f(x)] - g(y_0)| < \epsilon$$

Therefore (2-140) is proved and the first part of Theorem C is established. When f is continuous at x_0, $y_0 = f(x_0)$ and (2-140) gives $\lim F(x) = F(x_0)$, so that F is continuous at x_0. Thus all of Theorem C is proved.

PROOF OF THEOREM A. *Limit of Sum.* Let f and g be defined as above and let

$$\lim_{x \to x_0} f(x) = c, \qquad \lim_{x \to x_0} g(x) = k$$

Then given $\epsilon > 0$, we can choose $\delta_1 > 0$ so that $|f(x) - c| < \epsilon/2$ for $0 < |x - x_0| < \delta_1$, and $\delta_2 > 0$ so that $|g(x) - k| < \epsilon/2$ for $0 < |x - x_0| < \delta_2$. Hence, if δ is the smaller of δ_1, δ_2, then $0 < |x - x_0| < \delta$ implies that $|f(x) - c| < \epsilon/2$ and $|g(x) - k| < \epsilon/2$, so that
$$|\{f(x) + g(x)\} - \{c + k\}| =$$
$$|\{f(x) - c\} + \{g(x) - k\}| \le |f(x) - c| + |g(x) - k| < \frac{\epsilon}{2} + \frac{\epsilon}{2} = \epsilon$$

Therefore,

$$\lim_{x \to x_0} [f(x) + g(x)] = c + k = \lim_{x \to x_0} f(x) + \lim_{x \to x_0} g(x)$$

as asserted in Theorem A.

Next we prove the rule: $\lim[k\, f(x)] = k \lim f(x)$, where k is a constant. If

$k = 0$, the assertion is immediate. Hence, we assume $k \neq 0$. We are given $\lim_{x \to x_0} f(x) = c$ as above. Then

$$|kf(x) - kc| = |k||f(x) - c| < \epsilon$$

if $|f(x) - c| < \epsilon/|k|$. But we can choose δ so that $0 < |x - x_0| < \delta$ implies $|f(x) - c| < \epsilon/|k|$ and thus implies $|kf(x) - kc| < \epsilon$; that is,

$$\lim_{x \to x_0} kf(x) = kc = k \lim_{x \to x_0} f(x)$$

Consequently, we now also have, under the assumptions for the sum rule,

$$\begin{aligned}
\lim_{x \to x_0} [f(x) - g(x)] &= \lim_{x \to x_0} [f(x) + (-1)g(x)] \\
&= \lim_{x \to x_0} f(x) + \lim_{x \to x_0} (-1)g(x) \\
&= \lim_{x \to x_0} f(x) - \lim_{x \to x_0} g(x)
\end{aligned}$$

To obtain the rule for product and quotient, we use an auxiliary step or lemma.

LEMMA. *The function $g(y) = y^2$ is continuous for all y. The function $g(y) = 1/y$ is continuous for $y \neq 0$.*

The proof is left as an exercise (Problems 2 and 3 below). Assuming the lemma proved, we conclude by Theorem C that, if $\lim_{x \to x_0} f(x) = c$, then

$$\lim_{x \to x_0} [f(x)]^2 = [\lim f(x)]^2 = c^2$$

that is, limit of product equals product of limits, for the special case $f \cdot f$. For the general case of $f \cdot g$ we write

$$f(x) \cdot g(x) = \frac{1}{4} [\{f(x) + g(x)\}^2 + \{f(x) - g(x)\}^2]$$

and, therefore, by the rules for sum, difference, square, constant times function,

$$\begin{aligned}
\lim_{x \to x_0} f(x) \cdot g(x) &= \frac{1}{4} [\{\lim f + \lim g\}^2 + \{\lim f - \lim g\}^2] \\
&= \frac{1}{4} [(c + k)^2 + (c - k)^2] = ck \\
&= \lim_{x \to x_0} f(x) \cdot \lim_{x \to x_0} g(x)
\end{aligned}$$

as asserted.

For the quotient rule we observe first that, if $g(x) \neq 0$ for the values of x considered and $\lim_{x \to x_0} g(x) = k \neq 0$ then, by the lemma (using the function $1/y$) and Theorem C,

$$\lim_{x \to x_0} \frac{1}{g(x)} = \frac{1}{\lim_{x \to x_0} g(x)} = \frac{1}{k}$$

Hence, by the product rule

$$\lim_{x \to x_0} \frac{f(x)}{g(x)} = \lim_{x \to x_0} f(x) \cdot \frac{1}{g(x)}$$

$$= \lim_{x \to x_0} f(x) \cdot \lim_{x \to x_0} \frac{1}{g(x)} = c \cdot \frac{1}{k}$$

$$= \frac{\lim f(x)}{\lim g(x)}$$

Thus Theorem A is completely proved.

As we stated in earlier sections, the extensions of Theorem A to limits as $x \to x_0+$, $x \to x_0^-$, $x \to \infty$, $x \to -\infty$ are proved in exactly the same way. Some cases are left as exercises (Problem 5 below). Also the analogous theorems for sequences are valid and are again proved in the same way. As an illustration we prove the analogue of Theorem C:

$$\lim_{n \to \infty} g(x_n) = g(x_0) = g(\lim_{n \to \infty} x_n) \qquad (2\text{-}141)$$

if g is continuous at x_0, $\lim x_n = x_0$, and $g(x_n)$ is defined (at least for n sufficiently large).

Given $\epsilon > 0$, we choose δ so that $|g(x) - g(x_0)| < \epsilon$ for $|x - x_0| < \delta$. Then we choose N so that $|x_n - x_0| < \delta$ for $n > N$. Hence, for $n > N$, we have $|g(x_n) - g(x_0)| < \epsilon$, so that (2-141) is proved.

Before we consider the Intermediate Value Theorem, we shall make two remarks.

Remark 1. *If a set E has least upper bound $\bar{B}$, then every interval of form $\bar{B} - \epsilon < x \leq \bar{B}$ contains at least one number x in E.* For if, for some $\epsilon > 0$, such an interval contained no member of E, then every number in E would be less than $\bar{B} - \epsilon$, and $\bar{B}$ would not be the least upper bound. It should be noted that the only number of E in the interval may be $\bar{B}$ itself, as is illustrated by the set E consisting of two numbers: 0 and 1. Here $\bar{B} = 1$, and the interval $\frac{1}{2} < x \leq 1$ contains no member of E other than 1. There is a similar statement about the greatest lower bound: *if E has glb. $\bar{A}$, then every interval of form $\bar{A} \leq x < \bar{A} + \epsilon$ contains at least one member of E (perhaps $\bar{A}$ itself).*

Remark 2. If a function f is continuous at x_0 and $f(x_0) > k$, then for some $\delta > 0$ we have $f(x) > k$ for all x for which $|x - x_0| < \delta$. To show this, we write $f(x_0) = k + \epsilon$, where $\epsilon > 0$, and choose δ, as in the definition of continuity, so that $f(x_0) - \epsilon < f(x) < f(x_0) + \epsilon$ for $|x - x_0| < \delta$. Since $f(x_0) - \epsilon = k$, our assertion is proved. There is a similar remark for the case $f(x_0) < k$.

PROOF OF THEOREM D. *The Intermediate Value Theorem.* We are given a function $f(x)$ continuous for $x_1 \leq x \leq x_2$ and a value y_0 between $y_1 = f(x_1)$ and $y_2 = f(x_2)$, where $y_1 \neq y_2$. We want to show that $f(x_0) = y_0$ for some x_0, $x_1 < x_0 < x_2$. Let us assume $y_1 < y_0 < y_2$ (see Figure 2-34). Then $f(x_1) = y_1 < y_0$. Hence, by Remark 2 above, $f(x) < y_0$ in some interval $|x - x_1| < \delta$. Our idea is now to "edge over" in this way, by going as far as we can, keeping $f(x) < y_0$. Eventually, f has to get close to y_2, so that a transition must occur, that is, $f(x_0) = y_0$ for some x_0. To make this precise, we let E be the set of all x in the interval $x_1 \leq x \leq x_2$ such that $f(x) < y_0$. Then E is

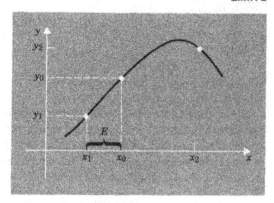

Figure 2-34 Intermediate Value Theorem.

bounded above by x_2 and, thus, has a least upper bound x_0, $x_1 \leq x_0 \leq x_2$. If $f(x_0) < y_0$, then $x_0 < x_2$ and, by Remark 2, we have $f(x) < y_0$ in an interval $|x - x_0| < \delta$, so that x_0 could not be the least upper bound. If $f(x_0) > y_0$, then again, by Remark 2, $f(x) > y_0$ in an interval $|x - x_0| < \delta$; but that means that the interval $x_0 - \delta < x \leq x_0$ contains no member of E, which is impossible by Remark 1. Hence, the only possibility is $f(x_0) = y_0$ and, thus, $x_1 < x_0 < x_2$. The theorem is proved.

PROOF OF THEOREM F. *Maximum and Minimum.* We are given a function $f(x)$ defined and continuous for $a \leq x < b$ and want to show that there is a value x_0 for which $f(x_0) = M$ but $f(x) \leq M$ for all x in the interval. If $f(x) \leq f(a)$ for all x or $f(x) \leq f(b)$ for all x, then we are done, so let us suppose that neither a nor b has this property. In particular, there is a value x so that $f(a) < f(x)$. In general, let E be the set of all numbers ξ such that there is a number x_1 with the property

$$f(x) < f(x_1) \quad \text{for} \quad a \leq x \leq \xi$$

(See Figure 2-35). Thus, a belongs to E, but b cannot belong to E. Let x_0 be lub E. If x_0 does not provide an absolute maximum for f, then $f(x_0) < f(x_2)$ for some x_2. Choose k so that $f(x_0) < k < f(x_2)$. Then by Remark 2 above, we can choose δ so that $f(x) < k$ for $|x - x_0| < \delta$. Thus, in particular,

$$f(x) < f(x_2) \quad \text{for} \quad x_0 - \frac{\delta}{2} \leq x \leq x_0 + \frac{\delta}{2}$$

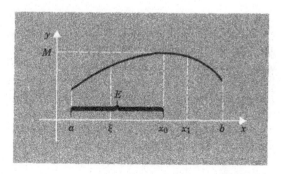

Figure 2-35 Existence of maximum.

By Remark 1 above, there is a member ξ of E so that $x_0 - (\delta/2) < \xi \leq x_0$. We choose such a ξ and then choose x_1 so that $f(x) < f(x_1)$ for $a < x \leq \xi$. Then, one of $f(x_1), f(x_2)$ is larger or equal to the other; let x' be the corresponding value of x. Hence, for $x_0 \neq b$,

$$f(x) < f(x') \qquad \text{for} \qquad a \leq x \leq x_0 + \frac{\delta}{2} \qquad (2\text{-}142)$$

This contradicts our choice of x_0 as lub E. Hence, x_0 must provide an absolute maximum of f, as asserted. If $x_0 = b$, then (2-142) becomes $f(x) < f(x')$ for $a \leq x \leq b$, which is absurd.

The proof of existence of an absolute minimum is similar (Problem 10 below).

PROBLEMS

1. Prove the following extension of Theorem C. Let $g(y)$ be defined for $c < y < d$ and continuous at y_0. Let $f(x)$ be defined for $a \leq x < \infty$ and let $\lim_{x \to \infty} f(x) = y_0$. Then

$$\lim_{x \to \infty} g[f(x)] = g(y_0) = g[\lim_{x \to \infty} f(x)]$$

2. Prove that the function $g(y) = y^2$ is continuous for all y. [*Hint.* Write

$$|g(y) - g(y_0)| = |y^2 - y_0^2| = |y + y_0||y - y_0|$$

and conclude that $|g(y) - g(y_0)| < \epsilon$ if $|y - y_0| < \delta$, provided that $0 < \delta < 1$ and $\delta < \epsilon/(1 + 2|y_0|)$. Notice that, if δ and y are so chosen, then $|y + y_0| < 1 + 2|y_0|$.]

3. Prove that the function $g(y) = 1/y$ is continuous for $y \neq 0$. [*Hint.* Write, for $y_0 \neq 0$,

$$|g(y) - g(y_0)| = \left| \frac{1}{y} - \frac{1}{y_0} \right| = \frac{|y - y_0|}{|y||y_0|}$$

and conclude that $|g(y) - g(y_0)| < \epsilon$ if $|y - y_0| < \delta$, provided that $0 < \delta < |y_0|/2$ and $\delta < \epsilon|y_0|^2/2$. Notice that, if δ and y are so chosen then $|y| > |y_0|/2$.]

4. (a) Prove the rule: $\lim_{x \to x_0} f \cdot g = \lim_{x \to x_0} f \cdot \lim_{x \to x_0} g$ for the special case $\lim f = 0$, $\lim g = 0$. [*Hint.* Given $\epsilon > 0$, choose $\delta > 0$ so small that $|f(x)| < \sqrt{\epsilon}$ and $|g(x)| < \sqrt{\epsilon}$ for $0 < |x - x_0| < \delta$.]

   (b) Prove the rule of part (a) for the general case: $\lim f = c$, $\lim g = k$. [*Hint.* Write

$$f(x) \cdot g(x) = [f(x) - c] \cdot [g(x) - k] + cg(x) + kf(x) - ck$$

   and apply the rules for limit of a sum, limit of a constant times a function, and the result of part (a).]

5. Formulate and prove the rule: $\lim(f + g) = \lim f + \lim g$ of Theorem A in the cases

   (a) $x \to x_0 +$ (b) $x \to -\infty$

6. Let $\{x_n\}$, $\{y_n\}$ be sequences such that $\lim_{n \to \infty} x_n = c$, $\lim_{n \to \infty} y_n = k$. Prove that

   (a) $\lim_{n \to \infty} (x_n + y_n) = c + k$ (b) $\lim_{n \to \infty} (C \cdot x_n) = C \cdot c$, $C = \text{const}$

   (c) $\lim_{n \to \infty} (x_n - y_n) = c - k$ (d) $\lim_{n \to \infty} x_n^2 = c^2$

(e) $\lim_{n \to \infty} x_n y_n = ck$ (f) $\lim_{n \to \infty} \dfrac{x_n}{y_n} = \dfrac{c}{k}$, if $k \neq 0$, $y_n \neq 0$ all n

7. Prove that, if $f(x)$ is continuous at x_0 and $f(x_0) < k$, then there is a $\delta > 0$ so that, wherever $f(x)$ is defined and $|x - x_0| < \delta$, we have $f(x) < k$.

8. Prove that, if $f(x)$ is continuous and nonconstant in the open interval (a, b), then the range of f is an interval. Show by examples that the range may be an open interval, a closed interval, a half-open interval, an infinite interval $c \leq y < \infty$, or an infinite interval $-\infty < y < \infty$.

9. Extend the result of Problem 8 by showing that, if f is nonconstant and continuous in an interval, then the range of f is an interval.

10. Prove the part of Theorem F relative to the existence of an absolute minimum.

11. (a) Prove that, if f is defined and continuous for all real x and $f(x) = 0$ whenever x is rational, then $f(x) \equiv 0$. [*Hint.* Suppose that $f(x_0) \neq 0$ and derive a contradiction.]
    (b) Prove that, if f and g are defined and continuous for all real x and if $f(x) = g(x)$ whenever x is rational, then $f(x) \equiv g(x)$.

12. Let a real function f have domain $[a, b]$. The function f is said to be *piecewise continuous* if f is continuous except perhaps for a finite number of jump discontinuities; that is, except perhaps for values $x_1, x_2, \ldots, x_n$ at each of which f has a limit to the right and to the left (excluding limits to the left at a and to the right at b). A function

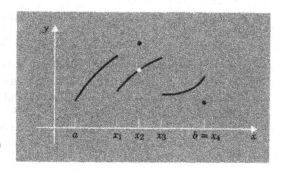

Figure 2-36 Piecewise continuous function.

with four jump discontinuities is shown in Figure 2-36. At x_2 the left and right limits are equal but disagree with $f(x_2)$.
(a) Let $f(x) = x/|x|$ for $0 < |x| \leq 1$, $f(0) = 0$. Show that f is piecewise continuous.
(b) Let $f(x) = \cos x (\tan^2 x)^{1/2}$ for $-\pi \leq x \leq \pi$, except $f(\pm \pi/2) = 0$. Graph the function and show that it is piecewise continuous.
(c) Prove that, if f and g are piecewise continuous in $[a, b]$, then fg is also.
(d) Prove that the piecewise continuous functions on $[a, b]$ form a vector space.

13. Let $f(x) = 1$ for x rational and $f(x) = 0$ for x irrational. Show that f is discontinuous for all x. Can we graph the function?

3
DIFFERENTIAL CALCULUS

3-1 MOTIVATION FOR THE DIFFERENTIAL CALCULUS

The differential calculus was invented at about the same time (1665 to 1675) by Leibnitz and Newton as a tool for problems in geometry and mechanics. It soon became an invaluable aid in many other branches of mathematics and physics. Many familiar experiences in daily life have at their basis the crucial concepts—the *derivative* and the *differential*. Thus the mathematical formalism of the calculus, at first appearance abstract and removed from reality, can be related to patterns of reasoning used by people generally. In this first section we present several examples of these patterns.

EXAMPLE 1. *Sensitivity*. The direction of the front wheels of an automobile is determined by the steering wheel. If we turn the steering wheel through x degrees from the neutral position, the front wheels turn through y degrees (Figure 3-1). In one type of car, a slight additional change in x

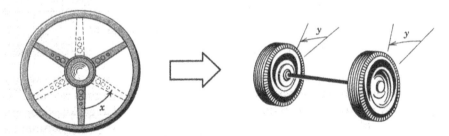

Figure 3-1 Sensitivity in steering.

—we denote such a change by Δx—leads to a small change Δy in y. When Δy is exceedingly small compared to Δx, for instance, $\Delta y = 0.01$ degrees for $\Delta x = 1$ degree (so that Δy is only one hundredth of Δx), we say that the direction of the wheels is only *slightly sensitive* to the direction of the steering wheel. When Δy is about the same size as Δx, we say that the direction of the wheels is *fairly sensitive* to the direction of the steering wheel. When Δy is very large compared to Δx, we say that the direction of the wheels is *highly sensitive* to the direction of the steering wheel. For a normal car, the ratio $\Delta y / \Delta x$ is about $\frac{1}{4}$ for most directions, so that the term slightly sensitive is appropriate. When we have turned the wheels as far as they can go, any further turning of the steering wheel produces no change in y; that is, the sensitivity drops to zero at the extreme position. Thus the sensitivity depends on the value of x at which we are working.

EXAMPLE 2. *Coefficient of expansion.* The word sensitivity is common, and we can cite many examples like that of steering a car. We consider one here: the sensitivity of the length of an iron bar (for example, a rail of a railway track) to temperature. This sensitivity is termed the coefficient of expansion of the bar—that is, the *coefficient of expansion* is defined as the ratio of the change Δy in length of the bar, having unit length at $0°$, to a change Δx in temperature x, in degrees C.[1] However, to be precise, we must require that Δx be very small, as we shall show clearly below. We tend to think of the coefficient of expansion as a fixed number; however, for very high reference temperatures (say approximately $900°C$), the ratio $\Delta y/\Delta x$ is found to be much larger than for very low temperatures. This can be clearly seen in Figure 3-2, which records the length of a typical bar as a function of temperature x. The ratio $\Delta y/\Delta x$ is about twice as large as $900°$ as at $100°$. It is also clear from the figure why Δx must be taken very small.

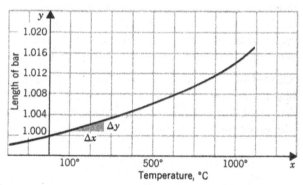

Figure 3-2   Coefficient of expansion of a bar.

EXAMPLE 3. *Marginal analysis in economics.* In economic theory we use terms such as "marginal efficiency," "marginal cost," "marginal revenue," in each case referring to the ratio of the change Δy to the change Δx, where y is a function of x. If x is the rate of production of an object manufactured for sale (for example, x = number of objects produced per day), and y is the production cost (total cost per day), then normally y will increase as x increases. Thus, if we are operating at $x = 1000$ (1000 units produced per day) and increase x to 1001, then y may increase by \$2,000; we would then say that the *marginal cost* of increasing production is \$2,000 per unit.

EXAMPLE 4. *Temperature gradient.* As we move upward through the earth's atmosphere, the temperature generally drops, although an "inversion" may occur—warm air lying above cool air. Typically the temperature drops by $1°C$ per 1000 feet. However, this is only a rough overall average. In a precise study of a given atmospheric condition, we would need to know, for each

[1] Strictly speaking, the coefficient is $(1/y)(\Delta y/\Delta x)$, for very small Δx, but y is approximately 1 here.

small change in altitude, how much the temperature T drops for a given very small increase in altitude h; thus, as above, we seek $\Delta T/\Delta h$. Since T is usually decreasing, ΔT should be considered as negative when Δh is positive, as suggested in Figure 3-3. The value of $\Delta T/\Delta h$, for "very small" Δh, is the *temperature gradient*.

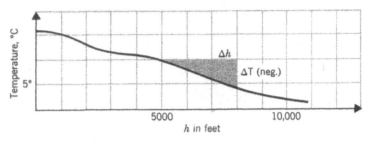

Figure 3-3   Temperature gradient.

EXAMPLE 5. *Gradient of a road.* The steepness or gradient of a road is usually measured as the ratio of the change in altitude h to change in horizontal distance x. For a straight road, as in Figure 3-4a, this ratio, $\Delta h/\Delta x$, is the slope of the straight line that represents the road in the xh-plane. For a general road (in a vertical plane), h is a function of x, and we must take Δx very small, as in Figure 3-4b. The gradient varies as we move along the road.

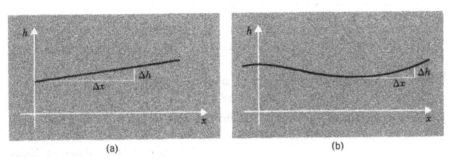

Figure 3-4   Gradient of a road.

EXAMPLE 6. *Slope of tangent to a curve.* This concept has been considered in a qualitative way in Section 2-2. A general function f has as graph a curve $y = f(x)$ in the xy-plane. At a particular point (x, y) on the curve, the tangent line should be the line that best approximates the curve near the point. Let the tangent line form angle α with the x-axis (Figure 3-5), so that the slope of the tangent line is $m = \tan \alpha$. Since the line is a close approximation to the curve, its slope is approximately equal to the ratio $\Delta y/\Delta x$, for very small Δx, as in Figure 3-5. The ratio $\Delta y/\Delta x$ can be interpreted as the slope of a *secant of the curve*, through the points (x, y) and $(x + \Delta x, y + \Delta y)$. As Δx approaches zero, the secant line approaches the tangent. We are clearly dealing with a limit process, as in Section 2-5.

EXAMPLE 7. *Velocity of a particle moving on a line.* This concept is introduced in Section 2-5. The position of a particle moving on a line, the x-axis, is given by a function $x = f(t)$ shown graphically in Figure 3-6. To find the

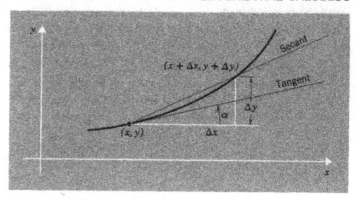

Figure 3-5 Tangents.

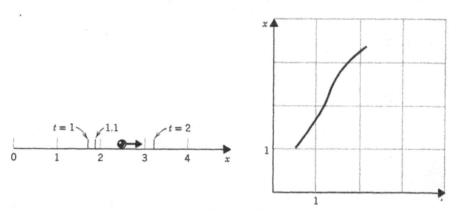

Figure 3-6 Velocity of a particle.

velocity at time $t = 1$, we observe that at times $t = 1$, $t = 2$, x has values 1.7, 3.2, respectively. Hence the ratio distance ÷ time has the value $1.5/1 = 1.5$ units of distance per unit time. This is, by definition, the average velocity over the chosen time interval, and it can be considered as a first approximation of the true velocity at time $t = 1$. To obtain a more accurate value, we take a time closer to $t = 1$, for instance, $t = 1.1$. We find that $x = 1.87$ at $t = 1.1$, so that the average velocity in the time interval $[1, 1.1]$ would be

$$\frac{\Delta x}{\Delta t} = \frac{1.87 - 1.7}{1.1 - 1.0} = \frac{0.17}{0.1} = 1.7$$

units of distance per unit time. As we pointed out in Section 2-5, to obtain the precise instantaneous velocity we must find the limiting value of the average velocity as the length of the time interval approaches 0. In the calculus and its applications, the word "velocity" normally means "instantaneous velocity."

If we have found the velocity v at time t, for all t of a certain interval, then v is itself a function of t: $v = g(t)$. The ratio of the change Δv in v to the corresponding change in t over an interval is the average acceleration over the interval. A limit process just like the one for the instantaneous velocity gives us the value of the *instantaneous acceleration*. Again the word "instantaneous" is normally omitted.

EXAMPLE 8. *Rate of change.* A recording barometer provides a graph like the one shown in Figure 3-7. Between Sunday noon and Monday noon the pressure reading drops from 31 inches to 29 inches (inches referring to length

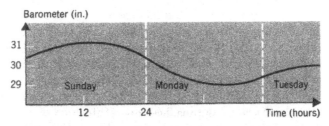

Barometer (in.)

31
30
29

Sunday   Monday   Tuesday

12   24   Time (hours)

Figure 3-7   Rate of change of pressure.

of a column of mercury). Since 24 hours have elapsed, we say that the average rate of change of pressure over this 24-hour period is $-\frac{2}{24}$ or $-\frac{1}{12}$ inches per hour (the minus sign indicates that the pressure is decreasing). From the graph we clearly observe that the pressure is not changing at a constant rate. When we choose the period from Sunday noon to Sunday midnight, we find that the average rate of change is $-0.75/12$ or $-\frac{1}{16}$ inches per hour. We can describe most accurately what is happening at Sunday noon by giving the instantaneous rate of change of the pressure; this would be found by taking shorter and shorter intervals—noon to 1 P.M., noon to 12:30 P.M., noon to 12:15 P.M., noon to 12:01 P.M., . . . —and seeking the limit of the average rates found:

instantaneous rate of change = limit of average rate of change as length of interval approaches zero.

We can give many other examples of rate of change. Usually the rate refers to a quantity that changes with time. However, we can also refer to rate of change with respect to distance, density, market value, and so on. Thus the cost per hour of heating a house depends on the outside temperature, as suggested in Figure 3-8. From the figure we notice that the cost drops by 1 cent per hour when the temperature rises from $20°$ to $30°$; hence, the average rate of change during this interval is -0.1 cents per hour per degree. The exact rate of change at $10°$ temperature (like the instantaneous rate considered

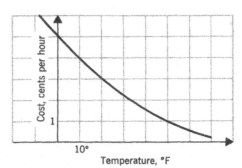

Cost, cents per hour

1

10°

Temperature, °F

Figure 3-8   Cost of heating a house.

above) is the limit of the average rate over intervals $10°$ to $10° + h$, as h approaches zero.

3-2 DEFINITION OF DERIVATIVE

From these examples we can extract one central idea. In each case, a function is given, for instance, $y = f(x)$, $a \leq x \leq b$, and we are interested in the ratio $\Delta y / \Delta x$; here Δx is the change in x (or increment in x) from a chosen value, and Δy is the corresponding change (or *increment*) in $y = f(x)$: $\Delta y = f(x + \Delta x) - f(x)$, as in Figure 3-9. It appears from the many examples considered that normally the ratio $\Delta y / \Delta x$ will have approximately the same values for all sufficiently small Δx. We now make our concept precise by defining the *derivative* of our function f, at a particular x, as the limit of the ratio $\Delta y / \Delta x$ as Δx approaches zero. If this limit does not exist, we say that f has no derivative (or is not differentiable) at this x. We denote the derivative at this x by $f'(x)$. Thus, by definition,

$$f'(x) = \lim_{\Delta x \to 0} \frac{\Delta y}{\Delta x} \tag{3-20}$$

If x is not an end point of the interval, Δx may be positive or negative, as in Figure 3-9. For $x = a$, Δx must be chosen positive [so that, to be precise, at

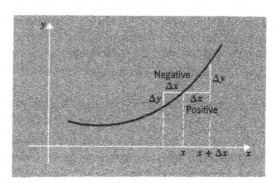

Figure 3-9 Formation of the derivative.

$x = a$ we are forming the "derivative to the right," sometimes denoted by $f'_+(a)$]. Similarly at $x = b$, Δx must be chosen as negative [derivative to the left, $f'_-(b)$]. That is, for a function given in the interval $[a, b]$,

$$f'_+(a) = \lim_{\Delta x \to 0+} \frac{\Delta y}{\Delta x}, \qquad f'_-(b) = \lim_{\Delta x \to 0-} \frac{\Delta y}{\Delta x}$$

For a function given in an open interval, the question about derivatives at the end points does not arise.

We can write our definition in other ways. From the meaning of Δy, we can write

$$f'(x) = \lim_{\Delta x \to 0} \frac{f(x + \Delta x) - f(x)}{\Delta x} \tag{3-20'}$$

The symbol Δx for the increment in x is arbitrary; we can use another letter, for example h, for this increment. Then (3-20') becomes

$$f'(x) = \lim_{h \to 0} \frac{f(x + h) - f(x)}{h} \qquad (3\text{-}20'')$$

We can denote the starting value of x by x_0 and let the displaced value be denoted by x. Then our definition becomes

$$f'(x_0) = \lim_{x \to x_0} \frac{f(x) - f(x_0)}{x - x_0} \qquad (3\text{-}20''')$$

In each case, a modification as above must be made at the end points.

Derivative as Slope of Tangent. Let $f'(x_1)$ exist and let P_1 be the point (x_1, y_1), where $y_1 = f(x_1)$. In view of Example 6 in Section 3-1, we are led to define the tangent line to the graph of f at P_1 to be the line through P_1 whose slope is

$$\lim_{\Delta x \to 0} \frac{\Delta y}{\Delta x} = \lim_{x \to x_1} \frac{f(x) - f(x_1)}{x - x_1}$$

Hence *the tangent line has slope* $m = f'(x_1)$, and the derivative can be interpreted as the slope of the tangent line.

Computation of the Derivative. A function f can be given by a table, by a graph, by a formula, or through various mathematical processes. We illustrate the determination of the derivative when f is given in some of these ways:

EXAMPLE 1. Here our function is given by Table 3-1.

Table 3-1

x	2.00	2.01	2.02	2.03	2.04	2.05	2.06	2.07	2.08	2.09	2.10
y	1.000	1.010	1.020	1.031	1.042	1.052	1.064	1.075	1.086	1.098	1.110

For a function defined over an interval, a complete table would require an infinite number of entries. However, in practical work, we are often forced to work with an "incomplete table," such as the one shown. From the table we can compute approximately the derivative $f'(x)$ for $x = 2.00$, $x = 2.01, \ldots$ For example, for $x = 2.05$, we find that

$$\frac{f(x + h) - f(x)}{h} = \frac{f(2.05 + h) - f(2.05)}{h} = \frac{f(2.05 + h) - 1.052}{h}$$

For $h = 0.03$, the quotient has the value $(1.086 - 1.052)/0.03 = 1.13$. Similarly, it has the value 1.15 for $h = 0.02$, the value 1.20 for $h = 0.01$, the value 1.00 for $h = -0.01$, the value 1.05 for $h = -0.02$, and the value

$$\frac{f(2.02) - 1.052}{-0.03} = \frac{1.020 - 1.052}{-0.03} = 1.07 \qquad \text{for} \qquad h = -0.03$$

From the information available, we say that

$$\lim_{h \to 0} \frac{f(2.05 + h) - f(2.05)}{h} = 1.20 \qquad \text{or} \qquad 1.00$$

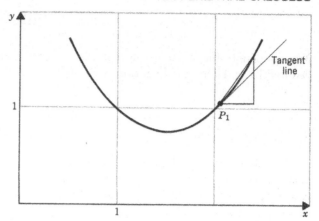

Figure 3-10 Function of Example 2.

and reasonably we take the average of these two values, 1.10, as our best estimate of the derivative $f'(2.05)$.

EXAMPLE 2. A function is given graphically in Figure 3-10 (this function is meant to agree with that of Example 1 at the points of Table 3-1). To find the derivative for $x = 2.05$, we try to measure the slope of the tangent line to the graph at the corresponding point P_1. To find the tangent line, we place a straight edge along a secant line through P_1 and a nearby point P on the graph, corresponding to a small increment Δx. We then observe the behavior of the secant lines as Δx approaches 0. The limiting position is the tangent line. By measuring slopes, we find that $f'(2.05) = 1.1$.

EXAMPLE 3. Finally, the function may be given by a formula. For the function of Examples 1 and 2, there is a formula (from which the table and the graph were constructed), namely, $f(x) = x^2 - 3x + 3$. From this formula we now reason that

$$f(2.05) = 2.05^2 - 3(2.05) + 3 = 1.0525$$
$$f(2.05 + h) = (2.05 + h)^2 - 3(2.05 + h) + 3$$
$$= 1.0525 + 1.1\,h + h^2$$
$$\frac{f(2.05 + h) - f(2.05)}{h} = \frac{1.0525 + 1.1\,h + h^2 - 1.0525}{h}$$
$$= 1.1 + h$$

Here h is not allowed to be zero. However

$$\lim_{h \to 0} \frac{f(2.05 + h) - f(2.05)}{h} = \lim_{h \to 0} (1.1 + h) = 1.1$$

Therefore, the derivative $f'(2.05) = 1.1$. Thus, from a precise formula, we obtain a precise value for the derivative.

EXAMPLE 4. Let f be given by the formula: $y = f(x) = 5x^3$. We find the derivative $f'(x)$ at a general value of x. With x fixed

$$f'(x) = \lim_{h \to 0} \frac{f(x + h) - f(x)}{h} = \lim_{h \to 0} \frac{5(x + h)^3 - 5x^3}{h}$$

$$= \lim_{h \to 0} \frac{5(x^3 + 3x^2h + 3xh^2 + h^3) - 5x^3}{h}$$

$$= \lim_{h \to 0} (15x^2 + 15xh + 5h^2) = 15x^2 + 0 + 0 = 15x^2$$

Thus, $f'(x) = 15x^2$ for exery x.

EXAMPLE 5. Let $f(x) = 1/x$ ($x \neq 0$). Then again for fixed x, not 0,

$$f'(x) = \lim_{h \to 0} \frac{[1/(x + h)] - (1/x)}{h} = \lim_{h \to 0} \frac{x - (x + h)}{hx(x + h)}$$

$$= \lim_{h \to 0} \frac{-h}{hx(x + h)} = \lim_{h \to 0} \frac{-1}{x(x + h)} = -\frac{1}{x^2}$$

Accordingly, $f'(x) = -1/x^2$ for every x other than 0. (In this example our function is defined in each of the two infinite intervals $-\infty < x < 0$ and $0 < x < \infty$.)

Example 5 illustrates the fact that the derivative of f cannot exist at a point of discontinuity of f. We state a general rule:

THEOREM 1. *If f has a derivative at x_0, then f is continuous at x_0.*

PROOF. For $x \neq x_0$, we can write

$$f(x) = f(x) - f(x_0) + f(x_0) = (x - x_0)\frac{f(x) - f(x_0)}{x - x_0} + f(x_0)$$

and therefore, by the basic theorem on limits,

$$\lim_{x \to x_0} f(x) = \lim_{x \to x_0} (x - x_0) \lim_{x \to x_0} \frac{f(x) - f(x_0)}{x - x_0} + f(x_0)$$

$$= 0 \cdot f'(x_0) + f(x_0) = f(x_0)$$

Accordingly, f is continuous at x_0.

EXAMPLE 6. Let $f(x)$ be given by the formula: $y = \sqrt{x} = x^{1/2}, x \geq 0$. Then for fixed x,

$$f'(x) = \lim_{h \to 0} \frac{f(x + h) - f(x)}{h} = \lim_{h \to 0} \frac{\sqrt{x + h} - \sqrt{x}}{h}$$

This limit is more difficult to evaluate. We multiply numerator and denominator by $\sqrt{x + h} + \sqrt{x}$:

$$\frac{\sqrt{x + h} - \sqrt{x}}{h} = \frac{\sqrt{x + h} - \sqrt{x}}{h} \frac{\sqrt{x + h} + \sqrt{x}}{\sqrt{x + h} + \sqrt{x}}$$

$$= \frac{x + h - x}{h(\sqrt{x - h} + \sqrt{x})} = \frac{1}{\sqrt{x + h} + \sqrt{x}}$$

Accordingly

$$\lim_{h \to 0} \frac{\sqrt{x + h} - \sqrt{x}}{h} = \lim_{h \to 0} \frac{1}{\sqrt{x + h} + \sqrt{x}} = \frac{1}{2\sqrt{x}}$$

provided that $x \neq 0$. At $x = 0$ [where we would seek $f'_+(0)$ and restrict h to be positive in our calculation], the procedure fails. From Figure 3-11 we observe the reason for the difficulty. If we start at $x = 0$, where $y = 0$, the

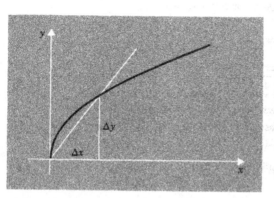

Figure 3-11 Derivative of $y = \sqrt{x}$.   **Figure 3-12** Discontinuity and corner.

ratio $\Delta y / \Delta x$ becomes larger and larger, approaching infinity as Δx approaches 0. The tangent line at $(0, 0)$ is, in fact, the y-axis, which has infinite slope.

Example 6 shows that the converse of Theorem 1 does not hold true: continuity of f at x_0 does not assure the existence of the derivative of f at x_0. Advanced books even give examples of functions continuous in an interval and having a derivative at no point of the interval.

The most common cases of failure of existence of $f'(x)$ at a point are the following:

(a) A discontinuity of f, as at x_0 in Figure 3-12.
(b) Infinite slope at the, point, as at $x = 0$ in Figure 3-11.
(c) A corner, as at x_1 in Figure 3-12.   (See Problem 13 below.)

At a corner, the left and right derivatives $f'_+(x_0)$ and $f'_-(x_0)$ exist but are unequal.

Other Notations for the Derivative. If $y = f(x)$, then we denote the derivative at x by $f'(x)$ or y', as above, so that $y' = f'(x)$ is a new function "derived" from $y = f(x)$. We also write $D_x y$ or $D_x f(x)$ for the derivative. For example, if $y = x^2$, then $D_x y = 2x$ or $D_x x^2 = 2x$. We may even write simply Dy or Df for the derivative: if $y = x^2$, then $Dy = 2x$. Furthermore, we write dy/dx for the derivative. Here, dy and dx are "differentials," whose meaning is explained in Section 3-22. Until differentials have been explained, it is best to think of the symbol dy/dx as a unit, not to be decomposed:

$$\text{if } y = x^2, \quad \text{then} \quad \frac{dy}{dx} = 2x$$

We can also write: if $y = f(x) = x^2$, then $df/dx = 2x$ or $(d/dx)(x^2) = 2x$. We use the term "differentiate a function" to mean "form the derivative of a function," and say that "f is differentiable" to mean that "f has a derivative."

In mechanics, for a function of time t, we often denote the derivative by a superposed dot. Thus $\dot{x}, \dot{y}$ stand for $dx/dt, dy/dt$.

PROBLEMS

1. Interpret each of the following as a derivative:
   (a) The rate of growth (in height) of a person.
   (b) The rate of cooling (say of a cup of coffee).
   (c) The rate of climb of an airplane.
   (d) The sensitivity of a loaded beam to additional loading.
   (e) The time it takes a kettle to boil as affected by the height of the gas flame.

2. Figure 3-13 shows a contour map of landscape, with a road leading from the river to the top of the hill. Evaluate (approximately) the gradient of the road at A, B, C and D. What would be the gradient at E for a road through that point (regardless of direction)?

3. Let $f(x) = (x^2 - 2)/3$.
   (a) Tabulate this function for $x = 0.97, 0.98, \ldots, 1.06$ and from the table evaluate approximately the derivatives $f'(1), f'(1.02), f'(1.05)$.

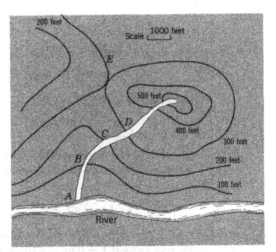

Figure 3-13   Gradient of a road.

   (b) Graph the function in the interval $0 \le x \le 2$, by using a scale of 2 in. per unit and, on the graph, show several secants through the point $(1, -\frac{1}{3})$ as well as the tangent line at the point. From the graph find the slopes of the secants and tangent and compare with $f'(1)$ as found in (a).
   (c) Show that $f(1 + h) - f(1) = \frac{2}{3}h + \frac{1}{3}h^2$ and, hence, that $f'(1) = \frac{2}{3}$.
   (d) Show that for a general x

$$f(x + h) - f(x) = \frac{2}{3}xh + \frac{h^2}{3}$$

   and, hence, that $f'(x) = (2x)/3$.

4. Carry out the steps of Problem 3 with $f(x) = 1/(x + 3)$ to find $f'(x)$ at the special values 1, 1.02, 1.05, and in general.

5. Find the derivative $f'(x)$ for general x for each of the following functions, and check
the result by graphing.

(a) $y = 13x + 3$ (b) $y = 7x + 17$ (c) $y = 4x^2 + 5x$
(d) $y = -x^2 - x$ (e) $y = x^3 - 2x^2$ (f) $y = 1/(x^2 + 1)$
(g) $y = x/(x + 2),\ x \neq -2$ (h) $y = \sqrt{x^2 - 1},\ x \geq 1$ (i) $y = \sqrt[3]{x}$

[*Hint.* For (i) use the algebraic identity $a - b = (a^{1/3} - b^{1/3})(a^{2/3} + a^{1/3}b^{1/3} + b^{2/3})$.]

6. (a) In Figure 3-14a, a function f is graphed, along with three other functions g_1, g_2,
g_3, of which one is the derivative of f. Determine which one is the derivative.
(b) Do the same, using Figure 3-14b.
(c) Do the same, using Figure 3-14c.

7. The functions $f(x) = \sin x$ and $g(x) = \cos x$ are tabulated herewith, for x in radians,
$0 \leq x \leq \pi/2$. From these tables compute approximately the derivatives f' and g'
and check by using the rule (see Chapter 5 and also Problem 11): $f'(x) = \cos x$,
$g'(x) = -\sin x$.

x	$\sin x$	$\cos x$	x	$\sin x$	$\cos x$
0	0.0000	1.0000	0.9	0.7833	0.6216
0.1	0.0998	0.9950	1.0	0.8415	0.5403
0.2	0.1987	0.9801	1.1	0.8912	0.4536
0.3	0.2955	0.9553	1.2	0.9320	0.3624
0.4	0.3894	0.9211	1.3	0.9636	0.2675
0.5	0.4794	0.8776	1.4	0.9855	0.1700
0.6	0.5646	0.8253	1.5	0.9975	0.0707
0.7	0.6442	0.7648	$\pi/2$	1.0000	0.0000
0.8	0.7174	0.6967			

8. *Falling body.* It is proved in physics that, for a body falling freely
under the influence of gravity near the earth's surface, we have the
relation

$$s = \frac{1}{2}gt^2 + v_0 t$$

where (Figure 3-15) s is the distance fallen from a given reference
point, t is the time elapsed ($t = 0$ for $s = 0$), v_0 is the initial velocity
—that is, the velocity at the instant $t = 0$, g is a constant—the
acceleration of gravity. If t is measured in seconds and s in feet,
then velocity is measured in feet per seconds, and acceleration in
feet per second per second; in these units $g = 32$ ft/sec² for a
typical point on the earth's surface.

Figure 3-15
Falling body.

(a) Consider the special case $v_0 = 0$ and show that the velocity at any instant is
$v = gt$, that the acceleration is always g, and that $v^2 = 2gs$.

(b) Consider the special case $v_0 = -10$ feet per sec (What is the meaning of the
minus sign?). Show that, in the units described, $v = 32t - 10$ and that the
acceleration is 32 ft per sec². At what instant is $v = 0$? What is the value of s at
that instant, and what is the physical meaning of the value of s?

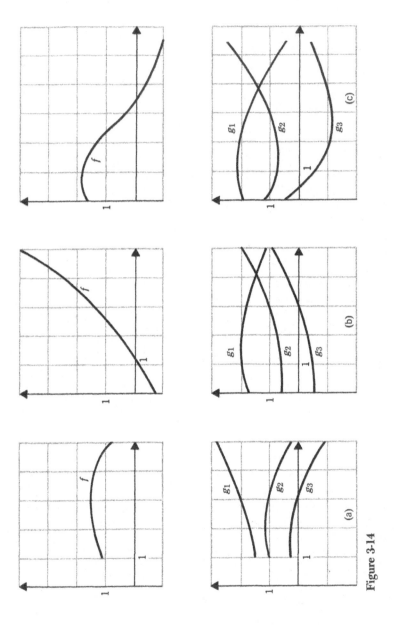

Figure 3-14

151

(c) Generalize the result of (b) by showing that, for each negative initial velocity v_0, the body rises to a certain maximum height and then falls toward the earth. Find the maximum height and the time it takes to reach it. Does the conclusion conflict with recent achievements with earth rockets?

9. The growth of money at compound interest can be described as follows. Let $x = f(t)$ be the amount of money at time t (in years); let t be the start of an interest period. Then x will be increased, a half-year later, by an amount Δx equal to $(r/2)x$, where r is the annual rate (for example, 0.04). We can regard this as the change Δx for a change $\Delta t = 1/2$ year in t. For compounding quarterly $\Delta x = (r/4)x$ and $\Delta t = 1/4$ year. We can conceive of taking Δt shorter (1 month, 1 week, 1 day, 1 hour, . . .) and passing to the limit to get *instantaneous compounding*. Show that if this is done then

$$\lim_{\Delta t \to 0} \frac{\Delta x}{\Delta t} = f'(t) = rf(t)$$

(The effect of instantaneous compounding for $r = 4\%$ is to increase the annual interest on \$100 from \$4 to \$4.08. By semiannual compounding it is \$4.04, and by quarterly compounding it is \$4.06.)

10. Figure 3-16 shows temperature measurements at certain points in the atmosphere in a vertical plane with coordinates x, y; x denotes the horizontal displacement from O in miles, y the altitude above O in miles. The temperature gradient described in the text, more accurately the *vertical temperature gradient*, is a measure of sensitivity of temperature to altitude y. From the figure, find the vertical temperature gradient approximately at the points A, B, C. Also evaluate the *horizontal*

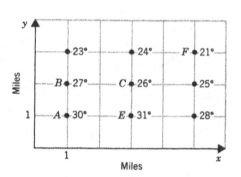

Figure 3-16   Temperature gradient.

temperature gradient (sensitivity to horizontal coordinate x) at C and E. We can also consider oblique displacements, as from C to F. The temperature gradient for such a direction is the limiting value of $\Delta T/\Delta s$, as Δs approaches zero, where Δs is the length of the displacement and ΔT is the corresponding increment in temperature. Evaluate this oblique gradient at C in the direction of F.

We are here dealing with T as a function of two variables x and y and, as our discussion indicates, the temperature gradient in a given direction is a "directional derivative."

11. Let $f(x) = \sin x$, for x in radians. Let $k = f'(0)$ (assuming that this derivative exists).
    (a) Prove that

$$k = \lim_{h \to 0} \frac{\sin h}{h}$$

We shall prove in Chapter 5 that $k = 1$. See also Sections 2-5 and 3-4.
    (b) Prove that $f'(x) = k \cos x$. [*Hint.* Show that

$$\frac{f(x + h) - f(x)}{h} = \frac{\sin h}{h} \left(\cos x - \sin x \frac{\sin h}{1 + \cos h} \right).]$$

12. Let a be a fixed positive number and let $f(x) = \log_a x$ for $x > 0$. Let $l = f'(1)$ (assuming that this derivative exists).

(a) Prove that $l = \lim\limits_{u \to 0} \dfrac{\log_a(1 + u)}{u}$ (b) Prove that $f'(x) = l \cdot \dfrac{1}{x}$

Remark: We shall prove in Chapter 5 that if a is chosen as $e = 2.718\ldots$, then $l = 1$, so that the function $\log_e x$ has derivative $1/x$.

13. Let $f(x)$ be defined as follows:

$$f(x) = 0 \quad \text{for} \quad x \le 0, \quad f(x) = x^2 \quad \text{for} \quad 0 \le x \le 1, \quad f(x) = 2 - x \quad \text{for} \quad x \ge 1$$

Show that f has a derivative for all x except $x = 1$, and explain why $f'(1)$ does not exist.

3-3 FUNDAMENTAL RULES OF DIFFERENTIATION

In Table 3-2, we present basic rules concerning derivatives. By means of these rules, derivatives of commonly occurring functions can be found by straightforward manipulation.

This table is useful for reference; it is reproduced as part of a larger table inside the front covers of this volume.

In Rules 1 to 5, we consider two functions f and g defined in a certain interval and both having a derivative at a particular x. Rule 1 then asserts that $f + g$ also has a derivative at x, equal to $f'(x) + g'(x)$.

EXAMPLE 1. By Rules 6 and 8, x^2 has derivative $2x$, $\sin x$ has derivative $\cos x$, so that $x^2 + \sin x$ has derivative $2x + \cos x$, that is,

$$(x^2 + \sin x)' = (x^2)' + (\sin x)' = 2x + \cos x$$

We can say: *a function that is a sum of several terms can be differentiated term by term.*

Rule 2 gives the derivative of a constant times a function.

EXAMPLE 2 $(5 \sin x)' = 5(\sin x)' = 5 \cos x$

Rules 1 and 2 can be combined to yield the statement: *the derivative of a linear combination of several differentiable functions equals the corresponding linear combination of the derivatives.* For example,

Table 3-2   Rules of Differentiation

1. $(f + g)' = f' + g'$	2. $(cf)' = cf'$ ($c = $ const)
3. $(fg)' = fg' + f'g$	4. $\left(\dfrac{f}{g}\right)' = \dfrac{gf' - fg'}{g^2}$
5. Derivative of constant function $\equiv 0$	7. $(f^n)' = nf^{n-1}f'$
6. $(x^n)' = nx^{n-1}$	
8. (a) $(\sin x)' = \cos x$	9. (a) $(\log_a x)' = \dfrac{1}{\log_e a} \cdot \dfrac{1}{x}$
(b) $(\cos x)' = -\sin x$	
(c) $(\tan x)' = \sec^2 x$	(b) $(\ln x)' = \dfrac{1}{x}$
(d) $(\cot x)' = -\csc^2 x$	
(e) $(\sec x)' = \sec x \tan x$	10. (a) $(a^x)' = a^x \log_e a$
(f) $(\csc x)' = -\csc x \cot x$	(b) $(e^x)' = e^x$

$$[c_1 f(x) + c_2 g(x)]' = c_1 f'(x) + c_2 g'(x)$$

The following is an illustration:

EXAMPLE 3 $(7x^2 + 4 \sin x)' = 14x + 4 \cos x$

By Rules 5 and 6 the derivative of a constant function is 0, the derivative of x^n is nx^{n-1}. Hence we can now find the derivative of a polynomial:

EXAMPLE 4 $(5x^3 + 3x^2 + 6x + 2)' = 15x^2 + 6x + 6$

Rule 3 is the *product rule*. We give an illustration:

EXAMPLE 5

$$(x^2 \sin x)' = x^2 (\sin x)' + (x^2)' \sin x = x^2 \cos x + 2x \sin x$$

More complicated products can be handled by repeated application of the rule:

EXAMPLE 6

$$\begin{aligned}
[x^3(2x^2 - 5) \sin x]' &= x^3[(2x^2 - 5) \sin x]' + (x^3)'(2x^2 - 5) \sin x \\
&= x^3[(2x^2 - 5)(\sin x)' + (2x^2 - 5)' \sin x] \\
&\qquad\qquad\qquad\qquad\qquad + 3x^2(2x^2 - 5) \sin x \\
&= x^3[(2x^2 - 5) \cos x + 4x \sin x] + 3x^2(2x^2 - 5) \sin x
\end{aligned}$$

Rule 4 is the *quotient rule*. Here we assume that $g(x) \neq 0$ at each x considered.

EXAMPLE 7 $y = x^2/(2x^3 + 1)$. With $f = x^2$, $g = 2x^3 + 1$, Rule 4 gives

$$y' = \frac{(2x^3 + 1)(2x) - x^2(6x^2)}{(2x^3 + 1)^2} = \frac{2x - 2x^4}{(2x^3 + 1)^2}, x \neq \sqrt[3]{-1/2}$$

Rule 5 requires no comment. In Rule 6 n is understood to be a positive integer, and x to be any real number. It will be seen below that the rule is valid also for $n = -1, -2, \ldots$, if we exclude the point $x = 0$. In fact, we shall show that the rule is valid for n any real number, rational or irrational, provided that we exclude the values of x for which the function is undefined (that is, exclude negative x for a function such as $x^{1/2}$) and exclude $x = 0$ for $n < 1$.

In Rule 7, n is again understood to be a positive integer, but the rule can also be extended as for Rule 6.

EXAMPLE 8. Let $y = (7x^3 + 3 \sin x)^5$. Then

$$y' = 5(7x^3 + 3 \sin x)^4 (21x^2 + 3 \cos x).$$

Rules 8(a) to 8(f) give the derivatives of the six trigonometric functions. For $\sin x$ and $\cos x$ the rules are valid for all x, but for each of the other functions we must exclude the values of x for which the function is discontinuous (as is implied by Theorem 1 above). For $\tan x$ and $\sec x$ the excluded values are the zeros of $\cos x$: $\pi/2, 3\pi/2, \ldots$ and, in general, $(\pi/2) + k\pi$ ($k = 0,$

$\pm 1, \dots$); for cot x and csc x the excluded values are the zeros of sin x: $k\pi$ ($k = 0, \pm 1, \dots$).

In Rule 9(a) a is a fixed positive number other than 1. The most important choice of a is the number $e = 2.71828 \dots$ (defined precisely in the next section). We denote $\log_e x$ by $\ln x$ (or occasionally simply by $\log x$) and call $\ln x$ the *natural* or *Napierian* logarithm of x. The derivative of $\ln x$ is $1/x$, as in Rule 9(b). In both rules, x must be *positive*.

In Rule 10(a), a is a fixed positive number; again the most important choice of a is e, for which Rule 10(b) applies. In both rules, x can be any real number.

Miscellaneous Examples. These examples further illustrate the rules and how they can be combined:

EXAMPLE 9 $(\tan^2 x)' = 2 \tan x (\tan x)' = 2 \tan x \sec^2 x$

EXAMPLE 10

$$\left(\frac{\sin x}{1 + 2 \cos x}\right)' = \frac{(1 + 2 \cos x) \cos x - \sin x(-2 \sin x)}{(1 + 2 \cos x)^2}$$

$$= \frac{1 + 2 \cos^2 x + 2 \sin^2 x}{(1 + 2 \cos x)^2} = \frac{3}{(1 + 2 \cos x)^2}$$

EXAMPLE 11 $(x \ln x)' = x \cdot \dfrac{1}{x} + 1 \cdot \ln x = 1 + \ln x$

EXAMPLE 12 $y = (\ln x)^2, \qquad y' = 2 \ln x \cdot \dfrac{1}{x} = \dfrac{2 \ln x}{x}$

EXAMPLE 13 $y = e^{2x} = (e^x)^2, \qquad D_x y = 2e^x \cdot e^x = 2e^{2x}$

EXAMPLE 14

$y = \csc^2 x - \cot^2 x, \qquad y' = 2 \csc x(-\csc x \cot x) - 2 \cot x(-\csc^2 x) = 0$

In Examples 9, 10, 11, 12, and 14 certain values of x must be excluded, for example, negative x and 0 for Example 12.

†3-4 PROOFS OF RULES FOR DERIVATIVES

The proofs are mainly a direct application of a theorem on limits: Theorem A of Section 2-7. Throughout we assume x to be fixed and that

$$\lim_{h \to 0} \frac{f(x + h) - f(x)}{h} = f'(x), \qquad \lim_{h \to 0} \frac{g(x + h) - g(x)}{h} = g'(x) \quad (3\text{-}40)$$

By Theorem 1 of Section 3-2, we then know that f and g are continuous at x. We can express this condition as follows:

$$\lim_{h \to 0} f(x + h) = f(x), \qquad \lim_{h \to 0} g(x + h) = g(x) \qquad (3\text{-}41)$$

For Rule 1, we seek the derivative of $F = f + g$:

$$F'(x) = \lim_{h \to 0} \frac{F(x + h) - F(x)}{h} = \lim_{h \to 0} \frac{f(x + h) + g(x + h) - [f(x) + g(x)]}{h}$$

$$= \lim_{h \to 0} \left[\frac{f(x + h) - f(x)}{h} + \frac{g(x + h) - g(x)}{h} \right]$$

$$= \lim_{h \to 0} \frac{f(x + h) - f(x)}{h} + \lim_{h \to 0} \frac{g(x + h) - g(x)}{h}$$

by (2-70). Accordingly,

$$F'(x) = f'(x) + g'(x)$$

as we asserted.

The proofs of Rules 2, 3, 5, and 7 are left as exercises (Problems 6 to 8 below).

For Rule 4 we have $H = f/g$, then

$$\frac{H(x + h) - H(x)}{h} = \frac{1}{h} \left[\frac{f(x + h)}{g(x + h)} - \frac{f(x)}{g(x)} \right]$$

$$= \frac{f(x + h)g(x) - f(x)g(x + h)}{hg(x)g(x + h)}$$

$$= \frac{[f(x + h) - f(x)]g(x) + f(x)[g(x) - g(x + h)]}{hg(x)g(x + h)}$$

$$= \frac{1}{g(x + h)} \frac{f(x + h) - f(x)}{h}$$

$$- \frac{f(x)}{g(x)g(x + h)} \frac{g(x + h) - g(x)}{h}$$

We now take the limit as $h \to 0$, by applying Theorem A of Section 2-7. We must, in particular, exclude each value of x for which $g(x) = 0$. By (3-40) and (3-41) we conclude that

$$H'(x) = \lim_{h \to 0} \frac{H(x + h) - H(x)}{h} = \frac{1}{g(x)} f'(x) - \frac{f(x)}{[g(x)]^2} g'(x)$$

$$= \frac{g(x)f'(x) - f(x)g'(x)}{[g(x)]^2}$$

For Rule 6 we must evaluate

$$\lim_{h \to 0} \frac{(x + h)^n - x^n}{h}$$

Now, from algebra (Problem 6 following Section 0-21),

$$a^n - b^n = (a - b)(a^{n-1} + a^{n-2}b + \cdots + ab^{n-2} + b^{n-1})$$

Hence

$$(x^n)' = \lim_{h \to 0} \frac{(x + h)^n - x^n}{h}$$

$$= \lim_{h \to 0} \frac{1}{h} h\{(x + h)^{n-1} + (x + h)^{n-2}x + \cdots + (x + h)x^{n-2} + x^{n-1}\}$$

$$= x^{n-1} + x^{n-2}x + \cdots + xx^{n-2} + x^{n-1}$$

$$= nx^{n-1}$$

by Theorem A of Section 2-7. Other proofs of Rule 6 are suggested in Problem 6 below.

A thorough discussion of the sine and cosine functions and their derivatives is given in Chapter 5. Here we give geometric justification for Rule 8(a). For $\sin x$ we find the derivative by seeking the limit, as $h \to 0$, of

$$\frac{\sin (x + h) - \sin x}{h} = \frac{\sin x \cos h + \cos x \sin h - \sin x}{h}$$

$$= \sin x \frac{(\cos h - 1)}{h} + \cos x \frac{\sin h}{h}$$

Let us assume $h > 0$, the case of negative h being similar. We construct a circle of radius 1 (Figure 3-17) and consider the sector of angle h as shown. The circle has area π and the sector has area $h/(2\pi)$ times the area of the circle; therefore, the sector has area $h/2$. We shall show that

$$\lim_{h \to 0+} \frac{\text{area of sector } ADC}{\text{area of triangle } ABC} = 1 \tag{3-42}$$

This implies that

$$\lim_{h \to 0+} \frac{h/2}{\frac{1}{2} \sin h \cos h} = 1$$

or, since $\cos h \to 1$ as $h \to 0$,

$$\lim_{h \to 0} \frac{h}{\sin h} = 1$$

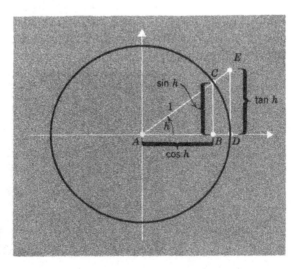

Figure 3-17 Derivative of sine function.

and, hence, also that

$$\lim_{h\to0+} \frac{\sin h}{h} = 1$$

Furthermore

$$\lim_{h\to0+} \frac{\cos h - 1}{h} = \lim_{h\to0+} \frac{-\sin^2 h}{h(\cos h + 1)}$$

$$= \lim_{h\to0+} \frac{\sin h}{h} \lim_{h\to0+} \frac{-\sin h}{\cos h + 1} = 1\cdot\frac{0}{2} = 0$$

Accordingly

$$\lim_{h\to0+} \frac{\sin (x + h) - \sin x}{h} = \sin x \lim_{h\to0+} \frac{\cos h - 1}{h} + \cos x \lim_{h\to0+} \frac{\sin h}{h}$$

$$= \sin x \cdot 0 + \cos x \cdot 1$$

$$= \cos x$$

It remains to prove (3-42). From the geometry we have

area of triangle ABC < area of sector ADC < area of triangle ADE

and thus

$$1 < \frac{\text{area of sector } ADC}{\text{area of triangle } ABC} < \frac{\text{area of triangle } ADE}{\text{area of triangle } ABC}$$

$$1 < \frac{\text{area of sector } ADC}{\text{area of triangle } ABC} < \frac{\frac{1}{2}\tan h}{\frac{1}{2}\sin h \cos h} = \frac{1}{\cos^2 h}$$

and, since $\cos^2 h \to 1$ as $h \to 0$, (3-42) follows. [See the remarks at the end of Section 2-5 and Figure 2-21.]

Remark 1. The preceding argument, although correct in principle, contains several steps that require further justification, in particular, that $\sin h \to 0$ and $\cos h \to 1$ as $h \to 0$, and that the expressions for areas are correct. A proof without these gaps is given in Chapter 5 (see also Problem 17, following Section 2-8).

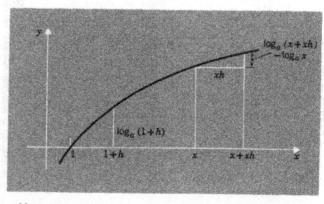

Figure 3-18   Derivative of $\log_a x$.

The proofs of the other rules for trigonometric functions are given in Problems 16 and 17 below.

Complete proofs of Rules 9(a), 9(b), 10(a), 10(b) are given in Chapter 5. Here we indicate in an informal way the basis for Rules 9(a) and 9(b). Rules 10(a) and 10(b) are deduced from Rules 9(a) and 9(b) in Section 3-6 below.

Let a fixed base a be chosen and, for simplicity, let $a > 1$, so that $y = f(x) = \log_a x \, (x > 0)$ has a graph as shown in Figure 3-18. We seek $f'(x)$ at a typical x. We take as increment Δx a value proportional to x: $\Delta x = hx$. Then our derivative is

$$f'(x) = \lim_{\Delta x \to 0} \frac{f(x + \Delta x) - f(x)}{\Delta x} = \lim_{h \to 0} \frac{\log_a(x + xh) - \log_a x}{xh}$$

$$= \lim_{h \to 0} \frac{1}{xh} \cdot \log_a\left(\frac{x + xh}{x}\right) = \lim_{h \to 0} \frac{1}{xh} \cdot \log_a(1 + h)$$

$$= \frac{1}{x} \lim_{h \to 0} \frac{\log_a(1 + h)}{h}$$

In particular

$$f'(1) = \lim_{h \to 0} \frac{\log_a(1 + h)}{h}$$

and, therefore, in general

$$f'(x) = \frac{1}{x} f'(1)$$

Thus, $\log_a x$ has a derivative inversely proportional to x. The constant of proportionality is $f'(1)$. By changing our base a, we can change this constant of proportionality. In fact, from properties of logarithms we easily observe that, for a suitable base, we can make $f'(1)$ precisely equal to 1 (see Problem 18 below). That base we denote by e. Thus, $y = \log_e x = f(x)$ has the property

$$y' = (\log_e x)' = \frac{f'(1)}{x} = \frac{1}{x}$$

By rules for logarithms (Section 0-19)

$$\log_a x = \frac{\log_e x}{\log_e a}$$

Hence, $f(x) = \log_a x$ has derivative

$$(\log_a x)' = \frac{1}{\log_e a} \cdot \frac{1}{x}$$

Thus, Rules 9(a) and 9(b) are established.

The number e is thus defined as the number for which $f(x) = \log_e x$ has derivative $f'(1) = 1$ or

$$\lim_{h \to 0} \frac{\log_e(1 + h)}{h} = 1$$

Accordingly

$$\lim_{h \to 0} \log_e(1 + h)^{1/h} = 1$$

and

$$\lim_{h \to 0} e^{\log_e(1+h)^{1/h}} = e$$

But $e^{\log_e u} = u$, so that our last equation can be written

$$\lim_{h \to 0} (1 + h)^{1/h} = e \tag{3-43}$$

This is the basic equation that defines the number e. We find that

$$(1 + 1)^1 = 2, \quad (1 + 0.5)^2 = 2.25, \quad (1 + 0.1)^{10} = 2.60$$
$$(1.01)^{100} = 2.705, \quad (1.001)^{1000} = 2.717, \ldots$$

and a detailed study of the limit gives

$$e = 2.7182818285 \ldots.$$

Remark 2. The rules for differentiating particular functions, such as x^2, x^3, $\sin x$, e^x, $\ln x$, apply to these functions in their complete domains. From the definition of the derivative, it follows that the rules continue to apply to these functions, restricted to smaller domains. For example, the function f for which $f(x) = 5x^3$ for $0 \le x \le 3$ has derivative $f'(x) = 15x^2$ in the same interval.

PROBLEMS

1. Differentiate (that is, find the derivative):

   (a) $y = 3x^7$

   (b) $y = 10x^3 - 2x^2 + 1$

   (c) $y = 3$

   (d) $y = 0$

   (e) $y = (x^2 + 1)^3$

   (f) $y = \dfrac{2x - 1}{x^2 + 5x + 4}$

   (g) $y = \dfrac{3x^2 + x - 1}{x}$

   (h) $y = (x^2 - 1)^3(2x^2 + 3x + 2)^2$

   (i) $y = (x^3 + x^2 + 1)(x^4 - 1)^3$

   (j) $y = \left(\dfrac{x^2 - 2}{x^3 + 1}\right)^4$

2. Prove that the rule $D_x(x^n) = nx^{n-1}$ is valid for $n = -1, -2, \ldots$ provided that $x \ne 0$. (*Hint.* For n negative we can write $n = -k$, where k is positive. Then $x^n = x^{-k} = 1/x^k$. Now differentiate by the rule for quotients.)

3. Differentiate (see Problem 2):

   (a) $y = x^2 + \dfrac{2}{x^2} - \dfrac{3}{x^4}$

   (b) $y = \dfrac{x^3 + 5x + 7}{x^4}$

   (c) $y = 5x^{-3} - 7x^{-4} - 3x^{-1}$

   (d) $y = \left(\dfrac{1}{x^2} - 2x\right)(x^3 + 2x + 4)^3$

4. (a) Prove that, on the basis of Rule 7, the rule $D_x[f(x)]^n = n[f(x)]^{n-1}f'(x)$ is valid for $n = -1, -2, \ldots$, in any interval in which f is differentiable and has no zeros. (*Hint.* See the hint for Problem 2.)

(b) Derive the quotient rule from the product rule by writing

$$\frac{f(x)}{g(x)} = f(x)[g(x)]^{-1}$$

and by using the result of part (a).

5. Differentiate (see Problem 4):

(a) $y = \dfrac{1}{(x^2 + 3x + 2)^3}$

(b) $y = \dfrac{1}{(x^4 - 16)^4}$

(c) $y = \dfrac{x^2}{(x^2 + 2)^3}$

(d) $y = \dfrac{(x^2 - 1)^2}{(2x^2 + x - 1)^3}$

6. (a) Prove Rule 6: $D_x(x^n) = nx^{n-1}$ for $n = 1, 2, \ldots$ by writing $f(x + h) - f(x) = (x + h)^n - x^n$ and by expanding $(x + h)^n$ by the binomial theorem (Section 0-21).

(b) Prove Rule 6 by mathematical induction (Section 0-20) with the aid of the product rule (Rule 3). (*Hint*. For the step from n to $n + 1$, write $x^{n+1} = x[x^n] = fg$.)

(c) Prove Rule 7 by mathematical induction.

7. Prove Rule 3. (*Hint*. Imitate the proof given for Rule 4.)

8. (a) Prove Rule 2.

(b) Prove Rule 5.

9. Let $f, g, p, q, \ldots, F, G, P, Q, \ldots$ be functions differentiable in a given interval, and let $F, G, \ldots$ have no zeros in the interval. Prove the rules:

(a) $(fgp)' = fgp' + fg'p + f'gp$

(b) $(fgpq)' = fgpq' + fgp'q + fg'pq + f'gpq$

(c) $\left(\dfrac{fg}{FG}\right)' = \dfrac{fg}{FG}\left(\dfrac{f'}{f} + \dfrac{g'}{g} - \dfrac{F'}{F} - \dfrac{G'}{G}\right)$

(d) $\left(\dfrac{fgpq}{FGPQ}\right)' = \dfrac{fgpq}{FGPQ}\left[\dfrac{f'}{f} + \dfrac{g'}{g} + \dfrac{p'}{p} + \dfrac{q'}{q} - \dfrac{F'}{F} - \dfrac{G'}{G} - \dfrac{P'}{P} - \dfrac{Q'}{Q}\right]$

Remark. In (c) and (d), if p, g, f or q has a zero in the interval, the right-hand side is to be evaluated by first cancelling with the corresponding factor in the numerator outside the parenthesis.

10. Differentiate the following functions:

(a) $y = 3 \sin x + 5 \cos x$

(b) $y = 2 \sin x \cos x = \sin 2x$

(c) $y = \tan^4 x$

(d) $y = 5 \sin^3 x$

(e) $y = \sin^4 x + 2 \sin^2 x \cos^2 x + \cos^4 x$

(f) $y = \dfrac{\sin x}{1 + \cos x}$

11. Differentiate:

(a) $x = t^2 \ln t + t^3$

(b) $y = \dfrac{t^2 + 2t + 3}{1 + \ln t}$

(c) $y = \dfrac{t}{\ln t}$

(d) $y = \ln(x^5) + (\ln x)^5$

(e) $y = \log_{10} x^4$

(f) $y = (1 + \ln x)^3 \ln x$

12. Differentiate:

(a) $y = e^{3x}$

(b) $y = x^2 e^x + 3x + 5$

(c) $y = \dfrac{e^x - e^{-x}}{2} = \sinh x$ (the *hyperbolic sine* of x)

(d) $y = \dfrac{e^x + e^{-x}}{2} = \cosh x$ (the *hyperbolic cosine* of x)

(e) $y = \dfrac{e^x - e^{-x}}{e^x + e^{-x}} = \tanh x = \dfrac{\sinh x}{\cosh x}$ (the hyperbolic tangent of x)

13. Let $f(x)$ be defined for $a < x < b$ and let x_0 be a point of this interval.
    (a) Prove that if $y = f(x)$ has a derivative, equal to m, at x_0, then there is a function $p(h)$, so that

    $$(°) \qquad f(x_0 + h) - f(x_0) = h[m + p(h)]$$

    and $p(h)$ is continuous at $h = 0$, with $p(0) = 0$. [*Hint.* Equation $(°)$ defines $p(h)$ near $h = 0$, except at $h = 0$. Require $p(0) = 0$ and then prove that $p(h)$ approaches zero as $h \to 0$.] Show also that Theorem 1 follows from the result established.
    (b) Prove that, if a function $p(h)$ exists, continuous at $h = 0$, with $p(0) = 0$, so that $(°)$ holds true, then f has a derivative, equal to m, at x_0.
    (c) Prove that the existence of the derivative at x_0, with value m, is equivalent to the existence of a function $q(x)$ continuous at x_0, such that

    $$(°°) \qquad f(x) - f(x_0) = (x - x_0)q(x), \qquad q(x_0) = m$$

    (d) Find $f'(0)$ if $f(x) = xe^x \cos x$.
    (e) Find $f'(1)$ if $f(x) = (x - 1)\sin[(x - 1)^2 e^x + x^2]$.
14. Verify that the functions

    $$f(x) = \frac{x^2 + 1}{x^2 + x + 1}, \qquad g(x) = \frac{-x}{x^2 + x + 1}$$

    have the same derivative. Why is this true?
15. Let $y = |x| = f(x)$, so that $y = x$ for $x \geq 0$, $y = -x$ for $x < 0$.
    (a) Show that $y' = 1$ for $x > 0$, $y' = -1$ for $x < 0$.
    (b) Show that at $x = 0$, f has a right-hand derivative, equal to $+1$, and a left-hand derivative equal to -1. Does $f'(0)$ exist?
16. (a) From the identity $\cos x = \sin[x + (\pi/2)]$, show that the derivative of the cosine function at x equals the derivative of the sine function at $x + (\pi/2)$.
    (b) From the result of part (a), prove Rule 8(b): $(\cos x)' = -\sin x$.
17. Prove Rules 8(c) to 8(f) by showing that, whenever the functions are defined,
    (a) $(\tan x)' = \sec^2 x$ (*Hint.* $\tan x = \sin x/\cos x$)
    (b) $(\cot x)' = -\csc^2 x$ (*Hint.* $\cot x = \cos x/\sin x$)
    (c) $(\sec x)' = \sec x \tan x$ (*Hint.* $\sec x = 1/\cos x$)
    (d) $(\csc x)' = -\csc x \cot x$ (*Hint.* $\csc x = 1/\sin x$)
18. Let $f(x) = \log_{10} x$ have derivative M for $x = 1$. Without using Rule 9, show that

    $$g(x) = \frac{1}{M} \log_{10} x$$

    has derivative $g'(1) = 1$, and that $g(x)$ can be written as $\log_e x$, where e is defined by the equation

    $$\log_{10} e = M$$

19. Let $f(x) = \sqrt{x}$, $x \geq 0$; let $g(x) = x - \sqrt{x + x^2}$, $x \geq 0$.
    (a) Show that neither $f'(0)$ nor $g'(0)$ exists (see Example 6 in Section 3-2).
    (b) Show that $f(x) + g(x)$ has derivative 1 for $x = 0$. [*Hint.* Show that for $h > 0$

$$\frac{f(h) + g(h) - f(0) - g(0)}{h} = 1 - \frac{\sqrt{h}}{1 + \sqrt{1 + h}} . \Big]$$

Remark. This problem shows that $f + g$ may have a derivative at x, even though $f'(x)$, $g'(x)$ do not exist.

20. Evaluate:

(a) $\lim\limits_{x \to 0} \dfrac{e^x - 1}{x}$,   (b) $\lim\limits_{x \to \pi} \dfrac{\cos x + 1}{x - \pi}$ (c) $\lim\limits_{x \to 1} \dfrac{\ln x}{x - 1}$,

(d) $\lim\limits_{x \to 0} \dfrac{\ln x - 1}{x - e}$,   (e) $\lim\limits_{x \to \infty} x \sin \dfrac{1}{x}$ (f) $\lim\limits_{x \to \infty} \dfrac{x}{x - 1} \ln x$

‡21. Let function f be defined for $a < x < b$ and let $a < x_0 < b$. A function g is said to be a better approximation to f at x_0 than is h if there is a $\delta > 0$ for which $|f(x) - g(x)| \le |f(x) - h(x)|$ for $|x - x_0| < \delta$. Prove that f has a derivative at x_0 if, and only if, f has a best linear approximation at x_0.

3-5 THE CHAIN RULE

A function such as

$$y = (x^2 + 3x + 7)^5 = F(x)$$

can be regarded as a composition of two functions:

$$y = u^5 = f(u), \qquad u = x^2 + 3x + 7 = g(x)$$

In general, a function $F = f \circ g$ is given by

$$y = F(x) = f[g(x)] \tag{3-50}$$

or by

$$y = f(u), \qquad u = g(x)$$

We shall learn that, for the function (3-50), the derivative is given by the *Chain Rule*

$$F'(x) = f'(u)g'(x) \tag{3-51}$$

where $u = g(x)$. Thus, in our example,

$$F'(x) = 5u^4(2x + 3) = 5(x^2 + 3x + 7)^4(2x + 3)$$

Notice that the same result is obtained by Rule 7. The rule (3-51) is easier to remember in the form

$$\frac{dy}{dx} = \frac{dy}{du}\frac{du}{dx} \tag{3-51'}$$

If we were allowed to cancel du against du, (3-51') would be a trivial identity. Most often the rule (3-51) or (3-51') is applied, as in our example, to a function built up with the aid of parentheses or with brackets. Thus we could write our example as follows:

$$y = (\quad)^5, \qquad (\quad) = x^2 + 3x + 7$$

so that, in a sense, ( ) stands for the auxiliary variable u. Then the rule can be stated: *differentiate with respect to the parenthesis, then multiply the result by the derivative of the function inside the parenthesis:*

$$y' = 5(\quad)^4(2x + 3)$$

We consider further examples:

EXAMPLE 1

$$y = \left(\frac{x}{x + 1}\right)^3 + 2\frac{x}{x + 1}$$

that is, $y = u^3 + 2u, u = x/(x + 1)$ or $y = (\quad)^3 + 2(\quad), (\quad) = x/(x + 1)$. Then

$$y' = (3u^2 + 2)\frac{du}{dx} = (3u^2 + 2)\frac{(x + 1) \cdot 1 - x \cdot 1}{(x + 1)^2}$$

$$= \left(3\left(\frac{x}{x + 1}\right)^2 + 2\right)\frac{1}{(x + 1)^2} = \frac{5x^2 + 4x + 2}{(x + 1)^4}$$

or we can write

$$y' = [3(\quad)^2 + 2]\left(\frac{x}{x + 1}\right)' = \left[3\left(\frac{x}{x + 1}\right)^2 + 2\right]\frac{1}{(x + 1)^2} \text{ etc.}$$

EXAMPLE 2. $y = \sqrt{x^2 + 1}$. Here, $y = \sqrt{u}, u = x^2 + 1$. In Section 3-2, Example 6, we found the derivative of $y = \sqrt{x} = x^{1/2}$ to be $y' = \frac{1}{2}x^{-1/2} = 1/(2\sqrt{x})$. Therefore, here

$$\frac{dy}{dx} = \frac{d}{du}\sqrt{u}\frac{du}{dx}, \qquad u = x^2 + 1$$

$$\frac{dy}{dx} = \frac{1}{2}u^{-1/2} \cdot 2x = \frac{x}{\sqrt{x^2 + 1}}$$

EXAMPLE 3. $y = \sin^2 x$. Here $y = u^2, u = \sin x$. By Rule 8(a)

$$\frac{dy}{dx} = \frac{dy}{du}\frac{du}{dx} = 2u \cos x = 2 \sin x \cos x = \sin 2x$$

EXAMPLE 4. $y = \sin(x^3 - 5x)$. Here, $y = \sin(\quad), (\quad) = x^3 - 5x$. Hence

$$\frac{dy}{dx} = \cos(\quad)\frac{d}{dx}(\quad) = [\cos(x^3 - 5x)](3x^2 - 5)$$

EXAMPLE 5. $y = 3[(x^2 + 2)^7 + (x^2 + 2)^5 + 3(x^2 + 2)^4]^3$. We could of course expand out and could ultimately express y as a polynomial of degree 42! It is much simpler to reason that

$$\frac{dy}{dx} = 9[\quad]^2\frac{d}{dx}[\quad]$$

Now $[\quad] = (x^2 + 2)^7 + (x^2 + 2)^5 + 3(x^2 + 2)^4 = (\quad)^7 + (\quad)^5 + 3(\quad)^4$, where $(\quad) = x^2 + 2$. Hence

$$\frac{d}{dx}[\ \ ] = \{7(\ \ )^6 + 5(\ \ )^4 + 12(\ \ )^3\}\frac{d}{dx}(\ .\ )$$

$$= \{7(\ \ )^6 + 5(\ \ )^4 + 12(\ \ )^3\}\, 2x$$

Accordingly

$$\frac{dy}{dx} = 9[\ \ ]^2\{7(\ \ )^6 + 5(\ \ )^4 + 12(\ \ )^3\}\, 2x$$

$$= 9[(x^2 + 2)^7 + (x^2 + 2)^5 + 3(x^2 + 2)^4]^2$$
$$\cdot\, \{7(x^2 + 2)^6 + 5(x^2 + 2)^4 + 12(x^2 + 2)^3\}\cdot 2x$$

We could also write

$$y = 3u^3, \qquad u = v^7 + v^5 + 3v^4, \qquad v = x^2 + 2$$

Then we apply the chain rule twice:

$$\frac{dy}{dx} = \frac{dy}{du}\frac{du}{dx} = \frac{dy}{du}\frac{du}{dv}\frac{dv}{dx} = 9u^2(7v^6 + 5v^4 + 12v^3)\, 2x$$

where u and v are as above.

The last example suggests a generalization. If $y = f(u)$, $u = g(v)$, $v = h(w)$, $w = p(x)$, so that ultimately, by composition, y can be expressed as a function of x, then

$$\frac{dy}{dx} = \frac{dy}{du}\frac{du}{dv}\frac{dv}{dw}\frac{dw}{dx} \tag{3-52}$$

and similarly for any number of such equations. Here y is "linked" to u by the equation $y = f(u)$, u is linked to v, v to w, w to x, so that y is connected to x by a "chain of relations" (Figure 3-19). It is for this reason that (3-51) and (3-52) are known as *chain rules*.

Figure 3-19 Chain rule for many variables.

Before presenting a precise formulation of the rule as a theorem, we give an informal justification. Suppose that y is a linear function of u, where u is a linear function of x:

$$y = m_1 u + b_1, \qquad u = m_2 x + b_2$$

Then y is also a linear function of x:

$$y = m_1(m_2 x + b_2) + b_1 = m_1 m_2 x + m_1 b_2 + b_1$$

The slopes m_1, m_2, which are the derivatives dy/du and du/dx, tell us how fast y changes with respect to u, and u with respect to x: y changes m_1 times as fast as u, and u changes m_2 times as fast as x. The composite function, expressing y in terms of x, has slope (or derivative) $m_1 m_2$, which is the product of the two slopes. We thus conclude that y changes $m_1 m_2$ times as fast as x. Now for general functions $y = f(u)$, $u = g(x)$, near a fixed x and the corresponding fixed $u = g(x)$, y and u are approximately linear functions—linear functions that are represented by the tangents to the graphs (Figure 3-20); the two slopes are precisely the derivatives at the two points. Accordingly,

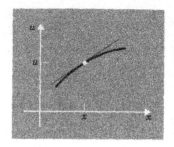
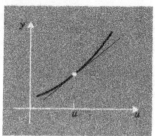

Figure 3-20 Chain rule.

near the fixed value of x, y is approximately a linear function of x with derivative equal to the product of the two derivatives. That is precisely the chain rule.

THEOREM 2. *Let $y = F(x)$, $a \leq x \leq b$, be defined as the composition of two functions:*

$$y = f(u), c \leq u \leq d, \qquad u = g(x), a \leq x \leq b$$

so that $F = f \circ g$. If at a particular x, $g'(x)$ exists, and if $f'(u)$ exists at the corresponding value $u = g(x)$, then $F'(x)$ exists and

$$F'(x) = f'(u)g'(x)$$

PROOF: For simplicity we assume that the chosen value of x and the corresponding $u = g(x)$ are interior to the intervals in which the functions are defined. Let $x + \Delta x$ be a value in the interval $[a, b]$ and let $u + \Delta u$ be the corresponding value of g (Figure 3-21a); let $y + \Delta y$ be the value of f at $u + \Delta u$ (Figure 3-21b). We now assume that, for $|\Delta x|$ sufficiently small and positive, Δu is never 0. This assumption simplifies our proof. A proof without any special assumption is given in Section 3-22 below. We now write, for Δx as stated,

$$\frac{\Delta y}{\Delta x} = \frac{\Delta y}{\Delta u} \frac{\Delta u}{\Delta x}$$

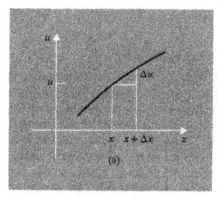
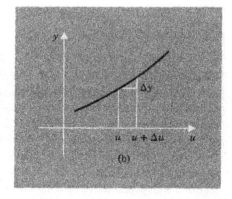

Figure 3-21 Proof of chain rule.

Since $g'(x)$ exists, $\Delta u/\Delta x$ has limit $g'(x)$ as $\Delta x \to 0$; also $\Delta u = \Delta x(\Delta u/\Delta x)$ has limit 0 as $\Delta x \to 0$ (this is just the continuity of g at x). Therefore, we can write

$$\lim_{\Delta x \to 0} \frac{\Delta y}{\Delta x} = \lim_{\Delta x \to 0} \frac{\Delta y}{\Delta u} \cdot \lim_{\Delta x \to 0} \frac{\Delta u}{\Delta x} = f'(u)g'(x)$$

that is, $F'(x) = f'(u)g'(x)$, as asserted.

The rule (3-52) is proved similarly and, by induction (Section 0-20), we can extend the rule to a chain of arbitrary length.

Derivative of x^α. We can write, for $x > 0$,

$$y = x^\alpha = e^{\alpha \ln x} = e^u, \qquad u = \alpha \ln x$$

Therefore, by the chain rule,

$$\frac{dy}{dx} = \frac{dy}{du}\frac{du}{dx} = e^u \cdot \frac{\alpha}{x} = e^{\alpha \ln x} \cdot \frac{\alpha}{x} = x^\alpha \cdot \frac{\alpha}{x} = \alpha x^{\alpha-1}$$

Therefore, the rule (Rule 6 in Table 3-2)

$$\frac{d}{dx} x^\alpha = \alpha x^{\alpha-1}, \qquad x > 0 \tag{3-53}$$

is valid for every constant α. For $\alpha > 1$, the rule is valid for $x = 0$ also (Problem 6 below). For $x < 0$, the rule remains valid for α rational and such that x^α is defined for $x < 0$ (Problem 7 below).

PROBLEMS

1. In each of the following examples find the derivative of the composite function by the chain rule and check by direct differentiation of the composite function:

   (a) $y = (x^2 - 5)^3 = u^3, \qquad u = x^2 - 5$

   (b) $y = \dfrac{1}{x^3 + 1} = \dfrac{1}{u}, \qquad u = x^3 + 1$

   (c) $x = (t^2 - t)^2 - (t^2 - t) = u^2 - u, \qquad u = t^2 - t$

   (d) $y = \dfrac{(x^3 - x)^2 + 3}{(x^3 - x)^4 + x^3 - x} = \dfrac{u^2 + 3}{u^4 + u}, \qquad u = x^3 - x$

2. Differentiate:

   (a) $y = \sqrt{x^2 - 1}$

   (b) $y = \dfrac{1}{(x^2 + 1)^{1/2}}$

   (c) $y = (x^2 + 5)^{7/3}$

   (d) $y = (2x^3 - 5x + \sqrt{2x^3 - 5x})^3$

   (e) $y = \sqrt{1 + \sqrt{x^2 - 1}}$

   (f) $y = \sqrt{1 + \sqrt{1 + \sqrt{x}}}$

   (g) $y = \left(\dfrac{x^2 - 1}{x^2 + 2x}\right)^5$

   (h) $y = \sqrt{1 + \left(\dfrac{x + 2}{x - 3}\right)^3}$

   (i) $y = \dfrac{1}{1 + (x^2 - 2\sqrt{x + 1})^5}$

3. Differentiate:

(a) $y = \cos^3 x$ (b) $y = \cos(2x - 1)$ (c) $r = \sin(\theta - \cos\theta)$

(d) $r = \cos^2[\theta + \sin(\theta \cos\theta)]$ (e) $x = 3\tan^4 t$ (f) $y = \sec^2\theta \tan^3\theta$

(g) $y = \cos^3(\cos^2 x)$ (h) $y = \sin[\sin(\sin x)]$

4. Differentiate:

(a) e^{x^2} (b) e^{e^x} (c) $\ln(x^2 - 1)$ (d) $\ln\dfrac{e^x + 1}{e^x - 1}$

(e) $\ln\dfrac{1}{x^2 - 1}$ (f) $\ln(\cos 2x)$ (g) $\ln\sin^2 x$ (h) $\ln\ln x$

(i) $\ln\ln\ln x$ (j) $e^{\cos^2 x}$ (k) $x^x = e^{x\ln x}$ (l) x^{x^2}

5. Figure 3-22 shows three gears of radii a, b, c. If the first gear turns through angle x, the second turns through angle u, and the third through angle y. If the second is turning m_1 times as fast as the first and the third is turning m_2 times as fast as the second, show

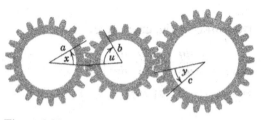

Figure 3-22

that the third is turning $m_1 m_2$ times as fast as the first. Interpret in terms of the chain rule. How is the conclusion modified if the radii are varying because of expansion under rising temperature?

6. Prove that, if $\alpha > 1$, then $y = x^\alpha = f(x)$, $x > 0$ has derivative 0 at $x = 0$. [*Hint.* Consider $[f(0 + h) - f(0)]/h$ for $h > 0$.] What happens for $\alpha = 1$?

7. Prove that, if $\alpha = p/q$, where p is an integer and q is a positive odd integer, then $y = x^\alpha$ is defined for $x < 0$ and has derivative $\alpha x^{\alpha-1}$. [*Hint.* Since q is odd and positive,

$$x^{1/q} = -(-x)^{1/q}$$

and, hence, $y = (-1)^p[(-x)^{p/q}] = (-1)^p v^{p/q}$, $v = -x$, where $v > 0$ for $x < 0$. Now use (3-53) and the chain rule.]

8. Prove that, if α is as in Problem 7 and $p > q$, then $y = x^\alpha$ has derivative $\alpha x^{\alpha-1}$ for all real x. [*Hint.* By (3-53) and Problems 6 and 7, we need only show that

$$\lim_{h \to 0} \frac{f(0 + h) - f(0)}{h} = 0.]$$

9. Find the slope of the tangent to the curve at the point requested:

(a) $y = e^{x^3}$ at $(0, 1)$ (b) $y = \sqrt{x^2 - 8}$ at $(3, 1)$

(c) $y = \ln(1 + \ln^2 x)$ at $(1, 0)$ (d) $y = [x^3 - (x^2 - 1)^4]^7$ at $(1, 1)$

10. Find a possible f if f' is   (a) $2xe^{x^2}$,   (b) $20x^4(x^5 + 1)^3$,   (c) xe^{x^2},   (d) $x^4(x^5 + 1)^3$,   (e) $x\sqrt{x^2 - 1}$,   (f) $(3x^2 - 1)\sin(x^3 - x)$.

3-6 DERIVATIVE OF INVERSE FUNCTIONS

We observed in Section 2-11 (Theorem D) that, if $y = f(x)$ is continuous and monotone strictly increasing (or decreasing) in an interval $a \le x \le b$, then there is an inverse function $x = f^{-1}(y)$, which is also continuous. If we start with a fixed pair of corresponding values x, y (see Figure 3-23) and add

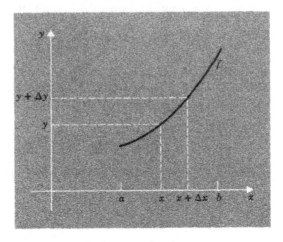

Figure 3-23 Derivative of inverse function.

increments Δx, Δy, so that $f(x + \Delta x) = y + \Delta y$ and $x + \Delta x = f^{-1}(y + \Delta y)$, then continuity of f and its inverse f^{-1} at the chosen point is equivalent to the two conditions:

$$\lim_{\Delta x \to 0} \Delta y = 0 \quad \text{and} \quad \lim_{\Delta y \to 0} \Delta x = 0 \tag{3-60}$$

In the first limit Δy is considered as a function of Δx; in the second limit Δx is considered as a function of Δy. Let us now suppose that f' exists and is not 0 at the value x considered, so that

$$\lim_{\Delta x \to 0} \frac{\Delta y}{\Delta x} = f'(x) \ne 0$$

and, hence, that

$$\lim_{\Delta x \to 0} \frac{\Delta x}{\Delta y} = \frac{1}{f'(x)} \tag{3-61}$$

We now consider Δx as a function of Δy. Then by the second part of (3-60), by choosing $|\Delta y|$ sufficiently small, we can make $|\Delta x|$ as small as desired and, consequently, by (3-61), can make $\Delta x / \Delta y$ as close as desired to $1/f'(x)$, that is,

$$\lim_{\Delta y \to 0} \frac{\Delta x}{\Delta y} = \frac{1}{f'(x)}$$

Thus, the derivative of the inverse function f^{-1} at y is the *reciprocal of the derivative of the function f at $x = f^{-1}(y)$*, provided that the derivative f exists and is not 0 at the point considered. We can write the conclusion as follows:

$$\frac{dx}{dy} = \frac{1}{dy/dx} \tag{3-62}$$

If we could rearrange the fraction on the right, the conclusion would be trivial. We state our result formally:

THEOREM 3. *Let $y = f(x)$ be defined, continuous, and monotone strictly increasing (or decreasing) for $a \leq x \leq b$, so that f has a continuous inverse f^{-1}. If $f'(x)$ exists and is not 0 at a particular x, then f^{-1} has a derivative at the corresponding value of y, given by (3-62), or by*

$$[(f^{-1})' \text{ at } y] = \frac{1}{[f' \text{ at } x]}, \qquad x = f^{-1}(y) \qquad (3\text{-}62')$$

EXAMPLE 1. $y = 2x - 3$, a linear function. The inverse function is $x = (1/2)y + (3/2)$ (Figure 3-24). The two derivatives are 2 and 1/2 and are reciprocals of each other:

$$\frac{dx}{dy} = \frac{1}{2} = \frac{1}{dy/dx}$$

Since forming the inverse function interchanges the roles of x and y and

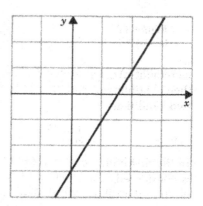

Figure 3-24   Example 1.

reflects the graph in the line $y = x$, it is clear that slopes of lines or curves will always be replaced by their reciprocals.

EXAMPLE 2. $y = x^3$. The inverse is $x = y^{1/3}$. We find that

$$\frac{dx}{dy} = \frac{1}{dy/dx} = \frac{1}{3x^2}$$

This gives the derivative in terms of x. To obtain the derivative in terms of y, we can write

$$\frac{dx}{dy} = \frac{1}{3x^2} = \frac{1}{3y^{2/3}} = \frac{1}{3}y^{-2/3}$$

We observe that the derivative $3x^2$ is 0 at $x = 0$, where $y = 0$. Correspondingly, the derivative dx/dy does not exist at this point (the tangent has infinite slope at the point, as Figure 3-25 shows).

EXAMPLE 3. $y = f(x) = x^3 + x$ (Figure 3-26). Since both terms increase as x increases, f is monotone strictly increasing for all x. We find, for the inverse

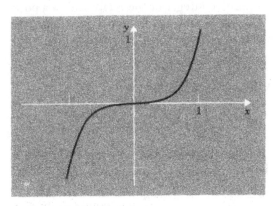

Figure 3-25 Example 2. **Figure 3-26** Example 3.

function $x = f^{-1}(y)$, that

$$\frac{dx}{dy} = \frac{1}{3x^2 + 1}$$

The derivative of $x = f^{-1}(y)$ is again given in terms of x, but in this example it is very difficult to express the result in terms of y, for this would require solving the cubic equation for x: $x^3 + x - y = 0$. It is convenient to leave our derivative dx/dy in terms of x. For any particular numerical value of y, we can find x by a calculation, or even roughly from the graph. For example, for $y = 1$, we find by a numerical method (Problem 3 below) that $x = 0.682$ and, hence, that

$$\frac{dx}{dy} = [3(0.682)^2 + 1]^{-1} = 0.418 \qquad \text{at} \qquad y = 1$$

At the end of Section 3-5 we showed, with the aid of the logarithmic function and the chain rule, that for all α, x^α has derivative $\alpha x^{\alpha-1}$. The next example leads to another way of proving this for α rational.

EXAMPLE 4. The function $y = x^n$ $(n = 1, 2, 3, \dots)$. These functions are all strictly increasing for $x \geq 0$, and we confine our attention to this interval, corresponding to the interval $y \geq 0$. We have $y' = nx^{n-1}$ and, therefore, for the inverse function we have

$$\frac{dx}{dy} = \frac{1}{nx^{n-1}}$$

We are forced to exclude the value $x = 0$ (except for $n = 1$). However, the inverse function is $x = y^{1/n}$ and, consequently, we can write

$$\frac{d}{dy}(y^{1/n}) = \frac{1}{ny^{(n-1)/n}} = \frac{1}{n}[y^{(1/n)-1}], \qquad y > 0$$

Our result shows that the general rule

$$\frac{d}{dx}x^\alpha = \alpha x^{\alpha-1} \tag{3-63}$$

is valid, at least for $x > 0$, for α a positive integer or the reciprocal of a positive integer. We can now extend the rule to arbitrary fractional powers. First, we consider $y = x^{m/n}$, $x > 0$, where m and n are positive integers. We write

$$y = (x^{1/n})^m = u^m, \qquad u = x^{/1n}$$

Hence, by the chain rule and by the result of Example 4,

$$y' = mu^{m-1} \frac{d}{dx} x^{1/n} = mx^{(m-1)/n} \left(\frac{1}{n}\right) x^{(1/n)-1} = \frac{m}{n} x^{(m/n)-1}$$

Thus, (3-62) is valid for positive fractional powers. For $\alpha = -m/n$, where m and n are positive integers, we have

$$y = x^{-m/n} = \frac{1}{u}, \qquad u = x^{m/n}$$

and therefore

$$y' = \frac{d}{du}\left(\frac{1}{u}\right) \cdot \frac{d}{dx} x^{m/n} = -\frac{1}{u^2} \frac{m}{n} x^{(m/n)-1} = -\frac{m}{n} x^{-2m/n} x^{(m/n)-1}$$

$$= -\frac{m}{n} x^{-(m/n)-1} = \alpha x^{\alpha-1}$$

Thus (3-63) is proved for fractional powers. A complete discussion is given in Chapter 5.

Remark. In all these examples the continuity and monotone character of the function has been easy to verify directly. Occasionally this verification is not so easy. The following theorem may then be helpful.

THEOREM 4. *Let $y = f(x)$ be defined and continuous in the interval $a \le x \le b$. If $f'(x)$ exists and $f'(x) > 0$ for $a < x < b$, then f is monotone strictly increasing; if $f'(x) < 0$ for $a < x < b$ then f is monotone strictly decreasing.*

A proof is given in Section 3-21 below. We can justify the theorem intuitively as follows: if $f'(x) > 0$ in the interval, then the tangent to the curve has positive slope, so that the curve must rise as x increases (Figure 3-23); that is, f must be monotone strictly increasing. We can also reason that if $f'(x) > 0$ in the interval then, at each x, $\Delta y/\Delta x$ must be positive for Δx sufficiently small and positive; hence, Δy is also positive for such Δx. This also indicates that y increases as x increases.

EXAMPLE 5. Let $y = 10x - x^2 - x^3$ for $0 \le x \le 1$. Then $y' = 10 - 2x - 3x^2 = 10 - (2x + 3x^2)$. Now for $0 \le x \le 1$ we have $2x + 3x^2 < 5$, so that $y' > 5$ in the interval. Accordingly, by Theorem 4, f is monotone strictly increasing and there is an inverse function $x = f^{-1}(y)$ in the corresponding interval $0 \le y \le 8$, with derivative

$$\frac{dx}{dy} = \frac{1}{10 - 2x - 3x^2}$$

Our theorems can be used to obtain *rules for derivatives of inverse trigo-nometric functions:*

$$(\operatorname{Sin}^{-1} x)' = \frac{1}{\sqrt{1 - x^2}}, \qquad -1 < x < 1 \qquad (3\text{-}64)$$

$$(\operatorname{Cos}^{-1} x)' = \frac{-1}{\sqrt{1 - x^2}}, \qquad -1 < x < 1 \qquad (3\text{-}65)$$

$$(\operatorname{Tan}^{-1} x)' = \frac{1}{1 + x^2}, \qquad -\infty < x < \infty \qquad (3\text{-}66)$$

We prove the first rule and leave the other two as exercises (Problem 5 below). The function $y = \sin x$ is continuous for all x and has derivative $y' = \cos x$; hence, $y' > 0$ for $-\frac{1}{2}\pi < x < \frac{1}{2}\pi$. Therefore, by Theorem 4, $y = \sin x$ is monotone strictly increasing for $-\frac{1}{2}\pi \leq x \leq \frac{1}{2}\pi$, and has an inverse function; that is precisely the function $\operatorname{Sin}^{-1} y$, $-1 \leq y \leq 1$ (Section 2-7, see Figure 3-27). By Theorem 3

$$(\operatorname{Sin}^{-1} y)' = \frac{1}{(\sin x)'} = \frac{1}{\cos x} = \frac{1}{\sqrt{1 - \sin^2 x}}$$

$$= \frac{1}{\sqrt{1 - y^2}}, \qquad -1 < y < 1$$

The plus sign is chosen before the square root, since $\cos x$ is positive for $-\frac{1}{2}\pi < x < \frac{1}{2}\pi$. If we now interchange x and y, we obtain the rule (3-63).

The inverse of the function $y = \ln x$ is the function $x = e^y$. Hence, from the rule $(\ln x)' = 1/x$, we deduce the rule for the derivative of the exponential function:

$$(e^y)' = \frac{1}{(\ln x)'} = \frac{1}{1/x} = x = e^y$$

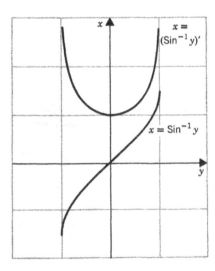

Figure 3-27 Derivative of inverse sine.

Similarly, the inverse of the function $y = e^x$ is the function $x = \ln y$. Thus, from the rule $(e^x)' = e^x$, we deduce the rule for the derivative of the logarithmic function:

$$(\ln y)' = \frac{1}{(e^x)'} = \frac{1}{e^x} = \frac{1}{y}$$

In the same way, either one of the two rules $(a^x)' = a^x \ln a$, $(\log_a x)' = 1/(x \ln a)$ can be deduced from the other.

PROBLEMS

1. In each case, find the derivative of the inverse function f^{-1} on the basis of Theorem 3. Also, check by finding the inverse function explicitly and differentiating it.

(a) $y = \dfrac{x}{x+1} = f(x)$, $x > -1$

(b) $y = \dfrac{x^2}{x+2} = f(x)$, $x > 0$

(c) $y = \sqrt{1-x} = f(x)$, $0 < x < 1$

(d) $y = e^{\sin x}$, $-\dfrac{1}{2}\pi < x < \dfrac{1}{2}\pi$

(e) $y = \ln(1 + e^x)$, $-\infty < x < \infty$

(f) $y = \tan^3 x$, $-\dfrac{1}{2}\pi < x < \dfrac{1}{2}\pi$

2. Show that each of the following functions f has an inverse f^{-1}, graph f and f^{-1}, and find the derivative of the inverse function:
   (a) $y = 2x^3 - 3x^2 + 6x - 2 = f(x)$, all x
   (b) $y = x^2/(x+1) = f(x)$, $x \geq 0$
   (c) $y = x + \sqrt{x+1} = f(x)$, $x \geq -1$.

3. In Example 3 in the text, it is asserted that, when $y = 1$, the derivative of the inverse

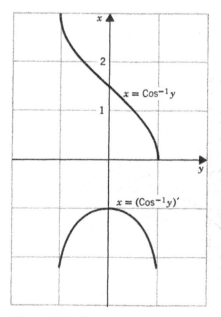

Figure 3-28   Derivative of inverse cosine.

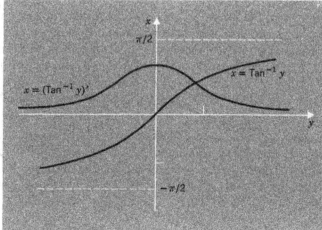

Figure 3-29   Derivative of inverse tangent.

$f^{-1}(y)$ of f, where $f(x) = x^3 + x$, has value 0.418. Verify all the steps leading to this value. (A table of squares and cubes will be helpful.)

4. Let $y = f(x)$ be differentiable and have a differentiable inverse f^{-1}. Show by the chain rule that

$$\frac{dy}{dx} \cdot \frac{dx}{dy} = 1$$

[*Hint*. If $F = f^{-1} \circ f$, then $F(x) = f^{-1}[f(x)] = x$.]

5. (a) Prove (3-65) (see Figure 3-28).

(b) Prove (3-66) (see Figure 3-29).

(c) Explain the relationship between the derivatives in (3-63) and (3-64).

6. Differentiate:

(a) $y = \text{Sin}^{-1}\dfrac{x}{2}$

(b) $y = \dfrac{1}{2}x\sqrt{1 - x^2} + \dfrac{1}{2}\text{Sin}^{-1} x$

(c) $y = \text{Sin}^{-1}\dfrac{3x + 1}{x}$

(d) $y = 5\,\text{Tan}^{-1} 5x$

(e) $y = \ln(\pi + \text{Tan}^{-1} x)$

(f) $y = \text{Sin}^{-1}\dfrac{x}{\sqrt{1 + x^2}}$

[Explain the result of (f)]

3-7 RELATED FUNCTIONS

On occasion we have several functions that we know are related by an equation. The calculus then provides some information about their derivatives.

EXAMPLE 1. Two differentiable functions f and g are known to be such that $f(1) = 3$, $f'(1) = -2$, and for $0 \le x \le 2$

$$2[f(x)]^2 - [g(x)]^3 + 9 = 0 \qquad\qquad (3\text{-}70)$$

Find $g(1)$ and $g'(1)$.

Solution. We find $g(1)$ by substituting $x = 1$ in Equation (3-70):

$$2[f(1)]^2 - [g(1)]^3 + 9 = 0, \qquad 2 \times 9 - [g(1)]^3 + 9 = 0$$
$$[g(1)]^3 = 27, \qquad g(1) = 3$$

Next we differentiate (3-70) to obtain

$$4f(x)f'(x) - 3[g(x)]^2 g'(x) = 0$$

By substituting $x = 1$ and by using the known values of f, f', g for $x = 1$, we obtain

$$4 \times 3 \times (-2) - 3 \times 9 \times g'(1) = 0$$

and, therefore, find that $g'(1) = -\frac{8}{9}$.

EXAMPLE 2. We show in mechanics that the potential energy U of a stretched spring is kx^2, where k is a constant and x is the distance the string has been stretched from equilibrium. Thus, $U = kx^2$ (see Figure 3-30). When

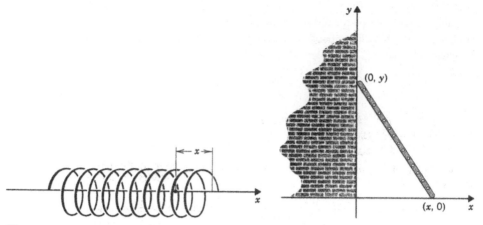

Figure 3-30   Stretched spring.   Figure 3-31   Ladder sliding.

the spring is oscillating, x becomes a function of time t, say $x = f(t)$. Then U also becomes a function of t. Since $U = kx^2$, we have two related functions of t and, by the chain rule,

$$\frac{dU}{dt} = 2kx\,\frac{dx}{dt}$$

This equation shows that $dU/dt = 0$ either when $x = 0$ (at equilibrium) or when $dx/dt = 0$ (at the point of maximum extension of the spring).

EXAMPLE. 3. A ladder 10 ft long is placed against a building, but starts to slide down (Figure 3-31). At a certain instant the bottom of the ladder is 6 ft from the wall and moving away from the wall at 7 ft per sec. How fast is the top of the ladder moving down at that instant?

Solution. We use coordinates (x, y), as in Figure 3-31, so that the top of the ladder is at $(0, y)$, and the bottom at $(x, 0)$. We measure x and y in feet and time t in seconds. We are given that at time t_0: $x = 6$, $dx/dt = 7$. Here x and y are related by the fact that the ladder has length 10 ft: $x^2 + y^2 = 100$. Therefore

$$2x\,\frac{dx}{dt} + 2y\,\frac{dy}{dt} = 0$$

Now at time t_0, $x = 6$ and $x^2 + y^2 = 100$, so that $y = 8$ (y is positive), and we know that $dx/dt = 7$. Hence, at time t_0,

$$2 \times 6 \times 7 + 2 \times 8 \times \frac{dy}{dt} = 0$$

Therefore, $dy/dt = -42/8 = -5.25$; that is, at time t_0 the top of the ladder is moving down at 5.25 ft per sec.

PROBLEMS

1. Let f and g be differentiable functions related by the given equation for each x of a certain interval. Find a relationship between f, g, f', g':

(a) $2f(x) + 3g(x) = 0$ (b) $3f(x) - 7g(x) = 0$
(c) $f(x)g(x) + 1 = 0$ (d) $f(x)[g(x)]^2 - g(x) + 1 = 0$
(e) $x^2f(x) - xg(x) + e^x f(x)g(x) = 0$ (f) $f(x) + \sin g(x) = 0$

2. A spherical balloon is being inflated so that its volume increases at the rate of 5 cu ft per sec. How fast is the radius increasing at the instant when the radius is 7 ft?

3. A rod of length 3 ft is attached at end A to a wheel that is turning at the rate of 0.3 radians per sec. The other end B is attached to a ring that is free to slide along a second rod (Figure 3-32). If the radius of the wheel is 2 ft, determine how fast the ring is moving at the instant when A is at its highest position. [*Hint.* Show first that, by the law of cosines, $9 = 4 + x^2 - 4x \cos \theta$.]

4. In physics, we find that the behavior of a gas in a container is described under normal conditions by Boyle's law: $pV = bT$, where p is the pressure, V the volume, T the absolute temperature, and b is a constant. When a gas is compressed, all three quantities p, V, T become functions of time t. Let us suppose that,

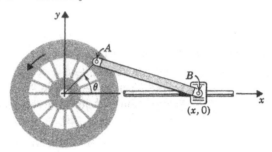

Figure 3-32

at a certain instant, $p = 10$ (lbs/in²), $V = 75$ cu in, $T = 300°$, the pressure is rising at the rate of 1 lb per sq in per minute and that the volume is decreasing at the rate of 5 cu in per minute. Find the rate of change of T at this instant.

5. The concept of a vector space of functions (Section 2-9) involves functions related to each other through addition, multiplication by constants and, in general, formation of linear combinations. Show that each of the following collections of functions forms a vector space:
   (a) V_a: all functions having a continuous derivative on the interval $[0, 1]$.
   (b) V_b: all functions having a derivative (not necessarily continuous) on the interval $[0, 1]$.
   (c) V_c: all functions having a derivative except at a finite number of points on $[0, 1]$.
   (d) V_d: all functions on $[0, 1]$ which are derivatives of other functions (that is, all f so that $g' = f$ for some g).

6. In the notation of Problem 5, each function f in V_b has a derivative f' in V_d. Show that if f, g are in V_b and are linearly dependent, then f' and g' are also linearly dependent. Does the converse hold true?

7. Show that formation of the derivative can be interpreted as a mapping of the vector space of all polynomials of degree at most n ($n \geq 1$) *onto* the vector space of all polynomials of degree $n - 1$. Is this mapping one-to-one?

3-8 IMPLICIT·FUNCTIONS

Let the following equation be given:

$$x^2 + xy + y^3 - 2x + 3y = 0 \tag{3-80}$$

We could try to solve this equation for y and, with much effort, might succeed in finding an expression for y in terms of x. Thus we conjecture that the equation defines a function $y = f(x)$, and we can seek the derivative $f'(x)$. The following reasoning simplifies this task greatly. We know that if f is the function in question and that if f is defined in an interval, then for each x of that interval we have

$$x^2 + xf(x) + [f(x)]^3 - 2x + 3f(x) = 0 \qquad (3\text{-}81)$$

Thus we have a situation much like that of the preceding section, in which several functions are related by an equation. If f has a derivative, then we can differentiate our equation (3-81):

$$2x + xf'(x) + f(x) + 3[f(x)]^2 f'(x) - 2 + 3f'(x) = 0 \qquad (3\text{-}82)$$

Consequently, we obtain a relation between x, $f(x)$ and $f'(x)$. If we give a particular x, we may be able to find $y = f(x)$ from Equation (3-80) and then can use (3-82) to find $f'(x)$. For example, for $x = 0$, Equation (3-80) becomes $y^3 + 3y = 0$, an equation with just one real solution: $y = 0$; that is, $f(0) = 0$. Hence, from (3-82), with $x = 0$, $f(0) = 0$, we have

$$2 \times 0 + 0 \times f'(0) + 0 + 3 \times 0 \times f'(0) - 2 + 3f'(0) = 0$$

so that $f'(0) = \frac{2}{3}$.

This procedure can normally be written more concisely as follows:

$$x^2 + xy + y^3 - 2x + 3y = 0, \qquad 2x + xy' + y + 3y^2 y' - 2 + 3y' = 0$$

For $x = 0$, $y^3 + 3y = 0$, so that $y = 0$, and the second equation gives $y' = \frac{2}{3}$ for $x = 0$.

EXAMPLE 1. $x + \cos x - e^y + xy^2 = 0$. We assume that the equation defines a differentiable function $y = f(x)$ and obtain

$$1 - \sin x - e^y y' + 2xyy' + y^2 = 0$$

We can solve this equation for y':

$$y' = \frac{\sin x - 1 - y^2}{-e^y + 2xy} \qquad (3\text{-}83)$$

This equation gives y' in terms of x and y. For any particular x, we can use the given equation to find y (if possible) and can then use (3-83) to find $y' = f'(x)$. For example, for $x = 0$, our given equation becomes $1 - e^y = 0$, so that $y = 0$; hence by (3-83)

$$y' = \frac{0 - 1 - 0}{-1 + 0} = 1 = f'(0)$$

Discussion. We are here dealing with functions that are "implied" by an equation or, as we say, are *implicit functions*. When we differentiate, we *assume* that there really is a differentiable function $y = f(x)$ satisfying the equation over an interval. The result of the differentiation is an equation relating x, y, and y'. It can be shown that we can always solve for y', as in Example 1, so that y' is expressed in terms of x and y. For any particular x, the given equation gives y (obtainable perhaps only after a difficult calculation) and the differentiated equation gives y'. The whole process is called *implicit differentiation.*

An equation such as $y = x^2 - e^x$ is said to give the function *explicitly*, in contrast to the implicit functions of our examples.

Our assumption that an equation in x and y defines a differentiable function $y = f(x)$ over some interval may fail to be justified. The equation $x^2 + y^2 = 0$ is satisfied only for $x = 0$, $y = 0$ and, therefore, there is no implicit function! The equation $x^2 + y^2 = 1$ leads to

$$y = \pm\sqrt{1 - x^2}$$

Hence, the equation "implies" two functions:

$$y = \sqrt{1 - x^2}, \qquad y = -\sqrt{1 - x^2}$$

each defined for $-1 \leq x \leq 1$ (but differentiable only for $-1 < x < 1$). We can give more complicated examples in which there are infinitely many differentiable functions that satisfy a given equation. Thus, our assumption is more a matter of hope than of confidence!

In practice, we usually have some good reasons for expecting the equation to define a single differentiable function. Frequently, we know one point (x_0, y_0) on the graph of the equation and are interested only in a differentiable function $y = f(x)$, defined in some small interval about x_0, with $f(x_0) = y_0$. An important theorem, the Implicit Function Theorem (proved in Chapter 12) provides a useful test as to whether there is such a function f, satisfying the equation. We can also give some information about the size of the interval in which f is defined.

A geometric interpretation of the problem is helpful. We are given an equation in x and y that has a certain graph in the xy-plane (Figure 3-33).

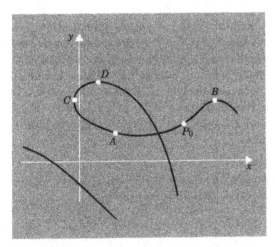

Figure 3-33 Graph of an equation in x and y.

The graph may be very complicated and may consist of many curves, crossing each other. However, given a point P_0: (x_0, y_0) on the graph, we may be able to find a piece of the graph, containing P_0, that is the graph of a continuous function f. In the figure, the curve from A to B is such a piece, and the function f appears to be differentiable. The process of implicit differentiation would enable us to find the tangent line at each point of this piece. If we try to use the piece from C to A to B, we expect difficulty at C, since the tangent line at C appears to be vertical. The piece from D to C to A cannot be the graph of a function, since there are two values of y for some values of x.

EXAMPLE 2. $xy^2 - y + x = 0$. Here we can solve explicitly, since we have a quadratic equation in y:

$$y = \frac{1 \pm \sqrt{1 - 4x^2}}{2x}$$

Observe that we have two functions defined:

$$y = \frac{1 + \sqrt{1 - 4x^2}}{2x} = f(x), \qquad y = \frac{1 - \sqrt{1 - 4x^2}}{2x} = g(x)$$

Both are defined in the intervals $-\frac{1}{2} \leq x < 0$ and $0 < x \leq \frac{1}{2}$. Both appear to be discontinuous at $x = 0$. However, we can show (Problem 4 below) that $g(x)$ has limit 0 as x approaches 0; in addition, from the given implicit equation we find that $y = 0$ for $x = 0$. Hence, if we define $g(0)$ to be 0, then $y = g(x)$ is a continuous function that satisfies the given equation for $-\frac{1}{2} \leq x \leq \frac{1}{2}$.

We now differentiate implicitly:

$$y^2 + 2xyy' - y' + 1 = 0$$

$$y' = \frac{y^2 + 1}{1 - 2xy}$$

This gives us one expression that can be used for the derivative of both functions f and g. For f', we must replace y by $f(x)$ on the right and, similarly for g', we replace y by $g(x)$.

Remark. In this example the equation gives two values of y for each x. In the interval $0 < x < \frac{1}{2}$, we could assign the values $f(x)$ to some values of x and the values $g(x)$ to the remaining values of x. Clearly, such a procedure would yield a function that satisfies the equation. However, it would not be continuous, let alone differentiable. Normally there is only one "natural" way to choose the functions implied by the equation, and this natural way usually leads to continuous functions that are differentiable except for some special points.

EXAMPLE 3. $y^3 + xy^2 + x^2y + 2x^3 = 0$. Here we have a difficult equation. We first experiment with a few x-values. For $x = 0$, $y = 0$. For $x = 1$, we have the cubic equation for y:

$$y^3 + y^2 + y + 2 = 0$$

If we let $g(y) = y^3 + y^2 + y + 2$, so that our equation is $g(y) = 0$, then we notice that

$$g'(y) = 3y^2 + 2y + 1$$

and that $g'(y)$ is *always positive*. For by completing the square

$$g'(y) = 3\left(y^2 + \frac{2}{3}y + \frac{1}{9}\right) + 1 - \frac{1}{3} = 3\left(y + \frac{1}{3}\right)^2 + \frac{2}{3} \geq \frac{2}{3}$$

Hence, $g(y)$ is monotone strictly increasing for all y. Furthermore (see Section 2-11)

$$\lim_{x \to +\infty} g(y) = +\infty, \qquad \lim_{x \to -\infty} g(y) = -\infty$$

Accordingly, $g(y)$ is positive for y sufficiently large and positive and $g(y)$ is negative for y sufficiently large and negative. Therefore, by the Intermediate Value Theorem, $g(y) = 0$ for at least one value of y and, since $g(y)$ is steadily increasing, for precisely one y. Accordingly, for $x = 1$, there is exactly one value of y defined. We can start to compute this value:

$$g(-2) = -4, \qquad g(-1) = 1, \qquad g(0) = 2, \ldots.$$

Thus, $y = -1$ for $x = 1$.

We can try to extend the reasoning just used to arbitrary x-values. Thus for x fixed, and not 0, the left-hand side of our given equation is always a function of y, which we denote by $g_x(y)$, and for all y:

$$D_y[g_x(y)] = 3y^2 + 2xy + x^2 = 3\left(y^2 + \frac{2}{3}xy + \frac{x^2}{9}\right) + \frac{2x^2}{3}$$

$$= 3\left(y + \frac{x}{3}\right)^2 + \frac{2x^2}{3} > 0$$

Accordingly, as above, $g_x(y)$ is monotone strictly increasing and, furthermore, $g_x(y)$ has limit $+\infty$ for $y \to +\infty$, g_x has limit $-\infty$ for $y \to -\infty$. Therefore, by the Intermediate Value Theorem, $g_x(y) = 0$ for exactly one value of y. Thus, for each $x \neq 0$ there is exactly one y such that x, y satisfy the given equation. For $x = 0$, we also have just one value, namely, $y = 0$.

We conclude that our given equation defines implicitly a unique function $y = f(x)$, $-\infty < x < \infty$. If we *assume* that $f(x)$ is differentiable, then we can obtain the derivative as before:

$$3y^2y' + 2xyy' + y^2 + x^2y' + 2xy + 6x^2 = 0$$

$$y' = \frac{-y^2 - 2xy - 6x^2}{3y^2 + 2xy + x^2}$$

We observe that the denominator here is the derivative $g_x'(y)$ of the left side of the given equation with x fixed and that, as was shown, $g_x'(y) > 0$ for all y (and x), provided that $x \neq 0$.

For this example we can prove that f is indeed differentiable for $x \neq 0$ by showing that the Implicit Function Theorem of Chapter 12 is applicable.

Logarithmic Differentiation. By taking logarithms of both sides of an equation, we may possibly shorten the process of finding the derivative. For example, if $y = x^3 e^{2x} \cos^2 x$, then

$$\ln y = 3 \ln x + 2x + 2 \ln \cos x$$

so that, by implicit differentiation,

$$\frac{y'}{y} = \frac{3}{x} + 2 - 2\frac{\sin x}{\cos x}$$

and

$$y' = y\left(\frac{3}{x} + 2 - 2\tan x\right)$$

$$= x^3 e^{2x} \cos^2 x \left(\frac{3}{x} + 2 - 2\tan x\right)$$

$$= x^2 e^{2x} \cos x \,(3\cos x + 2x\cos x - 2x\sin x)$$

This procedure is called *logarithmic differentiation*. We have ignored the fact that $\ln u$ is undefined for $u \leq 0$; nevertheless we can observe, by differentiating the original product, that the final result is correct for all x.

PROBLEMS

1. Find the indicated derivative by implicit differentiation and check by finding the explicit function and its derivative:
   (a) y' for $2x + 3y - 1 = 0$ (b) $D_t x$ for $t^2 x + tx - t^3 + 1 = 0$
   (c) $D_v u$ for $u^2 + v^2 = 4$ (d) $D_x y$ for $y^2 + x^2 y + 1 = 0$
2. Differentiate implicitly [that is find y' in terms of x and y under the assumption that $y = f(x)$ is a differentiable function satisfying the equation]:
   (a) $x^3 + x^2 y - 2xy^2 + y^3 - 1 = 0$ (b) $1 + x^2 y^3 - xy^2 + 2xy - 1 = 0$
   (c) $y^4 + (x^2 + 1)^3 y^3 - 5xy^2 - 4 = 0$ (d) $x + e^x + y^3 + y = 0$
   (e) $x\sin(xy) + \cos(xy) = 0$ (f) $x + y^2 + \ln(x + y) = 0$
3. Show that the equation $y^3 + y^2 + x^2 y + y + 1 = 0$ defines a single function $y = f(x)$, $-\infty < x < \infty$, and find an expression for $f'(x)$.
4. (a) For the equation of Example 2 in the text consider x as a function of y, $x = p(y)$, and show that p is monotone strictly increasing for $-1 \leq y \leq 1$. Now obtain the derivative y' for the inverse of $p(y)$ by Theorem 3, and compare with the results given in the text.
   (b) Show that, for the function $g(x)$ defined in Example 2,
   $$\lim_{x \to 0} g(x) = 0$$
5. Find the equation of the tangent line to the given curve at the specified point:
   (a) $x^3 + x^2 y - 2xy^2 + y^3 - 1 = 0$ at $(1, 0)$
   (b) $1 + x^2 y^3 - xy^2 + 2xy - 1 = 0$ at $(-1, 1)$
   (c) $\sin xy + y - x^2 = 0$ at $(0, 0)$
6. Use logarithmic differentiation to find the derivative:
   (a) $y = x^2 e^{3x} \tan^3 x$ (b) $y = x^3(1 + x)^2(1 - x)^{2/3}$
   (c) $y = \dfrac{x^2 \cos 5x}{(x^2 + 1)^3(x - 1)^2}$ (d) $y = \dfrac{x^3 \sin e^x}{e^{2x}(x^2 + 1)^2}$
   (e) $y = x^x$ (f) $y = x^{\sin x}$
   (g) $y = (\cos x)^{\sin x}$ (h) $y = x^{(x^x)}$
   (i) $y = f_1(x)f_2(x)\cdots f_n(x)$ (j) $y = \dfrac{f_1(x)f_2(x)\cdots f_n(x)}{g_1(x)g_2(x)\cdots g_m(x)}$
   (k) $y = \dfrac{[f_1(x)]^{k_1}[f_2(x)]^{k_2}\cdots[f_n(x)]^{k_n}}{[g_1(x)]^{h_1}[g_2(x)]^{h_2}\cdots[g_m(x)]^{h_m}}$

7. For functions of form $y = f(x)^{g(x)}$ we can find the derivative by logarithmic differentiation, as in Problem 5. We can also use the identity $a^b = e^b \ln a$ to write the function as

$$y = e^{g(x) \ln f(x)}$$

and then differentiate by the chain rule. Apply this method to find the derivative:

(a) $y = x^x$ (b) $y = x^{\sin x}$ (c) $y = (\cos x)^{\sin x}$ (d) $y = (e^x)^{e^x}$

3-9 PARAMETRIC EQUATIONS

In Section 1-15, we obtained parametric equations for a straight line. For example, the equations $x = 1 - t$, $y = 2 + t$ are parametric equations for the line through the point $(1, 2)$ with "velocity vector" $-\mathbf{i} + \mathbf{j}$. In general, a pair of equations

$$x = f(t), \qquad y = g(t) \tag{3-90}$$

where f and g are continuous in an interval, are said to be parametric equations (or a parametric representation) of a curve or *path* in the xy-plane. We can think of t as time and (x, y) as the coordinates of a point P moving in the xy-plane. Equations (3-90) give the position of P at each time t and, therefore, describe how P moves. Every conceivable motion is allowed here: P may stand still for an interval and then move along a curve; P may retrace a curve, back and forth; P may go around and around an oval. In each case, the nature of the motion is completely specified by the parametric equations.

EXAMPLE 1. $x = 2 \cos t$, $y = 2 \sin t$. For $t = 0$, $x = 2$, and $y = 0$; for $t = \pi/4$, $x = \sqrt{2}$, $y = \sqrt{2}$, and so on. In general, for every t

$$x^2 + y^2 = 4 \cos^2 t + 4 \sin^2 t = 4$$

Hence the point (x, y) is moving on a circle of radius 2 and center $(0, 0)$ (Figure 3-34). As t increases from 0 to π, x decreases from 2 to -2, while y rises from 0 to 2 and then decreases to 0; thus (x, y) traces the upper half of the circle from $(\tfrac{1}{2}, 0)$ to $(-2, 0)$. Similarly, as t increases from π to 2π, (x, y)

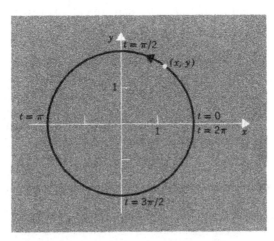

Figure 3-34 Path of Example 1.

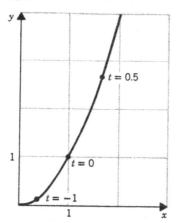

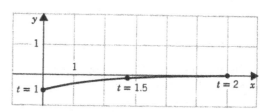

Figure 3-35 Path of Example 2.   **Figure 3-36** Parametric equations.

traces the lower half of the circle from $(-2, 0)$ to $(2, 0)$. [In fact, t is simply the polar coordinate angle of (x, y)]. If we allow t to increase beyond 2π, the circle is traced over and over again; there is a similar statement for t decreasing from 0 through negative values.

EXAMPLE 2. $x = e^t$, $y = e^{2t}$, $-\infty < t < \infty$. Here x and y both increase as t increases, approaching ∞ as $t \to \infty$ and approaching 0 as $t \to -\infty$. We notice that $e^{2t} = (e^t)^2$, so that $y = x^2$ for all t. Hence, we are following a parabola (Figure 3-35). However, since x is always positive, we remain in the first quadrant and, in fact, obtain the half of the parabola in that quadrant, with the vertex $(0, 0)$ as a limiting position for $t \to -\infty$.

In the general case of equations $x = f(t)$, $y = g(t)$, we may be able to solve one of the equations for t. For example, we may have $t = h(x)$ in an interval of x. This is, of course, a question of forming the inverse function as in Section 3-6 above, and $h = f^{-1}$. If, in particular, f is monotone strictly increasing for $a \leq t \leq b$, then we know that we have an inverse function $h = f^{-1}$ in the corresponding interval $c \leq x \leq d$, where $c = f(a)$, $d = f(b)$. If $g(t)$ is also defined for $a \leq t \leq b$, then we can express y in terms of x:

$$y = g[h(x)] = g[f^{-1}(x)] = \varphi(x), \qquad c \leq x \leq d$$

The process is suggested in Figure 3-36. The figure is based on the equations

$$x = t^2 + 3t - 4, \qquad y = \frac{t - 2}{t^2 + 1}, \qquad 1 \leq t \leq 2$$

We now seek the derivative $\varphi'(x)$. We assume f and g are differentiable functions of t. By the chain rule

$$\varphi'(x) = g'(t)h'(x), \qquad \text{where} \qquad t = h(x)$$

Now $h = f^{-1}$, so that by Theorem 3, wherever $f'(t) \neq 0$,

$$h'(x) = \frac{1}{f'(t)}$$

Thus we obtain the expression:

$$y' = \frac{g'(t)}{f'(t)} \tag{3-91}$$

This is easier to remember in the differential notation

$$\frac{dy}{dx} = \frac{dy/dt}{dx/dt} \tag{3-91'}$$

Equations (3-91) and (3-91') give y' in terms of t, but t is expressible in terms of x, so that y' can be expressed in terms of x if so desired; for most applications the expression in terms of t is sufficient.

We summarize our conclusions:

THEOREM 5. *Let functions $x = f(t)$, $y = g(t)$ be given for $a \le t \le b$ and differentiable in this interval, with either $f'(t) > 0$ (so that f is monotone strictly increasing) or $f'(t) < 0$ (so that f is monotone strictly decreasing) throughout the interval. Then y can be expressed in terms of x by the equation $y = g[f^{-1}(x)] = \varphi(x)$ and dy/dx is given by (3-91) or (3-91').*

In many cases, we cannot solve for t in terms of x over the whole interval considered, for the function f may be alternately increasing and decreasing. In that case, the conclusion (3-91) is still usable for each partial curve on which we can solve for t in terms of x. In some cases, it may be better to solve for t in terms of y, that is, to form $g^{-1}(y)$. Then the same reasoning as above leads us to the formula

$$\frac{dx}{dy} = \frac{dx/dt}{dy/dt} = \frac{f'(t)}{g'(t)} \tag{3-92}$$

This formula will be valid in an interval in which $g'(t) > 0$ or $g'(t) < 0$. Figure 3-37 shows a curve ABC for which it would be best to use (3-91') on

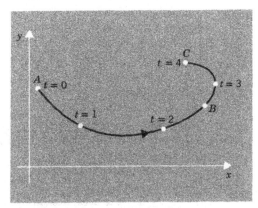

Figure 3-37 Functions defined by parametric equations.

the portion AB, and (3-92) on the portion BC. In this example, t-values are shown suggesting that x increases as t increases [that is, $f'(t) > 0$] for $0 \le t < 3$, but x decreases as t increases from 3 to 4; at the point for which

$t = 3$, the tangent should have infinite slope [in (3-91) we have a zero in the denominator at $t = 3$].

For the equations of Example 1, $dx/dt < 0$ for $0 < t < \pi$, so that in this interval we do obtain a function $y = \varphi(x)$ (the upper semicircle) and

$$\frac{dy}{dx} = \frac{dy/dt}{dx/dt} = \frac{2 \cos t}{-2 \sin t} = -\cot t$$

As $t \to 0+$ or $t \to \pi-$, dy/dx becomes infinite. There is a similar result for the interval $\pi < t < 2\pi$, corresponding to the lower semicircle.

For the equations of Example 2, $dx/dt > 0$ for all t, and therefore for the whole path

$$\frac{dy}{dx} = \frac{dy/dt}{dx/dt} = \frac{2e^{2t}}{e^t} = 2e^t$$

This agrees with the result of differentiating $y = x^2$: $y' = 2x = 2e^t$.

Parametric equations often provide a way of representing a curve when no simple representation $y = f(x)$ is available. An example is given by the equations $x = t + e^t$, $y = t \cos t$. Here elimination of t is difficult, to say the least.

PROBLEMS

1. Graph the curve with given parametric equations, find dy/dx, and check by solving for y in terms of x:
   (a) $x = 2t - 3$, $y = t^2 - 1$ (b) $x = t^2 + 1$, $y = t + t^3$, $t > 0$
   (c) $x = \sqrt{t - 1}$, $y = \sqrt{t + 1}$, $t > 1$ (d) $x = e^t + e^{-t}$, $y = e^t - e^{-t}$, $t > 0$
2. For each of the following curves given in parametric form, graph, find the slope of the tangent line at the point indicated, and graph the tangent line.
   (a) $x = t$, $y = t^3 - 1$ at $(1, 7)$
   (b) $x = t^2 + 2t + 3$, $y = 2t^2 - t + 2$
       at $(2, 5)$
   (c) $x = \cos^3 t$, $y = \sin^3 t$ at
       $(\sqrt{2}/4, \sqrt{2}/4)$
   (d) $x = t + e^t$, $y = t \cos t$ at $(1, 0)$
3. *The cycloid.* For a wheel that moves on a straight track, the path of a point P fixed on the wheel is known as a cycloid. Let the wheel have radius a and roll on the x-axis, starting with P at the origin (Figure 3-38). We can then show that P follows the path

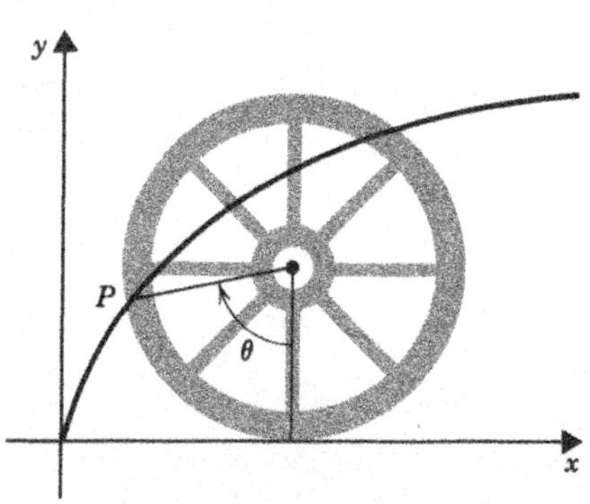

Figure 3-38   Cycloid.

$$(\circ) \quad x = a\theta - a \sin \theta, \qquad y = a - a \cos \theta$$

where θ is the angle through which the wheel has turned. Thus, here, θ is the parameter.
   (a) Graph a typical cycloid.   (b) Find dy/dx in terms of θ.
   (c) Derive equations $(\circ)$.

4. Show that for $t = 2$, the path $x = t^3 - t^2 - 2t + 4$, $y = 2t^3 - t^2 - 28t + 47$ meets the circle $x^2 + y^2 = 25$, and that the path is tangent to the circle at this point (that is, that the path and the circle have a common tangent).

3-10 VECTOR FUNCTIONS

In studying the straight line in Section 1-15, we observed that the parametric equations were equivalent to one vector equation $\overrightarrow{OP} = \overrightarrow{OP_1} + t\mathbf{u}$. There is a similar reasoning for general parametric equations: $x = f(t)$, $y = g(t)$. We can replace these equations by one vector equation:

$$\mathbf{u} = f(t)\mathbf{i} + g(t)\mathbf{j} = \mathbf{F}(t)$$

where $\mathbf{u} = \overrightarrow{OP} = x\mathbf{i} + y\mathbf{j}$ and $\mathbf{F}(t)$ is now a *vector function* of t (Figure 3-39).

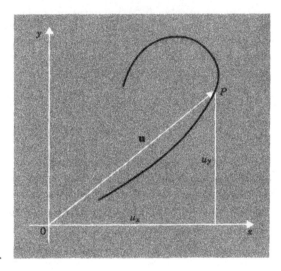

Figure 3-39   Vector function.

EXAMPLE 1. $x = 2 \cos t$, $y = 2 \sin t$ (Example 1 of Section 3-9). The corresponding vector function is

$$\mathbf{u} = 2 \cos t\,\mathbf{i} + 2 \sin t\,\mathbf{j}$$

For $t = 0$, $\mathbf{u} = 2\mathbf{i}$, for $t = \pi/4$, $\mathbf{u} = \sqrt{2}\mathbf{i} + \sqrt{2}\mathbf{j}$, and so on. This vector function can be graphed as in Figure 3-34.

In general, by a vector function $\mathbf{F}$ we mean a function that assigns a vector $\mathbf{u}$ to each number of an interval. These functions arise naturally in describing the motion of a point P in the plane, as in Example 1, with $\mathbf{u}$ as the *position vector* $\overrightarrow{OP}$ at time t. They also arise in many other ways. For example, the force applied to a moving particle is a vector, in general varying with time t (or with position). The velocity vector $\mathbf{v}$ of a moving point can also be considered as a vector function: $\mathbf{v} = \mathbf{G}(t)$. There are many other examples arising from physics and geometry.

Let a vector function $\mathbf{F}$ be defined in an interval $a \leq t \leq b$ except perhaps at t_0. Then we write

$$\lim_{t \to t_0} \mathbf{F}(t) = \mathbf{r}_0$$

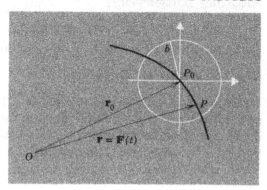

Figure 3-40   Limits for vector
functions.

if, for every number $b > 0$, we can choose an interval $|t - t_0| < q$ in which, except perhaps for $t = t_0$, $|\mathbf{F}(t) - \mathbf{r}_0| < b$. Thus, as in Figure 3-40, the moving point P, where $\overline{OP} = \mathbf{F}(t)$, must lie inside a circle with center P_0 and radius b for $0 < |t - t_0| < q$. In other words, we can make P as close to P_0 as desired by restricting t to be sufficiently close to, but not at t_0. (If $t_0 = a$, then we refer to a limit as $t \to t_0+$; if $t_0 = b$, we refer to a limit as $t \to t_0-$.)

We define $\mathbf{F}(t)$ to be continuous at t_0 if $\mathbf{F}(t_0)$ is defined and

$$\lim_{t \to t_0} \mathbf{F}(t) = \mathbf{F}(t_0)$$

The definitions parallel those for real-valued functions, and the expected properties (limit of sum, etc.) carry over, as long as they are meaningful (Problem 3 below). Ordinary multiplication and division do not carry over to vectors, but there is an inner product, and we can prove (Problem 3 below) that

$$\lim_{t \to t_0} [\mathbf{F}(t) \cdot \mathbf{G}(t)] = \lim_{t \to t_0} \mathbf{F}(t) \cdot \lim_{t \to t_0} \mathbf{G}(t)$$

provided that the limits on the right exist. Also, there is scalar multiplication —that is, formation of $\varphi(t)\mathbf{F}(t)$, where $\varphi(t)$ is a real valued function—and the expected theorem on limits holds true (Problem 3 below). All these theorems can be deduced from the following theorem which reduces the properties of vector functions to properties of real functions.

THEOREM 6. *Let* $\mathbf{F}(t) = f(t)\mathbf{i} + g(t)\mathbf{j}$ *be defined for* $a \leq t \leq b$ *except perhaps for* $t = t_0$. *Then*

$$\lim_{t \to t_0} \mathbf{F}(t) = \mathbf{r}_0 = x_0\mathbf{i} + y_0\mathbf{j} \qquad (3\text{-}100)$$

if, and only if,

$$\lim_{t \to t_0} f(t) = x_0 \qquad and \qquad \lim_{t \to t_0} g(t) = y_0 \qquad (3\text{-}101)$$

The function $\mathbf{F}$ *is continuous at* t_0, *if, and only if,* f *and* g *are continuous at* t_0.

PROOF. Let us suppose that (3-100) holds true. Then, given $b > 0$, we can choose an interval $|t - t_0| < q$, in which $|\mathbf{F}(t) - \mathbf{r}_0| < b$, except perhaps at $t = t_0$. But the condition $|\mathbf{F}(t) - \mathbf{r}_0| < b$ is the same as

$$\sqrt{(x - x_0)^2 + (y - y_0)^2} < b$$

where $\mathbf{F}(t) = x\mathbf{i} + y\mathbf{j}$, and the last inequality implies $|x - x_0| < b$ and $|y - y_0| < b$; that is, if the distance from P to P_0 is less than b, then the coordinates of P differ from those of P_0 by less than b, as suggested in Figure 3-40. Hence, for $|t - t_0| < q$, $t \neq t_0$, we have $|x - x_0| < b$ where $x = f(t)$; that is, $\lim_{t \to t_0} f(t) = x_0$. Similarly $\lim_{t \to t_0} g(t) = y_0$. Thus (3-101) holds true.

Conversely, let (3-101) hold true. Then, given $b > 0$, we can choose q so small that for $0 < |t - t_0| < q$ we have

$$|f(t) - x_0| < \frac{b}{\sqrt{2}}, \qquad |g(t) - y_0| < \frac{b}{\sqrt{2}}$$

Hence

$$[f(t) - x_0]^2 + [g(t) - y_0]^2 < \frac{b^2}{2} + \frac{b^2}{2} = b^2$$

so that $|\mathbf{F}(t) - \mathbf{r}_0| < b$. Hence (3-100) holds true.

If $\mathbf{F}(t)$ is continuous at t_0, then $\mathbf{F}(t)$ has limit $\mathbf{F}(t_0) = f(t_0)\mathbf{i} + g(t_0)\mathbf{j}$, as $t \to t_0$, so that, as was just proved,

$$\lim_{t \to t_0} f(t) = f(t_0), \qquad \lim_{t \to t_0} g(t) = g(t_0)$$

and, therefore, f and g are continuous at t_0. The converse is proved in the same way (Problem 7 below).

3-11 DIFFERENTIATION OF VECTOR FUNCTIONS

Derivatives of vector functions are defined just as for ordinary functions. Thus, if $\mathbf{u} = \mathbf{F}(t)$ is defined for $a \leq t \leq b$ and t_0 is a point of this interval, then the derivative of $\mathbf{F}$ at t_0 is the vector

$$\mathbf{F}'(t_0) = \lim_{\Delta t \to 0} \frac{\Delta \mathbf{u}}{\Delta t} = \lim_{h \to 0} \frac{\mathbf{F}(t_0 + h) - \mathbf{F}(t_0)}{h} \qquad (3\text{-}110)$$

(with limits to the right and left used at $t = a$ and $t = b$). As for ordinary real functions, if $\mathbf{F}'(t)$ exists throughout the given interval, it defines a new function $\mathbf{F}'(t)$. We also denote this function by $d\mathbf{u}/dt$, and so on.

The geometric meaning of the derivative of a vector function is suggested in Figure 3-41, in which $\mathbf{u} = \mathbf{F}(t)$ is represented, as above, by the motion of a point P on a curve, with t as time. The difference $\Delta \mathbf{F} = \Delta \mathbf{u}$ represents the displacement of the moving point as t changes from t_0 to $t_0 + \Delta t$. If we divide $\Delta \mathbf{F}$ by Δt (that is, multiply by $1/\Delta t$), we are evaluating the displacement per unit time. Since this is taken over an interval of length Δt, we can interpret it as an average velocity vector. The limiting value of $\Delta \mathbf{F}/\Delta t$ as $\Delta t \to 0$ is the *instantaneous velocity vector*. For Δt very small, our point P is moving approximately on a straight line at a definite speed. Our instantaneous velocity vector is simply the velocity vector associated with that motion

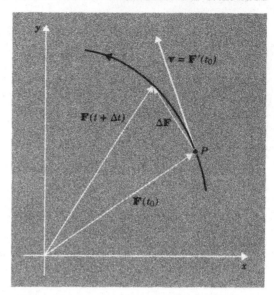

Figure 3-41   Derivative of vector function.

on a straight line at constant speed, as in Example 7, Section 3-1 above. We should think of a stone whirled on a string and flying off "on a tangent." Since $\Delta \mathbf{F}$ corresponds to a chord of the curve and $\Delta \mathbf{F}/\Delta t$ has the same direction as $\Delta \mathbf{F}$ (or the opposite direction for $\Delta t < 0$), the limiting direction is that of the tangent line to the path. Thus *the velocity vector is tangent to the path* as in Figure 3-41. [The tangent interpretation breaks down when $\mathbf{F}'(t_0) = \mathbf{0}$.]

To evaluate the derivative of a given vector function, we can introduce fixed rectangular coordinate axes (Figure 3-39), so that $\mathbf{u}$ is represented in terms of its orthogonal components $u_x = f(t)$, $u_y = g(t)$. We then have the simple rule: If $\mathbf{F}'(t)$ exists, then $f'(t)$ and $g'(t)$ exist and conversely. Furthermore

$$\frac{d\mathbf{u}}{dt} = \mathbf{F}'(t) = f'(t)\mathbf{i} + g'(t)\mathbf{j} \tag{3-111}$$

that is, *to differentiate a vector function, one differentiates its components separately*. To prove the rule, we observe that, if we write $\mathbf{u} = x\mathbf{i} + y\mathbf{j} = f(t)\mathbf{i} + g(t)\mathbf{j}$, then

$$\begin{aligned}
\Delta \mathbf{u} &= [f(t + \Delta t)\mathbf{i} + g(t + \Delta t)\mathbf{j}] - [f(t)\mathbf{i} + g(t)\mathbf{j}] \\
&= [f(t + \Delta t) - f(t)]\mathbf{i} + [g(t + \Delta t) - g(t)]\mathbf{j} \\
&= \Delta x\mathbf{i} + \Delta y\mathbf{j}
\end{aligned}$$

so that

$$\frac{\Delta \mathbf{u}}{\Delta t} = \frac{\Delta x}{\Delta t}\mathbf{i} + \frac{\Delta y}{\Delta t}\mathbf{j}$$

By Theorem 6, it now follows that $\Delta \mathbf{u}/\Delta t$ has a limit if, and only if, $\Delta x/\Delta t$ and $\Delta y/\Delta t$ have limits and, when these limits exist,

$$\lim_{\Delta t \to 0} \frac{\Delta \mathbf{u}}{\Delta t} = \lim_{\Delta t \to 0} \frac{\Delta x}{\Delta t}\mathbf{i} + \lim_{\Delta t \to 0} \frac{\Delta y}{\Delta t}\mathbf{j}$$

Thus (3-111) holds true.

EXAMPLE 1. Let $\mathbf{u} = \mathbf{F}(t) = t^2\mathbf{i} + 3t\mathbf{j}$. Then

$$\frac{d\mathbf{u}}{dt} = 2t\mathbf{i} + 3\mathbf{j}$$

EXAMPLE 2. Let $\mathbf{u} = t\mathbf{i} + t^2\mathbf{j}$. Then

$$\frac{d\mathbf{u}}{dt} = \mathbf{i} + 2t\mathbf{j}$$

If we represent $\mathbf{u}$ as $\overrightarrow{OP} = x\mathbf{i} + y\mathbf{j}$, then our path has parametric equations

$$x = t, \qquad y = t^2$$

so that (by eliminating t) the path is the parabola $y = x^2$ (Figure 3-42). The velocity vector $d\mathbf{u}/dt$ is shown at the point for which $t = 1$.

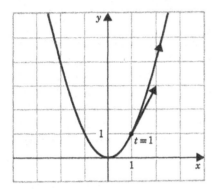

Figure 3-42 Tangent to parabola.

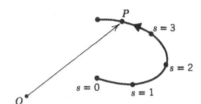

Figure 3-43 Arc length.

In general, if $\mathbf{u} = \mathbf{F}(t)$ and we represent $\mathbf{u}$ as $x\mathbf{i} + y\mathbf{j}$, then the derivative can be interpreted as the velocity vector $\mathbf{v}$ and

$$\mathbf{v} = \frac{d\mathbf{u}}{dt} = \frac{dx}{dt}\mathbf{i} + \frac{dy}{dt}\mathbf{j} \tag{3-112}$$

Thus the velocity vector has components dx/dt, dy/dt, with respect to the chosen axes. Since the velocity vector lies along the tangent line, that line has slope [see Equation (1-151) in Section 1-15]

$$m = \frac{b}{a} = \frac{dy/dt}{dx/dt} \tag{3-113}$$

(provided that the denominator is not zero). Equation (3-113) is in agreement with Equation (3-91′).

The magnitude of the velocity vector can be interpreted as *speed* (or *instantaneous speed*). If we imagine distance s measured along our curve (as on a tape measure), and increasing in the direction of motion, as in Figure 3-43, then s also becomes a function of t. Over a small interval the motion is approximately on a straight line at constant speed. Hence, as in Section 3-1, we expect that

$$\frac{ds}{dt} = |\mathbf{v}| = \left|\frac{d\mathbf{u}}{dt}\right|$$

or by (3-112) that

$$\frac{ds}{dt} = \sqrt{\left(\frac{dx}{dt}\right)^2 + \left(\frac{dy}{dt}\right)^2} \tag{3-114}$$

If the vector function $\mathbf{u} = \mathbf{F}(t) = x\mathbf{i} + y\mathbf{j}$ is given, Equation (3-114) gives us only the derivative of s with respect to t; finding s itself as a function of t requires "integration." In fact, a complete justification of (3-114) requires the integral calculus (see Chapter 4).

Remark. The tangent line to a curve at a point P has been defined as the limiting position of a secant line through P and a second point Q on the curve, as Q approaches P. When the curve is represented as the graph of a differentiable function $y = f(x)$, and P is (x_0, y_0), the tangent is found to be the line through P of slope $m = f'(x_0)$. When the curve is represented by parametric equations or, equivalently, by a vector equation $\mathbf{u} = \mathbf{F}(t)$, we find the tangent line to be the line through P determined by the velocity vector $\mathbf{F}'(t_0)$ (provided that this vector is not $\mathbf{0}$); we also observe that, when the same path is represented in the form $y = f(x)$, the velocity vector determines a line of slope $m = f'(x_0)$ (see Problem 13). Thus the tangent line, although obtained in different ways, is always the same line. We can say that the tangent line has a geometric meaning, independent of the manner in which the curve is presented. It is also independent of the choice of coordinates (see Problem 16 below).

3-12 DIFFERENTIAL CALCULUS RULES FOR VECTOR FUNCTIONS

We state the rules formally. The precise conditions are analogous to those for derivatives of ordinary functions:

$$\frac{d}{dt}[\mathbf{F}(t) + \mathbf{G}(t)] = \mathbf{F}'(t) + \mathbf{G}'(t) \tag{3-120}$$

$$\frac{d}{dt}[\mathbf{F}_0] = \mathbf{0}, \text{ if } \mathbf{F}_0 \text{ is a constant function} \tag{3-121}$$

$$\frac{d}{dt}[k\mathbf{F}(t)] = k\mathbf{F}'(t), \text{ if } k \text{ is a constant scalar} \tag{3-122}$$

$$\frac{d}{dt}[k(t)\mathbf{F}(t)] = k(t)\mathbf{F}'(t) + k'(t)\mathbf{F}(t) \tag{3-123}$$

$$\frac{d}{dt}[\mathbf{F}(t) \cdot \mathbf{G}(t)] = \mathbf{F}(t) \cdot \mathbf{G}'(t) + \mathbf{F}'(t) \cdot \mathbf{G}(t) \tag{3-124}$$

$$\frac{d}{dt}[\mathbf{F}(w)] = \frac{dw}{dt}\mathbf{F}'(w), \quad \text{if } w = g(t) \tag{3-125}$$

These rules can be proved in direct analogy with the proofs for ordinary functions. They can be deduced, in fact, from the rules for ordinary functions by Equation (3-111). We prove the rule (3-120) both ways as an illustration. Let $\mathbf{F}$ and $\mathbf{G}$ have derivatives at a particular t. Then at this t

$$(\mathbf{F} + \mathbf{G})' = \lim_{h \to 0} \frac{\mathbf{F}(t + h) + \mathbf{G}(t + h) - \mathbf{F}(t) - \mathbf{G}(t)}{h}$$

$$= \lim_{h \to 0} \left[\frac{\mathbf{F}(t + h) - \mathbf{F}(t)}{h} + \frac{\mathbf{G}(t + h) - \mathbf{G}(t)}{h} \right]$$

$$= \lim_{h \to 0} \frac{\mathbf{F}(t + h) - \mathbf{F}(t)}{h} + \lim_{h \to 0} \frac{\mathbf{G}(t + h) - \mathbf{G}(t)}{h}$$

$$= \mathbf{F}'(t) + \mathbf{G}'(t)$$

Thus (3-120) is proved. Alternatively, we can write $\mathbf{F}(t) = f(t)\mathbf{i} + g(t)\mathbf{j}$, $\mathbf{G}(t) = p(t)\mathbf{i} + q(t)\mathbf{j}$ and by (3-111) at t

$$\begin{aligned}(\mathbf{F} + \mathbf{G})' &= [(f + p)\mathbf{i} + (g + q)\mathbf{j}]' \\ &= [(f'(t) + p'(t)]\mathbf{i} + [g'(t) + q'(t)]\mathbf{j} \\ &= [f'(t)\mathbf{i} + g'(t)\mathbf{j}] + [p'(t)\mathbf{i} + q'(t)\mathbf{j}] \\ &= \mathbf{F}'(t) + \mathbf{G}'(t)\end{aligned}$$

The proofs of (3-121) . . . (3-125) are left as exercises (Problem 15 below).

An interesting special case of rule (3-124) is the following:

$$\frac{d}{dt} [\mathbf{F}(t) \cdot \mathbf{F}(t)] = 2\,\mathbf{F}(t) \cdot \mathbf{F}'(t) \tag{3-124'}$$

If now $\mathbf{F}(t)$ is such that $|\mathbf{F}(t)| \equiv$ const. over an interval, then $|\mathbf{F}(t)|^2 = \mathbf{F}(t) \cdot \mathbf{F}(t)$ is also constant, so that its derivative is 0. Hence by (3-124')

$$\mathbf{F}(t) \cdot \mathbf{F}'(t) = 0$$

that is, *if* $|\mathbf{F}(t)| \equiv$ const, *then* $\mathbf{F}(t)$ *is orthogonal to* $\mathbf{F}'(t)$.

The result has a simple geometric meaning: if $\overrightarrow{OP} = \mathbf{F}(t)$, where O is the origin, then P moves on a circle, since $|\overrightarrow{OP}|$ is constant. Hence the velocity vector of P, $\mathbf{F}'(t)$, must be tangent to the circle and perpendicular to $\overrightarrow{OP}$; that is, $\mathbf{F}(t) \cdot \mathbf{F}'(t) = 0$ (see Figure 3-44).

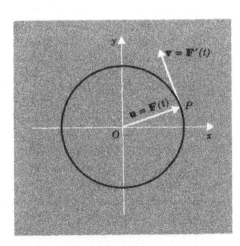

Figure 3-44   Motion on a circular path.

PROBLEMS

1. Graph the vector functions:
   (a) $\mathbf{r} = (1 - 2t)\mathbf{i} + (3 + t)\mathbf{j}, \quad -\infty < t < \infty$
   (b) $\mathbf{r} = t^2\mathbf{i} + 2t\mathbf{j}, \quad -\infty < t < \infty$
   (c) $\mathbf{r} = (1 + \cos \tau)\mathbf{i} + (2 + \sin \tau)\mathbf{j}, \quad 0 \leq \tau \leq 2\pi$
   (d) $\mathbf{r} = (\theta + \cos \theta)\mathbf{i} + \sin \theta\,\mathbf{j}, \quad 0 \leq \theta \leq 2\pi$
   (e) $\mathbf{r} = 2 \cos \theta\,\mathbf{i} + 3 \sin \theta\,\mathbf{j}, \quad 0 \leq \theta \leq 2\pi$

Figure 3-45

2. Figure 3-45 shows two vectors $\mathbf{u}$ and $\mathbf{v}$. Trace these on a sheet of paper and represent graphically the vector function given:
   (a) $\mathbf{r} = t\mathbf{u} + 2t\mathbf{v}, 0 \leq t \leq 1$ (b) $\mathbf{r} = t\mathbf{u} + (1 - t)\mathbf{v}, 0 \leq t \leq 1$
   (c) $\mathbf{r} = t^2\mathbf{u} + t\mathbf{v}, -1 \leq t \leq 1$ (d) $\mathbf{r} = \cos \theta\,\mathbf{u} + \sin \theta\,\mathbf{v}, 0 \leq \theta \leq 2\pi$

3. Prove, with the aid of Theorem 6, that if $\mathbf{F}_1(t) = f_1(t)\mathbf{i} + g_1(t)\mathbf{j}, \mathbf{F}_2(t) = f_2(t)\mathbf{i} + g_2(t)\mathbf{j}$ and $\varphi(t)$ are defined for $a \leq t \leq b$, except perhaps for t_0, and each has a limit as $t \to t_0$, then
   (a) $\lim\limits_{t \to t_0} [\mathbf{F}_1(t) + \mathbf{F}_2(t)] = \lim\limits_{t \to t_0} \mathbf{F}_1(t) + \lim\limits_{t \to t_0} \mathbf{F}_2(t)$
   (b) $\lim\limits_{t \to t_0} [\varphi(t)\mathbf{F}_2(t)] = \left[\lim\limits_{t \to t_0} \varphi(t)\right]\left[\lim\limits_{t \to t_0} \mathbf{F}_2(t)\right]$
   (c) $\lim\limits_{t \to t_0} \mathbf{F}_1(t) \cdot \mathbf{F}_2(t) = \left[\lim\limits_{t \to t_0} \mathbf{F}_1(t)\right] \cdot \left[\lim\limits_{t \to t_0} \mathbf{F}_2(t)\right]$
   (d) $\lim\limits_{u \to u_0} \mathbf{F}_1[\psi(u)] = \mathbf{F}_1\left[\lim\limits_{u \to u_0} \psi(u)\right]$, provided that $\psi(u)$ has limit t_0 as $u \to u_0$, $\mathbf{F}_1(t)$ is continuous at t_0, and $\mathbf{F}_1[\psi(u)]$ is defined in an interval containing u_0.

4. Prove the result of Problem 3(a) directly by showing that, if $\lim\limits_{t \to t_0} \mathbf{F}_1(t) = \mathbf{r}_1$ and $\lim\limits_{t \to t_0} \mathbf{F}_2(t) = \mathbf{r}_2$, then

$$|\mathbf{F}_1(t) + \mathbf{F}_2(t) - (\mathbf{r}_1 + \mathbf{r}_2)|$$

   can be made less than a prescribed positive number b by restricting t to an appropriate interval $|t - t_0| < q$, except for $t = t_0$.

5. From the results of Problem 3 show that, if $\mathbf{F}_1(t)$, $\mathbf{F}_2(t)$ and $\varphi(t)$ are continuous at t_0, then so is each of the following:
   (a) $\mathbf{F}_1(t) + \mathbf{F}_2(t)$ (b) $\varphi(t)\mathbf{F}_1(t)$ (c) $\mathbf{F}_1(t) \cdot \mathbf{F}_2(t)$
   (d) Show also that $\mathbf{F}[\psi(u)]$ is continuous at u_o if ψ is continuous at u_0 and $\psi(u_0) = t_0$.

6. Formulate and prove the analogue of Problem 3(a) for limits as $t \to \infty$.

7. Complete the proof of Theorem 6 by showing that, if f and g are continuous at t_0, then so is $\mathbf{F}(t)$.

8. Evaluate the limits:

   (a) $\lim\limits_{t \to 1} [(t^2 - 2)\mathbf{i} + 2t\mathbf{j}]$ (b) $\lim\limits_{t \to 0} \left[\dfrac{\sin^3 t}{t^3}\mathbf{i} + \cos t\,\mathbf{j}\right]$

   (c) $\lim\limits_{t \to 1} \left[\dfrac{1 - t^2}{1 - t}\mathbf{i} + \dfrac{1 + t^3}{1 + t}\mathbf{j}\right]$ (d) $\lim\limits_{t \to \infty} \left[\dfrac{t^2}{t^2 + 1}\mathbf{i} + \dfrac{t}{t^2 + 1}\mathbf{j}\right]$

9. Differentiate the following vector functions:

   (a) $\mathbf{u} = t\mathbf{i} + (3 - t)\mathbf{j}$ (b) $\mathbf{u} = \dfrac{1}{t - 1}\mathbf{i} + \dfrac{3}{t^2 - 2}\mathbf{j}$

   (c) $\mathbf{u} = (t - 2)^5\mathbf{i} + \dfrac{3t}{1 - t^3}\mathbf{j}$ (d) $\mathbf{u} = e^{2t}(\cos 3t\,\mathbf{i} + \sin 3t\,\mathbf{j})$

   Where is each function discontinuous?

10. Find the derivative of the vector function $\mathbf{u} = (t^2 - 3)\mathbf{i} + 5t\mathbf{j}$. Graph the function and show the derivative at $t = 1$ and at $t = 3$. Write parametric equations for the tangent line at each point.

11. As in Section 3-9, the equations $x = \cos t$, $y = \sin t$ (t in radians) are parametric equations for a circle, and $\mathbf{u} = \cos t\,\mathbf{i} + \sin t\,\mathbf{j}$ describes a motion at constant speed (1 unit of distance per unit time). Hence the velocity vector $\mathbf{v} = d\mathbf{u}/dt$ is a unit vector tangent to the path, as in Figure 3-44. Show that, at the point (x, y), this vector is $-y\mathbf{i} + x\mathbf{j} = \mathbf{u}^\perp$ and conclude that $\mathbf{v} = -\sin t\,\mathbf{i} + \cos t\,\mathbf{j}$. Thus

$$\mathbf{v} = \frac{dx}{dt}\mathbf{i} + \frac{dy}{dt}\mathbf{j} = -\sin t\,\mathbf{i} + \cos t\,\mathbf{j}$$

and accordingly

$$\frac{d}{dt}\cos t = -\sin t \qquad \frac{d}{dt}\sin t = \cos t$$

Remark. This "proof" of the rules for differentiating $\sin t$ and $\cos t$ assumes the relation between velocity and arc length has been established.

12. Obtain the velocity vector and the speed for each of the following motions:
    (a) $\mathbf{u} = a \cos t\,\mathbf{i} + b \sin t\,\mathbf{j}$ (ellipse)
    (b) $\mathbf{u} = a \cos 2t\,\mathbf{i} + a \sin 2t\,\mathbf{j}$ [circle, use (3-125)]

13. Vector methods can be applied to a curve given as $y = f(x)$. A parametrization is $x = t$, $y = f(t)$, so that one has the vector equation $\overrightarrow{OP} = t\mathbf{i} + f(t)\mathbf{j}$.
    (a) Show that the velocity vector is $\mathbf{i} + f'(t)\mathbf{j}$ and that the slope of the tangent line at (x, y) is $f'(x)$.
    (b) Show that the velocity vector has its shortest length where $f'(x) = 0$ (if such points occur).
    (c) Apply the results of (a) and (b) to the curve $y = x - x^3$, and indicate the velocity vector at the points $(1, 0)$ and $(2, -6)$.

14. If (x, y) moves on the circle $x^2 + y^2 = 1$, we can write $x = \cos\theta$, $y = \sin\theta$, where $\theta = h(t)$. Thus $\mathbf{u} = \overrightarrow{OP} = \cos\theta\,\mathbf{i} + \sin\theta\,\mathbf{j}$, $\theta = h(t)$. Use (3-125) to show that

$$\frac{d\mathbf{u}}{dt} = \frac{d}{dt}\overrightarrow{OP} = \frac{d\theta}{dt}\left[\cos\left(\theta + \frac{\pi}{2}\right)\mathbf{i} + \sin\left(\theta + \frac{\pi}{2}\right)\mathbf{j}\right] = \frac{d\theta}{dt}\mathbf{u}^\perp$$

Interpret this result geometrically.

15. (a) Prove (3-121).   (b) Prove (3-122).   (c) Prove (3-123).
    (d) Prove (3-124).   (e) Prove (3-125) with the aid of (3-111).

16. Let $\mathbf{u} = OP = \mathbf{F}(t)$, $a \le t \le b$, be a differentiable function that represents the path of the moving point P.
    (a) Show that the velocity vector of P is unaffected by a change of origin to O'; that is,

$$\frac{d}{dt}\overrightarrow{OP} = \frac{d}{dt}\overrightarrow{O'P}$$

    (b) Let a change of parameter be made from t to τ, where $\tau = \varphi(t)$ is a differentiable function for $a \le t \le b$, with $\varphi'(t) \ne 0$. Let φ have a differentiable inverse $t = \psi(\tau)$. Show that the velocity vector in terms of τ is a nonzero scalar times the velocity vector in terms of t. [*Hint.* Use (3-125).] Compare this result with that of Problem 14.

3-13 EQUATION OF TANGENT AND NORMAL LINES, ANGLE BETWEEN TWO CURVES

For a curve given in the form $y = f(x)$, the tangent line at a point (x_1, y_1) on the curve has slope $m = f'(x_1)$ (assuming that this derivative exists). Hence, by the point-slope formula (Section 0-7), *the tangent line has the equation*

$$y - y_1 = f'(x_1)(x - x_1) \qquad (3\text{-}130)$$

(see Figure 3-46). By the *normal line* to the curve at (x_1, y_1), we mean the line perpendicular to the tangent at (x_1, y_1). Since the tangent has slope $m = f'(x_1)$, *the normal line has slope* $-1/m = -1/f'(x_1)$ (see Section 0-7) *and the equation*

$$y - y_1 = -\frac{1}{f'(x_1)}(x - x_1) \qquad (3\text{-}131)$$

Let a second curve, with equation $y = g(x)$, also pass through $P_1:(x_1, y_1)$. It is natural to try to define the angle between the curves at P_1 as the angle QP_1R, where Q is on one curve, and R on the other, as in Figure 3-47. However, the size of this angle, in general, varies as we change Q and R. The variation is reduced if we stay close to P_1. Hence we are led to take a limit. As Q approaches P_1, the limiting direction of the line P_1Q is that of the tangent line to the curve $y = g(x)$ at P_1; the limiting direction of the line P_1R, as $R \rightarrow P$, is that of the tangent line to the curve $y = f(x)$ at P_1. We therefore define *the angle between the curves at P_1 to be the angle between the two tangent lines at the point.* A numerical value (in radians) for this angle can be given, but we have many choices: if α is one value, then $\alpha + 2\pi$, $\pi - \alpha$, etc., are also possible values (see Figure 3-47). However, there is but one α for which $0 \leq \alpha \leq \pi/2$.

If we write $m_1 = f'(x_1)$, $m_2 = g'(x_1)$, then one choice of the angle is given by φ, where

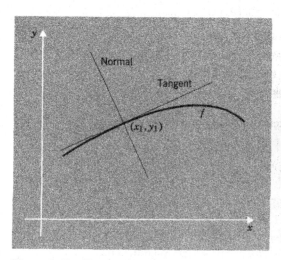

Figure 3-46   Tangent line.

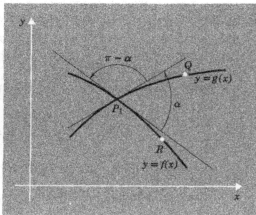

Figure 3-47   Angle between curves.

$$\tan \varphi = \frac{m_2 - m_1}{1 + m_1 m_2}, \qquad 0 \leq \varphi < \pi \qquad (3\text{-}132)$$

Equation (3-132) is a standard formula of analytic geometry, a consequence of the trigonometric formula for the tangent of the difference of two angles (identity No. 36, Appendix V). The two *curves* are said to be *orthogonal* at (x_1, y_1) if their tangent lines are orthogonal (perpendicular) to each other; in this case, $\varphi = \pi/2$ and in (3-132) $\tan \varphi$ is infinite, corresponding to a zero in the denominator: $m_1 m_2 = -1$.

   Vector Formulation. When the curves are given in parametric form, it is convenient to employ vectors. Let a curve be given by the equations $x = f(t)$, $y = g(t)$, or equivalently, by

$$\mathbf{u} = x\mathbf{i} + y\mathbf{j} = \mathbf{F}(t) \qquad (3\text{-}133)$$

Then, if $\overrightarrow{OP_1} = \mathbf{F}(t_1)$, the vector $\mathbf{v}_1 = \mathbf{F}'(t_1)$ is along the tangent line to the curve at P_1 (Figure 3-48); we assume $\mathbf{v}_1 \neq 0$. We can now give parametric equations for the tangent line:

$$\overrightarrow{OQ} = \overrightarrow{OP_1} + (t - t_1)\mathbf{F}'(t_1) \qquad (3\text{-}134)$$

(see Section 1-14). Equation (3-134) describes the motion of the point Q along the tangent line with constant velocity vector $\mathbf{v}_1 = \mathbf{F}'(t_1)$, Q passing through

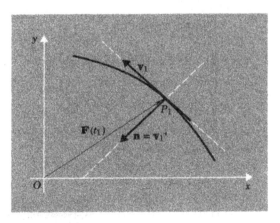

Figure 3-48   Tangent and normal vectors.

P_1 for $t = t_1$. If $\mathbf{v}_1 = a\mathbf{i} + b\mathbf{j}$, then $\mathbf{n} = -b\mathbf{i} + a\mathbf{j} = \mathbf{v}_1^\dashv$ represents a normal vector of the tangent line. Accordingly, the *equation of the tangent line* is

$$b(x - x_1) - a(y - y_1) = 0 \qquad (3\text{-}135)$$

(see Section 1-16), and that of the *normal line* is

$$a(x - x_1) + b(y - y_1) = 0 \qquad (3\text{-}136)$$

Since the slope of the tangent line is $m = b/a$, (3-135) is equivalent to (3-130) (provided that $a \neq 0$).

   If a second curve with equation $\overrightarrow{OP} = \mathbf{G}(t)$ passes through the point P_1 at time $t = t_1^*$ (which need not equal t_1), then the angle between the two curves at P_1 can be defined as the angle φ between the two tangent vectors

$v_1 = F'(t_1)$ and $v_1^* = G'(t_1^*)$; we assume both vectors to be different from 0. By Section 1-11, φ can then be computed by means of the inner product:

$$\cos \varphi = \frac{v_1 \cdot v_1^*}{|v_1||v_1^*|}, \qquad 0 \le \varphi \le \pi \qquad (3\text{-}137)$$

When the curves are orthogonal, $\varphi = \frac{1}{2}\pi$ and $v_1 \cdot v_1^* = 0$.

EXAMPLE. Show that the curves $\overrightarrow{OP} = ti + t^2j$ and $\overrightarrow{OP} = (2 + t)i + (10 - t^2)j$ meet at the point $(3, 9)$ and find the angle between the curves at this point.

The first curve passes through $(3, 9)$ when $t = 3$ and has derivative $i + 6j$ at the point. The second passes through $(3, 9)$ when $t = 1$ and has derivative $i - 2j$ at the point. The angle φ between the two tangent vectors $u = i + 6j$ and $v = i - 2j$ is determined by Equation (3-137):

$$\cos \varphi = \frac{u \cdot v}{|u||v|} = \frac{1 - 12}{\sqrt{37}\sqrt{5}} = \frac{-11}{\sqrt{185}}$$

After some arithmetic and the use of tables, we find that $\varphi = 2.51$ radians.

Remark. It can happen that two curves cross at one point Q for several values of t on one or both curves. The angle at crossing will then depend on which t-value is selected on each curve.

PROBLEMS

1. Write the equation of the tangent and normal lines at the point given and graph:

   (a) $y = 3x^2 - 5x + 2$ at $(2, 4)$ (b) $y = \frac{1}{x^2}$ at $(-1, 1)$

   (c) $y = \sqrt{1 - x^2}$ at $\left(\frac{3}{5}, \frac{4}{5}\right)$ (d) $y = \ln x$ at $(1, 0)$

2. Verify that the curves given meet at the point indicated and find the angle between the curves at the point:

   (a) $x^2 + y^2 = 25$ and $5x - 2y = 7$ at $(3, 4)$

   (b) $x^3 + xy^2 + y^3 = 3$ and $2x^3 + 3x^2y - y^3 = 4$ at $(1, 1)$

3. For the parabola $x^2 = 4py$ with focus at Q: $(0, p)$ and directrix: $y = -p$ (Section 0-15), show that the tangent line at a point P_1 on the parabola makes equal angles with the line P_1Q and with the line through P_1 parallel to the y-axis (Figure 3-49). (Thus, by geometric optics, a light ray from Q is reflected at P_1 in a direction parallel to the axis of the parabola.)

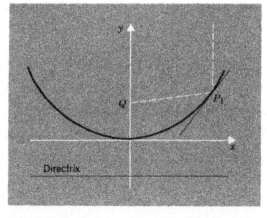

Figure 3-49   Reflection property of the parabola.

4. Find the equation of the tangent line at a general point (x_1, y_1) for each of the following curves:

(a) Circle: $x^2 + y^2 = a^2$

(b) Ellipse: $\dfrac{x^2}{a^2} + \dfrac{y^2}{b^2} = 1$

(c) Hyperbola: $\dfrac{x^2}{a^2} - \dfrac{y^2}{b^2} = 1$

(d) Circle: $Ax^2 + Ay^2 + Dx + Ey + F = 0$

(e) Conic section: $Ax^2 + Bxy + Cy^2 + Dx + Ey + F = 0$

‡5. *Geometry of a plane curve.*
Let a curve be given by $y = f(x)$, $a \le x \le b$, where f is differentiable. Let $P: (x, y)$ be a fixed point on the curve, at which $f'(x) = y' \ne 0$. Let the tangent line at P meet the x-axis at T, the y-axis at T_1. Let the normal line at P meet the x-axis at U, the y-axis at U_1. Let R be the foot of the Perpendicular from O to the tangent. Let M be the point $(x, 0)$ (see Figure 3-50). Justify these rules:

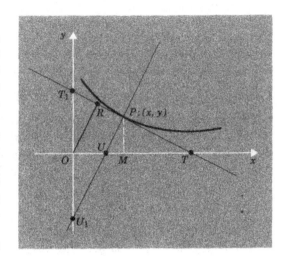

Figure 3-50   Geometry of plane curve.

(a) The distance MT (the "subtangent") equals $|y/y'|$.

(b) The distance PT equals $|y\sqrt{1 + y'^2}/y'|$.

(c) The x-intercept of the tangent is $x - (y/y')$.

(d) The y-intercept of the tangent is $y - xy'$.

(e) The distance OR from the origin O to the tangent is $|y - xy'|/\sqrt{1 + y'^2}$.

(f) The component of the radius vector $\overrightarrow{OP}$ on the tangent (the distance PR) equals $|x + yy'|/\sqrt{1 + y'^2}$.

(g) R has coordinates

$$\left(-\frac{y'(y - xy')}{1 + y'^2}, \quad \frac{y - xy'}{1 + y'^2} \right)$$

(h) The distance MU (the "subnormal") equals $|yy'|$.

(i) The distance PU is $|y|\sqrt{1 + y'^2}$.

(j) The x-intercept of the normal is $x + yy'$.

(k) The y-intercept of the normal is $y + (x/y')$.

6. Show that the curves $\overrightarrow{OP} = (2t - 1)\mathbf{i} + t^2\mathbf{j}$ and $\overrightarrow{OP} = (t^2 - 6)\mathbf{i} + (t + 1)\mathbf{j}$ cross at the point $(3, 4)$ and find the angle of intersection.

7. Show that the curve $x = t^2, y = t(1 - t^2)$ crosses itself at the point $(1, 0)$ and find the angle between the two tangent lines at the point.

8. Show, with the aid of vectors, that, if two curves $y = f_1(x)$ and $y = f_2(x)$ intersect at a point (x_0, y_0), then the tangent lines at the point form angle φ, where

$$\cos \varphi = \frac{1 + f_1'(x_0)f_2'(x_0)}{\sqrt{1 + f_1'(x_0)^2}\sqrt{1 + f_2'(x_0)^2}} \qquad \text{(see next page)}$$

(see Problem 13, following Section 3-12). Show that the curves are orthogonal at the point if, and only if, $f_1'(x_0)f_2'(x_0) = -1$.

3-14 SECOND DERIVATIVE, DERIVATIVES OF HIGHER ORDER

We have observed that, if a function $y = f(x)$ is differentiable over an interval, then its derivative $y' = f'(x)$ is a new function over that interval. Therefore, we can consider the derivative of f'; if this derivative exists at a particular x, we call its value the second derivative of f, at that x, and denote the value by $f''(x)$ or y''. If the second derivative exists throughout an interval, we get a new function f'' defined over that interval.

EXAMPLE 1. Let $y = f(x) = x^3$. Then $y' = f'(x) = 3x^2$, $y'' = f''(x) = 6x$.

We can repeat the process—if the second derivative of f exists throughout an interval, we can seek its derivative, the third derivative of f, which we denote by f''' or y'''. In the same way, we are led to fourth, fifth, . . . derivatives. The typical notations are

$$y', y'', y''', y^{iv}, y^v, \ldots, y^{(n)}, \ldots, \qquad f', f'', f''', f^{iv}, f^v, \ldots, f^{(n)}, \ldots,$$
$$\frac{dy}{dx}, \frac{d^2y}{dx^2}, \ldots, \qquad \frac{d^ny}{dx^n}, \ldots, \quad D_x y, D_x{}^2 y, D_x{}^3 y, \ldots, D_x{}^n y, \ldots$$
$$Dy, D^2 y, D^3 y, \ldots, \qquad D^n y, \ldots$$

We also refer to the first derivative as *derivative of first order*, to the second derivative as *derivative of second order*, and so on. We occasionally consider f itself to be $f^{(0)}$, the *derivative of order zero*.

Notice that the kth derivative $f^{(k)}(x)$ is defined at a point only if the $(k-1)$-st derivative is defined in an interval containing the point, so that also $f(x)$, $f'(x)$, $\ldots$, $f^{(k-2)}(x)$ are all defined in such an interval. Also, by Theorem 1 (Section 3-2), if $f^{(k)}(x)$ exists at x_1, then $f^{(k-1)}(x)$ must be continuous at x_1 and hence, by the preceding remark, $f(x)$, $f'(x)$, $\ldots$, $f^{(k-2)}(x)$ must be continuous in an interval containing x_1.

EXAMPLE 2. Let $y = x^n$, where n is a positive integer. Then $D_x y = nx^{n-1}$, $D_x{}^2 y = n(n-1)x^{n-2}$, $D_x{}^3 y = n(n-1)(n-2)x^{n-3}$, $\ldots$, $D_x{}^n y = n!$, $D_x{}^{n+1} y = 0$, $D_x{}^{n+2} y = 0$, $\ldots$; or equivalently, $D(x^n) = nx^{n-1}$, $D^2(x^n) = n(n-1)x^{n-2}, \ldots$

The rule for derivative of a product has an interesting generalization to derivatives of higher order:

THEOREM 7. (*Leibnitz's rule*). *Let $y = f(x)$ and $y = g(x)$ be two functions having derivatives through order n in the interval $a \leq x \leq b$. Then the product fg also has derivatives through order n in the interval and*

$$(fg)' = fg' + f'g,$$
$$(fg)'' = fg'' + 2f'g' + f''g$$
$$(fg)''' = fg''' + 3f'g'' + 3f''g' + f'''g$$

$$(fg)^{iv} = fg^{iv} + 4f'g''' + 6f''g'' + 4f'''g' + f^{iv}g,$$

$$\vdots$$

$$(fg)^{(n)} = fg^{(n)} + nf'g^{(n-1)} + \frac{n(n-1)}{2}f''g^{(n-2)} + \cdots + f^{(n)}g \qquad (3\text{-}140)$$

$$= \sum_{k=0}^{n} \binom{n}{k} f^{(k)}g^{(n-k)}.$$

The coefficients in the general rule (3-140) are the *binomial coefficients* (Section 0-21). The proof of the rule is left as an exercise (Problem 6 below, following Section 3-16).

For quotients f/h, we can deduce similar rules by writing $g = 1/h$, so that

$$\left(\frac{f}{h}\right)' = (fg)' = fg' + f'g, \qquad \left(\frac{f}{h}\right)'' = (fg)'' = fg'' + 2f'g' + f''g$$

and so on. Now $g(x) = [h(x)]^{-1}$, so that by the Chain Rule

$$g' = -h^{-2}h'$$
$$g'' = 2h^{-3}h'^2 - h^{-2}h''$$
$$g''' = -6h^{-4}h'^3 + 6h^{-3}h'h'' - h^{-2}h''' \qquad (3\text{-}141)$$
$$g^{(iv)} = 24h^{-5}h'^4 - 36h^{-4}h'^2h'' + 6h^{-3}h''^2 + 8h^{-3}h'h''' - h^{-2}h^{(iv)}$$
$$g^{(v)} = -120h^{-6}h'^5 + 240h^{-5}h'^3h'' - 90h^{-4}h'h''^2 - 60h^{-4}h'^2h'''$$
$$+ 20h^{-3}h''h''' + 10h^{-3}h'h^{(iv)} - h^{-2}h^{(v)}$$

[It is not easy to obtain a simple general rule for $g^{(n)}$.]

From (3-141) we then obtain our rules for derivatives of quotients:

$$\left(\frac{f}{h}\right)' = fg' + f'g = f(-h^{-2}h') + f'h^{-1} = h^{-2}(hf' - h'f)$$

$$\left(\frac{f}{h}\right)'' = fg'' + 2f'g' + f''g = f(2h^{-3}h'^2 - h^{-2}h'') + 2f'(-h^{-2}h') + f''h^{-1}$$

$$= h^{-3}(2fh'^2 - hfh'' - 2f'hh' + h^2f'')$$

and so on.

EXAMPLE 3. Let $y = F(x) = x/(x^2 + 1)$ (all x). Then we can write, as above, $F = f \cdot g$, $f(x) = x$, $g(x) = (x^2 + 1)^{-1} = [h(x)]^{-1}$, $h(x) = x^2 + 1$. We form the third derivative of F:

$$F'''(x) = (fg)''' = fg''' + 3f'g'' + 3f''g' + f'''g$$

and by (3-141)

$$g' = -2x(x^2 + 1)^{-2}, \qquad g'' = 2(x^2 + 1)^{-3}(2x)^2 - (x^2 + 1)^{-2}(2)$$
$$g''' = -48x^3(x^2 + 1)^{-4} + 24x(x^2 + 1)^{-3}$$

so that, after simplification,

$$F''' = (x^2 + 1)^{-4}(-54x^4 - 12x^2 - 6)$$

3-15 GEOMETRIC MEANING OF THE DERIVATIVES OF HIGHER ORDER

We have learned that the first derivative $f'(x)$ can be interpreted as the slope of the tangent line to the graph of the function at the point (x, y) on the graph. The second derivative $f''(x)$ is accordingly the rate of change, at x, of this slope. If, for example, $f''(x) > 0$, then the slope is increasing as x increases; if $f''(x) < 0$, the slope is decreasing as x increases. This is suggested in Figure 3-51, in which the second derivative is positive from a to c (but the

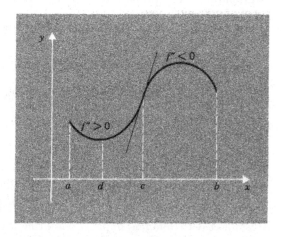

Figure 3-51 Sign of second derivative.

first derivative f' is negative from a to d and then positive from d to c), and the second derivative is negative from c to b. As the figure indicates, a positive second derivative is associated with a portion of the curve that is "concave upwards"; a negative derivative is associated with a portion of the curve that is "concave downwards." We can define *concave upwards* as meaning: near the point of tangency, the curve always lies above the tangent line. For *concave down* we replace "above" by "below." At $x = c$, in Figure 3-51, the curve crosses the tangent line and $f'' = 0$; such a point is called a *point of inflection.* A detailed discussion is given in Chapter 6.

The larger positive or negative f'' becomes, the more rapidly the slope f' is changing. We shall learn (see Section 6-5) that, roughly speaking, large values of f'' are associated with sharp curvature of the graph; where the graph is relatively flat (nearly straight), as near $x = c$ in the figure, f'' is close to 0. The curvature at x is actually given, not by f'', but by

$$\kappa = \frac{|f''(x)|}{[1 + f'(x)^2]^{3/2}}$$

as shown in Section 6-6. Thus, although values of f'' close to zero do indicate a very flat curve, large values of f'' do not necessarily indicate strong curvature, since f' may also be very large and thereby, because of the denominator, compensate for the large values of f''.

For the third derivative f''', we can similarly make an interpretation as rate of change of f'', or, with the qualification of the preceding paragraph, as rate of change of curvature. Thus, if f''' is very close to 0 over a portion of the graph, the curvature should remain nearly constant, and the curve should resemble a circular arc. If both f''' and f'' are close to 0, the curve should resemble a straight line—that is, should be extremely flat.

For higher derivatives, we can make similar interpretations. We gain insight into their significance by studying the graph of a polynomial $y = f(x)$ of degree n:

$$y = a_0 + a_1 x + a_2 x^2 + a_3 x^3 + \ldots a_n x^n = f(x) \qquad (a_n \neq 0)$$

near $x = 0$. When $x = 0$, $y = a_0$, so that $f(0) = a_0$. Next

$$f'(x) = a_1 + 2a_2 x + 3a_3 x^2 + \cdots + na_n x^{n-1}$$
$$f''(x) = 2a_2 + 6a_3 x + \cdots + n(n-1)a_n x^{n-2}$$
$$\vdots$$
$$f^{(k)}(x) = k(k-1) \ldots 1 a_k + \cdots + n(n-1) \ldots (n-k+1)a_n x^{n-k}$$

for $k = 1, 2, \ldots, n$. Accordingly

$$f'(0) = a_1, \qquad f''(0) = 2a_2, \ldots; \qquad f^{(k)}(0) = k!a_k \qquad \text{for} \qquad k = 1, \ldots, n$$

Also, $f^{(n)}(x)$ is constant, equal to $n!a_n$ and, accordingly, $f^{(n+1)}(x) = 0$, $f^{(n+2)}(x) = 0, \ldots$. We can now write:

$$a_0 = f(0), \qquad a_1 = f'(0), \qquad a_2 = \frac{f''(0)}{2!}, \qquad \ldots, \qquad a_n = \frac{f^{(n)}(0)}{n!}$$

Thus the *coefficient of x^k in the polynomial $a_0 + a_1 x + \cdots + a_n x^n$ is the kth derivative of the polynomial at $x = 0$, divided by $k!$.*

EXAMPLE. $y = 2 + 3x + 15x^2 + x^3 + 10x^5$. For $x = 1$, $y = 2 + 3 + 15 + 1 + 10 = 31$; for $x = 2$, $y = 2 + 5 + 60 + 8 + 320 = 396$; for $x = 3$, $y = 2 + 9 + 135 + 27 + 2430 = 2603$. Clearly, for large positive (or negative) x, the main contribution comes from the terms $10x^5$: the term of highest degree dominates for large positive or negative x. Now let us take x small. For $x = 0.01$,

$$y = 2 + 3 \times 0.01 + 15 \times 0.0001 + 0.000\,001 + 10 \times 0.000\,000\,000\,1$$
$$= 2 + 0.03 + 0.0015 + 0.000\,001 + 0.000\,000\,001$$
$$= 2.031501001,$$

where we have suggested the contributions of the terms of various degrees. As the degree increases, the relative significance decreases. For x sufficiently small only the constant term a_0 really matters. For example, for $x = 0.0001$, $y = 2.0003$ to extremely great accuracy. There is a similar analysis for negative x.

We can generalize what we have observed. The value of a polynomial $a_0 + a_1 x + \cdots + a_n x^n$, for x sufficiently close to 0 is given with great accuracy by $a_0 = f(0)$; as x moves away from 0, the term of first degree, $a_1 x$ [$a_1 = f'(0)$] gradually influences the value more and more; at a further stage, the term of second degree $a_2 x^2$ [$a_2 = f''(0)/2$] also has a significant effect, and so on. The further we recede from 0, the more significant become the terms of high degree. In Figure 3-52, we show this effect graphically. The

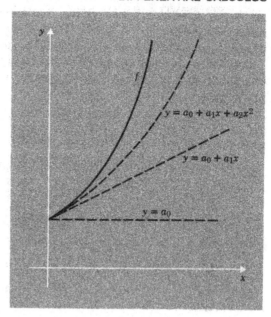

Figure 3-52 Significance of higher derivatives.

point at which a term of given degree begins to influence the value significantly depends on the size of the coefficient, relative to the size of the coefficients of the terms of lower degree—if the coefficient is very large, the influence can appear even for quite small x.

We have thus far concentrated on the meaning of the derivatives at the fixed value $x = 0$. A similar analysis holds true for a general value $x = c$. In particular, a polynomial $f(x)$ of degree n can always be written in the form

$$a_0 + a_1(x - c) + \cdots + a_n(x - c)^n$$

and we verify, as above, that $f(c) = a_0$, and

$$a_k = \frac{f^{(k)}(c)}{k!} \qquad \text{for} \qquad k = 1, \ldots, n$$

(see Problem 14 below). Thus the closer x is to c the less the relative importance of the higher derivatives at c.

We can prove that what has been shown by our study of polynomials is applicable to a general function (having continuous derivatives up to the order considered). In fact, we shall show in Section 6-12 that a general function $f(x)$ has, near $x = c$, a graph that closely resembles that of the polynomial $a_0 + a_1(x - c) + \cdots + a_n(x - c)^n$, where we use as coefficients the corresponding values

$$a_0 = f(c), \qquad a_1 = f'(c), \qquad a_2 = \frac{f''(c)}{2}, \ldots, \qquad a_n = \frac{f^{(n)}(c)}{n!}$$

obtained from the given function. In particular if, for instance, $f''(c) = 0$, $f'''(c) = 0, f^{(iv)}(c) = 0$ then, over a relatively large interval about $x = c$, the function is approximately the same as the linear function

$$y = a_0 + a_1(x - c) = f(c) + (x - c)f'(c)$$

which, of course, represents the tangent to the curve at $x = c$.

3-16 PHYSICAL MEANING OF THE HIGHER DERIVATIVES

We have observed in Section 3-1 that, for a particle moving on the x-axis, according to the equation $x = f(t)$, the derivative $f'(t)$ is interpreted as the velocity, and the derivative of the velocity, dv/dt, is interpreted as the *acceleration*. Accordingly, the acceleration a is the second derivative of x with respect to t:

$$a = \frac{dv}{dt} = f''(t) = \frac{d^2x}{dt^2}$$

According to *Newton's Second Law*, $F = ma$, where F is the force exerted on the particle (along the x-axis) and m is the mass of the particle. For a given particle the mass m is fixed, so that the acceleration is a measure of the force exerted. A very common example is the force of gravity of the earth. By *Newton's law of gravitation*, this force F is proportional to the mass m but inversely proportional to the distance from the center of the earth (Figure 3-53). Thus, we have the equation:

$$F = -km\,\frac{1}{x^2}, \qquad (x > 0)$$

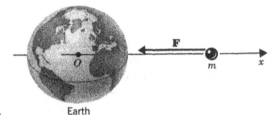

Figure 3-53 Gravitation.   Earth

where k is a positive constant; the minus sign indicates that the force has direction opposite to the x-axis. Near the surface of the earth, x is approximately constant (equal to the radius of the earth, 4,000 mi) and, hence, the force F is approximately constant. Therefore, the acceleration is also approximately constant:

$$a = \frac{F}{m} = \text{const}$$

and is the same for all objects. The numerical value of a is the familiar gravitational acceleration 32 ft/sec^2 observed for objects falling near the surface of the earth.

Third and higher order derivatives occur less often in basic physical theories, although they appear very frequently in the mathematical analysis of the theories. The third derivative $d^3x/dt^3 = d/dt(d^2x/dt^2)$ is a rate of change of acceleration. We can at least feel this derivative in an elevator. The change in speed (acceleration) because of a constant force is (within limits) not uncomfortable, whereas rapid changes in acceleration (large third derivative) can be distressing. The same applies to automobile travel or to airplane travel. We can verify this experimentally by accelerating a car, first steadily (so that the speedometer needle moves at a constant speed—zero third derivative) and then irregularly (fluctuating speed of the speedometer needle—third derivative not zero).

We can give other examples from physics and other sciences in which the higher derivatives play a significant role. The growth of money at compound interest, and generally the growth of population, and of some chemicals formed in a reaction, all follow laws for which all derivatives increase as time increases (Problem 13 below).

PROBLEMS

1. Find the first, second, and third derivatives:

   (a) $y = \sqrt{x} + \dfrac{7}{x^3}, x > 0$ (b) $y = (x^2 - 1)^5$ (c) $y = \sqrt{x + 1} \cdot (x^3 - 1)$

   (d) $y = \dfrac{x^2}{x + 5}$ (e) $y = \dfrac{x \sin x}{1 + \cos x}$ (f) $y = \cos 3x$

2. With the aid of Leibnitz's rule (3-140) prove the following:
   (a) $(f^2)'' = 2f'^2 + 2ff''$ (b) $(f^2)''' = 2ff''' + 6f'f''$
   (c) $[xf(x)]'' = xf'' + 2f'$ (d) $[xf(x)]''' = xf''' + 3f''$
   (e) $[xf(x)]^{(n)} = xf^{(n)} + nf^{(n-1)}$ (f) $[x^2f(x)]^n = x^2f^{(n)} + 2nxf^{(n-1)} + n(n-1)f^{(n-2)}$

3. Verify the following of the expressions (3-141) for the derivatives of $g(x) = 1/h(x)$ (in each case, assume that the previous rules have been verified):
   (a) for g' (b) for g'' (c) for g''' (d) for $g^{(iv)}$ (e) for $g^{(v)}$

4. Prove the rules (under appropriate hypotheses):
   (a) $D^2(f + g) = D^2f + D^2g$ (b) $D^2(cf) = cD^2f$, $c = \text{const.}$
   (c) $D^n(f + g) = D^nf + D^ng$, $n = 1, 2, 3, \ldots$
   (d) $D^n(cf) = cD^nf$, $c = \text{const.}$, $n = 1, 2, 3, \ldots$
   (e) $D^n(c) = 0$, $c = \text{const.}$, $n = 1, 2, 3, \ldots$
   [*Hint:* For (c), (d), and (e) use induction.]

5. Verify Leibnitz's rule (3-140) in the following cases (in each case assume that the previous rules have been verified):
   (a) for $(fg)''$ (b) for $(fg)'''$ [Use the result of (a)]
   (c) for $(fg)^{iv}$ (d) for $(fg)^{(5)}$

6. Prove Leibnitz's rule (Theorem 7) for general n. [*Hint:* Use mathematical induction. In the step from n to $n + 1$, write the coefficient of $f^{(r)} g^{(n-r)}$ in $(fg)^{(n)}$ as $n(n - 1) \ldots (n - r + 1)/r! \ldots$ Differentiate and show that the coefficient of $f^{(r)}g^{(n+1-r)}$ in $(fg)^{(n+1)}$ is

$$\frac{n(n - 1) \ldots (n - r + 1)}{r!} + \frac{n(n - 1) \ldots (n - r + 2)}{(r - 1)!}$$

and that this expression equals

$$\frac{(n + 1)(n) \cdots [(n + 1) - r + 1]}{r!} = \binom{n + 1}{r}$$

in agreement with (3-140) for $(fg)^{(n+1)}$.]

7. For each of the following functions (a) ... (d) given graphically (Figure 3-54), deter-

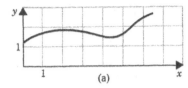
(a)

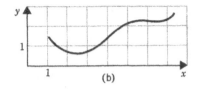
(b)

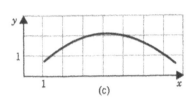
(c)

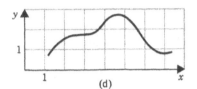
(d)

Figure 3-54

mine approximately from the figure the values of x for which $f''(x) > 0$, the values for which $f''(x) < 0$, the location of inflection points, and the value $f''(2)$.

8. Make a rough sketch of the graph of $y = f(x)$ for $-1 < x < 1$, where f is a polynomial and one is given only the following information:
   (a) $f(0) = 1, f'(0) = 2, f''(0) = -1$, higher derivatives are 0.
   (b) $f(0) = 1, f'(0) = 0, f''(0) = 1$, higher derivatives are 0.
   (c) $f(0) = 1, f'(0) = -1, f''(0) = 0, f'''(0) = 1$, higher derivatives are 0.

9. Each of the following functions is to be graphed, with accuracy of 3 decimal places, in the interval $0 \le x \le 0.01$. Determine in each case which terms must be considered:
   (a) $y = 3 + 2x + 5x^2 + x^3 + 200x^4$
   (b) $y = 1 - x + x^2 + 100x^3 + 2500x^4 + 10^8x^5$
   (c) $y = 0.0001 + 2203x^3 + 106x^4$

10. In each of the following cases a function $f(x)$ is known to be a polynomial of degree at most 3 and to satisfy the stated conditions. Find $f(x)$ in each case:
    (a) $f(0) = 1, f'(0) = 3, f''(0) = 10, f'''(0) = 6$
    (b) $f(0) = 0, f'(0) = 2, f''(0) + f'''(0) = 1, f(1) = 0$
    (c) $xf'(x) = f(x) - 2$ and $f'(0) = 5$
    (d) $f(2) = 1, f'(2) = 0, f''(2) = 0, f'''(2) = 1$
    (e) $f(0) = 1, f'(0) = 2, f(2) = 3, f'(2) = -1$

11. A meteor falls to the earth's surface and is found to weigh 10,000 lb. Find the force of gravity acting on the meteor when it was $3r$ miles out, where r is the radius of the earth. (Neglect air resistance and the possible loss of mass in falling through the atmosphere.)

12. A particle of mass 100 grams moves on a line, the x-axis, the position at time t being given by the equation $x = t^3 - t^2 + 3t$ (x in centimeters, t in seconds):
    (a) Find the velocity and acceleration when $t = 0$.
    (b) At what position is the acceleration 0?
    (c) Find the force applied when $t = 2$ seconds.

13. The "natural growth law" states that the instantaneous rate of growth dx/dt of a quantity measured by x is proportional to the value of x at that instant: $dx/dt = kx$ ($k = $ const., $k > 0$).
    (a) Show that the law is a reasonable one for the growth of bacteria in a favorable environment.

(b) Show that the law is a reasonable approximation to the way money grows at compound interest.

(c) Show that, if x is positive for $t = 0$, then x and all its derivatives are positive for $t \geq 0$.

(d) Let a second substance, measured by y, grow by the same law as x: $dy/dt = ky$. Show that $D_t(y/x) = 0$ (from which it follows that $y = \text{const} \cdot x$—see Section 3-21 below).

14. (a) Let $f(x) = b_0 + b_1 x + \cdots + b_n x^n$ be a polynomial of degree n. Show that, for given c, f can be written as follows:
$$f(x) = a_0 + a_1(x - c) + a_2(x - c)^2 + \cdots + a_n(x - c)^n$$
for appropriate constants $a_0, a_1, \ldots, a_n$, and express $a_0, a_1, \ldots, a_n$ in terms of $b_0, b_1, \ldots, b_n$. {*Hint.* Write each term $b_k x^k$ as $b_k[(x - c) + c]^k$ and expand by the binomial formula.}

   (b) Show that, in the expression for f in part (a), $f(c) = a_0, f'(c) = a_1, \ldots, f^{(k)}(c) = k! a_k$ for $k = 1, \ldots, n$.

15. Show that each of the following is a vector space of functions (Section 2-9):
   (a) All functions on $[0, 1]$ having a second derivative.
   (b) All functions on $[0, 1]$ having continuous derivatives of first, second, and third orders.
   (c) All functions f on $[0, 1]$ so that $f'' + f = 0$.
   (d) All functions f on $[0, 1]$ so that $f'' - 4f = 0$.

16. Show that the functions $\cos x$ and $\sin x$ are in the vector space of part (c) of Problem 15 and are linearly independent. (It can be shown that these two functions form a *basis* for the vector space, so that it has dimension 2.)

17. Show that the functions e^{2x} and e^{-2x} are in the vector space of Problem 15(d) and are linearly independent. (As in Problem 16, they form a basis for the vector space.)

3-17 HIGHER DERIVATIVES FOR COMPOSITE FUNCTIONS, INVERSE FUNCTIONS, FUNCTIONS DEFINED BY PARAMETRIC EQUATIONS

For each of the cases mentioned in this section heading, we can find second and higher derivatives by applying the rules already established. General formulas for these derivatives can also be derived.

In the following examples we proceed formally, assuming that the functions are defined in intervals in which the derivatives concerned are well defined.

EXAMPLE 1. Let $y = u^5$, where $u = x^3 - 1$. Then the chain rule gives

$$\frac{dy}{dx} = \frac{dy}{du} \cdot \frac{du}{dx} = 5u^4 \cdot 3x^2 = 15u^4 \cdot x^2$$

We leave our result in this form (not expressing u in terms of x) and proceed to form d^2y/dx^2:

$$\frac{d^2y}{dx^2} = 15(u^4 D_x x^2 + x^2 D_x u^4)$$

For the second term in the parenthesis we again apply the chain rule:

$$D_x u^4 = D_u u^4 \cdot D_x u = 4u^3 \cdot 3x^2 = 12u^3 \cdot x^2$$

Accordingly

$$\frac{d^2y}{dx^2} = 15(u^4 \cdot 2x + x^2 \cdot 12u^3 \cdot x^2) = 30(xu^4 + 6x^4u^3)$$

$$= 30x \cdot u^3 \cdot (u + 6x^3), \qquad u = x^3 - 1$$

EXAMPLE 2. Let $y = f(u)$, where $u = x/(x^2 + 1)$. Then

$$\frac{dy}{dx} = f'(u)\frac{du}{dx} = f'(u)\frac{1 - x^2}{(x^2 + 1)^2}$$

Next, by the product rule,

$$\frac{d^2y}{dx^2} = [D_x f'(u)]\frac{1 - x^2}{(x^2 + 1)^2} + f'(u)D_x\left[\frac{1 - x^2}{(x^2 + 1)^2}\right]$$

We evaluate $D_x f'(u)$ by the chain rule:

$$D_x f'(u) = D_u f'(u)\frac{du}{dx} = f''(u)\frac{1 - x^2}{(x^2 + 1)^2}$$

We thus find that

$$\frac{d^2y}{dx^2} = f''(u)\frac{(1 - x^2)^2}{(x^2 + 1)^4} + f'(u)\frac{2x^3 - 6x}{(x^2 + 1)^3}$$

EXAMPLE 3. Let $y = \sqrt[3]{u}$, where $u = g(x)$. Then

$$\frac{dy}{dx} = \frac{dy}{du} \cdot \frac{du}{dx} = D_u u^{1/3} \cdot g'(x) = \frac{1}{3}u^{-2/3}g'(x)$$

Next

$$\frac{d^2y}{dx^2} = \frac{1}{3}[D_x u^{-2/3}g'(x) + u^{-2/3}g''(x)]$$

$$= \frac{1}{3}\left[-\frac{2}{3}u^{-5/3}D_x u \cdot g'(x) + u^{-2/3}g''(x)\right]$$

$$= \frac{1}{9}\{-2u^{-5/3} \cdot [g'(x)]^2 + 3u^{-2/3} \cdot g''(x)\}$$

EXAMPLE 4. Let $y = f(u)$, $u = g(x)$. This is the general case for composition $f \circ g$:

$$\frac{dy}{dx} = f'(u)g'(x)$$

$$\frac{d^2y}{dx^2} = D_x f'(u) \cdot g'(x) + f'(u)D_x g'(x)$$

$$= D_u f'(u) \cdot D_x u \cdot g'(x) + f'(u) \cdot g''(x)$$

$$= f''(u) \cdot [g'(x)]^2 + f'(u) \cdot g''(x).$$

EXAMPLE 5. Let $y = f(x)$ have inverse $x = f^{-1}(y) = g(y)$. Then, by Theorem 3, $dx/dy = 1/f'(x)$, where x and y are paired by $y = f(x)$. Next, by the chain rule.

$$\frac{d^2x}{dy^2} = \frac{d}{dy}\frac{1}{f'(x)} = \frac{d}{dx}\left(\frac{1}{f'(x)}\right)\frac{dx}{dy}$$

$$= \frac{-f''(x)}{[f'(x)]^2}\cdot\frac{1}{f'(x)} = \frac{-f''(x)}{[f'(x)]^3} \tag{3-170}$$

Our conclusion shows that, in general, d^2x/dy^2 is not the reciprocal of d^2y/dx^2.

EXAMPLE 6. $y^3 + y - x = 0$ (implicit function). Here we notice that our equation gives x as a function of y:

$$x = y^3 + y = p(y), \qquad -\infty < y < \infty$$

where $p(y)$ is monotone strictly increasing. Hence, p has an inverse: $y = p^{-1}(x)$, which is the function $y = f(x)$ to be differentiated. We apply the conclusion of Example 5, by interchanging x and y and replacing f by p:

$$y'' = \frac{d^2y}{dx^2} = \frac{-p''(y)}{[p'(y)]^3} = \frac{-6y}{(3y^2 + 1)^3}$$

This gives y'' in terms of y.

EXAMPLE 7. $y^3 + y - x = 0$ (same equation as Example 6). We again seek y'', but regard the equation as an implicit equation for $y = f(x)$. As in Section 3-8 above,

$$3y^2y' + y' - 1 = 0 \tag{3-171}$$

The new equation is an implicit equation for y'. We differentiate again:

$$3y^2y'' + 6yy'^2 + y'' = 0$$

Accordingly

$$y'' = \frac{-6yy'^2}{3y^2 + 1}$$

This gives y'' in terms of y and y'. From Equation (3-171)

$$y' = \frac{1}{3y^2 + 1}$$

so that

$$y'' = \frac{-6y}{(3y^2 + 1)^3}$$

in agreement with the result of Example 6.

EXAMPLE 8. $x = t + t^3 = p(t)$, $y = 3t - 2t^3 = q(t)$. These are parametric equations defining a curve, expressible in the form $y = f(x)$ on portions for which p has an inverse: $t = p^{-1}(x)$ (since p is monotone strictly increasing and $p \to +\infty$ as $t \to +\infty$, $p \to -\infty$ as $t \to -\infty$, this holds true for all x and t). As in Section 3-9,

$$y' = \frac{dy}{dx} = \frac{q'(t)}{p'(t)} = \frac{3 - 6t^2}{1 + 3t^2}$$

Thus we have

$$x = p(t), \qquad y' = h(t) = \frac{q'(t)}{p'(t)}$$

These two equations are parametric equations for the graph of $y' = f'(x)$.
Consequently, we can differentiate in the same way:

$$y'' = f''(x) = \frac{dy'}{dx} = \frac{dy'/dt}{dx/dt} = \frac{h'(t)}{p'(t)}$$

$$= \frac{1}{1 + 3t^2} \frac{d}{dt}\left(\frac{3 - 6t^2}{1 + 3t^2}\right) = \left(\frac{1}{1 + 3t^2}\right)\left[\frac{-30t}{(1 + 3t^2)^2}\right]$$

$$= \frac{-30t}{(1 + 3t^2)^3}$$

In general,

$$y'' = \frac{1}{p'} \frac{d}{dt} \frac{q'(t)}{p'(t)} = \frac{p'(t)q''(t) - p''(t)q'(t)}{[p'(t)]^3} \tag{3-172}$$

Following Newton, we often denote the derivatives with respect to t as follows:

$$\frac{dx}{dt} = \dot{x}, \qquad \frac{dy}{dt} = \dot{y}, \qquad \frac{d^2x}{dt^2} = \ddot{x}, \qquad \frac{d^2y}{dt^2} = \ddot{y}$$

and so on.

Thus our expressions for y' and y'' become

$$y' = \frac{\dot{y}}{\dot{x}}, \qquad y'' = \frac{\dot{x}\ddot{y} - \ddot{x}\dot{y}}{\dot{x}^3} \tag{3-173}$$

3-18 HIGHER DERIVATIVES OF VECTOR FUNCTIONS

In Section 3-11 above, the derivative of a vector function $\mathbf{u} = \mathbf{F}(t)$ is defined [Equation 3-111], and it is shown that, if $\mathbf{u} = x\mathbf{i} + y\mathbf{i} = f(t)\mathbf{i} + g(t)\mathbf{j}$ then, under appropriate assumptions,

$$\frac{d\mathbf{u}}{dt} = \frac{dx}{dt}\mathbf{i} + \frac{dy}{dt}\mathbf{j} = f'(t)\mathbf{i} + g'(t)\mathbf{j} \tag{3-180}$$

The vector $d\mathbf{u}/dt$ can be interpreted as a *velocity vector* $\mathbf{v}$, and Equation (3-180) asserts that $\mathbf{v}$ has rectangular components dx/dt and dy/dt or, to use the notation of Example 8 in the preceding section, $v_x = \dot{x}$, $v_y = \dot{y}$. We can also write, in the same way.

$$\mathbf{v} = \dot{\mathbf{u}} = \frac{d\mathbf{u}}{dt} = \dot{x}\mathbf{i} + \dot{y}\mathbf{j}$$

Since $\mathbf{v} = d\mathbf{u}/dt = \mathbf{F}'(t)$ is again a vector function of t, we can seek its derivative, which we call the second derivative of the vector function $\mathbf{u} = \mathbf{F}(t)$, and which we denote by one of the following:

$$\frac{d^2u}{dt^2}, \qquad \mathbf{F}''(t), \qquad D_t{}^2\mathbf{u}, \qquad \ddot{\mathbf{u}}, \qquad \mathbf{u}''$$

Since

$$\mathbf{v} = \frac{d\mathbf{u}}{dt} = \frac{dx}{dt}\mathbf{i} + \frac{dy}{dt}\mathbf{j}$$

Equation (3-180) can be applied again:

$$\frac{d\mathbf{v}}{dt} = \frac{d^2\mathbf{u}}{dt^2} = \frac{d^2x}{dt^2}\mathbf{i} + \frac{d^2y}{dt^2}\mathbf{j} \qquad (3\text{-}181)$$

The vector $d\mathbf{v}/dt$, which describes the instantaneous rate of change of the velocity vector, can be interpreted as the *acceleration vector* $\mathbf{a}$ of the motion of the point P given by the equation $\overrightarrow{OP} = \mathbf{u} = \mathbf{F}(t)$. Equation (3-181) asserts that the acceleration vector has rectangular components d^2x/dt^2, d^2y/dt^2:

$$\mathbf{a} = \frac{d^2x}{dt^2}\mathbf{i} + \frac{d^2y}{dt^2}\mathbf{j} = \ddot{x}\mathbf{i} + \ddot{y}\mathbf{j} \qquad (3\text{-}182)$$

EXAMPLE. Let

$$\overrightarrow{OP} = x\mathbf{i} + y\mathbf{j} = 20t\mathbf{i} + (32t - 16t^2)\mathbf{j}$$

Then

$$\mathbf{v} = 20\mathbf{i} + (32 - 32t)\mathbf{j} \qquad \text{and} \qquad \mathbf{a} = \frac{d\mathbf{v}}{dt} = -32\mathbf{j}$$

In Figure 3-55 are shown the velocity vector and acceleration vector at the points for which $t = 0$, $t = 0.5$, $t = 1$. The acceleration vector is constant, of magnitude 32. (The motion can be interpreted as that of an object thrown up at an angle from the surface of the earth.)

In general, the magnitude of the acceleration vector is

$$|\mathbf{a}| = \sqrt{\left(\frac{d^2x}{dt^2}\right)^2 + \left(\frac{d^2y}{dt^2}\right)^2} \qquad (3\text{-}183)$$

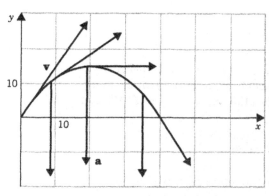

Figure 3-55 Velocity and acceleration.

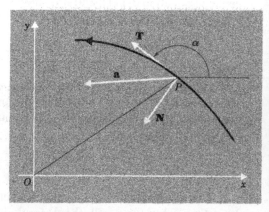

Figure 3-56 Tangential and normal components of acceleration.

This does not in general equal the derivative of the speed $|\mathbf{v}|$ (see Problem 7 below).

To analyze the relationship further, we write the velocity vector as $\mathbf{v} = |\mathbf{v}|\mathbf{T}$, where $\mathbf{T} = \mathbf{v}/|\mathbf{v}|$ is a *unit tangent vector* (Figure 3-56). Then by (3-123)

$$\mathbf{a} = \frac{d\mathbf{v}}{dt} = \frac{d|\mathbf{v}|}{dt}\mathbf{T} + |\mathbf{v}|\frac{d\mathbf{T}}{dt} \tag{3-184}$$

Since $|\mathbf{T}| = 1$, $\mathbf{T}\cdot\mathbf{T} = 1$ and we find, as in Section 3-12, that

$$\mathbf{T}\cdot\frac{d\mathbf{T}}{dt} = 0$$

Hence $d\mathbf{T}/dt$ is orthogonal to $\mathbf{T}$. Thus the acceleration vector $\mathbf{a}$ has a component, $\mathrm{comp}_{\mathbf{T}}\mathbf{a} = d|\mathbf{v}|/dt$, along the tangent, but it also has a component along the normal. A more detailed analysis leads to the interpretation of the normal component of $\mathbf{a}$ as $|\mathbf{v}|^2/\rho$, where ρ is the "radius of curvature" of the curve:

$$\mathrm{comp}_{\mathbf{T}}\mathbf{a} = \frac{d|\mathbf{v}|}{dt}, \qquad \mathrm{comp}_{\mathbf{N}}\mathbf{a} = \frac{|\mathbf{v}|^2}{\rho}$$

where $\mathbf{N}$ is an appropriately chosen unit normal vector, $\mathbf{N} = \pm\mathbf{T}^{\dashv}$ (see Section 6-7).

PROBLEMS

1. Find the second derivative of
   (a) $y = f(x)$, where $y = 3u^5 + u^3 + 1$, $u = 2x^7 + x^3 + 1$
   (b) $u = g(x)$, where $u = 3v^2 - v + 1$, $v = (1 + x^2)^{-1}$
   (c) $x = h(t)$, where $x = \sin y$, $y = \sqrt{t^2 - 1}$

   (d) $y = f(x)$, where $y = u^4 - u^2 + u$, $u = \dfrac{v}{2v + 3}$, $v = x^2 e^x$

2. Find the derivative requested:
   (a) d^2y/dx^2 for $y = f(u)$, $u = (x^3 + 2x^2 + 1)/x$
   (b) d^3y/dx^3 for $y = f(u)$, $u = (x^3 + 1)^4$
   (c) d^2y/dx^2 for $y = \sin u$, $u = g(x)$
   (d) d^3y/dx^3 for $y = e^u + e^{-u}$, $u = \sqrt[3]{2x + 5}$

3. Let $y = \varphi(u)$ be a function such that

$$(\circ) \quad \frac{d^2y}{du^2} - uy = u^2$$

and let $u = x^3 + x = g(x)$. Show that for the composite function $f = \varphi \circ g$, for which $y = f(x) = \varphi[g(x)]$, we have

$$\frac{d^2y}{dx^2} - 6x(3x^2 + 1)^{-1}\frac{dy}{dx} - (3x^2 + 1)^2(x^3 + x)y = (x^3 + x)^2(3x^2 + 1)^2$$

4. Let $u = g(x)$ be such that

$$(\circ\circ) \quad \frac{d^2u}{dx^2} + x^2\frac{du}{dx} = e^x$$

and let $y = \varphi(u) = u/(2u + 1)$. Show that, if y is expressed in terms of x (composite function), then

$$(1 - 2y)\frac{d^2y}{dx^2} + 4\left(\frac{dy}{dx}\right)^2 + x^2(1 - 2y)\frac{dy}{dx} = e^x(1 - 2y)^3$$

Remark. Equations (°) and (°°) are called *differential equations;* each describes an identity relating a function and its derivatives. Problems 3 and 4 illustrate ways of changing the variable in a differential equation.

5. (a) Show that, if $y = f(u)$, $u = g(x)$, then

$$\frac{d^3y}{dx^3} = f'(u)g'''(x) + 3f''(u)g'(x)g''(x) + f'''(u)[g'(x)]^3$$

   (b) Show that the chain rule, the result of Example 4 in Section 3-17, and the result of part (*a*) of this problem can be written, respectively, as follows:

$$(f \circ g)' = (f' \circ g)g'$$
$$(f \circ g)'' = (f'' \circ g)g'^2 + (f' \circ g)g''$$
$$(f \circ g)''' = (f' \circ g)g''' + 3(f'' \circ g)g'g'' + (f''' \circ g)g'^3$$

6. Graph the following paths and show the velocity and acceleration vectors at the points at which $t = 0$ and $t = 2$:

   (a) $x = t^2, y = t^3$ (b) $x = \dfrac{1 + t^2}{1 - t^2}, \quad y = \dfrac{2t}{1 - t^2}$

   (c) $x = \cos \pi t^2, y = \sin \pi t^2$ (d) $x = e^t \cos t, y = e^t \sin t$

7. Find the speed, the derivative of the speed, and the magnitude of the acceleration vector for the paths of Problem 6 at the point where $t = 0$.

3-19 MAXIMA AND MINIMA

In Sections 2-2 and 2-7, the concepts of maxima and minima are considered. Here, we relate this topic to derivatives. Further information is given in Sections 6-1 and 6-2, which can also be studied at this point.

Let a function f be defined in an interval. We say that f has a *local maximum* at the point x_0 of the interval if, for some $q > 0$, $f(x) \leq f(x_0)$ for all x in the interval such that $x_0 - q < x < x_0 + q$. A *local minimum* is defined similarly (with $\leq$ replaced by $\geq$) (see Figure 3-57).

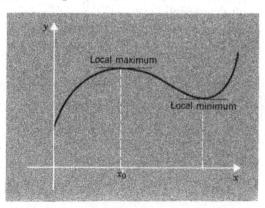

Figure 3-57 Local maximum and minimum.

With f as in the preceding paragraph, we say that f has *absolute maximum* M if $f(x_0) = M$ for, at least, one x_0 in the given interval and if $f(x) \leq M$ for all x in the interval. We notice that, if $f(x_0) = M$, then f also has a local maximum at x_0. The *absolute minimum* is defined similarly. We remark that a continuous function need not have an absolute maximum or minimum, but that it must have both if f is defined and continuous on a closed interval (Theorem F, Section 2-7).

THEOREM 8. *Let $y = f(x)$ be defined and differentiable in the interval $a \leq x \leq b$ and let f have a local maximum or local minimum at x_0, where $a < x_0 < b$. Then $f'(x_0) = 0$.*

Thus, for the graph of a differentiable function, the tangents at local maxima and minima *inside* the interval have slope zero, as suggested in Figure 3-57.

PROOF. We consider the case of a local maximum, the case of a local minimum being similar (see Problem 4 below).

Thus we are given that $a < x_0 < b$ and that $f(x) \leq f(x_0)$ for x sufficiently close to x_0. Accordingly, for h sufficiently small and positive, $f(x_0 + h) \leq f(x_0)$, so that (see Figure 3-57)

$$\frac{f(x_0 + h) - f(x_0)}{h} \leq 0$$

The limit of this expression, as $h \to 0+$, is the derivative $f'(x_0)$. Accordingly (see Remarks at the end of Section 2-5)

$$f'(x_0) \leq 0 \qquad\qquad (3\text{-}190)$$

Similarly, for h sufficiently small and *negative*, $f(x_0 + h) \leq f(x_0)$, so that

$$\frac{f(x_0 + h) - f(x_0)}{h} \geq 0$$

(negative divided by negative). Again, we take the limit, as $h \to 0-$, and obtain the derivative $f'(x_0)$. Accordingly

$$f'(x_0) \geq 0 \qquad\qquad (3\text{-}191)$$

The two inequalities (3-190), (3-191) imply that $f'(x_0) = 0$. Thus the theorem is proved.

Remark 1. In our proof we had to be able to choose h positive or negative; that is, all numbers $x_0 + h$ for h sufficiently small had to be in the interval. This is true for x_0 an interior point of the interval, but not for an end point. If f has a local maximum at a, we can use h positive and conclude that (3-190) holds true, that is, $f'(a) \leq 0$. Similarly, if f has a local maximum at b, then $f'(b) \geq 0$. For a local minimum these inequalities are reversed.

Remark 2. Our theorem states that, if f has a local maximum or minimum at x_0, where $a < x_0 < b$, then $f'(x_0) = 0$. The converse is not true; that is, if $f'(x_0) = 0$, we cannot be sure that f has a local maximum or minimum at

x_0. For, in particular, the graph may have a *horizontal inflection point* as shown in Figure 3-61 below.

COROLLARY OF THEOREM 8. *Let* $y = f(x)$ *be defined and differentiable in the interval* $a \leq x \leq b$ *and let* $f'(x) \neq 0$ *for* $a < x < b$. *Then the absolute maximum and minimum of* f *are taken on at* a *and* b *and only at these points.*

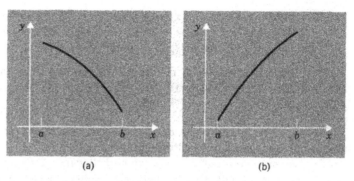

(a)   (b)

Figure 3-58   Absolute maximum and minimum at end points.

PROOF. In Figure 3-58 two typical cases are suggested. In case (*a*), the absolute maximum is $f(a)$, the absolute minimum is $f(b)$. They are reversed in case (*b*). If f had an absolute maximum, for example, at an interior point x_0, then it would be a local maximum, and, by Theorem 8, $f'(x_0)$ would have to be 0. This is forbidden by our assumption that $f'(x) \neq 0$ for $a < x < b$. Hence the absolute maximum must occur at a or b, and the same applies to the absolute minimum.

EXAMPLE 1. Find the absolute minimum of the function

$$y = 2x^3 + 6x^2 - 18x - 4 = f(x), \qquad 0 \leq x \leq 3$$

Solution. Since f is continuous on [0, 3], we know that f has an absolute minimum (Section 2-7) at some point of the interval. If the minimum occurs at an interior point x_0, then f has also a local minimum at x_0 and, by Theorem 8, $f'(x_0) = 0$. Now

$$f'(x) = 6x^2 + 12x - 18 = 6(x + 3)(x - 1)$$

Hence $f'(x) = 0$ at $x = 1$ and $x = -3$. However, only $x = 1$ falls in the interval considered and, thus, is the only interior point at which f can have its absolute minimum. Now

$$f(1) = -14, \qquad f(0) = -4, \qquad f(3) = 50$$

Therefore, the absolute minimum cannot occur at an end point ($x = 0$ or $x = 3$) and does occur at the interior point $x = 1$:

$$f(1) = -14, \qquad f(x) \geq -14, \qquad \text{for } 0 \leq x \leq 3$$

EXAMPLE 2. Find all local maxima and minima of the function

$$y = 3 + 2x^2 - x^4 = f(x), \qquad -2 \le x \le 2$$

Solution. We first examine the points where $f'(x) = 0$. Now

$$f'(x) = 4x - 4x^3 = 4x(1 - x)(1 + x)$$

Hence, $f'(x) = 0$ for $x = 0, x = 1, x = -1$, three values interior to the inter-

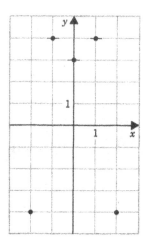

Figure 3-59 Analysis of Example 2.

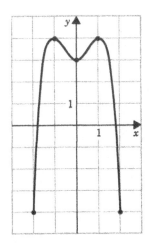

Figure 3-60 Graph of Example 2.

val $[-2, 2]$. We cannot tell from this information which of these are local maxima. However, we have

$$f(-2) = -5, \qquad f(-1) = 4, \qquad f(0) = 3, \qquad f(1) = 4, \qquad f(-2) = -5$$

In Figure 3-59 we show these five values and the horizontal tangents at $x = 0$, $x = \pm 1$. We can now reason as follows. In the interval $[-2, -1]$, $f'(x)$ is not 0 at an interior point. Consequently, by the Corollary of Theorem 8, the absolute maximum of f in that interval occurs at an end point; that end point must be -1, since $f(-1) = 4 > f(-2) = -5$. In the interval $[-1, 0]$ we find, in the same way, that the absolute maximum occurs at $x = -1$. Accordingly, f has a local maximum at $x = -1$, for $f(x) \le f(-1) = 4$ for $-2 \le x \le 0$. By similar reasoning, we conclude that f has a local minimum at $x = 0$ and another local maximum at $x = 1$. In addition, f has local minima at the end points $x = -2$, $x = +2$ (see Problem 3 below). The completed graph is shown in Figure 3-60.

EXAMPLE 3. Find the absolute maximum of the function

$$y = f(x) = x^3 - 3x^2 + 3x + 12, \qquad -2 \le x \le 2$$

Here we find that

$$f'(x) = 3x^2 - 6x + 3 = 3(x - 1)^2$$

Hence $f'(1) = 0$, and this is a point interior to the interval. Therefore, a local maximum may occur at $x = 1$. However, we find that $f(-2) = -14$, $f(2) = 14$, and $f(1) = 13$. Hence, by the Corollary of Theorem 8, in the interval $[-2, 1]$ the absolute maximum of f is $f(1) = 13$, but in the interval

[1, 2] the absolute minimum of f is $f(1) = 13$. Consequently, f has neither a local maximum nor a local minimum at $x = 1$, even though $f'(1) = 0$, so that the tangent is horizontal. This is a case of a *horizontal inflection point* (Section 2-2, Chapter 6). The function is graphed in Figure 3-61. The absolute maximum occurs at the end point $x = 2$, where f has the value 14.

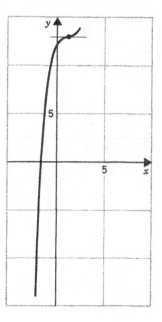

Figure 3-61 Graph of Example 3. Horizontal inflection point.

PROBLEMS

1. Each of the following functions is known to have exactly one local maximum at an interior point of the interval given. Locate the local maximum and find its value.

   (a) $y = 1 + 3x - x^2, 0 \leq x \leq 2$ (b) $y = \dfrac{x}{1 + x^2}, 0 \leq x \leq 2$

   (c) $y = 1 - \sqrt{x} - \dfrac{2}{\sqrt{x}}, 1 \leq x \leq 3$ (d) $y = e^{-x} \sin x, 0 \leq x \leq \pi$

2. For each of the following differentiable functions f in the interval [0, 4], the values at the end points of the interval of definition is given, as well as the location and value of all points at which $f'(x) = 0$. Use the information given to find all local maxima and minima, and the absolute maximum and minimum:

   (a) $f(0) = 1; f'(2) = 0, f(2) = 3; f(4) = 2$
   (b) $f(0) = 1; f'(2) = 0, f(2) = 2; f(4) = 0$
   (c) $f(0) = 1; f'(2) = 0, f(2) = 2; f(4) = 3$
   (d) $f(0) = 2; f'(3) = 0, f(3) = 1; f(4) = 2$
   (e) $f(0) = 1; f'(2) = 0, f(2) = 2; f'(3) = 0, f(3) = 0; f(4) = 1$
   (f) $f(0) = 0; f'(2) = 0, f(2) = 1; f'(3) = 0, f(3) = 2; f(4) = 3$
   (g) $f'(0) = 0, f(0) = 1; f(4) = 2$

3. Complete the discussion of Example 2 in the text.

4. Write out the proof of Theorem 8 for the case of a local minimum.

5. (a) Show that Theorem 8 remains valid if $f(x)$ is defined only in the *open* interval $a < x < b$.

(b) Show that Theorem 8 remains valid for arbitrary intervals (including infinite intervals) in the form: if f has a local maximum or minimum at an interior point x_0, then $f'(x_0) = 0$.

6. Find all local maxima and minima (see Problem 5).

(a) $y = \dfrac{1}{x-1} + \dfrac{1}{x}, 0 < x < 1$

(b) $y = \dfrac{x}{1+x^2}, -\infty < x < \infty$

(c) $y = \text{Tan}^{-1} x - \text{Tan}^{-1}(x-1), -\infty < x < \infty$

(d) $y = \sqrt{x(1-x)}, 0 < x < 1$

7. Find two positive numbers whose sum is 10 and whose product is a maximum.

8. A rectangle is inscribed in a circle of radius a. Find the sides of the rectangle of maximum area.

9. A metal cup is to be made of metal of given thickness and to hold exactly $\frac{1}{2}$ pt. What shape is least expensive? (*Hint.* Consider the surface area of a cylinder minus one base.)

10. A tin can is to be made of metal of given thickness and to hold exactly 1 qt. What shape is least expensive (see Problem 9)?

11. Prove that if f is differentiable in the interval $[a, b]$ and $f'(x) > 0$ for $a \le x \le b$, then f is monotone strictly increasing. [*Hint.* Let $a \le x_1 < x_2 \le b$. Apply the corollary of Theorem 8 to conclude that in the interval $[x_1, x_2]$, f has its absolute minimum at x_1 or x_2. Apply Remark 1 above to show that the absolute minimum is at x_1 and, hence, that $f(x_1) < f(x_2)$.]

12. Show on the basis of the proof given for Theorem 8 that the theorem can be restated as follows. Let $y = f(x)$ be defined in the interval $a \le x \le b$, let f have a local maximum or local minimum at x_0, where $a < x_0 < b$, and let $f'(x_0)$ exist. Then $f'(x_0) = 0$.

13. Let $f(x) = x^{2/3}, -1 \le x \le 1$. Show that f has a local minimum at a point at which f' does not exist.

3-20 ROLLE'S THEOREM

The following theorem has many applications in the calculus.

THEOREM 9 (*Rolle's Theorem*). *Let* $y = f(x)$ *be continuous in the closed interval* $[a, b]$, *and differentiable in the open interval* $a < x < b$, *and let* $f(a) = f(b)$. *Then* $f'(x)$ *has a zero at at least one point* x_0 *interior to the interval:*

$$f'(x_0) = 0, \qquad a < x_0 < b$$

PROOF. The hypotheses are illustrated in Figure 3-62. The theorem follows from the Corollary of Theorem 8. Thus, if $f'(x) \neq 0$ for $a < x < b$, then f has its absolute maximum and minimum at a and b. Since $f(a) = f(b)$, the only possibility is for f to be identically constant—but then $f'(x) \equiv 0$! Thus, f' must have a zero at an interior point.

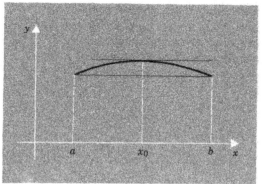

Figure 3-62   Rolle's theorem.   **Figure 3-63**   Mean Value Theorem.

EXAMPLE. Find the number of zeros of the derivative of

$$3(x - 1)(x - 2)(x - 4)(x - 7) = f(x) \text{ in the interval } 0 \leq x \leq 10$$

Solution. Instead of differentiating, we observe that $f(x)$ is a polynomial of degree 4: $f(x) = 3x^4 + \cdots$, and hence $f'(x)$ is a polynomial of degree 3. Next we notice that $f(x)$ has zeros at $x = 1, x = 2, x = 4, x = 7$. Accordingly, the hypotheses of Rolle's theorem are applicable to each of the intervals $[1, 2], [2, 4], [4, 7]$: $f(1) = f(2) = 0$, $f(2) = f(4) = 0$, $f(4) = f(7) = 0$. Hence $f'(x)$ has a zero x_1 between 1 and 2, a zero x_2 between 2 and 4, a zero x_3 between 4 and 7. Since f' is of degree 3, these are all the zeros of f', and there are exactly 3 zeros in the interval $0 \leq x \leq 10$.

3-21 MEAN VALUE THEOREM

Closely related to Rolle's theorem is the following:

THEOREM 10 (*Mean Value Theorem*). *Let* $y = f(x)$ *be continuous in the closed interval* $[a, b]$ *and differentiable in the open interval* (a, b). *Then there is at least one value* x_0, $a < x_0 < b$, *so that*

$$f(b) - f(a) = (b - a)f'(x_0) \tag{3-210}$$

Remark. Write the conclusion (3-210) in the form

$$f'(x_0) = \frac{f(b) - f(a)}{b - a} \tag{3-211}$$

Next observe that (3-211) is equivalent to the statement that there is at least one point (x_0, y_0) on the graph of the function (with $a < x_0 < b$) at which the tangent line is parallel to the chord through the end points of the graph (see Figure 3-63). In this form the statement has the same meaning as Rolle's theorem, for which the chord in question happens to have zero slope, as in Figure 3-62.

PROOF OF MEAN VALUE THEOREM. Let

$$g(x) = f(x) - \left[\frac{f(b) - f(a)}{b - a} \cdot (x - a) + f(a) \right]$$

Thus g is obtained from f by subtracting from f the linear function whose graph is the chord through the end points (Section 0-7). Now

$$g(a) = f(a) - f(a) = 0$$

$$g(b) = f(b) - \left[\frac{f(b) - f(a)}{b - a} \cdot (b - a) + f(a) \right] = 0$$

Accordingly, Rolle's theorem is applicable to $g(x)$, and we conclude that $g'(x_0) = 0$ for some x_0, $a < x_0 < b$. But

$$g'(x) = f'(x) - \frac{f(b) - f(a)}{b - a}$$

Hence the equation $g'(x_0) = 0$ is the same as (3-211) or (3-210), as asserted.

THEOREM 11. *Let $f(x)$ be defined and differentiable in the interval $[a, b]$ and let $f'(x) \equiv 0$ in the interval. Then f is identically constant on $[a, b]$.*

PROOF. We apply the Mean Value Theorem to two values x_1, x_2, with $x_1 < x_2$, in the given interval:

$$f(x_2) - f(x_1) = f'(x_0)(x_2 - x_1), \qquad x_1 < x_0 < x_2 \qquad (3\text{-}212)$$

Since $f'(x) \equiv 0$, the right-hand side is 0, and we conclude that $f(x_2) = f(x_1)$ for every pair x_1, x_2. Thus, f must be identically constant.

COROLLARY TO THEOREM 11. *If f and g are differentiable in the interval $[a, b]$ and $f'(x) \equiv g'(x)$ in this interval, then there is a constant C such that $f(x) \equiv g(x) + C$ in this interval.*

The proof is left as a problem (Problem 9 below).

THEOREM 12. *Let $f(x)$ be defined and differentiable in the interval $[a, b]$.*
   (a) *If $f'(x) > 0$ for $a < x < b$, then f is monotone strictly increasing.*
   (b) *If $f'(x) \geq 0$ for $a < x < b$, then f is monotone nondecreasing.*
   (c) *If $f'(x) < 0$ for $a < x < b$, then f is monotone strictly decreasing.*
   (d) *If $f'(x) \leq 0$ for $a < x < b$, then f is monotone nonincreasing.*

PROOF. We prove (a) as a sample, leaving the other parts as exercises (Problem 10 below). As in the proof of Theorem 11, we apply (3-212) to two arbitrary values x_1, x_2, with $x_1 < x_2$. Now $f'(x) > 0$ for $a < x < b$ and so in particular $f'(x_0) > 0$, thus the right-hand side is positive, and hence $f(x_2) - f(x_1) > 0$ or $f(x_2) > f(x_1)$; that is, f is monotone strictly increasing.

COROLLARY TO THEOREM 12. *Let $f(x)$ be defined and differentiable in the interval $[a, b]$. Let $f'(x)$ have at most a finite number of zeros in the interval.*
   (a) *If $f'(x) \geq 0$ for all x in the interval, then f is monotone strictly increasing.*
   (b) *If $f'(x) \leq 0$ for all x in the interval, then f is monotone strictly decreasing.*

PROOF. We prove (a), the proof of (b) being similar. By Theorem 12, f is monotone nondecreasing, that is, for $x_1 < x_2$, $f(x_1) \leq f(x_2)$. If we could choose x_1, x_2 such that $x_1 < x_2$ and $f(x_1) = f(x_2)$, then for every x between x_1 and x_2 we would have $f(x_1) = f(x) = f(x_2)$, so that $f(x)$ would remain constant over the whole interval $[x_1, x_2]$. But that would mean $f'(x) \equiv 0$ for $x_1 < x < x_2$, contrary to our assumption that $f'(x)$ has only a finite number of zeros. Hence $f(x_1) = f(x_2)$ cannot occur, and f is monotone strictly increasing.

EXAMPLE. A certain function $f(x)$ has the properties: $f(0) = 0$, $f'(x) = x^2/(1 + x^2)$ for all x. Show that

$$0 < f(x) < x \qquad \text{for} \qquad x > 0$$

Solution. We observe that $f'(x) \geq 0$ for all x, $f'(x) = 0$ only for $x = 0$. Hence, by the Corollary to Theorem 12, f is monotone strictly increasing in each interval $[0, b]$ for $b > 0$; that is, f is monotone strictly increasing for $x \geq 0$. Since $f(0) = 0$, $f(x) > 0$ for $x > 0$. Next

$$f'(x) = \frac{x^2}{1 + x^2} < \frac{1 + x^2}{1 + x^2} = 1$$

so that $f'(x) < 1$ for all x. Now let $x > 0$ and apply the Mean Value Theorem to the interval $[0, x]$:

$$f(x) - f(0) = f'(x_0)(x - 0)$$

or $f(x) = xf'(x_0)$, since $f(0) = 0$. We do not know the value x_0, but we can reason that

$$f(x) = xf'(x_0) < x \cdot 1 = x$$

since $f'(x_0) < 1$. Accordingly, for $x > 0$,

$$0 < f(x) < x$$

as asserted.

Remark. As the example illustrates, Theorem 11, Theorem 12, and the two corollaries can also be applied to general intervals (open, half-open, infinite). In each case, we are led to compare $f(x_1)$, $f(x_2)$ for x_1, x_2 in the interval and, hence, to a problem for the closed interval $[x_1, x_2]$.

THEOREM 13. *Let f be defined and continuous in the closed interval $a \leq x \leq b$ and let f be differentiable in the half-open interval $a < x \leq b$. If f' has a limit as x approaches a, equal to m, then f has a derivative at a and $f'(a) = m$.*

PROOF. By the Mean Value Theorem, for h positive (and less than $b - a$) there is an x_0 for which

$$\frac{f(a + h) - f(a)}{h} = f'(x_0), \qquad a < x_0 < a + h \qquad (3\text{-}213)$$

Now for h sufficiently small, $|f'(x) - m|$ can be made as small as we wish for $a < x < a + h$. Therefore by (3-213)

$$\left| \frac{f(a + h) - f(a)}{h} - m \right|$$

can also be made as small as we wish, that is,

$$\lim_{h \to 0+} \frac{f(a + h) - f(a)}{h} = m$$

and $f'(a) = m$.

Remark. There is a similar theorem concerning the derivative of f at b. If f is continuous for $a \leq x \leq b$ and differentiable except perhaps at c, where $a < c < b$, then existence of a limit of f' as $x \to c+$ assures the existence of a right-hand derivative at c; existence of a limit of f' as $x \to c-$ assures existence of a left-hand derivative at c. If both limits exist and are equal, f is differentiable at c.

PROBLEMS

1. Determine the number of zeros of $f'(x)$ in the interval stated:
   (a) $2 \leq x \leq 8$, for $f(x) = (x - 2)(x - 3)(x - 4)(x - 7)$
   (b) $-3 \leq x \leq 3$, for $f(x) = (x^2 - 1)(x^2 - 4)$
   (c) $0 \leq x \leq 5$, for $f(x) = \sin 2x$
2. If f is twice differentiable for $a \leq x \leq b$ and f has three zeros: $x_1 = a \leq x_2 < x_3 = b$, show that f'' has at least one zero in (a, b). Illustrate by an example.
3. (a) Prove that, if $f(x)$ and $g(x)$ are defined and differentiable for $a \leq x \leq b$ and $f(a) = g(a)$, $f(b) = g(b)$, then $f'(x_0) = g'(x_0)$ for some x_0, $a < x_0 < b$. (*Hint.* Use Rolle's theorem.)
   (b) Let $f(x) = x/(1 + x^3)$, $g(x) = x/(1 + x^4)$. Show that $f'(x) = g'(x)$ for at least one x, $0 < x < 1$.
   (c) For what choice of $g(x)$ does the rule of part (a) yield the Mean Value Theorem?
   (d) In part (a) take $g(x) = [f(b) - f(a)](b - a)^{-1}(x - a) + f(a) + (x - b)(x - a)$ and show that there is at least one x_0, $a < x_0 < b$, such that

   $$f'(x_0) - 2x_0 = \frac{f(b) - f(a)}{b - a} - a - b$$

4. (a) Show that for a quadratic function f: $f(x) = Ax^2 + Bx + C$, the value x_0 in the Mean Value Theorem is always halfway between a and b. Relate this information to the graph of f, a parabola.
   (b) Show that for a function f of form $f(x) = Ax^3 + Bx + C$ the value x_0 in the Mean Value Theorem satisfies the equation:

   $$x_0^2 = \frac{a^2 + ab + b^2}{3}$$

   Can there be two choices of x_0 for given f, a, and b?
5. Let $f(x)$ and $f'(x)$ be defined and differentiable [so that $f''(x)$ exists] for $a \leq x \leq b$. Let

   $$F(x) = f(b) - f(x) - (b - x)f'(x) - \left(\frac{b - x}{b - a}\right)^2[f(b) - f(a) - (b - a)f'(a)]$$

   Verify that $F(a) = 0$, $F(b) = 0$ and apply Rolle's Theorem to conclude that

$$(°) \quad f(b) = f(a) + (b - a)f'(a) + \frac{1}{2}f''(x_0)(b - a)^2$$

for some x_0, $a < x_0 < b$. The relation $(°)$ is a special case of *Taylor's Theorem with Remainder* (see Chapter 6).

6. Determine whether f is monotone strictly increasing, monotone strictly decreasing or neither in the interval given:

   (a) $f(x) = 3x^3 + x, -1 \le x \le 3$ (b) $f(x) = x^3 + 1, -1 \le x \le 1$
   (c) $f(x) = x \ln x - x, 5 \le x \le 7$ (d) $f = xe^{-x^2}, -1 \le x \le 1$
   (e) $f = x^{-1}, -1 \le x \le -0.1$ (f) $f = \cos^3 x, 0 \le x \le 0.1$

7. Show, by using the sign of $f'(x)$, that $f(x)$ has a local maximum at the point given. [*Hint*. Use Theorem 12.]

   (a) $f(x) = 2 + 2x - x^2, 0 \le x \le 2$, at $x = 1$
   (b) $f(x) = x^2/(1 + x), -3 \le x < -1$, at $x = -2$
   (c) $f(x) = \sin x + \cos x, 0 \le x \le \pi$, at $x = \pi/4$
   (d) $f(x) = 1 - x^{4/3}, 0 \le x \le 1$, at $x = 0$
   (e) $f(x)$ given, $-1 \le x \le 1$, with $f'(x) = -x/(x^2 + 1)^3$, at $x = 0$
   (f) $f(x)$ given, $0 \le x \le \pi$, with $f'(x) = \cos^3 x$, at $x = \pi/2$
   (g) $f(x) = \text{Sin}^{-1}(1 - x^2), -1 \le x \le 1$, at $x = 0$.

8. Let $f(x)$ have first and second derivatives for $0 \le x \le 1$.

   (a) If $f'(0) = 0$ and $f''(x) \ge 0$ for all x, show that f is monotone nondecreasing.
   (b) If $f'(0) = 1$ and $f''(x) \ge 0$ for all x, show that f is monotone strictly increasing.
   (c) If $f'(0) = 0$ and $f''(x) > 0$ for all x, show that f is monotone strictly increasing.

9. Prove the Corollary to Theorem 11.

10. Prove the following parts of Theorem 12:

    (a) Part (b) (b) Part (c) (c) Part (d)

11. Prove the following implications:

    (a) If $f'(x) = 1$ for all x, then $f(x) = x + c$ for some constant c. [*Hint*. Let $g(x) = x$ and apply the corollary to Theorem 11.]
    (b) If $f'(x) = x$ for all x, then $f = (x^2/2) + c$.
    (c) If $f'(x) = \sin x$ for all x, then $f(x) = -\cos x + c$.
    (d) If $f'(x) = x^n$ for all x $(n = 1, 2, 3, \ldots)$, then $f = x^{n+1}/(n + 1) + c$.
    (e) If $f'(x) = 2x/(1 + x^2)^2$ for all x, then $f = -1/(1 + x^2) + c$.
    (f) If $f''(x) \equiv 0$, for all x then $f(x) = c_1 x + c_2$ for some constants c_1, c_2.
    (g) If $f''(x) \equiv 1$, then $f(x) = x^2/2 + c_1 x + c_2$.

12. (a) Let $y = f(x)$ be such that $2yy' = 1$. Show that $y^2 = x + c$ for some constant c. [*Hint*. Let $u = y^2$. Then $du/dx = 2y(dy/dx) = 1$, and proceed as in Problem 11.]

    (b) If $yy' = x$, show that $y^2 = x^2 + c$.
    (c) If $2yy' = x^2$, show that $y^2 = (x^3/3) + c$.
    (d) If $3y^2y' = 1$, show that $y^3 = x + c$.
    (e) If $3y^2y' = x$, show that $y^3 = (x^2/2) + c$.

13. (a) Let $y = f(x)$ be such that $x^2y' + 2xy = 1$. Show that $x^2y = x + c$. (*Hint*. Let $u = x^2y$.)

    (b) If $x^3y' + 3x^2y = 1 + x$, show that $x^3y = x + (x^2/2) + c$.

   Remark. Problems 11–13 are examples of *differential equations*. In each case we are given an identity satisfied by $f(x)$ and its derivatives, and we seek expressions for $f(x)$. See Section 7-14 and Chapter 14 for a systematic discussion. The particular problem of finding $f(x)$ when $f'(x)$ is known is called *integration* (see Chapter 4).

14. *Kepler's equation.* An artificial satellite (or a natural one, such as the moon) moves, to first approximation, in an ellipse with parametric equations $x = a \cos \phi$, $y = b \sin \phi$. The relation between ϕ and time t is given by Kepler's equation

$$nt = \phi - e \sin \phi$$

Here e is the eccentricity of the ellipse, so that $0 < e < 1$, and n is a positive constant.
   (a) Show that t is a monotone strictly increasing function of ϕ and hence has an inverse function, say $\phi = g(t)$, $-\infty < t < \infty$.
   (b) Show that for all t, $\varphi[t + (2\pi/n)] = \varphi(t) + 2\pi$, so that x, y have period $2\pi/n$. The period $2\pi/n$ is the time it takes to complete one circuit of the ellipse.

†15. *Parachute.* A descending parachute is known to have acceleration $a = g - h(v)$, where h is a differentiable function for all $v \geq 0$, with $h(0) = 0$, $h(v) \to \infty$ as $v \to \infty$, $h'(v)$ continuous and positive for all v. It is also known that the motion is completely determined for all t (both less than t_0 and greater than t_0) by the velocity v_0 at time t_0. Show from this information that there is a *critical velocity* $v_1 > 0$ so that, if the initial velocity is greater than v_1, then v decreases steadily as t increases, approaching v_1 as $t \to \infty$; if the initial velocity is less than v_1, then v increases steadily as t increases, approaching v_1 as $t \to \infty$; if the initial velocity equals v_1, then $v \equiv v_1$ for all t.

16. Show that each of the following sets of functions on the interval $-\infty < x < \infty$ forms a vector space of functions and find its dimension (Section 2-9):
   (a) All f so that $f'' \equiv 0$ [see Problem 11(f)].
   (b) All functions f so that $f''' = 0$.

17. Determine which if any of the following is a vector space:
   (a) All functions f on $[0, 1]$ such that $f'(x) > 0$ on $[0, 1]$.
   (b) All functions f on $[0, 1]$ such that f can be expressed as $g - h$, where g and h are monotone strictly increasing.
   (c) All functions f on $[0, 1]$ such that $f''(x) \equiv \sin x$.
   (d) All functions f on $[0, 1]$ such that $f''(x) + \sin x \, f'(x) \equiv 0$.

3-22 THE DIFFERENTIAL

The calculus can be developed from an alternative to the derivative, namely, *the differential*. We illustrate this by an example.

Let $y = f(x) = x^2 + 3x$ and let us consider the relation between the increments Δx, Δy:

$$\Delta y = f(x + \Delta x) - f(x) = (x + \Delta x)^2 + 3(x + \Delta x) - (x^2 + 3x)$$
$$= \Delta x(2x + 3) + (\Delta x)^2$$

Thus, starting from a fixed x, Δy is given by the indicated function of Δx. If Δx is very small, the term in $(\Delta x)^2$ will be negligible compared to the term in Δx. For example, if $x = 1$ and $\Delta x = 0.01$, then $\Delta y = 5\Delta x + (\Delta x)^2 = 0.05 + 0.0001$. In general, if Δx is much smaller than 1, then $(\Delta x)^2$ is Δx times Δx, or

Δx times a quantity much smaller than 1, and hence $(\Delta x)^2$ will be much smaller than Δx.

It will be seen that our example is typical. For any (reasonable) function $f(x)$, the increment Δy can be split into two parts:

$$\Delta y = m \cdot \Delta x + q(\Delta x)$$

Here m is determined by the fixed x from which we started and is, therefore, a fixed number. The term $q(\Delta x)$ is a function of Δx, typically formed of terms of second, third, and higher degree in Δx; accordingly, for Δx sufficiently small, $q(\Delta x)$ represents a term much smaller than the first term. *The linear term $m \cdot \Delta x$ is the differential of y, or of the function f, at x and is denoted by dy or df:*

$$dy = df = m\,\Delta x$$

In our particular example, $m = 2x + 3$ and $q(\Delta x) = (\Delta x)^2$, $dy = df = (2x + 3)\,\Delta x$.

Instead of trying to describe $q(\Delta x)$ in terms of the degrees of Δx which appear, it is more convenient to say that $q(\Delta x)$ should have Δx as a factor

$$q(\Delta x) = \Delta x \cdot p(\Delta x)$$

and that the other factor, $p(\Delta x)$, should approach zero as Δx approaches 0. This leads us to our general definition:

The function $y = f(x)$ has a differential $dy = m\,\Delta x$ at a particular x if, for that x,

$$\Delta y = \Delta x[m + p(\Delta x)] \qquad (3\text{-}220)$$

where m does not depend on Δx and $p(\Delta x)$ is continuous at $\Delta x = 0$ with $p(0) = 0$. (Our function f is assumed defined in an interval containing x, and Δx is restricted so that $x + \Delta x$ is in that interval.)

EXAMPLE 1. Let $y = 1/x$ for $x > 0$. Then for fixed $x > 0$

$$\Delta y = \frac{1}{x + \Delta x} - \frac{1}{x} = \frac{-\Delta x}{x(x + \Delta x)}$$

$$= \Delta x \left\{ -\frac{1}{x^2} + \left[\frac{1}{x^2} - \frac{1}{x(x + \Delta x)} \right] \right\}$$

$$= \Delta x \left[-\frac{1}{x^2} + \frac{\Delta x}{x^2(x + \Delta x)} \right]$$

Hence we can take

$$p(\Delta x) = \frac{\Delta x}{x^2(x + \Delta x)}$$

and $p(\Delta x)$ is continuous at $\Delta x = 0$, with $p(0) = 0$. Accordingly,

$$dy = -\frac{\Delta x}{x^2}$$

so that

$$m = \frac{-1}{x^2}$$

THEOREM 14. *The function* $y = f(x)$ *has a derivative* $f'(x)$ *at a particular* x *if, and only if,* f *has a differential* $dy = m \cdot \Delta x$ *at that* x, *and* $m = f'(x)$.

PROOF. Let f have a differential at x, so that (3-220) holds true. Then for $\Delta x \neq 0$

$$\frac{\Delta y}{\Delta x} = m + p(\Delta x)$$

$$\lim_{\Delta x \to 0} \frac{\Delta y}{\Delta x} = \lim_{\Delta x \to 0} [m + p(\Delta x)] = m + p(0) = m$$

by the conditions on p. Hence, $f'(x)$ exists and equals m.

Conversely, let f have a derivative at x. We then *define* $p(\Delta x)$ by the conditions

$$p(\Delta x) = \frac{\Delta y}{\Delta x} - m \qquad \text{for } \Delta x \neq 0, \qquad p(0) = 0$$

where $m = f'(x)$. It then follows that p is continuous at $\Delta x = 0$ and that $p(0) = 0$. Furthermore,

$$\frac{\Delta y}{\Delta x} = m + p(\Delta x), \qquad \Delta y = \Delta x[m + p(\Delta x)]$$

Thus (3-220) holds true, and f has a differential $m \, \Delta x$ at x.

Remark 1. Our theorem shows that the number m in the differential $m \, \Delta x$ is the derivative $f'(x)$. Equation (3-220) can be interpreted as follows. For Δx sufficiently small, Δy is approximately a constant, m, times Δx; that is, the point $(x + \Delta x, y + \Delta y)$ moves approximately on a line of slope m. Since $m = f'(x)$, this line is the tangent to the graph at (x, y), as in Figure 3-64. The value of dy for each Δx is the number $m \, \Delta x$, which is the change in y that results from following the tangent line exactly.

Remark 2. The differential for a given function f depends on the "fixed"

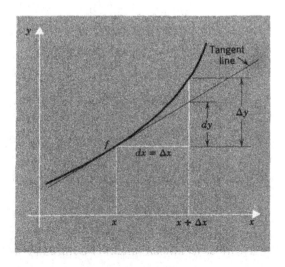

Figure 3-64 The differential.

x at which we start: $dy = m \cdot \Delta x$, where $m = f'(x)$. Thus, dy is a function of two variables, x and Δx. Other notations are often used for Δx—arbitrary letters such as h, k, ..., and the symbol dx. Thus we find

$$dy = f'(x)h, \qquad dy = f'(x)\,\Delta x, \qquad dy = f'(x)\,dx$$

all representing the same differential. The use of dx instead of Δx has a special significance which will be explained below. With this notation, we have

$$\frac{dy}{dx} = f'(x) \tag{3-221}$$

as used previously. [Technically, we should require $dx \neq 0$ in (3-221), but it is general practice to omit stating this restriction.]

3-23 RULES OF CALCULUS IN TERMS OF DIFFERENTIALS

The familiar rules for differentiation can be restated:

THEOREM 15. *Let f and g both be defined in an interval $[a, b]$ and let both have differentials at a point x of this interval. Then $f + g$, $f - g$, fg, f/g all have differentials at this x [provided that for f/g, $g(x) \neq 0$] and at x*

$$d(f + g) = df + dg$$
$$d(f - g) = df - dg$$
$$d(f \cdot g) = f\,dg + g\,df$$
$$d(c) = 0 \qquad (c = \text{constant})$$
$$d\left(\frac{f}{g}\right) = \frac{g\,df - f\,dg}{g^2}$$
$$d(f^n) = nf^{n-1}\,df \qquad (n = 1, 2, \ldots)$$

where f and g are evaluated at x. Furthermore

$$d(x^n) = nx^{n-1}\,dx \qquad (n = 1, 2, 3, \ldots)$$
$$d(\sin x) = \cos x\,dx, \qquad d(\cos x) = -\sin x\,dx$$
$$d(\tan x) = \sec^2 x\,dx, \qquad d(\cot x) = -\csc^2 x\,dx$$
$$d(\sec x) = \sec x \tan x\,dx \quad d(\csc x) = -\csc x \cot x\,dx$$
$$d(e^x) = e^x\,dx, \qquad\qquad d(a^x) = a^x \ln a\,dx$$
$$d(\ln x) = \frac{1}{x}\,dx, \qquad d\log_a x = \frac{1}{\ln a}\frac{1}{x}\,dx$$
$$d(\mathrm{Sin}^{-1} x) = \frac{1}{\sqrt{1 - x^2}}\,dx \qquad -1 < x < 1$$
$$d(\mathrm{Cos}^1 x) = \frac{-1}{\sqrt{1 - x^2}}\,dx, \qquad -1 < x < 1$$
$$d(\mathrm{Tan}^{-1} x) = \frac{1}{1 + x^2}\,dx, \qquad -\infty < x < \infty$$

PROOF. These rules can be derived from the corresponding rules for derivatives or proved directly by the same methods used previously (Section

3-4). For example, since f and g have differentials at x, $f'(x)$ and $g'(x)$ exist. Hence, $f + g$ has a derivative $f'(x) + f'(g)$ at x and, consequently by Theorem 14, $f + g$ has a differential $[f'(x) + g'(x)] \, \Delta x$. Thus

$$d(f + g) = [f'(x) + g'(x)] \, \Delta x = f'(x) \, \Delta x + g'(x) \, \Delta x$$
$$= df + dg$$

as asserted. We could have reasoned directly:

$$\Delta f = \Delta x[m_1 + p_1(\Delta x)], \qquad \Delta g = \Delta x[m_2 + p_2(\Delta x)] \qquad \text{(3-230)}$$
$$\Delta(f + g) = \Delta f + \Delta g = \Delta x[(m_1 + m_2) + p_1(\Delta x) + p_2(\Delta x)]$$
$$= \Delta x[m + p(\Delta x)]$$

where $p(\Delta x) = p_1(\Delta x) + p_2(\Delta x)$, $m = m_1 + m_2$. Since $p_1(\Delta x)$ and $p_2(\Delta x)$ are continuous and equal 0 at $\Delta x = 0$, $p(\Delta x)$, has these properties too. Hence, $f + g$ has a differential $m \, \Delta x = (m_1 + m_2) \, \Delta x = m_1 \, \Delta x + m_2 \, \Delta x = df + dg$.

The other rules are proved similarly (Problem 4 following Section 3-25).

The chain rule (Theorem 2, Section 3-5), can also be stated and proved in terms of differentials:

THEOREM 16. *Let $y = F(x)$, $a \le x \le b$, be defined as the composition of two functions: $y = f(u)$, $u = g(x)$, so that $F = f \circ g$. If at a particular x, g has a differential $du = n \, \Delta x$ and if, at the corresponding value $u = g(x)$, f has a differential $dy = m \, \Delta u$, then F has a differential dF at x:*

$$dF = mn \, \Delta x = f'(u)g'(x) \, \Delta x$$

PROOF. At the particular x and corresponding u considered,

$$\Delta u = \Delta x[n + p_1(\Delta x)], \qquad \Delta y = \Delta u[m + p_2(\Delta u)]$$

If we substitute the expression for Δu in that for Δy, we find that

$$\Delta y = \Delta x[n + p_1(\Delta x)] \cdot \{m + p_2[\Delta x(n + p_1(\Delta x))]\}$$
$$= \Delta x\{mn + mp_1(\Delta x) + np_2[\quad] + p_1(\Delta x)p_2[\quad]\}$$

where $p_2[\quad] = p_2[\Delta x(n + p_1(\Delta x))]$. Since $p_1(\Delta x)$ is continuous at $\Delta x = 0$, with value 0 at $\Delta x = 0$, $p_2[\quad]$ is continuous in Δx and also equals 0 for $\Delta x = 0$ (Theorem C, Section 2-7). Hence we can write

$$\Delta y = \Delta x[mn + p(\Delta x)]$$

as required, so that $dy = dF = mn \, \Delta x$.

Remark. The chain rule can be interpreted as follows: if $y = f(u)$ and $u = g(x)$ then, when y is expressed in terms of x,

$$dy = f'(u)g'(x) \, \Delta x = f'(u) \, du$$

Thus the relation between dy and du is the same as that between dy and Δu. This is the reason why Δu can be replaced by du in the differential symbol.

One can also interpret our conclusion as saying that, if one works with differentials, one can ignore the question of which variable depends on which; all variables are treated on the same basis. For example, if

$$x^2 + y^2 = 1$$

then we consider y as a function of x, or x as a function of y. Either way we are led to

$$2x \, dx + 2y \, dy = 0$$

and therefore

$$\frac{dy}{dx} = -\frac{x}{y} \quad \text{or} \quad \frac{dx}{dy} = -\frac{y}{x}$$

both formulas agree with the result of implicit differentiation, as in Section 3-8.

Similarly, if we have parametric equations

$$x = f(t) \qquad y = g(t)$$

then

$$dx = f'(t) \, dt, \qquad dy = g'(t) \, dt$$

so that

$$\frac{dy}{dx} = \frac{g'(t) \, dt}{f'(t) \, dt} = \frac{g'(t)}{f'(t)}$$

in agreement with the rule of Section 3-9. If $y = f(x)$, and we seek the derivative of the inverse function, we can reason that

$$dy = f'(x) \, dx, \qquad \frac{dx}{dy} = \frac{1}{f'(x)}$$

in agreement with the rule of Section 3-6.

In all these cases, we are dealing with a set of variables that are "chained" together. If one changes, then all change. We can regard the differentials $dx, dy, dt, du, \ldots$ as small changes in $x, y, t, u, \ldots$ from a fixed initial choice of these variables. The *ratio* of the differentials $dx : dy : dt : du : \ldots$ is then determined by the fixed initial choice, and *the quotient of any two is the corresponding derivative.*

3-24 NUMERICAL APPLICATIONS OF THE DIFFERENTIAL

Our basic formula (3-220) can be written

$$\Delta y = dy + \Delta x \cdot p(\Delta x)$$

where $dy = m \, \Delta x = f'(x) \, \Delta x$. As stressed in Section 3-22, the term $\Delta x \cdot p(\Delta x)$ is very small compared to the first degree term $dy = m \, \Delta x$, provided that Δx is sufficiently small. An exception must be made if $f'(x) = 0$ at the x considered, since then $dy = 0$ for every Δx, and it makes no sense to say that the second term is small compared to 0.

Provided that $f'(x) \neq 0$, we can then reason that, for sufficiently small Δx, Δy and dy are approximately equal, and we can obtain a good approximation for Δy by the formula

$$\Delta y \sim f'(x) \, \Delta x \tag{3-240}$$

This formula has had very wide application. (Here $\sim$ means "approximately equals.") The nature of the approximation can be seen in Figure 3-64. The smaller Δx is chosen, the better the approximation.

EXAMPLE 1. Evaluate $(1003)^2$. We take $y = f(x) = x^2$ and start at $x = 1000$, so that $y = 1,000,000$. We then take $\Delta x = 3$ and apply (3-240), with $f'(x) = 2x$, $f'(1000) = 2000$:

$$\Delta y \sim 2000\, \Delta x = 2000 \times 3 = 6000$$

Thus, $y = 1,000,000 + 6000 = 1,006,000$ (approximately). In this case, the exact value is 1,006,009, so that *the percentage error is negligible.*

EXAMPLE 2. Evaluate $\sqrt{50}$. We take $y = \sqrt{x}$ and start at $x = 49$, so that $y = 7$. Here $f'(x) = \frac{1}{2}x^{-1/2}$, $f'(49) = \frac{1}{14}$ and, with $\Delta x = 1$, $\Delta y \sim \frac{1}{14} \cdot 1 = \frac{1}{14} = 0.0714 \ldots$. Hence

$$\sqrt{50} \sim 7 + 0.0714 = 7.0714$$

The exact value is 7.071068 (to 6 decimal places).

EXAMPLE 3. Evaluate $\sin(\pi/8)$. Since $\pi/8 = 0.3927$, we take $y = \sin x$ and start at $x = 0$ (radians). Then $dy = \cos x\, dx$, so that here $\Delta y = \sin 0.3927 - \sin 0 \sim \cos 0 \cdot \Delta x = 1 \times 0.3927 = 0.3927$, and $y \sim 0 + 0.3927$. The exact value is 0.38268.

The preceding examples are typical of the applications of the formula (3-240). The accuracy of the result varies considerably, especially because we have given no information on how small Δx must be chosen to insure that the error in Δy is less than a prescribed amount. To obtain some estimate of our acuracy we can apply the Mean Value Theorem in the form

$$\Delta y = f(x + \Delta x) - f(x) = f'(x_0)\, \Delta x$$

where x_0 is between x and $x + \Delta x$. We do not know x_0 precisely, but we can reason: *if, on the interval* $[x, x + \Delta x]$, f' *has all its values between K and L, then* Δy *must lie betwen* $K\, \Delta x$ *and* $L\, \Delta y$.

Thus, in Example 1, $f'(x) = 2x$ and $x = 1000$, $x + \Delta x = 1003$. In this interval, f' varies between 2000 and 2006. Hence Δy lies between

$$2000 \times 3 = 6000 \quad \text{and} \quad 2006 \times 3 = 6018$$

Thus our error in using (3-240) is at most 18; as we observed, the error was 9.

In Example 2, $f'(x) = \frac{1}{2}x^{-1/2}$ and $x = 49$, $x + \Delta x = 50$. In this interval f' decreases from $\frac{1}{2}(49)^{-1/2} = \frac{1}{14}$ to $\frac{1}{2}(50)^{-1/2}$. However, we do not know $(50)^{-1/2}$—the number $\sqrt{50}$ is precisely what we are trying to find. But we can easily find a number K such that K is close to, but less than, $\frac{1}{2}(50)^{-1/2} = (200)^{-1/2}$. For example, since $15^2 > 200$, $\frac{1}{15} < (200)^{-1/2}$, and we can use $K = \frac{1}{15}$, $L = \frac{1}{14}$. Thus Δy lies between

$$\frac{1}{15} \times 1 = 0.0667 \quad \text{and} \quad \frac{1}{14} \times 1 = 0.0714$$

Hence our error in Δy is at most 0.0047. We observed that the error was actually 0.0003 (to four decimal places).

In Examples 2 and 3 we also are forced to rely on tables and perhaps on interpolation in order to check our results. Interpolation itself is based on a linear approximation to the function and, thus, is also related to the differential (see Problem 8 following Section 3-25).

A further discussion of the error in (3-240) is given in Chapter 6.

3-25 THE DIFFERENTIAL AND TANGENTS

Let (x_1, y_1) be a fixed point on the curve $y = f(x)$, assumed differentiable at $x = x_1$. Then as Figure 3-65 indicates, changing x_1 by $\Delta x = dx$ and y_1 by dy yields a point (x, y) on the tangent line. Since $dy = f'(x_1) \, dx$ and $y - y_1 = dy$, $x - x_1 = dx$, we conclude that

$$y - y_1 = f'(x_1)(x - x_1) \tag{3-250}$$

represents the equation of the tangent line. This is in agreement with our result in Section 3-13.

The formation of the equation of the tangent line (3-250) can be done quite mechanically by the following procedure:

(a) Take differentials.
(b) Evaluate the coefficients of dx, dy at the point (x_1, y_1) considered.
(c) Replace dx by $x - x_1$, dy by $y - y_1$.

The same procedure applies to implicit equations and parametric equations. For example, for the circle $x^2 + y^2 = 1$, we have

(a) $2x \, dx + 2y \, dy = 0$.
(b) $2x_1 \, dx + 2y_1 \, dy = 0$.
(c) $2x_1(x - x_1) + 2y_1(y - y_1) = 0$.

Thus, (c) gives the tangent line; if we divide by 2 and notice that $x_1^2 + y_1^2 = 1$, since (x_1, y_1) is a point of the circle, this simplifies to

$$x_1 x + y_1 y = 1$$

The justification for our procedure is that we know that (a) and (b) give the correct relation between the differentials dx, dy at the point considered and that, as above, $x = x_1 + dx$, $y = y_1 + dy$ locates a point (x, y) on the tan-

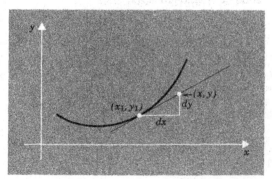

Figure 3-65   Tangent line and differentials.

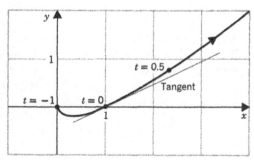

Figure 3-66   Tangent to curve $x = 1 + 2t + t^2$, $y = t + t^2$.

gent line, so that $x - x_1 = dx$ and $y - y_1 = dy$ are related, as in (c).

As an example of parametric equations we consider the curve

$$x = 1 + 2t + t^2, \qquad y = t + t^2 \tag{3-251}$$

Our procedure becomes

(a) $dx = (2 + 2t)\, dt, \qquad dy = (1 + 3t^2)\, dt$
(b) $dx = (2 + 2t_1)\, dt, \qquad dy = (1 + 3t_1{}^2)\, dt$
(c) $x - x_1 = (2 + 2t_1)(t - t_1), \qquad y - y_1 = (1 + 3t_1{}^2)(t - t_1)$

Equations (c) are *parametric equations for the tangent*. The justification is similar to that given for implicit equations. If we take $t_1 = 0$, our tangent becomes

$$x = 1 + 2t, \qquad y = t$$

These are precisely the terms through first degree in the given equations (3-251). The result is graphed in Figure 3-66.

In general, we observe that taking differentials is a process of *linearization*: curves are approximated by straight lines (the tangents), equations of general form become linear equations.

PROBLEMS

1. Evaluate Δy in terms of Δx, express in the form (3-220), and find dy:
   (a) $y = 2x + 3$ (b) $y = x^2$ (c) $y = x^3$ (d) $y = 3x^2 - x + 2$
2. For the function $y = x^2 + 3x$, considered in Section 3-22, tabulate dy and Δy for $\Delta x = 0.001, 0.01, 0.1, 0.5, 1$ for the following cases:
   (a) starting at $x = 1$ (b) starting at $x = 5$ (c) starting at $x = -1$
3. Find the differentials of the following functions:
   (a) $x^3 - 3x^2 + 5$ (b) $(x^2 - 1)^6$ (c) $\sqrt{2x - 3}$
   (d) $\sin^2 x$ (e) e^{2x} (f) $\ln \cos x$
   (g) $y = f(x)$, where $x^3 + 3xy^2 + y^3 = 1$ (h) $y = f(x)$, where $x \sin y - y = 1$
4. (a) Prove the rules: $d(fg) = f\, dg + g\, df$ and $d(f/g) = (g\, df - f\, dg)/g^2$ as applications of known rules for derivatives.
   (b) Prove the rule: $d(fg) = f\, dg + g\, df$ directly by differentials. {*Hint.* Verify first that $\Delta(fg) = f\, \Delta g + g\, \Delta f + \Delta f\, \Delta g$, where $f = f(x)$, $g = g(x)$. Express Δf, Δg as in (3-230), and hence express $\Delta(fg)$ in the form $\Delta x[f(x)m_2 + g(x)m_1] + \Delta x p(\Delta x)$}.
5. The variables x, u, y, v are "chained" by the following relations:

$$x^2 + xu + u^2 = 1, \qquad uy + y^3 = 1, \qquad y^2 + v^2 = 1$$

   (a) Find du/dx, dy/du, and dv/dy.
   (b) Find dx/dy and du/dv.
   (c) Find dx/dv.
6. Evaluate approximately with the aid of differentials (see Problem 7):
   (a) $\sqrt{99}$ (b) $1/1001$ (c) $(1.02)^{10}$ (d) $\sin 2\pi/7$
   (e) $\sin 9\pi/16$ (discuss the result)   (f) the solution of $y^3 + y = 0.3$

7. Discuss the error in evaluating by differentials the numbers described in parts (a) and (b) of Problem 6.

8. In *linear interpolation* we evaluate $f(x)$ by the rule

$$f(x) = \frac{f(b) - f(a)}{b - a} (x - a) + f(a)$$

where $a < x < b$.

(a) Show that this procedure is equivalent to replacing the curve $y = f(x)$ by the chord joining $[a, f(a)]$ and $[b, f(b)]$.

(b) Show that the procedure is equivalent to writing $\Delta f = f'(x_0) \, \Delta x$ for an appropriate choice of x_0, $a < x_0 < b$.

(c) Discuss the comparative accuracy of linear interpolation and the use of differentials. Use the function $f = x + cx^2$, $a = 0$, $b = 1$, to illustrate cases that can arise.

9. Find the equation (or equations) of the tangent line at the point requested:

(a) $x^3 + 2y^3 = 3$ at $(1, 1)$ (b) $x^3 + y^3 = 1$ at (x_1, y_1)

(c) $Ax^2 + Bxy + Cy^2 = 1$ at (x_1, y_1) (d) $x = 3t + t^2 + t^4$, $y = t + t^3$ at $(0, 0)$

(e) $xe^x + y \sin y + y^2 = 0$ at $(0, 0)$ (f) $x + xy \ln y + x^3 - 2 = 0$ at $(1, 1)$

MISCELLANEOUS PROBLEMS

1. Copy each of the following sentences and underline each word that is related to the concept of derivative:

(a) The parachute opened quickly and its speed slowly decreased—its vertical motion gradually changing to a partly horizontal one because of a strong wind.

(b) The annual inflation of 5% implied a corresponding rise in the cost of living and a steady drop in value of savings.

2. Evaluate the limits:

(a) $\displaystyle\lim_{x \to 2} \frac{e^x - e^2}{x - 2}$ (b) $\displaystyle\lim_{x \to e} \frac{\ln x - 1}{x - e}$

(c) $\displaystyle\lim_{x \to 0} \frac{\tan^2 x}{x}$ (d) $\displaystyle\lim_{x \to 0} \frac{e^{x^2} - 1}{x}$

(e) $\displaystyle\lim_{h \to 0} \frac{\sin^2(x + h) - \sin^2 x}{h}$ (f) $\displaystyle\lim_{h \to 0} \frac{\sqrt{(x + h)^3 + 1} - \sqrt{x^3 + 1}}{h}$

3. Differentiate:

(a) $y = 3x^7 - 2x^3 - 3$ (b) $y = 5x^4 - 2x^2 - 2$

(c) $y = 3x - \dfrac{4}{x^2}$ (d) $y = 2x^{-1} - 5x^{-2}$

(e) $y = \dfrac{\sin x}{1 + \cos x}$ (f) $y = \dfrac{e^x}{1 + 3e^{-x}}$

(g) $y = 5e^x \sin x$ (h) $y = x \ln x - x$

(i) $x = \sin(2u + 1)$ (j) $x = \cos(1 + e^v)$

(k) $r = \sin^2(1 + t^2)$ (l) $s = \ln \cos(t^3 + t)$

(m) $y = \mathrm{Tan}^{-1} \dfrac{x^2 + 1}{x}$ (n) $y = \mathrm{Sin}^{-1} \dfrac{1}{1 + x^2}$

(o) $y = \sin(\ln \cos x)$ (p) $y = \cos(\ln \sin x)$

(q) $y = \ln \sin \sqrt{1 + e^{3x}}$ (r) $y = e^{\sin\sqrt{1-x^2}}$

(s) $y = e^{x^2} \cos 2x \ln(1 - 2x)$ (t) $y = \dfrac{x^2 \operatorname{Tan}^{-1} x}{1 + (\operatorname{Sin}^{-1} x)^2}$

(u) $y = [3x^2 + 2(x^2 - 1)^{-3}]^{-2}$ (v) $y = [x^{1/2} + 5x^{-3/2}]^{-1}$

(w) $y = x(x^2 + 9)^{-1} + \operatorname{Tan}^{-1}(x/3)$ (x) $y = \ln\left(\dfrac{x - 1}{x - 2}\right) - \dfrac{1}{x - 2}$

(y) $y = (1 + x)^{x^2}$ (z) $y = (\ln x)^{\cos x}$

4. Show that each of the functions has a differentiable inverse function:

   (a) $y = x^3 + x$, $-1 \le x \le 1$ (b) $y = 2x - \cos x$, $-\pi \le x \le \pi$

   (c) $y = e^{-x^2}$, $1 \le x \le 2$ (d) $y = \ln(1 + \sin x)$, $0 \le x \le \dfrac{\pi}{4}$

5. (a) ... (d). Find the derivative of the inverse function for each of the functions of Problem 4.

6. Differentiate implicitly to find y'':

   (a) $x^2 y - xy^2 + x - 3y + 2 = 0$ (b) $xy^2 e^y + y \cos(xy) - 3x = 0$

   (c) $x \ln(1 + y) - 2y^3 - y \sin x = 0$ (d) $ye^y - x \cos y + x^3 \sin^2 3x = 0$

7. From the given parametric equations, find y' and y'':

   (a) $x = t^3 + t + 1$, $y = t^3 - t + 2$ (b) $x = te^{2t}$, $y = e^{2t} - t^2$

   (c) $x = \dfrac{1 - t^2}{1 + t^2}$, $y = \dfrac{2t}{1 + t^2}$ (d) $x = \cos^4 t$, $y = \sin^4 t$

8. Find the derivative of the vector function:

   (a) $\mathbf{F}(t) = \cos 2t\,\mathbf{i} + \sin 2t\,\mathbf{j}$ (b) $\mathbf{F}(t) = e^t(\cos t\,\mathbf{i} + \sin t\,\mathbf{j})$

   (c) $\mathbf{F}(t) = (1 + \sin t)^3(3\mathbf{i} - 5\mathbf{j})$ (d) $\mathbf{F}(t) = |t\,\mathbf{i} + t^2\mathbf{j}|(t\,\mathbf{i} + t^2\mathbf{j})$

9. Prove, under appropriate hypotheses:

   (a) If u and v are functions of x, then $d(u^v) = vu^{v-1}\,du + u^v \ln u\,dv$.

   (b) If u and v are functions of x, then $d(\log_u v) = (\ln u)^{-1}(-u^{-1}\log_u v\,du + v^{-1}\,dv)$.

10. Find the equation of the tangent line to the graph at the point requested:

   (a) The curve of Problem 3(a) at $(0, -3)$.

   (b) The curve of Problem 3(g) at $(0, 0)$.

   (c) The curve of Problem 6(a) at $(0, \frac{2}{3})$.

   (d) The curve of Problem 7(a) at the point where $t = 0$.

11. Find y'':   (a) $y = \cos^3 3x$ (b) $y = e^{-x^2}$ (c) $y = \sin \ln x$ (d) $y = \ln \cos x$

12. Let $f(x)$ be differentiable for $x \ge 0$, let $f(0) = 1$ and $f'(x) \ge 2x$ for $x \ge 0$. Show that $f(x) \ge x^2 + 1$ for $x \ge 0$, and that f has a local minimum only for $x = 0$.

‡13. Prove that, if f is differentiable in $[a, b]$ then, even though f' may not be continuous, f' has the Intermediate Value Property: if $f'(x_1) = m_1$, $f'(x_2) = m_2$ and $m_1 < m < m_2$, then $f'(x) = m$ for some x between x_1 and x_2.

4
INTEGRAL CALCULUS

4-1 INTRODUCTION

The integral calculus involves two concepts: *the indefinite integral,* symbolized by

$$\int f(x)\, dx$$

and *the definite integral,* symbolized by

$$\int_a^b f(x)\, dx$$

The indefinite integral represents a function whose derivative is f. The definite integral is a number and can be interpreted as an *area*. Both concepts have important applications in physical sciences. It will be seen that the two kinds of integral are closely related. The definite integral turns out to be the more fundamental idea, and is the starting point for important generalizations: double integrals, triple integrals, line integrals.

In this chapter, both the indefinite and the definite integral will be developed, and their relationship will be explained. Part I is a general introduction to the whole subject. In Parts II and III we consider the indefinite integral and definite integral in detail.

PART I. THE CONCEPTS OF INTEGRATION

4-2 THE INDEFINITE INTEGRAL

What functions have derivative $3x^2$? We answer at once: $y = x^3$ plus a constant. Similarly we respond to the following questions as indicated.

What is y if $y' = 2x + 1$?	Answer: $y = x^2 + x + C$
What is y if $y' = 4x^3$?	Answer: $y = x^4 + C$
What is y if $y' = 3x^2 + 4x + 2$?	Answer: $y = x^3 + 2x^2 + 2x + C$

In general, if $f(x)$ is a given function in a certain interval, the answer to the question: "What functions have derivative f?" is given by the *indefinite integral* of f, denoted by

$$\int f(x)\, dx$$

Thus we write:

$$\int 3x^2\, dx = x^3 + C, \qquad \int 4x^3\, dx = x^4 + C$$

We will see that our answer always has the form

$$\int f(x)\, dx = F(x) + C \qquad\qquad (4\text{-}20)$$

where C is an "arbitrary" constant; that is, F will be a specific function whose derivative is f and C can be any constant. The fact that adding a constant to F does not affect the derivative is clear, since

$$[F(x) + C]' = F'(x) + 0 = f(x)$$

Also, we know that we have all answers. For we proved in Section 3-21: If $F(x)$ is one function whose derivative is f, then every function whose derivative is f is representable as $F(x) + C$ for some constant C.

The ambiguity in the indefinite integral, in that we do not know the value of the constant C, is an unavoidable complication. We must always keep this in mind. In most applications, finding the indefinite integral is only an intermediary step, and we end with a definite value for C.

We can compare the situation with that of finding the inverse g^{-1} of a given function g. If g is monotone strictly increasing or decreasing in an interval, then there is an inverse function. However, for $y = g(x) = x^2$, $-1 \le x \le 1$, the inverse function cannot be defined: we can write $x = \pm \sqrt{y}$, but we do not know whether to use $+$ or $-$.

The comparison just made is in fact a very good analogy. For we can consider the assignment of derivative to a function as a mapping of one set into another (Figure 4-1). To make this precise, we consider the set X of all functions F having a continuous derivative $F' = DF$ on a certain interval. Then D itself denotes the mapping of X into the set Y of all continuous

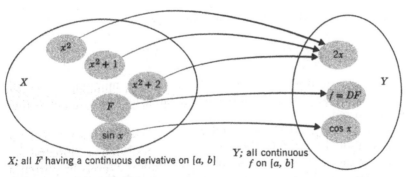

X; all F having a continuous derivative on $[a, b]$

Y; all continuous f on $[a, b]$

Figure 4-1   The mapping D.

functions on the interval. As indicated in Figure 4-1, the mapping D is not one-to-one; in particular, x^2, $x^2 + 1$, $x^2 + 2$, etc., all have the same derivative $2x$.

The problem of finding the indefinite integral is that of forming the inverse of the mapping D. However, since D is not one-to-one, there is no well-defined inverse mapping. The best we can do is to give, for each f in Y, all F such that $DF = f$; this amounts to finding one such F and then taking all functions $F(x) + C$.

Because finding the indefinite integral is, with the ambiguity noted, like forming the inverse of the operator D, it is sometimes denoted by D^{-1}:

$$D^{-1}f = \int f(x)\,dx = F(x) + C \qquad\qquad (4\text{-}21)$$

For the same reason $D^{-1}f$ is sometimes called the *antiderivative of f*. The term *primitive of f* is also used. We also say "integrate f" to mean "find the indefinite integral of f," although this term has additional meanings.

In the indefinite integral $\int f(x)\,dx$, we call $f(x)$ the *integrand*. Instead of thinking of integration as finding a function F whose derivative is our integrand, we can think of it as finding a function F whose differential is $f(x)\,dx$. For if $F' = f$, then

$$dF = F'(x)\,dx = f(x)\,dx$$

This explains the dx in the integration symbol. The differentials are important for the method of substitution (Section 4-7).

Having explained the meaning of the indefinite integral, we now consider how to evaluate it in particular cases. From the examples given above, we easily observe how to integrate polynomials. Thus

$$\int (x^2 + 3x + 5)\,dx = \frac{x^3}{3} + \frac{3x^2}{2} + 5x + C$$

as we can check by differentiating the right-hand side. We can obtain a long list of known indefinite integrals by observing that, for every function we have differentiated, our result gives a certain indefinite integral. Thus, since $(\sin x)' = \cos x$, we can write

$$\int \cos x\,dx = \sin x + C$$

and since $(\cos^3 x)' = -3\cos^2 x \sin x$, we can write

$$\int (-3\cos^2 x \sin x)\,dx = \cos^3 x + C$$

We can continue in this way, accumulating a vast number of known indefinite integrals. However, these results would be hard to use unless they were well organized.

The matter of organization is solved in two ways. First, a few general rules are established, related to those for derivatives, such as "integral of a sum = sum of the integrals." Second, certain key integrals are listed in systematic form in tables (see Appendix I). Through the rules and the tables, the known indefinite integrals are accessible to us in reasonably convenient form.

For most integrals occurring in practice, the rules and tables are adequate. However, in particular cases a long and time-consuming process may be required. This is true even for problems of innocent appearance, as the following example shows.

$$\int \frac{dx}{(1 + x^3)^3}$$

$$= \frac{1}{18} \left[\frac{x(8 + 5x^3)}{(1 + x^3)^2} + \frac{5}{3} \ln \frac{(1 + x)^2}{1 - x + x^2} + \frac{10}{\sqrt{3}} \operatorname{Tan}^{-1} \frac{2x - 1}{\sqrt{3}} \right] + C$$

(4-22)

The correctness of the result can be checked by differentiating the right-hand side.

Cases also arise in which the integral is not reducible to one in the tables. For example, it can be proved that

$$\int e^{-x^2} \, dx$$

is not expressible in terms of finitely many familiar functions! This integral is very important for probability and statistics. We must then ask: Does the function e^{-x^2} have an indefinite integral?

The difficulties of some examples and the uncertainty of being able to evaluate the integral at all, as in the example just mentioned, are discouraging. We might even be tempted to abandon the whole subject of integration as a tangled and frustrating one, with no definite and attainable goal. However, essentially all the serious mathematical problems that arise in physical sciences require integration! Thus, ignoring integration would be an ostrichlike attitude toward most of the problems of science.

Accordingly, the problem of integration is tantalizingly difficult but cannot be ignored! What is the way out of the dilemma?

To our great fortune there is a way out. That way is provided by the *definite integral*. In the next section we explain its meaning and how it answers our questions about indefinite integrals. In later sections we shall learn that the definite integral is in fact a basic tool for solving a great many problems of the sciences and is one of the central ideas of mathematics.

To conclude our first section on the indefinite integral, we simply state that, with the aid of the definite integral, we shall prove the theorem: *every continuous function, on an interval [a, b], has a well-defined indefinite integral* (not necessarily expressible in terms of familiar functions). Furthermore, the definite integral provides a practical way of calculating the indefinite integral in tabular form. The theorem mentioned can be restated as follows. The mapping D of Figure 4-1 has range equal to Y.

PROBLEMS

Where no interval is specified, the following problems refer to the natural interval on which each function is defined.

1. Verify the following integration formulas:

(a) $\int 1\, dx = x + C$

(b) $\int 0\, dx = C$

(c) $\int x\, dx = \frac{x^2}{2} + C$

(d) $\int (3x - 1)\, dx = \frac{3x^2}{2} - x + C$

(e) $\int (3x^2 + 6x + 5)\, dx = x^3 + 3x^2 + 5x + C$

(f) $\int (x^4 - x^3)\, dx = \frac{x^5}{5} - \frac{x^4}{4} + C$

(g) $\int \frac{1}{x}\, dx = \ln x + C, \, x > 0$

(h) $\int \frac{1}{x}\, dx = \ln(-x) + C, \; x < 0$

(i) $\int \sin x\, dx = -\cos x + C$

(j) $\int \sec^2 x\, dx = \tan x + C, \, -\frac{\pi}{2} < x < \frac{\pi}{2}$

(k) $\int e^{2x}\, dx = \frac{1}{2} e^{2x} + C$

2. A function F has derivative $3x^2 - 1$ and $F(0) = 2$. Find F. [*Hint.* F must have form $\int (3x^2 - 1)\, dx = x^3 - x + C$ for some particular C. Use the condition $F(0) = 2$ to find C.]

3. Find the function F in each of the following cases (cf. Problem 2).
   (a) $F'(x) = x$, $F(1) = 3$
   (b) $F'(x) = x^2 - x + 1$, $F(0) = -1$
   (c) $F'(x) = 1 - \sin x$, $F(0) = 2$
   (d) $F'(x) = m$, $F(0) = b$ (m, b are constants)
   (e) $F''(x) = 6x$, $F'(0) = 1$, $F(0) = 5$
   (f) $F''(x) = \sin x$, $F'(0) = 0$, $F(0) = 1$

4. A particle moves on a line, the x-axis, so that $x = x(t)$, where t is time, and its velocity $v(t)$ is $x'(t)$, its acceleration $a(t) = v'(t) = x''(t)$ (all in a chosen system of units). Use the method suggested in Problem 2 for the following exercises.
   (a) If $v(t) = t^2 - \sin t$ and $x(0) = 0$, find $x(t)$.
   (b) If $v(t) = t^2 + t - e^t$ and $x(1) = 0$, find $x(t)$.
   (c) If $v(t) = 1/(1 + t)$ and $x(0) = 0$, find $x(1)$.
   (d) If $a(t) \equiv 32$ and $v(0) = 0$, find $v(t)$.
   (e) If $a(t) = 1 + t^2$ and $v(0) = 0$, find $v(1)$.
   (f) If $a(t) = 10 - t$ and $x(0) = 0$, $v(0) = 0$, find $x(t)$ and $v(t)$.
   (g) If $a(t) \equiv 1$ and $x(0) = x(1) = 1$, find $x(t)$.
   (h) If $a(t) = 1/(1 + t)^2$ and $x(0) = 0$, $x(1) = 10$, find $x(t)$.
   (i) If $a(t) = 1 + \cos t$ and $x(0) = 0$, $v(\pi) = 0$, find $x(t)$.
   (j) If $v(t) = t^2 - 2t$ and $x(0) = 0$, when does x become 0 again for positive t?

5. The gradient of a road (in terms of horizontal distance x, in chosen units) is $0.12 + 0.009x^2$. What is the total rise in going from $x = 0$ to $x = 1$?

6. Verify Equation (4-22) by differentiation.

4-3 THE DEFINITE INTEGRAL

Let f be a given continuous function in the interval $[a, b]$. We seek a function F so that $F'(x) = f(x)$ in the interval; that is, we seek an indefinite integral of f. Does such an indefinite integral exist and, if so, how can we find it?

To answer this question, we first start with a known function $F(x)$, with known continuous derivative $f(x) = F'(x)$, and we seek a way of calculating the values of F from those of $F' = f$. We shall see that, if we know only $F(a)$, then all other values of F can be found from f alone. Our problem has a physical counterpart: An object moves on a straight line at a velocity v that is a known function of time t, $v = f(t)$. If the position of the object at time $t = a$ is known, how can we find the position at an arbitrary later time, by using only the known values of the velocity? We can imagine a passenger sitting in a car whose speedometer records speed but whose mileage gauge is defective. The passenger carefully records the varying speed and tries to determine how far the car has gone, for instance, after one hour.

We return to our function F with derivative f. Let us fix a value c, $a < c \leq b$, and try to relate $F(c)$ and $F(a)$. We can think of $F(c)$ as a value obtained from $F(a)$ by successive increments as x varies from a to c. In particular, we choose points $x_0, x_1, \ldots, x_n$ such that

$$a = x_0 < x_1 < \cdots < x_n = c \tag{4-30}$$

as in Figure 4-2. Then, as x goes from $a = x_0$ to x_1, the function F increases

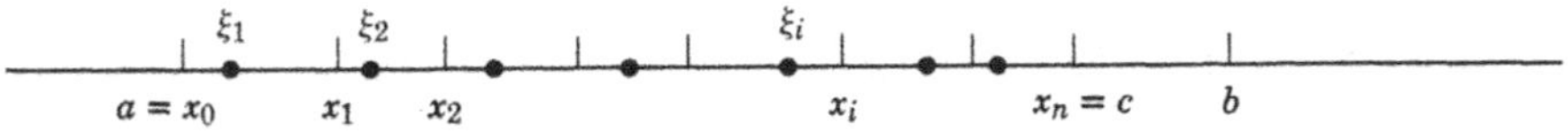

Figure 4-2 Subdivision for definite integral.

(or decreases) by adding the increment $F(x_1) - F(x_0)$:

$$F(x_1) = F(x_0) + [F(x_1) - F(x_0)]$$

Similarly, as x goes from x_0 to x_2, F increases by adding two increments:

$$F(x_2) = F(x_0) + [F(x_1) - F(x_0)] + [F(x_2) - F(x_1)]$$

and ultimately

$$F(c) = F(x_n) = F(x_0) + [F(x_1) - F(x_0)] + [F(x_2) - F(x_1)] + \cdots$$
$$+ [F(x_{n-1}) - F(x_{n-2})] + [F(x_n) - F(x_{n-1})] \tag{4-31}$$

The passenger in the car would be saying: the total distance can be found by totaling the distances traversed in successive intervals, say of 5 minutes.

Now to each term in brackets in Equation (4-31) we can apply the Mean Value Theorem:

$$F(x_1) - F(x_0) = (x_1 - x_0)F'(\xi_1), \qquad x_0 < \xi_1 < x_1$$
$$F(x_2) - F(x_1) = (x_2 - x_1)F'(\xi_2), \qquad x_1 < \xi_2 < x_2$$

and so on. If we replace F' by f everywhere, we thus find that

$$F(c) = F(x_0) + (x_1 - x_0)f(\xi_1) + (x_2 - x_1)f(\xi_2) + \cdots + (x_n - x_{n-1})f(\xi_n) \tag{4-32}$$

The passenger in the car would be saying that the distance traveled in each five minutes equals the time multiplied by an "average speed," somewhere between minimum and maximum for that interval of time.

In (4-32) we are close to our goal. If we knew the values $\xi_1, \ldots, \xi_n$ exactly, we could find $F(c)$ from $F(a)$ and the values of the derivative $F' = f$ at certain points. Now even though we do not know $\xi_1, \ldots, \xi_n$ exactly, we have them "cornered" in small intervals:

$$a < \xi_1 < x_1, \qquad x_1 < \xi_2 < x_2, \qquad \ldots, \qquad x_{n-1} < \xi_n < c \qquad (4\text{-}33)$$

If we make these intervals very small, it is plausible that, no matter what values $\xi_1, \ldots, \xi_n$ we choose in accordance with (4-33), they can be used in (4-32) to *approximate* the right side; the error in the approximation should approach 0 as $n \to \infty$ and the lengths of the intervals approach 0. Therefore, we are led to write

$$F(c) = F(a) + \lim \, [\, f(\xi_1)(x_1 - a) + f(\xi_2)(x_2 - x_1)$$
$$+ \cdots + f(\xi_n)(c - x_{n-1})] \quad (4\text{-}34)$$

where the limit refers to the process just described. We will see that, if we define this limit carefully, it exists and satisfies (4-34). The limit is called the *definite integral of f* over the interval $[a, c]$ and is symbolized by $\int_a^c f(x) \, dx$. Thus

$$\int_a^c f(x) \, dx = \lim \, [\, f(\xi_1)(x_1 - a) + f(\xi_2)(x_2 - x_1) +$$
$$\cdots + f(\xi_n)(c - x_{n-1})] \quad (4\text{-}35)$$

By combining (4-34) and (4-35), we can write

$$F(c) = F(a) + \int_a^c f(x) \, dx \qquad (4\text{-}36)$$

or, since $f = F'$

$$F(c) = F(a) + \int_a^c F'(x) \, dx \qquad (4\text{-}37)$$

Thus we have shown how the value of F can be found at any point c from knowledge of $F(a)$ and of the derivative of F.

We now return to our first question. Given a function f, continuous in $[a, b]$, can we find a function F whose derivative is f? To this end we choose c, $a < c \leq b$, and find the definite integral of f as in (4-35). Then we use Equation (4-36) to *define* $F(c)$. However, we do not know $F(a)$; this value we denote by C, so that

$$F(c) = \int_a^c f(x) \, dx + C \qquad (4\text{-}38)$$

If we now let c vary from a to b, Equation (4-38) defines $F(c)$ in this interval in terms of the arbitrary constant C. When $c = a$, (4-38) gives

$$F(a) = \int_a^a f(x) \, dx + C$$

Our limit definition (4-35) collapses for $c = a$ but clearly in this case we must assign the value 0 to that definite integral:

$$\int_a^a f(x) \, dx = 0$$

Accordingly, $F(a) = C$. Consequently, Equation (4-38) defines F at every c, $a \le c \le b$, and $F(a) = C$. It is a basic theorem, proved later in this chapter, that the function F thus defined has f as derivative: $F'(x) = f(x)$. Accordingly, *every continuous function has an indefinite integral.*

To recapitulate—given the function f, we choose subdivisions of $[a, c]$ (Figure 4-2), and for each we choose values $\xi_1, \xi_2, \ldots, \xi_n$, in the corresponding subintervals and form the sum on the right side of (4-34). The limit of the values of the sum is then the definite integral $\int_a^c f(x)\, dx$. We find this integral for each c in $[a, b]$ and then define $F(c)$ by (4-38), in terms of an arbitrary constant $C = F(a)$. The function F is the desired indefinite integral of f.

We have been somewhat vague about the limit process. This will be made precise in Section 4-15 below. For the present we observe that the evaluation of the limit is a systematic process, which in particular can be calculated as accurately as desired with the aid of a digital computer.

EXAMPLE 1. The following example illustrates the process. Let $f(x) = 2x$ on the interval $[0, 1]$. We choose c, $0 \le c \le 1$, and subdivide the interval $[0, c]$ by the equidistant points

$$0, \quad \frac{c}{n}, \quad \frac{2c}{n}, \quad \ldots, \quad \frac{(n-1)c}{n}, \quad c$$

We choose the ξ-value as the right-hand end point of each subinterval: $\xi_1 = c/n$, $\xi_2 = 2c/n, \ldots$ Then our sum becomes

$$f(\xi_1)(x_1 - a) + \cdots = f\left(\frac{c}{n}\right) \cdot \frac{c}{n} + f\left(\frac{2c}{n}\right) \cdot \frac{c}{n} + \cdots + f\left(\frac{nc}{n}\right) \cdot \frac{c}{n}$$

$$= \frac{c}{n}\left[\frac{2c}{n} + \frac{4c}{n} + \cdots + \frac{2nc}{n}\right]$$

$$= \frac{2c^2}{n^2}[1 + 2 + \cdots + n]$$

$$= \frac{2c^2}{n^2}\frac{n(n+1)}{2}$$

In the last step we used the rule for the sum of an arithmetic progression (Section 0-20). Our sum now reduces to

$$c^2 \cdot \frac{n+1}{n}$$

and as $n \to \infty$ this has limit c^2. Hence

$$\int_0^c 2x\, dx = c^2$$

and if we set

$$F(c) = \int_0^c 2x\, dx + C = c^2 + C$$

that is, $F(x) = x^2 + C$, then F should be the indefinite integral of $2x$, with $F(0) = C$. Since $\int 2x\, dx = x^2 + C$, we have perfect agreement!

Up to this point we have emphasized finding the indefinite integral of f and have used the definite integral as a means to that end. *If we already know the indefinite integral, we can use it to evaluate the definite integral and completely bypass the limit process.* We illustrate the process:

EXAMPLE 2. Evaluate

$$\int_0^{\pi/2} \cos x \, dx$$

An indefinite integral of $\cos x$ is $F(x) = \sin x$. Accordingly, by (4-37), with $F(x) = \sin x$, $F'(x) = \cos x$,

$$F\left(\frac{\pi}{2}\right) - F(0) = \sin \frac{\pi}{2} - \sin 0 = \int_0^{\pi/2} \cos x \, dx$$

so that

$$\int_0^{\pi/2} \cos x \, dx = 1$$

In general, we can write, by (4-37),

$$\int_a^b f(x) \, dx = F(b) - F(a) \tag{4-39}$$

where $F(x)$ is an indefinite integral of f. We might ask: What would be the result if $F(x)$ were replaced by $F(x) + C$? Clearly, it has no effect on (4-39), since

$$[F(b) + C] - [F(a) + C] = F(b) - F(a)$$

Hence the choice of arbitrary constant is irrelevant.

We thus observe that Equation (4-36) can be looked at in two different ways.

First, if f is given on $[a, b]$, then (4-36) gives $F(c)$ for every c on $[a, b]$, and F is an indefinite integral of f, with $F(a)$ as arbitrary constant.

Second, if f is given on $[a, c]$ and we know an indefinite integral F of f, then (4-36) allows us to compute the definite integral

$$\int_a^c f(x) \, dx$$

without resorting to a limit process.

PROBLEMS

1. For each of the following functions f write out the sum $f(\xi_1)(x_1 - a) + \cdots$ $f(\xi_n)(c - x_{n-1})$ for the interval $[a, c]$ given, the given subdivision points, and the choices of $\xi_1, \ldots$:
   (a) $f(x) = x^2$ on $[0, 1]$, with $x_0 = 0$, $x_1 = 1/n$, $x_2 = 2/n, \ldots$, and $\xi_1 = x_1$, $\xi_2 = x_2, \ldots, \xi_n = x_n$
   (b) $f(x) = x^2$ on $[0, 1]$ with $x_0 = 0, x_1 = 1/n, x_2 = 2/n, \ldots, \xi_1 = x_0, \xi_2 = x_1, \ldots$, $\xi_n = x_{n-1}$

(c) $f(x) = x^2$ on $[0, 1]$ with $x_0 = 0$, $x_1 = 1/n$, $x_2 = 2/n, \ldots, \xi_1 = (x_0 + x_1)/2$, $\xi_2 = (x_1 + x_2)/2, \ldots$ (midpoints)

(d) $f(x) = x^2$ on $[0, 1]$ with $x_0 = 0$, $x_1 = 1/n^2$, $x_2 = 4/n^2, \ldots, x_{n-1} = ([n-1]/n)^2$, $x_n = 1$, $\xi_1 = x_1$, $\xi_2 = x_2, \ldots, \xi_n = x_n$

2. (a) It can be shown by induction (Section 0-20) that

$$(°) \quad 1^2 + 2^2 + \cdots + n^2 = \frac{n^3}{3} + \frac{n^2}{2} + \frac{n}{6} \text{ for } n = 1, 2, \ldots$$

Use this result, Equation (4-35), and the answer given to Problem 1(a) to evaluate $\int_0^1 x^2 \, dx$ by a limit process.

(b) Prove $(°)$ by induction.

(c) Evaluate $\int_0^c x^2 \, dx$ without a limit process.

(d) Evaluate $\int_0^c x^2 \, dx$ by a limit process.

3. (a) Evaluate $\int_0^c (3x + 2) \, dx$ by a limit process and use the result to find $\int (3x + 2) \, dx$.

(b) Evaluate $\int_0^c (3x + 2) \, dx$ without a limit process.

4. Evaluate without a limit process:

(a) $\int_2^3 (x^4 - x) \, dx$ (b) $\int_0^{\pi/4} \sec^2 x \, dx$ (c) $\int_1^2 \frac{1}{x} \, dx$ (d) $\int_{-1}^1 e^x \, dx$

5. A passenger in a car observed speedometer readings as follows, in miles per hour:

Time	1:00	1:05	1:10	1:15	1:20	1:25	1:30	1:35	1:40	1:45	1:50	1:55	2:00
Speed	23	25	28	35	40	42	45	55	60	60	60	55	50

Estimate the distance traveled in the hour.

6. (a) Evaluate $\int_0^c e^x \, dx$, $c > 0$,   by a limit process.

[*Hint.* Subdivide and choose ξ-values as in Example 1 in the text. Simplify the sum by setting $r = e^{c/n}$ and by using the formula for the sum of a geometric progression (see Section 0-20). Also notice that $(e^{c/n} - 1)/(c/n)$ has limit 1, as $n \to \infty$, since the limit can be interpreted as the derivative of e^x at $x = 0$.]

(b) Evaluate $\int_0^c e^x \, dx$, $c > 0$,   without a limit process.

4-4 AREA

The discussion of the preceding section has a geometric interpretation in terms of area. Let f be defined on the interval $[a, b]$ and be continuous. For simplicity, we also assume that $f(x) > 0$ for $a \le x \le b$. Let $A(c)$ denote the area of the figure shaded in Figure 4-3; thus the figure is bounded by the interval $[a, c]$ on the x-axis, by segments on the lines $x = a$ and $x = c$, and by the portion of the graph of f for which $a \le x \le c$. (We call this the *area under the graph of f* from $x = a$ to $x = c$.) We here treat area intuitively, returning in Section 4-18 for a thorough discussion. We are interested in

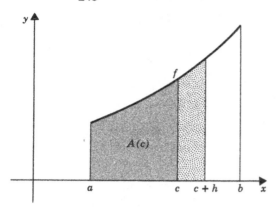

Figure 4-3 Area and integration.   **Figure 4-4** Area under $y = \cos x$.

the variation of $A(c)$ as c varies. Here $A(c)$ is a function of c defined for $a \leq c \leq b$.

Let us fix a particular c, $a < c < b$, and choose $h > 0$ so that $c < c + h < b$. Then $A(c + h) - A(c)$ represents the dotted area in Figure 4-3. This area is clearly included between two rectangles: one of base h and altitude m, the minimum of f in the interval $[c, c + h]$; the other of base h and altitude M, the maximum of f in the interval. Hence

$$mh < A(c + h) - A(c) < Mh$$

so that

$$m < \frac{A(c + h) - A(c)}{h} < M$$

Since f is continuous, if we let $h \to 0+$, m and M both approach the value of f at c. Accordingly (see Remarks at the end of Section 2-5),

$$\lim_{h \to 0+} \frac{A(c + h) - A(c)}{h} = f(c)$$

By a similar argument

$$\lim_{h \to 0-} \frac{A(c + h) - A(c)}{h} = f(c)$$

Accordingly, $A(c)$ is a function for which $A'(c) = f(c)$. Thus, if we replace c by x, $A(x)$ *is an indefinite integral of* $f(x)$. Since $A(a) = 0$, $A(x)$ is that indefinite integral having value 0 at $x = a$.

Our result can be used in two ways. First, it provides us with a way of evaluating certain areas. Second, it provides us with a new way of finding the indefinite integral of f—namely, by finding the area $A(c)$ for each c by some means. We can even graph f very carefully on high quality paper of constant known density, and then can measure $A(c)$ by cutting out the corresponding portion of the graph and weighing it!

EXAMPLE 1. Find the area under the graph of $y = x^2$ from $x = 0$ to $x = c$.

Solution. $A(x)$ is an indefinite integral of x^2; hence, $A(x) = (x^3/3) + C$. Also, $A(0) = 0$. Therefore, $A(x) = x^3/3$ and $A(c) = c^3/3$.

EXAMPLE 2. Find an indefinite integral of $y = \cos x$, $0 \leq x \leq \pi/2$.

Solution. We ignore the fact that we already know an indefinite integral $(\sin x)$ and solve the problem graphically. In Figure 4-4 we have graphed $y = \cos x$ on ruled paper. We then estimate the area by counting those squares entirely in the region—each square having area 0.04. The results are shown on the second line of Table 4-1. We can improve the accuracy of the approximation of $A(x)$ at these values [and also obtain approximations at more values] by using smaller squares. If we use squares of side 0.01, we obtain the values (to three decimal places) shown on the third line of Table 4-1. The table provides, in tabular form, an approximation to the desired indefinite integral. The exact indefinite integral is $A(x) = \sin x$, and we can verify that the tabulated values on the third line are very close to the true ones (given in the table in Appendix II).

Table 4-1

c	0	0.2	0.4	0.6	0.8	1.0	1.2	1.4	1.57
$A_1(c)$	0	0.16	0.32	0.48	0.60	0.68	0.72	0.72	0.72
$A_2(c)$	0	0.196	0.392	0.560	0.761	0.891	0.951	0.982	0.992

Thus far we have considered area only as an indefinite integral. Since $A(c)$ is an indefinite integral and $A(a) = 0$, we conclude from (4-36) that

$$A(c) = \int_a^c f(x)\, dx \qquad (4\text{-}40)$$

Consequently, *for a positive continuous function f, the definite integral of f from a to c equals the area under the graph of f from $x = a$ to $x = c$.* This result can also be obtained from the limit process. To that end we consider a subdivision of the interval $[a, c]$, as in Figure 4-5. A typical term in the sum for our definite integral is

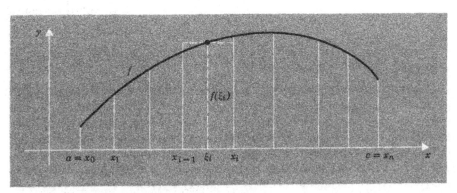

Figure 4-5 Area as a definite integral.

$$f(\xi_i)(x_i - x_{i-1}), \qquad x_{i-1} \leq \xi_i \leq x_i$$

We have allowed the ξ_i to be chosen freely in the subinterval. However, if ξ_i is properly chosen, the term $f(\xi_i)(x_i - x_{i-1})$ equals the area under the curve from x_{i-1} to x_i. To see this we notice that the area in question certainly lies between $m_i(x_i - x_{i-1})$ and $M_i(x_i - x_{i-1})$ where m_i and M_i are respectively the minimum and maximum of f in the subinterval. Now $m_i(x_i - x_{i-1})$ is the area of a rectangle included in the region whose area we seek, and $M_i(x_i - x_{i-1})$ is the area of a rectangle which includes that region. Therefore, for some number k_i between m_i and M_i, we have

$$\text{Area under the curve from } x_{i-1} \text{ to } x_i = k_i(x_i - x_{i-1})$$

Since f is continuous, $f(\xi_i) = k_i$ for some ξ_i in the subinterval (Intermediate Value Theorem, Section 2-7). It follows that if the ξ_i are properly chosen, our sum

$$f(\xi_1)(x_1 - a) + f(\xi_2)(x_2 - x_1) + \cdots$$

is precisely equal to the area. Hence, if these sums have a limit, as $n \to \infty$ and as the width of the subdivision intervals approaches 0, that limit must equal the area. Thus (4-40) is also obtained from the limit process.

The existence of the limit for the definite integral will be proved in Section 4-25 below. The area interpretation makes the existence plausible. For no matter how we choose the ξ_i, our sum is not less than $m_1(x_1 - a) + m_2(x_2 - x_1) + \cdots$ and is not greater than $M_1(x_1 - a) + M_2(x_2 - x_1) + \cdots$. The difference of these two sums is the sum of the areas of the shaded rectangles is Figure 4-6. It is reasonable to expect that, for a continuous

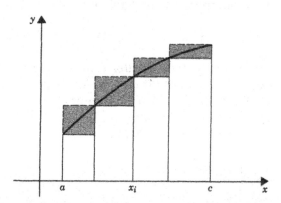

Figure 4-6 Limit defining definite integral.

function, as we let the length of our subdivision intervals shrink to 0, the sum of these areas also approaches 0. Thus the difference of the sum $m_1(x_1 + a) + \cdots$ and the sum $M_1(x_1 - a) + \cdots$ approaches 0. But the first is less than or equal to the area $A(c)$, and the second is greater than or equal to $A(c)$. Hence both have $A(c)$ as limit. Since our general sum $f(\xi_i)(x_1 - a) + f(\xi_2)(x_2 - x_1) + \cdots$ is squeezed between $m_1(x_1 - a) + m_2(x_2 - x_1) + \cdots$ and $M_1(x_1 - a) + M_2(x_2 - x_1) + \cdots$, it also has $A(c)$ as limit.

Throughout this section we have restricted attention to a function f which is ≥ 0. If we allow negative values, we can still give an area interpretation to the indefinite and definite integrals. $A(c)$ *becomes the area above the x-axis minus the area below the x-axis, from $x = a$ to $x = c$.* Thus the areas are counted *with signs*, as suggested in Figure 4-7.

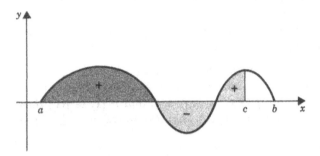

Figure 4-7   Signed area.

PROBLEMS

1. Find the area of the region bounded by the following curves. Graph the figure in each case and estimate the result graphically.
   (a) $y = 0$, $x = 1$, $y = x^2$, $x = 2$ (b) $y = 0$, $x = 0$, $y = e^x$, $x = 5$
   (c) $y = \frac{1}{2}$, $x = 1$, $y = 1/x$, $x = 2$ (d) $y = 1$, $x = 1$, $y = x^4$, $x = 2$
2. Find the area of the region bounded by the following curves: $y = x$, $y = x^2$. (*Hint.* Graph and notice that the area sought is the *difference* of two areas of the type considered in the text.)
3. Find the area bounded by the given curves (cf. Problem 2):
   (a) $y = x^3$, $y = x^3 + 1$, $x = 1$, $x = 2$ (Can you guess the result in advance?)
   (b) $y = \sin x$, $y = \cos x$, $x = 0$, with $0 \leq x \leq \pi/4$
   (c) $y = \sqrt{x}$, $y = 1/x$, $x = 2$
   (d) $y = x + \sin x$, $y = x$, $0 \leq x \leq \pi$
4. Let $A(c)$ be the area under the graph of $y = x^2 + \sin x$ from $x = 0$ to $x = c\ (c > 0)$. Find $A'(\pi/2)$.
5. Follow the procedure of Example 2 in the text to find an indefinite integral, in tabulated form, of $y = \sin x$ for $0 \leq x \leq \pi$, and compare it with the exact indefinite integral (Appendix II).

PART II.   THE INDEFINITE INTEGRAL

4-5 BASIC PROPERTIES OF THE INDEFINITE INTEGRAL

We now develop the theory of the indefinite integral in systematic fashion.

Equations in indefinite integrals have only a symbolic meaning and cannot be manipulated as ordinary equations. Thus

$$\int 3x^2\, dx = x^3 + C$$

expresses the meaning: the functions whose derivative is $3x^2$ are the functions of form $x^3 + C$, where C is a constant. It is also true that

$$\int 3x^2\, dx = x^3 + 1 + C$$

If we could combine the last two equations, we would conclude that $x^3 + C = x^3 + 1 + C$ or that $0 = 1!$

In practice, confusion arising from the equality sign is rare. However, we must always keep the ambiguity of the indefinite integral in mind.

THEOREM 1. *Let $f(x)$ be a continuous function on a certain interval. If $F'(x) = f(x)$ and $G'(x) = f(x)$ in the interval, then $G(x) = F(x) + C$ for some constant C. Conversely, if $G(x) = F(x) + C$ and $F'(x) = f(x)$, then $G'(x) = f(x)$. Accordingly, if $F'(x) = f(x)$, then the indefinite integral of f is given by*

$$\int f(x)\, dx = F(x) + C \qquad\qquad (4\text{-}50)$$

PROOF. This theorem is proved in Chapter 3 (Section 3-21).

Table 4-2 presents basic rules for indefinite integrals. Further rules are given in Table I in the Appendix, a part of which is reproduced inside the

Table 4-2   Rules for Indefinite Integrals

1. $\int [f(x) + g(x)]\, dx = \int f(x)\, dx + \int g(x)\, dx + C$

2. $\int kf(x)\, dx = k\int f(x)\, dx + C, \qquad k = \text{const.}$

3. (a) $\int x^n\, dx = \dfrac{x^{n+1}}{n+1} + C, n \neq -1$

   (b) $\int x^\alpha\, dx = \dfrac{x^{\alpha+1}}{\alpha+1} + C, \alpha \neq -1$

4. (a) $\int \dfrac{dx}{x} = \ln x + C, x > 0$

   (b) $\int \dfrac{dx}{x} = \ln(-x) + C, x < 0$

   (c) $\int \dfrac{dx}{x} = \ln |x| + C, x \neq 0$

5. (a) $\int e^x\, dx = e^x + C$

   (b) $\int a^x\, dx = \dfrac{a^x}{\ln a} + C, a > 0, a \neq 1$

6. $\int \sin x\, dx = -\cos x + C$

7. $\int \cos x\, dx = \sin x + C$

8. $\int \sec^2 x\, dx = \tan x + C$

9. $\int \csc^2 x\, dx = -\cot x + C$

10. $\int \sec x \tan x\, dx = \sec x + C$

11. $\int \csc x \cot x\, dx = -\csc x + C$

12. $\int \dfrac{dx}{\sqrt{a^2 - x^2}} = \mathrm{Sin}^{-1} \dfrac{x}{a} + C, \qquad a > 0$

13. $\int \dfrac{dx}{a^2 + x^2} = \dfrac{1}{a} \mathrm{Tan}^{-1} \dfrac{x}{a} + C, \qquad a > 0$

14. (a) $\int g[u(x)]u'(x)\, dx = \int g(u)\, du$

    (b) $\int f(x)\, dx = \int f[x(u)]x'(u)\, du$

15. $\int f(x)g'(x)\, dx = f(x)g(x) - \int g(x)f'(x)\, dx$

back covers of this volume. We discuss the rules 1–13 here, and the last three are discussed in Sections 4-7 and 4-9.

In Rules 1 and 2, f and g are assumed to be continuous in a certain interval and to have indefinite integrals. We write

$$\int f(x)\, dx = F(x) + C, \qquad \int g(x)\, dx = G(x) + C$$

Hence, $F'(x) = f(x)$, $G'(x) = g(x)$, so that $F + G$ has derivative $f + g$ and, therefore,

$$\int [f(x) + g(x)]\, dx = F(x) + G(x) + C = \int f(x)\, dx + \int g(x)\, dx + C$$

(only one arbitrary constant is needed). This proves Rule 1. Rule 2 is proved similarly by observing that kF has derivative kf.

Warning. Rule 2 states, in effect, that a constant factor can be moved outside the integral sign. A nonconstant factor cannot be moved outside! For example,

$$\int x^2\, dx \neq x \int x\, dx + C$$

since the left side is $(x^3/3) + C$, the right side is $(x^3/2) + C$.

Rules 3(a) to 13 are essentially restatements of the rules of differentiation (Table 3-2, reproduced inside the front covers of this volume). In each case x is restricted to an interval in which the integrand is continuous; this is indicated explicitly for Rules 4(a) and 4(b). For the other rules, such an interval is easy to recognize. For example, $-\pi/2 < x < \pi/2$ for Rule 8, and $-a < x < a$ for Rule 12.

In Rule 3(a), n is understood to be an integer other than -1; the case $n = -1$ is covered by Rules 4(a), 4(b), and 4(c). For $n = -2, -3, \ldots$, the integrand is discontinuous for $x = 0$ only. In Rule 3(b), α is allowed to be any fixed real number other than -1; here there is a discontinuity at $x = 0$ if α is negative, and all negative x must be excluded whenever x^α is not defined for $x < 0$ (for example, for $\alpha = \frac{1}{2}$).

Rule 4(a) corresponds to the familiar rule for $(\ln x)'$. Rule 4(b) is justified as follows: for $x < 0$, $-x$ is positive and $\ln(-x)$ is defined, with derivative

$$[\ln(-x)]' = \frac{1}{-x}(-1) = \frac{1}{x}$$

The Rules 4(a) and 4(b) are combined in the Rule 4(c):

$$\int \frac{dx}{x} = \ln|x| + C, \qquad x \neq 0$$

The two intervals are shown in Figure 4-8. In Rule 4(c), we are concealing the fact that the "constant" C can be given one value for $x < 0$ and another value for $x > 0$.

Rules 5 to 11 all correspond to familiar rules of differentiation. For $a = 1$, Rules 12 and 13 follow from the familiar rules

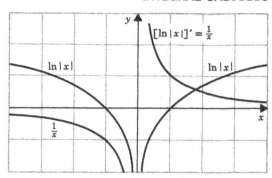

Figure 4-8 Integral of $1/x$.

$$(\text{Sin}^{-1} x)' = \frac{1}{\sqrt{1 - x^2}}, \qquad (\text{Tan}^{-1} x)' = \frac{1}{1 + x^2}$$

(Section 3-6). Here, the principal values are essential to avoid ambiguity and to give the correct sign. The Rules 12 and 13, as stated, with a a positive constant, follow from the special case mentioned and the chain rule. For example

$$\left(\text{Sin}^{-1} \frac{x}{a}\right)' = \frac{1}{\sqrt{1 - \dfrac{x^2}{a^2}}} \cdot \frac{1}{a} = \frac{1}{\sqrt{a^2 - x^2}}$$

The cancelation of a is correct for $a > 0$.

4-6 APPLICATIONS OF RULES OF INTEGRATION

The rules developed thus far permit us to integrate all linear combinations (Section 2-9) of the functions that occur in Rules 3(a) to 13. For each particular case, we must check the interval or intervals of validity. However, usually the restrictions on x are evident, and we often find the formulas stated (for example, in tables) without the explicit mention of the intervals of validity. The emphasis is on the formulas, or what we call a *formal theory*.

EXAMPLE 1. $\int(3x^4 + 5x^3 + 2x^2 + x + 5)\,dx$. This is a typical polynomial problem [linear combination of the functions x^4, x^3, x^2, x, 1 (constant function)]. Rule 1 applies to a sum of two functions. By induction we can extend it to give the rule "*integral of sum equals sum of integrals*" for any finite number of terms (Problem 3 below). Hence our integral equals

$$\int 3x^4\,dx + \int 5x^3\,dx + \int 2x^2\,dx + \int x\,dx + \int 5\,dx + C$$

Now Rule 2 allows us to factor out the constants and we get, with the aid of Rule 3(a),

$$3\int x^4\,dx + 5\int x^3\,dx + 2\int x^2\,dx + \int x\,dx + 5\int dx + C$$

$$= \frac{3x^5}{5} + \frac{5x^4}{4} + \frac{2x^3}{3} + \frac{x^2}{2} + 5x + C$$

The result is clearly valid for all x.

EXAMPLE 2. $$\int\left(4x^5 + \frac{2}{x} + \frac{3}{x^2}\right)dx$$

In the same way, we obtain, with the aid of Rules 1, 2, 3(a) and 4(c):

$$\int\left(4x^5 + \frac{2}{x} + \frac{3}{x^2}\right) = 4\frac{x^6}{6} + 2\ln|x| - \frac{3}{x} + C, \qquad x \neq 0$$

EXAMPLE 3 $$\int\frac{3x^2 + x + 1}{x^3}\,dx$$

We can replace the given integral by

$$\int\left(\frac{3}{x} + \frac{1}{x^2} + \frac{1}{x^3}\right)dx = 3\ln|x| - \frac{1}{x} - \frac{1}{2x^2} + C$$

and require $x \neq 0$.

EXAMPLE 4 $$\int(\sqrt{x^3} + \sqrt[3]{x^2})\,dx$$

We apply Rules 1 and 3(b):

$$\int\left(\sqrt{x^3} + \sqrt[3]{x^2}\right)dx = \int(x^{3/2} + x^{2/3})\,dx = \frac{2}{5}x^{5/2} + \frac{3}{5}x^{5/3} + C$$

Because of the first term, we must require $x \geq 0$.

EXAMPLE 5 $$\int(3\sin x - 2\cos x + 5e^x)\,dx$$

Here we obtain, by Rules 1, 2, 5(a), 6, and 7:

$$-3\cos x - 2\sin x + 5e^x + C$$

PROBLEMS

1. Evaluate and check by differentiation:

(a) $\int(2x^3 - x + 7)\,dx$

(b) $\int(3x^8 - 2x^4 + 1)\,dx$

(c) $\int(x - 1)^2\,dx$

(d) $\int(2x + 3)^3\,dx$

(e) $\int(x^{-2} + 2x^{-3})\,dx$

(f) $\int\left(\frac{1}{x^3} - \frac{1}{x^2}\right)dx$

(g) $\int\frac{2x^2 + x + 1}{x^3}\,dx$

(h) $\int\frac{x^3 - x + 5}{x^2}\,dx$

(i) $\int\frac{1 + \sqrt{x}}{x}\,dx$

(j) $\int(x^{3/4} + x^{7/4})\,dx$

(k) $\int\frac{x^2 - 1}{\sqrt[3]{x}}\,dx$

(l) $\int(1 + \sqrt{x} + \sqrt[3]{x})^2\,dx$

(m) $\int(\cos x - x)\,dx$

(n) $\int(3e^x + \sin x)\,dx$

(o) $\int \dfrac{1 - \cos^2 x}{\sin x}\, dx$ 　　　　　　　　　　(p) $\int e^{x+1}\, dx$

(q) $\int 0\, dx$ 　　　　　　　　　　　　　　　(r) $\int \dfrac{1}{|x|}\, dx$

(s) $\int \sin\left(x + \dfrac{\pi}{3}\right) dx$ 　　　　　　　　(t) $\int \cos\left(x - \dfrac{\pi}{4}\right) dx$

(u) $\int 2^x\, dx$ 　　　　　　　　　　　　　(v) $\int \tan^2 x\, dx$

(w) $\int \dfrac{\sqrt{1 - x^2}}{1 - x^2}\, dx$ 　　　　　　　　　(x) $\int \dfrac{3x^4 + 2x^2 + 5}{x^2 + 1}\, dx$

2. Is it correct to write Rule 2 in the form:

$$\int k\, f(x)\, dx = k\!\int f(x)\, dx? \qquad (\text{Try } k = 0.)$$

3. Prove by induction that, for every positive integer n, if $f_1(x), \ldots, f_n(x)$ are continuous in an interval and $\int f_1(x)\, dx = F_1(x) + C, \ldots, \int f_n(x)\, dx = F_n(x) + C$, then

$$\int [f_1(x) + \cdots + f_n(x)]\, dx = F_1(x) + \cdots + F_n(x) + C$$

4. Prove that, if $f(x) > 0$ for $a \le x \le b$ and $F(x) + C = \int f(x)\, dx$, then $F(x)$ is monotone strictly increasing for $a \le x \le b$.
5. Prove that, if $F(x)$ and $G(x)$ are defined and have second derivatives in an interval and if $F''(x) = G''(x)$, then $G(x) = F(x) + C_1 x + C_2$ for some constants C_1, C_2.
6. Find all functions $F(x)$ satisfying the given condition for all x (see Problem 5):

　　(a) $F''(x) = 0$ 　　(b) $F''(x) = x$ 　　(c) $F''(x) = \cos x$ 　　(d) $F''(x) = e^x$

7. Find the area under the graph of the function:

　　(a) $y = \dfrac{1}{x}, \quad e \le x \le e^2$ 　　(b) $y = \sin x, \quad 0 \le x \le \pi$

4-7 SUBSTITUTION IN INDEFINITE INTEGRALS

Very often the introduction of a new variable u by an equation $x = x(u)$, or $u = u(x)$, reduces an integral $\int f(x)\, dx$ to a simpler one $\int g(u)\, du$. In this section we consider the procedure in a purely formal way and shall give statements of precise theorems in the following section.

In considering substitutions, we think of $\int f(x)\, dx$ as describing a function $F(x)$ whose differential is $f(x)\, dx$:

$$\int f(x)\, dx = F(x) + C \qquad \text{if} \qquad dF \equiv F'(x)\, dx = f(x)\, dx$$

Accordingly, in introducing a new variable u, we must take into account both of the following:

　　(a) The expression of the integrand in terms of u.
　　(b) The expression of dx in terms of du (or of du in terms of dx).

EXAMPLE 1. Evaluate $\int e^{\sin x} \cos x\, dx$. We write

$$u = \sin x, \qquad du = \cos x\, dx$$

so that

$$e^{\sin x} \cos x \, dx = e^u \, du$$

and our given integral is replaced by $\int e^u \, du$:

$$\int e^{\sin x} \cos x \, dx = \int e^u \, du, \qquad (u = \sin x; \, du = \cos x \, dx)$$

The integral on the right can be evaluated as $e^u + C$, so that our given integral is $e^{\sin x} + C$.

We can indicate our whole procedure as follows:

$$\int e^{\sin x} \cos x \, dx = \int e^u \, du, \qquad (u = \sin x; \, du = \cos x \, dx)$$

$$= e^u + C$$
$$= e^{\sin x} + C$$

We can verify the correctness of our result with the aid of the chain rule:

$$(e^{\sin x})' = e^{\sin x}(\sin x)' = e^{\sin x} \cos x$$

We will see that the chain rule is the basis for all the substitution procedures.

EXAMPLE 2. $\int (x^2 + 1)^{5/2} x \, dx$. If we write $u = x^2 + 1$, then $du = 2x \, dx$. We take advantage of the fact that constants may be moved past the integral sign (Rule 2) to find the needed factor 2:

$$\int (x^2 + 1)^{5/2} x \, dx = \tfrac{1}{2} \int (x^2 + 1)^{5/2} 2x \, dx = \tfrac{1}{2} \int u^{5/2} \, du, \qquad \begin{bmatrix} u = x^2 + 1 \\ du = 2x \, dx \end{bmatrix}$$

$$= \frac{1}{2} \cdot \frac{2}{7} u^{7/2} + C$$

$$= \frac{1}{7} (x^2 + 1)^{7/2} + C$$

Check: $[\tfrac{1}{7}(x^2 + 1)^{7/2}]' = \tfrac{1}{7} \cdot \tfrac{7}{2}(x^2 + 1)^{5/2} \cdot 2x = x(x^2 + 1)^{5/2}$.

EXAMPLE 3

$$\int (x^3 + x + 1)^7 (3x^2 + 1) \, dx = \int u^7 \, du, \qquad \begin{bmatrix} u = x^3 + x + 1 \\ du = (3x^2 + 1) \, dx \end{bmatrix}$$

$$= \frac{u^8}{8} + C = \frac{(x^3 + x + 1)^8}{8} + C$$

We could have expanded the parenthesis to the 7th power and eventually could have obtained a polynomial of degree 23, which could then be integrated term by term. The substitution procedure is clearly shorter.

EXAMPLE 4

$$\int 2xe^{x^2} \cos(e^{x^2}) \, dx = \int e^u \cos e^u \, du, \qquad \begin{bmatrix} u = x^2 \\ du = 2x \, dx \end{bmatrix}$$

Our new integral is not familiar, but we can substitute again:

$$\int e^u \cos e^u \, du = \int \cos v \, dv, \qquad (v = e^u, \, dv = e^u \, du)$$
$$= \sin v + C$$
$$= \sin e^u + C$$
$$= \sin e^{x^2} + C$$

Sometimes three, four, or more substitutions may be used to reduce the problem to a familiar integral. In such cases, we can then redo the integration by one single substitution that is obtained by combining the previous ones. Thus, in the present example, we let $v = e^{x^2}$ and write

$$\int 2xe^{x^2} \cos(e^{x^2}) \, dx = \int (\cos e^{x^2})(\underbrace{e^{x^2} 2x \, dx})$$
$$= \int \cos v \, dv, \qquad (v = e^{x^2}, \, dv = e^{x^2} \, 2x \, dx)$$
$$= \sin v + C = \sin e^{x^2} + C$$

EXAMPLE 5

$$\int \frac{x \, dx}{\sqrt{1 + x^2}} = \frac{1}{2}\int \frac{du}{\sqrt{1 + u}}, \qquad (u = x^2, \, du = 2x \, dx)$$
$$= \frac{1}{2}\int \frac{dv}{\sqrt{v}}, \qquad (v = 1 + u, \, dv = du)$$
$$= \frac{1}{2}\int v^{-1/2} \, dv = v^{1/2} + C$$
$$= \sqrt{1 + u} + C = \sqrt{1 + x^2} + C$$

Here the substitution $v = 1 + x^2$, $dv = 2x \, dx$ would have given the result in one step.

EXAMPLE 6
$$\int \frac{dx}{2 + e^x}$$

Here we might try $u = e^x$, but $du = e^x \, dx$ is not visible. We force it into view by multiplying the numerator and denominator by e^x:

$$\int \frac{dx}{2 + e^x} = \int \frac{e^x \, dx}{2e^x + e^{2x}} = \int \frac{du}{2u + u^2}, \qquad (u = e^x, \, du = e^x \, dx)$$

We do not yet have a familiar integral. We observe that

$$\frac{du}{2u + u^2} = \frac{1}{(2/u) + 1} \cdot \frac{du}{u^2}$$

and, accordingly, set $v = 1/u$, $dv = -u^{-2} \, du$, to obtain

$$\int \frac{du}{2u + u^2} = \int \frac{1}{(2/u) + 1} \, du/u^2 = -\int \frac{1}{2v + 1} \, dv$$

Finally, we make one more substitution, $w = 2v + 1$, $dw = 2 \, dv$, and obtain

$$-\int \frac{1}{2v+1}\,dv = -\frac{1}{2}\int \frac{dw}{w}$$

$$= -\frac{1}{2}\ln|w| + C$$

$$= -\frac{1}{2}\ln|2v+1| + C$$

$$= -\frac{1}{2}\ln\left|\frac{2}{u}+1\right| + C$$

$$= -\frac{1}{2}\ln(2e^{-x}+1) + C$$

(The absolute value sign is dropped at the end, since $2e^{-x}+1 > 0$ for all x.)

This last example can be done in another way. Instead of forcing du into view, we express x in terms of u, $x = x(u)$, and then write $dx = x'(u)\,du$. Thus

$$f(x)\,dx = f[x(u)]x'(u)\,du$$

In this way we can make any desired substitution $x = x(u)$. At the end we must express u in terms of x. Thus this method requires that $u = u(x)$ have an inverse $x = x(u)$ and, for effective use of the method, we must be able to solve explicitly both ways—for x in terms of u and for u in terms of x. We carry out the program described for the preceding example:

$$\int \frac{dx}{2+e^x} = \int \frac{1}{2+u}\,x'(u)\,du, \qquad (u = e^x \text{ or } x = \ln u;\ dx = u^{-1}\,du)$$

$$= \int \frac{1}{2+u}\,\frac{1}{u}\,du$$

$$= \int \frac{1}{2+\dfrac{1}{v}}\,v\left(-\frac{1}{v^2}\right)dv, \qquad \begin{bmatrix} u = v^{-1} \text{ or } v = u^{-1} \\ du = -v^{-2}\,du \end{bmatrix}$$

$$= -\int \frac{dv}{2v+1}$$

$$= -\frac{1}{2}\int \frac{dw}{w}, \qquad \begin{bmatrix} w = 2v+1 \\ dw = 2\,dv \end{bmatrix}$$

$$= -\frac{1}{2}\ln|w| + C = -\frac{1}{2}\ln|2v+1| + C$$

$$= -\frac{1}{2}\ln\left|\frac{2}{u}+1\right| + C = -\frac{1}{2}\ln(2e^{-x}+1) + C$$

EXAMPLE 7
$$\int \frac{dx}{(4-x^2)\sqrt{4-x^2}}$$

Experimentation with substitutions $u = x^2$ or $u = \sqrt{4-x^2}$ or other similar ones proves fruitless. It developes that the substitution $x = 2\sin u$ is just what we need:

$$\int \frac{dx}{(4-x^2)\sqrt{4-x^2}} = \int \frac{2\cos u \, du}{(4-4\sin^2 u)\sqrt{1-\sin^2 u}} , \qquad \begin{bmatrix} x = 2\sin u \text{ or} \\ u = \operatorname{Sin}^{-1}(x/2), \\ dx = 2\cos u \, du \end{bmatrix}$$

$$= \frac{1}{2}\int \frac{\cos u \, du}{\cos^2 u \cos u} = \frac{1}{2}\int \sec^2 u \, du$$

$$= \frac{1}{2}\tan u + C = \frac{1}{2}\frac{\sin u}{\cos u} + C$$

$$= \frac{1}{2}\frac{\sin u}{\sqrt{1-\sin^2 u}} + C$$

$$= \frac{1}{2}\frac{x/2}{\sqrt{1-(x^2/4)}} + C = \frac{x}{2\sqrt{4-x^2}} + C$$

Several questions are raised by what we have done:

1. Why the substitution $x = 2\sin u$? It seems unrelated to the given integral. There are many cases where such "surprising" substitutions are called for, and a vast body of experience has been accumulated indicating when particular substitutions should be tried.

2. What are the functions $x(u)$, $u(x)$ in this example? We notice that our integrand is defined and continuous for $-2 < x < 2$. The function $x(u) = 2\sin u$ is defined for all u, but we must restrict u to have $-2 < x < 2$ or $-1 < \sin u < 1$, and further we want to have a differentiable inverse function $u(x)$. This suggests that we take $-\pi/2 < u < \pi/2$, as in Figure 4-9. In that interval, $x = 2\sin u$ is continuous and monotone strictly increasing, and there is a continuous inverse function $u = \operatorname{Sin}^{-1}(x/2)$, the principal value of the inverse sine (Section 3-6). Both $x = 2\sin u$ on $(-\pi/2, \pi/2)$ and its inverse function are differentiable (Section 3-6).

3. Have we made any error in sign? We observe that for $-\pi/2 < u < \pi/2$, $\cos u$ is positive, so that

$$\cos u = \sqrt{1 - \sin^2 u} \qquad (4\text{-}70)$$

(no minus sign), as in our derivation above. It follows that we can relate x

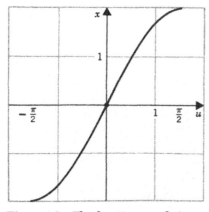

Figure 4-9 The function $x = 2\sin u$.

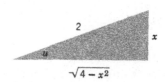

Figure 4-10 Diagram for substitution $x = 2\sin u$.

and trigonometric functions of u by the right triangle of Figure 4-10. We read off:

$$\sin u = \frac{x}{2}, \quad \cos u = \frac{\sqrt{4 - x^2}}{2}, \quad \tan u = \frac{x}{\sqrt{4 - x^2}},$$

$$\sec u = \frac{2}{\sqrt{4 - x^2}}, \quad \csc u = \frac{2}{x}, \quad \cot u = \frac{\sqrt{4 - x^2}}{x}.$$

Because of (4-70), these are valid whether x is positive or negative.

EXAMPLE 8

$$\int \sqrt{1 - x^2} \, dx = \int \cos^2 u \, du, \qquad (x = \sin u \text{ or } u = \text{Sin}^{-1} x; \, dx = \cos u \, du)$$

Now we use a trigonometric identity:

$$\cos^2 u = \frac{1 + \cos 2u}{2}$$

(Appendix V). Thus our integral equals

$$\frac{1}{2} \int du + \frac{1}{2} \int \cos 2u \, du = \frac{1}{2} u + \frac{1}{4} \int \cos v \, dv, \quad \begin{bmatrix} v = 2u \\ dv = 2 \, du \end{bmatrix}$$

$$= \frac{1}{2} u + \frac{1}{4} \sin v + C$$

$$= \frac{1}{2} u + \frac{1}{4} \sin 2u + C$$

$$= \frac{1}{2} u + \frac{1}{2} \sin u \cos u + C$$

$$= \frac{1}{2} \text{Sin}^{-1} x + \frac{1}{2} x \sqrt{1 - x^2} + C$$

EXAMPLE 9

$$\int \frac{dx}{(x^2 + 9)^2}$$

Here we set $x = 3 \tan u$, $u = \text{Tan}^{-1}(x/3)$, so that again $-\pi/2 < u < \pi/2$ (Section 3-6). Since $dx = 3 \sec^2 u \, du$, we obtain

$$\int \frac{3 \sec^2 u \, du}{(9 \tan^2 u + 9)^2} = \frac{1}{27} \int \frac{\sec^2 u \, du}{\sec^4 u} = \frac{1}{27} \int \cos^2 u \, du$$

$$= \frac{1}{54} (u + \sin u \cos u) + C$$

as in the previous example. We can again draw a triangle (Figure 4-11) from which we can read off all functions with correct signs. Thus, $\sin u = x/\sqrt{9 + x^2}$, $\cos u = 3/\sqrt{9 + x^2}$, ... and our integral equals

$$\frac{1}{54} \left(\text{Tan}^{-1} \frac{x}{3} + \frac{3x}{9 + x^2} \right) + C$$

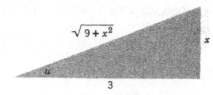

Figure 4-11   Diagram for substitution $x = 3 \tan u$.

†4-8 THEOREMS ON SUBSTITUTIONS

We now formulate and prove precise rules on substitution:

THEOREM 2. *Let the function f be a continuous function on a certain interval. Let $f(x)$ be expressible as $g[u(x)] \cdot u'(x)$, where $u = u(x)$ is also defined in the given interval, with continuous derivative $u'(x)$, and g is continuous in the appropriate interval. Then in the given interval*

$$\int f(x)\,dx \equiv \int g[u(x)]\,u'(x)\,dx = \int g(u)\,du \qquad (4\text{-}80)$$

where u is to be replaced by $u(x)$ after evaluating the last integral.

PROOF. Since $g(u)$ is continuous, g has an indefinite integral $G(u)$; this theorem, which we have discussed in Section 4-3, is proved in Section 4-16 below. We can now write $G'(u) = g(u)$ and

$$\int g(u)\,du = G(u) + C$$

We wish to show that $G[u(x)]$ has derivative $f(x)$. But the chain rule gives

$$\{G[u(x)]\}' = G'[u(x)]\,u'(x) = g[u(x)]\,u'(x) = f(x)$$

Hence, our rule is established.

We can combine Theorem 2 and the rules of Table 4-2 to obtain new general rules. Thus

$$\int [u(x)]^n\,u'(x)\,dx = \frac{[u(x)]^{n+1}}{n+1} + C \qquad (4\text{-}81)$$

$$\int \frac{u'(x)}{u(x)}\,dx = \ln|u(x)| + C \qquad (4\text{-}82)$$

$$\int e^{u(x)}\,u'(x)\,dx = e^{u(x)} + C \qquad (4\text{-}83)$$

and so on. We can write them more concisely:

$$\int u^n\,du = \frac{u^{n+1}}{n+1} + C \qquad (4\text{-}81')$$

$$\int \frac{du}{u} = \ln|u| + C \qquad (4\text{-}82')$$

$$\int e^u\,du = e^u + C \qquad (4\text{-}83')$$

where $u = u(x)$, and the conditions in Theorem 2 must be satisfied. For example, in (4-82') $u = u(x)$ must not be zero and must have a continuous derivative over the x-interval considered. Clearly, (4-81'), (4-82'), and (4-83'), respectively, are the same [except for notation and the significance of u as $u(x)$, du as $u'(x)\,dx$], as the previous rules for $\int x^n\,dx$, $\int (1/x)\,dx$, $\int e^x\,dx$. The other rules in Table 4-2 can be rewritten in the same way.

We observed in Examples 6 and 7 of the preceding section that we may not be able to express $f(x)$ in the form $g[u(x)]\,u'(x)$ [that is, $f(x)\,dx$ as $g(u)\,du$]. To cover those cases, we have the general rule:

THEOREM 3. *Let $x = x(u)$ be defined in a certain interval and let this function have an inverse $u = u(x)$; let each of the functions $x(u)$ and $u(x)$ have a continuous derivative in the corresponding interval. Let $f(x)$ be continuous in the same interval as $u(x)$. Then*

$$\int f(x)\,dx = \int f[x(u)]x'(u)\,du \qquad (4\text{-}84)$$

where, after the right side has been evaluated, u is to be replaced by $u(x)$.

PROOF. Under the hypotheses stated, $f[x(u)]x'(u)$ is continuous and has an indefinite integral $G(u)$, so that

$$\int f[x(u)]x'(u)\,du = G(u) + C$$

$$G'(u) = f[x(u)]x'(u)$$

We want to show that $D_x G[u(x)] = f(x)$. Now we can write

$$x'(u) = \frac{1}{u'(x)}$$

where x and u are corresponding values [$x = x(u)$ or $u = u(x)$]. Then, for such a pair of corresponding values,

$$G'(u) = f[x(u)] \cdot \frac{1}{u'(x)}$$

or

$$G'[u(x)] \cdot u'(x) = f(x)$$

Thus, by the chain rule,

$$D_x G[u(x)] = f(x)$$

This proves the Theorem.

PROBLEMS

In Problems 1–60 evaluate the indicated indefinite integral with the aid of one or more substitutions.

1. $\int (x + 3)^3\,dx$
2. $\int (x^2 + 1)^2 x\,dx$

3. $\int 2\sqrt{x^2 + 1} \cdot x \, dx$

4. $\int (2x - 1)^5 \, dx$

5. $\int \frac{dx}{5x - 1}$

6. $\int e^{2x} \, dx$

7. $\int e^{-3x} \, dx$

8. $\int \frac{dx}{(3x - 2)^3}$

9. $\int \frac{5x^4 - 1}{(x^5 - x - 7)^9} \, dx$

10. $\int (1 + \cos x)^5 \cdot \sin x \, dx$

11. $\int \frac{\cos x}{1 + \sin x} \, dx$

12. $\int \frac{x}{x^2 + 1} \, dx$

13. $\int \frac{2x + 1}{x^2 + x - 1} \, dx$

14. $\int \frac{x \, dx}{(x^2 + 1)^2}$

15. $\int \frac{dx}{4 + x}$

16. $\int \frac{dx}{\sqrt{1 + x}}$

17. $\int \frac{x^3}{x + 2} \, dx$

18. $\int \frac{x}{(2x + 1)^5} \, dx$

19. $\int \frac{x^4 - x + 1}{(3x + 2)^3} \, dx$

20. $\int \frac{e^x \, dx}{e^x + 1}$

21. $\int \frac{dx}{1 + e^{-x}}$

22. $\int \sin^3 x \, dx = \int (1 - \cos^2 x) \cdot \sin x \, dx$

23. $\int \cos^3 x \, dx$

24. $\int \sin^5 x \, dx$

25. $\int \cos^5 x \, dx$

26. $\int \tan x \, dx = \int \frac{\sin x}{\cos x} \, dx$

27. $\int \cot x \, dx$

28. $\int \sec^2 3x \, dx$

29. $\int \tan^4 x \, dx$

30. $\int \sec^4 x \, dx$

31. $\int \frac{dx}{x \ln x}$

32. $\int \frac{dx}{x \ln x \ln \ln x}$

33. $\int \frac{\sin (\ln x)}{x} \, dx$

34. $\int \frac{\ln 5x}{x} \, dx$

35. $\int \frac{dx}{x(\ln x)^2}$

36. $\int \frac{x^3 - x}{x^2 + 1} \, dx$

37. $\int \frac{dx}{\sqrt{4 - x^2}}$

38. $\int \frac{dx}{\sqrt{9 - x^2}}$

39. $\int \frac{dx}{\sqrt{1 - (x - 2)^2}}$

40. $\int \frac{dx}{\sqrt{4 - (x - 1)^2}}$

41. $\int \frac{dx}{\sqrt{1 - (2x - 3)^2}}$

42. $\int \frac{dx}{4 + x^2}$

43. $\int \frac{x^2 \, dx}{4 + x^2}$

44. $\int \frac{dx}{1 + 4x^2}$

45. $\int \frac{dx}{1 + (x + 2)^2}$

46. $\int \frac{dx}{9 + (x + 1)^2}$

47. $\int \frac{dx}{\sqrt{1 + x^2}} \left(\text{try } x = \frac{t^2 - 1}{2t}, t > 0 \right)$

48. $\int \frac{dx}{\sqrt{1 + (2x + 1)^2}}$

49. $\int \dfrac{dx}{\sqrt{9 + (2x + 3)^2}}$

50. $\int \dfrac{dx}{\sqrt{x^2 - 1}} \quad \begin{bmatrix} \text{try } x = (t^2 + 1)/2t \\ \text{and discuss intervals} \end{bmatrix}$

51. $\int \dfrac{2x + 3}{\sqrt{x^2 + 4}}\, dx$

52. $\int \dfrac{\cos \sqrt{x}\, dx}{\sqrt{x}}$

53. $\int \dfrac{x}{\sqrt[3]{1 + 2x}}\, dx$

54. $\int x(3 - x)^{1/3}\, dx$

55. $\int \dfrac{\cos(1 + x)^{1/3}}{(1 + x)^{2/3}}\, dx$

56. $\int \dfrac{e^{\sqrt{2 - x}}}{\sqrt{2 - x}}\, dx$

57. $\int (x + \sqrt{1 + x^2})^2\, dx$

58. $\int \dfrac{dx}{x + \sqrt{1 + x^2}}$

59. $\int \left(\dfrac{1 - \sqrt{1 - x^2}}{x} \right)^2 dx$

60. $\int (x + \sqrt{x^2 - 1})^3\, dx$

61. Explain the apparent paradoxes:

    (a) $\int (x + 1)\, dx = x^2/2 + x + C, \qquad \int (x + 1)\, dx = \int u\, du = \dfrac{(x + 1)^2}{2} + C$

    (b) $\int \sin x \cos x\, dx = \int \sin x\, d(\sin x) = \dfrac{\sin^2 x}{2} + C$

    $\int \sin x \cos x\, dx = -\int \cos x\, d(\cos x) = -\dfrac{\cos^2 x}{2} + C$

    In problems 62 to 67 let a and b denote nonzero constants. Obtain the formula given with the aid of an appropriate substitution.

62. $\int e^{ax}\, dx = \dfrac{e^{ax}}{a} + C$

63. $\int \sin ax\, dx = -\dfrac{1}{a} \cos ax + C$

64. $\int \cos ax\, dx = \dfrac{1}{a} \sin ax + C$

65. $\int \dfrac{dx}{\sqrt{a^2 - b^2x^2}} = \dfrac{1}{b} \operatorname{Sin}^{-1} \dfrac{bx}{a} + C, \qquad a > 0,\, b > 0$

66. $\int \dfrac{dx}{a^2 + b^2x^2} = \dfrac{1}{ab} \operatorname{Tan}^{-1} \dfrac{bx}{a} + C, \qquad a > 0,\, b > 0$

67. $\int \sqrt{a^2 - x^2}\, dx = \dfrac{1}{2} x\sqrt{a^2 - x^2} + \dfrac{1}{2} a^2 \operatorname{Sin}^{-1} \dfrac{x}{a} + C, \qquad a > 0$

68. *Separation of variables for first order differential equations.*
    (a) Find all functions $y(x)$ so that $y' = e^y \cos x$. [*Hint.* Let $u = e^{-y}$ and show that $u' = -\cos x$.]
    (b) Show that the solutions of part (a) can be obtained by "separating variables":

$$e^{-y}\, dy = \cos x\, dx, \text{ and hence } \int e^{-y}\, dy = \int \cos x\, dx$$

    (A general justification for this procedure is given in Problem 72 below.)

69. Solve by separating variables (cf. Problem 68):

    (a) $y' = \dfrac{x}{y}$ (b) $y' = \dfrac{x}{y^2}$ (c) $y' = \dfrac{\cos x}{\cos y}$

    (d) $y' = \dfrac{y^2 + 1}{y(x - 2)}$ (e) $y' = \dfrac{e^x e^{y^2}}{y}$

70. Show that, if y_0 is a zero of $h(y)$, then $y \equiv y_0$ satisfies the equation $y' = h(y)f(x)$.

71. Solve by separating variables and also by using the zeros of $h(y)$ as in Problem 70:

(a) $y' = xy$ (b) $y' = \dfrac{y}{x}$ (c) $y' = \dfrac{3y}{x}$

(d) $y' = \dfrac{y^2}{x}$ (e) $y' = \tan y \cos x$

72. (a) Show with the aid of Theorem 2 that, if $y = y(x)$ is a function for which $g[y(x)] \cdot y'(x) \equiv f(x)$, then under appropriate continuity conditions

$$(°) \int g(y)\, dy = \int f(x)\, dx$$

(b) Write the equation $(°)$ in the form

$$(°°)\ \ G(y) = F(x) + C$$

where $G' = g$, $F' = f$. Show that if $y(x)$ is a differentiable function defined implicitly by $(°°)$, then

$$g(y)y' = f(x), \qquad \text{where } y = y(x)$$

Remark. The results (a) and (b) justify the method of separation of variables of Problem 68(b). For an equation $y' = h(y)f(x)$, we separate by dividing by $h(y)$. This division is meaningless for y equal to a zero of h. The zeros of h provide constant solutions as in Problem 70. These constant solutions and those obtained by separation provide all the solutions of this differential equation.

73. Let $\int f(x)\, dx = F(x) + C$, and let a, b be constants, with $a \ne 0$. Show that

(a) $\int f(x + b)\, dx = F(x + b) + C$ (b) $\int f(ax)\, dx = (1/a)F(ax) + C$

(c) $\int f(ax + b)\, dx = (1/a)F(ax + b) + C$

(d) $\int f(x^2)x\, dx = \tfrac{1}{2}F(x^2) + C$ (e) $\int f(x)F(x)\, dx = \tfrac{1}{2}[F(x)]^2 + C$

74. Show that the substitution $u = \sin x$ can be applied to the integral $\int \sin^3 x \cos x\, dx$ on the interval $0 \le x \le 2\pi$ on the basis of Theorem 2, but not on the basis of Theorem 3.

75. Show that Theorem 3 is a special case of Theorem 2.

4-9 INTEGRATION BY PARTS

The rule for derivative of a product (Section 3-3) can be written:

$$(fg)' = f'g + g'f$$

Hence, by Theorem 1,

$$fg = \int (fg)'\, dx = \int gf'\, dx + \int fg'\, dx$$

or

$$fg = \int g\, df + \int f\, dg$$

provided that the integrals on the right exist.

Thus we can write

$$\int fg' \, dx = fg - \int gf' \, dx \qquad (4\text{-}90)$$

or

$$\int f \, dg = fg - \int g \, df \qquad (4\text{-}90')$$

Equations (4-90) and (4-90′) are both known as the formula for *integration by parts*. The reason for the name will be made clear below. We first state our conclusion formally:

THEOREM 4. *Let f and g both be defined in the same interval and let both have a continuous derivative. Then formulas (4-90) and (4-90′) are valid.*

Remark. In all the indefinite integrals that appear, there are arbitrary constants understood. By combining them, we have a single arbitrary constant, for instance, on the right side, in (4-90) and in (4-90′).

The formulas are useful whenever we are integrating a product $f(x) \cdot h(x)$ and can find a function g for which $g' = h$. Finding g means integrating h, which is one part of the product fh, and hence we have the term *integration by parts*. Having found g, we can use (4-90) to reduce integration of $fh = fg'$ to the new problem of integrating gf'. If the new problem is simpler, we have made progress.

EXAMPLE 1. $\int x \sin x \, dx$. Here we let $f(x) = x$, $g'(x) = \sin x$, so that $g(x)$ can be taken as $-\cos x$ and

$$\int x \sin x \, dx = -x \cos x + \int \cos x \, dx = -x \cos x + \sin x + C$$

EXAMPLE 2. $\int \ln x \, dx$. We take $f(x) = \ln x$, $g'(x) = 1$, $g(x) = x$ and find

$$\int \ln x \, dx = x \ln x - \int x \cdot \frac{1}{x} \, dx = x \ln x - x + C$$

EXAMPLE 3. $\int x^m \ln x \, dx$, $m \neq -1$. We let $f(x) = \ln x$, $g'(x) = x^m$, $g(x) = x^{m+1}/(m + 1)$ and find that

$$\int x^m \ln x \, dx = \frac{x^{m+1} \ln x}{m + 1} - \int \frac{x^m}{m + 1} \, dx$$

$$= \frac{x^{m+1} \ln x}{m + 1} - \frac{x^{m+1}}{(m + 1)^2} + C$$

EXAMPLE 4. $\int x^2 e^x \, dx$. We let $f(x) = x^2$, $g'(x) = e^x$, $g(x) = e^x$ and find that

$$\int x^2 e^x \, dx = x^2 e^x - \int e^x 2x \, dx$$

The new integral on the right is simpler than the one we started with, since we have lowered the power of x from 2 to 1. We integrate by parts again, letting $g' = e^x$, $f = 2x$, to obtain

$$\int e^x 2x \, dx = 2xe^x - \int e^x 2 \, dx = 2xe^x - 2e^x + C$$

Hence

$$\int x^2 e^x \, dx = x^2 e^x - 2x e^x + 2e^x + C$$

In some cases, we may have to integrate by parts several times to reach a familiar integral.

In the last example we might have tried another choice:

$$f(x) = e^x, \qquad g'(x) = x^2, \qquad g(x) = \frac{x^3}{3}$$

Then we would obtain

$$\int x^2 e^x \, dx = \frac{x^3}{3} e^x - \int \frac{x^3}{3} e^x \, dx$$

Now we have raised the power of x and have made matters worse! Sometimes, considerable experimentation is needed to discover a good choice of · f and g'. There is an art to using the method of integration by parts, which one masters only after much experimentation.

PROBLEMS

In Problems 1 to 15 evaluate the integral with the aid of integration by parts.

1. $\int x \cos x \, dx$ 2. $\int x \sin 2x \, dx$ 3. $\int x^2 \cos x \, dx$

4. $\int x^3 \cos x \, dx$ 5. $\int x e^{-x} \, dx$ 6. $\int x^2 e^{-x} \, dx$

7. $\int x^3 e^{-x^2} \, dx$ 8. $\int \mathrm{Tan}^{-1} x \, dx$ 9. $\int (\ln x)^2 \, dx$

10. $\int (\ln x)^3 \, dx$ 11. $\int \frac{\ln \ln x}{x} \, dx$ 12. $\int \ln \frac{x+1}{x-2} \, dx$

13. $\int \frac{\ln x^2}{x^2} \, dx$ 14. $\int \sin (\ln x) \, dx$ 15. $\int \sin(\ln x^2) \, dx$

16. Prove the reduction formulas:

  (a) $\int x^n e^{bx} \, dx = \frac{x^n e^{bx}}{b} - \frac{n}{b} \int e^{bx} x^{n-1} \, dx, \qquad b \neq 0$

  (b) $\int x^m (\ln x)^n \, dx = \frac{x^{m+1}(\ln x)^n}{m+1} - \frac{n}{m+1} \int x^m (\ln x)^{n-1} \, dx, \qquad m \neq -1$

  (c) $\int \sin^n x \, dx = -\frac{\sin^{n-1} x \cos x}{n} + \frac{n-1}{n} \int \sin^{n-2} x \, dx, \qquad n \geq 2$

  (d) $\int \cos^n x \, dx = \frac{\cos^{n-1} x \sin x}{n} + \frac{n-1}{n} \int \cos^{n-2} x \, dx, \qquad n \geq 2$

If n is a positive integer, one is reducing the integral to one with a smaller exponent. Repeated application of the formula in each case, leads us eventually to an easy integral.

17. Evaluate the following with the aid of the results of Problem 16:

  (a) $\int x^5 e^{2x} \, dx$ (b) $\int x^3 (\ln x)^2 \, dx$ (c) $\int \sin^5 x \, dx$ (d) $\int \cos^6 x \, dx$

18. Prove the formula (cf. Problem 16):

$$\int p(x)e^{-ax}\, dx = -e^{-ax}\left[\frac{p(x)}{a} + \frac{p'(x)}{a^2} + \cdots + \frac{p^{(n)}(x)}{a^{n+1}}\right] + C$$

where $p(x)$ is a polynomial of degree n and a is a constant, not 0.

19. In integrating by parts, the function g is determined up to an arbitrary constant. Show that the choice of the constant will not affect the difficulty of the new integral $\int gf'\, dx$.

20. Prove, under appropriate hypotheses, that

$$\int f'(x)g'(x)\, dx = \tfrac{1}{2}[f'(x)g(x) + f(x)g'(x)] - \tfrac{1}{2}\int [f(x)g''(x) + f''(x)g(x)]\, dx$$

21. Evaluate with the aid of Problem 20:

(a) $\int e^x \sin x\, dx$ (b) $\int e^x \cos x\, dx$ (c) $\int e^{ax} \sin bx\, dx$ (d) $\int e^{ax} \cos bx\, dx$

22. (a) To evaluate $\int e^{ax} \sin bx\, dx$ (see Problem 21), we integrate by parts twice, by taking first $f = e^{ax}$, $g' = \sin bx$, and then $f = e^{ax}$, $g' = \cos bx$. Show that this leads to the equation:

$$\int e^{ax} \sin bx\, dx = e^{ax}\left[\frac{a \sin bx - b \cos bx}{b^2}\right] - \frac{a^2}{b^2}\int e^{ax} \sin bx\, dx$$

Now solve this equation for the desired integral.

(b) Proceed as in part (a) to find $\int e^{ax} \cos bx\, dx$

4-10 PARTIAL FRACTION EXPANSION OF RATIONAL FUNCTIONS (CASE OF REAL ROOTS)

We recall that a rational function is one which is a ratio of two polynomials. For example,

$$y = f(x) = \frac{6x^2 + 6}{x^3 + 4x^2 + x - 6} \tag{4-100}$$

The rational function is called *proper* if the numerator has degree smaller than that of the denominator, as in the case of (4-100) ($2 < 3$). Integration of rational functions is made easier by a theorem of algebra which states that every proper rational function can be expressed as a sum of rational functions of a certain simple form. For the function (4-100), we have

$$\frac{6x^2 + 6}{x^3 + 4x^2 + x - 6} = \frac{1}{x - 1} + \frac{-10}{x + 2} + \frac{15}{x + 3} \tag{4-101}$$

The equality holds true whenever both sides have meaning and thus, in this case, except for $x = 1$, $x = -2$, $x = -3$, these being precisely the roots of the given denominator. From (4-101) integration is easy:

$$\int \frac{6x + 6}{x^3 + 4x^2 + x - 6}\, dx = \int \frac{dx}{x - 1} - 10\int \frac{dx}{x + 2} + 15\int \frac{dx}{x + 3}$$
$$= \ln|x - 1| - 10 \ln|x + 2| + 15 \ln|x + 3| + C$$

Equation (4-101) is called a *decomposition of the given rational function into partial fractions.* Each term on the right has a denominator that is a factor of the denominator on the left, since

$$x^3 + 4x^2 + x - 6 = (x - 1)(x + 2)(x + 3)$$

In general, if

$$f(x) = \frac{p(x)}{q(x)} = \frac{p(x)}{b(x - x_1)(x - x_2) \cdots (x - x_n)}$$

is a proper rational function and $x_1, \ldots, x_n$, the zeros of $q(x)$, are distinct real numbers, then f has a corresponding decomposition into partial fractions:

$$f(x) = \frac{A_1}{x - x_1} + \cdots + \frac{A_n}{x - x_n} \qquad (4\text{-}102)$$

where $A_1, \ldots, A_n$ are certain constants. When some of the zeros are repeated, we can write

$$f(x) = \frac{p(x)}{b(x - x_1)^{k_1}(x - x_2)^{k_2} \cdots (x - x_m)^{k_m}} \qquad (4\text{-}103)$$

where $k_1, k_2, \ldots, k_m$ are positive integers ($k_1 + k_2 + \cdots + k_m = n = $ degree of q) and $x_1, \ldots, x_m$ are distinct. The decomposition now takes the form:

$$f(x) = \frac{A_1}{x - x_1} + \frac{A_2}{(x - x_1)^2} + \cdots + \frac{A_{k_1}}{(x - x_1)^{k_1}} + \frac{A_{k_1+1}}{x - x_2} + \cdots$$

$$+ \frac{A_{n-k_m+1}}{x - x_m} + \cdots + \frac{A_n}{(x - x_m)^{k_m}} \qquad (4\text{-}104)$$

It is a theorem of algebra (Section 0-18) that every polynomial of degree n can be factored as is the denominator in (4-103), the factors corresponding to the zeros of the polynomial; $k_1, \ldots, k_m$ are the multiplicities of the zeros. However, the zeros may be complex numbers. We give our attention in this section to the case of real zeros, so that the denominator of $f(x)$ can be factored as in (4-103). We can then state our rule formally:

THEOREM 5. *Let $f(x)$ be a proper rational function whose denominator has only real zeros, so that f is expressible in form (4-103). Then f has a unique partial fraction decomposition of form (4-104).*

The proof is given in the next section. We here consider the question of finding the constants A_1, A_2, A_3, ... and of applying the theorem to integration.

Method I. *Comparison of terms of same degree.* We illustrate this for the function (4-100). We factor the denominator and write

$$\frac{6x^2 + 6}{(x - 1)(x + 2)(x + 3)} = \frac{A}{x - 1} + \frac{B}{x + 2} + \frac{C}{x + 3} \qquad (4\text{-}105)$$

We then clear denominators:

$$6x^2 + 6 = A(x + 2)(x + 3) + B(x - 1)(x + 3) + C(x - 1)(x + 2)$$
$$(4\text{-}106)$$

We reason that the left and right sides of (4-106) are polynomials of degree 2; since they are to agree for all values of x, except perhaps $x = 1$, $x = -2$, $x = -3$ and, since they are continuous, they agree for these values too. Hence they agree for all x, and therefore [see the discussion of linear independence of polynomials in Section 2-9] *the terms of the same degree must have equal coefficients.* The coefficient of x^2 on the left is 6, on the right is $A + B + C$, so that $A + B + C = 6$. Proceeding in this way for the term of first degree and the constant term, we obtain the three equations

$$
\begin{aligned}
A + \ B + \ \ C &= 6 \\
5A + 2B + \ \ C &= 0 \\
6A - 3B - 2C &= 6
\end{aligned}
$$

If we solve by elimination, we find $A = 1$, $B = -10$, $C = 15$, in agreement with (4-101).

It is clear from considerations of degree that, by this method, we always get n linear equations in n unknowns, and our Theorem 5 guarantees that they have a unique solution.

Method II. *Substitution of values.* We again consider Equation (4-105) and again multiply to obtain Equation (4-106). As we have observed, this last equation holds true for all x. Hence we can substitute $x = 1$ on both sides to obtain

$$6 + 6 = A \cdot 3 \cdot 4 + B \cdot 0 + C \cdot 0$$

so that $12 = 12A$ or $A = 1$. Similarly, by substituting $x = -2$ and $x = -3$, we obtain $30 = -3B$ and $60 = 4C$, so that $B = -10$ and $C = 15$, in agreement with (4-101). The three values of x which we used are the roots of the denominator of the given function.

When the roots are distinct, this method always gives all the unknown constants directly, as in the example (Problem 4 below). When the roots are not distinct, the method gives only some of the constants, for we have less roots to substitute than unknown constants. To find the other constants, we can substitute values of x other than the roots, to obtain the needed extra equations (Problem 4 below).

We illustrate the procedure by an example:

EXAMPLE 1

$$\frac{3x^2 - 1}{(x - 1)^2(x + 2)} = \frac{A}{x - 1} + \frac{B}{(x - 1)^2} + \frac{C}{x + 2}$$
$$3x^2 - 1 = A(x - 1)(x + 2) + B(x + 2) + C(x - 1)^2$$

$$
\begin{aligned}
x = 1: \quad & 2 = 3B, \quad B = \tfrac{2}{3} \\
x = -2: \quad & 11 = 9C, \quad C = \tfrac{11}{9} \\
x = 0: \quad & -1 = -2A + 2B + C
\end{aligned}
$$

From the three equations we find that $A = \frac{16}{9}$. Accordingly

$$\int \frac{3x^2 - 1}{(x - 1)^2(x + 2)}\, dx = \int \left[\frac{\frac{16}{9}}{x - 1} + \frac{\frac{2}{3}}{(x - 1)^2} + \frac{\frac{11}{9}}{x + 2} \right] dx$$

$$= \left(\frac{16}{9} \right) \ln |x - 1| - \frac{2}{3}\frac{1}{x - 1} + \frac{11}{9} \ln |x + 2| + C$$

To integrate a rational function that is not proper, we first carry out a long division. We illustrate by an example:

EXAMPLE 2

$$\int \frac{x^5 + x^4 - 2x^2 + 1}{x^3(x + 1)} = \int \left[x + \frac{-2x^2 + 1}{x^3(x + 1)} \right] dx$$

$$= \int \left(x - \frac{1}{x} - \frac{1}{x^2} + \frac{1}{x^3} + \frac{1}{x + 1} \right) dx$$

$$= \frac{x^2}{2} - \ln |x| + \frac{1}{x} - \frac{1}{2x^2} + \ln |x + 1| + C$$

The first step shows the result of long division. The second was carried out by writing

$$\frac{-2x^2 + 1}{x^3(x + 1)} = \frac{A_1}{x} + \frac{A_2}{x^2} + \frac{A_3}{x^3} + \frac{A_4}{x + 1}$$

and then by finding $A_1, \ldots, A_4$ as above.

‡4-11   PROOF OF PARTIAL FRACTION EXPANSION THEOREM FOR THE CASE OF REAL ROOTS

We prove Theorem 5 as an application of linear algebra. To make the writing simpler, we carry out the argument for a typical case of a proper rational function

$$f(x) = \frac{p(x)}{(x - x_1)^2(x - x_2)^3} = \frac{p(x)}{q(x)}, \qquad x_1 \neq x_2$$

A *partial fraction expansion* of f is an expression of f as a linear combination of certain other functions. This at once suggests that we consider a vector space—namely, the set V of all rational functions of the same type as f; that is, all rational functions of form:

$$\frac{C_4x^4 + C_3x^3 + C_2x^2 + C_1x + C_0}{(x - x_1)^2(x - x_2)^3}, \qquad x \neq x_1, \qquad x \neq x_2$$

These functions can be added and multiplied by scalars and, as in Section 2-9 we see that V is a vector space. Every member of V is a linear combination of the particular functions

$$f_0 = \frac{1}{q(x)}, \qquad f_1 = \frac{x}{q(x)}, \qquad f_2 = \frac{x^2}{q(x)}, \qquad f_3 = \frac{x^3}{q(x)}, \qquad f_4 = \frac{x^4}{q(x)}$$

The functions $f_0, \ldots, f_4$ are linearly independent, since a relation

$$c_0 f_0 + \cdots + c_4 f_4 = 0$$

is equivalent to the relation

$$c_0 + c_1 x + c_2 x^2 + c_3 x^3 + c_4 x^4 \equiv 0 \qquad (x \neq x_1, x \neq x_2)$$

But a polynomial of degree 0, 1, 2, 3 or 4 has, at most, 4 zeros. Thus the last relation is possible only if $c_0 = 0$, $c_1 = 0$, $c_2 = 0$, $c_3 = 0$, $c_4 = 0$. Consequently, f_0, f_1, f_2, f_3, f_4 are linearly independent and form a basis for V. Therefore, V is a vector space of dimension 5.

The vector space V also includes the functions

$$g_1 = \frac{1}{x - x_1}, \qquad g_2 = \frac{1}{(x - x_1)^2}, \qquad g_3 = \frac{1}{(x - x_2)}, \qquad g_4 = \frac{1}{(x - x_2)^2}$$

$$g_5 = \frac{1}{(x - x_2)^3}$$

For we can write, for example,

$$\frac{1}{x - x_1} = \frac{(x - x_1)(x - x_2)^3}{(x - x_1)^2 (x - x_2)^3}, \qquad (x \neq x_1, x \neq x_2)$$

The functions $g_1, \ldots, g_5$ are also linearly independent, as we now show. The relation,

$$c_1 g_1 + c_2 g_2 + \cdots + c_5 g_5 = 0$$

is the same as

$$\frac{c_1}{x - x_1} + \frac{c_2}{(x - x_1)^2} + \frac{c_3}{x - x_2} + \frac{c_4}{(x - x_2)^2} + \frac{c_5}{(x - x_2)^3} = 0 \quad (4\text{-}110)$$

If we clear of denominators, we obtain

$$c_1(x - x_1)(x - x_2)^3 + c_2(x - x_2)^3 + c_3(x - x_1)^2(x - x_2)^2$$
$$+ c_4(x - x_1)^2(x - x_2) + c_5(x - x_1)^2 = 0$$

This polynomial equation is valid first for $x \neq x_1$, $x \neq x_2$ and hence, by continuity, for all x. If we set $x = x_1$, all terms reduce to zero except one, and we obtain $c_2(x_1 - x_2)^3 = 0$; since $x_1 \neq x_2$, c_2 must be 0. In the same way, by setting $x = x_2$, we find that $c_5 = 0$. We can therefore delete the terms in c_2 and c_5 in (4-110). If we now clear denominators again and set $x = x_1$ and $x = x_2$ in turn, we find that $c_1 = 0$ and $c_4 = 0$, so that (4-110) reduces to

$$\frac{c_3}{x - x_2} = 0 \qquad (x \neq x_1, x \neq x_2)$$

and, consequently, $c_3 = 0$. Thus, $c_1 = c_2 = c_3 = c_4 = c_5 = 0$ and $g_1, \ldots, g_5$ are linearly independent. Since V has dimension 5, they form a basis for V (Theorem G of Section 2-9). Therefore, every member f of V can be represented uniquely as a linear combination of $g_1, \ldots, g_5$, that is,

$$f(x) = \frac{A_1}{x - x_1} + \frac{A_2}{(x - x_1)^2} + \frac{A_3}{x - x_2} + \frac{A_4}{(x - x_2)^2} + \frac{A_5}{(x - x_2)^3}$$

This is the desired partial fraction expansion.

This method of proof works in the same way for each proper rational function whose denominator has only real roots.

PROBLEMS

In Problems (1) to (21) integrate with the aid of partial fractions:

(1) $\int \dfrac{dx}{(x-1)(x-2)}$

(2) $\int \dfrac{x\,dx}{(x+3)(x-2)}$

(3) $\int \dfrac{(2x-1)}{x^2-4}\,dx$

(4) $\int \dfrac{(x^2+1)\,dx}{x(x^2-4)}$

(5) $\int \dfrac{dx}{(x-1)(x-2)^2}$

(6) $\int \dfrac{x^5+x^2+7x}{(x-1)(x-2)^2}\,dx$

(7) $\int \dfrac{(3x+5)\,dx}{2x(x+1)(x+5)}$

(8) $\int \dfrac{x^2\,dx}{(x-1)(2x-1)(2x+3)}$

(9) $\int \dfrac{x\,dx}{(x-1)(x-2)^3}$

(10) $\int \dfrac{x-1}{(x+2)^2}\,dx$

(11) $\int \dfrac{(x^2-5)\,dx}{x^3+2x^2-13x+10}$

(12) $\int \dfrac{(x+1)\,dx}{2x^3+3x^2-8x-12}$

(13) $\int \dfrac{x^2+1}{(4x^2-1)^2}\,dx$

(14) $\int \dfrac{dx}{(x-1)^2(x+2)^3}$

(15) $\int \dfrac{(x^2-x+1)}{x^3+3x^2-4}\,dx$

(16) $\int \dfrac{x^5+1}{x^4-x^2}\,dx$

(17) $\int \dfrac{x^3-x^2+1}{x^2+x}\,dx$

(18) $\int \dfrac{dx}{x^3+x^2(2\sqrt{2}+1)+x(2+2\sqrt{2})+2}$

(19) $\int \dfrac{2x^5+1}{2x^2+3x+1}\,dx$

(20) $\int \dfrac{(2x^3+1)\,dx}{4x^2+5x+1}$

(21) $\int \dfrac{x^7+x^3-2}{x^3-4x}\,dx$

22. Prove the rules:

(a) $\int \dfrac{dx}{(x-a)(x-b)} = \dfrac{1}{b-a}\ln\left|\dfrac{x-b}{x-a}\right| + C, \qquad a \neq b$

(b) $\int \dfrac{dx}{x^2-a^2} = \dfrac{1}{2a}\ln\left|\dfrac{x-a}{x+a}\right| + C, \qquad a \neq 0$

(c) $\int \dfrac{dx}{(x-a)(x-b)(x-c)} = \dfrac{-1}{(a-b)(b-c)(c-a)}\,[(b-c)\ln|x-a|$

$\qquad\qquad + (c-a)\ln|x-b| + (a-b)\ln|x-c|] + C \qquad a \neq b,\, b \neq c,\, a \neq c$

23. Let $f(x) = p(x)/q(x)$ be a proper rational function and let the roots of $q(x)$ be the distinct numbers $x_1, \ldots, x_n$. Let f have the partial fraction expansion

$$\frac{p(x)}{q(x)} = \frac{A_1}{x-x_1} + \cdots + \frac{A_n}{x-x_n}$$

(a) Prove that

$$A_k = \lim_{x \to x_k} \frac{(x-x_k)p(x)}{q(x)}, \qquad k = 1, \ldots, n$$

Remark. This result suggests a shortening of Method II. For example, consider

$$\frac{x}{(x-2)(x-3)} = \frac{A}{x-2} + \frac{B}{x-3}$$

To find A, simply cover up $(x-2)$ in the denominator on the left side:

$$\frac{x}{\boxed{}(x-3)}$$

and set $x = 2$ in what is left:

$$\frac{2}{2-3} = -2 = A$$

(b) Prove that

$$A_k = \frac{p(x_k)}{q'(x_k)}$$

[*Hint.* Use the result of (a) and notice that, since $q(x_k) = 0$,

$$\frac{(x-x_k)p(x)}{q(x)} = \frac{p(x)}{\dfrac{q(x)-q(x_k)}{x-x_k}}$$

24. (a) Show that, if $P(x)$ and $Q(x)$ are both polynomials of degree at most $n-1$ and if $P(x)$ and $Q(x)$ have the same value for n different values of x, then $P(x) \equiv Q(x)$.
    (b) Show that Method II always determines $A_1, \ldots, A_n$ uniquely.

25. Integrate

$$\int \frac{x^4 - x^3 + x + 1}{(x+2)^3}\, dx$$

(a) by partial fractions,   (b) by the substitution $u = x + 2$.

26. Integrate:

(a) $\int \sec x\, dx$ (*Hint:* Set $u = \sin x$.)   (b) $\int \csc x\, dx$

(c) $\int \sec^3 x\, dx$ (d) $\int \csc^3 x\, dx$

(e) $\int \frac{\sqrt{x^2+1}}{x}\, dx$ $\left(\text{Set } x = \dfrac{t^2-1}{2t}\right)$ (f) $\int \dfrac{dx}{x\sqrt{1-x^2}}$

†4-12 PARTIAL FRACTION EXPANSION (CASE OF COMPLEX ROOTS AND QUADRATIC FACTORS)

When some of the roots of the denominator q of the proper rational function p/q are complex, we can no longer factor q in the form $b(x-x_1)^{k_1} \ldots$ with real factors. It is shown in algebra (Section 0-18) that if the coefficients are real numbers, then the complex roots always come in conjugate pairs of form $a \pm bi$ ($i = \sqrt{-1}$) and that, corresponding to each pair of roots, there is a quadratic factor of q:

$$(x-a)^2 + b^2 = x^2 - 2ax + a^2 + b^2$$

When the roots are repeated, the quadratic factor is repeated a corresponding number of times.

We here assume that the factorization has been carried out, and consider our rational function given in the form:

$$f(x) = \frac{p(x)}{b(x - x_1)^{\alpha_1} \cdots (x - x_l)^{\alpha_l}(x^2 + r_1 x + s_1)^{\beta_1} \cdots (x^2 + r_m x + s_m)^{\beta_m}}$$

(4-120)

The degree of the denominator is

$$\alpha_1 + \cdots + \alpha_l + 2\beta_1 + \cdots + 2\beta_m = n \qquad (4\text{-}120')$$

and we assume that $p(x)$ has degree $\leq n - 1$, so that f is a proper rational function. What is the corresponding partial fraction expansion of f?

The answer resembles that for the case of real roots:

$$\frac{A_1}{x - x_1} + \cdots + \frac{A_{\alpha_1}}{(x - x_1)^{\alpha_1}} + \cdots + \frac{B_1 x + C_1}{x^2 + r_1 x + s_1} + \frac{B_2 x + C_2}{(x^2 + r_1 x + s_2)^2}$$

$$+ \cdots + \frac{B_{\beta_1} x + C_{\beta_1}}{(x^2 + r_1 x + s_1)^{\beta_1}} + \frac{B_{\beta_1+1} x + C_{\beta_1+1}}{x^2 + r_2 x + s_2} + \cdots \qquad (4\text{-}121)$$

Thus, for each parenthesis in the denominator of (4-120), the partial fraction expansion (4-121) has a term whose denominator is that parenthesis to the first power, a term with denominator that parenthesis to the second power, . . . and one with the highest power to which the parenthesis is raised. For the linear factors the numerator in each case is a constant: $A_1, A_2, \ldots$, and for the quadratic factors the numerator is a linear function $B_1 x + C_1, B_2 x + C_2, \ldots$ We state a formal theorem here. A proof can be given along the lines of Section 4-11.

THEOREM 6. *Let $f(x)$ be a proper rational function of form (4-120). Then f has a unique partial fraction expansion of form (4-121).*

To find the constants $A_1, \ldots, B_1, C_1, \ldots$, we have the same two methods available, as previously.

EXAMPLE 1

$$\frac{4x^2 + 4}{(x + 1)(x^2 + 2x + 5)} = \frac{A}{x + 1} + \frac{Bx + C}{x^2 + 2x + 5}$$

We clear of denominators and use Method I (comparison of coefficients). Thus, $4x^2 + 4 = A(x^2 + 2x + 5) + (Bx + C)(x + 1) = x^2(A + B) + x(2A + B + C) + (5A + C)$. Hence, $4 = A + B$, $0 = 2A + B + C$, and $4 = 5A + C$. We find by elimination that $A = 2$, $B = 2$, and $C = -6$. Accordingly

$$\int \frac{4x^2 + 4}{(x + 1)(x^2 + 2x + 5)}\, dx = \int \left(\frac{2}{x + 1} + \frac{2x - 6}{x^2 + 2x + 5} \right) dx$$

The first term is easy to integrate. For the second we write

$$\int \frac{2x - 6}{x^2 + 2x + 5}\, dx = \int \frac{(2x + 2) - 8}{x^2 + 2x + 5}\, dx$$

$$= \int \frac{2x + 2}{x^2 + 2x + 5}\, dx - 8\int \frac{dx}{(x + 1)^2 + 4}$$

$$= \int \frac{du}{u} - 4\int \frac{dv}{v^2 + 1} \qquad (u = x^2 + 2x + 5,\ 2v = x + 1)$$

$$= \ln |u| - 4 \operatorname{Tan}^{-1} v + C$$

$$= \ln|x^2 + 2x + 5| - 4 \operatorname{Tan}^{-1} \frac{x + 1}{2} + C$$

Accordingly

$$\int \frac{(4x^2 + 4)\, dx}{(x + 1)(x^2 + 2x + 5)}$$

$$= 2 \ln |x + 1| + \ln |x^2 + 2x + 5| - 4 \operatorname{Tan}^{-1} \frac{x + 1}{2} + C$$

This example illustrates that, for integration, time is saved if for each term with a quadratic (or power of a quadratic) in the denominator, the numerator is written in the form:

$$B \text{ (derivative of quadratic)} + C$$

and then B and C are found. In the example we would write

$$\frac{4x^2 + 4}{(x + 1)(x^2 + 2x + 5)} = \frac{A}{x + 1} + \frac{B(2x + 2) + C}{x^2 + 2x + 5}$$

and then find $A = 2$, $B = 1$, $C = -8$, so that we are ready to integrate the three terms, as above.

For the integrations that arise in these problems we have the following formulas:

$$\int \frac{dx}{x - x_1} = \ln |x - x_1| + C \tag{4-122}$$

$$\int \frac{(2x + b)\, dx}{x^2 + bx + c} = \ln x^2 + bx + c + C \tag{4-123}$$

$$\int \frac{(2x + b)\, dx}{(x^2 + bx + c)^m} = \frac{-1}{(m - 1)(x^2 + bx + c)^{m-1}} + C, \qquad m = 2, 3, \ldots \tag{4-124}$$

$$\int \frac{dx}{(x + a)^2 + b^2} = \frac{1}{b} \tan^{-1} \frac{x + a}{b} + C, \qquad b > 0 \tag{4-125}$$

$$\int \frac{dx}{[(x + a)^2 + b^2]^2} = \frac{x + a}{2b^2[(x + a)^2 + b^2]} + \frac{1}{2b^3} \operatorname{Tan}^{-1} \frac{x + a}{b} + C, \qquad b > 0 \tag{4-126}$$

$$\int \frac{dx}{[(x + a)^2 + b^2]^n} = \frac{x + a}{2(n - 1)b^2[(x + a)^2 + b^2]^{n-1}}$$

$$+ \frac{2n - 3}{2b^2(n - 1)} \int \frac{dx}{[(x + a)^2 + b^2]^{n-1}}$$

$$n = 2, 3, \ldots, \qquad b > 0 \tag{4-127}$$

The quadratic functions occuring are assumed to have only complex roots, so that, for example, $x^2 + bx + c$ can never change sign and therefore, must always be positive (Why ?); for that reason no absolute value sign is needed in (4-123). In (4-125) ... (4-127) the b must not be 0, and the corresponding complex roots are $-a \pm bi$.

The first three of these integration formulas are easily checked [use $u = x^2 + bx + c$ in (4-123) and (4-124)]. The last three are left to the exercises (Problem 4 below). Equation (4-127) is a recursion formula, which allows us to reduce our problem to one of lower and lower power in the denominator.

EXAMPLE 2

$$\int \frac{3x^7 + 6x^6 + 17x^5 + 19x^4 + 25x^3 + 14x^2 + 9x + 3}{(x^2 + 1)(x^2 + x + 1)^3} \, dx$$

We equate the integrand to

$$\frac{2Ax + B}{x^2 + 1} + \frac{C(2x + 1) + D}{x^2 + x + 1} + \frac{E(2x + 1) + F}{(x^2 + x + 1)^2} + \frac{G(2x + 1) + H}{(x^2 + x + 1)^3}$$

We cross-multiply and compare terms of same degree, finding that

$$x^7 \colon \ 3 = 2A + 2C$$
$$x^6 \colon \ 6 = 6A + B + 5C + D$$
$$x^5 \colon \ 17 = 12A + 3B + 10C + 2D + 2E$$
$$x^4 \colon \ 19 = 14A + 6B + 12C + 4D + 3E + F$$
$$x^3 \colon \ 25 = 12A + 7B + 12C + 4D + 5E + F + 2G$$
$$x^2 \colon \ 15 = 6A + 6B + 8C + 4D + 4E + 2F + G + H$$
$$x \colon \ 9 = 2A + 3B + 4C + 2D + 3E + F + 2G$$
$$1 \colon \ 3 = B + C + D + E + F + G + H$$

By solving (a long process!), we find that

$$A = \frac{1}{2}, \qquad B = 2, \qquad C = 1, \qquad D = -4,$$

$$E = \frac{3}{2}, \qquad F = -\frac{1}{2}, \qquad G = 1, \qquad H = 2$$

Accordingly, our integral equals

$$\int \left[\frac{\frac{1}{2} \cdot 2x}{x^2 + 1} + \frac{2}{x^2 + 1} + \frac{2x + 1}{x^2 + x + 1} + \frac{-4}{x^2 + x + 1} + \frac{\frac{3}{2}(2x + 1)}{(x^2 + x + 1)^2} \right.$$
$$\left. - \frac{\frac{1}{2}}{(x^2 + x + 1)^2} + \frac{2x + 1}{(x^2 + x + 1)^3} + \frac{2}{(x^2 + x + 1)^3} \right] dx$$

By (4-125), (4-126), and (4-127)

$$\int \frac{dx}{x^2 + x + 1} = \int \frac{dx}{(x + \frac{1}{2})^2 + \frac{3}{4}} = \frac{2}{\sqrt{3}} \, \text{Tan}^{-1} \frac{x + \frac{1}{2}}{\sqrt{3}/2} + C$$

$$\int \frac{dx}{(x^2 + x + 1)^2} = \frac{x + \frac{1}{2}}{\frac{3}{2}(x^2 + x + 1)} + \frac{4}{3\sqrt{3}} \, \text{Tan}^{-1} \frac{x + \frac{1}{2}}{\sqrt{3}/2} + C$$

$$\int \frac{dx}{(x^2 + x + 1)^3} = \frac{x + \frac{1}{2}}{3(x^2 + x + 1)^2} + \int \frac{dx}{(x^2 + x + 1)^2}$$

$$= \frac{x + \frac{1}{2}}{3(x^2 + x + 1)^2} + \frac{2x + 1}{3(x^2 + x + 1)}$$

$$+ \frac{4}{3\sqrt{3}} \operatorname{Tan}^{-1} \frac{2x + 1}{\sqrt{3}} + C$$

Thus finally our integral equals

$$\frac{1}{2} \ln(x^2 + 1) + 2 \operatorname{Tan}^{-1} x + \ln(x^2 + x + 1) - \frac{8}{\sqrt{3}} \operatorname{Tan}^{-1} \frac{2x + 1}{\sqrt{3}}$$

$$- \frac{3}{2} \frac{1}{x^2 + x + 1} - \frac{2x + 1}{6(x^2 + x + 1)} - \frac{2}{3\sqrt{3}} \operatorname{Tan}^{-1} \frac{2x + 1}{\sqrt{3}}$$

$$- \frac{1}{2(x^2 + x + 1)^2} + \frac{2x + 1}{3(x^2 + x + 1)^2} + \frac{4x + 2}{3(x^2 + x + 1)}$$

$$+ \frac{8}{3\sqrt{3}} \operatorname{Tan}^{-1} \frac{2x + 1}{\sqrt{3}} + C$$

PROBLEMS

1. Expand in partial fractions:

(a) $\dfrac{x^2 + x - 1}{(x - 1)(x^2 + 4)}$ (b) $\dfrac{1}{x(x^2 + 1)}$ (c) $\dfrac{x^2 - 2}{x(x^2 + 1)^2}$

(d) $\dfrac{1}{x(x^2 + x + 1)}$ (e) $\dfrac{1}{x(x^2 + x + 1)^2}$ (f) $\dfrac{1}{(x - 2)^2(x^2 + 2x + 10)}$

2. Integrate:

(a) $\displaystyle\int \frac{dx}{x^2 + 2x + 2}$ (b) $\displaystyle\int \frac{dx}{x^2 + 9}$ (c) $\displaystyle\int \frac{dx}{(x^2 + 9)^2}$

(d) $\displaystyle\int \frac{2x + 5}{9x^2 + 4} \, dx$ (e) $\displaystyle\int \frac{x + 5}{(x^2 + 1)^2} \, dx$ (f) $\displaystyle\int \frac{x^3 + 1}{x^2 + 1} \, dx$

3. Integrate:

(a) $\displaystyle\int \frac{dx}{(x - 2)(x^2 + 1)}$ (b) $\displaystyle\int \frac{dx}{x(x^2 + 2x + 2)}$

(c) $\displaystyle\int \frac{x - 1}{x^2(x^2 + 1)} \, dx$ (d) $\displaystyle\int \frac{x^5 - 1}{x(x^2 + 1)^2} \, dx$

(e) $\displaystyle\int \frac{5x^4 + 6x^2 + 1}{x^5 + 2x^3 + x} \, dx$ (Look for an easy way!)   (f) $\displaystyle\int \frac{dx}{x^3 + x^2 + 4x + 4}$

4. (a) Prove (4-125). (*Hint.* Set $x + a = b \tan u$.)

(b) Prove (4-126). {*Hint.* Write the integral as

$$\frac{1}{b^2} \int \frac{(x + a)^2 + b^2 - (x + a)^2}{[(x + a)^2 + b^2]^2} \, dx = \frac{1}{b^2} \int \frac{dx}{(x + a)^2 + b^2} - \frac{1}{b^2} \int \frac{(x + a)^2}{[\cdots]^2} \, dx$$

and integrate the last integral by parts, using $f = x + a$, $g' = (x + a)/[\cdots]^2$.}

(c) Prove (4-127). [See hint for part (b).]

5. Let V be the vector space of all functions of form:

$$\frac{C_3 x^3 + C_2 x^2 + C_1 x + C_0}{(x + 1)(x^2 + 1)} = \frac{p(x)}{q(x)}$$

(a) Show that V has dimension 4 and that the functions $f_1 = 1/q$, $f_2 = x/q$, $f_3 = x^2/q$, $f_4 = x^3/q$ are a basis for V.

(b) Show that the functions $g_1 = 1$, $g_2 = 1/(x + 1)$, $g_3 = x/(x^2 + 1)$, $g_4 = 1/(x^2 + 1)$ are also a basis for V.

6. (a) Verify the identity:

$$\frac{1}{(x - a)(x - b)} = \frac{1}{a - b}\frac{1}{x - a} + \frac{1}{b - a}\frac{1}{x - b}, \qquad x \neq a,\, x \neq b,\, a \neq b$$

(b) In the identity of part (a), consider x and b as fixed and a as variable. Differentiate both sides with respect to a to show that

$$\frac{1}{(x - a)^2(x - b)} = \frac{1}{a - b}\frac{1}{(x - a)^2} - \frac{1}{(a - b)^2}\frac{1}{x - a} + \frac{1}{(b - a)^2}\frac{1}{x - b}$$

7. (a) To solve Problem 1(a) write

$$\frac{x^2 + x - 1}{(x - 1)(x + 2i)(x - 2i)} = \frac{A}{x - 1} + \frac{B}{x - 2i} + \frac{C}{x + 2i}$$

and solve for A, B, and C, as in Section 4-10. Combine the last two terms to obtain the result in the usual form.

(b) Proceed as in part (a) for Problem 1(b).

(c) Proceed as in part (a) for Problem 1(c).

†4-13 INTEGRATION OF FUNCTIONS GIVEN BY DIFFERENT FORMULAS IN ADJOINING INTERVALS

We illustrate the problem to be considered by an example.

EXAMPLE 1. Let f be defined as follows:

$$f(x) = \begin{cases} 1 + x, & 0 \leq x \leq 1 \\ 2, & 1 \leq x \leq 2 \end{cases}$$

The function is graphed in Figure 4-12. We seek $\int f(x)\, dx = F(x) + C$. Thus we seek F so that

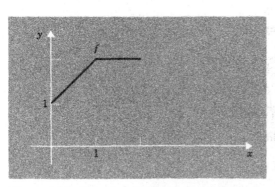

Figure 4-12   Function f of Example 1.

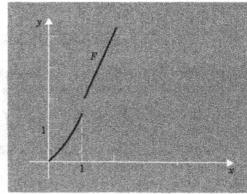

Figure 4-13   Function F with constants improperly adjusted.

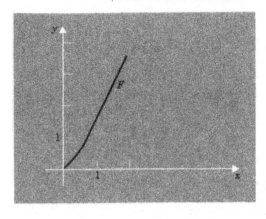

Figure 4-14 Function F whose derivative is function of Figure 4-12.

$$F'(x) = \begin{cases} 1 + x, & 0 \le x \le 1 \\ 2, & 1 < x \le 2 \end{cases}$$

Accordingly

$$F(x) = x + \frac{x^2}{2} + \text{const} \qquad \text{for} \qquad 0 \le x \le 1$$

$$F(x) = 2x + \text{const} \qquad \text{for} \qquad 1 < x \le 2$$

However, *the two constants are not necessarily the same*. If, for example, we took both equal to 0, then F would be as in Figure 4-13, with a discontinuity at $x = 1$. We choose the two constants so that the two pieces fit together. Thus, if $F(x) = x + (x^2/2)$ for $0 \le x \le 1$, then $F(1) = \frac{3}{2}$, and we must take $F(x) = 2x - (\frac{1}{2})$ for $1 < x \le 2$ in order that the two pieces fit together. Thus, finally,

$$F(x) = \begin{cases} x + \dfrac{x^2}{2}, & 0 \le x \le 1 \\ 2x - \dfrac{1}{2}, & 1 \le x \le 2 \end{cases}$$

defines a function F whose derivative is f. Notice in particular that at $x = 1$, F has left-hand derivative 2 at $x = 1$ and right-hand derivative 2, so that $F'(x) = f(x)$ at $x = 2$. By Theorem 1 (Section 4-5),

$$\int f(x)\, dx = F(x) + C = \begin{cases} x + \dfrac{x^2}{2} + C, & 0 \le x \le 1 \\ 2x - \dfrac{1}{2} + C, & 1 \le x \le 2 \end{cases}$$

The function $F(x)$ is graphed in Figure 4-14.

The example suggests a general theorem.

THEOREM 7. *Let* $y = f(x)$ *be defined and continuous for* $a \le x \le b$, *where*

$$f(x) = \begin{cases} f_1(x), & a \le x \le c \\ f_2(x), & c \le x \le b \end{cases}$$

with $a < c < b$. *Let*

$$\int f_1(x) \, dx = F_1(x) + C, \qquad \int f_2(x) \, dx = F_2(x) + C$$

Then

$$\int f(x) \, dx = F(x) + C$$

where

$$F(x) = \begin{cases} F_1(x) & a \le x \le c \\ F_2(x) + F_1(c) - F_2(c) & c \le x \le b \end{cases}$$

PROOF. We can imitate the example above. The function $F(x)$ is continuous for $a \le x \le b$, for the only point where there could be a discontinuity is at $x = c$. At $x = c$, $F(x) = F_1(c)$ by both formulas and, since $F_1(x)$ is continuous for $a \le x \le c$ and $F_2(x)$ is continuous for $c \le x \le b$, $F(x)$ is also continuous at $x = c$. Furthermore

$$\begin{aligned} F_1'(x) &= f_1(x), & a \le x \le c \\ F_2'(x) &= f_2(x), & c \le x \le b \end{aligned}$$

since f_1 has indefinite integral F_1, f_2 has indefinite integral F_2. Accordingly,

$$F'(x) = f(x), \qquad a \le x \le b$$

and $F(x)$ is an indefinite integral of f.

EXAMPLE 2. Let f be defined as follows

$$f(x) = \begin{cases} 1, & 0 \le x \le 1 \\ 2, & 1 < x \le 2 \end{cases}$$

as in Figure 4-15a. Thus f is discontinuous at $x = 1$, having a jump discontinuity there. We seek $\int f(x) \, dx = F(x) + C$. If $F' = f$, then $F'(x) = 1$ for $0 \le x \le 1$, and $F'(x) = 2$ for $1 < x \le 2$. Hence

$$\begin{aligned} F(x) &= x + \text{const}, & 0 \le x \le 1 \\ F(x) &= 2x + \text{const}, & 1 < x \le 2 \end{aligned}$$

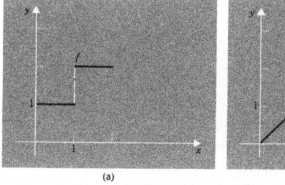
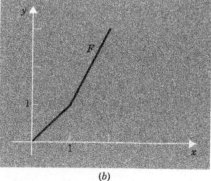

(a)   (b)

Figure 4-15   Integration of discontinuous function.

We can easily adjust the two constants so that the two pieces agree at $x = 1$. For example

$$F(x) = x, \qquad 0 \le x \le 1$$
$$F(x) = 2x - 1, \qquad 1 < x \le 2$$

as in Figure 4-15b. Thus we can make F continuous, but it is not differentiable on $[0, 2]$. For, as the graph shows, F has a corner at $x = 1$. The two slopes, to the left and to the right, are different:

$$F'_+(1) = 2, \qquad F'_-(1) = 1$$

This is unavoidable, because of the discontinuity in f at $x = 1$.

The procedure used in the example can be generalized to the case of an arbitrary function f with a jump discontinuity. A function F can be found that is continuous over the given interval and that has $F'(x) = f(x)$ except at the point of discontinuity, where F must have a corner. Strictly speaking, F is not an indefinite integral of f over the whole given interval, but conveniently we still consider it as an indefinite integral of f, and write

$$\int f(x)\, dx = F(x) + C$$

The addition of C can be justified as before (Problem 4 below). We call F a *continuous indefinite integral* of f.

The procedures of Examples 1 and 2 can further be generalized to a function f, given by different expressions over more than two intervals (Problem 2 below). If f is continuous throughout, we get a true indefinite integral. If f has several jump discontinuities, so that f is piecewise continuous, we get a continuous indefinite integral whose graph has corners at the points of discontinuity of f.

†4-14 APPROXIMATE METHODS FOR FINDING INDEFINITE INTEGRALS

Let f be a continuous function for $a \le x \le b$. When the known methods for finding $\int f(x)\, dx$ are unsuccessful, we consider approximate methods. We describe one such method here.

We subdivide the interval $[a, b]$ into several parts by points x_1, $x_2, \ldots, x_{n-1}$ (Figure 4-16), and take $x_0 = a$, $x_n = b$. In each subinterval we approximate f by a constant function, say equal to $f(x_i^*)$ on the interval $x_{i-1} \le x \le x_i$, where x_i^* is some point in this interval. Let $g(x)$ be the resulting approximation to f:

$$g(x) = \begin{cases} f(x_1^*), & a \le x \le x_1 \\ f(x_2^*), & x_1 < x \le x_2 \\ \vdots \\ f(x_n^*), & x_{n-1} < x \le b \end{cases} \qquad (4\text{-}140)$$

Thus, $g(x)$ has as graph a broken line, as in Figure 4-16. We call $g(x)$ a *step-function*. It is reasonable to expect that, if we subdivide very finely,

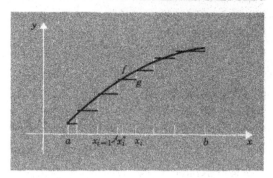

Figure 4-16 Approximation of continuous function f by a step-function.

then $g(x)$ will deviate very little from $f(x)$, and that, therefore, a continuous indefinite integral of g will be close to an indefinite integral of f. We make a precise statement in this regard below. First we consider some examples.

EXAMPLE 1. Let $f(x) = 3x^2$, $0 \leq x \leq 1$. Of course, we know that $x^3 = \int f(x)\, dx$, but let us use the approximate method and determine how well it works. We subdivide into two intervals, by using $x_0 = 0$, $x_1 = 0.5$, $x_2 = 1$. We take $x_1^* = 0.25$, $x_2^* = 0.75$ (the two midpoints). Thus we take

$$g(x) = \begin{cases} 3(0.25)^2 = 0.1875, & 0 \leq x \leq 0.5 \\ 3(0.75)^2 = 1.6875, & 0.5 < x \leq 1 \end{cases}$$

A continuous indefinite integral of $g(x)$ is

$$F_0(x) = \begin{cases} 0.1875x, & 0 \leq x \leq 0.5 \\ 1.6875x - 0.75, & 0.5 < x \leq 1 \end{cases}$$

Thus, as in the preceding section, $F_0(x)$ is chosen to be continuous, but has a corner at $x = 0.5$. In Figure 4-17, we graph $F_0(x)$ and $F(x) = x^3$, the true indefinite integral of f, with arbitrary constant chosen so that $F(0) = F_0(0)$. The agreement is strikingly good.

EXAMPLE 2. Let $f(x) = \cos x$, $0 \leq x \leq \pi/2$. We subdivide into four parts by

$$x_0 = 0, \qquad x_1 = \frac{\pi}{6}, \qquad x_2 = \frac{\pi}{4}, \qquad x_3 = \frac{\pi}{3}, \qquad x_4 = \frac{\pi}{2}$$

and take $x_1^* = 0$, $x_2^* = \pi/6$, $x_3^* = \pi/4$, $x_4^* = \pi/3$. (Thus we use the left-hand end points.) Our approximation to f is thus the function given by

$$g(x) = \begin{cases} \cos 0 = 1, & 0 \leq x \leq \dfrac{\pi}{6} \\[2ex] \cos \dfrac{\pi}{6} = 0.866, & \dfrac{\pi}{6} < x \leq \dfrac{\pi}{4} \\[2ex] \cos \dfrac{\pi}{4} = 0.707, & \dfrac{\pi}{4} < x \leq \dfrac{\pi}{3} \\[2ex] \cos \dfrac{\pi}{3} = 0.5, & \dfrac{\pi}{3} < x \leq \dfrac{\pi}{2} \end{cases}$$

A corresponding continuous indefinite integral of $g(x)$ is

$$F_0(x) = \begin{cases} x, & 0 \le x \le \dfrac{\pi}{6} \\[2ex] 0.866x + 0.069, & \dfrac{\pi}{6} < x \le \dfrac{\pi}{4} \\[2ex] 0.707x + 0.195, & \dfrac{\pi}{4} < x \le \dfrac{\pi}{3} \\[2ex] 0.5x + 0.415, & \dfrac{\pi}{3} < x \le \dfrac{\pi}{2} \end{cases}$$

Again we graph $F_0(x)$ against the true indefinite integral $F(x) = \sin x$ [with $F(0) = F_0(0)$] (see Figure 4-18). The agreement is even more striking.

We now return to the general question. Let a function f be given as above for $a \le x \le b$ and let the approximation $g(x)$ be chosen as in (4-140). Let $F_0(x)$ be the continuous indefinite integral of $g(x)$, with $F_0(a) = 0$. From $x = a$ to $x = x_1$, $F_0(x)$ has constant slope $f(x_1^*)$, so that $F_0(x)$ increases by

$$f(x_1^*)(x_1 - a)$$

Similarly, from $x = x_1$ to $x = x_2$, F_0 increases by

$$f(x_2^*)(x_2 - x_1)$$

and so on. Thus, in all, from $x = a$ to $x = b$, F_0 increases by

$$Q = f(x_1^*)(x_1 - a) + f(x_2^*)(x_2 - x_1) + \cdots + f(x_n^*)(b - x_{n-1})$$

Since $F_0(a) = 0$, Q is the value of F_0 at $x = b$. We now ask: how far is Q from the value $F(b)$ of the exact indefinite integral F of f chosen so that $F(a) = F_0(a) = 0$?

The answer is that we can make Q as close to $F(b)$ as desired by making all the subdivision intervals small enough. In fact, we are dealing here with

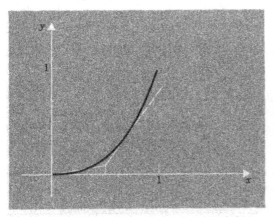

Figure 4-17 Exact versus approximate indefinite integral.

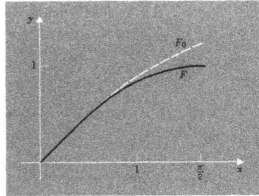

Figure 4-18 Approximate indefinite integral of $\cos x$.

a limit process—the very limit process discussed in Section 4-3—and our statement

$$F(b) = \lim \left[f(x_1^*)(x_1 - a) + f(x_2^*)(x_2 - x_1) + \cdots + f(x_n^*)(b - x_{n-1}) \right]$$

$$(4\text{-}141)$$

is equivalent [$F(a)$ being 0] to the statement already found in Section 4-3 (cf. Equation 4-36):

$$F(b) - F(a) = \int_a^b f(x)\, dx$$

A full justification of the statement will be given in Section 4-25 below.

We have asked only about the error at $x = b$. We can give a similar answer for the error at every x; that is

$$|F_0(x) - F(x)|$$

can be made as small as desired, for all x in $a \le x \le b$, by making all subdivision intervals small enough.

In all cases, our question leads us to the definite integral—the limit on the right of (4-141). The remainder of our chapter is devoted to the definite integral, the most fundamental concept of the integral calculus.

PROBLEMS

1. Find an indefinite integral:

   (a) $f(x) = \begin{cases} x, & 0 \le x \le 1 \\ 2 - x, & 1 < x \le 2 \end{cases}$ (b) $f(x) = \begin{cases} \sin x, & 0 \le x \le \pi/2 \\ 1, & \pi/2 < x \le \pi \end{cases}$

   (c) $f(x) = |x|, \; -\infty < x < \infty$ (d) $f(x) = \begin{cases} -x^2, & x \le 0 \\ x^2, & x > 0 \end{cases}$

2. Extend the statement and proof of Theorem 7 to a continuous function f, given as follows:

   (a) $f(x) = \begin{cases} f_1(x), & a \le x \le b \\ f_2(x), & b < x \le c \\ f_3(x), & c < x \le d \end{cases}$ (b) $f(x) = \begin{cases} f_1(x), & a \le x \le x_1 \\ f_2(x), & x_1 < x \le x_2 \\ \cdots \\ f_n(x), & x_{n-1} < x \le b \end{cases}$

3. Find a continuous indefinite integral and graph f and the integral:
   (a) $f(x) = 0$ for $0 \le x \le 1$, $f(x) = 1$ for $1 < x \le 2$
   (b) $f(x) = -1$ for $0 \le x \le 1$, $f(x) = 1$ for $1 < x \le 2$
   (c) $f(x) = x$ for $0 \le x \le 1$, $f(x) = 0$ for $1 < x \le 2$
   (d) $f(x) = -1$ for $x < 0$, $f(x) = 0$ for $x \ge 0$

4. Let f be piecewise continuous on $[a, b]$ and let F_1, F_2 be continuous indefinite integrals of f, so that $F_1'(x) = f(x)$ and $F_2'(x) = f(x)$ wherever f is continuous. Prove that, for some constant C, $F_2(x) = F_1(x) + C$.

5. Let f_1 and f_2 both be piecewise continuous on $[a, b]$. Let F_1, F_2 be continuous indefinite integrals of f_1, f_2, respectively; let k_1, k_2 be constants. Prove that $k_1 f_1 + k_2 f_2$ is piecewise continuous and that

   $$\int [k_1 f_1(x) + k f_2(x)]\, dx = k_1 F_1(x) + k_2 F_2(x) + C$$

6. Evaluate $\int f(x)\,dx$ approximately, where f is the function: $y = 3x^2$ in $[0, 1]$ of Example 1 in Section 4-14, by using as approximating function the step-function $q(x)$:

$$q(x) = \begin{cases} f(0), & 0 \le x \le \frac{1}{3} \\ f(\frac{1}{3}), & \frac{1}{3} < x \le \frac{2}{3} \\ f(\frac{2}{3}), & \frac{2}{3} < x \le 1 \end{cases}$$

Compare the result graphically with the exact indefinite integral, as in Figure 4-17.

7. Evaluate $\int f(x)\,dx$ approximately, where f is the function: $y = \cos x$ in $[0, \pi/2]$ of Example 2 in Section 4-14, by using the procedure of the text, with the same values $x_0, \ldots, x_4$, but by choosing $x_1^* = \pi/6$, $x_2^* = \pi/4$, $x_3^* = \pi/3$, $x_4^* = \pi/2$ (right-hand end points). Compare the result graphically with the exact indefinite integral, as in Figure 4-18.

PART III. THE DEFINITE INTEGRAL

4-15 THE DEFINITION OF THE DEFINITE INTEGRAL

We indicated in Part I of this chapter, and in Section 4-14, that the definite integral

$$\int_a^b f(x)\,dx$$

is given by

$$\int_a^b f(x)\,dx = \lim\left\{ f(\xi_1)(x_1 - x_0) + f(\xi_2)(x_2 - x_1) + \cdots + f(\xi_n)(x_n - x_{n-1}) \right\} \tag{4-150}$$

where the limit involves making the largest subinterval approach 0 in length; hence $n \to \infty$. In (4-150), $x_0 = a$, $x_n = b$ and $x_1, \ldots, x_{n-1}$ are chosen so that

$$x_0 = a < x_1 < x_2 < \cdots < x_{n-1} < b = x_n \tag{4-151}$$

so that the interval $[a, b]$ is subdivided as in Figure 4-19. The values $\xi_1, \ldots \xi_n$ are chosen in the subintervals as shown in Figure 4-19:

$$a \le \xi_1 \le x_1, \qquad x_1 \le \xi_2 \le x_2, \quad \cdots \tag{4-152}$$

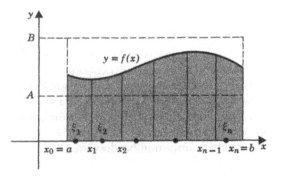

Figure 4-19 Definition of definite integral.

As we indicated in Section 4-4, for a function f whose values are *positive*, the integral can be interpreted as the area of the shaded region in Figure 4-19.

This presentation of the definite integral is essentially a descriptive one. Details must be filled in to give a precise meaning to what we are defining. In this section a new and precise form of the definition will be given. It will be made clear in the sections that follow that the new form is equivalent to the previous one.

Before stating our definition, we need two tools which simplify our work and are important in the definition.

The first tool is the Σ symbol for adding several terms. We define, for $n \geq 1$,

$$\sum_{i=1}^{n} a_i = a_1 + a_2 + \cdots + a_n$$

and, more generally,

$$\sum_{i=k}^{n} a_i = a_k + a_{k+1} + \cdots + a_n$$

where k and n are integers and $k \leq n$. We can also write, by using functional notation,

$$\sum_{i=1}^{n} g(i) = g(1) + g(2) + \cdots + g(n)$$

The following are examples:

$$\sum_{i=1}^{n} i = 1 + 2 + 3 + \cdots + n$$

$$\sum_{i=1}^{n} i^2 = 1^2 + 2^2 + 3^2 + \cdots + n^2 = 1 + 4 + 9 + \cdots + n^2$$

$$\sum_{k=2}^{5} \sin[(2k + 1)x] = \sin 5x + \sin 7x + \sin 9x + \sin 11x$$

and (4-150) itself can be written

$$\int_a^b f(x)\, dx = \lim \sum_{i=1}^{n} f(\xi_i)(x_i - x_{i-1}) \tag{4-153}$$

The second tool is the least upper bound axiom. We restate it and review it briefly here, but we recommend Section 2-13 for a detailed discussion. A set (or collection) of numbers is said to be *bounded above* if there is a number K for which $x \leq K$ for every x in the set; we call such a K an *upper bound* for the set. For example, the numbers $\sin(n\pi/4)$, for $n = 1, 2, 3, \ldots$, form a set that is bounded above; we can choose K to be 1, or any number larger than 1. In the example just given, 1 is the smallest K that can be used, so that 1 is the *least upper bound*. Generally, we say that a set that is bounded above has a least upper bound if there is a number K_0 such that $x \leq K_0$ for every number x in the set, and such that K_0 is the smallest number with this property (that is, any number less than K_0 is not an upper bound for the set).

LEAST UPPER BOUND AXIOM. *Every nonempty set of real numbers which is bounded above has a least upper bound.*

This axiom is an essential feature of the real number system, and we shall use it freely in what follows. It is clear from the very definition that in each case *the least upper bound is unique.* Hence, if E denotes a set that is bounded above, we can write

$$\text{lub } E$$

to denote the least upper bound of E. Thus

$$\text{lub}\left\{\sin\frac{\pi}{4}, \sin\frac{2\pi}{4}, \sin\frac{3\pi}{4}, \ldots\right\} = 1$$

We also write sup E for lub E (from the Latin word supremum).

There is a parallel discussion for a set E that is *bounded below*—that is, a set E for which there is a number L such that $x \geq L$ for all x in E. We could state a separate axiom, but as we pointed out in Section 2-13, such an axiom is a consequence of the least upper bound axiom and, therefore, we have the theorem:

GREATEST LOWER BOUND THEOREM. *Every nonempty set E that is bounded below has a greatest lower bound.*

We write glb E (or inf E, from infimum) for the unique greatest lower bound. Thus

$$\text{glb}\left\{\sin\frac{\pi}{4}, \sin\frac{2\pi}{4}, \sin\frac{3\pi}{4}, \ldots\right\} = -1$$

We can have lub E = glb E only when E consists of just one number. Otherwise

$$\text{glb } E < \text{lub } E$$

Now let f be a function defined and continuous on a closed interval $[a, b]$. Accordingly, f has a definite absolute minimum m and absolute maximum M, and $m \leq f(x) \leq M$ (see Section 2-7). We subdivide $[a, b]$ as in (4-151). In each interval $[x_{i-1}, x_i]$, f is continuous and, hence, has an absolute minimum m_i and an absolute maximum M_i. We form the sum

$$\sum_{i=1}^{n} M_i(x_i - x_{i-1}) \tag{4-154}$$

which we call an *upper sum* for f. We do the same for *all* such subdivisions of $[a, b]$ (the number n of subintervals will vary with the subdivision), and consider the set E of *all* values of the upper sums (4-154). We observe that every M_i is greater than or equal to m, and therefore every sum (4-154) is greater than or equal to

$$m(x_1 - x_0) + m(x_2 - x_1) + \cdots + m(x_n - x_{n-1})$$
$$= m(x_1 - x_0 + x_2 - x_1 + \cdots + x_n - x_{n-1})$$
$$= m(x_n - x_0) = m(b - a)$$

Thus, $L = m(b - a)$ serves as a lower bound for E. By the greatest lower bound theorem, E has a greatest lower bound. We define that greatest lower bound to be the definite integral of f from a to b.

Definition. *For a continuous function f on the closed interval $[a, b]$, $\int_a^b f(x)\, dx$ is the greatest lower bound of the set of all values of the upper sums (4-154) obtained from all subdivisions as in (4-151) of the interval $[a, b]$.*

We can abbreviate our definition, as follows:

$$\int_a^b f(x)\, dx = \text{glb} \left\{ \sum_{i=1}^{n} M_i(x_i - x_{i-1}) \right\} \tag{4-155}$$

or, by writing $\Delta_i x = x_i - x_{i-1}$,

$$\int_a^b f(x)\, dx = \text{glb} \left\{ \sum_{i=1}^{n} M_i \, \Delta_i x \right\} \tag{4-155'}$$

We shall see that we could just as well have defined the integral thus:

$$\int_a^b f(x)\, dx = \text{lub} \left\{ \sum_{i=1}^{n} m_i(x_i - x_{i-1}) \right\} \tag{4-156}$$

To justify this, we must show that the right side of (4-156) equals the right side of (4-155). Since $m_i \leq M_i$, clearly each sum in (4-156) is less than or equal to each sum in (4-155), so that

$$\sum_{i=1}^{n} m_i(x_i - x_{i-1}) \leq \sum_{i=1}^{n} M_i(x_i - x_{i-1}) \tag{4-157}$$

We shall show (Theorem 26 in Section 4-25 below) that, for a fine enough subdivision, the two members of (4-157) can be made to differ by as little as desired, from which it follows that the lub of the values of the left side must equal the glb of the values of the right side. If we replace each m_i or M_i by $f(\xi_i)$, where $x_{i-1} \leq \xi_i \leq x_i$ then, since $m_i \leq f(\xi_i) \leq M_i$, we get a sum that is squeezed between the two previous ones:

$$\sum_{i=1}^{n} m_i(x_i - x_{i-1}) \leq \sum_{i=1}^{n} f(\xi_1)(x_i - x_{i-1}) \leq \sum_{i=1}^{n} M_i(x_i - x_{i-1})$$

By what we have stated, it thus follows that, for sufficiently fine subdivisions, the middle sum will be as close as desired to $\int_a^b f(x)\, dx$; that is, (4-150) will be achieved. These statements will all be made precise and will be proved in later sections.

Terminology. As we indicated above, we refer to the sum $\Sigma M_i \Delta_i x$ in (4-154) as an *upper sum* for f. Similarly, the sums $\Sigma m_i \, \Delta_i x$ will be called *lower sums* for f. In $\int_a^b f(x)\, dx$, we call a the *lower limit* of integration, and b the *upper limit*.

Motivation for the Definition. If we think in terms of area, then it is clear that (for a positive function f), each upper sum $\sum_{i=1}^{n} M_i \, \Delta_i x$ is greater

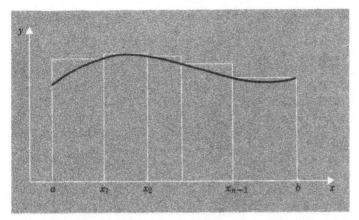

Figure 4-20 Geometric meaning of definition of definite integral.

than or equal to the "area under the curve." For the sum is the total of the areas of the rectangles as in Figure 4-20. However, as we subdivide more and more finely, we can make the total area of the rectangles smaller, and as close as we wish to the area sought. Hence, it is reasonable that the greatest lower bound should exactly equal the area.

The Dummy Variable. In the symbol $\int_a^b f(x)\,dx$ the letter x can be replaced by any desired letter. For example

$$\int_a^b f(x)\,dx = \int_a^b f(t)\,dt = \int_a^b f(u)\,du$$

and so on. The value in each case is the same, and in some books the integral is even written $\int_a^b f$. In general, we call the variable x in $\int_a^b f(x)\,dx$ a *dummy variable;* the integral is in no sense a function of x. A similar discussion applies to sums:

$$\sum_{i=1}^n a_i = \sum_{k=1}^n a_k = a_1 + a_2 + \cdots + a_n$$

Here i or k is the "dummy variable" or "dummy index."

PROBLEMS

1. Evaluate:

    (a) $\displaystyle\sum_{n=1}^3 (n^2 - 1)$ (b) $\displaystyle\sum_{n=1}^1 \sin\frac{n\pi}{2}$ (c) $\displaystyle\sum_{i=1}^3 1$

2. Simplify:

    (a) $\displaystyle\sum_{i=1}^n [a + (i - 1)d]$ (Sum of arithmetic progression)

    (b) $\displaystyle\sum_{i=1}^n ar^{i-1}$ (Sum of geometric progression);   (c) $\displaystyle\sum_{i=1}^n \ln\frac{i + 1}{i}$

3. Let $f(i)$ and $g(i)$ be functions defined for the appropriate integer values of i. Let α, β be constants. Prove the rules:

(a) $\sum\limits_{i=k}^{n} f(i) + \sum\limits_{i=n+1}^{m} f(i) = \sum\limits_{i=k}^{m} f(i)$

(b) If $A \leq f(i) \leq B$, then $A(n - k + 1) \leq \sum\limits_{i=k}^{n} f(i) \leq B(n - k + 1)$

(c) $\sum\limits_{i=k}^{n} [\alpha f(i) + \beta g(i)] = \alpha \sum\limits_{i=k}^{n} f(i) + \beta \sum\limits_{i=k}^{n} g(i)$

(d) Let $F(i) = \sum\limits_{j=k}^{i} f(j)$, $i = k, k + 1, \ldots, n$, $F(k - 1) = 0$. Then $F(i) - F(i - 1) = f(i)$ for $i = k, k + 1, \ldots, n$

(e) If $f(i) = F(i) - F(i - 1)$ for $i = k, k + 1, \ldots, n$, then $\sum\limits_{i=k}^{n} f(i) = F(n) - F(k - 1)$

4. Use the result of Problem 3(e) to simplify:

(a) $\sum\limits_{i=k}^{n} (2i - 1) = \sum\limits_{i=k}^{n} [i^2 - (i - 1)^2]$

(b) $\sum\limits_{i=k}^{n} \dfrac{1}{i(1 - i)} = \sum\limits_{i=k}^{n} \left(\dfrac{1}{i} - \dfrac{1}{i - 1}\right), k \geq 2$

(c) $\sum\limits_{i=1}^{n} (i - 1)[(i - 1)!] = \sum\limits_{i=1}^{n} [i! - (i - 1)!]$

5. Find the glb of the given sets of numbers:
   (a) $2/1, 3/2, 4/3, \ldots, (n + 1)/n, \ldots$.
   (b) Range of the function $y = 2^{-x}$, $x \geq 0$.
   (c) The set of areas of all convex polygons enclosing a circle of radius 1.
6. Prove directly from the definition of the definite integral that, if k is a positive constant, then

$$\int_a^b k\, dx = k(b - a)$$

7. Prove directly from the definition of the definite integral that

$$\int_0^b x\, dx = \frac{b^2}{2} \qquad (b > 0)$$

[*Hint.* Show that here $\Sigma M_i \Delta_i x$ can be written as

$$\frac{1}{2}\Sigma(x_i + x_{i-1})(x_i - x_{i-1}) + \frac{1}{2}\Sigma(x_i - x_{i-1})^2$$

and that the first term equals $b^2/2$. Conclude that $b^2/2$ is a lower bound for all sums. Take $x_1 = b/n$, $x_2 = 2b/n$, $\ldots$, $x_i = ib/n$, $\ldots$ to show that, for n sufficiently large, the sum $\Sigma M_i \Delta_i x$ can be made as close to $b^2/2$ as desired, so that $b^2/2$ must be the glb.]

4-16 PROPERTIES OF THE DEFINITE INTEGRAL

We now determine some properties that follow from our definition of the definite integral.

THEOREM 8. *Let f be a continuous function on the closed interval* $[a, b]$. *Then the definite integral of f from a to b,* $\int_a^b f(x)\, dx$, *exists.*

PROOF. Since f is continuous, f has an absolute minimum m. As we observed above, the set E of upper sums for f is bounded below by $m(b - a)$. Since E is surely nonempty, the Greatest Lower Bound Theorem applies, and E has a unique greatest lower bound. That lower bound is, by definition, the definite integral of f from a to b. Therefore, the integral exists.

THEOREM 9. *Let f be a continuous function on the closed interval* $[a, b]$ *and let* $A \leq f(x) \leq B$ *for* $a \leq x \leq b$. *Then*

$$A(b - a) \leq \int_a^b f(x)\, dx \leq B(b - a) \qquad (4\text{-}160)$$

PROOF. Since $A \leq f(x) \leq B$, we have also $A \leq M_i \leq B$ and hence

$$\sum_{i=1}^{n} A(x_i - x_{i-1}) \leq \sum_{i=1}^{n} M_i(x_i - x_{i-1}) \leq \sum_{i=1}^{n} B(x_i - x_{i-1})$$

or

$$A(b - a) \leq \sum_{i=1}^{n} M_i(x_i - x_{i-1}) \leq B(b - a)$$

It thus follows from the definition of glb, that

$$A(b - a) \leq \mathrm{glb} \sum_{i=1}^{n} M_i(x_i - x_{i-1}) \leq B(b - a)$$

so that (4-160) follows.

We can interpret the inequalities (4-160) geometrically for a positive function f defined on $[a, b]$. Since $A \leq f(x) \leq B$, the area under the graph of f is at least equal to that of a rectangle of base $b - a$ and altitude A, and is at most equal to that of a rectangle of base $b - a$ and altitude B. These rectangles are shown in Figure 4-19 (page 285).

THEOREM 10. *Let* $a < b < c$ *and let f be continuous on the closed interval* $[a, c]$. *Then*

$$\int_a^b f(x)\, dx + \int_b^c f(x)\, dx = \int_a^c f(x)\, dx \qquad (4\text{-}161)$$

PROOF. Let us subdivide the interval $[a, c]$ as usual and form the corresponding upper sum for f. If b happens to be a subdivision point, say $b = x_k$, then the upper sum can be written as

$$\sum_{i=1}^{k} M_i\, \Delta_i x + \sum_{i=k+1}^{n} M_i\, \Delta_i x$$

The first term here is an upper sum for f for the interval $[a, b]$; the second term is an upper sum for f for the interval $[b, c]$. Hence

$$\sum_{i=1}^{k} M_i\, \Delta_i x \geq \int_a^b f(x)\, dx, \qquad \sum_{i=k+1}^{n} M_i\, \Delta_i x \geq \int_b^c f(x)\, dx$$

Upon adding these two inequalities, we obtain

$$\sum_{i=1}^{n} M_i \, \Delta_i x \geq \int_a^b f(x) \, dx + \int_b^c f(x) \, dx \qquad (4\text{-}162)$$

If b is not a subdivision point, then we have $x_{k-1} < b < x_k$ for some k. Let M_k' be the absolute maximum of f in the interval $[x_{k-1}, b]$, M_k'' the absolute maximum of f in the interval $[b, x_k]$; also let $\Delta_k' x = b - x_{k-1}$, $\Delta_k'' x = x_k - b$, so that $\Delta_k x = \Delta_k' x + \Delta_k'' x$. (See Figure 4-21.) Since both of the small intervals are contained in the interval $[x_{k-1}, x_k]$, we have $M_k' \leq M_k$, $M_k'' \leq M_k$. Now we can write

$$\sum_{i=1}^{n} M_i \, \Delta_i x = \sum_{i=1}^{k-1} M_i \, \Delta_i x + M_k \, \Delta_k x + \sum_{i=k+1}^{n} M_i \, \Delta_i x$$

$$= \sum_{i=1}^{k-1} M_i \, \Delta_i x + M_k(\Delta_k' x + \Delta_k'' x) + \sum_{i=k+1}^{n} M_i \, \Delta_i x$$

$$\geq \left[\sum_{i=1}^{k-1} M_i \, \Delta_i x + M_k' \, \Delta_k' x \right] + \left[M_k'' \, \Delta_k'' x + \sum_{i=k+1}^{n} M_i \, \Delta_i x \right]$$

The terms in the first bracket are an upper sum for f for the interval $[a, b]$, and those in the second bracket are an upper sum for $[b, c]$. Therefore, each is greater than or equal to the corresponding integral, and (4-162) again follows. Thus the inequality (4-162) is true for every subdivision of $[a, c]$; whence its right side is a lower bound for all upper sums for f for the interval $[a, c]$. Accordingly

$$\int_a^c f(x) \, dx \geq \int_a^b f(x) \, dx + \int_b^c f(x) \, dx \qquad (4\text{-}163)$$

Next we subdivide $[a, b]$ and form the corresponding upper sum for f, subdivide $[b, c]$, and form the corresponding upper sum for f. If we add these two upper sums, we obtain an upper sum for f for the interval $[a, c]$. Therefore, we conclude successively that

$$(\text{upper sum for } [a, b]) + (\text{upper sum for } [b, c]) \geq \int_a^c f(x) \, dx$$

or $$(\text{upper sum for } [a, b]) \geq \int_a^c f(x) \, dx - (\text{upper sum for } [b, c])$$

The right side of the last inequality is thus a lower bound for all upper sums for f for $[a, b]$. Therefore we can conclude that

$$\int_a^b f(x) \, dx \geq \int_a^c f(x) \, dx - (\text{upper sum for } [b, c])$$

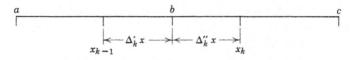

Figure 4-21 Proof of Theorem 10.

whence $\quad$ (upper sum for $[b, c]$) $\geq \int_a^c f(x)\, dx - \int_a^b f(x)\, dx$

so that $\qquad \int_b^c f(x)\, dx \geq \int_a^c f(x)\, dx - \int_a^b f(x)\, dx$

and, thus, $\qquad \int_a^c f(x)\, dx \leq \int_a^b f(x)\, dx + \int_b^c f(x)\, dx \qquad$ (4-164)

By (4-163) and (4-164), we observe that only the equals sign is valid; that is, (4-161) is proved.

$\qquad$ **An Extended Definition of** $\int_a^b f(x)\, dx$. When we defined the definite integral from a to b, we assumed $a < b$. It will be very convenient to remove this requirement. Since $\int_a^b f(x)\, dx$, with $a \geq b$, is thus far undefined, we are free to define it in any way we desire. We choose to define it so that (4-161) is valid for any three points a, b, c in an interval in which f is continuous. If we take $b = c$, this forces us to have

$$\int_a^c f(x)\, dx + \int_c^c f(x)\, dx = \int_a^c f(x)\, dx$$

and, consequently, we must take

$$\int_c^c f(x)\, dx = 0, \qquad \text{for every } c \qquad\qquad (4\text{-}165)$$

But then we would have

$$\int_a^c f(x)\, dx + \int_c^a f(x)\, dx = \int_a^a f(x)\, dx = 0$$

Thus we are forced to take

$$\int_c^a f(x)\, dx = -\int_a^c f(x)\, dx \qquad\qquad (4\text{-}166)$$

Equation (4-165) defines $\int_c^c f(x)\, dx$. Equation (4-166) defines $\int_c^a f(x)\, dx$ for $c > a$ in terms of the known value $\int_a^c f(x)\, dx$. By these definitions $\int_a^b f(x)\, dx$ is defined for every pair of numbers a, b in the interval in which f is defined and continuous.

$\qquad$ **COROLLARY OF THEOREM 10.** *If $f(x)$ is continuous for $\alpha \leq x \leq \beta$ and a, b, c are any 3 points in the interval $[\alpha, \beta]$, then*

$$\int_a^b f(x)\, dx + \int_b^c f(x)\, dx = \int_a^c f(x)\, dx$$

The proof is left as an exercise (Problem 3 below).

4-17 THE FUNDAMENTAL THEOREM OF CALCULUS

$\qquad$ We can now answer affirmatively the main question raised in Section 4-2: Does every continuous function have an indefinite integral?

THEOREM 11. *Let $f(x)$ be continuous for $a \leq x \leq b$. Let c be a fixed number in $[a, b]$. Let*

$$F(x) = \int_c^x f(t) \, dt, \qquad a \leq x \leq b \qquad (4\text{-}170)$$

Then

$$F'(x) = f(x) \qquad and \qquad F(c) = 0 \qquad (4\text{-}171)$$

Remarks. In (4-171), $F'(b)$ is understood as a left-hand derivative, and $F'(a)$ as a right-hand derivative, since the function is given on the closed interval $[a, b]$. The statement: $F' = f$ can be written

$$\frac{d}{dx}\left(\int_c^x f(t) \, dt\right) = f(x) \qquad (4\text{-}171')$$

This clearly indicates that the operations of integration (definite integral from c to x) and subsequent differentiation cancel each other. The rule (4-171') is one of the most important rules of the calculus, and is often termed *the fundamental theorem of the calculus.*

PROOF OF THEOREM 11. Let x be a fixed point for which $a \leq x < b$. Let $h > 0$ be such that $x + h \leq b$. Then

$$F(x + h) = \int_c^{x+h} f(t) \, dt = \int_c^x f(t) \, dt + \int_x^{x+h} f(t) \, dt$$

$$= F(x) + \int_x^{x+h} f(t) \, dt$$

and, hence,

$$F(x + h) - F(x) = \int_x^{x+h} f(t) \, dt$$

Let m be the minimum of f on $[x, x + h]$, and M the maximum. Then $m \leq f(x) \leq M$ on $[x, x + h]$ and hence, by Theorem 9,

$$mh \leq \int_x^{x+h} f(t) \, dt \leq Mh$$

so that

$$mh \leq F(x + h) - F(x) \leq Mh$$

and, h being positive,

$$m \leq \frac{F(x + h) - F(x)}{h} \leq M$$

But f is continuous at x, and therefore as h tends to 0, both m and M tend to $f(x)$ (Section 2-6). Hence we obtain a limit from the right (Section 2-5):

$$\lim_{h \to 0+} \frac{F(x + h) - F(x)}{h} = f(x), \qquad a \leq x < b \qquad (4\text{-}172)$$

Similarly, if x is a fixed point for which $a < x \leq b$ and $h < 0$, $x + h \geq a$, then

$$F(x) = \int_c^{x+h} f(t)\,dt + \int_{x+h}^x f(t)\,dt$$

$$= F(x+h) + \int_{x+h}^x f(t)\,dt$$

accordingly

$$F(x) - F(x+h) = \int_{x+h}^x f(t)\,dt$$

Now if m, M are the minimum and maximum, respectively of $f(x)$ on $[x+h, x]$, then

$$m \cdot [x - (x+h)] \le \int_{x+h}^x f(t)\,dt \le M \cdot [x - (x+h)]$$

so that

$$-mh \le \int_{x+h}^x f(t)\,dt \le -Mh$$

and hence

$$mh \ge -\int_{x+h}^x f(t)\,dt \ge Mh$$

Consequently,

$$mh \ge F(x+h) - F(x) \ge Mh$$

On dividing by h, which is negative, we obtain

$$m \le \frac{F(x+h) - F(x)}{h} \le M$$

Thus, as above,

$$\lim_{h \to 0-} \frac{F(x+h) - F(x)}{h} = f(x), \qquad a < x \le b \qquad (4\text{-}172')$$

By combining (4-172) and (4-172'), we have

$$\lim_{h \to 0} \frac{F(x+h) - F(x)}{h} = f(x), \qquad a < x < b$$

or $F'(x) = f(x)$ for $a < x < b$. In (4-172), we can take $x = a$ and conclude that $F'(a) = f(a)$, the derivative being a right-hand one. In the same way, from (4-172') with $x = b$, $F'(b) = f(b)$, the derivative being a left-hand one.

Finally, we observe that

$$F(c) = \int_c^c f(t)\,dt = 0$$

Thus, Theorem 11 is completely proved.

THEOREM 12. Let f_1, f_2 both be functions that are continuous for $a \le x \le b$, and let k_1, k_2 be fixed real numbers. Then

$$\int_a^b [k_1 f_1(x) + k_2 f_2(x)]\,dx = k_1 \int_a^b f_1(x)\,dx + k_2 \int_a^b f_2(x)\,dx$$

PROOF. Notice that, if f_1, f_2 are defined and continuous on $a \leq x \leq b$, so is $k_1f_1(x) + k_2f_2(x)$. For $a \leq x \leq b$, let

$$F_1(x) = \int_a^x f_1(t) \, dt,$$

$$F_2(x) = \int_a^x f_2(t) \, dt,$$

$$F(x) = \int_a^x [k_1f_1(t) + k_2f_2(t)] \, dt$$

Then, by Theorem 11,

$$F_1'(x) = f_1(x), \qquad F_2'(x) = f_2(x), \qquad F'(x) = k_1f_1(x) + k_2f_2(x)$$
$$F_1(a) = 0, \qquad\quad F_2(a) = 0, \qquad\quad F(a) = 0$$

Accordingly

$$F'(x) = k_1F_1'(x) + k_2F_2'(x) = [k_1F_1(x) + k_2F_2(x)]'$$

Hence, by Theorem 1 (Section 4-5),

$$F(x) = k_1F_1(x) + k_2F_2(x) + C, \qquad a \leq x \leq b$$

for some constant C. If we set $x = a$, we obtain $0 = 0 + 0 + C$, so that $C = 0$ and, thus,

$$F(x) = k_1F_1(x) + k_2F_2(x)$$

or

$$\int_a^x [k_1f_1(t) + k_2f_2(t)] \, dt = k_1 \int_a^x f_1(t) \, dt + k_2 \int_a^x f_2(t) \, dt$$

If we now set $x = b$, our theorem follows.

COROLLARY OF THEOREM 12. *If f is continuous on $[a, b]$ and k is constant, then*

$$\int_a^b kf(x) \, dx = k \int_a^b f(x) \, dx$$

PROOF. Let the constant k_2 in Theorem 12 be 0.

THEOREM 13. *Let $f(x)$ be continuous on $[a, b]$, then*

$$\int_a^b f(x) \, dx = \text{lub} \left\{ \sum_{i=1}^n m_i \, \Delta_i x \right\} \tag{4-173}$$

where the least upper bound is over the set of all values of the sum for all subdivisions (4-161) of $[a, b]$ and m_i is the minimum of f on $[x_{i-1}, x_i]$.

PROOF. If we had used (4-173) to define the integral, we would have had a development completely parallel to the one using (4-155), except that certain inequality signs would be reversed. Thus, we would again have Theorems 8, 9, 10, 11, 12 for the new integral. But Theorem 11 shows that $\int_c^x f(t) \, dt$ is $F(x)$, an indefinite integral of f, with $F(c) = 0$. There can be only one such indefinite integral by Theorem 1! Hence (4-173) must agree with (4-155).

COROLLARY OF THEOREM 13. *If f is continuous on* $[a, b]$, *then the number* $J = \int_a^b f(x) \, dx$ *is the one and only number such that*

$$\sum_{i=1}^n m_i \, \Delta_i x \le J \le \sum_{i=1}^n M_i \, \Delta_i x$$

for all subdivisions (4-151) *of* $[a, b]$, *where as usual* m_i *and* M_i *are the minimum and maximum values respectively of* $f(x)$ *for* $x_{i-1} \le x \le x_i$.

PROOF. If $J \le \Sigma M_i \, \Delta_i x$ for all subdivisions of $[a, b]$, then J is a lower bound for all such upper sums, and hence cannot exceed the greatest lower bound. Therefore

$$J \le \int_a^b f(x) \, dx = \text{glb}\{\Sigma M_i \, \Delta_i x\}$$

Similarly, if $J \ge \Sigma m_i \, \Delta_i x$ for all subdivisions, then

$$\int_a^b f(x) \, dx = \text{lub}\{\Sigma m_i \, \Delta_i x\} \le J$$

Hence

$$\int_a^b f(x) \, dx \le J \le \int_a^b f(x) \, dx$$

so that J equals the integral.

On the other hand, since $\int_a^b f(x) \, dx$ is both $\text{glb}\{\Sigma M_i \, \Delta_i x\}$ and $\text{lub}\{\Sigma m_i \, \Delta_i x\}$, we have for every subdivision

$$\Sigma m_i \, \Delta_i x \le \int_a^b f(x) \, dx \le \Sigma M_i \, \Delta_i x$$

THEOREM 14. *Let f be continuous on* $[a, b]$ *and let* $F'(x) = f(x)$ *for* $a \le x \le b$. *Then*

$$\int_a^b f(x) \, dx = F(b) - F(a) \tag{4-174}$$

PROOF. Let $G(x) = \int_a^x f(t) \, dt$. Then by Theorem 11, $G'(x) = f(x)$, $a \le x \le b$. Since $F'(x) = f(x)$ also, we have that $F(x) = G(x) + C$ for some constant C. But $F(a) = G(a) + C = 0 + C = C$, by Theorem 11. Hence, $F(x) = G(x) + F(a)$ or

$$F(x) = \int_a^x f(t) \, dt + F(a)$$

If we put $x = b$, then (4-174) follows.

Remarks. We can write our conclusion (4-174) as follows:

$$F(b) - F(a) = \int_a^b F'(x) \, dx$$

It is general practice to write $F(x) \Big|_a^b$ for $F(b) - F(a)$. Then the rule can be written

$$\int_a^b F'(x)\, dx = F(x) \Big|_a^b$$

Since in Theorem 11, C is arbitrary, there are many functions $F(x)$ for which $F'(x) = f(x)$, but any two differ from each other by a constant. If F_1 and F_2 are two such functions, then $F_1 = F_2 + C$ and $F_1(b) - F_1(a) = [F_2(b) + C] - [F_2(a) + C] = F_2(b) - F_2(a)$. Hence, the same result is obtained in (4-174), no matter which indefinite integral of f is chosen.

Theorem 14 is a basic result. It provides a very simple way of evaluating a definite integral, whenever we can find an indefinite integral (antiderivative) of the integrand.

EXAMPLE 1 $\qquad \displaystyle\int_0^3 x\, dx = \frac{x^2}{2} \Big|_0^3 = \frac{9}{2} - \frac{0}{2} = \frac{9}{2}$

EXAMPLE 2 $\qquad \displaystyle\int_1^2 \frac{1}{x}\, dx = \ln x \Big|_1^2 = \ln 2 - \ln 1 = \ln 2$

EXAMPLE 3 $\qquad \displaystyle\int_0^{\pi/4} \sec x\, dx = \ln(\sec x + \tan x) \Big|_0^{\pi/4} = \ln(\sqrt{2} + 1)$

EXAMPLE 4 $\qquad \displaystyle\int_0^1 2\sqrt{1 - x^2}\, dx = (x\sqrt{1 - x^2} + \text{Sin}^{-1} x) \Big|_0^1 = \frac{\pi}{2}$

For Examples 3 and 4 we can use the methods of Part II of this chapter, or we can use a table of integrals (Table I in the Appendix). Throughout the remainder of this text such a table will be found helpful. For a more extensive list of integrals, see Mathematical Tables from *Handbook of Chemistry and Physics* (published, with frequent new editions, by Chemical Rubber Publishing Co., Cleveland, Ohio) or B. O. Peirce, A *Short Table of Integrals* (published by Ginn and Co., Boston, 1929).

Remark. The formula (4-174) remains valid if $b \leq a$ [Problem 4(a) below]. A similar statement applies to the conclusion of Theorem 12 and its Corollary [Problem 4(b) below].

PROBLEMS

1. Evaluate:

(a) $\displaystyle\int_0^1 (2x - 1)^2\, dx$ (b) $\displaystyle\int_{-\pi/2}^{\pi/2} \sin 2x\, dx$ (c) $\displaystyle\int_0^1 e^{3x}\, dx$

(d) $\displaystyle\int_0^5 e^{x^2} x\, dx$ (e) $\displaystyle\int_0^1 \frac{dx}{x^2 + 4}$ (f) $\displaystyle\int_{-1}^1 \frac{dx}{x^2 - 4}$

(g) $\displaystyle\int_{-3}^{-2} \frac{dx}{x}$ (h) $\displaystyle\int_1^2 \frac{dx}{x(x^2 + 4)}$ (i) $\displaystyle\int_0^{0.5} \frac{dx}{\sqrt{1 - x^2}}$

(j) $\displaystyle\int_0^{\pi/4} \tan^4 x \sec^2 x\, dx$ (k) $\displaystyle\int_0^\pi \frac{\sin 2x}{3 + \cos 2x}\, dx$ (l) $\displaystyle\int_0^{\pi/4} e^{\tan x} \sec^2 x\, dx$

(m) $\int_0^1 (\sin \sqrt{x^4 + 1})' \, dx$ (n) $\int_0^{\pi/4} (\tan^3 x)' \, dx$ (o) $\int_0^1 (\sec^2 x)'' \, dx$

(p) $\int_{-1}^1 [(x^2 - 1)f(x)]' \, dx$

2. In each of the following cases, evaluate $\int_a^b f(x) \, dx$ and then verify by geometry that it equals the area under the curve $y = f(x)$ from $x = a$ to $x = b$.

(a) $\int_a^b m(x - a) \, dx$, $m > 0$, a, b, m constant (triangle)

(b) $\int_a^b (Ax + B) \, dx$, A, B, constant, $Ax + B > 0$ on $[a, b]$ (trapezoid)

(c) $\int_0^b \sqrt{1 - x^2} \, dx$, $0 < b \leq 1$ (portion of a circle)

(d) $\int_b^1 \sqrt{1 - x^2} \, dx$, $-1 \leq b < 1$ (portion of circle)

3. Prove the Corollary to Theorem 10. [*Hint.* Show that the equation to be proved can be written

$$\int_a^b f(x) \, dx + \int_b^c f(x) \, dx + \int_c^a f(x) \, dx = 0$$

Prove this from Theorem 10 for the two cases $a \leq b \leq c$ and $c \leq b \leq a$, and show that all other cases follow from these two.]

4. Let a, b be arbitrary numbers in the interval $[\alpha, \beta]$, so that possibly $b < a$ or $b = a$.
   (a) Show that (4-174) holds true, if f is continuous in $[\alpha, \beta]$ and $F' = f$ in $[\alpha, \beta]$.
   (b) Show that the conclusion of Theorem 12 and that of its corollary hold true, if $f_1(x), f_2(x)$ are continuous in $[\alpha, \beta]$.

5. Prove from the definition: Let $x_0, x_1, \ldots, x_n$ be a subdivision of $[a, b]$, as in (4-151), and let $x_0', \ldots x_k'$ be another. Then

$$\sum_{i=1}^n m_i(x_i - x_{i-1}) \leq \sum_{i=1}^k M_i(x_i' - x_{i-1}')$$

[*Hint.* Consider the subdivision consisting of all the points in either subdivision.]

6. (a) Prove that, if a subdivision (4-151) is such that $M_i - m_i < \dfrac{1}{10^k}$ for $i = 1, \ldots, n$, then

$$\sum_{i=1}^n M_i \, \Delta_i x - \frac{b - a}{10^k} \leq \int_a^b f(x) \, dx \leq \sum_{i=1}^n M_i \, \Delta_i x$$

(b) Evaluate $\ln 2 = \int_1^2 dt/t$ with an error less than 0.05.

4-18 AREA

From our experience in geometry we are led to an intuitive feeling for the area of a figure—a triangle, a circle, an ellipse, in general a figure enclosed by one or several boundaries (Figure 4-22). Geometry gives us a precise formula for the area of a triangle. From it, we find the area of a polygonal figure by cutting it up into triangles, as in the first figure in Figure 4-22; it

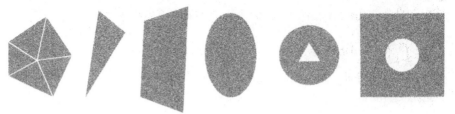

Figure 4-22   Figures having area.

is by no means a simple exercise in geometry to prove that the result does not depend on how we cut up the figure.

From these examples, we might be led to expect that area can be defined for *every figure* in the plane. If by "every figure" we mean every set of points in the plane, then we have gone too far. For the points in the first quadrant, in terms of xy-coordinates, form a set to which we can hardly assign a finite area. Other examples can be given of strange bounded sets of points for which the assignment of area in a reasonable way is in doubt.

Modern researches have shown that the best we can attain is a certain class of figures having area. Furthermore, for these figures, we have the following results (which themselves allow us to say we have defined area):

Rule I. If one figure is congruent to another and the first figure has area, then the second figure has area and the two areas are equal.

Rule II. If two figures have area and the first figure includes the second, then the area of the first figure is at least as large as that of the second; furthermore, the figure obtained by removing the second figure from the first has area, equal to that of the first figure minus that of the second figure.

Rule III. If a given figure is formed of a finite number of figures, no two of which have points in common, and each of which has area, then the given figure has area, equal to the sum of the areas of the several figures of which it is composed.

Rule IV. The area of a rectangle of sides a, b is ab. The area of the empty set is 0.

Rule V. If a given figure is included in other figures having area and if the given figure includes other figures having area, and if the greatest lower bound of the areas of the including figures equals the least upper bound of the area of the included figures, then the given figure has area, equal to the common value of the greatest lower bound and the least upper bound.

In Rule I the word "congruent" is used as in geometry. Hence two figures are congruent if, and only if, there is a one-to-one correspondence between the figures so that each pair of points of the first figure corresponds to an equidistant pair of points of the second figure. In the presence of the other rules, the first part of Rule IV is equivalent to giving the usual rule for the area of a triangle. Rule V is suggested by the familiar process of obtaining the area of a circle by considering the areas of inscribed and circumscribed regular polygons; in fact, the area of the circle is equal to the lub of the former set of numbers and the glb of the latter. A similar process of "exhaustion" was used by Archimedes to find the areas of figures such as a segment of a parabola.

Rule III allows us to find the area of a polygonal figure by cutting it up into triangles. At first, there seems to be a difficulty here, since the triangles will overlap at edges and vertexes. However, we can count the interiors of the triangles, the edges without end points, and the vertexes as separate figures. Then, we assign area 0 to each edge and to each vertex, and obtain the expected result.

That a line segment or point has zero area can also be justified by Rules IV and V. For every line segment or point can be enclosed in a rectangle of area as small as desired; it also encloses a set of zero area, namely, the empty set. Hence, by Rule V, its area must be 0. We might also be tempted to say that every rectangle can be formed by putting together many line segments (Figure 4-23). Since each segment has zero area, Rule III seems to imply that the rectangle has zero area. However, in Rule III, our figure is to be composed of a finite number of figures. By similar reasoning, we can show that the graph of a continuous function f in $[a, b]$ has zero area.

We shall not attempt to describe the class of allowed figures here or to prove that there is a class for which area can be defined in such a way as to satisfy the rules. (This is carried out in books on "measure theory.") Instead we shall assume that such a proof has been given and shall then deduce that certain figures do have area and shall find a value for that area.

4-19 AREA UNDER A CURVE

Let f be a continuous function on the interval $[a, b]$, and let $f(x) > 0$ for $a < x < b$. We seek the area of the figure or *region*, as we prefer to call it, bounded by the graph of $y = f(x)$, and the segments: $x = a, 0 \leq y \leq f(a)$; $x = b, 0 \leq y \leq f(b)$; $y = 0, a \leq x \leq b$. That is, we seek the area "beneath the curve $y = f(x)$ from $x = a$ to $x = b$" (see Figure 4-24). To this end, we subdivide our interval $[a, b]$ by points $x_0, x_1, \ldots, x_n$ as in construction of the definite integral. For each i, we construct the rectangle with the interval $[x_{i-1}, x_i]$ as base and with altitude m_i, the minimum of $f(x)$ on

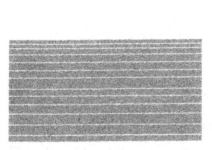

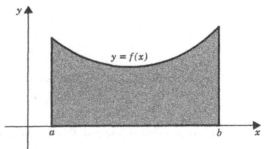

Figure 4-23 Rectangle as union of line segments.

Figure 4-24 Area beneath a curve.

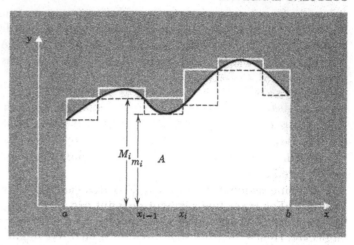

Figure 4-25   Definite integral and area.

$[x_{i-1}, x_i]$. The union of these rectangles is a polygon inscribed in the region under the curve (see Figure 4-25). By Rules III and IV, this inscribed polygon has as area

$$\sum_{i=1}^{n} m_i(x_i - x_{i-1})$$

Strictly speaking, the several figures overlap to the extent of having a line segment in common but, as in the case of cutting a polygonal figure into triangles, the edges contribute nothing.

Similarly, for each i, we construct the rectangle with $[x_{i-1}, x_i]$ as base and with altitude M_i, the maximum of $f(x)$ on $[x_{i-1}, x_i]$. The union of these rectangles is a polygon containing the region under the curve (see Figure 4-25) and this circumscribing polygon has as area

$$\sum_{i=1}^{n} M_i(x_i - x_{i-1})$$

We may now apply Rule V. For our region is included in all the circumscribed polygons and includes all the inscribed polygons, and the polygons all have area, by Rules III and IV. Furthermore, the glb of the areas of the circumscribed polygons equals the lub of the areas of the inscribed polygons. In fact, by Theorem 13 and the definition of the definite integral,

$$\text{lub} \sum_{i=1}^{n} m_i \, \Delta_i x = \text{glb} \sum_{i=1}^{n} M_i \, \Delta_i x = \int_a^b f(x) \, dx$$

Thus, by Rule V, our "region under the curve" has area, and it is given by

$$A = \int_a^b f(x) \, dx \qquad (4\text{-}190)$$

If f has indefinite integral F, we can now evaluate the area by (4-174):

$$A = \int_a^b f(x) \, dx = F(b) - F(a)$$

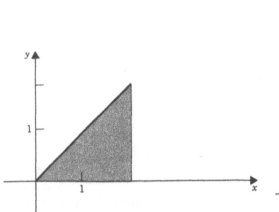

Figure 4-26 Example 1.   Figure 4-27 Example 2.

EXAMPLE 1. We take $y = x$, $0 \leq x \leq 2$. Then

$$A = \int_0^2 x \, dx = \frac{x^2}{2} \bigg|_0^2 = 2 \text{ [sq units]}$$

In this case, the figure is a right triangle (Figure 4-26) of legs 2 and 2. Accordingly, the result can be found easily by geometry.

EXAMPLE 2. $y = x^2$, $0 \leq x \leq 3$. The curve is a parabola. We find that the area beneath this parabola from $x = 0$ to $x = 3$ is

$$A = \int_0^3 x^2 \, dx = \frac{x^3}{3} \bigg|_0^3 = 9 \text{ [sq units]}$$

The area of the symmetric figure (Figure 4-27) is also

$$\int_{-3}^0 x^2 \, dx = \frac{x^3}{3} \bigg|_{-3}^0 = 9 \text{ [sq units]}$$

and hence by Rule II the area of the "parabolic segment" QOP is $6 \times 9 - 18 = 36$ sq units.

EXAMPLE 3. Let $y = \sqrt{a^2 - x^2}$, $-a \leq x \leq a$. The curve is a semicircle (Figure 4-28). We find that, with the aid of entry no. 30 of Table I in the Appendix,

$$A = \int_{-a}^a \sqrt{a^2 - x^2} \, dx = \frac{1}{2} \left(x\sqrt{a^2 - x^2} + a^2 \, \text{Sin}^{-1} \frac{x}{a} \right) \bigg|_{-a}^a$$

$$= \frac{1}{2} [a^2 \, \text{Sin}^{-1} 1 - a^2 \, \text{Sin}^{-1}(-1)]$$

$$= \frac{1}{2} \left(\frac{\pi a^2}{2} + \frac{\pi a^2}{2} \right) = \frac{\pi a^2}{2}$$

Thus the area of the whole circle is πa^2. This basic formula of geometry is thus proved by calculus. (Notice, however, that the sine function, used in

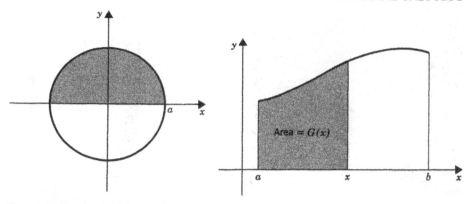

Figure 4-28   Example 3.   Figure 4-29   Area under curve from a to x.

our evaluation, has been defined on the basis of geometric properties of angle and circles. In Chapter 5 a definition of the sine function, wholly within the calculus, will be presented.)

THEOREM 15. *Let f be a continuous function on* $[a, b]$*, let* $f(x) > 0$ *for* $a < x < b$*. Let* $G(x)$ *denote the area under the curve* $y = f(x)$ *from a to x. Then* $G'(x) = f(x)$*,* $G(a) = 0$*.*

The proof is given as an exercise (Problem 4 below). The theorem states that an indefinite integral of f can be found as the area under the curve from a to a variable x (Figure 4-29). Thus we obtain a graphical way of finding an indefinite integral (see Section 4-4).

Remark. The value given above for "area under a curve" might seem at first to depend on the coordinates chosen. However, the area has true geometric meaning, and its value does not depend on the choice of axes. This is observed most easily from the fact that the area is the glb of the areas of certain circumscribed polygons; the area of each of these polygons is determined by geometry, by the rules of Section 4-18, and therefore is unrelated to the choice of axes. The only way the coordinate axes enter is through our choice of a figure for which the axes can be chosen so that the area in question is the area under a curve. In Chapter 7, other expressions for area are given, in which the coordinates do not appear.

PROBLEMS

1. For each of the functions given, graph and find the area under the curve:

    (a) $y = \frac{1}{x}$ (hyperbola), $1 \leq x \leq 2$ (b) $y = b\sqrt{1 - \frac{x^2}{a^2}}$ (ellipse), $-a \leq x \leq a$

    (c) $y = 2 + \sin x$, $c \leq x \leq c + 2\pi$ (d) $y = \ln x$, $1 \leq x \leq 2$

2. Find the area bounded by the x-axis and the curve given:

    (a) $y = 1 - x^2$ (b) $y = 2 + x - x^2$ (c) $y = \frac{1 - x^2}{1 + x^2}$

    (d) $y = -e^{2x} + 3e^x - 2$

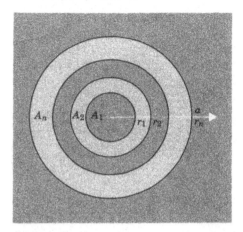

Figure 4-30

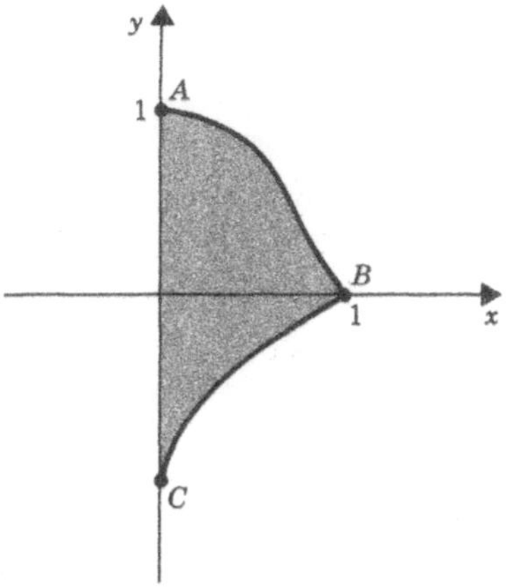

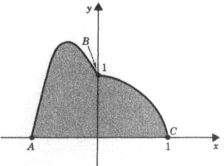

Figure 4-31

3. Find the area bounded by the y-axis and the curve given:
   (a) $y^2 = 1 - x$ (b) $2x + e^y + e^{-y} - 4 = 0$
   (c) $x - \sqrt{4y^2 - y^4} = 0$ (d) $xy^4 + 5xy^2 + 4x + y^2 - 4 = 0$

4. Prove Theorem 15.

5. To find the area of a circle of radius a, we might reason as follows. Subdivide the interval $0 \le r \le a$ by values $r_0 = 0 < r_1 < r_2 < \cdots < r_n = a$. The whole area A is then the sum of the areas $A_1, \ldots, A_n$ of n rings (Figure 4-30). Let the circumference of a circle of radius r be $g(r)$. Then "clearly" A_i lies between $m_i(r_i - r_{i-1})$ and $M_i(r_i - r_{i-1})$, where m_i and M_i are the minimum and maximum of $g(r)$ in $[r_{i-1}, r_i]$. Hence, as in the derivation of (4-191),

$$(\alpha) \quad A = \int_0^a g(r)\, dr$$

Also if $G(r)$ is the area of the circle of radius r then, by Theorem 11,
$$(\beta) \quad G'(r) = g(r)$$

   (a) Show by geometry that (α) and (β) are correct.
   (b) Deduce an analogous set of relations for the area of a square and its perimeter.
   (c) Obtain similar relations for the volume and surface area of a sphere.

6. (a) In Figure 4-31, curve AB has the equation $(1 - x)e^x - y = 0$, curve BC has the equation $(1 + y)e^y - x = 0$.
   Find the shaded area.

   (b) In Figure 4-32, curve AB has the equation $y = 1 - x + 2x^3$, curve BC has the equation $x = \cos(\pi y/2)$. Find the shaded area.

7. (a) In Figure 4-33, OAB is an isosceles right triangle, $|\overrightarrow{OA}| = 1$ and the curve shown is such that at each point P on the curve $|\overrightarrow{PQ}| = |\overrightarrow{OQ}| - |\overrightarrow{OQ}|^2$, where Q is the foot of the perpendicular from P to OA. Find the shaded area.

Figure 4-32

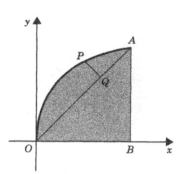

Figure 4-33   Figure 4-34

(b) In Figure 4-34, $ABCD$ is a rectangle, the curve DC is such that $|\overrightarrow{P_1Q_1}| = 1 + \cos|\overrightarrow{D_1Q_1}|$, the curve to the right of BC is such that $|\overrightarrow{P_2Q_2}| = \sin|\overrightarrow{CQ_2}|$; here P_1, P_2 are typical points on the curves and Q_1, Q_2 are the feet of the perpendiculars on DC and BC, respectively. Find the shaded area.

8. Find the area:

(a) Outside the circle $4(x - 1)^2 + 4(y - 1)^2 = 1$ and inside the ellipse $9x^2 + 16y^2 = 144$. [*Hint.* The area inside ellipse $(x^2/a^2) + (y^2/b^2) = 1$ is πab, as in Problem 1(b).].

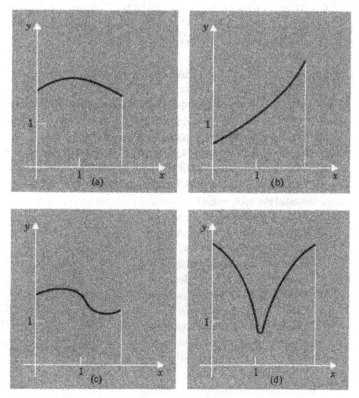

Figure 4-35

(b) Outside the triangle of vertexes $(1, 1)$, $(3, 2)$, $(2, 3)$ and inside the circle $x^2 + y^2 = 36$.

(c) Above the curve $y = \sin x$ and below the curve $y = 3 - \sin x$, $0 \leq x \leq \pi$.

(d) Above the curve $y = -1/x$ and below the curve $y = e^{-x}$, $-2 \leq x \leq -1$.

9. (a) to (d) In Figure 4-35, several functions are given graphically. Trace each graph and draw circumscribed and inscribed polygons as in Figure 4-25 and Figure 4-6. Then, by measurement, find the area of the polygons and estimate the area under the curve.

10. (a) For a discontinuous function such as

$$f(x) = \begin{cases} xe^x, & 0 \leq x \leq 1 \\ e^x - 1, & 1 < x \leq 2 \end{cases}$$

we interpret the region below the curve as the set of all (x, y) for which $0 \leq y \leq f(x)$, $0 \leq x \leq 2$. Graph the function and show by circumscribed and inscribed polygons that the area equals

$$\int_0^1 xe^x \, dx + \int_1^2 (e^x - 1) \, dx$$

Determine this area.

(b) Determine the area under the curve for the function

$$f(x) = \begin{cases} \sin^3 x, & 0 \leq x \leq \pi \\ 2 + \cos x, & \pi < x \leq 2\pi \end{cases}$$

11. Let f be the function defined on $[0, 1]$, for which $f(x) = 1$ if x is irrational and $f(x) = 0$ if x is rational. Let "circumscribed and inscribed polygons" be constructed for finding the area below the curve, as in Section 4-19. Show that the area of each circumscribed polygon is 1, and that the area of each inscribed polygon is 0. (Despite this paradox, modern measure theory would assign the value 1 to the area.)

4-20 THE INTEGRAL AS AN ACCUMULATOR

The formula (4-174) can be interpreted physically as follows. If $F(x)$ is a measure of how much of a certain quantity is present at index value x, then the change in F in going from $x = a$ to $x = b$ is the integral of the rate of change of F from a to b, that is, the integral of $F'(x)$ from a to b.

If we replace x by t and think of t as time, then $F(t)$ is a measure of how much is present at time t, and the formula states that *the net gain in F from time t = a to time t = b is the integral of the rate of production F'(t)*.

EXAMPLE 1. Water is flowing through a pipe at a variable rate. If the rate of flow at time t at the point A (Figure 4-36) is y cu ft per sec, then y is a function of t, $y = f(t)$ (with, say, t in seconds). But the rate of flow can be interpreted as the derivative of $V = F(t)$, where V is the total amount of fluid that has passed A, starting at a certain initial time. Therefore, the total volume of fluid passing A from $t = a$ to $t = b$ is

$$F(b) - F(a) = \int_a^b f(t) \, dt$$

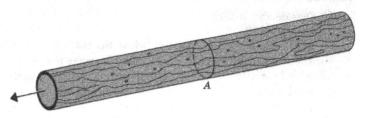

Figure 4-36 Flow of water in a pipe.

We can also justify this formula by reasoning similar to the reasoning that leads to the area formula. We subdivide the interval $[a, b]$ by values $t_0 = a < t_1 < \cdots < t_n = b$. In the time from $t = t_{i-1}$ to $t = t_i$ the amount of fluid passing A is at least equal to $m_i(t_i - t_{i-1})$, where m_i is the minimum rate in this time interval—that is, m_i is the minimum of f (assumed continuous) in the interval. Similarly, the amount is at most $M_i(t_i - t_{i-1})$, where M_i is the maximum rate. Hence, for each subdivision, the total amount lies between

$$\sum_{i=1}^{n} m_i(t_i - t_{i-1}) \quad \text{and} \quad \sum_{i=1}^{n} M_i(t_i - t_{i-1})$$

so that, by the Corollary to Theorem 13, it equals $\int_a^b f(t)\, dt$.

EXAMPLE 2. A point moves on a line, the y-axis. Its velocity at time t is $v(t)$ (for instance, in feet per second, t in sec, y in ft). The total displacement from time $t = \alpha$ to time $t = \beta$ is

$$y(\beta) - y(\alpha) = \int_\alpha^\beta v(t)\, dt \qquad (4\text{-}200)$$

Here v may be positive or negative at different times, corresponding to increasing and decreasing y. However, the same reasoning applies. The distance moved from time t_{i-1} to time t_i is between $m_i(t_i - t_{i-1})$ and $M_i(t_i - t_{i-1})$, where m_i and M_i are the minimum and maximum velocities. The rest follows as before. We can interpret (4-200) as saying that *total distance accumulated is the integral of velocity* (the rate of producing distance—as in the phrase "eating up mileage").

EXAMPLE 3. In a chemical reaction, a certain substance is being precipitated at the rate of y grams per sec, where y varies with time t. What is the total precipitated from $t = 0$ to $t = 10$ (secs)? Again, we use a similar *reasoning* and are led to the formula:

$$\text{Total precip.} = \int_0^{10} y(t)\, dt \qquad \text{(grams)}$$

Thus, in effect, we are again finding the area under a curve [Figure 4-37(a)]. Figure 4-37(b) shows the total accumulated at $t = 1, 2, \ldots, 10$ sec. The amount accumulated between $t = t_{i-1}$ and $t = t_i$ (for instance, between 1 and 2 sec, as in the figure) is shown as a layer of material. This corresponds to one term, lying between $m_i \Delta_i t$ and $M_i \Delta_i t$, as before.

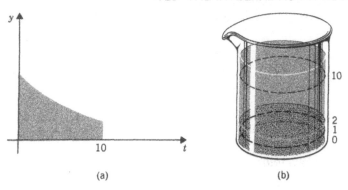

(a)

(b)

Figure 4-37 Precipitation of a chemical.

Clearly, from these examples, the integration process is present in countless practical situations. Whenever something is being accumulated (or lost) at a constant or variable rate, the total accumulated or lost is given by the integral of the function that represents the "rate of accumulation." The integral also appears in many other contexts, which are illustrated in the next example and in later sections.

EXAMPLE 4. *Work.* Let a particle of mass m move from $x = a$ to $x = b$ along the x-axis subject to a force F, directed along the x-axis (Figure 4-38). Let the force F vary with position, so that $F = F(x)$. The work done by a constant force is force times distance, or more precisely, the force component along the line of motion times the distance moved (Section 1-10). This component here corresponds to $F(x)$, which may be positive, negative of zero. Since $F(x)$ varies, we define the total work done by adding the work done in moving from x_{i-1} to x_i for $i = 1, 2, \ldots, n$, where we have subdivided the interval $[a, b]$, as before. The familiar argument then leads us to the equation:

$$\int_a^b F(x)\, dx = \text{work} \tag{4-201}$$

Here, a may be less than, greater than, or equal to b. Equation (4-201) can be considered to be a *definition* of work for the case considered.

Newton's Second Law states that the force equals the mass times the acceleration; this applies to the force and acceleration vectors, or to their components in a fixed direction. Consequently, here, if x is the position at time t,

$$F(x) = m\frac{d^2x}{dt^2} = m\frac{dv}{dt}$$

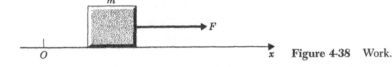

Figure 4-38 Work.

Now if we assume the velocity v is expressible in terms of position x, then by the chain rule

$$\frac{dv}{dt} = \frac{dv}{dx}\frac{dx}{dt} = v\frac{dv}{dx}$$

Accordingly,

$$F(x) = mv\frac{dv}{dx} = \frac{d}{dx}\left(\frac{1}{2}mv^2\right)$$

(We assume that the functions concerned have continuous derivatives.) Accordingly, by Theorem 14,

$$\text{Work} = \int_a^b \frac{d}{dx}\left(\frac{1}{2}mv^2\right)dx = \frac{1}{2}mv^2\Big|_a^b$$

The quantity $\frac{1}{2}mv^2$ is the *kinetic energy* of the particle. Therefore, we have proved that *for rectilinear motion, the work done by the force equals the gain in kinetic energy of the particle*. This is a basic law of physics. The law shows that the work integral measures the accumulation of kinetic energy by the particle.

PROBLEMS

1. Water is flowing through a pipe at a linearly increasing rate, observed to be 10 cu ft per sec at 8 A.M., and to be 12 cu ft per sec at 2 P.M. of the same day. How much water passed the point of observation in this time interval?

2. (a) Justify the rule: if a particle moves on a line, the y-axis, with acceleration a as a known function of time t, then the change in velocity v from time $t = \alpha$ to time t is given by $v(t) - v(\alpha) = \int_\alpha^t a(u)\, du$.

   (b) Use the result of part (a) to find $v(t)$ for a particle for which $a(t) = 3\sin 2t$ ft per sec^2, if the particle starts with $y = 0$ and $v = 0$ for $t = 0$. Also find $y(5)$.

   (c) For a falling body near the surface of the earth, if the y-axis is vertical and pointing up, we have $a = -g$. Find the velocity and the position at time t for a motion starting at $t = 0$ at a distance y_0 ft above the earth and with velocity v_0 ft per sec.

   (d) For the falling body motion of part (c), let y become y_0 again at a positive time t_1, at which $v = v_1$. Find the velocity $v(t_1)$. [*Hint.* Use the relation between work and energy as described for Example 4.]

3. (a) Write out in detail the justification of the integral representation of total precipitation for Example 3.

   (b) The rate of precipitation for a certain reaction is $2e^{-bt}$ gr per sec, where b can be adjusted by changing concentrations. How should b be chosen so that 50 gr are precipitated between $t = 0$ and $t = 100$ sec?

4. In a certain experiment water droplets are formed and grow in spherical shape. If for a single droplet mass can be added at a constant rate of k gr per sec, find the radius at time $t = 100$ sec for a droplet that has radius 0.1 cm at $t = 0$. (Take the density of water to be 1 gr per cc).

5. Introduce reasonable functions to show that each of the following can be represented by a definite integral:
   (a) Total rainfall
   (b) Erosion
   (c) The storage of electric charge in a battery
   (d) The heating of an object
   (e) Energy radiated
   (f) The growth of money in a bank
   (g) The growth of a population
   (h) The gain in weight of a person
   (i) The inventory of a manufacturer
   (j) The reading of a gas meter

6. The rate r of consumption of gasoline by a certain car is known to depend on the speed v by the equation:

$$r = \frac{1}{2000}(v^2 - 80v + 1800)$$

   Here, v is in miles per hour, and r is in gallons per mile.
   (a) What is the most efficient speed?
   ‡(b) On a certain trip the speed is observed to vary according to the equation $v = 40 + 5\sin 40\pi t$ (t in hours). Find the gasoline consumed in one hour.

7. The records of snowfall for many years in a certain northern region have the following average values: on Nov. 15, 0.49 in.; on Dec. 15, 1.3 in.; on Jan. 15, 1.6 in.; on Feb. 15, 1.3 in.; on Mar. 15, 0.47 in.
   (a) Show that those data are reasonably well represented by the equation $r = 1.6\sin(\pi t/150)$, where r is the rate of snowfall in inches per day, t is measured in days starting at the beginning of November, and each month is assumed to have 30 days.
   (b) Use the formula of part (a) to estimate the total snow precipitation during the following periods: (i) November 1 to March 30, and (ii) January 5 to January 20.

4-21   INTEGRATION BY PARTS AND SUBSTITUTION

The formula for integration by parts has a counterpart for definite integrals:

THEOREM 16. *Let f and g be functions having continuous first derivatives on $[a, b]$. Then*

$$\int_a^b f(x)g'(x)\,dx = [f(x)g(x)]\Big|_a^b - \int_a^b g(x)f'(x)\,dx \qquad (4\text{-}210)$$

PROOF. With $F(x) = f(x)g(x)$ we have

$$F(b) - F(a) = \int_a^b F'(x)\,dx = \int_a^b [f(x)g'(x) + g(x)f'(x)]\,dx$$

and hence

$$[f(x)g(x)]\Big|_a^b = \int_a^b f(x)g'(x)\,dx + \int_a^b g(x)f'(x)\,dx$$

Thus (4-210) follows.

EXAMPLE 1 $\quad \int_0^\pi \sin^4 x \, dx = \int_0^\pi \sin^3 x \sin x \, dx$

$$= [\sin^3 x(-\cos x)] \Big|_0^\pi + 3 \int_0^\pi \sin^2 x \cos^2 x \, dx$$

$$= 3 \int_0^\pi \sin^2 x \cos^2 x \, dx$$

[The term in brackets equals 0 at 0 and at π and, therefore, drops out.] We can now write

$$\int_0^\pi \sin^4 x \, dx = 3 \int_0^\pi \sin^2 x(1 - \sin^2 x) \, dx = 3 \int_0^\pi \sin^2 x \, dx - 3 \int_0^\pi \sin^4 x \, dx$$

so that

$$\int_0^\pi \sin^4 x \, dx = \frac{3}{4} \int_0^\pi \sin^2 x \, dx = \frac{3}{4} \int_0^\pi \frac{1 - \cos 2x}{2} \, dx =$$

$$\frac{3}{8} \left(x - \frac{\sin 2x}{2} \right) \Big|_0^\pi = \frac{3\pi}{8}$$

For substitution, the theory is simpler than for indefinite integrals (Section 4-8) above:

THEOREM 17. *Let $x(u)$ have a continuous derivative on $[\alpha, \beta]$ (or $[\beta, \alpha]$), let $x(\alpha) = a$, $x(\beta) = b$ and let f be a function so that $f(x)$ is continuous for $a \le x \le b$, and $f[x(u)]$ is defined and continuous for $\alpha \le u \le \beta$ (or $\beta \le u \le \alpha$). Then*

$$\int_a^b f(x) \, dx = \int_\alpha^\beta f[x(u)]x'(u) \, du \tag{4-211}$$

Remark. In contrast to Theorem 3 (Section 4-8), $x(u)$ is not required to have an inverse function. The values of $x(u)$ may partly fall outside the interval $[a, b]$, as long as $f[x(u)]$ is defined and continuous.

PROOF OF THEOREM 17. Let F be an indefinite integral of f. Then $F[x(u)]$ is an indefinite integral of $f[x(u)]x'(u)$. For by the chain rule

$$\frac{d}{du} F[x(u)] = \frac{dF}{dx} \cdot x'(u) = f(x)x'(u)$$

where x is expressed in terms of u; that is

$$\frac{d}{du} F[x(u)] = f[x(u)]x'(u)$$

Hence

$$\int_\alpha^\beta f[x(u)]x'(u) \, du = F[x(u)] \Big|_{u=\alpha}^{u=\beta}$$

$$= F[x(\beta)] - F[x(\alpha)] = F(b) - F(a) = \int_a^b f(x) \, dx$$

EXAMPLE 2. $\int_0^3 x\sqrt{1+x} \, dx$. We try $\sqrt{1+x} = u$, that is, $x = u^2 - 1$, so that $dx/du = 2u$; $x = 0$ corresponds to $u = \pm 1$, $x = 3$ to $u = \pm 2$. However, if we want $u = \sqrt{1 + x}$, we must use positive values of u. Consequently, we use the limits 1, 2 for u. By (4-211) our integral equals

$$\int_1^2 (u^2 - 1)u \cdot 2u \, du = \int_1^2 (2u^4 - 2u^2) \, du = \left(\frac{2u^5}{5} - \frac{2u^3}{3} \right) \Big|_1^2 = 116/15$$

EXAMPLE 3. $\int_0^{\pi/2} e^{\sin x} \cos x \, dx$. Here we set $\sin x = u$, so that $x = 0$ corresponds to $u = 0$, $x = \pi/2$ to $u = 1$, and thus

$$\int_0^{\pi/2} e^{\sin x} \cos x \, dx = \int_0^1 e^u \, du = e^u \Big|_0^1 = e - 1$$

Notice that, in this example, we have applied (4-211) as a way of evaluating the *right-hand side*. We observe this more clearly if we interchange u and x in (4-211):

$$\int_a^b f(u) \, du = \int_\alpha^\beta f[u(x)] \frac{du}{dx} \, dx \qquad (4\text{-}211')$$

If we take $u(x) = \sin x$, $f(u) = e^u$, $\alpha = 0$, $\beta = \pi/2$, the right side of (4-211') becomes the given integral $\int_0^{\pi/2} e^{\sin x} \cos x \, dx$, and the left side of (4-211') becomes $\int_0^1 e^u \, du$.

When we apply Theorem 17 in this way to evaluate the right-hand side in (4-211), it is important to notice that (4-211) remains valid for $a \geq b$ (Problem 10 below).

EXAMPLE 4. Transform the integral $\int_0^1 (1 - x^2)^3 \, dx$ by the substitution $x = \sin u$. We must choose α, β so that $\sin \alpha = 0$, $\sin \beta = 1$. We can use $\alpha = 0$, $\beta = \pi/2$; but we can equally well take, for example, $\alpha = \pi$, $\beta = -3\pi/2$. In the latter case, our integral becomes

$$\int_\pi^{-3\pi/2} (1 - \sin^2 u)^3 \cos u \, du = \int_\pi^{-3\pi/2} \cos^7 u \, du$$

4-22 ODD AND EVEN FUNCTIONS

We can take advantage of symmetry in evaluating definite integrals. The two most common cases are suggested in Figure 4-39. A function f defined in an interval $[-b, b]$ (or generally in an interval that is symmetric with respect to $x = 0$) is called *even* if $f(x) = f(-x)$ for all x, and *odd* if $f(x) = -f(-x)$ for all x. Figure 4-39(a) shows an even function, and Figure 4-39(b), an odd function. The names are used because the even and odd powers of x (that is, the functions $f(x) = x^n$ with n even or odd) are, respectively, even and odd. For example, x^4 is even, since $x^4 = (-x)^4$.

THEOREM 18. *Let f be continuous on the interval $[-b, b]$. If f is even then*

$$\int_{-b}^b f(x) \, dx = 2 \int_0^b f(x) \, dx \qquad (4\text{-}220)$$

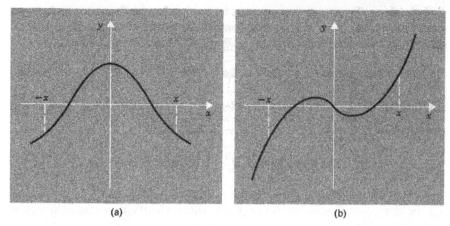

Figure 4-39   Even function and odd function.

If f is odd, then

$$\int_{-b}^{b} f(x)\, dx = 0 \tag{4-221}$$

PROOF. By the substitution, $x = -u$,

$$\int_{-b}^{0} f(x)\, dx = \int_{b}^{0} f(-u)(-1)\, du = \int_{0}^{b} f(-u)\, du = \int_{0}^{b} f(-x)\, dx$$

Accordingly

$$\int_{-b}^{b} f(x)\, dx = \int_{-b}^{0} f(x)\, dx + \int_{0}^{b} f(x)\, dx = \int_{0}^{b} f(-x)\, dx + \int_{0}^{b} f(x)\, dx$$

$$= \int_{0}^{b} [f(x) + f(-x)]\, dx$$

Now if f is even, we can set $f(-x) = f(x)$ and obtain (4-220). If f is odd, we can set $f(x) = -f(-x)$ and obtain (4-221).

EXAMPLE
$$\int_{-1}^{1} \frac{x^2 + \sin x}{1 + x^2}\, dx$$

The integrand is neither odd nor even. However, we can write our integral as the sum of two integrals:

$$\int_{-1}^{1} \frac{x^2}{1 + x^2}\, dx + \int_{-1}^{1} \frac{\sin x}{1 + x^2}\, dx$$

The second has an odd integrand, since

$$\frac{\sin(-x)}{1 + (-x)^2} = \frac{-\sin x}{1 + x^2}$$

so that the second term is 0. The first has an even integrand and, hence, equals

$$2 \int_0^1 \frac{x^2}{1+x^2} \, dx = 2 \int_0^1 \left[1 - \frac{1}{1+x^2} \right] dx = 2(x - \mathrm{Tan}^{-1} x) \Big|_0^1 = 2 - (\pi/2)$$

Thus the given integral equals $2 - (\pi/2)$.

PROBLEMS

1. Evaluate with the aid of integration by parts:

   (a) $\int_0^\pi x^2 \sin x \, dx$ (b) $\int_0^\pi \sin^5 x \, dx$

   (c) $\int_0^1 x^2 e^{2x} \, dx$ (d) $\int_1^2 x \ln x \, dx$

2. Let f and g be defined and continuous on the interval $[a, b]$, and let f have a continuous derivative. Prove that

   $$\int_a^b f(x)g(x) \, dx = f(b) \cdot \int_a^b g(x) \, dx - \int_a^b f'(x) \left\{ \int_a^x g(t) \, dt \right\} dx$$

3. Evaluate with the aid of the indicated substitution:

   (a) $\int_0^1 \frac{dx}{1+x^2}, \quad x = \tan u$ (b) $\int_0^1 (1 - x^2)^{3/2} \, dx, \quad x = \sin u$

   (c) $\int_0^7 \frac{dx}{1 + \sqrt[3]{1+x}}, \quad x = u^3 - 1$ (d) $\int_0^{\pi/2} \cos x \cos(\sin x) \, dx, \quad u = \sin x$

4. Transform with the aid of the indicated substitution but do not evaluate:

   (a) $\int_0^2 \frac{x^2}{x^3 + 1} \, dx, \quad u = x^3 + 1$ (b) $\int_0^{\pi/4} \frac{\cos 2x}{(2 + \sin 2x)^2} \, dx, \quad u = \sin 2x$

   (c) $\int_1^{\sqrt{5}} \sin x^2 \, dx, \quad x = \sqrt{u}$

5. Let f be defined and continuous in $[a, b]$. Justify the rules:

   (a) $\int_a^b f(x) \, dx = \int_{a-c}^{b-c} f(x + c) \, dx$ $(c = \text{const})$

   (b) $\int_a^b f(x) \, dx = \frac{1}{k} \int_{ka}^{kb} f\left(\frac{x}{k}\right) dx$ $(k = \text{const} \neq 0)$

   (c) $\int_a^b f(x) \, dx = (b - a) \int_0^1 f[a + (b - a)x] \, dx$

†6. Let f be defined and continuous in $[a, b]$, let f have a continuous derivative, and let f be monotone strictly increasing, so that f has an inverse g in the interval $[f(a), f(b)]$. Prove that

   $$(°) \quad \int_a^b f(x) \, dx = [x f(x)] \Big|_a^b - \int_{f(a)}^{f(b)} g(y) \, dy$$

   [*Hint.* Integrate by parts and then make a substitution $x = g(y)$ in the final integral.] Interpret the result graphically in terms of areas. The formula $(°)$ can be used to find the indefinite integral of the inverse of a given function (see Problem 8).

†7. Prove the result of Problem 6 without the assumption that f has a continuous derivative. [*Hint.* Each subdivision of the interval $[a, b]$ by points x_i determines a

subdivision of the interval $[f(a), f(b)]$ by the points $y_i = f(x_i)$; and every subdivision of $[f(a), f(b)]$ is so obtainable. For each subdivision, let m_i, M_i be the minimum and maximum of f in the ith subinterval, let $\bar{m}_i$, $\bar{M}_i$ be the minimum and maximum of g in the corresponding y-interval. Show, with the aid of a graph, that

$$\sum m_i \, \Delta_i x + \sum \bar{M}_i \, \Delta_i y = \sum M_i \, \Delta_i x + \sum \bar{m}_i \, \Delta_i y = [x \, f(x)] \Big|_a^b$$

Now use these equations and the fact that

$$\sum m_i \, \Delta_i x \leq \int_a^b f(x) \, dx \leq \sum M_i \, \Delta_i x$$

to show that

$$\sum \bar{m}_i \, \Delta_i y \leq [x \, f(x)] \Big|_a^b - \int_a^b f(x) \, dx \leq \sum \bar{M}_i \, \Delta_i y$$

and conclude that the middle expression equals the integral of $g(y)$ from $f(a)$ to $f(b)$.]

8. Apply the result of Problem 6 to evaluate:

(a) $\displaystyle\int_0^b \mathrm{Sin}^{-1} x \, dx$ (b) $\displaystyle\int_1^b \ln x \, dx$

(c) $\displaystyle\int_0^2 f(x) \, dx$, where f is defined implicitly by the equation:

$$x - y - y^3 = 0, \qquad 0 \leq y \leq 1$$

(d) $\displaystyle\int_\pi^{2\pi} f(x) \, dx$, where f is defined implicitly by the equation:

$$x = y + \tfrac{1}{2} \sin y, \qquad \pi \leq y \leq 2\pi$$

9. Justify, with the aid of symmetry:

(a) $\displaystyle\int_{-\pi}^\pi \sin^n x \, dx = 0$, n odd   (b) $\displaystyle\int_{-1}^1 \frac{x^5 + x^2 + 4x}{x^4 + 1} \, dx = 2 \int_0^1 \frac{x^2}{x^4 + 1} \, dx$

(c) $\displaystyle\int_{-1}^1 (e^x + e^{-x}) \, dx = 2 \int_0^1 (e^x + e^{-x}) \, dx$ (d) $\displaystyle\int_{-1}^1 (e^x - e^{-x}) \, dx = 0$

(e) $\displaystyle\int_a^b f(x) \, dx = 0$ if f is continuous and $f(x) = -f(a + b - x)$ for all x

10. Prove that (4-211) remains valid if $b \leq a$.

11. Evaluate with the aid of the indicated substitution (see Equation (4-211') and Problem 10).

(a) $\displaystyle\int_0^{2\pi} \cos x \ln(3 + 2 \sin x) \, dx$, $u = 3 + 2 \sin x$

(b) $\displaystyle\int_{-1}^1 (3x^2 - 1) \sin (x^3 - x) \, dx$, $u = x^3 - x$

12. Let f be a continuous function on the interval $[-a, a]$. Prove that there exist unique continuous functions $f_e(x)$ and $f_o(x)$ on $[-a, a]$ so that f_e is even, f_o is odd, and $f = f_e + f_o$. Show also that $\displaystyle\int_{-a}^a f(x) \, dx = 2 \int_0^a f_e(x) \, dx = \int_0^a [f(x) + f(-x)] \, dx$.

4-23 INEQUALITIES FOR INTEGRALS

THEOREM 19. *Let $f(x)$ be continuous and nonnegative: $f(x) \geq 0$ on the interval $[a, b]$. Then*

$$F(x) = \int_a^x f(t)\, dt \tag{4-230}$$

is monotone nondecreasing in $[a, b]$. Furthermore

$$F(b) = \int_a^b f(x)\, dx \geq 0 \tag{4-231}$$

and equality holds true only for f identically 0 on $[a, b]$.

PROOF. Since $F'(x) = f(x) \geq 0$, F is monotone nondecreasing (Section 3-21). Also $F(a) = 0$, so that $F(x) \geq 0$ for $a \leq x \leq b$ and, in particular, $F(b) \geq 0$, so that (4-231) holds true. If equality holds true in (4-231), then $F(b) = 0$ and, since F is monotone nondecreasing, $F(x) \equiv 0$. Thus $f(x) = F'(x) \equiv 0$.

THEOREM 20. *If f and g are continuous for $a \leq x \leq b$ and $f(x) \leq g(x)$ on $[a, b]$, then*

$$\int_a^b f(x)\, dx \leq \int_a^b g(x)\, dx \tag{4-232}$$

and the equality holds true only if $f(x) \equiv g(x)$ on $[a, b]$.

PROOF. The theorem follows from Theorem 19, if we replace f in Theorem 19 by $g - f$, and then apply Theorem 12.

THEOREM 21. *Let f be continuous on $[a, b]$. Then*

$$\left| \int_a^b f(x)\, dx \right| \leq \int_a^b |f(x)|\, dx \tag{4-233}$$

PROOF. We first remark that, since $|u|$ is continuous for all u, $|f(x)|$ is a continuous function of a continuous function and, therefore, is continuous in $[a, b]$. Since $f(x) \leq |f(x)|$ and $-f(x) \leq |f(x)|$, we have, by Theorem 20,

$$\int_a^b f(x)\, dx \leq \int_a^b |f(x)|\, dx$$

$$-\int_a^b f(x)\, dx \leq \int_a^b |f(x)|\, dx$$

These two inequalities imply (4-233).

THEOREM 22. *Let f be continuous on $[a, b]$ and let $|f(x)| \leq K = $ const. for $a \leq x \leq b$. Then*

$$\left| \int_a^b f(x)\, dx \right| \leq K(b - a) \tag{4-234}$$

PROOF. This follows from Theorems 20 and 21 (Problem 3 below).

4-24 MEAN VALUE THEOREM FOR INTEGRALS

The Mean Value Theorem for derivatives states that

$$F(b) - F(a) = F'(\xi)(b - a) \qquad (4\text{-}240)$$

provided that F is defined on the interval $[a, b]$ or $[b, a]$ and has a derivative inside this interval, and that ξ is appropriately chosen inside the interval (Section 3-21). There is a parallel theorem for integrals.

THEOREM 23. *Let f be continuous on the interval $[a, b]$ or $[b, a]$. Then there exists a value ξ inside the interval so that*

$$\int_a^b f(x)\, dx = f(\xi)(b - a) \qquad (4\text{-}241)$$

PROOF. The conclusion follows at once if we set

$$F(x) = \int_a^x f(t)\, dt$$

and apply Theorem 11: $F'(x) = f(x)$, Theorem 14 [integral $= F(b) - F(a)$], and the Mean Value Theorem for derivatives (4-240). For then the left side of (4-240) becomes the left side of (4-241), and $F'(\xi)$ becomes $f(\xi)$, so that the right side of (4-240) becomes the right side of (4-241).

The relation (4-241) is known as *the first mean value theorem for integrals* (or law of the mean for integrals). If f is positive and $a < b$, we can interpret the left side of (4-241) as the area under the curve. The equation states that the area equals the base times an average or mean altitude $f(\xi)$; that is, the area equals that of a rectangle of the same base and of altitude $f(\xi)$, which is an average of the vertical distances $f(x)$ (see Figure 4-40). The interpretation of $f(\xi)$ as an *average value* in general can be justified by writing

$$f(\xi) = \frac{1}{b - a} \int_a^b f(x)\, dx \qquad (4\text{-}242)$$

and then by replacing the right side by a sum, say $\Sigma M_i\, \Delta_i x$, which is close to the integral. If we subdivide into n equal parts, then $\Delta_i x = (b - a)/n$. Also we can write $M_i = f(\xi_i)$, where ξ_i is a suitable value between x_{i-1} and x_i. Then we have

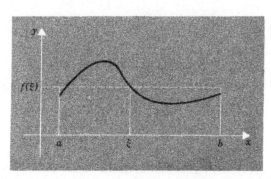

Figure 4-40 Mean Value Theorem.

$$f(\xi) = \frac{1}{b-a} \int_a^b f(x)\, dx \sim \frac{1}{b-a} \sum_{i=1}^{n} f(\xi_i) \frac{(b-a)}{n}$$

$$= \frac{f(\xi_1) + f(\xi_2) + \cdots + f(\xi_n)}{n}$$

The last expression is the ordinary average of the n values $f(\xi_1), \ldots, f(\xi_n)$. Hence we are justified in considering the right side of (4-242) as the limit of such an average, and make a general definition:

Definition. *Let f be continuous on the interval* $[a, b]$. *Then the average value* (*or arithmetic mean*) *of f over this interval is the number*

$$\frac{1}{b-a} \int_a^b f(x)\, dx$$

[If $a > b$, we can compute the average over the interval $[b, a]$ by the same expression (Problem 5 below).]

The law of the mean then says: *a function continuous over an interval equals its average value at at least one point ξ inside the interval.*

Instead of averaging f itself, it is often more suitable (for example, in statistics) to average the *square* of f, that is, for $a < b$, to form

$$\frac{1}{b-a} \int_a^b [f(x)]^2\, dx$$

The square root of this average is called the *root mean square* (rms) of f. We can show by linear algebra (see Chapter 11) that the *absolute value of the average of f is always less than or equal to the root mean square of f.*

We now have a number of ways to estimate integrals or to compare integrals by inequalities—namely, (4-231), (4-232), (4-233), (4-234), and (4-241). In addition, we have the earlier rule: if $A \le f(x) \le B$ for $a \le x \le b$, then

$$A(b-a) \le \int_a^b f(x)\, dx \le B(b-a)$$

Here, A can be taken as the absolute minimum of f, and B the absolute maximum. We can also go back to the definition of the integral, choose a subdivision of $[a, b]$, and write

$$\sum m_i \, \Delta_i x \le \int_a^b f(x)\, dx \le \sum M_i \, \Delta_i x \qquad (4\text{-}243)$$

Observe that this basic inequality also follows from (4-232). For we can write

$$\int_a^b f(x)\, dx = \int_a^{x_1} f(x)\, dx + \int_{x_1}^{x_2} f(x)\, dx + \cdots + \int_{x_{n-1}}^b f(x)\, dx$$

Then, in general, by (4-232)

$$m_i \, \Delta_i x = \int_{x_{i-1}}^{x_i} m_i \, dx \le \int_{x_{i-1}}^{x_i} f(x)\, dx \le \int_{x_{i-1}}^{x_i} M_i \, dx = M_i \, \Delta_i x \qquad (4\text{-}244)$$

If we add these relations for $i = 1, \ldots, n$, we obtain 4-243). We also observe that this argument and Theorem 20 show that equality can hold true in one or the other place in (4-244) only when f is constant over the subinterval. Hence, in (4-243), we have strict inequality in both places unless f is identically constant, in which case both equality signs hold true.

EXAMPLE. Let $J = \int_0^{\pi/2} e^{\sin x}\, dx$. Then, by the Mean Value Theorem, $J = (\pi/2)e^{\sin \xi}, \ 0 < \xi < \pi/2$. Since $e^{\sin x}$ has minimum 1 and maximum e in the given interval, we have

$$\frac{\pi}{2} < J < \frac{\pi}{2}e = 4.27$$

Since $x - \sin x$ has derivative $1 - \cos x \geq 0$, we conclude that this function has minimum value 0 for $x = 0$, so that $x - \sin x \geq 0$ or $\sin x \leq x$ for $x \geq 0$. Hence

$$J \leq \int_0^{\pi/2} e^x\, dx = e^{\pi/2} - 1 = 3.81$$

PROBLEMS

1. With the aid of theorems developed in the text, justify the following inequalities:

(a) $\int_0^1 e^{x^2}\, dx > 0$

(b) $\int_1^{-1} \cos^8 x\, dx < 0$

(c) $\int_0^1 x^2 \sin x^2\, dx < \int_0^1 x^2\, dx$

(d) $\int_0^1 x^2 \sin x^2\, dx < \int_0^1 \sin x^2\, dx < \int_0^1 x^2\, dx$

(e) $-4.9 \ln 10 < \int_{0.1}^5 \ln x\, dx < 4.9 \ln 5$

(f) $\frac{2}{17} < \int_{-1}^2 \frac{x}{1 + x^4}\, dx < \frac{1}{2}$ (Hint. Use the fact that the integrand is odd.)

2. Prove that, if f is continuous and $f(x) \leq 0$ for $a \leq x \leq b$, then $\int_a^b f(x)\, dx \leq 0$, with equality holding only for $f(x) \equiv 0$.

3. Prove Theorem 22.

4. Find the average and the root mean square for each of the following functions, and verify that the former is smaller in each case.
   (a) $f(x) = x^3, \ 1 \leq x \leq 2$ (b) $f(x) = e^x, \ -1 \leq x \leq 1$
   (c) $f(x) = \cos x, \ -\pi \leq x \leq \pi$ (d) $f(x) = \ln x, \ \frac{1}{2} \leq x \leq 2$

5. Show that if $b < a$, the right side of (4-242) can be interpreted as the average of f over the interval $[b, a]$.

4-25 THE DEFINITE INTEGRAL AS A LIMIT

At several points in this chapter, we have suggested that the definite integral can be defined as a limit of an expression

$$\sum_{i=1}^{n} f(\xi_i) \, \Delta_i x$$

as $n \to \infty$ and the widths of the subdivision intervals $\to 0$. We now make this precise, and are led to the usual definition of the Riemann integral.

By the *mesh* of a subdivision $x_0 = a \le x_1 < \cdots < x_n = b$ of $[a, b]$ we mean the largest of the values $\Delta_i x = x_i - x_{1-1}$. When we subdivide into n equal parts, the mesh is $(b - a)/n$. In general, to say that a subdivision has mesh less than δ, is equivalent to saying that $\Delta_i x < \delta$ for $i = 1, \ldots, n$.

THEOREM 24. *Let f be continuous on $[a, b]$, and let f have a continuous derivative f'. Let ϵ be a positive number, let K be a positive constant such that $|f'(x)| \le K$ on $[a, b]$ and let*

$$\delta = \frac{\epsilon}{K(b - a)} \tag{4-250}$$

Then for every subdivision of $[a, b]$ of mesh less than δ, we have

$$0 \le \sum_{i=1}^{n} M_i \, \Delta_i x - \sum_{i=1}^{n} m_i \, \Delta_i x < \epsilon \tag{4-251}$$

Hence, in particular,

$$\sum_{i=1}^{n} M_i \, \Delta_i x - \epsilon < \int_a^b f(x) \, dx \le \sum_{i=1}^{n} M_i \, \Delta_i x \tag{4-252}$$

$$\sum_{i=1}^{n} m_i \, \Delta_i x \le \int_a^b f(x) \, dx < \sum_{i=1}^{n} m_i \, \Delta_i x + \epsilon \tag{4-253}$$

PROOF. Since a function continuous on a closed interval has an absolute maximum and minimum, we can write $M_i = f(\xi_i)$ and $m_i = f(\eta_i)$, where $x_{i-1} \le \xi_i \le x_i$, $x_{i-1} \le \eta_i \le x_i$. Then, by the Mean Value Theorem for derivatives,

$$M_i - m_i = f(\xi_i) - f(\eta_i) = (\xi_i - \eta_i) f'(x_i^*)$$

where x_i^* lies between ξ_i and η_i. Since $M_i \ge m_i$, $|f'(x)| \le K$ and our mesh is less than δ, it follows that

$$M_i - m_i = |M_i - m_i| = |\xi_i - \eta_i| \cdot |f'(x_i^*)| < \delta K = \frac{\epsilon}{b - a}$$

Hence

$$0 \le \sum_{i=1}^{n} M_i \, \Delta_i x - \sum_{i=1}^{n} m_i \, \Delta_i x = \sum_{i=1}^{n} (M_i - m_i) \, \Delta_i x \le \frac{\epsilon}{(b - a)} \sum_{i=1}^{n} \Delta_i x$$

$$= \frac{\epsilon}{(b - a)} \cdot (b - a) = \epsilon$$

Therefore (4-251) is proved. From (4-251) we obtain

$$\sum_{i=1}^{n} m_i \, \Delta_i x > \sum_{i=1}^{n} M_i \, \Delta_i x - \epsilon$$

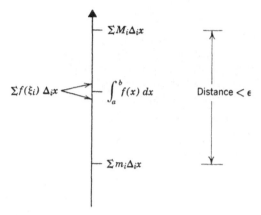

Figure 4-41   Sums approximating the definite integral.

and, since

$$\sum m_i \, \Delta_i x \leq \int_a^b f(x) \, dx \leq \sum M_i \, \Delta_i x$$

the assertion (4-252) follows. The assertion (4-253) follows in the same way. The relationships are shown graphically in Figure 4-41.

The theorem shows that, to estimate the value of the integral with error less than ϵ (where ϵ can be chosen as small as desired), one need only choose any subdivision of mesh $< \delta$, where δ is given by (4-250), and then use either the upper or lower sum to approximate the integral.

Thus, in the case considered (f' continuous), the definite integral of f can be computed *as accurately as desired* by subdividing finely enough and evaluating $\Sigma m_i \, \Delta_i x$ or $\Sigma M_i \, \Delta x$. There is still a computing problem: How do we determine M_i and m_i? Fortunately, as the next theorem shows, we do not need to find them!

THEOREM 25. *Under the hypotheses of Theorem 24, for every subdivision of mesh $< \delta$, and every choice of points $\xi_i, \ldots, \xi_n$ in the subintervals: $x_{i-1} \leq \xi_i \leq x_i$, one has*

$$\left| \int_a^b f(x) \, dx - \sum_{i=1}^{n} f(\xi_i) \, \Delta_i x \right| < \epsilon \qquad (4\text{-}254)$$

PROOF. We have for each $i = 1, \ldots, n$

$$m_i \leq f(\xi_i) \leq M_i$$

Hence

$$\sum m_i \, \Delta_i x \leq \sum f(\xi_i) \, \Delta_i x \leq \sum M_i \, \Delta_i x$$

But we also have

$$\sum m_i \, \Delta_i x \leq \int_a^b f(x) \, dx \leq \sum M_i \, \Delta_i x$$

and by (4-251) the outer members differ by at most ϵ. Therefore, the two middle expressions must be at most ϵ apart (see Figure 4-41).

EXAMPLE. $\int_0^{0.5} \cos x^4 \, dx$. Here $f'(x) = -4x^3 \sin x^4$. Since x^3, x^4 and $\sin x^4$ increase as x increases from 0 to 0.5, f' is monotone strictly decreasing, from 0 at $x = 0$, to $-0.5 \sin(1/16) = -0.032$ at $x = 0.5$. Hence we can take $K = 0.032$, and

$$\delta = \frac{\epsilon}{0.032 \times 0.5} = 62.5 \, \epsilon$$

Accordingly, for $\epsilon = 0.005$, $\delta = 0.315$. If we divide the interval $[0, 0.5]$ into two equal intervals, our mesh is $0.25 < \delta$ and the error will be less than ϵ. The value to be computed is

$$\cos(\xi_1{}^4) \times 0.25 + \cos(\xi_2{}^4) \times 0.25$$

where $0 \le \xi_1 \le 0.25$, $0.25 \le \xi_2 \le 0.5$. If we use the points $\xi_1 = 0.2$, $\xi_2 = 0.4$, we obtain $0.25[\cos(0.0016) + \cos(0.0256)] = 0.4996$, and our integral equals this value, with error at most 0.005. (By advanced methods the value can be shown to be 0.49989, to 5 decimal places.)

Even though we are not dealing with an ordinary limit, it is customary to use the terminology of limits here and to say that

$$\lim_{\text{mesh} \to 0} \sum_{i=1}^{n} f(\xi_i) \, \Delta_i x = \int_a^b f(x) \, dx \tag{4-255}$$

The meaning of the symbolism is just what is stated in the theorem: given $\epsilon > 0$, we can choose $\delta > 0$ so that, for every subdivision of mesh $< \delta$ and every choice of the points ξ_i $(x_{i-1} \le \xi_i \le x_i)$, the sum $\Sigma f(\xi_i) \, \Delta_i x$ differs from $\int_a^b f(x) \, dx$ by less than ϵ.

More generally, we can consider an arbitrary function defined on $[a, b]$ and form the sums $\Sigma f(\xi_i) \, \Delta_i x$ for all allowed subdivisions and choices of the ξ_i. We say that

$$\lim_{\text{mesh} \to 0} \sum_{i=1}^{n} f(\xi_i) \, \Delta_i x = I$$

if, for each $\epsilon > 0$, we can find $\delta > 0$ so that, for all subdivisions of mesh less than ϵ and all choices of the ξ_i, we have

$$\left| \sum_{i=1}^{n} f(\xi_i) \, \Delta_i x - I \right| < \epsilon$$

This limit, when it exists, is the Riemann integral of f. As for all limits in the calculus, the limit, when it exists, is unique (Problem 3 below).

Definition. *The Riemann integral of a function f defined on $[a, b]$ is the number I, where*

$$\lim_{\text{mesh} \to 0} \sum_{i=1}^{n} f(\xi_i) \, \Delta_i x = I$$

provided that this limit exists.

Theorem 25 above asserts that, if f has a continuous derivative, then f has a Riemann integral, which is just our definite integral. We shall see that this is true generally for every continuous f (Theorem 26 below): *every continuous function f has a Riemann integral, namely, our definite integral;* equivalently, (4-255) is valid for every continuous f.

If f is not continuous, our definite integral is not defined (although for certain important types of discontinuous functions we can easily modify our definition to get an integral—see Section 4-30). However, the sums $\Sigma f(\xi_i)\, \Delta_i x$ are always meaningful and in many cases, even though f is not continuous, the Riemann integral exists. Hence the Riemann integral is a more general concept than is our definite integral. For continuous functions, they are the same. Since in every case when our definite integral and the Riemann integral have meaning, the values are the same, we denote both by $\int_a^b f(x)\, dx$.

We now state the basic theorem for continuous functions.

THEOREM 26. *Let f be defined and continuous on $[a, b]$. Then Equation (4-255) is valid; that is, f has a Riemann integral, which equals the definite integral of f from a to b.*

The proof is given in the next section. We state here an important consequence of the theorem:

COROLLARY OF THEOREM 26. *Let f be continuous on $[a, b]$. Then*

$$\lim_{\text{mesh}\to 0} \sum_{i=1}^{n} |f(\xi_i) - f(\eta_i)|\, \Delta_i x = 0 \tag{4-256}$$

and hence

$$\lim_{\text{mesh}\to 0} \sum_{i=1}^{n} \{ f(\xi_i) - f(\eta_i) \}\, \Delta_i x = 0 \tag{4-257}$$

that is, given $\epsilon > 0$, one can choose $\delta > 0$ so that for every subdivision of mesh less than δ and every choice of ξ_i, η_i $(i = 1, \ldots, n)$ so that $x_{i-1} \le \xi_i \le x_i$, $x_{i-1} \le \eta_i \le x_i$, each sum differs from 0 by less than ϵ.

PROOF. Let I denote $\int_a^b f(x)\, dx$. Then by Theorem 26

$$\lim_{\text{mesh}\to 0} \sum_{i=1}^{n} M_i\, \Delta_i x = I, \qquad \lim_{\text{mesh}\to 0} \sum_{i=1}^{n} m_i\, \Delta_i x = I$$

since $M_i = f(\xi_i)$ for appropriate choice of ξ_i in the interval, and the same holds true for m_i. Hence, given $\epsilon > 0$, we can choose $\delta > 0$ so that for every subdivision of mesh less than δ,

$$\left| \sum M_i\, \Delta_i x - I \right| < \frac{\epsilon}{2}, \qquad \left| \sum m_i\, \Delta_i x - I \right| < \frac{\epsilon}{2}$$

and hence

$$\left| \sum (M_i - m_i)\, \Delta_i x \right| = \left| \left(\sum M_i\, \Delta_i x - I \right) - \left(\sum m_i\, \Delta_i x - I \right) \right|$$

$$\le \left| \sum M_i\, \Delta_i x - I \right| + \left| \sum m_i\, \Delta_i x - I \right| < \frac{\epsilon}{2} + \frac{\epsilon}{2} = \epsilon$$

Then, since $M_i - m_i \geq 0$ and $\Delta_i x > 0$ for each i, we have

$$\sum_{i=1}^{n} (M_i - m_i) \Delta_i x < \epsilon$$

Now for every choice of ξ_i, η_i in $[x_{i-1}, x_i]$, $|f(\xi_i) - f(\eta_i)| \leq M_i - m_i$, since M_i is the absolute maximum of f, m_i the absolute minimum, in the interval. Thus, for a subdivision of mesh less than δ,

$$\sum |f(\xi_i) - f(\eta_i)| \Delta_i x \leq \sum (M_i - m_i) \Delta_i x < \epsilon$$

and, therefore, (4-256) is proved. Since

$$\left| \sum (f(\xi_i) - f(\eta_i)) \Delta_i x \right| \leq \sum |f(\xi_i) - f(\eta_i)| \Delta_i x$$

(4-257) also follows.

‡4-26 PROOF OF EXISTENCE OF RIEMANN INTEGRAL OF A CONTINUOUS FUNCTION

We first state a lemma:

LEMMA. *Let f be continuous on $[a, b]$ and have absolute maximum M and minimum m on $[a, b]$. Let $a = x_0 < x_1 < \cdots < x_n = b$ be a subdivision of mesh less than δ_1 and form the corresponding upper sum of f. Then introduction of one additional subdivision point $\bar{x}$ either reduces the upper sum by an amount less than $\delta_1(M - m)$, or else leaves it unchanged.*

PROOF. Let $x_{k-1} < \bar{x} < x_k$, let $\Delta'_k x = \bar{x} - x_{k-1}$, $\Delta''_k x = x_k - \bar{x}$, let M'_k, M''_k be the absolute maxima of f on the intervals $[x_{k-1}, \bar{x}]$, $[\bar{x}, x_k]$, respectively (cf. the proof of Theorem 10 in Section 4-16). The introduction of $\bar{x}$ affects our upper sum by replacing $M_k \Delta_k x$ by $M'_k \Delta'_k x + M''_k \Delta''_k x$ and hence

(old upper sum) − (new upper sum)

$$\begin{aligned} &= M_k \Delta_k x - (M'_k \Delta'_k x + M''_k \Delta''_k x) \\ &= M_k(\Delta'_k x + \Delta''_k x) - (M'_k \Delta'_k x + M''_k \Delta''_k x) \\ &= (M_k - M'_k) \Delta'_k x + (M_k - M''_k) \Delta''_k x \end{aligned}$$

Since $M_k \geq M'_k$ and $M_k \geq M''_k$, the new upper sum does not exceed the old one. Since $M_k \leq M$, $M'_k \geq m$, $M''_k \geq m$, we have also

(old upper sum) − (new upper sum)

$$\begin{aligned} &\leq (M - m) \Delta'_k x + (M - m) \Delta''_k x = (M - m) \Delta_k x \\ &< (M - m)\delta_1 \end{aligned}$$

Thus the lemma is proved.

We proceed to prove Theorem 26. Let m be the absolute minimum of f, M the absolute maximum of f on $[a, b]$. If $m = M$, f is identically constant and the conclusion is immediate (Problem 4 below). Let us therefore assume that $m < M$. Let J denote the definite integral of f from a to b and let ϵ be a given positive number. By the definition of the definite integral, J is the glb of all upper sums $\Sigma M_i \, \Delta_i x$ for all subdivisions of $[a, b]$. Hence we can choose a subdivision of $[a, b]$ for which the upper sum is less than $J + \frac{1}{2}\epsilon$. Let us fix this subdivision and denote it as follows:

$$\bar{x}_0 = a < \bar{x}_1 < \bar{x}_2 < \cdots < \bar{x}_p = b$$

We denote the corresponding maxima by $\bar{M}_i$, so that we have

$$\sum_{i=1}^{p} \bar{M}_i(\bar{x}_i - \bar{x}_{i-1}) < J + \frac{1}{2}\epsilon \tag{4-260}$$

Let h be the mesh of this subdivision (largest of the numbers $\bar{x}_i - \bar{x}_{i-1}$). We now choose δ_1 to be the smaller of the two numbers $\epsilon/[2(p - 1)(M - m)]$ and h, so that

$$0 < \delta_1 \leq h, \qquad 0 < \delta_1 \leq \frac{\epsilon}{2(p - 1)(M - m)} \tag{4-261}$$

We now assert that, for every subdivision of mesh less than δ_1, we have

$$J \leq \Sigma M_i \, \Delta_i x < J + \epsilon \tag{4-262}$$

To prove this, we consider such a subdivision: $a = x_0 \leq x_1 < \cdots < x_n = b$, with corresponding upper sum $\Sigma M_i \, \Delta_i x$. Since our mesh is less than δ_1 and $\delta_1 \leq h$, there will be at least one point of this subdivision in each interval $[\bar{x}_{i-1}, \bar{x}_i]$. Now the points $\bar{x}_1, \ldots, \bar{x}_{p-1}$ are not necessarily points of the subdivision: $x_0, \ldots, x_n$. However, we can introduce these points one by one—first $\bar{x}_1$, then $\bar{x}_2$, and so on. By the lemma, for each one we introduce we decrease the upper sum by less than $(M - m)\delta_1$. Hence, the introduction of all of $\bar{x}_1, \ldots, \bar{x}_{p-1}$ decreases the sum by less than $(M - m)(p - 1)\delta_1 \leq \epsilon/2$. At the end of the process we have a subdivision using all the points x_i and $\bar{x}_i$; we can regard this as a subdivision obtained from $\bar{x}_0, \ldots, \bar{x}_p$ by adding some of the points $x_1, \ldots, x_n$. By the lemma again, the corresponding upper sum is at most equal to $\Sigma \bar{M}_i(\bar{x}_i - \bar{x}_{i-1})$ and by (4-260) is hence less than $J + \frac{1}{2}\epsilon$. Thus, by starting with the subdivision $x_0, \ldots, x_n$ and by introducing all the $\bar{x}_i$, we have reduced our upper sum to less than $J + \frac{1}{2}\epsilon$; but we know that we reduced it by less than $\frac{1}{2}\epsilon$. Therefore

$$J \leq \sum M_i(x_i - x_{i-1}) < J + \epsilon$$

and (4-262) is proved.

We can write our conclusion thus:

$$\lim_{\text{mesh}\to 0} \sum_{i=1}^{n} M_i \, \Delta_i x = \int_a^b f(x) \, dx \tag{4-263}$$

In exactly the same way we prove, by considering lower sums, that

$$\lim_{\text{mesh}\to 0} \sum_{i=1}^{n} m_i \, \Delta_i x = \int_a^b f(x) \, dx \tag{4-264}$$

Now finally, for arbitrary sums $\Sigma f(\xi_i) \, \Delta_i x$, we have

$$\sum m_i \, \Delta_i x \leq \sum f(\xi_i) \, \Delta_i x \leq \sum M_i \, \Delta_i x$$

and hence, from (4-263) and (4-264),

$$\lim_{\text{mesh}\to 0} \sum_{i=1}^{n} f(\xi_i)\, \Delta_i x = \int_a^b f(x)\, dx \qquad (4\text{-}265)$$

The details of these last steps are left as an exercise (Problem 6 below).

The Integral as a Mapping. Let us consider a fixed interval $[a, b]$ and the vector space V of all continuous functions on $[a, b]$. The definite integral of each function f in V is a number $I(f)$. We can regard I as a mapping of V into the real numbers R. For example, $I(e^x) = e^b - e^a$, $I(x) = (b^2 - a^2)/2$ and, in general, if $f_k(x) = x^k$ for $k = 0, 1, 2, \ldots$,

$$I(f_k) = \frac{b^{k+1} - a^{k+1}}{k+1}, \qquad k = 0, 1, 2, \ldots \qquad (4\text{-}266)$$

If c_1, c_2 are arbitrary scalars and f, g are in V, then by Theorem 12

$$I(c_1 f + c_2 g) = c_1 I(f) + c_2 I(g) \qquad (4\text{-}267)$$

We describe the rule (4-267) by saying that I is a *linear mapping* of V into R. We notice another property of the mapping I. For each f in V, we denote by $M(f)$ the absolute maximum of $|f(x)|$ on $[a, b]$. Then *there is a constant C such that*

$$|I(f)| \leqq CM(f) \qquad \text{for all } f \text{ in } V \qquad (4\text{-}268)$$

For

$$\left| I(f) \right| = \left| \int_a^b f(x)\, dx \right| \leq \int_a^b \left| f(x) \right| dx \leq M(f) \cdot (b - a)$$

Hence (4-268) holds true with $C = b - a$. The rule (4-268) is described by saying that I is a *bounded linear mapping*.

It is natural to surmise that the only linear mapping of our chosen vector space V into R is given by the integral I. However, there are many others, and in particular there are many bounded linear mappings of V into R. However, I is the only bounded linear mapping of V into R that satisfies (4-266). This theorem is proved in advanced texts in real analysis.

A study of the linear mappings other than I leads one naturally to other types of integrals: in particular, the *Stieltjes integral* and the *Lebesgue integral*. These are also studied in advanced texts.

PROBLEMS

1. For each of the following integrals, choose δ so that $\left| \Sigma f(\xi_i)\, \Delta_i x - \int_a^b f(x)\, dx \right| < \epsilon$ for all subdivisions of mesh $< \delta$, and evaluate a sum $\Sigma f(\xi_i)\, \Delta_i x$ for such a subdivision:

(a) $\int_0^1 \left(1 - \frac{x^2}{4}\right) dx$, $\quad \epsilon = 0.5$ $\qquad$ (b) $\int_1^2 \frac{1}{1 + x^2}\, dx$, $\quad \epsilon = 0.2$

(c) $\int_0^1 \cos x^2\, dx$, $\quad \epsilon = 0.3$

2. For each of the following integrals, give a formula for δ in terms of ϵ, so that for

each subdivision of mesh less than δ, each sum $\sum f(\xi_i) \Delta_i x$ differs from $\int_a^b f(x)\, dx$ by less than ϵ.

(a) $\int_0^1 e^{-x^2}\, dx$

(b) $\int_0^2 \sqrt{1 + x^3}\, dx$

(c) $\int_0^\pi \sin(1 + x^2)\, dx$

(d) $\int_1^2 \ln(1 + \ln x)\, dx$

3. Prove that the Riemann integral is unique; that is, if

$$\lim_{\text{mesh} \to 0} \sum f(\xi_i) \Delta_i x = I_1 \qquad \text{and} \qquad \lim_{\text{mesh} \to 0} \sum f(\xi_i) \Delta_i x = I_2$$

then $I_1 = I_2$.

4. Prove that, if $f \equiv$ constant on $[a, b]$, then Equation (4-255) holds true.

5. Let f and g be continuous in $[a, b]$. Show that each of the following is true:

(a) $\displaystyle \lim_{\text{mesh} \to 0} \sum_{i=1}^n [f(\xi_i) + g(\eta_i)] \Delta_i x = \int_a^b [f(x) + g(x)]\, dx$

(b) $\displaystyle \lim_{\text{mesh} \to 0} \sum_{i=1}^n f(\xi_i) g(\eta_i) \Delta_i x = \int_a^b f(x)g(x)\, dx$

that is, given $\epsilon > 0$, there is a $\delta > 0$ so that for every subdivision of mesh less than δ, and for every choice of ξ_i, η_i between x_{i-1} and x_i, the sum differs from the corresponding integral by less than ϵ. [*Hint.* For (a) we write the sum as

$$\sum [f(\xi_i) - f(\eta_i)] \Delta_i x + \sum [f(\eta_i) + g(\eta_i)] \Delta_i x$$

and apply the Corollary to Theorem 26 and Theorem 26 itself. For (b) write

$$f(\xi_i)g(\eta_i) = [f(\xi_i) - f(\eta_i)]g(\eta_i) + f(\eta_i)g(\eta_i)$$

and split the sum into two sums; for the first one, use the Corollary to Theorem 26 and the fact that $|g(x)| \leq K$, for some constant K, in $[a, b]$; for the second, use Theorem 26.]

6. (a) Prove (4-264) and show in detail how (4-263) and (4-264) give (4-265).

7. Let f be continuous in $[0, 1]$. Justify the statements:

(a) $\displaystyle \lim_{n \to \infty} \left[\frac{f(0) + f\left(\dfrac{1}{n}\right) + \cdots + f\left(\dfrac{n-1}{n}\right)}{n} \right] = \int_0^1 f(x)\, dx$

(b) $\displaystyle \lim_{n \to \infty} \left[\frac{f\left(\dfrac{1}{2^{n+1}}\right) + f\left(\dfrac{3}{2^{n+1}}\right) + f\left(\dfrac{2^{n+1} - 1}{2^{n+1}}\right)}{2^n} \right] = \int_0^1 f(x)\, dx$

8. Evaluate the limits (cf. Problem 7):

(a) $\displaystyle \lim_{n \to \infty} \frac{1}{n^3}(1 + 4 + 9 + \cdots + (n - 1)^2)$

(b) $\displaystyle \lim_{n \to \infty} \left(\frac{1}{n} + \frac{1}{n + 1} + \cdots + \frac{1}{2n - 1} \right)$

9. Let V be the vector space of all continuous functions on $[0, 1]$. Show that each of the following is a linear mapping of V into R and determine whether the mapping is bounded:

(a) $A(f) = \int_0^1 xf(x)\, dx$ (b) $B(f) = f(0)$

(c) $C(f) = f(0) + f(1)$ (d) $E(f) = f(0) + \int_0^1 \sin x\, f(x)\, dx$

4-27 ARC LENGTH

In geometry one easily assigns a length to each line segment and to each broken line (for example, the perimeter of a rectangle). With more difficulty one also finds a reasonable way of assigning length to a circular arc. We here develop a systematic way of assigning lengths to curves: for example, graphs of functions $y = f(x)$, or of paths given by parametric equations: $x = f(t), y = g(t)$.

The parametric case turns out to be the general one and we proceed with it. Let a path be given by the equations:

$$x = f(t), \qquad y = g(t), \qquad a \le t \le b \tag{4-270}$$

Here f and g will be assumed continuous and to have *continuous first derivatives* in $[a, b]$. We wish to define the length of this path; we think of the length as measuring the total distance traversed by the point $P:(x, y)$ as t goes from a to b.

Guided by our experience with the circle, we think of "inscribed polygons." Something like an inscribed polygon is obtained if we form the broken line $P_0 P_1 P_2 \ldots P_n$, where $P_0, P_1, P_2, \ldots$ are the positions for $t = t_0$, $t = t_1, t = t_2, \ldots$, where $a = t_0 < t_1 < t_2 < \cdots < t_n = b$ (Figure 4-42). The length of this broken line is

$$|\overrightarrow{P_0 P_1}| + |\overrightarrow{P_1 P_2}| + \cdots + |\overrightarrow{P_{n-1} P_n}| = \sum_{i=1}^{n} |\overrightarrow{P_{i-1} P_i}|$$

Now by the distance formula

$$|\overrightarrow{P_{i-1} P_i}| = \sqrt{(x_i - x_{i-1})^2 + (y_i - y_{i-1})^2}$$
$$= \sqrt{[f(t_i) - f(t_{i-1})]^2 + [g(t_i) - g(t_{i-1})]^2}$$

We now apply the Mean Value Theorem for derivatives:

$$f(t_i) - f(t_{i-1}) = f'(\xi_i)(t_i - t_{i-1}), \qquad g(t_i) - g(t_{i-1}) = g'(\eta_i)(t_i - t_{i-1})$$

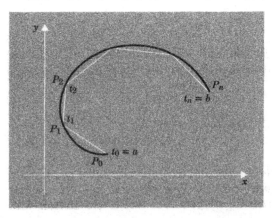

Figure 4-42 Arc length.

where $t_{i-1} < \xi_i < t_i$, $t_{i-1} < \eta_i < t_i$, and ξ_i, η_i are in general unequal. We now find that

$$|\overline{P_{i-1}P_i}| = \sqrt{([f'(\xi_i)]^2 + [g'(\eta_i)]^2) \cdot (t_i - t_{i-1})^2}$$
$$= \sqrt{[f'(\xi_i)]^2 + [g'(\eta_i)]^2}\,(t_i - t_{i-1})$$

$(t_i - t_{i-1}$ being positive). Hence the length of our broken line is

$$\sum_{i=1}^{n} \sqrt{[f'(\xi_i)]^2 + [g'(\eta_i)^2]} \cdot (t_i - t_{i-1})$$

This suggests the Riemann integral of the function:

$$F(t) = \sqrt{[f'(t)]^2 + [g'(t)]^2}$$

However, because ξ_i and η_i are in general unequal, the sum is not of the right form [as in (4-255)]. The difficulty is not serious, since we can apply the inequality $|\sqrt{a^2 + b^2} - \sqrt{c^2 + b^2}| \le |a - c|$ (Figure 4-43, Problem 7 following Section 4-29) to conclude that

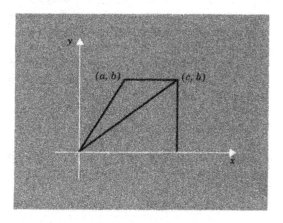

Figure 4-43   Proof of inequality.

$$\left| \sum_{i=1}^{n} \sqrt{[f'(\xi_i)]^2 + [g'(\eta_i)]^2}\,\Delta_i t - \sum_{i=1}^{n} \sqrt{[f'(\eta_i)]^2 + [g'(\eta_i)]^2}\,\Delta_i t \right|$$
$$\le \sum_{i=1}^{n} \left| \sqrt{[f'(\xi_i)]^2 + [g'(\eta_i)]^2} - \sqrt{[f'(\eta_i)]^2 + [g'(\eta_i)]^2} \right| \Delta_i t$$
$$\le \sum_{i=1}^{n} |f'(\xi_i) - f'(\eta_i)|\,\Delta_i t$$

By the Corollary to Theorem 26 (Section 4-25) this last sum can be made less than ϵ by taking the mesh less than δ. Thus we conclude (cf. Problem 5, following Section 4-26) that

$$\lim_{\text{mesh}\to 0} \sum \sqrt{[f'(\xi_i)]^2 + [g'(\eta_i)]^2}\,\Delta_i t = \lim_{\text{mesh}\to 0} \sum \sqrt{[f'(\eta_i)]^2 + [g'(\eta_i)]^2}\,\Delta_i t$$
$$= \int_a^b \sqrt{[f'(t)]^2 + [g'(t)]^2}\,dt$$

Accordingly, we are led to *define the length of the path to be the limit of the lengths of "inscribed polygons"* $P_0P_1 \ldots P_n$ *as the mesh* $\to 0$. It is important to observe that the inscribed polygons and mesh are chosen with reference to the parameter t; P_0, P_1, ..., P_n correspond to parameter values $t_0 = a < t_1 < \ldots < t_n = b$, and the mesh is the maximum of $t_i - t_{i-1}$. However, the lengths of the polygons have meaning independent of the coordinate system chosen and, hence, so does the length of the path. In terms of coordinates the length is given by the following theorem:

THEOREM 27. *Let f and g be continuous, with continuous first derivatives, in $[a, b]$. Then the length of the path: $x = f(t)$, $y = g(t)$, $a \leq t \leq b$, is*

$$L = \int_a^b \sqrt{[f'(t)]^2 + [g'(t)]^2}\, dt \qquad (4\text{-}271)$$

A curve $y = f(x)$, $a \leq x \leq b$, can be regarded as a path, with x as parameter:

$$x = t, \qquad y = f(t), \qquad a \leq x \leq b \qquad (4\text{-}272)$$

Accordingly, if f' is continuous, we can assign a length:

THEOREM 28. *Let f be defined and continuous, with continuous first derivative, on $[a, b]$. Then the graph of $y = f(x)$ over the interval $[a, b]$— that is, the path (4-272)—has length*

$$L = \int_a^b \sqrt{[f'(x)]^2 + 1}\, dx \qquad (4\text{-}273)$$

This follows at once from (4-271) (Problem 6 below).

EXAMPLE 1. $y = 3x + 5$, $0 \leq x \leq 2$. The path is a line segment. By (4-273) its length is

$$\int_0^2 \sqrt{9 + 1}\, dx = 2\sqrt{10} \text{ units}$$

This is precisely the distance between the end points $(0, 5)$, $(2, 11)$.

EXAMPLE 2. $x = 3t^2$, $y = t^3 - 3t$, $0 \leq t \leq 1$. By (4-271) the length is

$$\int_0^1 \sqrt{36t^2 + (3t^2 - 3)^2}\, dt = \int_0^1 \sqrt{9(t^2 + 1)^2}\, dt$$
$$= 3\int_0^1 (t^2 + 1)\, dt = 4 \text{ units}$$

EXAMPLE 3. $x = \cos^2 t$, $y = \sin^2 t$, $0 \leq t \leq \pi$. Here (4-271) gives the path length as

$$\int_0^\pi \sqrt{4\cos^2 t \sin^2 t + 4\sin^2 t \cos^2 t}\, dt = \int_0^\pi \sqrt{8\sin^2 t \cos^2 t}\, dt$$
$$= \int_0^{\pi/2} \sqrt{8\sin^2 t \cos^2 t}\, dt + \int_{\pi/2}^\pi \sqrt{8\sin^2 t \cos^2 t}\, dt$$

We have split the integral in two parts because we must take the positive square root and we know that $\cos t$ is negative in the second quadrant. Therefore our length is

$$\int_0^{\pi/2} 2\sqrt{2}\sin t \cos t\, dt - \int_{\pi/2}^\pi 2\sqrt{2}\sin t \cos t\, dt$$
$$= \sqrt{2}\sin^2 t \Big|_0^{\pi/2} - \sqrt{2}\sin^2 t \Big|_{\pi/2}^\pi = 2\sqrt{2}$$

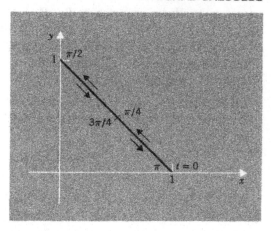

Figure 4-44   Path of Example 3.

Now on the path, $x + y = \cos^2 t + \sin^2 t = 1$, so that our path follows a line segment (Figure 4-44). As t increases from 0 to $\pi/2$, x decreases steadily from 1 to 0, and we move from $(1, 0)$ to $(0, 1)$; as t increases from $\pi/2$ to π, x increases from 0 to 1, and we retrace the segment from $(0, 1)$ to $(1, 0)$. The integral measures the total distance covered, which is $2\sqrt{2}$.

As the example illustrates, the arc length integral measures distance in the same manner as the mileage gauge in an automobile. At the end of a day the gauge reading may have increased by 27 miles; that does not mean one ends the day 27 miles from the starting point, but rather that one has moved, perhaps back and forth, a total of 27 miles.

Vector interpretation of the arc length. If we write $\overrightarrow{OP} = x\mathbf{i} + y\mathbf{j}$, then our path $x = f(t)$, $y = g(t)$, is equivalent to a vector function

$$\overrightarrow{OP} = \mathbf{F}(t), \qquad a \leq t \leq b$$

and can be interpreted as describing the motion of a point in the plane, as in Section 3-10. The velocity vector is

$$\mathbf{v} = \frac{d\,\overrightarrow{OP}}{dt} = \frac{dx}{dt}\mathbf{i} + \frac{dy}{dt}\mathbf{j}$$

and its magnitude, the speed, is

$$|\mathbf{v}| = \sqrt{\left(\frac{dx}{dt}\right)^2 + \left(\frac{dy}{dt}\right)^2} = \sqrt{[f'(t)]^2 + [g'(t)]^2}$$

Hence, by (4-271), the length of the path is

$$L = \int_a^b |\mathbf{v}|\, dt$$

The distance traversed is the integral of the speed $|\mathbf{v}|$ *with respect to time t.* This result again shows that path length has geometric meaning, independent of coordinates.

4-28 THE ARC LENGTH FUNCTION

With reference to our smooth path (4-270), we let s be the distance traversed as t increases from $t = a$ to a general t in $[a, b]$, so that s becomes a function of t. Explicitly, by Theorem 27,

$$s(t) = \int_a^t \sqrt{[f'(u)]^2 + [g'(u)]^2} \, du \qquad (4\text{-}280)$$

We call $s(t)$ the *arc length function* for the given path. Now f' and g' are continuous. Therefore, by the Fundamental Theorem of Calculus (Theorem 11, Section 4-17), $s(t)$ has a derivative:

$$\frac{ds}{dt} = \sqrt{[f'(t)]^2 + [g'(t)]^2} = \sqrt{\left(\frac{dx}{dt}\right)^2 + \left(\frac{dy}{dt}\right)^2} \qquad (4\text{-}281)$$

[Equation (4-281) is one of the basic formulas of calculus.]

In terms of the vector interpretation of the preceding section, (4-281) shows that

$$\frac{ds}{dt} = |\mathbf{v}|$$

Thus the speed $|\mathbf{v}|$, as expected, is a measure of distance traversed per unit time, where distance is measured along the path.

By our definition of the arc length function, $s = 0$ for $t = a$ and s increases (or at least does not decrease) as t increases. Occasionally, we have reason to measure s in some other way—for example, starting at some value s_0 other than 0 at $t = a$ and increasing to $s_0 + L$ as t goes from a to b; or we might measure s backwards, by starting with 0 at $t = b$ and by increasing s as t decreases. In this last case

$$s = \int_t^b \sqrt{[f'(u)]^2 + [g'(u)]^2} \, du$$

Another possibility is a curve for which the parameter interval is infinite, say $-\infty < t < \infty$. Then we can measure s by starting at any chosen value t_0 and, if s increases with t, we have

$$s = \int_{t_0}^t \sqrt{[f'(u)]^2 + [g'(u)]^2} \, du$$

In all cases, we are in effect introducing an s-scale along our path, just as the number scale is introduced on the x-axis; the unit of length is fixed, but we are free to choose any point as the origin ($s = 0$) and to choose either direction—that of increasing or decreasing t—as that of increasing s. No matter what choice we make, the length of the portion of the path traversed between $t = t_1$ and $t = t_2$, corresponding to $s = s_1$ and $s = s_2$, respectively, is

$$|s_2 - s_1| = \left| \int_{t_1}^{t_2} \sqrt{[f'(u)]^2 + [g'(u)]^2} \, du \right| \qquad (4\text{-}282)$$

[Problem 5(b) below].

†4-29 CHANGE OF PARAMETER

The two sets of equations:

$$x = t, \qquad y = t^2, \qquad 0 \le t \le 2; \qquad x = 2t, \qquad y = 4t^2, \qquad 0 \le t \le 1$$

both represent the same path, except for the t-values assigned (Figure 4-45). In both cases the arc of the parabola $y = x^2$ from $(0, 0)$ to $(2, 4)$ is traced once, with increasing t. We can compute the length of the path either way:

$$L_1 = \int_0^2 \sqrt{1 + 4t^2} \, dt; \qquad L_2 = \int_0^1 \sqrt{4 + 64t^2} \, dt$$

If we set $u = 2t$ in the second integral, it becomes

$$L_2 = \int_0^2 \sqrt{4 + 16u^2} \, \tfrac{1}{2} \, du = \int_0^2 \sqrt{1 + 4u^2} \, du$$

so that $L_1 = L_2$ as expected.

In general, we can introduce a new parameter on a given path: $x = f(t)$, $y = g(t)$, $a \le t \le b$, by a substitution: $t = \varphi(\tau)$, $\alpha \le \tau \le \beta$, where φ has range $[a, b]$. Then

$$x = f(t) = f[\varphi(\tau)] = F(\tau) \qquad y = g(t) = g[\varphi(\tau)] = G(\tau), \qquad \alpha \le \tau \le \beta$$

We say the new path is *equivalent* to the old one if φ has a continuous derivative and either $\varphi'(\tau) > 0$ on $[\alpha, \beta]$ or $\varphi'(\tau) < 0$ on $[\alpha, \beta]$; thus t may increase with τ, or decrease with τ. Under the conditions stated, φ defines a one-to-one correspondence, and we can solve for τ in terms of t: $\tau = \psi(t)$, $a \le t \le b$, and

$$F(\tau) = F[\psi(t)] = f(t), \qquad G(\tau) = G[\psi(t)] = g(t), \qquad a \le t \le b$$

Thus the old path is equivalent to the new one. The changes of variable are suggested in Figure 4-46. Because of the symmetry of the relationship, we speak simply of equivalent paths. We notice that every path is equivalent to itself, and that two paths equivalent to a third one are equivalent to each other (Problem 10 below).

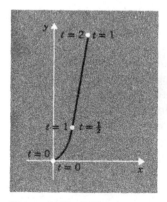

Figure 4-45 Different parametrizations.

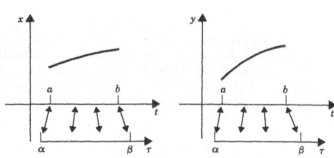

Figure 4-46 Change of parameter.

The main property of equivalent paths is given by the following theorem, suggested by our example:

THEOREM 29. *Let f and g have continuous first derivatives on [a, b]; let F and G have continuous first derivatives on [α, β]. If the paths*

$$x = f(t), \qquad y = g(t), \qquad a \le t \le b$$
$$x = F(\tau), \qquad y = G(\tau), \qquad \alpha \le \tau \le \beta$$

are equivalent, then both paths have the same length.

PROOF. The length of the first path is

$$L = \int_a^b \sqrt{[f'(t)]^2 + [g'(t)]^2} \, dt$$

Since the paths are equivalent, there is a function $t = \varphi(\tau)$ as above. Let us assume first that $\varphi'(\tau) > 0$, so that t increases with τ, and hence $\varphi(\alpha) = a$, $\varphi(\beta) = b$. Then by the rule for substitution (Section 4-21)

$$L = \int_\alpha^\beta \sqrt{\{f'[\varphi(\tau)]\}^2 + [g'(\varphi(\tau))]^2} \, \varphi'(\tau) \, d\tau$$
$$= \int_\alpha^\beta \sqrt{\{f'[\varphi(\tau)]\varphi'(\tau)\}^2 + \{g'[\varphi(\tau)]\varphi'(\tau)\}^2} \, d\tau$$
$$= \int_\alpha^\beta \sqrt{[F'(\tau)]^2 + [G'(\tau)]^2} \, d\tau$$

In the last step the chain rule for derivatives was used. The last expression is the length of the second curve, so that the two lengths are equal. The proof for the case $\varphi'(\tau) < 0$ is left as an exercise (Problem 11 below).

Arc Length as Parameter. Let us assume that we have introduced arc length s along our path, as in the preceding section, and that either $ds/dt > 0$ along the path, or $ds/dt < 0$ along the path. Then there is a one-to-one correspondence between t-values and s-values: $t = \varphi(s)$ and $s = \psi(t)$. Thus we can regard s itself as the new parameter. Under the assumptions made, the path is then given by equations

$$x = f[\varphi(s)] = x(s), \qquad y = g[\varphi(s)] = y(s)$$

where s is in an appropriate interval (perhaps infinite); $x(s)$ and $y(s)$ have continuous first derivatives and (assuming that $ds/dt > 0$)

$$\frac{dx}{ds} = \frac{dx}{dt}\frac{dt}{ds} = \frac{dx/dt}{ds/dt} = \frac{f'(t)}{\sqrt{[f'(t)]^2 + [g'(t)]^2}}$$
$$\frac{dy}{ds} = \frac{dy}{dt}\frac{dt}{ds} = \frac{g'(t)}{\sqrt{[f'(t)]^2 + [g'(t)]^2}} \tag{4-290}$$

so that

$$\left(\frac{dx}{ds}\right)^2 + \left(\frac{dy}{ds}\right)^2 = 1 \tag{4-291}$$

Thus the vector $x'(s)\mathbf{i} + y'(s)\mathbf{j}$ is a unit vector. In fact, by (4-290),

$$x'(s)\mathbf{i} + y'(s)\mathbf{j} = \frac{f'(t)\mathbf{i} + g'(t)\mathbf{j}}{\sqrt{[f'(t)]^2 + [g'(t)]^2}} = \frac{\mathbf{v}}{|\mathbf{v}|}$$

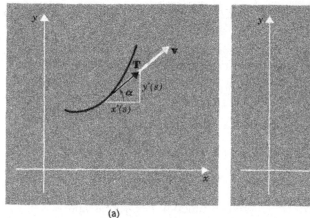

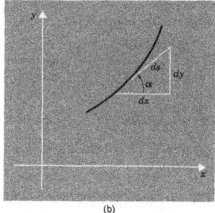

(a)   (b)

Figure 4-47 Arc length as parameter.

Thus $x'(s)\mathbf{i} + y'(s)\mathbf{j}$ is a unit vector in the direction of the velocity vector $\mathbf{v}$. Since $\mathbf{v}$ is tangent to the path (Section 3-11), $x'(s)\mathbf{i} + y'(s)\mathbf{j}$ is a unit tangent vector $\mathbf{T}$, as in Figure 4-47. Hence we can write

$$x'(s) = \cos \alpha, \qquad y'(s) = \sin \alpha \qquad\qquad (4\text{-}292)$$

where α is the angle from the positive x-direction to $\mathbf{v}$ or $\mathbf{T}$. If $ds/dt < 0$, the vector $\mathbf{T}$ has direction opposite to that of $\mathbf{v}$. From (4-291) we also conclude that

$$(dx)^2 + (dy)^2 = ds^2 \qquad\qquad (4\text{-}291')$$

where dx, dy, ds are the differentials of x, y, s in terms of t; hence they represent, approximately, the changes in x, y, s for a small change in t [Figure 4-47(b)].

PROBLEMS

1. Graph and find the length of the path:
   (a) $y = x^{3/2}$, $0 \le x \le 1$
   (b) $y = (x^3/6) + 1/(2x)$, $1 \le x \le 2$
   (c) $x = t^2, y = t^3$, $0 \le t \le 1$
   (d) $x = 8t^3, y = 6t^4 - 3t^2$, $-1 \le t \le 1$
   (e) $x = \cos^3 t, y = \sin^3 t$, $0 \le t \le 2\pi$
   (f) $x = \cos t, y = \cos t$, $0 \le t \le 2\pi$
   (g) $x = 3t^4, y = 3t^2 - t^6$, $0 \le t \le 1$
   (h) $x = t^2, y = 1 + t^2$, $-1 \le t \le 1$
   (i) $y = (e^x + e^{-x})/2$, $0 \le x \le b$
2. Estimate the length of each of the paths (a) to (d) of Figure 4-48, given graphically.
3. For each of the following paths, find s, as an integral, in terms of t but do not evaluate:
   (a) $x = t^2, y = t^5, 0 \le t \le 1$; $s = 0$ for $t = 0$; s increases with t.
   (b) $x = e^t, y = \cos t, 0 \le t \le 2\pi$; $s = 0$ for $t = \pi$; s increases with t.

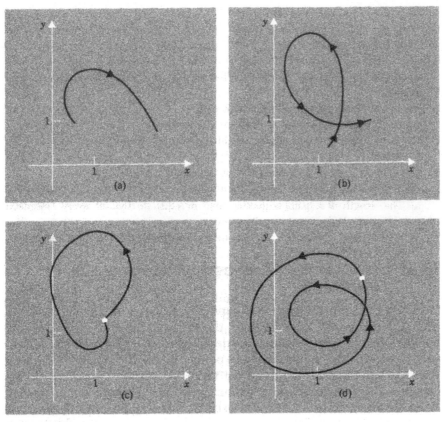

Figure 4-48

(c) $x = \ln t$, $y = 1/t$, $0 < t < \infty$; $s = 0$ for $t = 1$, s increases with t.

(d) $x = e^t$, $y = e^{3t}$, $-\infty < t < \infty$; $s = 0$ for $t = 1$, s decreases as t increases.

4. Find s in terms of t, and t in terms of s, for each of the following paths. Also express x and y in terms of s.

    (a) $x = t^2$, $y = t^3$, $0 \leq t \leq 1$; $s = 0$ for $t = 0$; s increases with t.

    (b) $x = \cos^3 t$, $y = \sin^3 t$, $0 \leq t \leq \frac{1}{2}\pi$; $s = 0$ for $t = 0$; s increases as t increases.

5. (a) Show that, if arc length s is introduced on a path $x = f(t)$, $y = g(t)$ as in Section 4-28, and $s = s_0$ for $t = t_0$, then

$$s = \pm \int_{t_0}^{t} \sqrt{[f'(u)]^2 + [g'(u)]^2}\, du + s_0$$

    with $+$ for s increasing with t, $-$ for s decreasing as t increases.

    (b) Deduce (4-282) from the result of (a).

6. Prove Theorem 28.

7. Prove the inequality $|\sqrt{a^2 + b^2} - \sqrt{c^2 + b^2}| \leq |a - c|$:

    (a) With the aid of vectors $a\mathbf{i} + b\mathbf{j}$, $c\mathbf{i} + b\mathbf{j}$.

    (b) By algebra.

8. Show that each of the following pairs of paths are equivalent, and give the function $t = \varphi(\tau)$ in each case:

    (a) $x = t^2$, $y = t^4$, $1 \leq t \leq 2$; $x = \tau$, $y = \tau^2$, $1 \leq \tau \leq 4$

    (b) $x = t$, $y = \sqrt{1 + t^2}$, $0 \leq t \leq 1$; $x = \tan \tau$, $y = \sec \tau$, $0 \leq \tau \leq \pi/4$

(c) $x = t^2 - 1$, $y = t^3 + t$, $0 \le t \le 1$; $x = \tau^2 - 2\tau$, $y = 2 - 4\tau + 3\tau^2 - \tau^3$, $0 \le \tau \le 1$

9. Show that the paths

$$x = t, y = t^2, \quad 0 \le t \le 1 \quad \text{and} \quad x = \sin^2 \tau, \quad y = \sin^4 \tau, \quad 0 \le \tau \le 3\pi/2$$

cover the same points in the xy-plane but are not equivalent. How are the lengths related?

‡ 10. Show from the definition of equivalent paths:
    (a) Each path is equivalent to itself.
    (b) If two paths are equivalent to a third path, then they are equivalent to each other.

‡11. Carry out the proof of Theorem 29 for the case $\varphi'(\tau) < 0$.

‡12. Show from the definition of length of a path:
    (a) The length of a path is greater than or equal to that of every "inscribed polygon." (When does equality occur?)
    (b) The length of a path is the lub of the lengths of all "inscribed polygons."

†4-30 INTEGRATION OF PIECEWISE CONTINUOUS FUNCTIONS

Up to this point, we have considered the definite integral only for continuous functions. We now consider the case of a function f on $[a, b]$ with a finite number of jump discontinuities [Figure 4-49 (See Section 2-2)]. At each discontinuity point (such as the point $x = c$ in the figure), f has a limit from the left and from the right, but these are usually unequal, as at c. In any case, the value at the point differs from one of these limits; the point d in the figure illustrates a case where the limits are equal, but different from the value at the point. The discontinuity can also occur at an end point: a or b. At a, a limit to the right would then exist, different from $f(a)$; at b, a limit to the left would exist, different from $f(b)$. A function of the type considered is called a *piecewise continuous function*.

If we think of the integral as "area under a curve," then we have no difficulty in assigning a value to the integral. In the case of a function f for which the only discontinuities are at the ends, we can reason that the values at the ends a, b can have no effect on the area (since the area of a line segment is 0). Hence we replace these values by the limiting values in

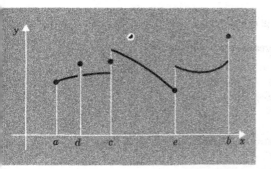

Figure 4-49   Piecewise continuous function.

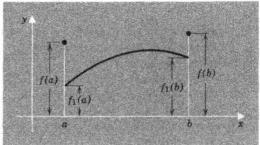

Figure 4-50   Piecewise continuous function discontinuous only at end points.

each case, thereby replacing f by a continuous function f_1 in $[a, b]$; see Figure 4-50. The functions f and f_1 agree except at the end points. We then define the integral of f to be that of f_1:

$$\int_a^b f(x)\, dx = \int_a^b f_1(x)\, dx$$

For a function as in Figure 4-49, we simply add the integrals for the separate parts of $[a, b]$; in the case of the figure

$$\int_a^b f(x)\, dx = \int_a^d f(x)\, dx + \int_d^c f(x)\, dx + \int_c^e f(x)\, dx + \int_e^b f(x)\, dx$$

Each of the separate integrals is then evaluated by the process of Figure 4-50; that is, by first changing the end values to get a continuous function.

It is of interest to note that the definition just given is also obtainable from the Riemann integral; that is

$$\int_a^b f(x)\, dx = \lim_{\text{mesh}\to 0} \sum_{i=1}^n f(\xi_i)\, \Delta_i x \qquad (4\text{-}300)$$

We do not write out a formal proof, but simply remark that, for a function as in Figure 4-50, the values assigned at a and b have no effect on the right side of (4-300). For the function f is surely bounded: $|f(x)| \le K$ for some K, and the values at the end points can affect at most the terms

$$f(\xi_1)\, \Delta_1 x + f(\xi_n)\, \Delta_n x$$

Their sum is in absolute value at most $K(\Delta_1 x + \Delta_n x)$, which approaches 0 as the mesh approaches 0. Hence, in the limit, we get the same result. A similar argument applies to the general case suggested in Figure 4-49, and we conclude that *every piecewise continuous function f has a Riemann integral* $\int_a^b f(x)\, dx$.

A linear combination of two piecewise continuous functions is piecewise continuous, and we can verify (Problem 5 below) that for such functions

$$\int_a^b [k_1 f_1(x) + k_2 f_2(x)]\, dx = k_1 \int_a^b f_1(x)\, dx + k_2 \int_a^b f_2(x)\, dx$$

Also, if f is piecewise continuous on an interval containing a, b, and c then

$$\int_a^b f(x)\, dx + \int_b^c f(x)\, dx = \int_a^c f(x)\, dx$$

Here we can allow a to be greater than or equal to b, $b \ge c$, $a \ge c$ by the usual procedure:

$$\int_a^b f(x)\, dx = -\int_b^a f(x)\, dx, \qquad \int_a^a f(x)\, dx = 0$$

(See Problem 6 below.)

Other properties can be deduced from the fact that the integral is in fact the sum of integrals of continuous functions (see Problems 5 and 7 below). The most important new feature is a modification of the Fundamental Theorem of Calculus:

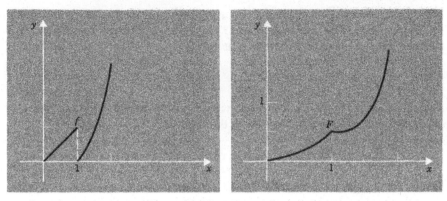

Figure 4-51 Indefinite integral F of piecewise continuous function f.

THEOREM 30. *Let f be defined and piecewise continuous on $[a, b]$ and let*

$$F(x) = \int_a^x f(t)\, dt$$

Then F is continuous on $[a, b]$ and $F'(x) = f(x)$ at every point of continuity of f. At each discontinuity x_0, F has a derivative to the right (except at b) equal to the right-hand limit of f at x_0, and a derivative to the left (except at a) equal to the left-hand limit of f.

Before proceeding to the proof we consider an example.

EXAMPLE 1. Let $f(x) = x$ for $0 \le x \le 1$, $f(x) = x^2 - 1$ for $1 < x \le 2$ (see Figure 4-51). Then, for $0 \le x \le 1$,

$$F(x) = \int_0^x t\, dt = \frac{x^2}{2}$$

For $x > 1$, by our definition of the integral.

$$F(x) = \int_0^1 t\, dt + \int_1^x (t^2 - 1)\, dt = \frac{1}{2} + \frac{x^3}{3} - x + \frac{2}{3} = \frac{x^3}{3} - x + \frac{7}{6}$$

At $x = 1$, F is continuous, with value $\frac{1}{2}$, and F is continuous elsewhere in $[0, 2]$. At $x = 1$, the derivative of F to the left is the derivative of $x^2/2$, hence equals 1; the derivative to the right is the derivative of $(x^3/3) - x + (7/6)$, hence equals 0. Thus, F has a corner at $x = 1$. For other values of x, we find that

$$F'(x) = \begin{cases} x, & 0 \le x < 1 \\ x^2 - 1, & 1 < x \le 2 \end{cases}$$

thus $F'(x) = f(x)$, except at $x = 1$.

PROOF OF THEOREM 30. The example suggests the reasoning. If x is not a discontinuity point of f, then

$$F(x) = \int_a^x f(t)\, dt = \int_a^{b_1} f(t)\, dt + \int_{b_1}^{b_2} f(t)\, dt + \cdots + \int_{b_k}^{x} f(t)\, dt \qquad (4\text{-}301)$$

where $b_1, b_2, \ldots, b_k$ (and possible a) are the discontinuity points of f to the left of x. In the last integral f is to be replaced by a function f_1, which is continuous in $[b_k, x]$. Hence F is continuous at x and $F'(x) = f(x)$. To study the behavior of F at b_k, we use the same equation (4-301) for $x \geq b_k$. Thus

$$F(x) = \text{const} + \int_{b_k}^{x} f_1(t)\, dt \qquad (4\text{-}302)$$

where f_1 is continuous in $[b_k, b_{k+1}]$, $b_{k+1} > x$. Hence, by Theorem 11 (Section 4-17), $F'(x) = f_1(x)$ in $[b_k, b_{k+1}]$, where the derivative at b_k is a derivative to the right, and hence equals

$$f_1(b_k) = \lim_{x \to b_k+} f(x)$$

Equation (4-302) also shows that F is continuous to the right at b_k. Similar arguments show that F is continuous to the left at b_k and that F has a derivative to the left, equal to the limit of f at b_k. (At the end points a and b, only one side can be considered.) Thus the theorem is proved.

The theorem shows that F is the "continuous indefinite integral" of Section 4-13, so that one can write

$$\int f(x)\, dx = F(x) + C$$

as in that section.

Application to Arc Length. In discussing arc length (Sections 4-27 to 4-29), we considered only paths $x = f(t)$, $y = g(t)$, where f and g have continuous first derivatives in the interval considered. For many common paths, f' and g' are only piecewise continuous—for example, for a broken line path. Such a path is called *piecewise smooth* (see Figure 4-52). More precisely, we define a path $x = f(t)$, $y = g(t)$, $a \leq t \leq b$, to be piecewise smooth if the interval $[a, b]$ can be subdivided by points $a = b_0, b_1, \ldots,$ $b_n = b$ into intervals $[b_{i-1}, b_i]$ in each of which f and g are continuous, with continuous first derivatives. (This implies that at each b_k, f and g have

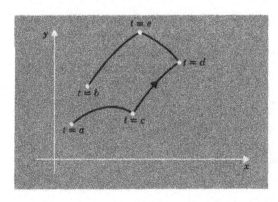

Figure 4-52   Piecewise smooth path.

derivatives to the left and to the right. By Theorem 13 of Section 3-21, it would be equivalent to require: f and g are continuous, f' and g' are continuous except perhaps at $b_1, \ldots, b_n$, where f' and g' have limits to the left and to the right.)

The length of such a piecewise smooth path can then be defined as the sum of the lengths of the pieces corresponding to the intervals $b_{i-1} \leq t \leq b_i$, or equivalently as the integral

$$\int_a^b \sqrt{[f'(t)]^2 + [g'(t)]^2}\, dt \qquad (4\text{-}303)$$

For, under the hypotheses stated, f' and g' are piecewise continuous. Strictly speaking, these derivatives are not defined at the subdivision points b_i, but we can always use the left-hand or right-hand derivative at these points. It follows that the integrand in (4-303) is piecewise continuous, and therefore the integral has meaning either as a Riemann integral or as a sum of the integrals for the intervals $[b_{i-1}, b_i]$. We can also verify that the integral equals the lub of the lengths of all inscribed polygons.

We can also introduce an arc length function on a piecewise smooth path: for example, as

$$s = \int_a^t \sqrt{[f'(u)]^2 + [g'(u)]^2}\, du \qquad (4\text{-}304)$$

By Theorem 30, s is then a continuous nondecreasing function of t, and

$$\frac{ds}{dt} = \sqrt{[f'(t)]^2 + [g'(t)]^2} = \sqrt{\left(\frac{dx}{dt}\right)^2 + \left(\frac{dy}{dt}\right)^2}$$

except at the subdivision points b_i. At these points $s(t)$ has right-hand and left-hand derivatives, equal to $(f'^2 + g'^2)^{\frac{1}{2}}$, with f' and g' taken as right-hand and left-hand derivatives, respectively.

We can also define equivalent paths as in Section 4-29 above, but can now allow a change of parameter $t = \varphi(\tau)$ where $\varphi(\tau)$ is continuous but φ' is only piecewise continuous, with left and right derivatives existing at the discontinuity points. The condition $\varphi'(\tau) > 0$ [or < 0] must then hold true

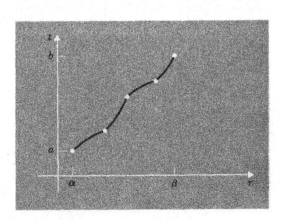

Figure 4-53 Change of parameter.

at continuity points, and both left and right-hand derivatives must be > 0 [or < 0] at the discontinuity points. Under these conditions $\varphi(\tau)$ is monotone strictly increasing [or decreasing] and has an inverse $\tau = \varphi^{-1}(t)$ with properties like those given for $\varphi(\tau)$ (see Figure 4-53). We can verify as above that equivalent paths have the same length.

We can in particular use arc length s as parameter on a piecewise smooth path and obtain an equivalent path. We can also verify that, with s as parameter, as in (4-304),

$$\frac{dx}{ds} = \frac{f'(t)}{\sqrt{[f'(t)]^2 + [g'(t)]^2}}, \qquad \frac{dy}{ds} = \frac{g'(t)}{\sqrt{[f'(t)]^2 + [g'(t)]^2}} \qquad (4\text{-}305)$$

(Problem 12 below). Here at the discontinuity points we must use right-hand derivatives on both sides of the equation, or left-hand derivatives on both sides.

PROBLEMS

1. Evaluate the integral $\int_0^3 f(x)\, dx$ in each of the following cases; and check by graphing and finding the corresponding area.
   (a) $f(x) = 1$ for $0 \le x \le 2$, $f(x) = 0$ for $2 < x \le 3$
   (b) $f(x) = x$ for $0 \le x \le 1$, $f(x) = x - 1$ for $1 < x \le 2$, $f(x) = x - 2$ for $2 < x \le 3$
   (c) $f(x) = \dfrac{\sin 2x}{|\sin 2x|}$ except for $f(0) = 0$, $f(\pi/2) = 0$

2. Find the function $F(x) = \int_1^x f(u)\, du$ in each of the following cases and verify that $F'(x) = f(x)$ where f is continuous:
   (a) $f(x) = 1$ for $0 \le x \le 1$, $f(x) = 0$ for $1 < x \le 2$
   (b) $f(x) = x^2$ for $0 \le x \le 1$, $f(x) = 2x^2 + 1$ for $1 < x \le 2$
   (c) $f(x) = e^x$ for $0 \le x \le 1$, $f(x) = e^{-x}$ for $1 < x \le 2$

3. A car travels at 50 mph for 2 hours, then at 60 mph for 1 hour, then at 30 mph for 1 hour. Represent the total distance traveled by an integral, and evaluate.

4. The rate of production of cars at a certain factory is maintained steadily over intervals as follows:

January 1–February 15	320 cars per day
February 16–February 28	300 cars per day
March 1–March 18	350 cars per day
March 19–March 31	310 cars per day

Represent the total production from January through March as an integral and evaluate.

5. Let $f_1(x)$ and $f_2(x)$ both be defined and piecewise continuous for $a \le x \le b$.
   (a) Show that $kf_1(x)$ is also piecewise continuous for each constant k (can kf_1 be continuous?) and that
   $$\int_a^b kf_1(x)\, dx = k \int_a^b f_1(x)\, dx$$
   (b) Show that $f_1(x) + f_2(x)$ is piecewise continuous (can it be continuous?) and
   $$\int_a^b [f_1(x) + f_2(x)]\, dx = \int_a^b f_1(x)\, dx + \int_a^b f_2(x)\, dx$$

(c) Prove:

$$\int_a^b [k_1 f_1(x) + k_2 f_2(x)]\, dx = k_1 \int_a^b f_1(x)\, dx + k_2 \int_a^b f_2(x)\, dx$$

6. Prove: if f is piecewise continuous in $[a, c]$ and $a < b < c$, then

$$\int_a^b f(x)\, dx + \int_b^c f(x)\, dx = \int_a^c f(x)\, dx$$

7. (a) Prove: if f is piecewise continuous in $[a, b]$ and $A \le f(x) \le B$ for all x, then

$$A(b - a) \le \int_a^b f(x)\, dx \le B(b - a)$$

(b) Show that the same conclusion holds true if we assume only that $A \le f(x) \le B$ at points of continuity of f.

8. Let $f(x)$ be piecewise continuous on $[a, b]$. Let $F(x)$ be continuous on $[a, b]$ with derivative $F'(x)$ equal to $f(x)$ at points of continuity of f. Show that

$$\int_a^b f(x)\, dx = F(b) - F(a)$$

9. Let $f(x)$ and $g(x)$ be continuous on $[a, b]$, with derivatives $f'(x)$ and $g'(x)$ existing at all but a finite number of points, at which left and right-hand derivatives exist. Let f_0' and g_0' be piecewise continuous on $[a, b]$ and suppose that $f_0'(x) = f'(x)$ and $g_0'(x) = g'(x)$ at the points where f' and g' are both defined. Show that

$$\int_a^b f(x) g_0'(x)\, dx = [f(x)g(x)]\, \Big|_a^b - \int_a^b g(x) f_0'(x)\, dx$$

10. Show that each of the following is a set of parametric equations for a piecewise smooth path, graph, and find the length of the path:
(a) $x = t^2 - 1, 0 \le t \le 1, x = t - 1, 1 \le t \le 2$
     $y = t^2, 0 \le t \le 1, y = 2 - t, 1 \le t \le 2$
(b) $x = t^2, 0 \le t \le 1, x = 4t - 3, 1 \le t \le 2$
     $y = t^3, 0 \le t \le 1, y = 6t - 5, 1 \le t \le 2$

11. (a) For the path of Problem 10(a), let arc length s be measured starting at $t = 0$ and increasing with t. Find $s(t), x(s), y(s)$ and discuss the continuity properties of $s(t), dx/ds, dy/ds$.
(b) Proceed as in (a) for the path of Problem 10(b).

12. Prove that the formulas (4-305) are valid as stated for a piecewise smooth path $x = f(t), y = g(t), a \le t \le b$.

4-31 INTEGRATION OF VECTOR FUNCTIONS

In Section 3-11, we have considered differentiation of a vector function $\mathbf{F}(t), a \le t \le b$. We now consider integration of such a function.

For the definite integral it is easier to use the Riemann integral. We say that $\mathbf{F}$ has a Riemann integral, over the interval $[a, b]$, if there is a vector $\mathbf{I}$ so that

$$\lim_{\text{mesh} \to 0} \sum_{i=1}^n \mathbf{F}(\xi_i)(t_i - t_{i-1}) = \mathbf{I} \tag{4-310}$$

Here, as usual, we are considering subdivisions of $[a, b]$:

$$a = t_0 < t_1 < \cdots < t_n = b$$

and choices of the ξ_i so that $t_{i-1} \leq \xi_i \leq t_i$. Each term of the summation is a vector in the plane, since it is a scalar $(t_i - t_{i-1})$ times a vector $\mathbf{F}(\xi_i)$; hence, the sum is a vector. The limit is understood as follows: Given a number $\epsilon > 0$, it must be possible to choose $\delta > 0$ so that

$$\left| \sum_{i=1}^{n} \mathbf{F}(\xi_i)(t_i - t_{i-1}) - \mathbf{I} \right| < \epsilon$$

for all subdivisions of mesh less than δ and all choices of the points ξ_i. The vector $\mathbf{I}$ is the Riemann integral of $\mathbf{F}$ over $[a, b]$ and we write $\mathbf{I} = \int_a^b \mathbf{F}(t)\,dt$.

Here we have a limit process resembling that for a vector function (Section 2-19), and we have a similar result:

THEOREM 31. *Let a vector function*

$$\mathbf{F}(t) = f(t)\mathbf{i} + g(t)\mathbf{j}, \qquad a \leq t \leq b$$

be given, in terms of the orthonormal basis $\mathbf{i}, \mathbf{j}$*. Then* $\mathbf{F}$ *has a Riemann integral*

$$\mathbf{I} = \int_a^b \mathbf{F}(t)\,dt$$

if, and only if, f *and* g *have Riemann integrals, and when these integrals exist*

$$\int_a^b \mathbf{F}(t)\,dt = \left[\int_a^b f(t)\,dt \right]\mathbf{i} + \left[\int_a^b g(t)\,dt \right]\mathbf{j}$$

PROOF: The proof of the theorem parallels that for Theorem 6 in Section 3-10. If we write $\mathbf{I} = A\mathbf{i} + B\mathbf{j}$, then the condition $|\Sigma \mathbf{F}(\xi_i)\,\Delta_i t - \mathbf{I}| < \epsilon$ is the same as

$$\left| \left[\sum f(\xi_i)\,\Delta_i t \right]\mathbf{i} + \left[\sum g(\xi_i)\,\Delta_i t \right]\mathbf{j} - (A\mathbf{i} + B\mathbf{j}) \right| < \epsilon$$

or

$$\left| \left[\sum f(\xi_i)\,\Delta_i t - A \right]\mathbf{i} + \left[\sum g(\xi_i)\,\Delta_i t - B \right]\mathbf{j} \right| < \epsilon \qquad (4\text{-}311)$$

If this condition holds true, then, since $|a\mathbf{i} + b\mathbf{j}| = (a^2 + b^2)^{\frac{1}{2}} \geq |a|$ and $\geq |b|$, we have

$$\left| \sum f(\xi_i)\,\Delta_i t - A \right| < \epsilon, \qquad \left| \sum g(\xi_i)\,\Delta_i t - B \right| < \epsilon \qquad (4\text{-}312)$$

This shows that the existence of the Riemann integral $\mathbf{I}$ for $\mathbf{F}$ implies the existence of the integrals for f and g, and that

$$\int_a^b f(t)\,dt = A, \qquad \int_a^b g(t)\,dt = B$$

Conversely, if these conditions hold true, then we can achieve (4-312) with

ϵ replaced by $\epsilon/2$ for all subdivisions of mesh less than some δ. The vector rule (Problem 6 following Section 1-9):

$$|a\mathbf{i} + b\mathbf{j}| \leq |a| + |b|$$

then gives (4-311), so that $\mathbf{F}$ has integral $\mathbf{I} = A\mathbf{i} + B\mathbf{j}$. Thus the theorem is proved.

Because of Theorem 31, questions concerning vector integrals can be reduced to questions for ordinary integrals, and we can verify that the expected properties of integrals carry over. In particular, *if $\mathbf{F}(t)$ is continuous or, more generally, piecewise continuous, then the integral exists. If $\mathbf{F}(t)$ is continuous and $\mathbf{F}(t) = \mathbf{G}'(t)$, then*

$$\int_a^b \mathbf{F}(t)\, dt = \int_a^b \mathbf{G}'(t)\, dt = \mathbf{G}(t) \Big|_a^b = \mathbf{G}(b) - \mathbf{G}(a) \qquad (4\text{-}313)$$

[(Problem 6(f) below].

EXAMPLE 1

$$\int_0^1 (t^2\mathbf{i} + t^3\mathbf{j})\, dt = \int_0^1 t^2\, dt\, \mathbf{i} + \int_0^1 t^3\, dt\, \mathbf{j} = \frac{1}{3}\mathbf{i} + \frac{1}{4}\mathbf{j}$$

The integral as total displacement. Let a point P move in the xy-plane with velocity vector $\mathbf{v}(t)$, $a \leq t \leq b$; we assume that $\mathbf{v}(t)$ is continuous. Then

$$\int_a^b \mathbf{v}(t)\, dt = \textit{total displacement of P over the given interval of time.}$$

For let $\overrightarrow{OP} = \mathbf{G}(t)$, so that $\mathbf{v}(t) = \mathbf{G}'(t)$. Then by (4-313) above

$$\int_a^b \mathbf{v}(t)\, dt = \int_a^b \mathbf{G}'(t)\, dt = \mathbf{G}(b) - \mathbf{G}(a) = \overrightarrow{OP_b} - \overrightarrow{OP_a} = \overrightarrow{P_a P_b}$$

where P_a, P_b are the positions for $t = a$, $t = b$, respectively (Figure 4-54).

Indefinite Integrals. By an indefinite integral of a vector function $\mathbf{F}(t)$, we mean a function $\mathbf{G}(t)$ whose derivative is $\mathbf{F}(t)$. From the fact that $\mathbf{G}'(t) \equiv 0$ over an interval implies: $\mathbf{G}(t) \equiv$ const [Problem 8(a) below], we

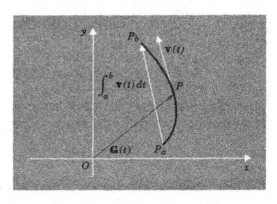

Figure 4-54 Integral of velocity.

conclude that, if $\mathbf{F}$ has an indefinite integral $\mathbf{G}(t)$, then all indefinite integrals are given by $\mathbf{G}(t) + \mathbf{C}$, where $\mathbf{C}$ is a *constant vector*. Hence we write

$$\int \mathbf{F}(t)\, dt = \mathbf{G}(t) + \mathbf{C}$$

as for ordinary functions. We also have: *if* $\mathbf{F}$ *is continuous in an interval containing the point* $t = c$, *then*

$$\int_c^t \mathbf{F}(u)\, du$$

defines an indefinite integral of $\mathbf{F}$ *in that interval* [Problem 6(e) below].

From the rules for derivatives of vector functions (Section 3-11), or from Theorem 31, we conclude that, if $\mathbf{F}(t) = f(t)\mathbf{i} + g(t)\mathbf{j}$ in an interval, where $\mathbf{F}$ is continuous (and, hence, so are f and g), then

$$\int \mathbf{F}(t)\, dt = \left(\int f(t)\, dt \right)\mathbf{i} + \left(\int g(t)\, dt \right)\mathbf{j} + \mathbf{C} \tag{4-314}$$

Thus, all the techniques for integrating ordinary functions are available. Also (4-314) shows that the usual rules for integral of sum and constant times function, integration by parts, and substitution have their counterparts for vector functions (Problem 8 below).

EXAMPLE 2. If $\mathbf{a}$ is a constant vector, then for $t > 0$,

$$\int t^n \mathbf{a}\, dt = \frac{t^{n+1}}{n+1}\mathbf{a} + \mathbf{C}, \qquad n \neq -1$$

$$\int \frac{1}{t}\mathbf{a}\, dt = \ln t\, \mathbf{a} + \mathbf{C}$$

These are verified most easily by differentiation.

EXAMPLE 3. Let a point P move in the xy-plane with constant nonzero acceleration $\mathbf{a}$, starting at time $t = 0$ at position P_0 with initial velocity $\mathbf{v}_0$. Find $\mathbf{v}(t)$ and $\mathbf{r}(t)$, where $\overrightarrow{OP} = \mathbf{r}$.

Solution. Since $\mathbf{v}'(t) = \mathbf{a}$,

$$\mathbf{v}(t) - \mathbf{v}(0) = \int_0^t \mathbf{a}\, du = u\,\mathbf{a} \Big|_{u=0}^{u=t} = t\,\mathbf{a}$$

so that $\mathbf{v}(t) = \mathbf{v}_0 + t\,\mathbf{a}$. Next $\mathbf{r}'(t) = \mathbf{v}(t)$, so that

$$\mathbf{r}(t) - \mathbf{r}(0) = \int_0^t (\mathbf{v}_0 + u\,\mathbf{a})\, du = \left\{ u\,\mathbf{v}_0 + \frac{u^2}{2}\mathbf{a} \right\}\Big|_0^t$$

and hence, with $\mathbf{r}_0 = \overrightarrow{OP_0} = \mathbf{r}(0)$,

$$\mathbf{r}(t) = \mathbf{r}_0 + t\,\mathbf{v}_0 + \frac{t^2}{2}\mathbf{a}$$

PROBLEMS

1. Evaluate the integrals:

   (a) $\int_1^2 [(t^2 - 1)\mathbf{i} + 2t^3\mathbf{j}]\, dt$ (b) $\int_0^\pi (\cos t\, \mathbf{i} + \sin t\, \mathbf{j})\, dt$

   (c) $\int_0^1 (t\, \mathbf{a} + t^2\, \mathbf{b})\, dt$, where **a** and **b** are constant vectors

   (d) $\int_0^1 [\mathbf{a} \cdot (t^2\mathbf{i} + (1 - t)\mathbf{j}]\, dt$, **a** = constant vector

2. Find the total displacement over the given time interval of a point with given velocity vector:

   (a) $\mathbf{v}(t) = e^{t^2}t\, \mathbf{i} + (1 + t)^{1/3}\, \mathbf{j}$, $0 \leq t \leq 1$

   (b) $\mathbf{v}(t) = \dfrac{1}{1 - t^2}\, \mathbf{a} + \dfrac{t}{1 - t^2}\, \mathbf{b}$, $2 \leq t \leq 3$,   **a**, **b**, constant vectors

3. A particle of mass 10 grams moves in the xy-plane with acceleration vector $\mathbf{a}(t) = 3t^2\mathbf{i} - e^t\mathbf{j}$ cm per sec^2 and velocity **i** at $t = 0$. Find the increase or decrease in kinetic energy of the particle from $t = 0$ to $t = 1$ (sec).

4. Find the indefinite integrals:

   (a) $\int \left(\dfrac{t^2}{1 + t^2}\, \mathbf{i} + \dfrac{2t}{1 + t^2}\, \mathbf{j} \right) dt$ (b) $\int (te^t\, \mathbf{i} + \cos 2t\, \mathbf{j})\, dt$

5. In Example 3 of Section 4-31, let Cartesian coordinates be introduced so that P_0 is the origin O and $\mathbf{a} = a\mathbf{j}$; also let $\mathbf{v}_0 = v_0(\cos \alpha\mathbf{i} + \sin \alpha\mathbf{j})$. Describe the path followed in each of the cases: (a) $\alpha = \pi/2$ (b) $0 \leq \alpha < \pi/2$.

6. Prove, under appropriate hypotheses:

   (a) $\int_a^b [\mathbf{F}(t) + \mathbf{G}(t)]\, dt = \int_a^b \mathbf{F}(t)\, dt + \int_a^b \mathbf{G}(t)\, dt$

   (b) $\int_a^b \varphi(t)\mathbf{a}\, dt = \mathbf{a} \int_a^b \varphi(t)\, dt$, **a** = constant vector

   (c) $\int_a^b k\mathbf{F}(t)\, dt = k \int_a^b \mathbf{F}(t)\, dt$, k = constant

   (d) $\int_a^b \mathbf{F}(t)\, dt + \int_b^c \mathbf{F}(t)\, dt = \int_a^c \mathbf{F}(t)\, dt$

   (e) $\dfrac{d}{dt} \int_a^t \mathbf{F}(u)\, du = \mathbf{F}(t)$, $a \leq t \leq b$

   (f) $\int_a^b \mathbf{G}'(t)\, dt = \mathbf{G}(b) - \mathbf{G}(a)$

† 7. (a) Prove directly from the definition of the integral that if $\mathbf{F}(t)$ is continuous for $a \leq t \leq b$, then

   $$\left| \int_a^b \mathbf{F}(t)\, dt \right| \leq \int_a^b |\mathbf{F}(t)|\, dt$$

   (b) From the result of part (a) prove that, if $f(t)$ and $g(t)$ are continuous for $a \leq t \leq b$, then

   $$\left\{ \left[\int_a^b f(t)\, dt \right]^2 + \left[\int_a^b g(t)\, dt \right]^2 \right\}^{1/2} \leq \int_a^b ([f(t)]^2 + [g(t)]^2)^{1/2}\, dt$$

8. Prove, under appropriate hypotheses:

   (a) If $\mathbf{G}'(t) \equiv 0$, then $\mathbf{G}(t)$ = constant

(b) $\int [\mathbf{F}(t) + \mathbf{G}(t)] \, dt = \int \mathbf{F}(t) \, dt + \int \mathbf{G}(t) \, dt + \mathbf{C}$

(c) $\int k\mathbf{F}(t) \, dt = k \int \mathbf{F}(t) \, dt + C$

(d) $\int \varphi(t)\mathbf{a} \, dt = \mathbf{a} \int \varphi(t) \, dt + C$

(e) $\int \mathbf{F}(t) \cdot \mathbf{G}'(t) \, dt = \mathbf{F}(t) \cdot \mathbf{G}(t) - \int \mathbf{F}'(t) \cdot \mathbf{G}(t) \, dt + C$

(f) $\int \varphi(t)\mathbf{F}'(t) \, dt = \varphi(t)\mathbf{F}(t) - \int \varphi'(t)\mathbf{F}(t) \, dt + C$

(g) $\int \mathbf{F}(t) \, dt = \int \mathbf{F}[\varphi(u)]\varphi'(u) \, du + \mathbf{C}$, where $t = \varphi(u)$

5

THE ELEMENTARY
TRANSCENDENTAL FUNCTIONS

In the previous chapters the trigonometric functions (sine, cosine, tangent, cosecant, secant, cotangent), the exponential function, the logarithm function, and those obtained from them by simple operations have all been used freely. Their definitions have been given in an informal fashion, and many properties have been stated without proof. In this chapter, precise definitions are given, and the crucial properties are proved. The procedures used in the chapter have more value than that of merely adding precision. They can be applied widely as a way of creating new functions and determining their properties.

†5-1 THE SINE AND COSINE FUNCTIONS

We now put ourselves in the frame of mind of one who has never heard of the sine or cosine or of any of the concepts of trigonometry, of one who has never even heard of the number π. With the aid of the calculus, we shall then proceed to develop the theory afresh. It will be seen that all desired properties can be obtained with relative ease.

We start with the semicircle $x^2 + y^2 = 1$, $x \geq 0$, shown in Figure 5-1, and introduce a special parametrization:

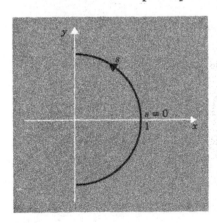

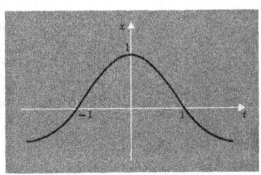

Figure 5-1   Basic semicircle.   Figure 5-2 $x = \dfrac{1 - t^2}{1 + t^2}$

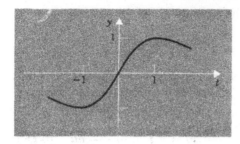

Figure 5-3 $y = \dfrac{2t}{1 + t^2}$.

$$x = \frac{1 - t^2}{1 + t^2}, \qquad y = \frac{2t}{1 + t^2}, \qquad -1 \le t \le 1 \qquad (5\text{-}10)$$

Other parametrizations can be used; (5-10) happens to be a simple and convenient one (see Problem 1 below).

The functions $x = (1 - t^2)/(1 + t^2), y = (2t)/(1 + t^2)$ appearing in (5-10) are graphed in Figures 5-2 and 5-3 over the whole interval $-\infty < t < \infty$. We notice that both functions are continuous for all t and have continuous first derivatives:

$$\frac{dx}{dt} = \frac{-4t}{(1 + t^2)^2}, \qquad \frac{dy}{dt} = \frac{2(1 - t^2)}{(1 + t^2)^2} \qquad (5\text{-}11)$$

Accordingly, on the interval $-1 \le t \le 1$, y is monotone strictly increasing, going from the value -1 at $t = -1$ to the value $+1$ at $t = 1$; on the same interval x remains positive, going from 0 to 1 and back to 0. Furthermore,

$$[x(t)]^2 + [y(t)]^2 = \frac{(1 - t^2)^2 + 4t^2}{(1 + t^2)^2} = \frac{(1 + t^2)^2}{(1 + t^2)^2} = 1$$

so that our path does follow the circle, and for $-1 \le t \le 1$ we have a parametrization of the semicircle as described. We observe also that $x(t)$ is an *even* function, $y(t)$ is an *odd* function.

Now we can introduce arc length:

$$s = s(t) = \int_0^t \sqrt{[x'(u)]^2 + [y'(u)]^2} \, du, \qquad -1 \le t \le 1 \qquad (5\text{-}12)$$

as in Section 4-28. Equation (5-12) implies that $s = 0$ at $t = 0$ [at (1, 0)], and that s increases as t increases (hence as y increases). From (5-12) we thus have

$$s = s(t) = \int_0^t \sqrt{\frac{16u^2}{(1 + u^2)^4} + \frac{4(1 - u^2)^2}{(1 + u^2)^4}} \, du = \int_0^t \frac{2}{1 + u^2} \, du \qquad (5\text{-}13)$$

We shall not use this formula to compute the arc length. We remark that it does *define* $s(t)$ precisely, as the integral of a continuous function, for $-1 \le t \le 1$ and that

$$\frac{ds}{dt} = \frac{2}{1 + t^2} > 0 \qquad (5\text{-}14)$$

Also, in (5-13) the integrand is even, so that $s(-t) = -s(t)$; that is, *s is an odd function of t* (Problem 3 below).

We now *define* the number π by the equation

$$\pi = 2s(1) = 2 \int_0^1 \frac{2}{1 + u^2}\, du \tag{5-15}$$

Thus π is a positive real number which is defined as twice the length of the quarter-circle from $(1, 0)$ to $(0, 1)$. We notice that, since $s(t)$ is odd, we have also

$$\pi = -2s(-1) = -2 \int_0^{-1} \frac{2}{1 + u^2}\, du = 2 \int_{-1}^0 \frac{2}{1 + u^2}\, du \tag{5-15'}$$

Thus $s(-1) = -\pi/2$, $s(0) = 0$, $s(1) = \pi/2$ and, because of (5-14), as t increases from -1 to 1, $s(t)$ increases steadily from $-\pi/2$ to $\pi/2$. Furthermore $s(t)$ has the same sign as t.

Since arc length was obtained by inscribing polygons in the path (Section 4-27), Equation (5-15) is simply the precise calculus version of the definition of π in plane geometry. Equation (5-15) can be used to calculate π to any desired number of decimal places (Problem 9 below).

Because of (5-14), we can now use s as parameter along our semicircle; that is, we obtain an equivalent path by expressing x and y in terms of s instead of t (Section 4-29). These functions are *defined* to be the cosine and sine functions for the interval considered:

$$x = x(s) = \cos s, \qquad y = y(s) = \sin s, \qquad -\frac{\pi}{2} \le s \le \frac{\pi}{2} \tag{5-16}$$

[We here write $x(s)$ for $x[t(s)]$, $y(s)$ for $y[t(s)]$; the context will always make clear which functions are being considered.] By the theory of Sections 4-28 and 4-29, the functions (5-16) are continuous and even have continuous derivatives. By the basic formula (4-290) we have:

$$\frac{dx}{ds} = \frac{x'(t)}{\sqrt{x'(t)^2 + y'(t)^2}} = \frac{x'(t)}{s'(t)}$$

$$\frac{dy}{ds} = \frac{y'(t)}{\sqrt{[x'(t)]^2 + [y'(t)]^2}} = \frac{y'(t)}{s'(t)}$$

Accordingly, by (5-11) and (5-14),

$$\frac{d}{ds} \cos s = \frac{-4t/(1 + t^2)^2}{2/(1 + t^2)} = \frac{-2t}{1 + t^2}$$

But the last expression is minus the value of y at that t, or at the corresponding s; that is,

$$\frac{dx}{ds} = \frac{d}{ds} \cos s = -y(s) = -\sin s$$

Similarly,

$$\frac{dy}{ds} = \frac{d}{ds} \sin s = \frac{2(1 - t^2)/(1 + t^2)^2}{2/(1 + t^2)} = \frac{1 - t^2}{1 + t^2} = x(s) = \cos s$$

Accordingly, we have proved the basic rules

$$\frac{d}{ds}\cos s = -\sin s, \qquad \frac{d}{ds}\sin s = \cos s \qquad (5\text{-}17)$$

for $-\frac{1}{2}\pi \le s \le \frac{1}{2}\pi$. At the same time, we have proved that $x(s) = \cos s$ and $y(s) = \sin s$ have continuous first derivatives in the interval. By differentiating again, we find that $\cos s$ and $\sin s$ have continuous second derivatives:

$$(\cos s)'' = (-\sin s)' = -\cos s$$
$$(\sin s)'' = (\cos s)' = -\sin s$$

and, by induction, they have continuous derivatives of every order.

Since $x^2 + y^2 = 1$ for every s, we have the identity

$$\cos^2 s + \sin^2 s = 1 \qquad \text{for} \qquad -\frac{\pi}{2} \le s \le \frac{\pi}{2} \qquad (5\text{-}18)$$

Along our semicircle x is nonnegative, reducing to zero only for $t = \pm 1$, that is, for $s = \pm(\pi/2)$. On the semicircle, y has the same sign as t, hence the same sign as s, and $y = 0$ only for $s = 0$. Accordingly, we have

$$\cos s \ge 0, \qquad -\frac{1}{2}\pi \le s \le \frac{1}{2}\pi$$

$$\sin s > 0 \text{ for } 0 < s < \frac{1}{2}\pi, \qquad \sin s < 0 \text{ for } -\frac{1}{2}\pi < s < 0 \qquad (5\text{-}19)$$

In $\left[-\frac{1}{2}\pi, \frac{1}{2}\pi\right]$, $\cos s = 0$ only for $s = \pm\frac{1}{2}\pi$, $\sin s = 0$ only for $s = 0$.

Since $(\sin s)' = \cos s$, we conclude from (5-19) that $\sin s$ is monotone strictly increasing over the interval considered (see Section 3-21). For $s = -\frac{1}{2}\pi$, $y = \sin s$ has the value given by (5-10) for $t = -1$, so that $\sin(-\frac{1}{2}\pi) = -1$; similarly, $\sin \frac{1}{2}\pi = 1$. Thus $\sin s$ increases from -1 to 1 as s goes from $-\frac{1}{2}\pi$ to $\frac{1}{2}\pi$. Similarly, since $(\cos s)' = -\sin s$, $\cos s$ increases from 0 to 1 as s increases from $-\frac{1}{2}\pi$ to 0, and $\cos s$ decreases from 1 to 0 as s increases from 0 to $\frac{1}{2}\pi$. Accordingly, we obtain the graphs of Figure 5-4 and Figure 5-5. Since $s(t)$ is an odd function and x and y are, respectively, even and odd functions of t, we conclude that $x(-s) = x(s)$, $y(-s) = -y(s)$; that is, $\cos s$ is an *even* function, $\sin s$ is an *odd* function, as seen in the symmetry of the graphs.

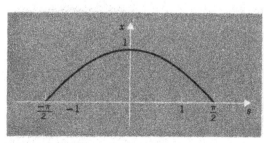

Figure 5-4 The cosine function.

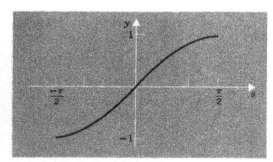

Figure 5-5 The sine function.

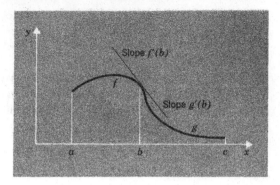

Figure 5-6 Extension of a function.

†5-2 EXTENSION OF COS s AND SIN s TO THE INFINITE INTERVAL

Before proceeding to our construction, we point out some basic facts on functions obtained by piecing together two functions on adjacent intervals. Let the functions

$$y = f(x), \qquad a \le x \le b, \qquad y = g(x), \qquad b \le x \le c$$

be given. Let both functions be continuous and let $f(b) = g(b)$. We can then piece these functions together to form a single function F (Figure 5-6):

$$F(x) = \begin{cases} f(x), & a \le x \le b \\ g(x), & b \le x \le c \end{cases}$$

We notice that $F(b) = f(b) = g(b)$. From the continuity of f and g, it follows that F *is continuous in* $[a, c]$. At b a special argument is needed; at this point the continuity of f and g make F continuous both to the left and to the right, and hence continuous. If f and g also have continuous first derivatives and if $f'(b) = g'(b)$, then F *has a continuous first derivative in* $[a, c]$. Again a special argument is needed at b: We first observe that, at $x = b$, F has a derivative to the left equal to $f'(b)$ and a derivative to the right equal to $g'(b)$. Since $f'(b) = g'(b)$, the derivatives of F to the left and right are equal, so that F has a derivative at b: $F'(b) = f'(b) = g'(b)$. The continuity of F' in $[a, c]$ now follows by the reasoning given above for the continuity of F itself.

We can think of the function F as being obtained from the function f by

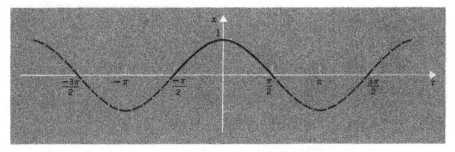

Figure 5-7 Extension of cosine function to the infinite interval.

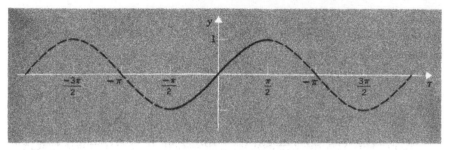

Figure 5-8   Extension of the sine function to the infinite interval.

extending the domain of the function from $[a, b]$ to the larger interval $[a, c]$. We call F an *extension* of f and refer to the continuity rules just derived as *extension rules.*

In Section 5-1 we have defined the cosine and sine functions for the interval $-\tfrac{1}{2}\pi \leq s \leq \tfrac{1}{2}\pi$. We now use these functions as building blocks to define the functions for $-\infty < s < \infty$. For clarity, we temporarily write $x = \cos\tau$, $y = \sin\tau$ for the functions to be defined over the interval $-\infty < \tau < \infty$; however, we shall see that τ can be interpreted as arc length along the circle.

The way we build our functions over the infinite interval is suggested graphically in Figures 5-7 and 5-8. In Figure 5-7, the heavy line is our building block, the cosine function of Figure 5-4. We turn this upside down (that is, reflect it in the τ-axis) and slide it π units to the right to obtain the dashed curve, providing an extension of the function to the interval $[-\pi/2, 3\pi/2]$; since $\cos(\pm\tfrac{1}{2}\pi) = 0$, the extension rules apply and the two pieces fit together to form a continuous function on the interval $[-\pi/2, 3\pi/2]$. By sliding our original building block 2π units to the right, we obtain an extension to the interval $[-\pi/2, 5\pi/2]$. By reflecting it in the τ-axis and sliding it π units to the left, we obtain an extension to the interval $[-3\pi/2, 5\pi/2]$, and so on. It is clear that in this way we can build a continuous function on the whole τ-axis. For the sine function we proceed in analogous fashion. Since $\sin(\pm\tfrac{1}{2}\pi)=\pm 1$, we again obtain a continuous function for $-\infty < \tau < \infty$ (Figure 5-8).

We can carry out our extension in one step by the following procedure. For each number τ we let n be the largest integer so that $-\tfrac{1}{2}\pi + n\pi \leq \tau$. Hence

$$-\frac{1}{2}\pi + n\pi \leq \tau < -\frac{1}{2}\pi + (n+1)\pi$$

Then we set

$$\cos\tau = (-1)^n \cos(\tau - n\pi), \qquad \sin\tau = (-1)^n \sin(\tau - n\pi) \quad (5\text{-}20)$$

Since $\tau - n\pi$ lies in the interval $[-\pi/2, \pi/2]$, the values of $\cos(\tau - n\pi)$ and $\sin(\tau - n\pi)$ are known, and hence $\cos\tau$, $\sin\tau$ are defined for $-\infty < \tau < \infty$. From the continuity of $\cos\tau$, $\sin\tau$ for $-\pi/2 \leq \tau \leq \pi/2$ and the fact that $\cos(-\pi/2) = \cos(\pi/2) = 0$, $\sin(-\pi/2) = -\sin(\pi/2) = -1$, we conclude as above that $\cos\tau$, $\sin\tau$ are continuous for $-\infty < \tau < \infty$. From (5-20) we also verify that

$$\cos(\tau + \pi) = -\cos\tau \qquad \sin(\tau + \pi) = -\sin\tau$$
$$\cos(\tau + 2\pi) = \cos\tau \qquad \sin(\tau + 2\pi) = \sin\tau \qquad (5\text{-}21)$$

for all τ (Problem 5 below).

The identity

$$\cos^2\tau + \sin^2\tau = 1 \qquad (5\text{-}22)$$

also holds true for all τ. For by (5-20),

$$\cos^2\tau + \sin^2\tau = \cos^2(\tau - n\pi) + \sin^2(\tau - n\pi) = 1$$

since $\tau - n\pi$ is in the interval $[-\pi/2, \pi/2]$, and we know that (5-22) holds true in the interval $[-\pi/2, \pi/2]$.

From (5-22) we conclude that

$$x = \cos\tau, \qquad y = \sin\tau, \qquad -\infty < \tau < \infty \qquad (5\text{-}23)$$

is a parametrization of a path following the circle. From the properties of the cosine and sine in the interval $[-\pi/2, \pi/2]$ and the extension formulas (5-20), we verify that, as τ increases from 0, the circle is traced in a counterclockwise direction, a complete circuit being completed each time τ increases by 2π; similar remarks apply to negative τ.

For all τ we have

$$\frac{d}{d\tau}\cos\tau = -\sin\tau, \qquad \frac{d}{d\tau}\sin\tau = \cos\tau \qquad (5\text{-}24)$$

This we know first for the interval $[-\pi/2, \pi/2]$. Then by (5-20), for $\pi/2 \le \tau \le 3\pi/2$ (that is, $n = 1$),

$$\frac{d}{d\tau}\cos\tau = -\frac{d}{d\tau}\cos(\tau - \pi) = \sin(\tau - \pi)$$

since $\tau - \pi$ is in the interval $[-\pi/2, \pi/2]$. By (5-20), $\sin(\tau - \pi) = -\sin\tau$, so that

$$\frac{d}{d\tau}\cos\tau = -\sin\tau, \qquad \frac{\pi}{2} \le \tau \le \frac{3\pi}{2}$$

At $\tau = \pi/2$ this equation gives a right-hand derivative of -1, the same as the left-hand derivative of $\cos\tau$ at $\pi/2$ (obtained from the interval $[-\pi/2, \pi/2]$). Hence by the extension rules $\cos\tau$ has a continuous derivative, equal to $-\sin\tau$, in the whole interval $[-\pi/2, 3\pi/2]$. Repetition of the argument for the general case of (5-20) gives $(\cos\tau)' = -\sin\tau$ for all τ. Similarly, we show that $(\sin\tau)' = \cos\tau$ for all τ.

Accordingly, if we define arc length s along our path (5-23) by the equation

$$s = \int_0^\tau \sqrt{[x'(u)]^2 + [y'(u)]^2}\, du,$$

we find that

$$s = \int_0^\tau \sqrt{\sin^2 u + \cos^2 u}\, du = \int_0^\tau du = \tau$$

Thus $s = \tau$, as stated, and our parametrization (5-23) could be written:

7. To prove the law of cosines, consider a triangle ABC in the plane and introduce Cartesian coordinates so that C is at the origin and A is on the positive x-axis at $(b, 0)$. Show that our definition of the sine and cosine implies that B has coordinates $(a \cos \theta, a \sin \theta)$ (see Figure 5-9). Now find c^2 by the distance formula of analytic geometry.

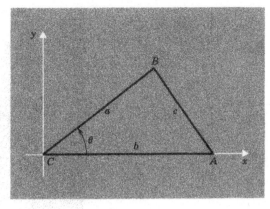

Figure 5-9 Law of cosines.

8. The ellipse

$$x^2 + \frac{y^2}{4} = 1$$

(with foci along the y-axis) can be parametrized by the equations

$$x = \cos t, \qquad y = 2 \sin t, \qquad -\infty < t < \infty$$

just as for the circle. Show that, if arc length is measured in the counterclockwise direction, starting with $s = 0$ for $t = 0$, then

$$s = 2 \int_0^t \sqrt{1 - \tfrac{3}{4} \sin^2 u}\, du$$

The integral $\int_0^t \sqrt{1 - k^2 \sin^2 u}\, du$ is known as an incomplete elliptic integral of second kind, and is denoted by $E(k, t)$. From tables[1] we find that $E[\sqrt{3}/2,\, \pi/2] = 1.2111$. Use this information to find the circumference of the ellipse, and also to sketch rough graphs of $x(s)$, $y(s)$.

9. Show from (5-15) that for each positive integer n

$$4n \left(\frac{1}{n^2 + 1} + \frac{1}{n^2 + 4} + \cdots + \frac{1}{n^2 + n^2} \right) < \pi$$

$$< 4n \left[\frac{1}{n^2} + \frac{1}{n^2 + 1} + \cdots + \frac{1}{n^2 + (n - 1)^2} \right]$$

and that the outer members of the inequality have π as a limit. Take $n = 4$ and evaluate the outer members and also their average, and compare with $\pi = 3.14159$.

†5-4 ANGLE FUNCTIONS

In the calculus and its applications, one is often led to assign an angle to a direction that is varying continuously. The fact that angles are determined only up to multiples of 2π leads to ambiguity. In this section we show how the ambiguity can be resolved in a satisfactory fashion. As a side result, we obtain a method for evaluating certain difficult definite integrals.

Let a path

$$x = f(t), \qquad y = g(t), \qquad a \le t \le b \qquad (5\text{-}40)$$

---

[1] See *Tables of Functions*, by E. Jahnke and F. Emde, Berlin: Teubner, 1938, page 72.

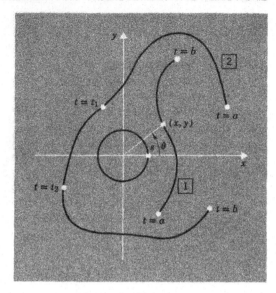

Figure 5-10 Angle functions.

be given in the xy-plane, with f and g continuous, and let the path *not pass through the origin,* so that $f^2 + g^2$ has no zeros. Two such paths are shown in Figure 5-10.

At each point (x, y) on such a path, we can assign a value of the angle θ as in polar coordinates, or as in trigonometry. However, θ is determined only up to multiples of 2π; that is, if θ_0 is one value, then so is $\theta_0 + 2k\pi$ for $k = \pm 1$, $\pm 2, \ldots$.

We now ask: If a value α has been chosen for θ at the point for which $t = a$ [the point $(f(a), g(a))$], can we give a value for θ for each t in such a fashion that $\theta(t)$ is a continuous function for $a \leq t \leq b$? Thus we seek a continuous function $\theta(t)$, $a \leq t \leq b$, so that $\theta(a) = \alpha$ and for each t

$$x = f(t) = r \cos \theta, \qquad y = g(t) = r \sin \theta, \qquad \text{with } r = \sqrt{x^2 + y^2}$$

or, explicitly,

$$f(t) = \sqrt{[f(t)]^2 + [g(t)]^2} \cos \theta(t)$$
$$g(t) = \sqrt{[f(t)]^2 + [g(t)]^2} \sin \theta(t)$$

Let us first consider the case of a path such as No. 1 in Figure 5-10. For $t = a$, we assign a value α, approximately $-\pi/3$. As t increases from a, we can clearly choose values for θ, moving up to 0, and then varying between 0 and $\pi/2$, ending up at about $\pi/3$ at $t = b$. Thus we have no doubt about which value to assign at each point, and it appears obvious that we get a continuous function $\theta(t)$ as required.

For a path such as No. 1 *on which x is greater than* 0, we can in fact give a formula that answers our question. We set

$$\theta = \operatorname{Tan}^{-1} \frac{y}{x} = \operatorname{Tan}^{-1} \frac{g(t)}{f(t)} \tag{5-41}$$

Now the function $\operatorname{Tan}^{-1} u$ (graphed in Figure 5-11) is continuous for

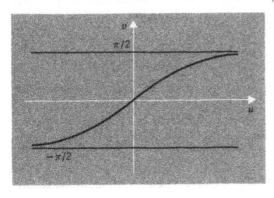

Figure 5-11 $v = \text{Tan}^{-1} u$.

$-\infty < u < \infty$; since $f(t) \neq 0$, $g(t)/f(t)$ is also continuous, so that (5-41) defines $\theta(t)$ as a continuous function. The values of $\text{Tan}^{-1} u$ lie between $-\pi/2$ and $\pi/2$ and, hence, we do get an allowable θ for each point. In particular, the value α at $t = a$ will be between $-\pi/2$ and $\pi/2$. If another initial value α is required, it must differ from $\text{Tan}^{-1}[g(a)/f(a)]$ by $2k\pi$ for some integer k. We must then modify our answer (5-41) by setting

$$\theta = \text{Tan}^{-1} \frac{y}{x} + \alpha - \text{Tan}^{-1} \frac{g(a)}{f(a)} = \text{Tan}^{-1} \frac{g(t)}{f(t)} + 2k\pi \qquad (5\text{-}41')$$

Similar formulas can be given for paths on which $x < 0$, or $y > 0$, or $y < 0$ (Problem 2 below). However, for a path such as No. 2 in Figure 5-10 one such formula will not suffice, and we are forced to give one formula for a part of the curve on which $y > 0$, then one for a part on which $x < 0$, then one for a part on which $y < 0$; that is, one for each of the intervals $a \leq t \leq t_1, t_1 \leq t \leq t_2$, $t_2 \leq t \leq b$, where t_1, t_2 are as shown in Figure 5-10.

Thus the assignment of formulas becomes clumsy, and in some cases we might even need infinitely many formulas. Also there are many ways of choosing the points $t_1, t_2, \ldots$ This raises doubts about our original question. However, the answer is always affirmative, and for the case of a *smooth curve* (f' and g' continuous) we can give *one formula*:

THEOREM 1. *Let the functions f and g be continuous on $[a, b]$, let $f^2 + g^2$ have no zeros, and let α be such that*

$$f(a) = r(a) \cos \alpha, \qquad g(a) = r(a) \sin \alpha$$

where $r(t) = \{[f(t)]^2 + [g(t)]^2\}^{1/2}$. Then there is a unique function: $\theta = \psi(t)$, $a \leq t \leq b$ so that $\psi(a) = \alpha$,

$$f(t) = r(t) \cos \psi(t), \qquad g(t) = r(t) \sin \psi(t)$$

and $\psi(t)$ is continuous for $a \leq t \leq b$.

If f and g have continuous first derivatives, then

$$\psi(t) = \int_a^t \left(\frac{f(u)g'(u) - g(u)f'(u)}{[f(u)]^2 + [g(u)]^2} \right) du + \alpha \qquad (5\text{-}42)$$

Remark. Equation (5-42) is obtainable from (5-41) by noting that, for

$\psi(t) = \text{Tan}^{-1}[g(t)/f(t)]$, we have

$$\psi'(t) = \frac{[f(t)g'(t) - g(t)f'(t)]/[f(t)]^2}{1 + [g(t)/f(t)]^2} = \frac{f(t)g'(t) - g(t)f'(t)}{[f(t)]^2 + [g(t)]^2} \quad (5\text{-}43)$$

PROOF OF VALIDITY OF (5-42). Under the hypotheses stated, the integrand of (5-42) is continuous in the interval $[a, b]$. Thus (5-42) does define ψ as a continuous function of t and, by the Fundamental Theorem of Calculus (Section 4-17), $\psi'(t)$ is given by (5-43). Also from (5-42), $\psi(a) = 0 + \alpha = \alpha$.

We now let

$$u(t) = \frac{f(t)}{r(t)} - \cos \psi(t), \qquad v(t) = \frac{g(t)}{r(t)} - \sin \psi(t)$$

Then straightforward differentiation, with the aid of the expression (5-43) for $\psi'(t)$, gives (Problem 7 below):

$$u'(t) = -\left(\frac{g(t)}{r(t)} - \sin \psi(t)\right) \cdot \psi'(t)$$

$$v'(t) = \left(\frac{f(t)}{r(t)} - \cos \psi(t)\right) \cdot \psi'(t) \qquad (5\text{-}44)$$

that is

$$u' = -v\psi', \qquad v' = u\psi'$$

Accordingly

$$uu' + vv' = -uv\psi' + uv\psi' = 0$$

so that

$$\frac{d}{dt}(u^2 + v^2) = 0, \qquad a \le t \le b$$

Accordingly, $u^2 + v^2 = \text{const.}$ As noted above, $\psi(a) = \alpha$, so that at $t = a$,

$$u^2 + v^2 = \left(\frac{f(a)}{r(a)} - \cos \alpha\right)^2 + \left(\frac{g(a)}{r(a)} - \sin \alpha\right)^2 = 0$$

by hypothesis. Hence $u^2 + v^2 \equiv 0$, so that $u \equiv 0$, $v \equiv 0$ and, hence, $f(t) = r(t) \cos \psi(t)$, $g(t) = r(t) \sin \psi(t)$ as asserted.

The proof of uniqueness of $\psi(t)$ and the proof of existence of ψ for the general case when f and g are merely continuous are given in the next section.

We observe that (5-42) also solves the problem for the case of a piecewise smooth path. The discussion is just like that for arc length (Section 4-30).

EXAMPLE 1. Find $\psi(t)$ for the path

$$x = e^{0.1t} \cos 3t, \qquad y = e^{0.1t} \sin 3t, \qquad 0 \le t \le \pi$$

with $\psi(0) = 0$. The path is shown in Figure 5-12. Equation (5-42) gives

$$\psi(t) = \int_0^t \frac{e^{0.1u} \cos 3u[3e^{0.1u} \cos 3u + 0.1e^{0.1u} \sin 3u] - \cdots}{e^{0.2u}(\cos^2 3u + \sin^2 3u)} \, du$$

$$= \int_0^t 3 \, du = 3t$$

The result could be predicted!

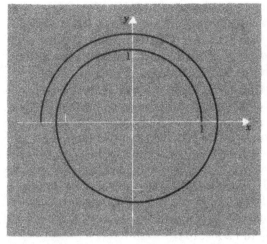

Figure 5-12 Path of Example 1.   **Figure 5-13** Path of Example 2.

EXAMPLE 2. Evaluate the integral

$$\int_0^{2\pi} \frac{f(t)g'(t) - g(t)f'(t)}{[f(t)]^2 + [g(t)]^2} \, dt, \qquad f(t) = 2\cos t + 1, \qquad g(t) = \sin t$$

On substitution, we are led to

$$\int_0^{2\pi} \frac{2 + \cos t}{3\cos^2 t + 4\cos t + 2} \, dt$$

a not particularly easy integral. Instead of struggling with this, we see from the original expression that our integral is related to the angle function $\psi(t)$ for the path $x = 2\cos t + 1$, $y = \sin t$, $0 \leq t \leq 2\pi$. In fact, by (5-42), the integral equals $\int_0^{2\pi} \psi'(t) \, dt = \psi(2\pi) - \psi(0)$, the total change in the angle. Now the path is easily graphed as in Figure 5-13, and we verify that the path encircles the origin once in the counterclockwise direction. *Hence, the integral equals $\psi(2\pi) - \psi(0) = 2\pi$.*

Remark. For a general path (5-40), not passing through the origin, for which $f(a) = f(b)$, $g(a) = g(b)$, $\psi(b) - \psi(a)$ must be an integral multiple of 2π, say $n \cdot 2\pi$. The number n is called *the winding number of the path.*

‡5-5 EXISTENCE AND UNIQUENESS OF THE ANGLE FUNCTION

Let f and g be given as in Theorem 1 of the preceding section. At each t, $f^2 + g^2 > 0$, so that at least one of f, g must differ from 0. If, for example, $f > 0$ at this t, then by continuity $f > 0$ in an interval including t (see Section 2-14, Remark 2), and ψ can be defined as a continuous function in that interval by a formula as above:

$$\psi = \text{Tan}^{-1} \frac{g}{f} + 2k\pi \tag{5-50}$$

Similar formulas can be given (Problem 2 below) if $f < 0$ at t, if $g > 0$ at t, and if $g < 0$ at t. In particular, ψ can be defined as a continuous function in some interval $a \leq t \leq c$, with $\psi(a) = \alpha$. Let E be the set of all numbers c for which this is possible. If E does not include b, then E has a lub t_0, $a < t_0 \leq b$. If $t_0 < b$, ψ can be defined by a formula such as (5-50) in an interval $t_0 - \delta < t < t_0 + \delta$. Since $t_0 = \text{lub } E$, we can choose a number c in E for which $t_0 - \delta < c \leq t_0$. Now we have ψ defined as a continuous function for $a \leq t \leq c$ and also for $c \leq t < t_0 + \delta$. By adjusting the multiple of 2π in (5-50) (or the analogous formula) we can make the two functions agree at c, and hence together give a continuous $\psi(t)$ for $a \leq t < t_0 + \delta$, with $\psi(a) = \alpha$. This contradicts the fact that $t_0 = \text{lub } E$ with $t_0 < b$. Hence, $t_0 = b$. The argument just given then shows that $t_0 = b$ is also in E. Thus ψ can be defined to be continuous for $a \leq t \leq b$, with $\psi(a) = \alpha$.

If there were a second such function $\psi_1(t)$, the difference $\psi_1(t) - \psi(t)$ would also be continuous for $a \leq t \leq b$, and equal to 0 for $t = a$. But any two choices of the angle differ by an integral multiple of 2π. If $\psi_1(t) - \psi(t) = 2k\pi$, $k \neq 0$, for some t, then by the Intermediate Value Theorem $\psi_1(t) - \psi(t)$ must take on all values between 0 and $2k\pi$. This is impossible. Hence, $\psi_1(t) - \psi(t) = 0 \cdot 2\pi = 0$ for $a \leq t \leq b$; that is, ψ_1 and ψ are the same function.

PROBLEMS

1. For each of the paths in Figure 5-14, given graphically, give the values of the continuous angle function $\psi(t)$ for all the values $t_1, t_2, \ldots$ indicated, noting that $\psi(a) = \alpha$ is given:

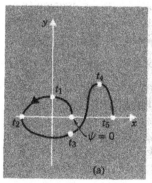 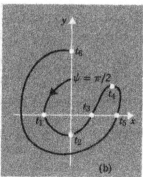 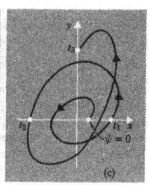

(a)   (b)   (c)

Figure 5-14

2. Let a path $x = f(t)$, $y = g(t)$, $a \leq t \leq b$, be given. Show that in each of the following cases $\psi(t)$ is defined as a continuous angle function (see Section 2-4).
   (a) $g(t) > 0$ for $a \leq t \leq b$; $\psi(t) = \text{Cot}^{-1}[f(t)/g(t)]$
   (b) $f(t) < 0$ for $a \leq t \leq b$; $\psi(t) = \pi + \text{Tan}^{-1}[g(t)/f(t)]$
   (c) $g(t) < 0$ for $a \leq t \leq b$; $\psi(t) = \pi + \text{Cot}^{-1}[f(t)/g(t)]$

3. Describe the solution to the following part of Problem 1 with the aid of several formulas like those of Problem 2:
   (a) Part (a) of Problem 1   (b) Part (b) of Problem 1
   (c) Part (c) of Problem 1

4. Find $\psi(t)$ by (5-42) for the path: $x = 1$, $y = t$, $-1 \le t \le 1$, with $\psi(-1) = -\pi/4$.
5. Find an integral expression for $\psi(t)$ for the elliptical path: $x = 2 \cos t$, $y = \sin t$, $0 \le t \le 2\pi$, and evaluate $\psi(2\pi) - \psi(0)$.
6. Evaluate the integral

$$\int_0^{2\pi} \frac{f(t)g'(t) - g(t)f'(t)}{[f(t)]^2 + [g(t)]^2} \, dt$$

(a) with $f(t) = 2 \cos 2t - \cos t$, $g(t) = 2 \sin 2t - \sin t$, and (b) with $f(t) = e^{\cos^3 t}$, $g(t) = \sin^4 t$.

7. Prove the correctness of the equations (5-44), in the proof of Theorem 1.
‡8. Prove that for a path as in Theorem 1 for a closed interval $a \le t \le b$, the angle function can always be defined by a finite number of formulas such as those in Problem 2, but this is not always true for an open interval: $a < t < b$.
‡9. Let f, g, f', g' be continuous in $[0, 2\pi]$, $f(0) = f(2\pi)$, $g(0) = g(2\pi)$, $f^2 + g^2 \ne 0$ in $[0, 2\pi]$. Show that

$$\int_0^{2\pi} \frac{f(t)}{\sqrt{[f(t)^2 + [g(t)]^2}} \frac{f(t)g'(t) - g(t)f'(t)}{[f(t)]^2 + [g(t)]^2} \, dt = 0$$

†5-6 INTEGRAL OF A RATIONAL FUNCTION OF SIN X AND COS X

We consider here indefinite integrals such as

$$\int \frac{\sin x}{4 + 5 \sin x} \, dx, \qquad \int \frac{\cos^2 x}{3 \cos x + 5 \sin x} \, dx, \qquad \int \frac{\sin x \cos x}{2 \cos x + 1} \, dx$$

or, in general, an integral of form

$$\int R(\cos x, \sin x) \, dx \tag{5-60}$$

where $R(u, v)$ is a rational function of u and v, that is, a ratio of two polynomials in u and v (Section 0-12). Many of these integrals can be handled by special devices. It is important to know that every one can be reduced to the integral of a rational function of one variable and, hence, can always be integrated by reduction to partial fractions as in Section 4-10 and 4-12.

Throughout we shall assume that x is restricted to the interval $-\pi < x < \pi$. The method is easily modified to take care of other intervals. We make the substitution

$$t = \tan \frac{x}{2}, \qquad \text{or} \qquad x = 2 \, \text{Tan}^{-1} t, \qquad -\infty < t < \infty \tag{5-61}$$

Thus we have a one-to-one correspondence between the infinite t-interval and the interval $-\pi < x < \pi$. By the substitution procedure of Section 4-7, we must express our integrand in terms of t and replace dx by $(dx/dt) \, dt$. Now, by trigonometry,

$$\cos x = 2 \cos^2 \frac{x}{2} - 1 = \frac{2}{\sec^2(x/2)} - 1 = \frac{2}{\tan^2(x/2) + 1} - 1$$

$$= \frac{2}{t^2 + 1} - 1 = \frac{1 - t^2}{1 + t^2}$$

$$\sin x = 2 \sin \frac{x}{2} \cos \frac{x}{2} = \frac{2 \tan(x/2)}{\sec^2(x/2)} = \frac{2t}{1 + t^2}$$

and by calculus

$$dx = \frac{2}{1 + t^2} \, dt$$

Thus the crucial formulas are

$$x = 2 \operatorname{Tan}^{-1} t, \qquad t = \tan \frac{x}{2}$$

$$\cos x = \frac{1 - t^2}{1 + t^2}, \qquad \sin x = \frac{2t}{1 + t^2}, \qquad dx = \frac{2 \, dt}{1 + t^2} \tag{5-62}$$

Through these relations an integral of the form (5-60) becomes an integral of the form $\int S(t) \, dt$, where S is a rational function. It is worth noting that, if x is intepreted as arc length s as in Section 5-1, then t is the parameter used in that section.

EXAMPLE 1 $$\int \frac{\sin x}{4 + 5 \sin x} \, dx$$

becomes

$$\int \frac{2t}{1 + t^2} \cdot \frac{1}{4 + [(10t)/(1 + t^2)]} \cdot \frac{2 \, dt}{1 + t^2}$$

$$= \int \frac{4t \, dt}{(t^2 + 1)(4t^2 + 10t + 4)} = \int \frac{t \, dt}{(t^2 + 1)(t + \frac{1}{2})(t + 2)}$$

$$= \int \left(\frac{4}{15} \cdot \frac{1}{t + 2} - \frac{4}{15} \cdot \frac{1}{t + \frac{1}{2}} + \frac{2}{5} \cdot \frac{1}{t^2 + 1} \right) dt$$

$$= \frac{4}{15} \ln \left| \frac{t + 2}{t + \frac{1}{2}} \right| + \frac{2}{5} \operatorname{Tan}^{-1} t + C$$

$$= \frac{4}{15} \ln \left| \frac{2 + \tan(x/2)}{\frac{1}{2} + \tan(x/2)} \right| + \frac{x}{5} + C$$

We can also write

$$t = \tan \frac{x}{2} = \frac{1 - \cos x}{\sin x} = \frac{\sin x}{1 + \cos x}$$

and get other forms of the answer.

EXAMPLE 2

$$\int \sec x \, dx = \int \frac{1}{\cos x} \, dx = \int \frac{1+t^2}{1-t^2} \frac{2 \, dt}{1+t^2} = 2\int \frac{dt}{1-t^2}$$

$$= -2\int \frac{dt}{(t+1)(t-1)} = -2\int \left(\frac{1}{2} \cdot \frac{1}{t-1} - \frac{1}{2} \cdot \frac{1}{t+1}\right) dt$$

$$= \ln \left|\frac{t+1}{t-1}\right| + C$$

$$= \ln \left|\frac{\tan (x/2) + 1}{\tan (x/2) - 1}\right| + C$$

EXAMPLE 3

$$\int \frac{dx}{1 + \cos x} = \int \frac{1}{1 + [(1-t^2)/(1+t^2)]} \, 2 \, \frac{dt}{1+t^2} = \int dt = t + C$$

$$= \tan \frac{x}{2} + C.$$

In each example, the result is valid in each interval containing no discontinuity.

Remark. The formulas (5-62) are helpful in proving trigonometric identities. In particular, if both sides are rational functions of sin x and cos x (or more generally rational functions of sin x, cos x, tan x, csc x, sec x, cot x), then by (5-62) both sides can be expressed as rational functions of t; the identity is then easily verified.

PROBLEMS

1. Integrate

(a) $\int \dfrac{1}{3 \sin x + 4 \cos x} \, dx$

(b) $\int \dfrac{1}{1 + 2 \sin x} \, dx$

(c) $\int \dfrac{1}{\sin x - \cos x - 1} \, dx$

(d) $\int \dfrac{1}{1 + \sin x + \cos x} \, dx$

(e) $\int \csc x \, dx$

(f) $\int \dfrac{(1 + \cos x) \, dx}{(\sin x - \cos x - 1)(\sin x - 2 \cos x - 2)}$

(g) $\int \dfrac{\sin^2 x \, dx}{(1 + \cos x)^3}$

(h) $\int \dfrac{dx}{\sqrt{1 - x^2} \, (4 + 5x)}$

2. Use the substitution $t = \tan x$ to integrate:

(a) $\int \dfrac{dx}{4 \tan x - \tan^3 x}$

(b) $\int \dfrac{dx}{1 + 2 \tan x + \tan^2 x}$

3. (a) Let $I_n = \int \cos^n x \, dx$, $n = 0, 1, 2, \ldots$, so that $I_0 = x + C$, $I_1 = \sin x + C$. Show by integration by parts that one has the recursion formula

$$I_n = \frac{1}{n} \cos^{n-1} x \sin x + \frac{n-1}{n} I_{n-2}; \qquad n \geq 2$$

and find I_2, I_3, I_4.

(b) Find a recursion formula for $J_n = \int \sin^n x \, dx$.

(c) Show from the result of (a) that

$$\int_0^{\pi/2} \cos^n x \, dx = \begin{cases} \dfrac{\pi}{2} \cdot \dfrac{1}{2} \cdot \dfrac{3}{4} \cdot \dfrac{5}{6} \cdots \dfrac{n-1}{n}, & (n \text{ even}) \\[3mm] \dfrac{2}{3} \cdot \dfrac{4}{5} \cdot \dfrac{6}{7} \cdots \dfrac{n-1}{n}, & (n \text{ odd}) \end{cases}$$

(d) Show, without evaluating the integrals, that

$$\int_0^{\pi/2} \cos^n x \, dx = \int_0^{\pi/2} \sin^n x \, dx, \qquad n = 0, 1, 2, \ldots$$

4. Let $K_n = \int (\sin nx / \cos x) \, dx$, $n = 1, 2, 3, \ldots$.

(a) Prove: $K_n = [2/(1-n)] \cos(n-1) x - K_{n-2}$, $\qquad n = 3, 4, \ldots$.

(b) Find K_1, K_2, K_3, K_4, K_5.

5. Verify

(a) $\displaystyle \int \frac{\sin(5/2) \, x}{\sin(x/2)} \, dx = \int \frac{\sin(5/2) \, x - \sin(3/2) \, x}{\sin(x/2)} \, dx$

$$+ \int \frac{\sin(3/2) \, x - \sin(x/2)}{\sin(x/2)} \, dx + \int \frac{\sin(x/2)}{\sin(x/2)} \, dx$$

$$= 2 \int \cos 2x \, dx + 2 \int \cos x \, dx + \int dx = \sin 2x + 2 \sin x + x + C.$$

(b) $\displaystyle \int \frac{\sin[(2n+1)/2] \, x}{\sin(x/2)} \, dx = x + 2 \left(\sin x + \frac{1}{2} \sin 2x \cdots + \frac{1}{n} \sin nx \right) + C$

(c) $\displaystyle \int \frac{\cos[(2n+1)/2] \, x \, dx}{\sin(x/2)} = 2 \ln \left| \sin \frac{x}{2} \right| + 2 \left(\cos x + \frac{1}{2} \cos 2x \right.$

$$\left. + \cdots + \frac{1}{n} \cos nx \right) + C$$

‡5-7 THE EXPONENTIAL AND LOGARITHMIC FUNCTIONS

We wish here to define precisely a^x and $\log_a x$ and to establish the basic properties of these functions. As in Section 5-1, we put ourselves in the frame of mind of one who has never heard of logarithms and of e or e^x. Accordingly, we return to the definition of a^x from algebra. For $a > 0$, we set

$$a^1 = a, \quad a^2 = a \cdot a, \quad \ldots, \quad a^n = a \cdot a \ldots a \quad (n \text{ factors})$$

$$a^{-n} = \frac{1}{a^n} \ (n = 1, 2, 3, \ldots), \qquad a^0 = 1$$

$$a^{1/n} = \sqrt[n]{a}, \qquad a^{m/n} = [a^{(1/n)}]^m, \qquad n = 1, 2, \ldots, \qquad m = 0, \pm 1, \ldots$$

The existence of the nth root (positive nth root) follows from the fact that $y = x^n$ describes a one-to-one correspondence between the intervals $x > 0$ and $y > 0$. It is proved in elementary algebra that $a^{m/n}$ is unambiguously defined for all rational numbers m/n.

Thus we know a^x for $a > 0$ and x rational, and $a^x > 0$ for all such values.

We further know from algebra that

$$a^{s+t} = a^s a^t, \qquad a^{s-t} = a^s/a^t$$
$$(a^s)^t = a^{st}, \qquad a^s b^s = (ab)^s, \qquad 1^s = 1 \qquad (5\text{-}70)$$

for all rational numbers s, t. We should like to define a^x for all real numbers so that the properties (5-70) continue to hold true.

We are led to our definition in the following way. Instead of considering a^x, we consider the function

$$y = u^x$$

a function of two variables x and u. We want to define this function for all $u > 0$ and all x. We already know the values for each rational x and each $u > 0$; hence, we know u^x on the half-lines $u > 0$, $x = \text{const} = \text{rational}$ number, as shown in Figure 5-15. These half-lines fill up "most" of the half-plane $u > 0$, but not all; there are gaps that consist of lines $x = \text{const} =$ irrational number. It is these gaps in the domain of the function, that we wish to fill by defining the function properly.

We can fill some of the gaps easily, by setting

$$1^x = 1 \qquad \text{for all } x \qquad (5\text{-}71)$$

This defines our function on the line $u = 1$, perpendicular to the previous lines (Figure 5-15).

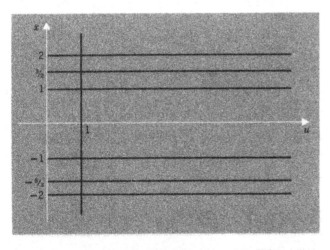

Figure 5-15 Domain for u^x.

Now let us take a fixed irrational x and try to define u^x for all $u > 0$. It is natural to require that, for x fixed, u^x should have a derivative with respect to u given by

$$\frac{dy}{du} = xu^{x-1} \qquad (5\text{-}72)$$

For this holds true when x is rational (Section 3-6). Since $y = u^x$, we more conveniently write (5-72) in the form:

$$\frac{dy}{du} = x\frac{y}{u} \qquad (5\text{-}72')$$

We seek a function $y(u)$ (x being fixed) that satisfies (5-72′) and, because of (5-71), for which

$$y(1) = 1$$

We shall find such a function and thereby define $y = u^x$ on the missing lines parallel to the u-axis in Figure 5-15.

From Equation (5-72′) we proceed formally and obtain

$$\frac{dy}{y} = x \frac{du}{u}, \qquad \int \frac{dy}{y} = x \int \frac{du}{u}$$

Since $y = 1$ for $u = 1$, we can relate y and u in terms of definite integrals:

$$\int_1^y \frac{dt}{t} = x \int_1^u \frac{ds}{s} \tag{5-73}$$

[We have in fact solved the differential equation (5-72′) by separating variables, as in Problem 68 following Section 4-8.] We shall see that (5-73) really gives just the definition we seek. To see this more clearly, we define a "new" function

$$v = \ln y = \int_1^y \frac{dt}{t}, \qquad 0 < y < \infty \tag{5-74}$$

As stated above, we here ignore all previous knowledge of logarithms. By the Fundamental Theorem of Calculus, $\ln y$ is continuous for $y > 0$ and has derivative

$$\frac{d}{dy} \ln y = \frac{1}{y} > 0$$

Hence, $\ln y$ is monotone strictly increasing. Furthermore, it is easily shown from (5-74) that

$$\lim_{y \to \infty} \ln y = \infty, \qquad \lim_{y \to 0+} \ln y = -\infty$$

(Problem 2 below.) Hence, $\ln y$ has as range the whole v-axis, and there is a continuous inverse function, which we denote by $\exp v$:

$$v = \ln y \quad \text{if, and only if,} \quad y = \exp v \tag{5-75}$$

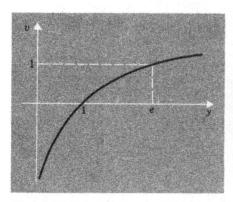 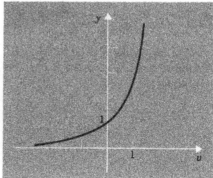

Figure 5-16 $\ln y$ and $\exp v$.

The two functions are graphed in Figure 5-16. Since ln y is monotone strictly increasing, exp v is also and

$$\lim_{v \to -\infty} \exp v = 0, \qquad \lim_{v \to \infty} \exp v = \infty$$

Now we can rewrite (5-73):

$$\ln y = x \ln u$$

or, by (5-75),

$$y = \exp[x \ln u] \tag{5-76}$$

Thus, finally, we have:

Definition. For $u > 0$ and all real x, u^x is defined by the equation

$$u^x = \exp[x \ln u] \tag{5-76'}$$

where

$$\ln u = \int_1^u \frac{ds}{s}$$

and $y = \exp v$ is the inverse of the function $v = \ln y$.

We shall see that for x rational the new functions u^x agrees with u to the power x, as defined in algebra. Hence the function can be consistently de-noted by u^x for all x.

THEOREM 2. *The function* ln x *has derivative* $1/x$ $(x > 0)$; *the function* exp x *has derivative* exp x $(-\infty < x < \infty)$. *For fixed $u > 0$, the function* u^x *is continuous in x for all x. For fixed x, the function u^x is continuous in u for $u > 0$. For $u > 0$ and x rational, u^x agrees with u to the power x, as defined in algebra. The function u^x has the properties:*

(a) $u^{x_1 + x_2} = u^{x_1} u^{x_2}$ (b) $u^{x_1 - x_2} = u^{x_1}/u^{x_2}$

(c) $(u^{x_1})^{x_2} = u^{x_1 x_2}$ (d) $(u_1 u_2)^x = u_1{}^x u_2{}^x$

(e) $1^x = 1$ (f) $\dfrac{d}{du} u^x = x u^{x-1}$ (x fixed)

(g) $\dfrac{d}{dx} u^x = u^x \ln u$ (u fixed)

PROOF. As noted above, the definition (5-74) implies that ln x is con-tinuous for $x > 0$ and has derivative $1/x$. The function $x = \exp y$ is the corre-sponding inverse function and, therefore, has derivative

$$\frac{d}{dy}(\exp y) = \frac{dx}{dy} = \frac{1}{dy/dx} = \frac{1}{1/x} = x = \exp y$$

Thus the function exp x has derivative exp x. The continuity properties of u^x, for fixed x or u, now follow from those of the functions ln u and exp v.

By the chain rule, for fixed x,

$$\frac{d}{du}[\exp(x \ln u)] = \exp(x \ln u) \cdot \frac{x}{u}$$

that is, $y = \exp(x \ln u)$ defines a function of u (x being fixed) for which

$$\frac{dy}{du} = x\frac{y}{u} \qquad (5\text{-}77)$$

(as expected, since that is the differential equation we used to find y). Now, if x is rational, if $y = \exp(x \ln u)$ and if u^x denotes u to the power x as defined in algebra, we have

$$\frac{d}{du}(u^{-x}y) = u^{-x}\frac{dy}{du} - xu^{-x-1}y = u^{-x}\cdot x\frac{y}{u} - xu^{-x-1}y = 0$$

Therefore, $u^{-x}y = \text{const}$ (for fixed x). But for $u = 1$, $y = \exp(x \ln 1) = \exp 0 = 1$, from the definition of ln and exp. Hence

$$u^{-x}y = \text{const} = u^{-x}y|_{u=1} = 1^{-x}\cdot 1 = 1\cdot 1 = 1$$

Therefore, for x rational

$$y = u^x$$

as defined by algebra.

We now know that $y = \exp(x \ln u)$ defines a function, continuous in x and u (for all x and $u > 0$) and, for x rational, this function coincides with u^x as defined by algebra. Hence it makes sense to denote this function by u^x for all real x and for all $u > 0$. The continuity implies that the laws (a), (b), ... hold true for all real x. For example, for fixed u the equation

$$u^{x_1}u^x = u^{x_1+x}$$

is satisfied for fixed rational x_1 and all rational x. But both sides are continuous functions of x for all real x. Hence they coincide for all real x (Problem 3 below). Therefore, the equation

$$u^x u^{x_2} = u^{x+x_2}$$

is true, for each fixed real x_2, for all rational x. Again both sides are continuous in x, so that equality holds true for all real x. But that is precisely rule (a). The proofs of (b), (c), and (d) are left as exercises; (e) follows from

$$1^x = \exp(x \ln 1) = \exp 0 = 1$$

as above. In (5-77) we have $y = \exp(x \ln u) = u^x$, so that by (b)

$$\frac{dy}{dx} = x\frac{u^x}{u} = xu^{x-1}$$

and (f) is proved. For (g) we have

$$\frac{d}{dx}(u^x) = \frac{d}{dx}\exp(x \ln u) = \exp(x \ln u)\cdot \ln u = u^x \ln u$$

by the chain rule.

THEOREM 3. *For fixed $u > 1$, $y = u^x$ is monotone strictly increasing in x, with range the interval $0 < y < \infty$ and $\lim u^x = 0$ as $x \to -\infty$, $\lim u^x = \infty$ as $x \to +\infty$. The inverse function, denoted by $x = \log_u y$,* *is defined and continuous for $y > 0$, is monotone strictly increasing in y, has range $(-\infty, \infty)$ and $\lim \log_u y = -\infty$ as $y \to 0$, $\lim \log_u y = \infty$ as $y \to \infty$. Furthermore*

(h) $\log_u y_1 y_2 = \log_u y_1 + \log_u y_2$

(i) $\log_u \dfrac{y_1}{y_2} = \log_u y_1 - \log_u y_2$

(j) $\log_u y^x = x \log_u y$

(k) $\dfrac{d}{dy} \log_u y = \dfrac{1}{\ln u} \cdot \dfrac{1}{y}$ *(for fixed u)*

(l) $\log_u y \cdot \log_y v = \log_u v$ *for* $y > 1, v > 0$

The number

$$e = \exp 1 = 2.7182818285 \ldots$$

has the properties:

(m) $1 = \displaystyle\int_1^e \dfrac{dt}{t} = \ln e$

(n) $\exp x = e^x$

(o) $\ln x = \log_e x$

(p) $e = \displaystyle\lim_{h \to 0} (1 + h)^{1/h}$

PROOF. From (g), for fixed $u > 1$, u^x has a positive derivative with respect to x, so that $y = u^x$ is monotone strictly increasing and has an inverse function: $x = \log_u y$. In fact, from $y = \exp(x \ln u)$, we have

$$\ln y = x \ln u$$

so that

$$\log_u y = \frac{\ln y}{\ln u}, \qquad y > 0 \tag{5-78}$$

Thus, $\log_u y$ is continuous and monotone increasing as asserted, with derivative as in (k). The proofs of the statements about limits and of (h), (i), (j), (l) are left as exercises.

With $e = \exp 1$, (m) follows, since $\exp$ is the inverse of $\ln$. Now

$$e^x = \exp(x \ln e) = \exp(x)$$

which is (n) and, as above,

$$\log_e x = \frac{\ln x}{\ln e} = \ln x$$

which is (o).

Only (p) remains. To prove this, we use the rule $(\ln x)' = 1/x$ at $x = 1$:

$$\frac{d}{dx} \ln x \bigg|_{x=1} = \lim_{h \to 0} \frac{\ln(1 + h) - \ln 1}{h} = 1$$

Hence

$$\lim_{h \to 0} \frac{1}{h} \ln(1 + h) = \lim_{h \to 0} \ln(1 + h)^{1/h} = 1$$

Since e^x is continuous for all x,

$$\lim_{h \to 0} e^{\ln(1+h)1/h} = e^{\lim \ln(1+h)1/h} = e^1 = e$$

But $e^{\ln u} = \exp(\ln u) = u$, so that our equation can be written

$$\lim_{h \to 0} (1 + h)^{1/h} = e$$

The Numerical Value of e. We can verify (Problem 1 below) that the function

$$f(t) = \begin{cases} (1 + t)^{1/t}, & -1 < t < 0 \quad \text{and} \quad 0 < t < \infty \\ e, & t = 0 \end{cases}$$

is continuous and monotone strictly decreasing. Hence, for $0 < t < 1$, we have

$$(1 + t)^{1/t} < e < (1 - t)^{-1/t}$$

On letting $t = 1/n$, where n is an integer greater than 1, we obtain

$$\left(1 + \frac{1}{n}\right)^n < e < \left(1 - \frac{1}{n}\right)^{-n} \qquad (n = 2, 3, \dots)$$

and the sequence $\{[1 + (1/n)]^n\}$ is monotone strictly increasing with limit e, whereas the sequence $\{[1 - (1/n)]^{-n}\}$ is monotone strictly decreasing with limit e. On taking $n = 10$, we obtain $(1.1)^{10} < e < (.9)^{-10}$ or $2.59 < e < 2.87$.

It is shown in Section 6-12 that

$$e = \lim_{n \to \infty} \left(1 + \frac{1}{1!} + \frac{1}{2!} + \frac{1}{3!} + \cdots + \frac{1}{n!}\right)$$

Again we have a monotone increasing sequence, but one much more rapidly approaching its limit. For $n = 10$, we obtain a value correct to 7 decimal places!

PROBLEMS

1. Let f be the function defined by the following: $x = (1 + t)^{1/t}$ for $t > -1$, $x = e$ for $t = 0$.
   (a) Prove that f is continuous for $t > -1$.
   (b) Prove that $f'(t) = f(t)t^{-2}[t(1 + t)^{-1} - \ln(1 + t)]$ for $t \neq 0$, and hence that $f'(t) < 0$ for $-1 < t < 0$ and for $0 < t < \infty$. (*Hint.* Show that the quantity in square brackets has its maximum value, 0, for $t = 0$.)
   (c) Conclude from (a) and (b) that, for $0 < t < 1$,

   $$(1 + t)^{1/t} < e < (1 - t)^{-1/t}$$

   (d) Take $t = 0.001$ in the inequalities of part (c) to get estimates for e.

2. (a) Use the change of variable $y = u^2$ to show that

$$\int_1^{c^2} \frac{dy}{y} = 2 \int_1^c \frac{du}{u}, \qquad c > 0$$

(b) From the result of (a), show that $\ln c^2 = 2 \ln c$
(c) From the result of (b) and the fact that $\ln y$ is monotone increasing for $y > 0$, conclude that

$$\lim_{y \to \infty} \ln y = \infty, \qquad \lim_{y \to 0+} \ln y = -\infty$$

(d) From the fact that $y = \exp v$ is the inverse of $v = \ln y$ and from the results of part (c), show that $y = \exp v$ is monotone strictly increasing and that

$$\lim_{v \to -\infty} \exp v = 0, \qquad \lim_{v \to \infty} \exp v = \infty$$

(e) From the results of part (d), show that for fixed $u > 1$, the function $y = u^x = \exp(x \ln u)$, $-\infty < x < \infty$, has range $(0, \infty)$ and that

$$\lim_{x \to -\infty} u^x = 0, \qquad \lim_{x \to \infty} u^x = \infty$$

(f) From the results of part (e) and the fact that $x = \log_u y$ is the inverse of $y = u^x$ (u fixed, $u > 1$), show that $x = \log_u y$ has range $(-\infty, \infty)$ and that

$$\lim_{u \to 0+} \log_u y = -\infty, \qquad \lim_{u \to \infty} \log_u y = \infty$$

(g) How are the results in parts (e) and (f) modified if $0 < u < 1$?
3. (a) Prove that, if $f(x)$ is defined and continuous for $a \le x \le b$ and $f(x) = 0$ whenever x is rational, then $f(x) \equiv 0$. {*Hint.* If, for example, $f(x_0) > 0$ for some x_0, then we can find $\delta > 0$ so that $f(x) > 0$ in the interval $(x_0 - \delta, x_0 + \delta)$; now choose a rational number x in this interval and in $[a, b]$ to obtain a contradiction.}
(b) Prove that, if $p(x)$ and $q(x)$ are continuous in $[a, b]$ and $p(x) = q(x)$ whenever x is rational, then $p(x) \equiv q(x)$ in $[a, b]$.
4. Prove the following parts of Theorems 2 and 3.
(a) Part (b) by the method used to prove part (a).
(b) Part (b) as a consequence of part (a).
(c) Part (c) by the method used to prove part (a).
(d) Part (d) by the method used to prove part (a).
(e) Part (h) as a consequence of part (a) and the fact that $u^{\log_u x} = x$, $\log_u u^x = x$.
(f) Part (i).
(g) Part (j).
(h) Part (l) with the aid of (5-78).
5. Find the inverse function and give its domain:
(a) $y = \ln(x + 1)$ (b) $y = \log_2(e^x + 1)$
(c) $y = e^{3 \ln x}$ (d) $y = \log_3 e^{(\ln x)^3}$

†5-8 THE COMPLEX EXPONENTIAL FUNCTION

The trigonometric functions and the exponential function e^x appear to be quite unrelated. However, there is a deep connection between them that is revealed when we consider complex numbers $x + iy$, $i = \sqrt{-1}$.

By a complex-valued function F of a real variable t, we mean a function that assigns a complex number to each t in some interval. Thus, $F(t)=x+iy$, where x and y are real numbers that vary with t; in particular, x and y are real functions of t, so that we can write

$$F(t) = f(t) + ig(t)$$

where f and g are real functions defined on the given interval. We can represent the function F graphically as in Figure 5-17; as t varies, $x + iy$ traces a

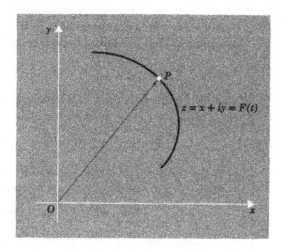

Figure 5-17   Complex function of t.

path in the plane of complex numbers. If we associate with each point P: (x, y) the vector $\overrightarrow{OP}$, then our complex function F is simply another way of describing the vector function $f(t)\mathbf{i} + g(t)\mathbf{j}$.

We can define limits, continuity, derivative, and integral for these complex functions. The development is the same as for vector functions (Sections 3-10, 3-11, and 4-31), and we conclude in particular that $F(t)$ has a limit as $t \to c$ if, and only if, $f(t)$ and $g(t)$ have limits and

$$\lim_{t \to c} F(t) = \lim_{t \to c} f(t) + i \lim_{t \to c} g(t) \tag{5-80}$$

Similarly, F is continuous if, and only if, f and g are continuous; F has a derivative if, and only if, f and g have derivatives. When F is continuous, F has a definite integral. Furthermore

$$F'(t) = f'(t) + ig'(t), \qquad \int_a^b F(t) \, dt = \int_a^b f(t) \, dt + i \int_a^b g(t) \, dt \tag{5-81}$$

For complex functions we have two operations that are not available for vector functions—multiplication and division. We verify that the usual theorems on limits, continuity; and derivatives for products and quotients hold true for complex functions (Problem 8 below).

Now let $F(t) = \cos t + i \sin t$, $-\infty < t < \infty$. This function has the property

$$F(t_1 + t_2) = F(t_1)F(t_2) \tag{5-82}$$

For we verify by multiplication and trigonometry that

$$\cos(t_1 + t_2) + i \sin(t_1 + t_2) = (\cos t_1 + i \sin t_1)(\cos t_2 + i \sin t_2)$$

Now the function e^{at} also satisfies (5-82):

$$e^{a(t_1+t_2)} = e^{at_1}e^{at_2}$$

This suggests that $F(t) = e^{at}$ for some constant a. To find the value of a, we observe that

$$F'(t) = (\cos t)' + i(\sin t)' = -\sin t + i \cos t = i(\cos t + i \sin t) = iF(t)$$

But e^{at} should have derivative ae^{at}. Consequently, we take $a = i$ and *define* the *complex exponential function* of t by the equation:

$$e^{it} = \cos t + i \sin t \tag{5-83}$$

This relation, called the *Euler identity*, can be justified in many other ways.

The complex exponential function has very many applications in mathematics. In particular, many problems concerning trigonometric functions are greatly simplified when expressed in terms of the exponential function. For this purpose we notice that

$$e^{it} = \cos t + i \sin t, \qquad e^{-it} = \cos t - i \sin t$$

so that

$$\cos t = \frac{e^{it} + e^{-it}}{2} = \frac{1}{2}\left(w + \frac{1}{w}\right), \qquad w = e^{it}$$

$$\sin t = \frac{e^{it} - e^{-it}}{2i} = \frac{1}{2i}\left(w - \frac{1}{w}\right), \qquad w = e^{it} \tag{5-84}$$

EXAMPLE 1

$$\int \sin^5 x \, dx = \int \frac{1}{32i}(e^{ix} - e^{-ix})^5 \, dx$$

$$= \int \frac{1}{32i}(e^{5ix} - 5e^{3ix} + 10e^{ix} - 10e^{-ix} + 5e^{-3ix} - e^{-5ix}) \, dx$$

$$= \int \frac{1}{32i}[(e^{5ix} - e^{-5ix}) - 5(e^{3ix} - e^{-3ix}) + 10(e^{ix} - e^{-ix})] \, dx$$

$$= \frac{1}{16}\int (\sin 5x - 5 \sin 3x + 10 \sin x) \, dx$$

$$= -\frac{\cos 5x}{80} + \frac{5 \cos 3x}{48} - \frac{5 \cos x}{8} + C$$

We more generally define

$$e^{x+iy} = e^x e^{iy} = e^x(\cos y + i \sin y) \tag{5-85}$$

These equations define the general *complex exponential function*

$$\exp z = e^z, \qquad z = x + iy$$

a complex-valued function of a complex variable, defined for all complex numbers z. We verify (Problem 3) that

$$e^{z_1 + z_2} = e^{z_1} e^{z_2}, \qquad e^{z_1 - z_2} = e^{z_1}/e^{z_2}, \qquad (e^z)^n = e^{nz}, \; n = 0, \pm 1, \pm 2, \ldots$$

EXAMPLE 2. We verify (Problem 6 below) that, if a and b are real, then

$$\frac{d}{dx}[e^{(a+bi)x}] = \frac{d}{dx}[e^{ax} \cos bx + ie^{ax} \sin bx] = (a + bi)e^{(a+bi)x}$$

and, therefore, that

$$\int e^{(a+bi)x} dx = \frac{e^{(a+bi)x}}{a + bi} + C \qquad (a + bi \neq 0)$$

(C = complex constant). We can compare real and imaginary parts here (Section 0-17). The left side equals

$$\int [e^{ax} \cos bx + ie^{ax} \sin bx]dx = \int e^{ax} \cos bx \, dx + i \int e^{ax} \sin bx \, dx$$

Therefore, on comparing real parts,

$$\int e^{ax} \cos bx \, dx = \text{Re}\left(\frac{e^{(a+b)ix}}{a + bi}\right) + \text{const}$$

$$= \text{Re}\left(\frac{e^{ax}(\cos bx + i \sin bx)(a - bi)}{(a + bi)(a - bi)}\right) + \text{const}$$

$$= \frac{e^{ax}(a \cos bx + b \sin bx)}{a^2 + b^2} + \text{const}$$

Similarly, on comparing imaginary parts,

$$\int e^{ax} \sin bx \, dx = \text{Im}\left(\frac{e^{(a+bi)x}}{a + bi}\right) + \text{const} = \frac{e^{ax}(a \sin bx - b \cos bx)}{a^2 + b^2} + \text{const}$$

These are useful integration rules (Nos. 95 and 96 in Table I in the Appendix).

EXAMPLE 3. The equation $z = \rho e^{i\omega t}$, where ρ and ω are real constants and t is real, is equivalent to the equations

$$x = \rho \cos \omega t, \qquad y = \rho \sin \omega t$$

describing motion on a circular path: $x^2 + y^2 = \rho^2$, and is an alternative to a vector representation $\overrightarrow{OP} = \rho \cos \omega t \, \mathbf{i} + \rho \sin \omega t \, \mathbf{j}$. The derivatives

$$\frac{dz}{dt} = \rho i \omega e^{i\omega t}, \qquad \frac{d^2 z}{dt^2} = -\rho \omega^2 e^{i\omega t}$$

represent the velocity and acceleration vectors, respectively.

†5-9 HYPERBOLIC FUNCTIONS

The hyperbolic cosine: cosh x, hyperbolic sine: sinh x, and the hyperbolic tangent: tanh x are defined first as real functions of a real variable by the equations:

$$\cosh x = \frac{e^x + e^{-x}}{2}, \qquad \sinh x = \frac{e^x - e^{-x}}{2}$$

$$(5\text{-}90)$$

$$\tanh x = \frac{e^x - e^{-x}}{e^x + e^{-x}} = \frac{\sinh x}{\cosh x}$$

All three functions are defined for $-\infty < x < \infty$ and are clearly continuous and differentiable:

$$\frac{d}{dx} \cosh x = \sinh x, \qquad \frac{d}{dx} \sinh x = \cosh x$$

$$(5\text{-}91)$$

$$\frac{d}{dx} \tanh x = \frac{1}{\cosh^2 x}$$

(In the last case we can also write $\operatorname{sech}^2 x$, in terms of a hyperbolic secant, but the sech and cosech are rarely used.)

From (5-90) and (5-91) the graphs of these functions are readily sketched (Figure 5-18). A detailed discussion is left to the problems (Problems 10, 11, and 12 below). The hyperbolic functions satisfy a set of identities, for example,

$$\cosh^2 x - \sinh^2 x = 1 \qquad (5\text{-}92)$$
$$\cosh(x + y) = \cosh x \cosh y + \sinh x \sinh y \qquad (5\text{-}93)$$
$$\sinh(x + y) = \sinh x \cosh y + \cosh x \sinh y \qquad (5\text{-}94)$$
$$\cosh(-x) = \cosh x, \qquad \sinh(-x) = -\sinh x \qquad (5\text{-}95)$$

The proofs are left as exercises (Problem 13 below).

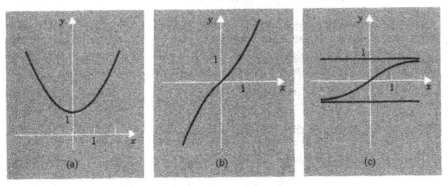

(a) (b) (c)

Figure 5-18 Hyperbolic functions: (a) cosh x, (b) sinh x, (c) tanh x.

Since $\sinh x$ and $\tanh x$ are monotone strictly increasing for all x, and $\cosh x$ is monotone strictly increasing for $x \geq 0$, we can define inverses of these functions, and we have then the functions

$$y = \sinh^{-1} x, \qquad -\infty < x < \infty$$
$$y = \operatorname{Cosh}^{-1} x, \qquad 1 \leq x < \infty \qquad (5\text{-}96)$$
$$y = \tanh^{-1} x, \qquad -1 < x < 1$$

Here $\operatorname{Cosh}^{-1} x$ is the *principal value* and $\operatorname{Cosh}^{-1} x \geq 0$ for $1 \leq x < \infty$. The domain in each case is the range of the original function (Problem 14 below).

Since $y = \sinh^{-1} x$ if $x = \sinh y$, we have

$$\frac{dy}{dx} = \frac{1}{dx/dy} = \frac{1}{\cosh y} = \frac{1}{\sqrt{1 + \sinh^2 y}} = \frac{1}{\sqrt{1 + x^2}}$$

Thus

$$\frac{d}{dx}(\sinh^{-1} x) = \frac{1}{\sqrt{1 + x^2}}, \qquad -\infty < x < \infty \qquad (5\text{-}97)$$

Similarly, we find that (Problem 15 below)

$$\frac{d}{dx} \operatorname{Cosh}^{-1} x = \frac{1}{\sqrt{x^2 - 1}}, \qquad 1 < x < \infty$$

$$\frac{d}{dx} \tanh^{-1} x = \frac{1}{1 - x^2}, \qquad -1 < x < 1 \qquad (5\text{-}98)$$

These relations are useful in integration:

$$\int \frac{dx}{\sqrt{1 + x^2}} = \sinh^{-1} x + C, \qquad \int \frac{dx}{\sqrt{x^2 - 1}} = \operatorname{Cosh}^{-1} x + C,$$

$$\int \frac{dx}{1 - x^2} = \tanh^{-1} x + C$$

[For $x < -1$, $1/\sqrt{x^2 - 1}$ has indefinite integral $-\operatorname{Cosh}^{-1}(-x)$; for $|x| > 1$, $1/(1 - x^2)$ has indefinite integral $\tanh^{-1}(1/x)$.]

†5-10 RELATION BETWEEN HYPERBOLIC FUNCTIONS AND TRIGONOMETRIC FUNCTIONS

The striking resemblance between the identities for trigonometric functions and those for hyperbolic functions leads one to suspect that the two types of functions are closely related. The relationship is found through complex functions. In Section 5-8 we observed [Equation (5-84)] that

$$\cos y = \frac{e^{iy} + e^{-iy}}{2}, \qquad \sin y = \frac{e^{iy} - e^{-iy}}{2i}$$

This suggests *defining, for every complex number* z,

$$\cos z = \frac{e^{iz} + e^{-iz}}{2}, \qquad \sin z = \frac{e^{iz} - e^{-iz}}{2i} \qquad (5\text{-}100)$$

These definitions are meaningful because [by Equation (5-85)] we already have defined e^{a+bi}. Similarly, from (5-90), we are led to write, for every complex z,

$$\cosh z = \frac{e^z + e^{-z}}{2}, \qquad \sinh z = \frac{e^z - e^{-z}}{2} \qquad (5\text{-}101)$$

The definitions (5-100) and (5-101) are the standard ones for the *complex trigonometric functions and complex hyperbolic functions.*

From the definitions we notice at once that

$$\cos z = \cosh iz, \qquad \sin z = \frac{1}{i} \sinh iz \qquad (5\text{-}102)$$

These are the desired relations. Now we can verify that the identity

$$\cos^2 z + \sin^2 z = 1 \tag{5-103}$$

is valid for all complex z. Hence from (5-102)

$$\cosh^2 iz - \sinh^2 iz = 1$$

or, since every complex number can be written in the form iz,

$$\cosh^2 z - \sinh^2 z = 1 \tag{5-104}$$

for all complex z. We have thus shown how a trigonometric identity becomes a hyperbolic identity. The other familiar identities can be treated in the same way.

PROBLEMS

1. Evaluate: (a) $e^{\pi i}$ (b) e^{2i} (c) $e^{2\pi i}$ (d) $e^{1+\pi i}$ (e) e^{-2-i}
2. Verify that the definition (5-85) of e^z agrees with the meaning of e^z when z is real $(z = x = x + 0i)$.
3. Prove that the complex exponential function satisfies the laws:
   (a) $e^{z_1+z_2} = e^{z_1} \cdot e^{z_2}$ (Hint. Write $z_1 = x_1 + iy_1$, $z_2 = x_2 + iy_2$.)
   (b) $e^{z_1-z_2} = e^{z_1}/e^{z_2}$
   (c) $(e^z)^n = e^{nz}$, $n = 0, 1, 2, \ldots$ (Hint. Use induction)
   (d) $(e^z)^n = e^{nz}$, $n = -1, -2, \ldots$
   (e) $e^{z+2\pi i} = e^z$
4. Prove that $e^z \neq 0$ for all complex z.
5. Find all z such that
   (a) $\cos z = 0$ (b) $\sin z = 0$
6. Prove, on the basis of the rule: $[f(x) + ig(x)]' = f'(x) + ig'(x)$, that $[e^{(a+bi)x}]' = (a + bi)e^{(a+bi)x}$, $(a + bi = \text{constant})$.
7. Let a, b, c be real constants, $a + bi \neq 0$; explain how each of the integrals can be found:
   (a) $\int xe^{ax} \cos bx \, dx$ (b) $\int xe^{ax} \sin bx \, dx$
   (c) $\int x^2 e^{ax} \cos bx \, dx$ (d) $\int x^2 e^{ax} \sin bx \, dx$
   (e) $\int e^{ax} \sin bx \cos x \, dx$ (f) $\int \cos ax \cos bx \cos cx \, dx$, $(a \neq b, b \neq c, a \neq c)$
   (g) $\int \cos^2 ax \cos^2 bx \, dx$, $a \neq b$ (h) $\int e^{ax} \sin bx \cos^2 cx \, dx$
8. Let $F(t) = f(t) + ig(t)$, $G(t) = p(t) + iq(t)$ be complex functions defined for $a \leq t \leq b$, and let $a < c < b$. Prove the following:
   (a) If F and G have limits as $t \to c$, then so does FG and

$$\lim_{t \to c} F(t)G(t) = \{\lim_{t \to c} F(t)\}\{\lim_{t \to c} G(t)\}$$

   [Hint. Multiply out $FG = (f + ig)(p + iq)$ and use the rule (5-80).]
   (b) If $F'(c)$ and $G'(c)$ exist, then FG has a derivative at c, equal to $F(c)G'(c) + F'(c)G(c)$. [Hint. Use (5-81).]
9. Graph the complex functions of t as paths in the z-plane:
   (a) $z = 2e^{3it}$, $-\pi \leq t \leq \pi$ (b) $z = te^{it}$, $0 \leq t \leq 4\pi$
   (c) $z = e^{(1+2i)t}$, $0 \leq t \leq 2\pi$ (d) $z = e^{(-1+2i)t}$, $0 \leq t \leq 2\pi$
   (e) $z = e^{(-1-i)t}$, $0 \leq t \leq 2\pi$

10. Establish the following properties of the function $y = \cosh x$:

(a) It is monotone strictly decreasing for $x \leq 0$, monotone strictly increasing for $x \geq 0$.

(b) Its graph is symmetric with respect to the y-axis.

(c) $\lim\limits_{x \to -\infty} \cosh x = \infty$, $\qquad \lim\limits_{x \to \infty} \cosh x = \infty$.

(d) $e^x/2 < \cosh x < e^x$ for $x > 0$.

(e) Its absolute minimum is 1, taken only for $x = 0$.

(f) Its range is the interval $[1, \infty)$.

11. Establish the following properties of the function $y = \sinh x$.

(a) It is monotone strictly increasing for $-\infty < x < \infty$.

(b) Its graph is symmetric with respect to the origin.

(c) $\lim\limits_{x \to \infty} \sinh x = \infty$, $\qquad \lim\limits_{x \to -\infty} \sinh x = -\infty$.

(d) $(e^x - 1)/2 < \sinh x < e^x/2$ for $x > 0$.

(e) Its range is the interval $(-\infty, \infty)$.

12. With the aid of the properties listed in Problems 10 and 11, establish the following properties of the function $y = \tanh x$:

(a) It is monotone strictly increasing for $-\infty < x < \infty$.

(b) Its graph is symmetric with respect to the origin.

(c) $\lim\limits_{x \to \infty} \tanh x = 1$, $\qquad \lim\limits_{x \to -\infty} \tanh x = -1$.

(d) Its range is the interval $(-1, 1)$.

13. Prove the identities:

(a) (5-92)　　　(b) (5-93)　　　(c) (5-94)　　　(d) (5-95)

(e) $\cosh 2x = \cosh^2 x + \sinh^2 x$　　　(f) $\sinh 2x = 2 \sinh x \cosh x$

(g) $\dfrac{1}{2} + \cosh x + \cdots + \cosh nx = \dfrac{e^{(n+1)x} - e^{-nx}}{2(e^x - 1)}$, $\qquad x \neq 0$

(h) $\dfrac{1}{2} + \cosh x + \cdots + \cosh nx = \dfrac{\sinh(n + \frac{1}{2})x}{2 \sinh \frac{1}{2}x}$, $\qquad x \neq 0$

14. (a) Graph the inverse hyperbolic functions (5-96), with the aid of the results of Problems 10, 11, and 12.

(b) Prove that $\sinh^{-1} x = \ln(x + \sqrt{x^2 + 1})$, $\qquad -\infty < x < \infty$

(c) Prove that $\cosh^{-1} x = \ln(x + \sqrt{x^2 - 1})$, $\qquad x \geq 1$

(d) Prove that $\tanh^{-1} x = \frac{1}{2} \ln[(1 + x)/(1 - x)]$, $\qquad -1 < x < 1$

15. Prove:(a) $\dfrac{d}{dx} \cosh^{-1} x = \dfrac{1}{\sqrt{x^2 - 1}}$, $\quad 1 < x < \infty$

(b) $\dfrac{d}{dx} \cosh^{-1}(-x) = \dfrac{-1}{\sqrt{x^2 - 1}}$, $\quad x < -1$

(c) $\dfrac{d}{dx} \tanh^{-1} x = \dfrac{1}{1 - x^2}$, $\quad -1 < x < 1$

(d) $\dfrac{d}{dx} \tanh^{-1}\left(\dfrac{1}{x}\right) = \dfrac{1}{1 - x^2}$, $\quad |x| > 1$

16. Integrate:

(a) $\int \dfrac{dx}{(x^2 - 1)^{3/2}}$, $x > 1$.　　　(Hint. Set $x = \cosh t$.)

(b) $\int \dfrac{dx}{(1 - x^2)^2}$, $-1 < x < 1$.   (*Hint.* Set $x = \tanh t$.)

(c) $\int (1 + x^2)^{3/2}\, dx$.   (*Hint.* Set $x = \sinh t$.)

17. Prove from (5-100) and the results of Problem 3 that for all complex z, z_1, z_2:

    (a) $\cos^2 z + \sin^2 z = 1$

    (b) $\cos(z_1 + z_2) = \cos z_1 \cos z_2 - \sin z_1 \sin z_2$

    (c) $\sin(z_1 + z_2) = \sin z_1 \cos z_2 + \cos z_1 \sin z_2$

    (d) $\sin(z + 2\pi) = \sin z$, $\cos(z + 2\pi) = \cos z$

    (e) $\sin[(\pi/2) - z] = \cos z$

18. Use (5-102) and the identities of Problem 17(b) and (c) to prove that

    (a) $\cosh(z_1 + z_2) = \cosh z_1 \cosh z_2 + \sinh z_1 \sinh z_2$

    (b) $\sinh(z_1 + z_2) = \sinh z_1 \cosh z_2 + \cosh z_1 \sinh z_2$

19. Evaluate: (a) $\cos(1 + i)$ (b) $\sin 2i$ (c) $\cosh(\pi i)$ (d) $\tanh(\pi i/2)$

20. Show that $x = a \cosh t$, $y = b \sinh t$, $-\infty < t < \infty$, is a set of parametric equations for one half of a hyperbola. (Hence the name "hyperbolic functions.")

‡5-11 CLASSIFICATION OF FUNCTIONS

For real functions of a real variable we can introduce a classification based on the operations used to define the function. The *polynomial functions* are obtained from the functions 1 (a constant function) and x (the identity function) by repeated applications of the operations of addition, multiplication, and multiplication by constants. The *rational functions* are obtained in the same way from the functions 1 and x, by the operations mentioned plus the operation of division. An *algebraic function* is a continuous function on an interval defined implicitly by a polynomial equation in x and y:

$$a_1 x^{n_1} y^{m_1} + a_2 x^{n_2} y^{m_2} + \cdots + a_k x^{n_k} y^{m_k} = 0 \tag{5-110}$$

$(n_1, n_2, \ldots$, and $m_1, m_2, \ldots$ being nonnegative integers). Thus every rational function and every continuous inverse of a rational function is algebraic. For example, $y = x^{1/3}$ is an algebraic function defined implicitly by the equation $y^3 - x = 0$.

The functions that are not algebraic are called *transcendental*. It can be verified that the trigonometric functions, the function e^x, and the function $\ln x$ are transcendental; that is, for each of these functions $y = f(x)$ satisfies no algebraic equation of form (5-110). For example, in the case of e^x, if we write

$$f_1(x) = x^{n_1} e^{m_1 x}, \qquad f_2(x) = x^{n_2} e^{m_2 x}, \ldots, \qquad f_k(x) = x^{n_k} e^{m_k x},$$
$$-\infty < x < \infty \tag{5-111}$$

where $n_i \neq n_j$ whenever $m_i = m_j$, then there are no constants $a_1, \ldots, a_k$ other than $0, \ldots, 0$ so that

$$a_1 f_1(x) + a_2 f_2(x) + \cdots + a_k f_k(x) \equiv 0, \qquad -\infty < x < \infty$$

that is, the functions (5-111) are linearly independent, no matter how the n's and m's are chosen as nonnegative integers. In general, the fact that the function $y(x)$ is transcendental is equivalent to the linear independence of certain sets of functions of form $x^{n_i}[y(x)]^{m_j}$ on the appropriate interval. For e^x, $\ln x$ and the trigonometric functions, the proof is left to Problems 2 and 3 below.

The trigonometric functions e^x, $\ln x$, and the nonalgebraic functions obtained from them by addition, subtraction, multiplication, division, multiplication by constants, composition, and formation of inverse functions are all called *elementary transcendental functions*. Thus, in particular,

$$\text{Sin}^{-1} x, \qquad \ln \tan x, \qquad e^{\cos x}$$

are elementary transcendental functions (one has to verify that they are not algebraic). Most of the functions encountered in the calculus and its applications are either algebraic or elementary transcendental.

In going beyond algebraic functions to create transcendental functions, the crucial step is the *limit process*. In our definition of $\sin x$, $\cos x$, and e^x this has appeared mainly through the definite integral. Thus the sine and cosine arose naturally through studying the arc length along a circle, and the arc length is the limit of the lengths of certain inscribed polygons—hence, finally a definite integral. The definition of e^x and $\ln x$ depends on the integral:

$$\int_1^x \frac{1}{u}\, du$$

There are other limit processes which define transcendental functions. For example, as shown in Section 6-11,

$$e^x = \lim_{n \to \infty} \left(1 + x + \frac{x^2}{2!} + \cdots + \frac{x^n}{n!}\right)$$

or, as we usually write,

$$e^x = 1 + x + \frac{x^2}{2!} + \cdots + \frac{x^n}{n!} + \cdots$$

This is a representation of e^x as sum of an infinite series, a power series.

There are also many other transcendental functions definable by integration: for example, elliptic functions, Bessel functions, and the Gamma function.

All the functions thus far mentioned in this section can be represented in suitable intervals by power series like that given for e^x. We shall study such functions in detail in Chapter 8. Functions so representable are called *analytic functions*. The power series also serves to give meaning to the function for complex values. For example, we can obtain e^z, for z complex, by the equation:

$$e^z = 1 + \frac{z}{1!} + \frac{z^2}{2!} + \cdots + \frac{z^n}{n!} + \cdots = \lim_{n \to \infty} \left(1 + \frac{z}{1!} + \frac{z^2}{2!} + \cdots + \frac{z^n}{n!}\right)$$

The study of such analytic functions of a complex variable is a major branch of mathematics.

PROBLEMS

1. Show that the following functions are algebraic:

   (a) $y = x/(x^2 + 1)$, $-\infty < x < \infty$ (b) $y = x/(x^3 - 1)$, $x \neq 1$

   (c) $y = \sqrt{x^2 + 1}$, $-\infty < x < \infty$ (d) $y = \sqrt[3]{x^3 - x}$

   (e) $y = x + \sqrt{x}, x > 0$ (f) $y = |x| = \sqrt{x^2}$

   (g) $y = \sqrt{x + 1} + \sqrt{x - 1}$, $x > 1$ (h) $y = x - \sqrt[3]{x}$

2. Consider the following functions as members of the vector space of all functions continuous on a given interval.

   (a) Show that $f_1(x) = e^x$ and $f_2(x) = e^{2x}$ are linearly independent: that is, that
   $$c_1 e^x + c_2 e^{2x} \equiv 0$$
   is possible, for constant c_1, c_2, only if $c_1 = 0$, $c_2 = 0$.

   (b) Show that $f_1(x) = xe^x$, $f_2(x) = x^2 e^x$, $f_3(x) = e^{2x}$ are linearly independent.

   (c) Let $f_1(x) = p_1(x)e^x$, $f_2(x) = p_2(x)e^{2x}$, where $p_1(x)$ and $p_2(x)$ are polynomials. Show that $p_1(x)e^x + p_2(x)e^{2x} \equiv 0$ is possible only if $p_1(x) \equiv 0$, $p_2(x) \equiv 0$. What does this imply in terms of linear independence? {*Hint.* Divide the given equation by e^x and differentiate repeatedly to obtain an equation of the form $q(x)e^x \equiv 0$, where $q(x)$ is a polynomial of the same degree as $p_2(x)$.]

   (d) Prove that, if $p_0(x)$, $p_1(x)$, $\dots$, $p_k(x)$ are polynomials, then
   $$(^\circ) \quad p_0(x) + p_1(x)e^x + \cdots + p_k(x)e^{kx} \equiv 0$$
   is possible only if $p_0(x) \equiv 0$, $p_1(x) \equiv 0$, $\dots$, $p_k(x) \equiv 0$, and show that this implies that e^x is transcendental. [*Hint.* Use induction. If the identity $(^\circ)$ holds true for a particular k, differentiate repeatedly to eliminate $p_0(x)$, but show that differentiation does not affect the degrees of the coefficients of $e^x, \dots, e^{kx}$; then divide by e^x and apply the induction hypothesis.]

   (e) Prove that, if $p_0(x)$, $p_1(x)$, $\dots$, $p_k(x)$ are polynomials and α_0, α_1, $\dots$, α_k are distinct constants, then
   $$(^{\circ\circ}) \quad p_0(x)e^{\alpha_0 x} + p_1(x)e^{\alpha_1 x} + \cdots + p_k(x)e^{\alpha_k x} \equiv 0$$
   is possible only if $p_0(x) \equiv 0$, $p_1(x) \equiv 0$, $\dots$, $p_k(x) \equiv 0$ and state implications of this result for linear independence and transcendentality of functions.

   (f) Show that the result and proof for part (e) remain valid when $p_0(x)$, $\dots$, $p_k(x)$ are polynomials with *complex* coefficients and α_0, α_1, $\dots$, α_k are *complex* constants.

   (g) Prove that the functions $\sin x$ and $\cos x$ are transcendental. [*Hint.* See part (f).]

   (h) Prove that the functions 1, $\cos x$, $\cos 2x$, $\dots$, $\cos nx$ are linearly independent. [*Hint.* As in (5-84), write $\cos kx = \frac{1}{2}(w^k + w^{-k})$, $w = e^{ix}$, and rewrite a relation $c_0 + c_1 \cos x + \cdots + c_n \cos nx = 0$ as an algebraic equation in w.]

   (i) Prove that the functions 1, $\cos x$, $\sin x$, $\cos 2x$, $\sin 2x$, $\dots$, $\cos(m + n)x$, $\sin(m + n)x$ form a basis for the vector space of all linear combinations of the functions $\sin^p x \cos^q x$, where p, q are integers and $0 \leq p \leq m$, $0 \leq q \leq n$. [*Hint.* See hint for part (h).]

3. Prove that the function $\ln x$ is transcendental. [*Hint.* If $a_1 x^{n_1}(\ln x)^{m_1} + \cdots + a_k x^{n_k}(\ln x)^{m_k} \equiv 0$, let $t = \ln x$ and consider the resulting identity in functions of t.]

4. Prove that (a) a continuous inverse of an algebraic function is an algebraic function, and that (b) a continuous inverse of a transcendental function is a transcendental function.

6

APPLICATIONS OF
DIFFERENTIAL CALCULUS

Applications of the derivative are considered at various points in Chapters 3 and 5—in particular, for finding tangent lines (Section 3-13), for finding maxima and minima (Section 3-19), and for computation of functional values (Section 3-24). In this chapter these and other applications are considered more thoroughly. Specifically, we present new techniques for graphing a function.

6-1 TESTS FOR MAXIMA AND MINIMA

Let the function f be differentiable in the interval $[a, b]$. To find the absolute maximum M and minimum m of f, we first find all points x for which $f'(x) = 0$; each such point is called a *critical point* of f. We know from the theory of Section 3-19 that the absolute maximum M and minimum m must be taken on either at a critical point or at an end point; also that every *local* maximum or minimum of f inside the interval must occur at a critical point. A local maximum or minimum is also called a *relative* maximum or minimum. We shall use both terms.

If x_0 is a critical point inside the interval and if, for h sufficiently small,

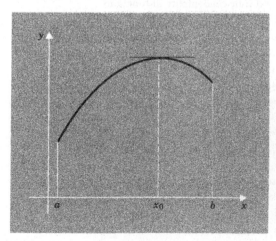

Figure 6-1   Local maximum at x_0.

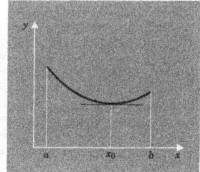

Figure 6-2   Local minimum at x_0.

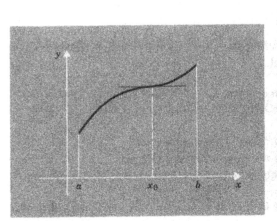

Figure 6-3 Horizontal inflection point at x_0.   **Figure 6-4** Function of Example 1.

f' is positive for $x_0 - h < x < x_0$ and negative for $x_0 < x < x_0 + h$, then f must first increase and then decrease. Thus f has a local maximum at x_0, as in Figure 6-1. Similarly, if f' is negative for $x_0 - h < x < x_0$ and positive for $x_0 < x < x_0 + h$, then f has a local minimum at x_0 (Figure 6-2). If f' is positive in both intervals or negative in both, then f has a *horizontal inflection point* at x_0. The case of f' positive is illustrated in Figure 6-3.

In many cases, there is exactly one point x_0 inside $[a, b]$ at which f has a local maximum or minimum. In such a case, f takes on its *absolute* maximum or minimum at x_0. For simplicity, we state the rule for the case of a maximum:

THEOREM 1. *Let f be continuous in $[a, b]$ and let f have a local maximum at x_0, $a < x_0 < b$. Let f have no other local maxima or minima inside $[a, b]$. Then $f(x_0) = M$, the absolute maximum of f, and $f(x) < M$ for $x \neq x_0$ in $[a, b]$.*

PROOF. Let x_1 be in $[a, b]$, $x_1 > x_0$. Then f has an absolute minimum in the interval $[x_0, x_1]$. This could occur at x_0 only if f were constant near x_0, but this cannot happen since f has no local maxima or minima in (a, b) other than x_0. The absolute minimum cannot occur between x_0 and x_1 for the same reason. Hence it occurs only at x_1. Therefore $f(x_1) < f(x_0)$. Similarly, $f(x_1) < f(x_0)$ for $x_1 < x_0$ and hence $f(x) < f(x_0)$ for $x \neq x_0$ in $[a, b]$. Therefore $f(x_0) = M$, the absolute maximum of f.

EXAMPLE 1. Let $f(x) = -2x^2 + 5x - 4$, $1 \leq x \leq 2$. Here $f'(x) = -4x + 5 = -4[x - (\frac{5}{4})]$ and the sole critical point is $x = \frac{5}{4}$, which is inside our interval. Now $f'(x) > 0$ for $x < \frac{5}{4}$, $f'(x) < 0$ for $x < \frac{5}{4}$. Consequently, there is a local maximum at $\frac{5}{4}$. By Theorem 1, $f(x_0) = -\frac{7}{8} = M$, the absolute maximum of f. The function is graphed in Figure 6-4.

EXAMPLE 2. $f(x) = e^{2x} + e^{-3x}$, $-\infty < x < \infty$. Here $f'(x) = 2e^{2x} -$

$3e^{-3x} = e^{-3x}(2e^{5x} - 3)$. Thus there is one critical point, where $2e^{5x} = 3$ or $x = x_0 = (\frac{1}{5}) \ln (\frac{3}{2}) = 0.081$. Also $f'(x) < 0$ for $x < x_0$, $f'(x) > 0$ for $x > x_0$. Hence, in each interval $[a, b]$ containing x_0, $f(x) > f(x_0)$ for $x \neq x_0$. Since this holds true for *all* such intervals $[a, b]$, $f(x_0)$ must be the absolute minimum of f on the given *infinite interval*.

Remark 1. Example 2 shows that we can apply the principle of Theorem 1 to an arbitrary type of interval.

THEOREM 2. (*Second derivative test*). *Let f be defined and differentiable in the interval* $[a, b]$ *and let* $a < x_0 < b$. *Let* $f'(x_0) = 0$ *and* $f''(x_0)$ *exist. Then*

(a) *if* $f''(x_0) < 0$, *f has a relative maximum at* x_0,
(b) *if* $f''(x_0) > 0$, *f has a relative minimum at* x_0,
(c) *if* $f''(x_0) = 0$, *f may have a relative maximum or a relative minimum or neither at* x_0.

In cases (a) *and* (b) *there is an interval* $(x_0 - \delta, x_0 + \delta)$ *containing no critical point of f other than* x_0.

PROOF. In case (a) $f''(x_0) < 0$. But, since $f'(x_0) = 0$,

$$f''(x_0) = \lim_{h \to 0} \frac{f'(x_0 + h) - f'(x_0)}{h} = \lim_{h \to 0} \frac{f'(x_0 + h)}{h}$$

Hence for $|h|$ sufficiently small and not 0, say $0 < |h| < \delta$, the fraction $f'(x_0 + h)/h$ must be negative. Thus

$$f'(x_0 + h) < 0, \text{ for } 0 < h < \delta$$
$$f'(x_0 + h) > 0, \text{ for } -\delta < h < 0$$

Accordingly, f' changes sign at x_0 and there is a local maximum at x_0.

The discussion of case (b) is similar (Problem 17 below). For case (c) we consider the three examples of Figure 6-5: $y = x^3$, $y = x^4$, $y = -x^4$. In each case $f(0) = 0$, $f'(0) = 0$, and $f''(0) = 0$. The first has a horizontal inflection point at $x = 0$, the second a relative minimum, and the third a relative maximum. Hence, all three can arise when $f''(x_0) = 0$.

In case (c) further information can be obtained from higher derivatives (Problems 18 and 19 below).

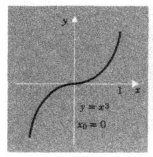 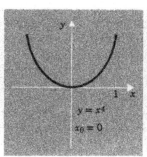 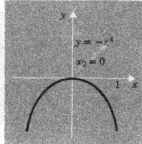

Figure 6-5   The functions $y = x^3$, $y = x^4$, $y = -x^4$.

EXAMPLE 3. (*Intermolecular potential*). The potential energy U of two molecules at distance r is often represented as follows:

$$U = \frac{a}{r^n} - \frac{b}{r^m} = f(r), \qquad 0 < r < \infty$$

where a, b, n, and m are positive constants and $n > m$. Here

$$f'(r) = \frac{-na}{r^{n+1}} + \frac{mb}{r^{m+1}} = \frac{-na + mbr^{n-m}}{r^{n+1}}$$

$$f''(r) = \frac{n(n+1)a}{r^{n+2}} - \frac{m(m+1)b}{r^{m+2}} = \frac{n(n+1)a - m(m+1)br^{n-m}}{r^{n+2}}$$

Hence there is but one critical point r_0:

$$r_0 = \left(\frac{na}{mb}\right)^{1/(n-m)} > 0$$

We now find that

$$f''(r_0) = na(n-m)\left(\frac{mb}{na}\right)^{(n+2)/(n-m)} > 0$$

Therefore by Theorem 2 the function f has a local minimum at r_0. By Theorem 1, f has its absolute minimum at r_0 alone:

$$f(r_0) = \frac{a(m-n)}{mr_0^n} < 0$$

Since $f'(r)$ is continuous and f' has but the one zero r_0, we conclude also that $f'(r) < 0$ for $0 < r < r_0$, $f'(r) > 0$ for $r > r_0$. Consequently, f is monotone strictly decreasing for $r < r_0$ and strictly increasing for $r > r_0$. We verify also (Problem 8 below) that

$$\lim_{r \to 0+} f(r) = \infty, \qquad \lim_{r \to \infty} f(r) = 0$$

so that f has a graph, as in Figure 6-6.

EXAMPLE 4. $y = \sin x + \sin 3x = f(x)$, $0 \le x \le 2\pi$. Here $f'(x) = \cos x + 3 \cos 3x$. By trigonometry (Appendix, Table V), $\cos 3x = 4 \cos^3 x - 3 \cos x$.

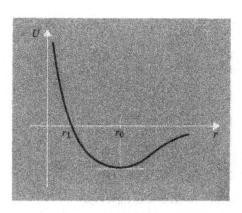

Figure 6-6 Intermolecular potential.

Hence

$$f'(x) = 12 \cos^3 x - 8 \cos x = 4 \cos x (3 \cos^2 x - 2)$$

and $f'(x) = 0$ when $\cos x = 0$—that is, $x = x_1 = \pi/2$, $x = x_2 = 3\pi/2$—and when $\cos x = \pm\sqrt{2/3}$. For the latter case we obtain 4 values of x, which we number as x_3, x_4, x_5, x_6, with $0 < x_3 < \pi/2 < x_4 < \pi < x_5 < 3\pi/2 < x_6 < 2\pi$. Now

$$f''(x) = -36 \cos^2 x \sin x + 8 \sin x = 4 \sin x \, (2 - 9 \cos^2 x)$$

Consequently, $f''(x_1) = f''(\pi/2) = 8$, $f''(x_2) = -8$ and, at each of the points x_3, x_4, x_5, x_6,

$$f''(x) = 4(\sin x)\left(2 - 9 \cdot \frac{2}{3}\right) = -16 \sin x$$

Hence, $f''(x_3) < 0$, $f''(x_4) < 0$, $f''(x_5) > 0$, $f''(x_6) > 0$. Therefore, f has a relative maximum at each of x_2, x_3, x_4 and relative minima at the other three critical points (Figure 6-7).

EXAMPLE 5 (*Law of reflection*). The path of a ray of light upon reflection from a mirror (Figure 6-8) is known to be such that the total time is a minimum. Since the speed of light (in a given medium) is constant, the path is also such that the total distance is a minimum. If we introduce an x-axis along the mirror, as in Figure 6-8 (which shows a view in a plane perpendicular to the surface of the mirror), then the total distance y depends on the location x of the point of reflection:

$$y = \sqrt{h_1^2 + x^2} + \sqrt{h_2^2 + (c - x)^2} = f(x)$$

and we consider f in the interval $0 \le x \le c$. Now

$$f'(x) = \frac{x}{\sqrt{h_1^2 + x^2}} - \frac{c - x}{\sqrt{h_2^2 + (c - x)^2}}$$

The equation for the critical point is complicated. Rather than studying this equation, we notice that, if x_0 is a critical point, then

$$\frac{x_0}{\sqrt{h_1^2 + x_0^2}} = \frac{c - x_0}{\sqrt{h_2^2 + (c - x_0)^2}}$$

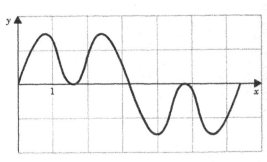

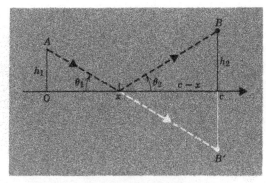

Figure 6-7 $y = \sin x + \sin 3x,\ 0 \le x \le 2\pi.$ **Figure 6-8**   Reflection of light ray.

or $\cos \theta_1 = \cos \theta_2$, where θ_1, θ_2 are the angles shown in Figure 6-8. Since θ_1, θ_2 are acute angles, our condition is equivalent to $\theta_1 = \theta_2$. Now as x increases from 0 to c, θ_1 decreases monotonically from $\pi/2$ to a certain value α_1, while θ_2 increases monotonically, from α_2 to $\pi/2$ ($\cos \alpha_1 = c/\sqrt{h_1{}^2 + c^2}$, $\cos \alpha_2 = c/\sqrt{h_2{}^2 + c^2}$). Accordingly, θ_1 can equal θ_2 only for one value x_0. Thus, there is a unique critical point, at which $\theta_1 = \theta_2$. Now

$$f''(x) = \frac{h_1{}^2}{(h_1{}^2 + x^2)^{3/2}} + \frac{h_2{}^2}{[h_2{}^2 + (c-x)^2]^{3/2}}$$

so that $f''(x_0) > 0$, and we have a relative minimum. Since x_0 is the only critical point in the interval, it provides the absolute minimum for f. We conclude: *the path that provides the minimum distance is the one for which the angle of incidence equals the angle of reflection.* According to geometric optics, that is the path followed by the light ray.

Remark 2. If a continuous function on a closed interval fails to be differentiable at one or more points of the interval, then those points must also be considered in seeking maxima and minima. The most common such cases are vertical tangents (derivative infinite) and corners (derivative to the left different from derivative to the right). Figure 6-9(a) shows a function having its absolute minimum at $x = a$, where the tangent is vertical; a relative minimum at $x = x_1$, where the tangent is again vertical and there is a cusp; its absolute maximum at $x = x_2$, where there is a corner. In such cases, as remarked above, we can still apply Theorem 1 and a knowledge of the sign of f' to determine the behavior at the critical point.

Another complication is suggested in Figure 6-9(b). Here the function is differentiable in the interval $[0, 1]$ but has infinitely many critical points (including $x = 0$). The absolute minimum is 0, taken on at infinitely many values of x. The absolute maximum is taken on at just one x (see Problem 21 below).

If the function f is not continuous, we are not sure that f has a maximum and minimum in each closed interval and a special study is needed. We return to discontinuous functions in Section 6-4 below.

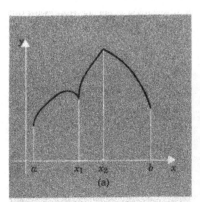
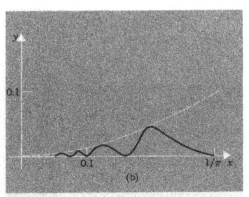

Figure 6-9 (a) Function with discontinuous derivative. (b) Function with infinitely many critical points.

PROBLEMS

1. Show that the function has exactly one critical point x_0 in the interval given and determine whether there is a relative maximum or minimum. Also determine the absolute maximum M and the absolute minimum m (if they exist) and graph.
   (a) $y = x^2 - 2x$, $0 \le x \le 2$
   (b) $y = x^2 - 4x + 5$, $0 \le x \le 3$
   (c) $y = x^3 - 6x^2 + 9x - 3$, $0 \le x \le 2$
   (d) $y = x^3 - 3x + 1$, $-2 \le x \le 0$
   (e) $y = x^2 + (1/x)$, $0 < x < \infty$
   (f) $y = x^4 + 6x^2 + 1$, $-\infty < x < \infty$
   (g) $y = e^x + e^{-2x}$, $-\infty < x < \infty$
   (h) $y = \sqrt{x(2 - x)}$, $0 \le x \le 2$
   (i) $y = \sin x + \cos x$, $0 \le x \le \pi$
   (j) $y = e^{-x} \sin 2x$, $0 \le x \le \pi/2$
2. Locate all critical points, determine whether there is a relative maximum or minimum at each and graph.
   (a) $y = 2x^3 - 3x^2 + 1$, $-\infty < x < \infty$
   (b) $y = 4x^3 - 6x^2 + 3x + 1$, $-\infty < x < \infty$
   (c) $y = x^4 - 2x^2 + 2$, $-\infty < x < \infty$
   (d) $y = 6x^5 - 15x^4 + 10x^3 - 30x^2$, $-\infty < x < \infty$
   (e) $y = \sin 2x - 2 \sin x$, $-\pi \le x \le \pi$
   (f) $y = e^{3x} - 3e^x + 1$, $-\infty < x < \infty$
   (g) $y = 2 \ln x - x$, $0 < x < \infty$
3. (a) What rectangle of given perimeter has maximum area?
   (b) What rectangle of given area has minimum perimeter?
   (c) What rectangle of given area has maximum perimeter?
4. The potential energy of a particle P of mass m attracted by the gravitational pull of two bodies of masses m_1, m_2 is

$$U = -k\left(\frac{m_1 m}{r_1} + \frac{m_2 m}{r_2}\right)$$

where k is a positive universal constant and r_1, r_2 are the distances from P to the two bodies (Figure 6-10). If P lies on the line segment joining m_1, m_2, at what point does U have a maximum? (This is the point at which the two attractive forces cancel each other.)

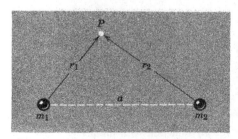

Figure 6-10   Problem 4.

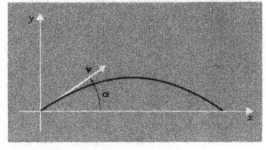

Figure 6-11   Problem 5.

5. At what angle of elevation should a baseball be thrown to travel the maximum horizontal distance? (See Figure 6-11.) Ignoring air resistance it follows from mechanics that the ball follows a parabola

$$y = x \tan \alpha - \frac{16x^2}{v^2 \cos^2 \alpha}$$

where x and y are in feet and v is the initial speed: $v = |\mathbf{v}|$, in feet per second.

6. *A problem in economics.* A shipping center to serve cities A, B, and C is to be located at P on a highway. If, in terms of the coordinates shown in Figure 6-12, A is $(-1, 0)$, B is $(1, 0)$, C is $(1, 1)$, and the highway is the y-axis, show that the total distance of P from A, B, and C has a unique minimum. [*Hint.* Let u denote the sum of the three distances, so that $u = f(y)$, $-\infty < y < \infty$. Show that $f'(y)$ changes sign exactly once, between $y = 0$ and $y = 1$. Do not attempt to find the exact location of the critical point.]

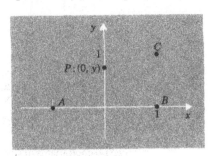

Figure 6-12   Problem 6.

7. *A problem in travel economy.* A car is to be driven by a chauffeur on a 300-mile trip. The cost of operation of the car is $5 + 0.1(v - 50)$ cents per mile, where v is the speed in mph. The chauffeur charges \$5.00 per hour. At what speed is the total cost a minimum?

8. Let $f(r)$ be the intermolecular potential in Example 3. Show:
   (a) $\lim\limits_{r \to 0+} f(r) = \infty$.   (b) $\lim\limits_{r \to \infty} f(r) = 0$.

9. In triangle ABC (Figure 6-13) choose point P on side AC to maximize the area of the inscribed rectangle $PQRS$. [*Hint.* Choose axes so that A is $(-a, 0)$, B is $(b, 0)$, and C is $(0, c)$. Let P be (x, y) and express the area in terms of x.]

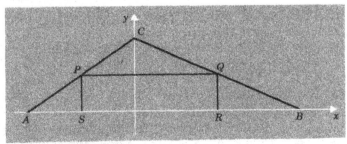

Figure 6-13   Problem 9.

10. In a circle of diameter PS a chord RQ is drawn perpendicular to PS. How should R be chosen to maximize the area of triangle PRQ? [*Hint.* As in Figure 6-14, let the circle have equation $x^2 + y^2 = a^2$, with P at $(-a, 0)$, R at (x, y). Express the area (*a*) in terms of x or (*b*) in terms of $\theta = \angle SPR$.] (Figure is on next page.)

11. From a rectangular piece of cardboard of sides a and b, a box is made by cutting out squares of side x from the corners (Figure 6-15). Choose x to obtain the box of maximum volume.

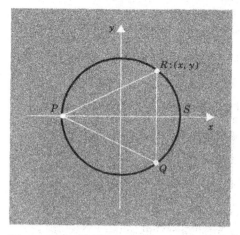

Figure 6-14 Problem 10.   **Figure 6-15** Problem 11.

12. Let x, y, and z be coordinates in space and let the yz-plane be illuminated by light from a source at P: $(x, y, 0)$, $x > 0$ (see Figure 6-16). It is shown in physics that, at the origin, the light has intensity $kx(x^2 + y^2)^{-3/2}$, where k is a positive constant. If P is known to be on the line $x + 3y = 4$ in the xy-plane, determine where P should be placed to give maximum illumination at the origin.

13. A right circular cone is to be inscribed in a sphere of radius a. What should be the radius of the base of the cone if it is to have maximum volume?

14. A right circular cylinder is to be in-   **Figure 6-16** Problem 12.
    scribed in a sphere of radius a. What
    should be the radius of the cylinder if it is to have maximum volume?

15. Let $y = f(x) = x^{4/5}$, $-1 \leq x \leq 1$.
    (a) Graph the function.
    (b) Show that $f'(x)$ exists except at $x = 0$.
    (c) Show that f has its absolute minimum at $x = 0$.

16. Carry out steps (a), (b), and (c) of Problem 15 for the function $y = f(x) = |x|$, $-1 \leq x \leq 1$.

17. Carry out the proof for case (b) of Theorem 2.

18. With reference to case (c) of Theorem 2, show that, if $f'''(x_0)$ exists, then
    (a) If $f'(x_0) = 0, f''(x_0) = 0, f'''(x_0) = 0$, f may have a relative maximum or a relative minimum or neither at x_0.
    (b) If $f'(x_0) = 0, f''(x_0) = 0, f'''(x_0) \neq 0$, f has neither minimum nor maximum at x_0. [*Hint.* Here show that if, for example, $f'''(x_0 > 0$, then $f''(x) > 0$ for $x_0 < x < x + \delta, f''(x) < 0$ for $x_0 - \delta < x < x_0$ for δ sufficiently small, so that $f'(x)$ is positive for x sufficiently close to x_0, $x \neq x_0$.]

19. Extend the result of Problem 19 to show that if, for some positive integer n, $f'(x_0) = 0, f''(x_0) = 0, \ldots, f^{(n)}(x_0) = 0, f^{(n+1)}(x_0) \neq 0$, then

   (a) If n is odd, f has a relative minimum at x_0 if $f^{(n+1)}(x_0) > 0$ and a relative maximum at x_0 if $f^{(n+1)}(x_0) < 0$.

   (b) If n is even, f has neither relative maximum nor relative minimum at x_0.

20. *Method of least squares.* Experiments are carried out to determine the value of the constant m in an equation $y = mx$ governing two physical quantities x, y. Let the measured values be $(x_1, y_1), \ldots, (x_n, y_n)$. Generally, they will not fall on a straight

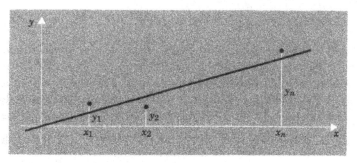

Figure 6-17   Method of least squares.

line through the origin (Figure 6-17). In order to find such a line, which best fits the data, we try to choose m to minimize the sum of the squares of the errors:

$$E = (y_1 - mx_1)^2 + \cdots + (y_n - mx_n)^2$$

Thus, E becomes a function of m whose minimum is sought.

   (a) Carry out the procedure described for the data $(1, 2.1), (2, 3.8), (3, 6.3)$.

   (b) Show that in general E is a quadratic function of m, with a unique minimum.

‡21. Let function f be defined by the equations $y = x^2 \sin^2(1/x)$ for $0 < x \leq 1/\pi$, $y = 0$ for $x = 0$.

   (a) Verify that $f'(x)$ exists for $0 < x \leq 1/\pi$ and that

$$\left| \frac{f(0 + h) - f(0)}{h} \right| \leq |h| \qquad (h > 0)$$

   and hence $f'(0) = 0$.

   (b) Show that the relative minima of f occur at $x = (n\pi)^{-1}$ $(n = 1, 2, 3, \ldots)$ and at $x = 0$, and that the absolute minimum of f is 0.

   (c) Show that f has infinitely many relative maxima, one between each pair of relative minima.

   (d) Show that the graph of f is as in Figure 6-9(b).

22. For what integer x does $f(x) = -3x^2 + 10x - 1$ have its maximum value?

6-2   MAXIMA AND MINIMA WITH SIDE CONDITIONS. LAGRANGE MULTIPLIER

We sometimes encounter the problem of finding the absolute maximum or minimum of z, where z is expressed in terms of x and y, and where x and y are related by an equation, called a *side condition*. From the side condition, y may be determined as an implicit function of x, so that z itself becomes a function of x alone. It may be inconvenient to eliminate y and express z in terms of x. However, dz/dx can be found and, hence, the critical points can be located.

EXAMPLE 1. Find the absolute minimum of $z = x + 2y$, where $xy = 2$ and $x > 0$.

Here we could eliminate y, but we illustrate the general method by not doing so. From the two given equations, we obtain

$$\frac{dz}{dx} = 1 + 2\frac{dy}{dx} \quad \text{and} \quad x\frac{dy}{dx} + y = 0 \qquad (6\text{-}20)$$

Therefore $dz/dx = 0$ when $dy/dx = -\frac{1}{2}$ and, consequently, from the second equation (6-20), when $y = \frac{1}{2}x$. The side condition $xy = 2$ now gives $x^2 = 4$ and, since $x > 0$, our only critical point is $x = 2$. For $x = 1$, $y = 2$ and $z = 5$; for $x = 2$, $y = 1$ and $z = 4$; for $x = 4$, $y = \frac{1}{2}$ and $z = 5$. These values show that z, as a function of x for $x>0$, has its absolute minimum, 4, at $x=2$.

In this example, elimination of y gives $z = x + (4/x)$, $x > 0$, and we find directly that

$$\frac{dz}{dx} = 1 - \frac{4}{x^2}, \qquad \frac{d^2z}{dx^2} = \frac{8}{x^3}$$

The critical point is $x = 2$ and, since $z''(2) > 0$, z has a local minimum at $x = 2$. Hence, by Theorem 1, z has its absolute minimum, 4, at $x = 2$.

When elimination is difficult, we can follow the procedure first illustrated. In each particular case, a proof that one has indeed found the absolute maximum or minimum may be very difficult. However, often a simple physical or geometrical argument indicates what is happening.

EXAMPLE 2. $z = 3x + 4y$, where $x^2 + y^2 = 1$, $y \geq 0$. We proceed formally as for Example 1:

$$\frac{dz}{dx} = 3 + 4\frac{dy}{dx}, \qquad 2x + 2y\frac{dy}{dx} = 0 \qquad (6\text{-}21)$$

Now $dz/dx = 0$ when $dy/dx = -\frac{3}{4}$, whence $2x + 2y(-\frac{3}{4}) = 0$. But then $y = \frac{4}{3}x$ and hence $x^2 + \frac{16}{9}x^2 = 1$, or $x^2 = \frac{9}{25}$ and $x = \pm\frac{3}{5}$. When $x = \pm\frac{3}{5}$, we have $y = \frac{4}{3}x = \pm\frac{4}{5}$. Since $y \geq 0$, we obtain only one critical point: $x = \frac{3}{5}$, for which $y = \frac{4}{5}$ and $z = 5$. We can apply the second derivative test. From (6-21)

$$\frac{d^2z}{dx^2} = 4\frac{d^2y}{dx^2}, \qquad 2 + 2\left(\frac{dy}{dx}\right)^2 + 2y\frac{d^2y}{dx^2} = 0$$

If we substitute $x = \frac{3}{5}$, $y = \frac{4}{5}$, $dy/dx = -\frac{3}{4}$ in the second equation, we find that $d^2y/dx^2 = -\frac{385}{128}$, so that d^2z/dx^2 is negative. Thus, z has an absolute maximum of 5 at $x = \frac{3}{5}$.

In this example, we could again eliminate y:

$$z = 3x + 4\sqrt{1 - x^2}, \qquad -1 \leq x \leq 1 \qquad (6\text{-}22)$$

We then verify directly that the absolute maximum is at $x = \frac{3}{5}$ (Problem 7 below).

If we had not required $y \geq 0$ in Example 2, then we would have to write:

$$z = 3x \pm 4\sqrt{1 - x^2}, \qquad -1 \leq x \leq 1$$

Thus, there would be *two* values of z for each x (except ± 1), and z is not defined as a function of x! However, it still makes sense to ask for the largest and smallest value of $z = 3x + 4y$ for (x, y) on the circle $x^2 + y^2 = 1$. To reduce this question to a familiar form, we represent the circle parametrically:

$$x = \cos t, \qquad y = \sin t, \qquad 0 \leq t \leq 2\pi \tag{6-23}$$

Then $z = 3 \cos t + 4 \sin t, 0 \leq t \leq 2\pi$, and we have an ordinary maximum-minimum problem. Now

$$\frac{dz}{dt} = -3 \sin t + 4 \cos t = 0 \qquad \text{when} \qquad \tan t = \frac{4}{3}$$

We thus obtain two critical points: $t = \alpha$ and $t = \alpha + \pi$, where $\alpha = \text{Tan}^{-1} \frac{4}{3}$. We find that z has a local maximum at $t = \alpha$ and a local minimum at $t = \alpha + \pi$; we verify also that z has its absolute maximum and minimum at these points (Problem 8 below). Hence, z has its absolute maximum at the point $(\cos \alpha, \sin \alpha)$ or $(\frac{3}{5}, \frac{4}{5})$ on the circle $x^2 + y^2 = 1$; its absolute minimum is at $[\cos(\alpha + \pi), \sin(\alpha + \pi)]$ or $(-\frac{3}{5}, -\frac{4}{5})$.

We pursue this problem further, to develop a general method for such problems. We first remark that we can obtain the results sought without explicitly using the parameter. We write

$$z = 3x + 4y, \qquad x^2 + y^2 = 1 \tag{6-24}$$

and differentiate with respect to t:

$$\frac{dz}{dt} = 3\frac{dx}{dt} + 4\frac{dy}{dt}, \qquad 2x\frac{dx}{dt} + 2y\frac{dy}{dt} = 0$$

At the critical points, $dz/dt = 0$. Hence we obtain the equations

$$3\frac{dx}{dt} + 4\frac{dy}{dt} = 0, \qquad 2x\frac{dx}{dt} + 2y\frac{dy}{dt} = 0$$

These equations can be written in vector language:

$$(3\mathbf{i} + 4\mathbf{j}) \cdot \left(\frac{dx}{dt}\mathbf{i} + \frac{dy}{dt}\mathbf{j} \right) = 0, \qquad (2x\mathbf{i} + 2y\mathbf{j}) \cdot \left(\frac{dx}{dt}\mathbf{i} + \frac{dy}{dt}\mathbf{j} \right) = 0$$

Therefore, both vectors $3\mathbf{i} + 4\mathbf{j}$, $2x\mathbf{i} + 2y\mathbf{j}$ are orthogonal to the *velocity vector*: $(dx/dt)\mathbf{i} + (dy/dt)\mathbf{j}$. Since we know from the parametrization (6-23), that the velocity vector is never $\mathbf{0}$, the vectors $3\mathbf{i} + 4\mathbf{j}$, $2x\mathbf{i} + 2y\mathbf{j}$ must be *linearly dependent*. Hence, one vector is a scalar λ times the other:

$$3\mathbf{i} + 4\mathbf{j} = \lambda(2x\mathbf{i} + 2y\mathbf{j}) \tag{6-25}$$

[We could also have put λ as a factor on the other side; we are justified in putting it on the right, since the vector on the right cannot be $\mathbf{0}$ at a point (x, y) on the circle $x^2 + y^2 = 1$.]

From (6-25) we obtain

$$3 = 2x\lambda, \qquad 4 = 2y\lambda \tag{6-26}$$

From the side condition in (6-24) we thus obtain

$$\frac{9}{4\lambda^2} + \frac{16}{4\lambda^2} = 1 \quad \text{or} \quad 25 = 4\lambda^2, \quad \lambda = \pm\frac{5}{2}$$

and hence from (6-26)

$$x = \frac{3}{2\lambda} = \pm\frac{3}{5}, \qquad y = \frac{4}{2\lambda} = \pm\frac{4}{5}$$

Consequently, we obtain the two desired critical points $(\pm\frac{3}{5}, \pm\frac{4}{5})$.

The method illustrated can be applied whenever the side condition describes a curve that can be parametrized by differentiable functions $x = x(t)$, $y = y(t)$. The vector argument and the introduction of the extra variable λ (*called Lagrange multiplier*) are also general; we must be sure that the velocity vector is not $\mathbf{0}$, at least at the critical points. In particular cases, complications may arise in applying the method. However, it has proved to be of value in many important problems.

EXAMPLE 3. $z = x^2 + y^2$, where $Ax^2 + 2Bxy + Cy^2 = 1$ and $B^2 - AC < 0$ and $A + C > 0$. It is shown in Section 6-5 that the side condition describes an ellipse with center at the origin. Thus $z = x^2 + y^2$ is the square of the distance from the origin to a point (x, y) on the ellipse (Figure 6-18).

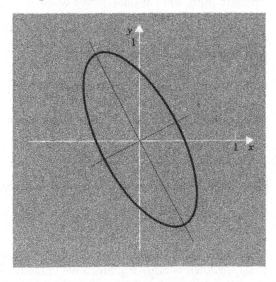

Figure 6-18   Ellipse $Ax^2 + 2Bxy + Cy^2 = 1$.

Hence, it is clear that z has its maximum value at the ends of the major axis and its minimum value at the ends of the minor axis. We proceed to find these points by the method of the Lagrange multiplier. Differentiation gives

$$\frac{dz}{dt} = 2x\frac{dx}{dt} + 2y\frac{dy}{dt} = 0, \qquad (2Ax + 2By)\frac{dx}{dt} + (2Bx + 2Cy)\frac{dy}{dt} = 0$$

To simplify, we divide by 2 throughout. As above, the vectors $x\mathbf{i} + y\mathbf{j}$ and $(Ax + By)\mathbf{i} + (Bx + Cy)\mathbf{j}$ are linearly dependent at each critical point. Therefore,

$$Ax + By = \lambda x, \qquad Bx + Cy = \lambda y \tag{6-27}$$

or

$$\begin{array}{c} (A - \lambda)x + By = 0 \\ Bx + (C - \lambda)y = 0 \end{array} \tag{6-27'}$$

For fixed λ, these are two homogeneous linear equations for x, y, as in Section 0-8. If the determinant of coefficients is not 0, the only solution is $x = 0$, $y = 0$. The point $(0, 0)$ is clearly excluded as a critical point. Hence, we get a critical point only when the determinant is 0:

$$\begin{vmatrix} A - \lambda & B \\ B & C - \lambda \end{vmatrix} = 0 \tag{6-28}$$

Expanding, we obtain a quadratic equation for λ:

$$\lambda^2 - \lambda(A + C) + AC - B^2 = 0 \tag{6-28'}$$

The discriminant of this quadratic is

$$(A + C)^2 + 4(B^2 - AC) = (A - C)^2 + 4B^2 \geq 0$$

The discriminant can equal 0 only when $A = C$ and $B = 0$. In this case, our ellipse becomes a circle, and $z \equiv$ constant. Therefore, let us assume that the discriminant is positive. Then (6-28') has two distinct roots λ_1, λ_2, say $\lambda_1 < \lambda_2$. Corresponding to each root, we obtain solutions (x, y) of (6-27'), forming a line through the origin, as in Section 0-8. Each line meets the given ellipse at a critical point of z—hence, at the ends of the major or minor axis. We observe further from (6-27) that, at each critical point,

$$\lambda x^2 + \lambda y^2 = x(Ax + By) + y(Bx + Cy) = Ax^2 + 2Bxy + Cy^2 = 1$$

Consequently, $\lambda(x^2 + y^2) = 1$, or $z = 1/\lambda$, and the equation $\lambda(x^2 + y^2) = 1$ shows that $\lambda > 0$ at each critical point. Therefore, the two values λ_1, λ_2 give the absolute minimum and maximum of z:

$$\text{minimum of } z = \frac{1}{\lambda_2} < \text{maximum of } z = \frac{1}{\lambda_1}$$

Therefore the major axis of the ellipse is $2/\sqrt{\lambda_1}$, the minor axis is $2/\sqrt{\lambda_2}$. As a numerical example, we take the equation

$$5x^2 + 4xy + 2y^2 = 1$$

The equations (6-27') become

$$(5 - \lambda)x + 2y = 0, \qquad 2x + (2 - \lambda)y = 0 \tag{6-27''}$$

The equation (6-28') becomes

$$\lambda^2 - 7\lambda + 6 = 0 \tag{6-28''}$$

The roots are $\lambda_1 = 1$, $\lambda_2 = 6$. For $\lambda = \lambda_1 = 1$, the equations (6-27'') become

$$4x + 2y = 0, \qquad 2x + y = 0$$

and hence represent one line: $y = -2x$. The line meets the given ellipse where

$$5x^2 - 8x^2 + 8x^2 = 1$$

that is, at $(\pm 1/\sqrt{5}, \mp 2/\sqrt{5})$. These are the end points of the major axis. Similarly, for $\lambda = \lambda_2 = 6$, Equations (6-27'') yield one line: $x - 2y = 0$, meeting the ellipse at $(\pm 2/\sqrt{30}, \pm 1/\sqrt{30})$. The lengths of the major and minor axes are $2/\sqrt{\lambda_1} = 2$ and $2/\sqrt{\lambda_2} = 2/\sqrt{6}$. The ellipse is shown in Figure 6-18.

PROBLEMS

1. In each case find the absolute maximum and minimum of z (if they exist), where the given side condition holds true:
   (a) $z = x^2 + y^2$, $x - y = 1$ (b) $z = x^2 + y^2$, $2x - 3y = 1$
   (c) $z = x^2 + y^2$, $x^2 - y^2 = 1$, $x > 0$ (d) $z = x^2 + y^2$, $x^2 - 4y^2 = 1$, $x > 0$
   (e) $z = x^2 + y^2$, $2x^2 - y = 1$ (f) $z = 2x^2 - xy + y^2$, $x^2 + y^2 = 1$, $y \geq 0$
   (g) $z = x^2 + y^2$, $xy = 1$, $x > 0$ (h) $z = xy$, $x^2 + y^2 = 1$, $y \geq 0$
   (i) $z = x - y$, $x^2 + y^2 = 1$, $y \geq 0$ (j) $z = 2x + y$, $x^2 + y^2 = 1$, $y \geq 0$
   (k) $z = x + 8y$, $x^4 + y^4 = 1$, $y \geq 0$ (l) $z = 3x - y$, $x^{2/3} + y^{2/3} = 1$, $y \geq 0$

2. Find the ends of the major and minor axes of the given ellipse:
   (a) $x^2 + xy + y^2 = 1$ (b) $x^2 + 3xy + 5y^2 = 1$
   (c) $5x^2 + 12xy + 10y^2 = 1$ (d) $13x^2 + 5xy + y^2 = 1$

3. Find the minimum in Example 5 of Section 6-1 by writing $y = h_1 \cos \theta_1 + h_2 \cos \theta_2$, with the side condition $h_1 \cot \theta_1 + h_2 \cot \theta_2 = c$.

4. Establish the result of Example 5 in Section 6-1 geometrically by showing that the distance $y = |\overrightarrow{AP}| + |\overrightarrow{PB'}|$, where B' is the image of B for a reflection in the x-axis (Figure 6-8), and that hence y is a minimum when all three points A, P, and B' are on a straight line.

5. Find the area of the rectangle of maximum area inscribed in the ellipse: $(x^2/a^2) + (y^2/b^2) = 1$.

6. Find the perimeter of the rectangle of maximum perimeter inscribed in the ellipse of Problem 5.

7. Show that the function of Equation (6-22) has its absolute maximum for $x = \frac{2}{3}$.

8. Show, with the aid of Equations (6-23), that $z = 3x + 4y$ has its absolute maximum and minimum on the circle $x^2 + y^2 = 1$ at the points corresponding to $t = \alpha$, $t = \alpha + \pi$, respectively. Here $\alpha = \text{Tan}^{-1} \frac{4}{3}$.

6-3 CONCAVITY AND CONVEXITY, INFLECTION POINTS

Let a function $y = f(x)$ be given on the interval $a \leq x \leq b$, and let f have a unique local minimum at x_0, $a < x_0 < b$. Then, as we have observed, f has its absolute minimum $f(x_0)$ at x_0 and remains above that value otherwise. If $f'(x) = 0$ only at x_0 and $f''(x_0) > 0$, we are sure that x_0 does provide an absolute minimum, so that $f(x)$ is above the minimum value elsewhere in the interval. The condition $f'(x_0) = 0$ states that the tangent is horizontal at x_0, and the additional condition $f''(x_0) > 0$ insures that the curve lies above this tangent. What can we say about a point x_1, where $f'(x_1) \neq 0$, but $f''(x_1) > 0$? It is natural to conjecture that again the curve must lie above the tangent at x_1 even though the tangent is not horizontal. Such is indeed the case.

THEROEM 3. *Let $y = f(x)$ be differentiable on the interval $a \leq x \leq b$. If $f''(x_1)$ exists and $f''(x_1) > 0$, where $a < x_1 < b$, then there is an interval $[x_1 - h, x_1 + h]$ in which the curve $y = f(x)$ lies above the tangent line at x_1, that is,*

$$f(x) \geq f'(x_1)(x - x_1) + f(x_1), \qquad x_1 - h \leq x \leq x_1 + h \quad (6\text{-}30)$$

with equality only at x_1. If $f''(x)$ exists for $a < x < b$ and $f''(x) > 0$ for $a < x < b$, then the curve lies above the tangent line at x_1 throughout:

$$f(x) \geq f'(x_1)(x - x_1) + f(x_1), \qquad a \leq x \leq b \quad (6\text{-}31)$$

with equality only at x_1.

PROOF. We let $g(x)$ be the "vertical distance" of each point of the curve from the tangent line at x_1:

$$g(x) = f(x) - [f'(x_1)(x - x_1) + f(x_1)]$$

(as in Figure 6-19), and wish to show that $g(x)$ has its unique minimum, namely 0, at x_1 only. Now

$$g'(x) = f'(x) - f'(x_1), \qquad g''(x) = f''(x)$$

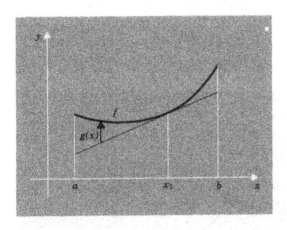

Figure 6-19 $f''(x) > 0$.

Hence, $g'(x_1) = 0$, $g''(x_1) = f''(x_1) > 0$. Therefore, g has a relative minimum at x_1 and $g(x) \geq g(x_1) = 0$ at least in an interval $[x_1 - h, x_1 + h]$, with equality only at x_1; that is, (6-30) holds true. If we assume that $f''(x)$ exists and that $f''(x) > 0$ for $a < x < b$, then also $g''(x) > 0$ for $a < x < b$. This implies that $g(x)$ has only the one critical point x_1 inside the interval. For, if there were another x_2, we would have

$$g'(x_1) = 0, \qquad g'(x_2) = 0$$

and hence, by Rolle's theorem (Section 3-20), $g''(x_3) = 0$ for some x_3 between x_1 and x_2. This is impossible, since we are given $g''(x) > 0$ for $a < x < b$. Therefore, $g(x)$ has its absolute minimum, 0, at x_1, and at x_1 alone. Accordingly, (6-31) holds true, and we have the situation illustrated in Figure 6-19.

We can formulate a similar theorem for the condition: $f''(x_1) < 0$. In this case, the curve lies below the tangent line, as in Figure 6-20.

For the case of Figure 6-19, in which the curve lies above each of its tangents, we say that the graph is *concave upward* or that the function f is a *convex function* in the interval. For the case of Figure 6-20, we say that the graph is *concave downward* and call f a *concave function*.

It often happens that a graph is concave upward in certain intervals and concave downward in others, as in Figure 6-21. The transition points are called *inflection points*. At each such point the curve lies above the tangent on one side and below the tangent on the other. If $f''(x)$ exists at these points,

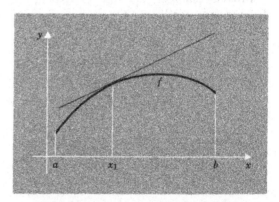

Figure 6-20 $f''(x) < 0$.

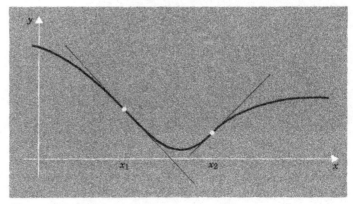

Figure 6-21   Inflection points.

then by Theorem 3 (and its counterpart for curves concave downward), we can have neither $f''(x) > 0$ nor $f''(x) < 0$; accordingly, $f''(x) = 0$ at each inflection point. However, just as $f'(x_1) = 0$ does not guarantee a local maximum or minimum at x_1, the condition $f''(x_1) = 0$ does not guarantee that there is an inflection point at x_1; consider the example $y = x$ for which $f''(x) \equiv 0$. A simple condition for an inflection point is

$$f''(x_1) = 0, \qquad f'''(x_1) \neq 0 \tag{6-32}$$

The proof is left as a problem (Problem 3 below).

Remark. At an inflection point the condition $f''(x) = 0$ implies that the rate of change of the slope is 0. Hence, near the inflection point, the graph is approximately straight, following the tangent line at the inflection point. These qualitative statements can be made precise (see Section 6-6 below).

PROBLEMS

1. Determine the intervals in which the graph is concave up or concave down, and locate all inflection points:

   (a) $y = x^3$ (b) $y = x^4$

   (c) $y = x^5$ (d) $y = x^3 - 2x^2 + x - 2$

   (e) $y = x^4 - 2x^3 + x^2 + 3$ (f) $y = x^5 + 10x^2 + 5x - 1$

   (g) $y = 3x^5 - 10x^3 + x + 7$ (h) $y = \sec x$

   (i) $y = \tan x$ (j) $y = \ln x$ (k) $y = x^2 \ln x$

2. On graph paper make a careful graph of $y = \sin x$ by drawing the tangent lines at critical points and inflection points and fitting a smooth curve to them. (As in the Remark at the end of Section 6-3, the graph is very close to the tangent lines at inflection points.) Read the values of $\sin \pi/6$, $\sin \pi/4$, $\sin \pi/6$, and $\sin 3\pi/8$ from the graph and compare with the precise values.

3. Prove: If $f(x)$ is defined for $a < x < b$, and $f''(x_1) = 0, f'''(x_1) > 0$, where $a < x_1 < b$, then f has an inflection point at $x = x_1$. [*Hint.* Define $g(x)$ as in the proof of Theorem 3 and show that, by Theorem 2 applied to g', $g'(x) > 0$ for $|x - x_1| < h$, $x \neq x_1$, for h sufficiently small. Thus g is monotone strictly increasing for $|x - x_1| < h$.]

4. Show that the function has an inflection point at the point stated (cf. Problem 3):

   (a) $y = e^x \sin x$, $(\pi/2, e^{\pi/2})$ (b) $y = x^{1/3} - x^{1/5}$, $\left[\left(\dfrac{18}{25} \right)^{15/2}, \dfrac{-378\sqrt{2}}{3125} \right]$

6-4 REMARKS ON GRAPHING

To graph a particular function $y = f(x)$ given by an equation, we should first determine:

(a) The values of x for which f is defined (domain of f) and the values of y which occur (range of f). Finding the range may be difficult and often some partial information suffices. For example, y is always positive, or $-B < y < B$ (f is bounded).

We should next observe:

(b) The x- and y-intercepts. The x-intercepts are the zeros of f; the y-intercept is $f(0)$.

(c) Symmetry properties. If f is even $[f(-x) = f(x)]$, the graph is symmetric in the y-axis. If f is odd $[f(-x) = -f(x)]$, the graph for negative x can be obtained from that for positive x by reflecting first in the y-axis and then in the x-axis.

(d) Periodicity. If $f(x + c) = f(x)$ for all x, for some nonzero constant c, f is said to be periodic and to have period c. For example, $\sin x$ has period 2π. A function of period c repeats in intervals of length c $[f(x) = f(x \pm c) = f(x \pm 2c) = \cdots]$ and, hence, the whole graph can be obtained from the graph in one interval of length c.

We can then proceed to apply the calculus to determine, as far as possible, the following:

(e) The critical points, the corresponding relative minima and maxima, and the intervals in which $f'(x) > 0$ (f increasing), and those in which $f'(x) < 0$ (f decreasing);

(f) The inflection points, and the corresponding intervals of concavity upward or downward.

In addition, we can observe the following:

(g) The discontinuities of f and behavior of f near each discontinuity;

(h) The discontinuities of f';

(i) The vertical asymptotes and horizontal asymptotes.

Recall that f is discontinuous at x_0 if f is defined at all points in some open interval having x_0 as an endpoint but either f is not defined at x_0, or the limit of f as x approaches x_0 does not exist, or the limit exists and does not equal $f(x_0)$. Here we shall restrict our attention for the most part to cases in which f has a limit, finite or infinite, as the point is approached from left or right. The possibilities are suggested in Figure 6-22.

At x_0 both limits are finite, but different. At x_1 both the limits are $+\infty$. At x_2 the left-hand limit is $+\infty$, the right-hand limit is $-\infty$. At x_4 the left-

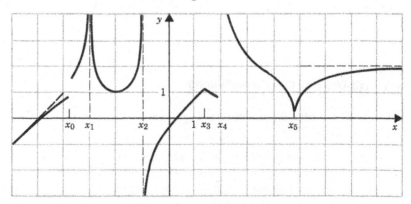

Figure 6-22 Discontinuities and asymptotes.

hand limit is finite, the right-hand limit is $+\infty$. One type of discontinuity of f' is suggested at x_3; here there is a *corner*, with different slopes to the left and to the right. Another type, a *cusp*, occurs at x_5.

Vertical asymptotes appear at the discontinuities at which one or both of the limiting values is infinite. Thus, in Figure 6-22, the lines $x = x_1$ and $x = x_2$ are vertical asymptotes. The line $x = x_4$ is a vertical asymptote relative to the portion of the graph to the right of $x = x_4$.

Horizontal asymptotes may appear if f is defined in an infinite interval. Thus, in Figure 6-22, f has limit 2 as $x \to +\infty$, so that the line $y = 2$ is a horizontal asymptote for the corresponding portion of the graph. As $x \to -\infty$, the graph suggests that $f \to -\infty$ and there is no horizontal asymptote for this portion. However, the graph indicates that

$$\lim_{x \to \infty} [f(x) - (x + 5)] = 0$$

that is, that the line $y = x + 5$ is an asymptote. In some cases, we may find a nonlinear function $g(x)$ so that $f(x) - g(x) \to 0$ as $x \to +\infty$ or as $x \to -\infty$; then the curve $y = g(x)$ serves as an "asymptotic curve" for a portion of the graph.

The possibilities and their combinations are so varied that we confine our attention here to some simple examples which indicate the typical procedures and computational problems.

EXAMPLE 1

$$y = x^2 + (2/x) - 2 = f(x), \qquad -\infty < x < \infty, x \neq 0$$

Here f is continuous at all points except at $x = 0$. We have

$$f'(x) = 2x - \frac{2}{x^2} = \frac{2(x^3 - 1)}{x^2}, \qquad f''(x) = 2 + \frac{4}{x^3} = \frac{2(x^3 + 2)}{x^3}$$

Hence $f'(x) = 0$ for $x = 1$, and this is the only critical point; the function f' can change sign only at a crtical point or at a discontinuity of f. Here we verify that the only change of sign is at $x = 1$ and that

$$f'(x) > 0 \text{ for } x > 1, \qquad f'(x) < 0 \text{ for } x < 0 \text{ or } 0 < x < 1$$

$f'(x)$ is undefined at $x = 0$. Consequently, f is steadily increasing for $x > 1$, steadily decreasing for $x < 0$ and for $0 < x < 1$. At the critical point $x = 1$, f has thus a local minimum, with value $f(1) = 1$. We notice that $f''(1) = 6 > 0$, in agreement with the existence of a local minimum.

A similar analysis of $f''(x)$ shows that $f''(x) > 0$ for $-\infty < x < -\sqrt[3]{2}$ and for $x > 0$, consequently on these intervals the graph is concave up. For $-\sqrt[3]{2} < x < 0$, $f''(x) > 0$, and the graph is concave down. At $x = -\sqrt[3]{2}$ there is an inflection point. At this point, $y = -2, y' = -3\sqrt[3]{2}$. At $x = 0, f''$ is undefined.

At the discontinuity $x = 0$ (as in Section 2-11),

$$\lim_{x \to 0+} f(x) = \lim_{x \to 0+} \left(x^2 + \frac{2}{x} - 2\right) = +\infty, \qquad \lim_{x \to 0-} \left(x^2 + \frac{1}{x} - 2\right) = -\infty$$

Therefore, the y-axis is a vertical asymptote.

As x becomes infinite, we have (see Section 2-11)

$$\lim_{x \to +\infty} f(x) = \lim_{x \to +\infty} \left(x^2 + \frac{2}{x} - 2\right) = +\infty$$

$$\lim_{x \to -\infty} f(x) = \lim_{x \to -\infty} \left(x^2 + \frac{2}{x} - 2\right) = +\infty$$

Hence there is no horizontal asymptote. We can use $y = x^2 - 2 = g(x)$ as an asymptotic curve on both sides, since $f(x) - g(x) = 2/x \to 0$ as $x \to \pm\infty$.

We also observe the signs of f itself: for $x > 0$, $f(x) > 0$; f is steadily decreasing for $x < 0$ and $f(-2) = 1 > 0$, $f(-1) = -3 < 0$, so that f has one zero between -2 and -1, at which f changes sign; f also changes sign at the discontinuity $x = 0$.

From the information obtained, we can sketch the graph as in Figure 6-23.

EXAMPLE 2

$$y = \frac{e^x - e^{-2x}}{e^x + e^{-2x}} = f(x), \qquad -\infty < x < \infty$$

Since the numerator and denominator are continuous and the denominator is never 0, f is continuous for all x. Before considering derivatives, we seek horizontal asymptotes. Since (see Section 5-7)

$$\lim_{x \to \infty} \frac{e^x - e^{-2x}}{e^x + e^{-2x}} = \lim_{x \to \infty} \frac{1 - e^{-3x}}{1 + e^{-3x}} = 1$$

$$\lim_{x \to -\infty} \frac{e^x - e^{-2x}}{e^x + e^{-2x}} = \lim_{x \to -\infty} \frac{e^{3x} - 1}{e^{3x} + 1} = -1$$

we conclude that the graph has asymptotes $y = 1$ (for large positive x) and $y = -1$ (for large negative x). We find that

$$f' = \frac{6e^{3x}}{(1 + e^{3x})^2}, \qquad f'' = \frac{18e^{3x}(1 - e^{3x})}{(1 + e^{3x})^3}$$

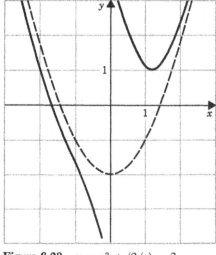

Figure 6-23 $y = x^2 + (2/x) - 2$.

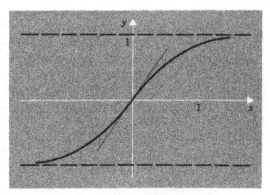

Figure 6-24 $y = (e^x - e^{-2x})/(e^x + e^{-2x})$.

Accordingly, $f'(x) > 0$ for all x, $f''(x) > 0$ for $x < 0$, $f''(x) < 0$ for $x > 0$. There is an inflection point at $x = 0$, where $y = 0$ and $y' = \frac{3}{2}$. Since f is steadily increasing, with the horizontal asymptotes found above, we readily obtain the graph of Figure 6-24.

EXAMPLE 3

$$y^3 + (x^4 - 4x + 4)(y + 2) = 0 \tag{6-40}$$

Here a function $y = f(x)$ is defined only implicitly. By the methods of Section 3-8, we verify that the equation does assign a unique y to each x, $-\infty < x < \infty$. We can show further that f' and f'' are continuous. We shall assume this here and obtain these derivatives by "implicit differentiation." We find successively that

$$3y^2y' + (x^4 - 4x + 4)y' + (4x^3 - 4)(y + 2) = 0 \tag{6-41}$$

$$3y^2y'' + 6yy'^2 + (x^4 - 4x + 4)y'' + 2(4x^3 - 4)y' + 12x^2(y + 2) = 0 \tag{6-42}$$

From the first equation,

$$y' = \frac{-4(x^3 - 1)(y + 2)}{3y^2 + x^4 - 4x + 4}$$

so that $y' = 0$ when $x = 1$ or $y = -2$. However, from our given equation (6-40), $y = -2$ implies $-8 = 0$, which is absurd. Hence, $y = -2$ cannot arise. Accordingly, $x = 1$ gives the only critical point; we find easily that $f(1) = -1$. If we substitute $x = 1$, $y = -1$, $y' = 0$ in (6-42), we find that $4y'' + 12 = 0$, or $y'' = -3$. Consequently, f has a relative maximum at $x = 1$. Since this is the only critical point, we conclude that f has its absolute maximum at $x = 1$. Since f is continuous and $f(x) \neq -2$, we have by the Intermediate Value Theorem:

$$-2 < f(x) \le -1 \qquad \text{for all } x \tag{6-43}$$

A discussion of the sign of $f''(x)$ is complicated. However, we can determine the horizontal asymptotes easily. If we write our equation (6-40) as follows

$$x^4 - 4x + 4 = -\frac{y^3}{y + 2}$$

we notice that as $x \to \pm\infty$, the left-hand side approaches $+\infty$. Hence the right-hand side must also approach $+\infty$. By (6-43) the numerator y^3 is bounded between -8 and -1. Therefore, the denominator must approach 0; that is, $y \to -2$. Thus the line $y = -2$ serves as a horizontal asymptote in both directions.

The function is graphed in Figure 6-25.

EXAMPLE 4

$$y = f(x) = e^{1/x}, \qquad -\infty < x < \infty, x \neq 0$$

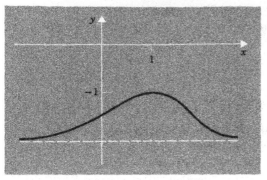

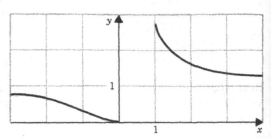

Figure 6-25 $y^3 + (x^4 - 4x + 4)(y + 2) = 0$.   **Figure 6-26** $y = e^{1/x}$

Here there is a discontinuity at $x = 0$. We find that

$$f'(x) = \frac{-e^{1/x}}{x^2}, \qquad f''(x) = e^{1/x}\frac{2x + 1}{x^4}, \qquad x \neq 0$$

Hence there is no critical point. There is an inflection point at $x = -\frac{1}{2}$: we find that $f''(x) < 0$ for $x < -\frac{1}{2}$ and find that $f''(x) > 0$ for $-\frac{1}{2} < x < 0$ and for $x > 0$. At the discontinuity at $x = 0$, we have limits from left and right:

$$\lim_{x \to 0+} e^{1/x} = \lim_{t \to +\infty} e^t = \infty \qquad \lim_{x \to 0-} e^{1/x} = \lim_{t \to -\infty} e^t = 0$$

Hence the y-axis is a vertical asymptote, for positive values of x. Also

$$\lim_{x \to \infty} e^{1/x} = \lim_{t \to 0+} e^t = 1, \qquad \lim_{x \to -\infty} e^{1/x} = \lim_{t \to 0-} e^t = 1$$

Thus the line $y = 1$ is a horizontal asymptote for large positive or negative x. The graph is sketched in Figure 6-26.

EXAMPLE 5

$$x = t^3 + t, \, y = t^3 - 3t, \qquad -\infty < t < \infty$$

Here our curve is given in parametric form. However, x is a steadily increasing function of t, since $x'(t) = 3t^2 + 1 \geq 1 > 0$; as $t \to +\infty$, $x \to +\infty$ and as $t \to -\infty$, $x \to -\infty$. Hence, there is an inverse function $t(x)$, $-\infty < x < \infty$. Thus we can write $y = y(t) = y[t(x)] = f(x)$, and the function f is defined for $-\infty < x < \infty$. Furthermore

$$y' = f'(x) = \frac{dy}{dx} = \frac{dy/dt}{dx/dt} = \frac{3t^2 - 3}{3t^2 + 1}$$

$$f''(x) = \frac{dy'}{dx} = \frac{dy'/dt}{dx/dt} = \frac{24t}{(3t^2 + 1)^3}$$

Accordingly, $f'(x) = 0$ for $t = \pm 1$; that is, for $(2, -2)$, $(-2, 2)$ and at these points $y'' = \pm 3/8$. Thus f has a relative minimum at $x = 2$, with value -2, and a relative maximum at $x = -2$, with value 2. Furthermore, $f''(x) > 0$ for $t > 0$, that is for $x > 0$, and $f''(x) < 0$ for $t < 0$, that is for $x < 0$. Accordingly, there is an inflection point at $t = 0$, where $x = 0$, $y = 0$, $y' = -3$.

There are no discontinuities—hence, no vertical asymptotes. As $x \to \pm\infty$, $t \to \pm\infty$ and $y \to \pm\infty$; consequently, there are no horizontal asymptotes. We notice that

$$\frac{y}{x} = \frac{t^3 - 3t}{t^3 + t} \to 1$$

as $t \to \pm\infty$; therefore, we might expect the line $y = x$ to serve as an asymptote. However, $y - x = -4t \nrightarrow 0$ as $t \to \pm\infty$, so that this line is not an asymptote. We can obtain an *asymptotic curve* as follows:

$$t^3 = x - t, \qquad t = \sqrt[3]{x - t} = \sqrt[3]{x} \, \sqrt[3]{1 - t/x}$$

which suggests that $t - \sqrt[3]{x} \to 0$ as $x \to \pm\infty$. This can indeed be verified (Problem 11 below). Hence

$$y = t^3 - 3t = t^3 + t - 4t = x - 4\sqrt[3]{x} + h(x)$$

where $h(x) \to 0$ as $x \to \pm\infty$. Accordingly,

$$y = g(x) = x - 4\sqrt[3]{x}$$

is the desired asymptotic curve for large positive or negative x.

Our function is graphed in Figure 6-27. The graph suggests symmetry with respect to the origin; this can be verified (Problem 10 below).

EXAMPLE 6.

$$y = f(x) = x^4 - 5x^3 + 6x^2 = x^2(x - 2)(x - 3)$$

Here $f(0) = 0$, $f(2) = 0$, and $f(3) = 0$. Hence, by Rolle's theorem, f' has a zero between 0 and 2 and one between 2 and 3. In addition, $f'(0) = 0$ because f has a double zero at $x = 0$. But f' is a polynomial of degree 3 and, consequently, has at most 3 real roots. Therefore, we have accounted for all of them: one at 0, one between 0 and 2, and one between 2 and 3. We find that

$$f'(x) = 4x^3 - 15x^2 + 12x = x(4x^2 - 15x + 12)$$
$$f''(x) = 12x^2 - 30x + 12$$

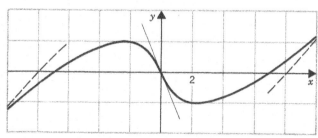

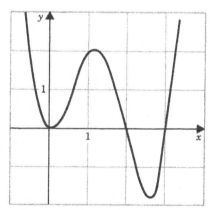

Figure 6-27 $x = t^3 + t, \, y = t^3 - 3t$

Figure 6-28 $y = x^2(x - 2)(x - 3)$.

We find the zeros of f' to be 0, 1.2, and 2.6 (to the nearest tenth). Since $f''(0) = 12 > 0$, $f''(1.2) = -6.7 < 0$, $f''(2.6) = 15.1 > 0$, we have successively a relative minimum, a relative maximum, and a relative minimum at these points. These conclusions can also be reached as follows. Since $f(x) = x^2(x - 2)(x - 3)$, the sign of f is determined by the sign of the 3 factors. For $x > 3$ all factors are positive, hence, so is f. As x decreases through 3, the factor $x - 3$ changes sign, thus, f becomes negative. Therefore, f is negative for x between 2 and 3, and we know f has just one critical point between 2 and 3; at the critical point f takes on its absolute minimum for the interval $[2, 3]$. Similarly, f becomes positive between $x = 0$ and $x = 2$, and the critical point must provide a local maximum. At $x = 0$, $f = 0$, but f does not change sign because of the factor x^2; since f is positive to the left and right of $x = 0$, f has a local minimum at this point.

The function is graphed in Figure 6-28.

Clearly, the reasoning above can be generalized to apply to a function

$$f(x) = a(x - x_1)^{n_1}(x - x_2)^{n_2} \cdots (x - x_k)^{n_k}$$

where $n_1, n_2, \ldots, n_k$ are positive integers; that is to any polynomial all of whose zeros are real. If the roots are arranged in order: $x_1 < x_2 < \cdots < x_k$, then f has precisely one critical point between each successive pair of zeros, and that critical point is either a local minimum or a local maximum, depending on whether f is negative or positive between the zeros. In general, f changes sign at each zero for which the corresponding factor is raised to an odd power (zero of odd multiplicity). In addition to the critical points between the zeros, f has a critical point at each multiple zero. By the previous remark, if the multiplicity is odd, this is neither a local maximum nor a local minimum and, therefore, is an inflection point. If the multiplicity is even, there is a local minimum or local maximum at the point.

EXAMPLE 7. $y^2 = x^4 - 5x^3 + 6x^2$. We can write

$$y = \pm \sqrt{x^4 - 5x^3 + 6x^2} = \pm \sqrt{f(x)}$$

where f is the function of Example 6. Having graphed f, we can at once graph $y = \sqrt{f(x)}$ and $y = -\sqrt{f(x)}$; wherever f is positive or zero, both functions are defined; wherever f is negative neither is defined; where f has a maximum (and is positive), $\sqrt{f}$ has a maximum and $-\sqrt{f}$ has a minimum. Hence, we obtain the graph of Figure 6-29 which is the union of the graphs $y = \sqrt{f(x)}$ and $y = -\sqrt{f(x)}$. We clearly have symmetry with respect to the x-axis.

Figure 6-29 indicates a *corner* in the graph of $y = \sqrt{f}$ (and of $y = -\sqrt{f}$) at $x = 0$. To see this we can write $y = \sqrt{x^2(6 - 5x + x^2)}$ or

$$y = \sqrt{x^2}\sqrt{6 - 5x + x^2}$$

Now for $x \geq 0$, $\sqrt{x^2} = x$ so that

$$y = x\sqrt{6 - 5x + x^2}, \qquad x \geq 0$$

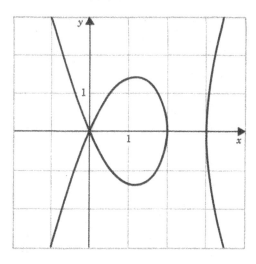

Figure 6-29 $y^2 = x^4 - 5x^3 + 6x^2$.

Since $y = 0$ for $x = 0$, we can also interpret y as Δy, x as Δx and write the last equation as

$$\Delta y = \Delta x \sqrt{6 - 5\,\Delta x + \overline{\Delta x^2}}$$

so that, as $\Delta x \to 0+$,

$$\frac{\Delta y}{\Delta x} = \sqrt{6 - 5\,\Delta x + \overline{\Delta x^2}} \to \sqrt{6}$$

Similarly, $y = -x\sqrt{6 - 5x + x^2}$, for $x \le 0$ and $\Delta y / \Delta x \to -\sqrt{6}$ as $\Delta x \to 0-$. Hence the right- and left-hand derivatives at $x = 0$ are $\sqrt{6}$ and $-\sqrt{6}$, respectively, and the graph has a corner.

At $x = 2$ and at $x = 3$, a similar analysis shows that the derivative is infinite at these points (Problem 12 below).

PROBLEMS

1. Graph, determining wherever possible the features described at the beginning of Section 6-4:

(a) $y = x^3 - 3x^2 - 9x + 11$ (b) $y = x^3 + 6x^2 + 12x + 8$

(c) $y = x^4 + 2x^3 + 6x^2$ (d) $y = 3x^5 + 5x^3 + 15x$

(e) $y = x^{10} + 2x^5 + 1$ (f) $y = x^7 + x + 1$

(g) $y = \dfrac{1}{x} - \dfrac{2}{x^2}$ (h) $y = \dfrac{x}{x + 2}$

(i) $y = \dfrac{x^2}{x - 1}$ (j) $y = \dfrac{x}{x^2 + 4}$

(k) $y = \dfrac{x}{x^2 - 4}$ (l) $y = \dfrac{x^2}{x^2 - 1}$

(m) $y = 2 \sin x + \sin^2 x$ (n) $y = 8 \cos x + \cos 2x$

(o) $y = \text{Tan}^{-1} x$ (p) $y = \ln(\tan^2 x)$

(q) $y = \dfrac{e^x - 1}{e^x + 1}$ (r) $y = e^{-x^2}$

(s) $y = xe^{-x^2}$ (t) $y = x^2 e^{-x^2}$

(u) $y = \sqrt{x} + \dfrac{1}{\sqrt{x}}$ (v) $y = 3x^{2/3} + x^2$

2. Graph the function $y = f(x)$, $-\infty < x < \infty$, with the aid of the table given. Assume f, f', f'' to be continuous for all x and that all zeros of f' and f'' are given in the table.

(a)

x	f	f'	f''
0	0	0	1
1	1	5	0
2	3	0	-2
5	0	-2	-2

(b)

x	f	f'	f''
1	0	2	-1
2	1	0	0
3	2	2	0
4	5	0	-1
5	4	-2	0
6	2	0	1

3. Graph the function $y = f(x)$, $-\infty < x < \infty$, with the aid of the diagram given. Assume f, f', f'' to be continuous for all x and that all changes of sign of these three functions are given by the diagram.

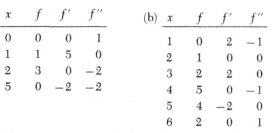

(a)

Value of f	0	3	2	1	3
Sign of f'	+	+	-	-	+	+
Sign of f''	-	-	-	+	+	-

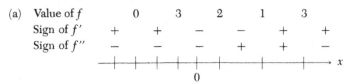

(b)

Value of f	1	2	3	2	1
Sign of f'	-	+	+	-	-	+
Sign of f''	+	+	-	-	+	+

4. Let $y = f(x)$ be defined for $x \ge a$ and have a continuous second derivative. Furthermore, let $\lim\limits_{x \to \infty} f(x) = b$, so that $y = b$ is a horizontal asymptote for f.

(a) Show that if $f''(x) > 0$ for $x \ge a$, then $f(x) > b$ for all x.

(b) Show that if $f(x) > b$ for all x and $f''(x) \ne 0$, then $f''(x) > 0$ for all x.

5. For each of the following equations verify that a function $y = f(x)$ is defined implicitly, and graph:

(a) $y^3 + (x^2 + 1)y - 2 = 0$, $-\infty < x < \infty$

(b) $y - \frac{1}{2}\sin y - x = 0$, $-2\pi \le x \le 2\pi$

6. Show that Equation (6-40) does define a function $y = f(x)$, $-\infty < x < \infty$. (See Section 3-8.)

7. Graph, observing discontinuities:

(a) $y = \dfrac{1}{\ln \dfrac{1}{x^2}}$ (b) $y = \text{Tan}^{-1}\dfrac{1}{x}$

8. Show that the parametric equations define $y = f(x)$, $-\infty < x < \infty$, and graph:

(a) $x = t + e^t$, $y = te^t$ (b) $x = 2t + \sin t$, $y = 2t + \cos t$

9. Graph:

(a) $y = (x - 1)(x + 2)$

(b) $y = (x + 3)(x - 3)$

(c) $y = x(x + 1)(x - 2)$

(d) $y = -(x - 2)(x - 3)(x + 4)$

(e) $y = (x - 1)^2(x + 2)$

(f) $y = (x + 3)^2(x - 3)$

(g) $y = (x - 1)^2(x + 2)^2$

(h) $y = x^3(x + 1)(x - 2)$

(i) $y^2 = x^2 - 4$

(j) $y^2 = x(x^2 - 9)$

(k) $y^2 = (x + 3)(x - 1)^2(x - 2)$

(l) $y^2 = x/(x - 1)$

(m) $y^2 = (x + 2)/x$

(n) $y^2 = \sin x$

(o) $y^3 = (x + 2)(x - 3)$

(p) $y^3 = x^2(x^2 - 1)$

(q) $y^2 = -(x - 1)^2(x - 2)^2(x - 3)^2$

10. For Example 5 of Section 6-4 prove that f is odd by considering the effect of replacing t by $-t$.

11. In the discussion of Example 5 above, it is asserted that

(1)
$$\lim_{x \to +\infty} (t - \sqrt[3]{x}) = 0$$

where $t(x)$ is the inverse of the function $x(t)$ defined by $x = t^3 + t$. Verify the correctness of the assertion by showing first that it is equivalent to

(2)
$$\lim_{t \to +\infty} (t - \sqrt[3]{t^3 + t}) = 0$$

and then to

(3)
$$\lim_{t \to +\infty} \frac{-t}{t^2 + t(t^3 + t)^{1/3} + (t^3 + t)^{2/3}} = 0$$

[To obtain (3) from (2), use the identity $(a - b)(a^2 + ab + b^2) = a^3 - b^3$. Now divide numerator and denominator by t.]

12. Let $g(x)$ be the function $y = \sqrt{x^4 - 5x^3 + 6x^2}$ considered in Example 7.

(a) Evaluate the (right-hand) derivative of $g(x)$ at $x = 3$ by showing that

$$\lim_{x \to 3+} \frac{g(x) - g(3)}{x - 3} = \lim_{x \to 3+} \frac{\sqrt{x^3 - 2x^2}}{\sqrt{x - 3}} = \infty$$

(b) Show that $g'(2) = \infty$ (left-hand derivative).

13. The graph of Example 7 (Figure 6-29) above can be regarded as the union of the graphs of

$$y = F(x) = \begin{cases} \sqrt{x^4 - 5x^3 + 6x^2}, & \text{if } 0 \le x \le 2 \text{ or } x > 3 \\ -\sqrt{x^4 - 5x^3 + 6x^2}, & \text{if } x < 0 \end{cases}$$

and

$$y = G(x) = \begin{cases} -\sqrt{x^4 - 5x^3 + 6x^2}, & \text{if } 0 \le x \le 2 \text{ or } x > 3 \\ \sqrt{x^4 - 5x^3 + 6x^2}, & \text{if } x < 0 \end{cases}$$

With the aid of Figure 6-29, sketch the graphs of F and G and show, with the aid of the results in the text, that F and G are differentiable for $-\infty < x < 2$ (including $x = 0$). Compare these functions and $y = \sqrt{x^4 - 5x^3 + 6x^2}$.

6-5 CHANGE OF COORDINATES

Very often graphing an equation is simplified by choosing new coordinate axes.

EXAMPLE 1. $x^2 + 2y^2 - 6x + 8y + 16 = 0$. As in Section 0-14, we complete the square and obtain the equation

$$(x - 3)^2 + 2(y + 2)^2 = 1$$

which we recognize as the equation of an ellipse with center at $x = 3$, $y = -2$. If we choose new axes parallel to the old ones with origin at $(3, -2)$, as in Figure 6-30, our equation becomes

$$x'^2 + 2y'^2 = 1$$

Here the new coordinates are related to the old coordinates by the equations

$$x' = x - 3, \qquad y' = y + 2$$

In general, a choice of new axes parallel to the old ones, as in the example, is called a *translation of axes*. If the new origin is (h, k) (in old coordinates), then new and old coordinates are related by the equation

$$x' = x - h, \qquad y' = y - k \qquad (6\text{-}50)$$

The signs can be checked by observing that, when $x' = 0$, $x = h$ and, when $y' = 0$, $y = k$.

EXAMPLE 2. Translate axes to eliminate the terms of first degree in the equation

$$x^2 + 2xy + 4y^2 + 4x - 8y + 5 = 0$$

For a general translation (6-50), $x = x' + h$, $y = y' + k$ and, in new coordinates, our equation becomes

$$(x' + h)^2 + 2(x' + h)(y' + k) + 4(y' + k)^2$$
$$+ 4(x' + h) - 8(y' + k) + 5 = 0$$

On expanding and collecting terms, the equation becomes

$$x'^2 + 2x'y' + 4y'^2 + x'(2h + 2k + 4) + y'(2h + 8k - 8)$$
$$+ h^2 + 2hk + 4k^2 + 4h - 8k + 5 = 0$$

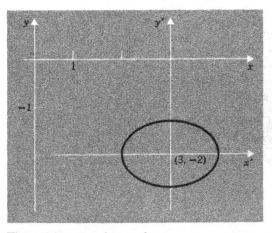

Figure 6-30   Translation of axes.

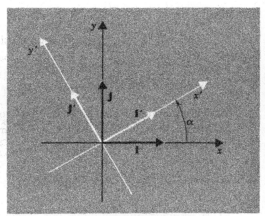

Figure 6-31   Rotation of axes.

If the first degree terms are to disappear, we must have

$$2h + 2k + 4 = 0, \qquad 2h + 8k - 8 = 0$$

Solving, we find $h = -4$, $k = 2$. In new coordinates, the equation becomes

$$x'^2 + 2x'y' + 4y'^2 - 11 = 0$$

The equation can be simplified further before graphing, as described below.

Now let the origin be unchanged, and let new axes be chosen in new directions, as in Figure 6-31. The directions can be given by unit vectors $\mathbf{i'}$, $\mathbf{j'}$, as shown. In the case of the figure, $\mathbf{i'}$, $\mathbf{j'}$ are oriented similarly to $\mathbf{i}$, $\mathbf{j}$ (Section 1-12), and we speak of a *rotation of axes* through angle α. Here α is the directed angle from $\mathbf{i}$ to $\mathbf{i'}$ which is the same as that from $\mathbf{j}$ to $\mathbf{j'}$. Hence

$$\mathbf{i'} = \cos \alpha \, \mathbf{i} + \sin \alpha \, \mathbf{j}, \qquad \mathbf{j'} = \mathbf{i'^{\dashv}} = -\sin \alpha \, \mathbf{i} + \cos \alpha \, \mathbf{j} \qquad (6\text{-}51)$$

For a general point P we have

$$\overrightarrow{OP} = x\mathbf{i} + y\mathbf{j} = x'\mathbf{i'} + y'\mathbf{j'}$$

and hence, by Equations (6-51),

$$x\mathbf{i} + y\mathbf{j} = x'(\cos \alpha \, \mathbf{i} + \sin \alpha \, \mathbf{j}) + y'(-\sin \alpha \, \mathbf{i} + \cos \alpha \, \mathbf{j})$$
$$= (x' \cos \alpha - y' \sin \alpha)\mathbf{i} + (x' \sin \alpha + y' \cos \alpha)\mathbf{j}$$

Accordingly,

$$x = x' \cos \alpha - y' \sin \alpha, \qquad y = x' \sin \alpha + y' \cos \alpha \qquad (6\text{-}52)$$

These equations relate old coordinates and new. We solve these equations for x', y' and find that

$$x' = x \cos \alpha + y \sin \alpha, \qquad y' = -x \sin \alpha + y \cos \alpha \qquad (6\text{-}52').$$

EXAMPLE 3. Simplify by rotating axes: $3x^2 + xy + 3y^2 = 1$. Because of the symmetry of the equation, we try rotating through $\pi/4$. Equations (6-52) give (with $\alpha = \pi/4$):

$$x = \frac{x' - y'}{\sqrt{2}}, \qquad y = \frac{x' + y'}{\sqrt{2}}$$

If we substitute in the given equation and simplify, we find that

$$7x'^2 + 5y'^2 = 2$$

The curve is an ellipse and is easily graphed as in Figure 6-32.

For a general equation $Ax^2 + Bxy + Cy^2 + F = 0$, where A, B, C, and F are constants and $B \neq 0$, a rotation of axes yields another equation of the same form. By proper choice of the rotation angle α, the equation has no term in $x'y'$; that is, it has the form:

$$A'x'^2 + Cy'^2 + F' = 0$$

Therefore, it can be graphed as in Section 0-14. The curve is an ellipse or hyperbola (possibly degenerate) or a degenerate parabola. The details are left to Problem 5 below.

For the *general equation of second degree*

$$Ax^2 + Bxy + Cy^2 + Dx + Ey + F = 0 \qquad (6\text{-}53)$$

we can proceed similarly to eliminate the term in xy and to obtain an equation

$$A'x'^2 + C'y^2 + D'x' + E'y' + F' = 0$$

This can be graphed as in Sections 0-13 and 0-14, or with the aid of a translation of axes. Hence we conclude that *the graph of a second degree equation (6-53) is always an ellipse, a hyperbola, or a parabola (possibly degenerate).*

If the new axes are chosen with the same origin but with the *opposite orientation*, as in Figure 6-33, then we have

$$\mathbf{i}' = \cos \alpha \, \mathbf{i} + \sin \alpha \, \mathbf{j}, \qquad \mathbf{j}' = -\mathbf{i}'^{\dashv} = \sin \alpha \, \mathbf{i} - \cos \alpha \, \mathbf{j} \qquad (6\text{-}54)$$

The new axes can be obtained from the old ones in two stages: by first rotating through angle α, to obtain unit vectors $\mathbf{i}''$, $\mathbf{j}''$, and then *reflecting* in the axis through O in the direction of $\mathbf{i}''$. The new coordinates (x', y') can be related, as above, to the old coordinates:

$$x = x' \cos \alpha + y' \sin \alpha, \qquad y = x' \sin \alpha - y' \cos \alpha \qquad (6\text{-}55)$$

and

$$x' = x \cos \alpha + y \sin \alpha, \qquad y' = x \sin \alpha - y \cos \alpha \qquad (6\text{-}55')$$

If new axes are chosen with both a new origin at (h, k) (old coordinates) and new directions (Figure 6-34), then we obtain new coordinates (x', y') that are related to old coordinates (x, y) by the following reasoning:

$$\overrightarrow{OP} = x\mathbf{i} + y\mathbf{j} = \overrightarrow{OO'} + \overrightarrow{O'P} = h\mathbf{i} + k\mathbf{j} + x'\mathbf{i}' + y'\mathbf{j}'$$
$$= h\mathbf{i} + k\mathbf{j} + x'(\cos \alpha \, \mathbf{i} + \sin \alpha \, \mathbf{j}) + y'(\mp\sin \alpha \, \mathbf{i} \pm \cos \alpha \, \mathbf{j})$$

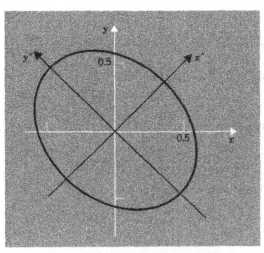

Figure 6-32   Ellipse of Example 3.

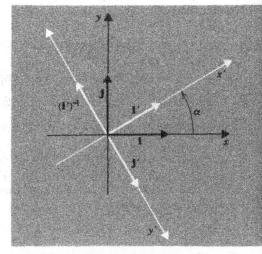

Figure 6-33   Opposite orientation of axes.

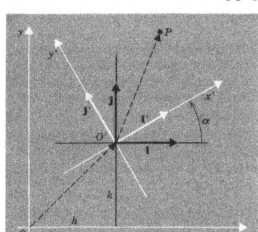

Figure 6-34   General Euclidean change of coordinates.

Hence

$$x = h + x' \cos \alpha \mp y' \sin \alpha, \qquad y = k + x' \sin \alpha \pm y' \cos \alpha \qquad (6\text{-}56)$$

Here the upper signs hold true for a similar orientation and the lower signs hold true for the opposite orientation. We can consider the new coordinates to be obtained from the old by first translating axes to O': (h, k), then rotating through angle α, and then, if the orientation is opposite, reflecting in the axis through O' with direction $\mathbf{i}'$.

The equations (6-56) are called the *equations for Euclidean coordinate transformations*. They describe how all Cartesian coordinate systems in the plane can be obtained from one given system. For a given graph in the plane, we can use whichever coordinate system we choose. Certain features of the graph will be the same in all coordinate systems; they are the basic *geometric features*. An example is the *degree* of an equation. It is easily seen that a graph given by a linear (first degree) equation $ax + by + c = 0$ in one coordinate system is given by a linear equation in all other Cartesian coordinate systems. The underlying geometric feature here is that the graph is a *straight line* (Section 0-7). Similarly, a second degree equation (6-53) remains of second degree under all changes of coordinates (6-56); here we are dealing with the geometric property of being a *conic section*. Other geometric features are the angle between two lines, the eccentricity of an ellipse, the distance from a point to a line, and the area of a triangle. Each can be computed in one coordinate system; a change of coordinates does not affect the result. However, the *slope* of a line is not a geometric feature; it is changed when the axes are rotated. The features that are not changed are often called *geometric invariants*, or simply *invariants*.

For the second degree equation (6-53), the quantity $B^2 - 4AC$ can also be shown to be an invariant (Problem 5 below). For an ellipse $(x^2/a^2) + (y^2/b^2) = 1$, we have $B^2 - 4AC = -4/(a^2 b^2) < 0$. Hence $B^2 - 4AC$ *must be negative for all ellipses*. Similarly, we find that $B^2 - 4AC > 0$ for hyperbolas, $B^2 - 4AC = 0$ for parabolas. (Here we include the degenerate cases as in Sections 0-13, and 0-14.)

When the calculus is applied, it is important to know which quantities are invariants. The derivative, like the slope of a line, is not an invariant; however, the angle between two curves (Section 3-13) is an invariant. The definite integral represents area, and hence is an invariant (for the specific figure involved).

EXAMPLE 4. Graph the curve $x - y = e^{-(x+y)^2}$. The equation is difficult as it stands. As in Example 3, a rotation of axes by $\pi/4$ is suggested. We denote the new coordinates by u, v, rather than x', y', to avoid confusion with derivatives. As in Example 3, we have

$$x = \frac{u - v}{\sqrt{2}}, \qquad y = \frac{u + v}{\sqrt{2}}$$

so that $x + y = \sqrt{2}\, u$, $x - y = -\sqrt{2}\, v$ and our equation becomes

$$-\sqrt{2}\, v = e^{-2u^2} \qquad \text{or} \qquad v = -(1/\sqrt{2})e^{-2u^2}$$

Hence

$$\frac{dv}{du} = 2\sqrt{2}\, u e^{-2u^2}, \qquad \frac{d^2v}{du^2} = 2\sqrt{2}\, e^{-2u^2}(1 - 4u^2)$$

Consequently, there is a critical point at $u = 0$, and there are inflection points at $u = \pm 1/2$. The graph is easily completed as in Figure 6-35. The curve is symmetric with respect to the v-axis; that is, with respect to the line $x + y = 0$. The equation of the line of symmetry depends on the coordinate system, but the property of *being symmetric* is a geometric feature, independent of the coordinates.

Non-Cartesian Coordinate Systems. At times it is of advantage to use different scales on the two axes, or to use oblique axes. We may even use special irregular scales on the axes, such as logarithmic scales. We may also

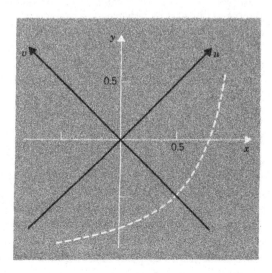

Figure 6-35   Curve of Example 4.

use coordinate systems based on certain curves, rather than on straight lines; we call them *curvilinear* coordinate systems. The most important example is *polar coordinates* (Section 0-16 and Section 6-8 below).

PROBLEMS

1. The axes are translated so that the new origin is $(3, -2)$.
   (a) Find the equations relating new and old coordinates.
   (b) Find the new coordinates of the points whose old coordinates are $(0, 0)$, $(1, 3)$ and $(-2, 4)$, and check graphically.
   (c) Find the new equation of the curve whose old equation is $x - y + 2 = 0$, $x^2 + y^2 = 1$, $xy = 2$.

2. The axes are rotated through an angle $\pi/3$.
   (a) Find the equations for (x, y) in terms of (x', y') and those for (x', y') in terms of (x, y).
   (b) Find the new coordinates of the points whose old coordinates are $(1, 0)$, $(0, 1)$, $(1, 1)$, and $(-3, 2)$.
   (c) Find the new equation for each of the curves $y - \sqrt{3}\,x = 0$, $x^2 + y^2 = 1$, $xy = 1$.

3. Reduce to standard form by translation of axes and graph:
   (a) $3x^2 + 3y^2 - 6x + 12y - 16 = 0$ (b) $x^2 + y^2 - 2x + 4y = 0$
   (c) $4x^2 + y^2 + 8x - 6y + 9 = 0$ (d) $4y^2 - 32x + 4y - 63 = 0$

4. (a) Show that a translation of axes does not affect the second degree terms in the equation of second degree (6-53).
   (b) Choose a translation to make the constant term zero for the equation $x^2 + xy + 3x - 2y + 5 = 0$.

5. (a) Show that a rotation of axes through angle α replaces the second degree equation (6-53) by a similar equation $A'x'^2 + B'x'y' + C'y'^2 + \cdots = 0$ in which
$$A' = A\cos^2\alpha + B\cos\alpha\sin\alpha + C\sin^2\alpha$$
$$B' = B(\cos^2\alpha - \sin^2\alpha) - 2(A - C)\sin\alpha\cos\alpha$$
$$C' = A\sin^2\alpha - B\cos\alpha\sin\alpha + C\cos^2\alpha$$
   (b) Show from the result of (a) that $B^2 - 4AC = B'^2 - 4A'C'$, also that $A + C = A' + C'$.
   (c) Show that $B' = 0$ if α is chosen so that
$$\tan 2\alpha = \frac{B}{A - C}$$
   How should α be chosen if $A = C$?
   (d) Show that a transformation $x'' = x$, $y'' = -y$ transforms the Equation (6-53) into a similar one
$$A''x''^2 + B''x''y'' + C''y''^2 + \cdots = 0$$
   in which $A'' = A$, $B'' = -B$, $C'' = C$, and that, hence, $B''^2 - 4A''C'' = B^2 - 4AC$, $A'' + C'' = A + C$. Show that this transformation corresponds to a change in the orientation.
   (e) Conclude from the results of parts (a), (b) and (d) and Problem 4(a) that both $B^2 - 4AC$ and $A + C$ represent geometric invariants of the second degree Equation (6-53).

(f) Show from the result of part (e) that the degree of Equation (6-53) is a geometric invariant. (*Hint.* If a change of coordinates is made, the new equation has form $A'x'^2 + B'x'y' + C'y'^2 + \cdots$ and, hence, is of degree at most 2. Show that $A' = 0, B' = 0, C' = 0$ is possible only if $B^2 - 4AC = 0$ and $A + C = 0$ and therefore only if $A = 0$, $B = 0$, and $C = 0$.)

6. *Elimination of term in xy.* As in Problem 5(c) we can always eliminate the term in xy in (6-53) by a suitable rotation, with α chosen so that $\tan 2\alpha = B/(A - C)$. We can always choose 2α between 0 and π to satisfy this condition. If $B > 0$, we then consider the triangle with sides B, $A - C$ to read off $\cos 2\alpha$, $\sin 2\alpha$; $\sin 2\alpha = B/\lambda$, $\cos 2\alpha = (A - C)/\lambda$, where $\lambda = [B^2 + (A - C)^2]^{1/2}$. If $B < 0$, we consider triangle with sides $-B$, $C - A$ and have $\sin 2\alpha = -B/\lambda$, and $\cos 2\alpha = (C - A)/\lambda$ (See Figure 6-36).

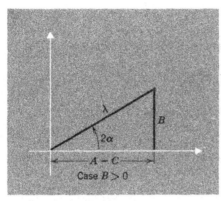 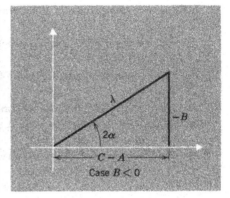

Figure 6-36

(a) Show that a rotation has no effect on the constant term in (6-53) and that, by the result of Problem 5(a),

$$A' = \frac{1}{2}[(A + C) + (A - C)\cos 2\alpha + B \sin 2\alpha]$$

(b) Find $\cos 2\alpha$, $\sin 2\alpha$ for a rotation to eliminate the xy-term for the equation

$$x^2 - 5xy + 7y^2 - 4 = 0$$

and find the new equation. [*Hint.* A' is given in part (a) and by Problem 5(b) $A' + C' = A + C$. By (a) the constant term is unchanged.]

7. Determine the type of curve and graph, with the aid of the procedures of Problem 6:
   (a) $3x^2 - 8xy - 3y^2 - 4 = 0$
   (b) $15x^2 + 2\sqrt{3}\,xy + 5y^2 - 1 = 0$
   (c) $8x^2 - 4xy + 5y^2 - 36x + 18y + 75 = 0$
   (d) $2x^2 - 4xy - y^2 + 7x - 2y + 3 = 0$
   [*Hint.* For (c) and (d) eliminate the terms in x and y first, by a suitable translation.]

8. Let T be the locus of a polynomial equation in x and y. Let d be the lowest possible degree of an equation for T in the given Cartesian coordinates. Show that d is a geometric invariant. [*Hint.* Show that if (x', y') are new Cartesian coordinates, then an equation $p(x, y) = 0$ becomes an equation $q(x', y') = 0$, where the degree of q is at most equal to the degree of p. Now reverse the roles of (x, y) and (x', y').]

9. Graph with the aid of a suitable rotation of axes:

(a) $x + y = \sin^2(x - y)$ (b) $x + y = (x - y) - (x - y)^3$

(c) $3x - 4y = 1/(4x + 3y)^2$ (d) $x^3 + xy + y^3 = 0$

10. Reduce to the form $v = A \sin u$ by a suitable translation of axes and graph:

(a) $y = \sin x + \cos x$ (b) $y = 3 \sin x + 4 \cos x$

(c) $y = 2 - 5 \sin x + 12 \cos x$ (d) $y = 3 + \sin x + \sqrt{3} \cos x$

11. An equation in x, y is said to be unchanged by a particular change of coordinates if the new equation is the same except for the replacement of x by x', y by y'. For example, the equation $x^2 + y^2 = 1$ is unchanged by every rotation of axes. For each of the following changes of coordinates, what special features does a graph have if its equation is unchanged by this change (or changes) of coordinates?

(a) $x = x' + 2\pi, y = y'$ (b) $x = x' + 1, y = y' + 1$

(c) $x = -x', y = y'$ (d) $x = -x', y = -y'$

(e) rotation of axes by π (f) rotation of axes by $\pi/3$

(g) $x = (x' + y')/\sqrt{2}, y = (x' - y')/\sqrt{2}$

(h) $x = x' + 1, y = y'$ and $x = x', y = y' + 1$

(i) all translations of the form $x = x' + h, y' = y$

(j) all translations

6-6 PLANE CURVES: VECTOR EQUATIONS, CURVATURE

In this section we study a plane curve given in parametric form:

$$x = f(t), \qquad y = g(t), \qquad a \le t \le b \tag{6-60}$$

or, equivalently, by a vector equation

$$\mathbf{r} = \overrightarrow{OP} = \mathbf{F}(t) = f(t)\mathbf{i} + g(t)\mathbf{j}, \qquad a \le t \le b \tag{6-61}$$

(See Sections 3-10, 3-11, and 4-27 to 4-31.) Throughout f and g will be assumed to be continuous.

Now let a path (6-60) be given that does not pass through the origin. Then we can assign polar coordinates r, θ to each point on the path. The angle θ is determined only up to addition of multiples of 2π. However, once we have chosen θ for $t = a$—that is, at the point $[f(a), g(a)]$, we can choose θ to be a continuous function of t on the path, as suggested in Figure 6-37. We call

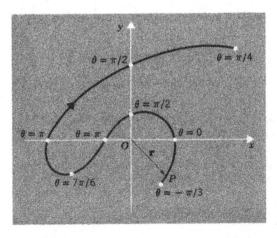

Figure 6-37 Angle function.

such a choice of θ an *angle function* for the path. If f and g have continuous derivatives f', g', then we can write $\tan \theta = y/x = g(t)/f(t)$, except where $f(t) = 0$. Hence

$$\sec^2\theta \frac{d\theta}{dt} = \frac{f(t)g'(t) - g(t)f'(t)}{f(t)^2}$$

Since $\sec^2\theta = 1 + \tan^2\theta = 1 + (y/x)^2 = 1 + [g(t)/f(t)]^2$, this equation can be written:

$$\frac{d\theta}{dt} = \frac{xy' - yx'}{x^2 + y^2} = \frac{f(t)g'(t) - g(t)f'(t)}{[f(t)]^2 + [g(t)]^2} \tag{6-62}$$

This equation is also valid where $f(t) = 0$. For a proof and a more complete discussion of angle functions, see Section 5-4.

A vector function (6-61) can always be interpreted as a path (6-60), and hence also has an angle function (provided always that $\mathbf{F}(t) \neq \mathbf{0}$ for $a \leq t \leq b$). At times the angle associated with the vector must be distinguished from a polar angle θ; for that reason, we often use another notation, for instance, $\varphi(t)$.

THEOREM 4. *Let the vector function* $\mathbf{F}(t)$ *be differentiable for* $a \leq t \leq b$ *and let* $|\mathbf{F}(t)| \equiv 1$ *for* $a \leq t \leq b$, *so that* $\mathbf{F}(t)$ *is a unit vector. Let* $\varphi(t)$ *be an angle function for* $\mathbf{F}(t)$, *so that*

$$\mathbf{F}(t) = \cos \varphi(t)\mathbf{i} + \sin \varphi(t)\mathbf{j}$$

Then

$$\mathbf{F}'(t) = \frac{d\varphi}{dt}[-\sin \varphi(t)\mathbf{i} + \cos \varphi(t)\mathbf{j}] = \frac{d\varphi}{dt} \mathbf{F}^{\dashv}$$

where $\mathbf{F}^{\dashv} = -\sin \varphi \, \mathbf{i} + \cos \varphi \, \mathbf{j}$ *is obtained from* $\mathbf{F}$ *by rotating through* $\pi/2$ *in the counterclockwise direction.*

PROOF. By rules of calculus (Section 3-12)

$$\frac{d\mathbf{F}}{dt} = \frac{d}{dt}[\cos \varphi(t)]\mathbf{i} + \frac{d}{dt}[\sin \varphi(t)]\mathbf{j} = -\frac{d\varphi}{dt}\sin \varphi(t)\mathbf{i} + \frac{d\varphi}{dt}\cos \varphi(t)\mathbf{j}$$

$$= \frac{d\varphi}{dt}[-\sin \varphi(t)\mathbf{i} + \cos \varphi(t)\mathbf{j}] = \frac{d\varphi}{dt}\left\{\cos\left[\varphi(t) + \frac{\pi}{2}\right]\mathbf{i} + \sin\left[\varphi(t) + \frac{\pi}{2}\right]\mathbf{j}\right\}$$

$$= \frac{d\varphi}{dt} \mathbf{F}^{\dashv}$$

If we interpret $\mathbf{F}(t)$ as $\overrightarrow{OP}$, as in Figure 6-38, then P is moving on a circle of radius 1 and center O, φ is the polar coordinate angle of P and $d\varphi/dt$ is the *angular velocity* of P. The derivative $\mathbf{F}'(t)$ is the velocity vector of P and is tangent to the circle at P, with magnitude equal to $|d\varphi/dt|$.

We now return to a general path (6-60) and assume that f and g have continuous first and second derivatives. The velocity vector

$$\mathbf{v} = \frac{d\,\overrightarrow{OP}}{dt} = \mathbf{F}'(t) = f'(t)\mathbf{i} + g'(t)\mathbf{j}$$

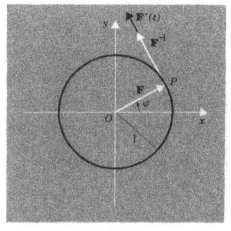

Figure 6-38 Derivative of unit vector.

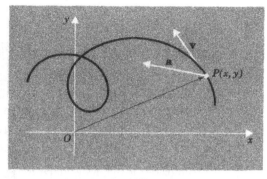

Figure 6-39 Velocity and acceleration vectors.

and the acceleration vector

$$\mathbf{a} = \frac{d\mathbf{v}}{dt} = \mathbf{F}''(t) = f''(t)\mathbf{i} + g''(t)\mathbf{j}$$

are then also defined for $a \le t \le b$ (see Figure 6-39).

In Section 4-28, it is shown that arc length s can be defined along our path. The value of s for each t gives the distance the point P has moved from time a to time t. Thus, for $t = a$, $s = 0$, and for $t = b$, $s = l =$ total distance along the path. (Occasionally one may measure s backward starting at $t = b$, but it is usually more convenient to have s increase with t.) Accordingly, s is a function of t for $a \le t \le b$ and, as shown in Section 4-28,

$$\frac{ds}{dt} = |\mathbf{v}| = \sqrt{\left(\frac{dx}{dt}\right)^2 + \left(\frac{dy}{dt}\right)^2} \qquad (6\text{-}63)$$

In particular, $ds/dt \ge 0$, as is to be expected, since s is a nondecreasing function of t. Usually, $ds/dt = 0$ only at exceptional points and, for simplicity, we shall here assume $ds/dt > 0$, so that s is a strictly increasing function of t. It follows that the inverse function $t(s)$ is also well defined and continuous for $0 \le x \le l$ and

$$\frac{dt}{ds} = \frac{1}{|\mathbf{v}|} = \frac{1}{ds/dt}$$

In fact we can change from t to s as parameter by the equations:

$$x = f[t(s)] = \alpha(s), \qquad y = g[t(s)] = \beta(s)$$

or

$$\mathbf{r} = \alpha(s)\mathbf{i} + \beta(s)\mathbf{j}, \qquad 0 \le s \le l$$

As in Section 4-29, we have thereby replaced our path by an equivalent one. The new "velocity vector"

$$\frac{d\mathbf{r}}{ds} = \frac{dx}{ds}\mathbf{i} + \frac{dy}{ds}\mathbf{j}$$

has components

$$\frac{dx}{ds} = \frac{dx}{dt}\frac{dt}{ds} = \frac{1}{|\mathbf{v}|}\frac{dx}{dt}$$
$$\frac{dy}{ds} = \frac{dy}{dt}\frac{dt}{ds} = \frac{1}{|\mathbf{v}|}\frac{dy}{dt} \qquad (6\text{-}64)$$

which are proportional to the old ones and since

$$\left(\frac{dx}{ds}\right)^2 + \left(\frac{dy}{ds}\right)^2 = \frac{1}{|\mathbf{v}|^2}\cdot\left[\left(\frac{dx}{dt}\right)^2 + \left(\frac{dy}{dt}\right)^2\right] = 1$$

the new "velocity vector" is a unit vector. We denote this vector by $\mathbf{T}$:

$$\mathbf{T} = \frac{dx}{ds}\mathbf{i} + \frac{dy}{ds}\mathbf{j} \qquad (6\text{-}65)$$

From (6-64) it follows at once that

$$\mathbf{T} = \frac{1}{|\mathbf{v}|}\mathbf{v} \qquad \text{or} \qquad \mathbf{v} = |\mathbf{v}|\,\mathbf{T} \qquad (6\text{-}66)$$

so that $\mathbf{T}$ is simply the unit tangent vector in the direction of $\mathbf{v}$, hence in the direction of increasing s.

Curvature. Since $\mathbf{T}$ is a unit vector, we can introduce an angle function $\varphi(t)$ as in Theorem 4 and write

$$\mathbf{T} = \cos\varphi\,\mathbf{i} + \sin\varphi\,\mathbf{j}$$

The angle φ, as shown in Figure 6-40, is the angle from the direction of the positive x-axis to $\mathbf{T}$ (or, equivalently, to $\mathbf{v}$). Thus φ is simply the angle of inclination (Section 0-7) of the tangent line. As P moves along the path, φ varies and, consequently, φ becomes a function of t or s. Any two choices of the angle function $\varphi(t)$ differ by the constant function $2k\pi$ for some integer k.

Regardless of which function we choose for φ, the derivative $d\varphi/ds$ will then be uniquely determined as a continuous function of s; the absolute value of this derivative is defined as the *curvature* κ:

$$\kappa = \kappa(s) = \left|\frac{d\varphi}{ds}\right| \qquad (6\text{-}67a)$$

The reciprocal of the curvature is the *radius of curvature* ρ:

$$\rho = \rho(s) = \frac{1}{\kappa(s)} \qquad (6\text{-}67b)$$

The geometric meaning of these quantities is as follows: In general, $d\varphi/ds$ measures the rate of change of direction along the path. For example, in riding in an automobile, we could observe the variation in φ by holding a compass; if we are traveling at 60 mph and φ changes by $10°$ or $\pi/18$ radian over 1 minute (distance 1 mile), then

$$\left|\frac{d\varphi}{ds}\right| \sim \frac{\pi/18}{1} = .174 \text{ radians per mile}$$

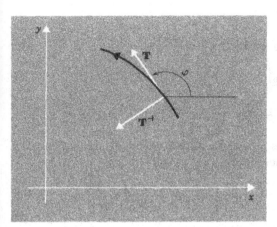

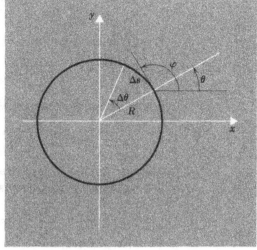

Figure 6-40 Unit tangent vector.

Figure 6-41 Radius of curvature as radius of circular path.

The more rapidly we are changing direction, the larger the value of $\kappa = |d\varphi/ds|$, the more curved the path. In the case of travel along a straight line, $d\varphi/ds = 0$, and we have the extreme case $\kappa = 0$; ρ would have to be defined to be infinite (a straight line is regarded as a "circle of infinite radius"). In the case of travel along a circular path $x^2 + y^2 = R^2$, ρ is simply the radius R. Let the polar angle θ increase along the path. Then, as in Figure 6-41, we have $\varphi = \theta + \pi/2$; but $\Delta s = R\,\Delta\theta$, so that $d\theta/ds = 1/R$ and

$$\frac{1}{\rho} = \kappa = \left|\frac{d\varphi}{ds}\right| = \left|\frac{d\theta}{ds}\right| = \frac{1}{R}$$

so that $\rho = R$. If θ decreases with s we still have $\rho = R$.

For the general path, ρ can be interpreted as a radius of a "circle of curvature" that approximates the path near a particular point P (see Section 6-7 below).

THEOREM 5. *For the path (6-60), let f and g have continuous first and second derivatives on $[a, b]$, let $\mathbf{a} = f''(t)\mathbf{i} + g''(t)\mathbf{j}$, and let arc length s be introduced so that $ds/dt = |\mathbf{v}| \neq 0$. Then the curvature κ is given at each point by each of the formulas:*

$$\kappa = \left|\frac{d\mathbf{T}}{ds}\right| \tag{6-68a}$$

$$\kappa = \left|\frac{x'y'' - y'x''}{(x'^2 + y'^2)^{3/2}}\right| \tag{6-68b}$$

$$\kappa = \frac{|\mathbf{v}^{\perp} \cdot \mathbf{a}|}{|\mathbf{v}|^3} \tag{6-68c}$$

PROOF. Since $\mathbf{T}$ is a unit vector, we can apply Theorem 4, with $\mathbf{T} = \mathbf{T}(s)$, so that s is the parameter instead of t. Accordingly, at each s

$$\frac{d\mathbf{T}}{ds} = \frac{d\varphi}{ds} \mathbf{T}^{\dashv}$$

Now $\mathbf{T}^{\dashv}$ is also a unit vector (see Figure 6-40). Accordingly

$$\left|\frac{d\mathbf{T}}{ds}\right| = \left|\frac{d\varphi}{ds}\right| |\mathbf{T}^{\dashv}| = \left|\frac{d\varphi}{ds}\right| = \kappa$$

so that (6-68a) is proved. Now

$$\frac{d\varphi}{ds} = \frac{d\varphi}{dt}\frac{dt}{ds} = \frac{1}{|\mathbf{v}|}\frac{d\varphi}{dt}$$

Also $\varphi(t)$ can be regarded as the angle function for $\mathbf{v}(t) = x'(t)\mathbf{i} + y'(t)\mathbf{j}$. Hence by (6-62), with θ replaced by φ, x by x', y by y'

$$\frac{d\varphi}{ds} = \frac{1}{|\mathbf{v}|}\frac{d\varphi}{dt} = \frac{1}{|\mathbf{v}|}\frac{x'y'' - y'x''}{x'^2 + y'^2} = \frac{x'y'' - y'x''}{(x'^2 + y'^2)^{3/2}} \qquad (6\text{-}69)$$

If we take absolute values, we obtain (6-68b). Finally, $\mathbf{v} = x'\mathbf{i} + y'\mathbf{j}$, $\mathbf{v}^{\dashv} = -y'\mathbf{i} + x'\mathbf{j}$, $\mathbf{a} = x''\mathbf{i} + y''\mathbf{j}$, so that (6-69) can be written

$$\frac{d\varphi}{ds} = \frac{\mathbf{v}^{\dashv} \cdot \mathbf{a}}{|\mathbf{v}|^3} \qquad (6\text{-}69')$$

If we take absolute values, we obtain (6-68c).

COROLLARY OF THEOREM 5. *For a curve given in rectangular coordinates as $y = f(x)$, $a \le x \le b$, where f has continuous first and second derivatives,*

$$\kappa = \frac{|y''|}{(1 + y'^2)^{3/2}} \qquad (6\text{-}68\text{d})$$

PROOF. We can regard our curve as the path $x = t$, $y = f(t)$, $a \le t \le b$. In (6-68b) we then have

$$x' = \frac{dx}{dt} = 1, \qquad x'' = \frac{d^2x}{dt^2} = 0$$

$$y' = f'(t) = \frac{dy}{dx}, \qquad y'' = f''(t) = \frac{d^2y}{dx^2}$$

and (6-68d) follows.

EXAMPLE 1. $x = 3\cos t$, $y = 2\sin t$, $0 \le t \le 2\pi$. By eliminating t, we observe that these are parametric equations of the ellipse

$$\frac{x^2}{9} + \frac{y^2}{4} = 1$$

(see Figure 6-42). From (6-68b) we find at once that

$$\kappa = \frac{1}{\rho} = \frac{6}{(9\sin^2 t + 4\cos^2 t)^{3/2}}$$

Thus, for $t = 0$, $\kappa = \frac{3}{4}$, and $\rho = \frac{4}{3}$; for $t = \pi/2$, $\kappa = \frac{2}{9}$, and $\rho = \frac{9}{2}$; for $t =$

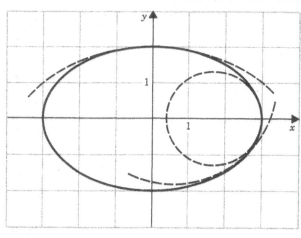

Figure 6-42 Curvature of ellipse.

$7\pi/4, \kappa = (\tfrac{12}{13})\sqrt{\tfrac{2}{13}} = 0.36$, and $\rho = 2.8$. The corresponding circles of curvature are shown in Figure 6-42. As the figure suggests, the circles of curvature very closely approximate the path near each point and, hence, are a considerable aid in graphing.

EXAMPLE 2. $y = x^2$, $-\infty < x < \infty$. Here Equation (6-58d) gives

$$\kappa = \frac{2}{(1 + 4x^2)^{3/2}}, \qquad -\infty < x < \infty$$

The x-interval is infinite here. The discussion above applies to each finite interval and hence, in effect, to all x. At $x = 0$, $\kappa = 2$, $\rho = \tfrac{1}{2}$; at $x = 1$, $\kappa = 2 \times 5^{3/2} = 0.18$, $\rho = 5.6$, as shown in Figure 6-43.

6-7 TANGENTIAL AND NORMAL COMPONENTS OF ACCELERATION. CIRCLE OF CURVATURE

With reference to our path of Section 6-6, we now introduce a *unit normal vector* **N**, as suggested in Figure 6-44. Thus **N** is obtained from **T** by rotating through $90°$. However, we wish to choose **N** always on the "concave" side of

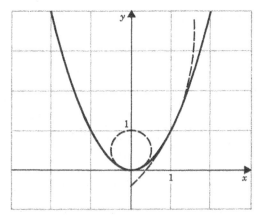

Figure 6-43 Curvature of parabola.

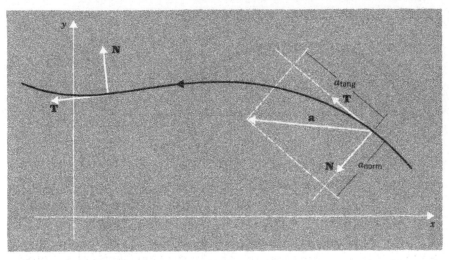

Figure 6-44   Tangential and normal components of acceleration.

the path. This means that when $d\varphi/ds \geq 0$, as at P_1 in Figure 6-45, we choose

$$\mathbf{N} = \cos\left(\varphi + \frac{\pi}{2}\right)\mathbf{i} + \sin\left(\varphi + \frac{\pi}{2}\right)\mathbf{j} = -\sin\varphi\,\mathbf{i} + \cos\varphi\,\mathbf{j} = \mathbf{T}^\dashv \quad (6\text{-}70a)$$

and when $d\varphi/ds < 0$, as at P_2 in Figure 6-45, we choose

$$\mathbf{N} = \cos\left(\varphi - \frac{\pi}{2}\right)\mathbf{i} + \sin\left(\varphi - \frac{\pi}{2}\right)\mathbf{j} = \sin\varphi\,\mathbf{i} - \cos\varphi\,\mathbf{j} = -\mathbf{T}^\dashv \quad (6\text{-}70b)$$

In each case, $\mathbf{T}$ and $\mathbf{N}$ are a pair of perpendicular unit vectors, and we can write the acceleration vector as follows:

$$\mathbf{a} = a_{\text{tang}}\mathbf{T} + a_{\text{norm}}\mathbf{N} \quad (6\text{-}71)$$

where a_{tang} and a_{norm} are the tangential and normal components of $\mathbf{a}$, as suggested in Figure 6-44. We shall observe that, with our choice of $\mathbf{N}$, a_{norm} is always positive or zero, so that $\mathbf{a}$ also points to the concave side of the curve (see Problem 7 below).

THEOREM 6. *For the path of Theorem 5, let the normal vector $\mathbf{N}$ be chosen as $\mathbf{T}^\dashv$ when $d\varphi/ds \geq 0$ and as $-\mathbf{T}^\dashv$ when $d\varphi/ds < 0$. Then the acceleration vector $\mathbf{a}$ has corresponding components*

$$a_{\text{tang}} = \frac{d|\mathbf{v}|}{dt} = \frac{d^2s}{dt^2}, \qquad a_{\text{norm}} = \frac{|\mathbf{v}|^2}{\rho} = \kappa|\mathbf{v}|^2 \quad (6\text{-}72)$$

PROOF. We write

$$\mathbf{v} = |\mathbf{v}|\mathbf{T}$$

$$\mathbf{a} = \frac{d\mathbf{v}}{dt} = |\mathbf{v}|\frac{d\mathbf{T}}{dt} + \frac{d|\mathbf{v}|}{dt}\mathbf{T} = |\mathbf{v}|\frac{ds}{dt}\frac{d\mathbf{T}}{ds} + \frac{d}{dt}\left(\frac{ds}{dt}\right)\mathbf{T}$$

As in Theorem 4 and in the proof of Theorem 5,

$$\frac{d\mathbf{T}}{ds} = \frac{d\varphi}{ds}\mathbf{T}^\dashv$$

Accordingly

$$\mathbf{a} = |\mathbf{v}|^2 \frac{d\varphi}{ds} \mathbf{T}^{\dashv} + \frac{d^2s}{dt^2} \mathbf{T} = \pm |\mathbf{v}|^2 \frac{d\varphi}{ds} \mathbf{N} + \frac{d^2s}{dt^2} \mathbf{T}$$

By our choice of $\mathbf{N}$, the $+$ sign holds true when $d\varphi/ds \geq 0$, the $-$ sign when $d\varphi/ds \leq 0$. Accordingly, in either case,

$$\mathbf{a} = |\mathbf{v}|^2 \left| \frac{d\varphi}{ds} \right| \mathbf{N} + \frac{d^2s}{dt^2} \mathbf{T} = (|\mathbf{v}|^2\kappa)\mathbf{N} + \frac{d^2s}{dt^2} \mathbf{T}$$

Thus the components of $\mathbf{a}$ are as in (6-72).

Remark. By (6-69)

$$\frac{d\varphi}{ds} = \frac{x'y'' - y'x''}{(x'^2 + y'^2)^{3/2}}$$

From this equation, we can determine whether $d\varphi/ds$ is positive or negative and, hence, whether the curve is concave "to the left" or "to the right" as at P_1, P_2, respectively, in Figure 6-45. As the figure suggests, the transition from $+$ to $-$ occurs at inflection points (Problem 5 below). A skater executing such a pattern (or a football player) would be very well aware of passing through these points, since his acceleration must swing from one side of the path to the other.

Circle of Curvature. This is defined, at each point P along the path, as the circle of radius ρ passing through P with center Q along the normal line at P in the direction of $\mathbf{N}$. Thus $\overrightarrow{OQ} = \overrightarrow{OP} + \rho\mathbf{N}$, as in Figure 6-46. This circle is in a sense the *best circular approximation* to the path near P (Problem 4 below); in the same sense the tangent is the best straight line approximation of the path at P.

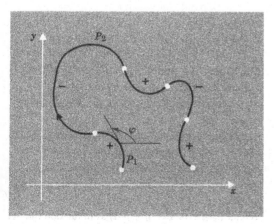

Figure 6-45 Variation of sign of $d\varphi/ds$.

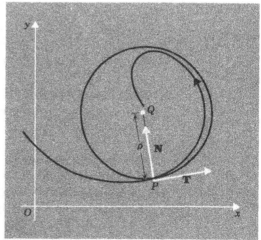

Figure 6-46 Circle of curvature.

The point Q is called the *center of curvature*. From the definition we find that Q is given by the equations:

$$\overrightarrow{OQ} = \overrightarrow{OP} + \left|\frac{ds}{d\varphi}\right|\mathbf{N} = \overrightarrow{OP} + \frac{x'^2 + y'^2}{x'y'' - y'x''}(-y'\mathbf{i} + x'\mathbf{j}) \quad (6\text{-}73)$$

Hence Q has coordinates (x_c, y_c) given by

$$x_c = x - y'\frac{x'^2 + y'^2}{x'y'' - y'x''} = f(t) - g'(t)(\ldots)$$

$$(6\text{-}74)$$

$$y_c = y + x'\frac{x'^2 + y'^2}{x'y'' - y'x''} = g(t) + f'(t)(\ldots)$$

As t varies, (x_c, y_c) traces a path called the *evolute* of the given path.

From the vector formulas given, we observe that curvature, radius of curvature, the tangent vector $\mathbf{T}$, and the normal vector $\mathbf{N}$ all have geometric meaning, independent of the coordinates chosen. The sign of $d\varphi/ds$ depends on the choice of positive direction for angles (or, equivalently, on the orientation of the basis $\mathbf{i}, \mathbf{j}$). The direction of $\mathbf{T}$ depends on the direction of motion on the path, but $\mathbf{N}$ always points to the concave side and so is independent of the direction of the motion.

EXAMPLE. A particle P of mass m is moving on the path $y = x^2$ in the xy-plane at constant speed v_0. Find the force exerted when P passes through the origin O.

Solution. Since the speed is constant, $a_T = 0$ and $a_N = v_0^2/\rho$. At O, $\rho = \frac{1}{2}$, as in Example 2 of Section 6-6, and $\mathbf{N} = \mathbf{j}$, so that $a_N = 2v_0^2$ (in the given units) and $\mathbf{a} = 2v_0^2\mathbf{j}$. The force $\mathbf{F}$ exerted is $m\mathbf{a} = 2mv_0^2\mathbf{j}$.

PROBLEMS

1. Graph, evaluate κ for general t, and show the circle of curvature at the points stated:
   (a) $x = 2t + 1$, $y = t^3 - 3t + 1$, $-2 \leq t \leq 2$, points for which $t = -1, 0, +1$
   (b) $x = 3t - 2$, $y = t^4 - 4t^3 + 4t^2$, $-3 \leq t \leq 3$, points for which $t = -1, t = 0$, $t = 1.5$
   (c) $x = \cos 2t$, $y = 2\sin 2t$, $0 \leq t \leq \pi$, points for which $t = 0, \pi/8, \pi/4$
   (d) $x = \cosh t$, $y = \sinh t$, $-2 \leq t \leq 2$, points for which $t = -1, 0, 1$
   (e) $y = \sin x$, $0 \leq x \leq 2\pi$, points for which $x = 0, \pi/4, \pi/2$
   (f) $y = \sec x$, $-\pi/2 < x < \pi/2$, points for which $x = 0, \pi/4$
   (g) $x = (1 - t^2)/(1 + t^2)$, $y = 2t/(1 + t^2)$, $-\infty < t < \infty$, points for which $t = 0, 1, 2$

2. From the paths given graphically in Figure 6-47, estimate the t-intervals in which $d\varphi/ds > 0$ and $d\varphi/ds < 0$, and estimate ρ at the points P_1, P_2 shown.

3. Complete a sketch of each of the paths of Figure 6-48, given the tangents at inflection points as shown, given that the path never crosses itself, and given that the points $P_1, P_2, \ldots$ are on the path in that order.

4. *Meaning of circle of curvature.* Given a path $x = x(t)$, $y = y(t)$ as in Section 6-6, let t_1, t_0, t_2 be successive parameter values, and let

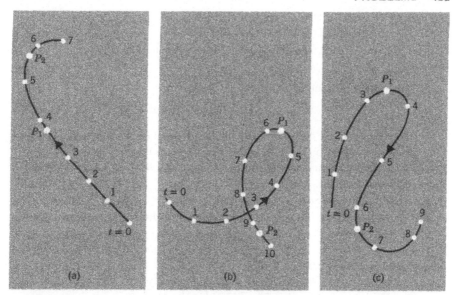

Figure 6-47

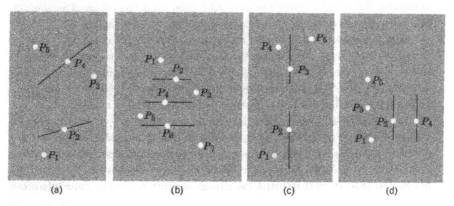

Figure 6-48

$$F(t) = [x(t) - \alpha]^2 + [y(t) - \beta]^2 - R^2$$

where α, β, R are constants.

(a) Show that if the circle of center (α, β) and radius R passes through the points $(x[t_i], y[t_i])$ for $i = 0, 1, 2$, then

$$F(t_1) = 0, \qquad F(t_0) = 0, \qquad F(t_2) = 0$$

(b) Show from the result of (a) that $F'(\xi) = 0, F'(\xi') = 0, F''(\zeta) = 0$ for some ξ, ξ', ζ with $t_1 < \xi < t_0$, $t_0 < \xi' < t_2$, $\xi < \zeta < \xi'$.

(c) Let $t_1 \to t_0$, $t_2 \to t_0$ to show that for the limiting form of the circle, α, β, R must satisfy the three equations

$$F(t_0) = 0, \qquad F'(t_0) = 0, \qquad F''(t_0) = 0$$

that is

$$(°)\begin{cases}(x - \alpha)^2 + (y - \beta)^2 = R^2 \\ (x - \alpha)x' + (y - \beta)y' = 0 \\ (x - \alpha)x'' + (y - \beta)y'' + (x')^2 + (y')^2 = 0\end{cases}$$

where x, y, x', y', x'', y'' are evaluated at t_0.

(d) Eliminate α, β from the equations ($°$) to show that

$$R = \rho = [(x'^2 + y'^2)^{3/2}/|x'y'' - x''y'|].$$

Thus the radius of curvature is the limit of the radius of a circle through three successive points on the path as these points come together.

(e) Show from the equations ($°$) that the limiting center (α, β) is on the normal to the curve for $t = t_0$.

5. For a path, as given in Section 6-6, let $d\varphi/ds < 0$ for $t < t_1$, $d\varphi/ds > 0$ for $t > t_1$, so that (by continuity) $d\varphi/ds = 0$ for $t = t_1$. Show that there is an inflection point at $t = t_1$. [*Hint.* We are given that $\mathbf{v}(t) \neq \mathbf{0}$, hence $x'(t)$ and $y'(t)$ are not both 0 at any t. If $x'(t_1) \neq 0$, show that x can be used as parameter near t_1 and that, when y is expressed in terms of x,

$$\frac{d^2y}{dx^2} = \frac{d\varphi}{ds} \cdot \left(\frac{|\mathbf{v}|}{x'}\right)^3$$

so that d^2y/dx^2 changes sign at $x = x_1 = x(t_1)$. Now apply the results of Section 6-3. If $x'(t_1) = 0$, proceed similarly by using y as parameter and by finding d^2x/dy^2.]

6. The gravitational force of attraction of a planet by the sun has magnitude kmm'/r^2, where k is a universal constant, m is the mass of the sun, m' is the mass of the planet, and r is the distance between them. If the planet moves in an ellipse with one focus at the sun, find the radius of curvature of the ellipse at one end of a major axis, given that at this point, the planet is at distance r_0 from the sun and has speed v_0.

‡ 7. For the path of Theorem 6, let P_0 be the position at a fixed time t_0 at which $d\varphi/ds \neq 0$ and let $\mathbf{T}_0$, $\mathbf{N}_0$ be the corresponding choices of $\mathbf{T}$, $\mathbf{N}$ at the point. Show that for t sufficiently close to t_0 the path lies on the side of the tangent line to which $\mathbf{N}_0$ points. [*Hint.* Let $h = \overrightarrow{P_0P} \cdot \mathbf{N}_0$. Show with the aid of Theorem 6 that for $t = t_0$, $h = 0$, $dh/dt = 0$, $d^2h/dt^2 > 0$, so that h has a local minimum for $t = t_0$. Now interpret h geometrically.]

6-8 CURVES IN POLAR COORDINATES

We consider again a path $x = x(t)$, $y = y(t)$, $a \leq t \leq b$, as in the preceding sections, and *assume that the path does not pass through the origin.* Then, as in Section 6-6, we can define both polar coordinates r, θ along the path so that r and θ become continuous functions of t: $r = r(t)$, $\theta = \theta(t)$, and

$$\begin{aligned}x(t) &= r \cos \theta = r(t) \cos \theta(t) \\ y(t) &= r \sin \theta = r(t) \sin \theta(t) \\ r(t) &= \sqrt{[x(t)]^2 + [y(t)]^2}\end{aligned} \tag{6-80}$$

We assume, as in Section 6-6, that $x(t)$ and $y(t)$ have continuous first derivatives, so that the same holds true for $r(t)$ and $\theta(t)$, and

$$r' = \frac{xx' + yy'}{r}, \qquad \theta' = \frac{xy' - yx'}{r^2}$$

To study these derivatives, we introduce the vectors

$$\mathbf{R} = \frac{x}{r}\mathbf{i} + \frac{y}{r}\mathbf{j} = \cos\theta\,\mathbf{i} + \sin\theta\,\mathbf{j}$$

$$\mathbf{B} = -\sin\theta\,\mathbf{i} + \cos\theta\,\mathbf{j} = \mathbf{R}^\dashv \tag{6-81}$$

Thus, $\mathbf{R}$ is a unit vector in the direction of $\overrightarrow{OP} \stackrel{!}{=} x\mathbf{i} + y\mathbf{j}$ and, in fact,

$$\overrightarrow{OP} = r\mathbf{R} \tag{6-82}$$

We call $\mathbf{R}$ the *radial unit vector*.

The vector $\mathbf{B} = \mathbf{R}^\dashv$, so that $\mathbf{B}$ is obtained by rotating $\mathbf{R}$ through $90°$ in the counterclockwise direction, as in Figure 6-49. We call $\mathbf{B}$ the *transverse unit vector*.

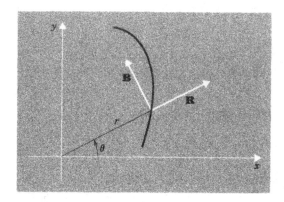

Figure 6-49 Radial and transverse unit vectors.

Since $\mathbf{R}$ and $\mathbf{B}$ form an orthonormal basis for vectors in the plane (Section 1-9), we can express an arbitrary vector as a linear combination of $\mathbf{R}$ and $\mathbf{B}$; that is, we can find the *radial* and *transverse components* of each vector. For the position vector $\mathbf{r} = \overrightarrow{OP}$, Equation (6-82) provides the answer: $\mathbf{r}$ has radial component r, transverse component 0.

THEOREM 7. *Let a path* $x = f(t)$, $y = g(t)$, $a \le t \le b$, *be given in the plane, not passing through the origin. Let f and g have continuous first derivatives, and let $r(t)$, $\theta(t)$ be defined as continuous functions on the path, so that Equations (6-80) hold true. Then the velocity vector $\mathbf{v} = f'(t)\mathbf{i} + g'(t)\mathbf{j}$ can be expressed as follows:*

$$\mathbf{v} = \frac{dr}{dt}\mathbf{R} + r\frac{d\theta}{dt}\mathbf{B}$$

so that

$$v_{\text{rad}} = \frac{dr}{dt}, \qquad v_{\text{trans}} = r\frac{d\theta}{dt}$$

PROOF. We first note that $\mathbf{R}$ is a unit vector having angle function $\theta(t)$, $\mathbf{B}$ is a unit vector having angle function $\theta(t) + (\pi/2)$. Hence by Theorem 4 (Section 6-6)

$$\frac{d\mathbf{R}}{dt} = \frac{d\theta}{dt}\,\mathbf{R}^{\dashv} = \frac{d\theta}{dt}\,\mathbf{B}$$

$$\frac{d\mathbf{B}}{dt} = \frac{d}{dt}\left[\theta(t) + \frac{\pi}{2}\right]\mathbf{B}^{\dashv} = \frac{d\theta}{dt}\,\mathbf{B}^{\dashv} = -\frac{d\theta}{dt}\,\mathbf{R}$$

(6-83)

since $\mathbf{B}^{\dashv} = -\mathbf{R}$. Now we can write

$$\mathbf{v} = \frac{d\mathbf{r}}{dt} = \frac{d}{dt}(r\mathbf{R}) = \frac{dr}{dt}\,\mathbf{R} + r\frac{d\mathbf{R}}{dt} = \frac{dr}{dt}\,\mathbf{R} + r\frac{d\theta}{dt}\,\mathbf{B}$$

as asserted.

COROLLARY 1. *For the path of Theorem 7, if arc length s is measured to increase with increasing t, then*

$$\frac{ds}{dt} = \sqrt{\left(\frac{dr}{dt}\right)^2 + r^2\left(\frac{d\theta}{dt}\right)^2}$$

so that the length of the path is

$$l = \int_a^b \sqrt{\left(\frac{dr}{dt}\right)^2 + r^2\left(\frac{d\theta}{dt}\right)^2}\;dt$$

PROOF. Since $\mathbf{R}$, $\mathbf{B}$ form an orthonormal basis,

$$|\mathbf{v}| = \sqrt{v^2{}_{rad} + v^2{}_{trans}} = \sqrt{\left(\frac{dr}{dt}\right)^2 + r^2\left(\frac{d\theta}{dt}\right)^2}$$

The Corollary now follows from the fact that $ds/dt = |\mathbf{v}|$.

COROLLARY 2. *Let a path be defined by an equation in polar coordinates: $r = G(\theta)$, $\alpha \le \theta \le \beta$, where $G(\theta)$ has a continuous first derivative and $G(\theta) > 0$ for $\alpha \le \theta \le \beta$. Then for each θ the vector*

$$\mathbf{u} = \frac{dr}{d\theta}\,\mathbf{R} + r\mathbf{B} = G'(\theta)\mathbf{R} + G(\theta)\mathbf{B}$$

is tangent to the path at the point $(G(\theta), \theta)$. The length of the path is given by

$$l = \int_\alpha^\beta \sqrt{\left(\frac{dr}{d\theta}\right)^2 + r^2}\;d\theta = \int_\alpha^\beta \sqrt{[G'(\theta)]^2 + [G(\theta)]^2}\;d\theta$$

PROOF. We can regard θ as the parameter t, so that Equations (6-80) become

$$x(\theta) = r\cos\theta = G(\theta)\cos\theta$$
$$y(\theta) = r\sin\theta = G(\theta)\sin\theta \qquad \alpha \le \theta \le \beta \qquad (6\text{-}84)$$

Since $\theta \equiv t$, $dr/dt = dr/d\theta$, $d\theta/dt = 1$ and the velocity vector $\mathbf{v}$ of Theorem 7 becomes $\mathbf{u} = (dr/d\theta)\mathbf{R} + r\mathbf{B}$, the expression for l of Corollary 1 becomes the one given in Corollary 2.

The two expressions for arc length can be stated concisely in differential form:

$$(ds)^2 = (dr)^2 + r^2(d\theta)^2 \qquad (6\text{-}85)$$

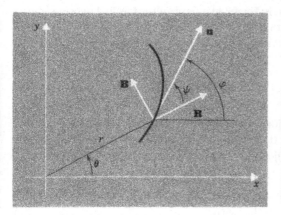

Figure 6-50 Tangent to path for path given in polar coordinates.

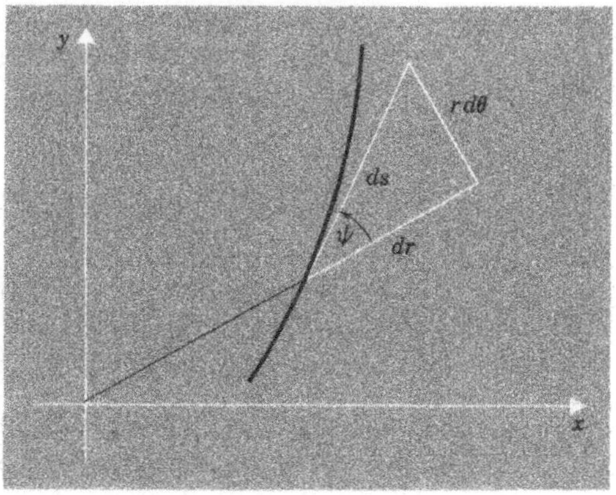

Figure 6-51 Differential of arc in polar coordinates

Angle ψ Between Radial and Tangent Vectors. The directed angle from **R** to the tangential vector **u** is denoted by ψ (see Figure 6-50). Hence, by Corollary 2,

$$\mathbf{u} = |\mathbf{u}| \cos \psi\, \mathbf{R} + |\mathbf{u}| \sin \psi\, \mathbf{B} = \frac{dr}{d\theta}\, \mathbf{R} + r\mathbf{B}$$

so that

$$|\mathbf{u}| \cos \psi = \frac{dr}{d\theta}, \qquad |\mathbf{u}| \sin \psi = r$$

and

$$\cot \psi = \frac{1}{r}\frac{dr}{d\theta} \tag{6-86}$$

We observe that, since $r > 0$, $\sin \psi > 0$ and ψ must be an angle in the first or second quadrant: it can be chosen uniquely so that $0 < \psi < \pi$. Other choices are obtained by adding multiples of 2π. The knowledge of ψ enables us to draw the tangent line at a given point on the path.

If arc length s is introduced along the path, increasing with θ, then Corollary 2 shows that

$$|\mathbf{u}| = \frac{ds}{d\theta}$$

so that

$$\frac{ds}{d\theta} \cos \psi = \frac{dr}{d\theta}, \qquad \frac{ds}{d\theta} \sin \psi = r$$

If s is taken as parameter, these relations become

$$\cos \psi = \frac{dr}{ds}, \qquad \sin \psi = r\frac{d\theta}{ds} \tag{6-87}$$

as suggested in Figure 6-51.

We observe that, for a path given in terms of a parameter t as in Theorem 7, the angle ψ can be defined as the angle from $\mathbf{R}$ to $\mathbf{v}$, provided that $d\theta/dt > 0$ at the point; if $d\theta/dt < 0$, ψ is the angle from $\mathbf{R}$ to $-\mathbf{v}$. However, in either case,

$$\cot \psi = \frac{v_{\text{rad}}}{v_{\text{trans}}} = \frac{dr/dt}{r(d\theta/dt)} = \frac{1}{r}\frac{dr}{d\theta} \tag{6-86'}$$

We should also notice that Equations (6-84) remain valid for a path passing through the origin [that is $G(\theta) = 0$ for some θ], and can be used to find the tangent vector $(dx/d\theta)\mathbf{i} + (dy/d\theta)\mathbf{j}$ at such a point (see Problem 7 below).

EXAMPLE 1. $r = 2 \cos \theta$, $-\pi/2 \le \theta \le \pi/2$. From the familiar properties of the cosine function we easily obtain the graph of Figure 6-52, which appears to be a circle. Indeed, along the path,

$$r^2 = 2r \cos \theta \qquad \text{or} \qquad x^2 + y^2 = 2x$$

so that we see that we have a circle of radius 1 with center at $x = 1$, $y = 0$. From (6-86) we obtain

$$\cot \psi = \frac{-2 \sin \theta}{2 \cos \theta} = -\tan \theta$$

Thus, for $\theta = \pi/4$, $\cot \psi = -1$ and $\psi = 3\pi/4$, as in Figure 6-52. In this case, we can in fact show by geometry that for all θ, $\psi = \theta + \pi/2$ (see Problem 8 below).

In this example, $G(-\pi/2) = G(\pi/2) = 0$, so that the path passes through the origin. Our equations (6-84) give

$$x = 2 \cos^2 \theta, \qquad\qquad y = 2 \sin^2 \theta$$

$$\frac{dx}{d\theta} = -4 \cos \theta \sin \theta, \qquad \frac{dy}{d\theta} = 4 \sin \theta \cos \theta$$

Hence our "velocity vector" $(dx/d\theta)\mathbf{i} + (dy/d\theta)\mathbf{j}$ reduces to $\mathbf{0}$ for $\theta = \pm \pi/2$, and does not help in finding the tangent vector. However, the very fact that $G(\pm \pi/2) = 0$ implies that the rays $\theta = \pm\pi/2$—that is, the y-axis—are tangent to the path (Problem 7 below).

EXAMPLE 2. $x = 2 \cos t$, $y = \sin t$, $0 \le t \le 2\pi$. These are parametric equations for an ellipse:

$$\frac{x^2}{4} + y^2 = 1$$

We can study the motion in polar coordinates:

$$r = \sqrt{x^2 + y^2} = \sqrt{4 \cos^2 t + \sin^2 t}$$

$$\theta = \text{Tan}^{-1} \frac{y}{x} = \text{Tan}^{-1}\left(\frac{\tan t}{2}\right)$$

The equation for θ requires specification of the quadrant, and of the values when $y/x = \infty$. As in Section 5-4, we can also write

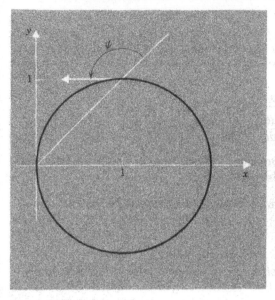

Figure 6-52 $r = 2 \cos \theta$.

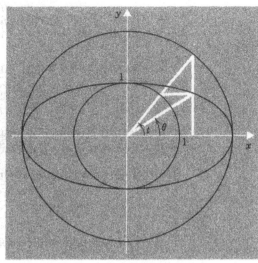

Figure 6-53 Motion on an ellipse.

$$\theta = \int_0^t \frac{x(u)y'(u) - x'(u)y(u)}{[x(u)]^2 + [y(u)]^2}\, du = 2 \int_0^t \frac{1}{4 \cos^2 u + \sin^2 u}\, du$$

since $\theta = 0$ when $t = 0$. The relationship between θ and t is suggested in Figure 6-53. The angle t is called the *eccentric angle*. From the formulas given we conclude that

$$v_{\text{rad}} = -\frac{3 \sin t \cos t}{r}, \qquad v_{\text{trans}} = \frac{2}{r}, \qquad \cot \psi = -\frac{3}{2} \sin t \cos t$$

6-9 ACCELERATION AND CURVATURE IN POLAR COORDINATES

We now seek the radial and transverse components of acceleration:

THEOREM 8. *For the path of Theorem 7, let f and g have continuous second derivatives for $a \leq t \leq b$. Then the acceleration vector $\mathbf{a} = f''(t)\mathbf{i} + g''(t)\mathbf{j}$ can be expressed as follows:*

$$\mathbf{a} = \left[\frac{d^2 r}{dt^2} - r\left(\frac{d\theta}{dt}\right)^2\right]\mathbf{R} + \left[2 \frac{dr}{dt}\frac{d\theta}{dt} + r\frac{d^2\theta}{dt^2}\right]\mathbf{B} \qquad (6\text{-}90)$$

so that

$$a_{\text{rad}} = \frac{d^2 r}{dt^2} - r\left(\frac{d\theta}{dt}\right)^2, \qquad a_{\text{trans}} = 2\frac{dr}{dt}\frac{d\theta}{dt} + r\frac{d^2\theta}{dt^2} \qquad (6\text{-}91)$$

PROOF. By Theorem 7

$$\mathbf{v} = \frac{dr}{dt}\mathbf{R} + r\frac{d\theta}{dt}\mathbf{B}$$

We differentiate this relation to obtain

$$\mathbf{a} = \frac{d\mathbf{v}}{dt} = \frac{d^2r}{dt^2}\mathbf{R} + \frac{dr}{dt}\frac{d\mathbf{R}}{dt} + \left(r\frac{d^2\theta}{dt^2} + \frac{dr}{dt}\frac{d\theta}{dt}\right)\mathbf{B} + r\frac{d\theta}{dt}\frac{d\mathbf{B}}{dt}$$

$$(6\text{-}92)$$

Now by Equation (6-83)

$$\frac{d\mathbf{R}}{dt} = \frac{d\theta}{dt}\mathbf{B}, \qquad \frac{d\mathbf{B}}{dt} = -\frac{d\theta}{dt}\mathbf{R}$$

If we substitute these two expressions in (6-92) and collect terms, we obtain (6-90), so that (6-91) also holds true.

The derivatives appearing in (6-90) are often abbreviated as follows:

$$\frac{dr}{dt} = \dot{r}, \qquad \frac{d^2r}{dt^2} = \ddot{r}$$

$$\frac{d\theta}{dt} = \omega = angular\ velocity, \qquad \frac{d^2\theta}{dt^2} = \alpha = angular\ acceleration$$

Thus we can write $v_{rad} = \dot{r}$, $v_{trans} = r\omega$. Then by (6-90) and (6-91)

$$\mathbf{a} = (\ddot{r} - r\omega^2)\mathbf{R} + (2\dot{r}\omega + r\alpha)\mathbf{B} \qquad (6\text{-}90')$$

and

$$a_{rad} = \ddot{r} - r\omega^2, \qquad a_{trans} = 2\dot{r}\omega + r\alpha \qquad (6\text{-}91')$$

EXAMPLE 1. Let the path follow a circle $r = \rho = $ constant. Then $\dot{r} = 0$, $\ddot{r} = 0$ and

$$a_{rad} = -\rho\omega^2, \qquad a_{trans} = \rho\alpha$$

In particular, if the angular velocity ω is constant, then $\alpha = 0$, so that $a_{trans} = 0$, but $a_{rad} = -\rho\omega^2$. *For motion of a particle on a circular path at constant angular velocity, there is always an acceleration towards the center* (centripetal). By Newton's second law (force = mass × acceleration), there is accordingly a corresponding force pulling the particle towards the center. The familiar example is that of gravitation attracting satellites; the paths are only approximately circular, but in principle can be made extremely close to perfect circular motion at constant angular velocity.

Kepler's Laws. In 1609, on the basis of observations, Kepler formulated three laws on the motion of the planets in the solar system:

1. *Each planet moves in a plane and traces an ellipse with the sun as one focus.*
2. *The radius vector from the sun to the planet sweeps out area at a constant rate.*
3. *The squares of the periods (years) of the planets are proportional to the cubes of the major axes of the corresponding ellipses.*

We here assume it has been proved that each planet moves in a plane, and we show that the path is a conic section. Let the sun (of mass M) be at the origin, and let P be the position of a planet of mass m; let r, θ be the polar coordinates of P. The gravitational force of attraction of the sun then acts on P. Its magnitude is $(\gamma Mm)/r^2$, where γ is the constant of gravitation, and its direction is toward the sun. Hence, the force is $-(\gamma Mm/r^2)\mathbf{R}$, and by Newton's second law

$$m\mathbf{a} = m\frac{d\mathbf{v}}{dt} = -\gamma\frac{Mm}{r^2}\mathbf{R}$$

or, with $k = \gamma M$,

$$\mathbf{a} = \frac{d\mathbf{v}}{dt} = -\frac{k}{r^2}\mathbf{R} \tag{6-93}$$

Since $\mathbf{a}$ is proportional to $\mathbf{R}$, $a_{\text{trans}} = 0$ or, by (6-91'), $2\dot{r}\omega + r\alpha = 0$. We multiply this equation by r and obtain $2r\dot{r}\omega + r^2\alpha = 0$ or

$$\frac{d}{dt}(r^2\omega) = 0$$

Hence, $r^2\omega$ is constant throughout the motion. We write

$$r^2\omega = h \tag{6-94}$$

We notice that $h = 0$ leads to $\theta = \text{const.}$ or motion on a line. We do not consider this case further and assume that $h \neq 0$. We can also assume axes chosen so that $h > 0$, so that $d\theta/dt > 0$, and θ increases with t. By (6-94), Equation (6-93) can be written

$$\frac{d\mathbf{v}}{dt} = -\frac{k}{h}\omega\mathbf{R} \tag{6-93'}$$

Now for any differentiable vector function $\mathbf{u} = \mathbf{F}(t)$

$$\frac{d}{dt}\mathbf{u}^\dashv = \frac{d}{dt}(-u_y\mathbf{i} + u_x\mathbf{j}) = -\frac{du_y}{dt}\mathbf{i} + \frac{du_x}{dt}\mathbf{j} = \left(\frac{d\mathbf{u}}{dt}\right)^\dashv$$

Thus, from (6-93'), we obtain

$$\frac{d\mathbf{v}^\dashv}{dt} = \left(-\frac{k}{h}\omega\mathbf{R}\right)^\dashv = -\frac{k}{h}\omega\mathbf{R}^\dashv$$

But $d\mathbf{R}/dt = \omega\mathbf{R}^\dashv$, by (6-83), so that we can write

$$\frac{d\mathbf{v}^\dashv}{dt} = -\frac{k}{h}\frac{d\mathbf{R}}{dt}$$

and, hence, $\mathbf{v}^\dashv = -(k/h)(\mathbf{R} + \mathbf{b})$, where $\mathbf{b}$ is a constant vector. Since $\mathbf{R}$, $\mathbf{B}$ form an orthonormal basis oriented similarly to $\mathbf{i}$, $\mathbf{j}$, $\mathbf{v}^\dashv = (\dot{r}\mathbf{R} + r\omega\mathbf{B})^\dashv = -r\omega\mathbf{R} + \dot{r}\mathbf{B}$. Therefore,

$$-r\omega\mathbf{R} + \dot{r}\mathbf{B} = -\frac{k}{h}(\mathbf{R} + \mathbf{b})$$

We form the inner product of both sides with the unit vector $\mathbf{R}$ and obtain

$$-r\omega = -\frac{k}{h}(1 + e\cos\theta), \quad \text{where } e = |\mathbf{b}|, \text{ and } \theta = \sphericalangle(\mathbf{b}, \mathbf{R})$$

By (6-94), our equation can be written

$$r = \frac{h^2/k}{1 + e \cos \theta} \tag{6-95}$$

This last equation can be written as a second degree equation in x and y:

$$x^2 + y^2 = (h^2/k - ex)^2$$

Thus (6-95) represents a conic section (Section 6-4). We can show directly that (6-95) represents a conic section of eccentricity, e, with focus at the origin and major axis along the direction given by $\mathbf{b}$. When $\mathbf{b} = \mathbf{0}$, the path is a circle.

A special case of Kepler's third law is proved in Problem 4 below. The second law is contained in (6-94), as shown in Section 7-2.

Curvature in Polar Coordinates. By Theorem 5, Equation (6-68c),

$$\kappa = \frac{|\mathbf{v}^{\dashv} \cdot \mathbf{a}|}{|\mathbf{v}|^3}$$

Now since $\mathbf{R}$, $\mathbf{B}$ form a basis oriented similarly to $\mathbf{i}$, $\mathbf{j}$ (because $\mathbf{B} = \mathbf{R}^{\dashv}$), we can write, as in Section 1-12,

$$\mathbf{v} = \dot{r}\mathbf{R} + r\omega\mathbf{B}, \qquad \mathbf{v}^{\dashv} = -r\omega\mathbf{R} + \dot{r}\mathbf{B}$$

Accordingly, by (6-90'),

$$\kappa = \frac{|\mathbf{v}^{\dashv} \cdot \mathbf{a}|}{|\mathbf{v}|^3} = \frac{|-r\omega(\ddot{r} - r\omega^2) + \dot{r}(2\dot{r}\omega + r\alpha)|}{(\dot{r}^2 + r^2\omega^2)^{3/2}} \tag{6-96}$$

For a path $r = G(\theta)$, as in Corollary 2 of Theorem 7, $\theta \equiv t$, $\omega = 1$, $\alpha = 0$, $\dot{r}$ becomes $r' = G'(\theta)$. $\ddot{r}$ becomes $r'' = G''(\theta)$, and (6-96) becomes

$$\kappa = \frac{|-rr'' + r^2 + 2r'^2|}{(r'^2 + r^2)^{3/2}} \tag{6-97}$$

For the path of Example 1 of Section 6-8, $r = 2 \cos \theta$, $r' = -2 \sin \theta$, $r'' = -2 \cos \theta$, and (6-97) gives

$$\kappa = \frac{|4 \cos^2 \theta + 4 \cos^2 \theta + 8 \sin^2 \theta|}{(4 \sin^2 \theta + 4 \cos^2 \theta)^{3/2}} = 1$$

as expected.

Concavity can also be studied in polar coordinates. From (6-69')

$$\frac{d\varphi}{ds} = \frac{\mathbf{v}^{\dashv} \cdot \mathbf{a}}{|\mathbf{v}|^3}$$

and, as in the preceding paragraph, we find that for a path $r = G(\theta)$

$$\frac{d\varphi}{ds} = \frac{-rr'' + r^2 + 2r'^2}{(r'^2 + r^2)^{3/2}} \tag{6-98}$$

The sign of the numerator gives the sign of $d\varphi/ds$ and, hence, tells whether the curve is concave to the left (as s and θ increase together) or to the right.

EXAMPLE 2. $r = 3 - \cos\theta$, $0 \leq \theta \leq 2\pi$. Here $r' = \sin\theta$, $r'' = \cos\theta$. Thus r' has zeros at $\theta = 0$, π, and 2π. At these points r'' has the values 1, -1, and 1, respectively, and we conclude (even though 0 and 2π are end points of the given interval) that r has its absolute minimum, 2, for $\theta = 0$, 2π, and its absolute maximum, 4, for $\theta = \pi$. From (6-98), we find that

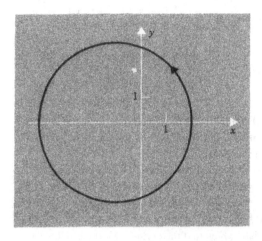

Figure 6-54 $r = 3 - \cos\theta$.

$$\frac{d\varphi}{ds} = \frac{11 - 9\cos\theta}{(10 - 6\cos\theta)^{3/2}}$$

Hence, $d\varphi/ds$ is always positive, so that the path is always concave to the left. Thus concavity is unrelated to maxima and minima. The path is graphed in Figure 6-54.

PROBLEMS

1. Given the path $x = 1 + 2t$, $y = 2 - t$, $-\infty < t < \infty$:
   (a) Show that the path is a straight line not passing through the origin and graph.
   (b) Find $r'(t)$, $\theta'(t)$, v_{rad} and v_{trans}.
   (c) Show that $\theta(t)$ is monotone strictly decreasing and, assuming $\theta(2) = 0$, find $\lim\limits_{t \to \infty} \theta(t)$ and $\lim\limits_{t \to -\infty} \theta(t)$, and interpret on the graph. [*Hint.* Observe the way $\tan\theta$ behaves as $t \to +\infty$ or $-\infty$.]
   (d) Show that r' has one zero, and interpret geometrically.
2. Let the path $x = 2\cos t + \cos 2t$, $y = 2\sin t + \sin 2t$, $0 \leq t \leq 2\pi$, be given.
   (a) Show that the path does not pass through the origin.
   (b) Find $r'(t)$ and $\theta'(t)$, and show that $0 \leq \theta'(t) \leq \frac{4}{3}$.
   (c) Show that $\theta(t)$ is monotone strictly increasing. Show from the equations for x and y that θ must increase by a multiple of 2π as t goes from 0 to 2π. Use the result of (b) to show that $\theta(t)$ increases by exactly 2π.
   (d) Show that $r(t)$ has its absolute minimum for $t = \pi$, its absolute maximum for $t = 0$ and 2π, and $r'(t) = 0$ only at these values of t.
   (e) Graph.
   (f) Show that v_{trans} and v_{rad} are both 0 only for $t = \pi$, so that $v = 0$ at this point.

‡(g) Show that at $t = \pi$, the graph has a cusp. [*Hint.* Show that y is expressible in terms of x in each of the intervals $2\pi/3 \leq t \leq \pi$, $\pi \leq t \leq 4\pi/3$, and that both functions are defined for $-\frac{3}{2} \leq x \leq -1$ and have derivative $y'(-1)$ equal to 0.]

3. With the aid of the derivative $dr/d\theta$, graph the following curves given in polar coordinates:

(a) $r = 2, 0 \leq \theta \leq 2\pi$ (b) $r = 2 \sin \theta, 0 \leq \theta \leq \pi$
(c) $r = 2 \cos [\theta - (\pi/3)], \pi/6 \leq \theta \leq 5\pi/6$ (d) $r = 4 \sin 2\theta, 0 \leq \theta \leq \pi/2$
(e) $r = \sin^2 \theta, 0 \leq \theta \leq 2\pi$ (f) $r = 2 \cos^2 2\theta, 0 \leq \theta \leq 2\pi$
(g) $r = 3 + \cos \theta, 0 \leq \theta \leq 2\pi$ (h) $r = 1 - \sin \theta, 0 \leq \theta \leq 2\pi$
(i) $r = 1 - 2 \cos \theta, \pi/3 \leq \theta \leq 5\pi/3$ (j) $r = e^\theta, -\infty < \theta < \infty$
(k) $r = \theta, 0 \leq \theta < \infty$ (l) $r = \sec \theta, -\pi/2 < \theta < \pi/2$
(m) $r = \tan \theta, -\pi/2 < \theta < \pi/2$ (n) $r = 2/(1 + \cos \theta), -\pi < \theta < \pi$

In graphing we can take advantage of *symmetry*. If the equation is unchanged when θ is replaced by $-\theta$, the graph is symmetric in the x-axis; if it is unchanged when θ is replaced by $\pi - \theta$, it is symmetric in the y-axis.

4. For a satellite moving in a circular orbit about the earth, we can introduce polar coordinates in the plane of the path, with the origin at the center of the earth. If the radius of the circle is ρ, then the gravitational force acting on the satellite is directed toward the origin and has magnitude $\gamma M m/\rho^2$, where γ is a constant, M is the mass of the earth, and m is that of the satellite. Show that the period T for the satellite (time for one revolution) is proportional to $\rho^{3/2}$; that is, for two satellites at distances ρ_1, ρ_2, we have

$$\frac{T_1}{T_2} = \frac{\rho_1^{3/2}}{\rho_2^{3/2}}$$

5. A ball of mass m is rolling in a horizontal circular track of radius ρ subject to a constant frictional force of magnitude k. If the ball has angular velocity ω_0 at time $t = 0$, how long will it take for the ball to come to rest? (*Hint.* By Newton's Second Law, applied to transverse components, $ma_{\text{trans}} = -k$.)

6. (a) Show that for a curve given, as in Section 6-8, by a polar coordinate equation $r = r(\theta)$, we can choose the angle functions ψ and φ as differentiable functions of θ so that $\varphi(\theta) = \psi(\theta) + \theta$ (see Figure 6-50).
   (b) Show from the result of (a) that the curvature κ is given by

$$\kappa = \frac{|rr'' - 2r'^2 - r^2|}{(r^2 + r'^2)^{3/2}}$$

and that, with s an increasing function of θ,

$$\frac{d\varphi}{ds} = \frac{r^2 + 2r'^2 - rr''}{(r^2 + r'^2)^{3/2}}$$

7. Let a path, given in polar coordinates by $r = f(\theta), a \leq \theta \leq b$, be such that $f(\theta) > 0$ for $a \leq \theta < b$, $f(b) = 0$, and f is continuous for $a \leq \theta \leq b$. Show that the line $\theta = \text{const.} = b$ is tangent to the path at O. [Proof: Consider the limiting direction, as $\theta \to b-$, of a chord through the origin and the point $[\theta, f(\theta)]$.]

8. Show that, for the circle in Example 1 in Section 6-8, one can choose $\psi(\theta)$ to be $\theta + (\pi/2)$.

9. The conic sections can be defined by the following locus condition: a point moves so that its distance from a fixed point (the focus) and its distance from a fixed line L (the directrix, not containing the focus) are in a given ratio e (the eccentricity).

Choose polar coordinates with origin O at the focus, and so that L is perpendicular to the ray $\theta = 0$ at $(d, 0)$, where $d > 0$. Show that $P: (r, \theta)$ satisfies the locus condition precisely when

$$(*) \qquad r = \frac{de}{1 + e \cos \theta}$$

In the derivation, notice that for $e > 1$ (*) gives negative values of r for some θ and that each such point must be graphed as $(-r, \theta + \pi)$.

6-10 NEWTON'S METHOD

A basic problem of mathematics and its applications is the solution of equations. This problem occurs, as we have seen, in finding critical points and inflection points. In algebra one learns how to solve linear and quadratic equations. For the cubic (third degree) and biquadratic (fourth degree) equations, there are formulas for the roots in terms of the coefficients (see Dickson, *Theory of Equations*), but they are complicated. For higher degree polynomials there are no comparable formulas (see Dickson, cited above), and we are forced to use some method of approximating the zeros. The problem is equally difficult for most nonalgebraic functions (often called transcendental functions) such as

$$x^2 - \cos x, \qquad e^x + \sin x \tag{6-100}$$

Here graphical procedures, as developed earlier in this chapter, can indicate approximate location of zeros. In addition, we can split one equation into two equations and then graph. We illustrate this for the first function of (6-100), which we replace by

$$y = x^2, \qquad y = \cos x$$

The x-coordinates of the points of intersection of these two curves are the zeros of the function $x^2 - \cos x$. If we graph the two functions as in Figure 6-55, we observe that the graphs must intersect at precisely two points, at which x equals approximately ± 0.9. The graphical procedures are not precise, but they do indicate where zeros may exist. To prove that zeros exist in certain intervals, we can use the Intermediate Value Theorem, properties of

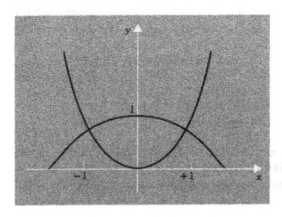

Figure 6-55 Graphical solution of $x^2 - \cos x = 0$.

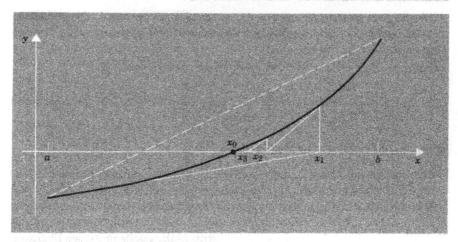

Figure 6-56   Idea of Newton's method.

increasing and decreasing functions, boundedness of functions, and other tools of the calculus. For example, between $x = 0$ and $x = \frac{1}{2}\pi$, both functions $y = x^2$ and $y = -\cos x$ are strictly increasing, so that their sum $x^2 - \cos x$ is also strictly increasing; the sum equals -1 at $x = 0$, equals $\frac{1}{4}\pi^2$ at $x = \frac{1}{2}\pi$, hence, there is exactly one zero between $x = 0$ and $x = \frac{1}{2}\pi$. The same reasoning can be used to restrict the zero to successively smaller intervals: $[\pi/4, \pi/3]$, $[0.8, 0.9]$, ... For $|x| > \frac{1}{2}\pi$, we reason otherwise: x^2 is at least equal to $\pi^2/4 = 2.5\ldots$, while $-\cos x$ is bounded between -1 and $+1$, so that the sum cannot be zero.

We now ask how we can systematically locate the zero as precisely as desired. The typical starting point of the process is suggested by our example. We have a differentiable function $y = f(x)$ on an interval $[a, b]$; f is strictly monotone and $f(a)$, $f(b)$ have opposite signs, so that by the Intermediate Value Theorem f has exactly one zero x_0 between a and b. To find x_0, we can replace our function f by a linear function—that is, replace our graph by a straight line. One such linear function is the chord joining $[a, f(a)]$ and $[b, f(b)]$. A second one is a tangent line, for instance, at $[a, f(a)]$, as in Figure 6-56. The figure suggests that the tangent line crosses the x-axis closer to x_0 than the chord does, and it can be shown that the tangent line is the better choice. Let x_1 be the point where the tangent crosses. We then draw the tangent line at $[x_1, f(x_1)]$; it crosses at x_2, closer to x_0 than x_1. We draw the tangent at $[x_2, f(x_2)]$ and so on. The process described is *Newton's method*. It provides a sequence $x_1, x_2, x_3, \ldots, x_n, \ldots$; and x_n appears to approach our zero x_0 as $n \to \infty$:

$$\lim_{n \to \infty} x_n = x_0$$

The process does not always succeed and, in fact, for appropriately chosen $f(x)$, we may even be taken out of the interval $[a, b]$ by our successive approximations x_n. Below we state precise conditions under which success is guaranteed.

We can describe the process by a formula. For we can write the equation of the tangent line at a general point $[x_n, f(x_n)]$:

$$y - f(x_n) = f'(x_n)(x - x_n)$$

To find our next approximation, x_{n+1}, we set $y = 0$ and solve for x: We find that

$$x_{n+1} = x_n - \frac{f(x_n)}{f'(x_n)}, \qquad \text{provided that } f'(x_n) \neq 0 \qquad (6\text{-}101)$$

EXAMPLE 1. Let $f(x) = x^2 - 5$. Then $f(2) = -1 < 0$, $f(3) = 4 > 0$, so that f has a zero x_0 in the interval $[2, 3]$. Since $f'(x) = 2x > 0$ in this interval, f is monotone strictly increasing. We choose $x_1 = 3$ as a first approximation. Now, for $f(x) = x^2 - 5$, (6-101) becomes

$$x_{n+1} = x_n - \frac{x_n^2 - 5}{2x_n} = \frac{1}{2}\left(x_n + \frac{5}{x_n}\right) \qquad (6\text{-}102)$$

With $n = 1$ and $x_1 = 3$, (6-102) gives $x_2 = \frac{1}{2}[3 + (5/3)] = 7/3$. With $n = 2$ and $x_2 = 7/3$, (6-102) gives $x_3 = 47/21$. Similarly we find that $x_4 = 2207/987 = 2.236069$, to six decimal places. If we look at our original function, we see that we have found $\sqrt{5} = 2.236068 \ldots$ with an error of about 0.000001, or one part in $2,000,000$. For work with a computer, it is better to use decimals throughout and round off as needed.

The example we have chosen is an important one, for it suggests *a general method for finding square roots*. For the general case we take $f(x) = x^2 - c$ ($c > 0$) and replace (6-102) by

$$x_{n+1} = \frac{1}{2}\left(x_n + \frac{c}{x_n}\right) \qquad (6\text{-}103)$$

In words, (6-103) says that to find $\sqrt{c}$ we improve our trial value x_n by averaging it with c/x_n. Thus we can find $\sqrt{2}$ by taking $x_1 = 1$, averaging with $2/1 = 2$ to get $x_2 = 3/2$, average $3/2$ with $2 \div (3/2) = 4/3$ to obtain $x_3 = 17/12$. Since $17/12 = 1.417$, we already have $\sqrt{2}$ with an error less than 0.003, by a computation that we can do in our heads!

EXAMPLE 2. Let $f(x) = x^2 - \cos x$. We noticed above that f is monotone and has a zero in the interval $[0.8, 0.9]$. We take $x_1 = 0.9$. Here (6-101) reads

$$x_{n+1} = x_n - \frac{x_n^2 - \cos x_n}{2x_n + \sin x_n}$$

We do not combine terms, because the second term represents the correction to x_n; as n increases this correction becomes small, and it is easier to control accuracy if we observe it separately. With $x_1 = 0.9$, we find (with the aid of tables) that

$$x_2 = 0.9 - 0.07 = 0.83$$

to two decimal places. With $x_2 = 0.83$, we find that

$$x_3 = 0.83 - \frac{0.01402}{2.2979} = 0.82411$$

The numerator in the above fraction is our error $x_2^2 - \cos x_2$. The denominator is $f'(x_2)$. Typically, this denominator changes only slightly from one point to the next, and we can hold one value, not recalculating at each step. The slight loss in accuracy may require more steps, but each step takes less time. In order to check the accuracy of x_3, we need an even larger table, and we soon exhaust existing tables. To proceed beyond the tables, other tools are needed, especially *infinite series* (see Chapter 8).

THEOREM 9. *Let the function f be defined and differentiable on the interval $[a, b]$ and let f'' also exist in the interval. Let f' and f'' each preserve the same sign throughout the interval; in particular, let f' and f'' have no zeros. Let $f(a)$ and $f(b)$ have opposite signs. Then f has a single zero x_0 in $[a, b]$.*

Case 1. *Let $f(a)$ and $f''(a)$ have the same sign. Then x_1 can be chosen as a, and Equation (6-101) defines $x_2, x_3, \ldots, x_n, \ldots$ in $[a, b]$ so that $x_n < x_{n+1}$ for $n = 1, 2, \ldots$ and $\lim_{n \to \infty} x_n = x_0$.*

Case 2. *Let $f(a), f''(a)$ have opposite signs. Then x_1 can be chosen as b, and Equation (6-101) defines $x_2, x_3, \ldots, x_n, \ldots$ in $[a, b]$ so that $x_{n+1} < x_n$ for $n = 1, 2, \ldots$ and $\lim_{n \to \infty} x_n = x_0$.*

The two cases are illustrated in Figure 6-57.

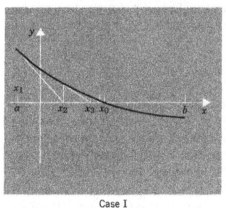
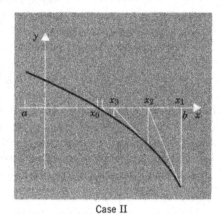

Case I   Case II

Figure 6-57   Newton's method.

Remarks. The figures suggest why in each case the sequence $\{x_n\}$ is monotone and bounded between x_1 and x_0. Accordingly, in each case, the sequence has a limit, and we can deduce from (6-101) that the limit must be a zero of f.

We notice that, if in the Case 1 we used b as a starting value x_1, then (6-101) gives a value x_2 that is less than x_0 and may even be less than a (Problem 4 below). Similar remarks apply to Case 2.

PROOF OF THEOREM 9. Since $f(a)$, $f(b)$ have opposite signs, the Intermediate Value Theorem implies that f has a zero between a and b. Since f'

has no zeros in the interval, Rolle's theorem implies that f has at most one zero in the interval. Hence, there is a unique zero, x_0, $a < x_0 < b$.

Now let Case 1 hold true and let $f(a) < 0$, $f''(a) < 0$. Then $f(b)$ must be > 0, and since f' does not change sign, $f'(x)$ must be > 0 for all x in the interval. By the Mean Value Theorem, there is a ξ so that

$$f(x_0) - f(a) = f'(\xi)(x_0 - a), \qquad a < \xi < x_0$$

Since f'' does not change sign and $f''(a) < 0$, f'' must be negative in the interval, so that f' is monotone strictly decreasing. Therefore, $f'(a) > f'(\xi)$ and $f(x_0) - f(a) < f'(a)(x_0 - a)$ and, since $f(x_0) = 0$, $f'(a) > 0$, we have

$$x_0 > a - \frac{f(a)}{f'(a)}$$

But, with $x_1 = a$, the right side of this inequality is x_2, as given by (6-101). Therefore, $x_0 > x_2$. Also, since $f(a) < 0$ and $f'(a) > 0$,

$$x_2 = a - \frac{f(a)}{f'(a)} > a$$

Accordingly, $a = x_1 < x_2 < x_0$. Since f cannot change sign between a and x_0, we must have $f(x_2) < 0$. Thus we can repeat the previous reasoning with x_2 as starting value and can conclude that $a = x_1 < x_2 < x_3 < x_0$. By induction (Section 0-20) we conclude that the sequence $\{x_n\}$ is well defined and is a bounded monotone increasing sequence in the interval $[a, x_0]$. Therefore, by Theorem H of Section 2-12, the sequence has a limit x°, $a < x^\circ \le x_0$. Since f and f' are continuous,

$$\lim_{n \to \infty} f(x_n) = f(x^\circ), \qquad \lim_{n \to \infty} f'(x_n) = f'(x^\circ)$$

(see Section 2-12). From (6-101), we now obtain (on letting $n \to \infty$ on both sides)

$$x^\circ = x^\circ - \frac{f(x^\circ)}{f'(x^\circ)}$$

This equation is possible only if $f(x^\circ) = 0$. Therefore $x^\circ = x_0$.

The proof for the other cases is similar (Problem 5 below).

The hypotheses are illustrated by Example 2, considered above. Here $f(x) = x^2 - \cos x$, $a = 0$, $b = \pi/2$, and $f'(x) = 2x + \sin x > 0$, $f''(x) = 2 + \cos x > 0$ in the interval. Also $f(a) = -1$, $f(b) = \pi^2/4$. Thus, all conditions are fulfilled, and Case 2 applies.

In Section 6-13 below we discuss the problem of estimating the error, after n steps, in finding a zero by Newton's method.

PROBLEMS

1. Using Newton's method, find all solutions to the accuracy indicated by underlining:
   (a) $x^2 - 3 = 0$ $(\pm \underline{1.732})$ (b) $x^2 - 10 = 0$ $(\pm\underline{3.162})$

(c) $x^3 - 2 = 0$　　(1.25922)　　　　　　　　(d) $x^3 + x + 1 = 0$　　(−0.6823)

(e) $x^3 + x^2 - x + 1 = 0$　　(−1.8393)　　　(f) $e^x + x = 0$　　(−0.5671)

(g) $\tan x - 2x = 0,\ -\pi/2 < x < \pi/2$　　(0, ±1.166)

(h) $\ln x - \cos x = 0$　　(1.3030)

2. Find the critical points of the functions to the nearest hundredth:

   (a) $y = x^4 - 14x^2 + 28x + 3$　　(1.3569, 1.6920, −3.0489)

   (b) $y = x^4 + 4x^3 - 4x^2 - 20x + 1$　　(−1.201639, 1.330058, −3.128419)

3. Find the absolute maximum of the function to the accuracy given:

   (a) $y = (1 - x)(e^x - 1),\ 0 \le x \le 1$　　(0.3305)

   (b) $y = (1 - x)\sin x,\ 0 \le x \le 1$　　(0.240)

4. Let $f(x) = \cos x - 0.7,\ 0.1 \le x \le 1$. Show that in this interval $f'(x) < 0, f''(x) < 0$. Use $x_1 = 0.1$ as starting value for Newton's method and show that the method fails. Show graphically what happens.

5. (a) Carry out the proof of Theorem 9 for the case $f(a) > 0, f''(a) > 0$.

   (b) Show by a geometric argument that all cases can be reduced to the one for which $f(a) < 0, f''(a) < 0$.

6-11　ESTIMATION OF ERROR

Throughout this chapter we have described processes that are in general *approximations*. Thus our graphs are "sketches," and we can claim a certain accuracy only in the points that we have computed numerically. We now show how the calculus can be used to estimate the accuracy of our results.

To this end we first consider the accuracy of the graph of the function f over an interval $a \le x \le b$. Let c be a point of this interval and let $f(c)$ be known exactly; let it also be known that f' is continuous and that

$$A \le f'(x) \le B$$

over the interval, where A and B are certain constants. Then for $c \le x \le b$ we have

$$f(x) - f(c) = \int_c^x f'(u)\,du$$

so that, as in Section 4-23,

$$A(x - c) \le f(x) - f(c) \le B(x - c)$$

and

$$f(c) + A(x - c) \le f(x) \le f(c) + B(x - c) \tag{6-110}$$

For $a \le x \le c$ we have

$$f(c) - f(x) = \int_x^c f'(u)\,du$$

so that

$$A(c - x) \le f(c) - f(x) \le B(c - x)$$

or

$$f(c) + A(x - c) \ge f(x) \ge f(c) + B(x - c) \tag{6-110'}$$

We can combine the two double inequalities (6-110) and (6-110') by stating

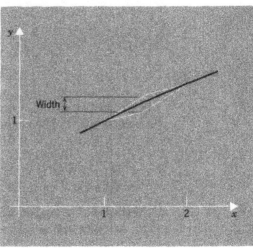

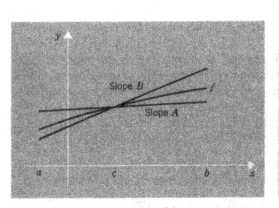

Figure 6-58 Linear upper and lower estimates for f.

Figure 6-59 Parallelogram estimate for f.

that *the graph of $y = f(x)$ lies between the graphs of the two straight lines* $y = f(c) + A(x - c)$, $y = f(c) + B(x - c)$ (see Figure 6-58).

EXAMPLE 1. In Figure 6-59 we have graphed the portion of the curve $y = \sqrt{x}$ between $x = 1$ and $x = 2$. In this interval $y' = 1/(2\sqrt{x})$ lies between $A = 1/(2\sqrt{2}) = 0.35$ and $B = \frac{1}{2}$. Hence, since $y = 1$ for $x = 1$, the graph lies between the lines of slopes A and B through $(1, 1)$. It also lies between the lines of slopes A and B through $(2, \sqrt{2})$. This restricts the graph to a thin parallelogram. The figure shows that the curve lies inside the parallelogram and, therefore, the error in the graph is at most the "width" of the parallelogram. In our example, the width is 0.064.

Second Derivative Estimates. When we know $f(c)$ and the derivative $f'(c)$ we can draw a tangent to the graph. How much can the graph of f deviate from the tangent? To answer this, we assume that f'' is continuous and that we have found numbers A, B so that

$$A \le f''(x) \le B$$

over the given interval $[a, b]$. Then for $c \le x \le b$ as above [with (6-110) applied to f' instead of to f]

$$f'(c) + A(x - c) \le f'(x) \le f'(c) + B(x - c)$$

so that, for $c \le x \le b$,

$$\int_c^x [f'(c) + A(u - c)]\, du \le \int_c^x f'(u)\, du \le \int_c^x [f'(c) + B(u - c)]\, du$$

$$f'(c)(x - c) + A\frac{(x - c)^2}{2} \le f(x) - f(c) \le f'(c)(x - c) + B\frac{(x - c)^2}{2}$$

Hence

$$f(c) + f'(c)(x - c) + A\frac{(x - c)^2}{2} \leq f(x)$$

$$\leq f(c) + f'(c)(x - c) + B\frac{(x - c)^2}{2} \qquad (6\text{-}111)$$

If we take $a \leq x \leq c$ and proceed as above, we again obtain (6-111) (Problem 6 below). Hence the graph of f lies above the graph of the parabola $y = f(c) + f'(c)(x - c) + A[(x - c)^2/2]$, and below the parabola $y = f(c) + f'(c)(x - c) + B[(x - c)^2/2]$. The tangent line L to $y = f(x)$ at $[c, f(c)]$ is $y = f(c) + f'(c)(x - c)$. Hence the curve differs from the tangent L by a vertical distance at most $K(x - c)^2/2$, where K is the larger of $|A|$, $|B|$.

EXAMPLE 2. Figure 6-60 shows a graph of $y = e^x, 0 \leq x \leq 1$. In this interval $y'' = e^x$ lies between 1 and $e = 2.72$. Hence, by (6-111), e^x satisfies

$$1 + x + \frac{x^2}{2} \leq e^x \leq 1 + x + \frac{ex^2}{2}$$

Here we have taken $c = 0$, and we obtain the two parabolas shown. We could also use $c = 1$ and obtain the estimates:

$$e + e(x - 1) + \frac{(x - 1)^2}{2} \leq e^x \leq e + e(x - 1) + \frac{e(x - 1)^2}{2}$$

From either or both we can estimate the error in graphing e^x. As remarked above, both estimates tell us how far the graph is from a tangent line. Thus at x the graph is at most $ex^2/2$ units above the tangent line at $(0, 1)$.

Estimates of the type discussed can be used to locate zeros of functions with known accuracy. Thus, with f squeezed between an "A" and "B" parabola (Figure 6-61), its zero is located in a definite interval.

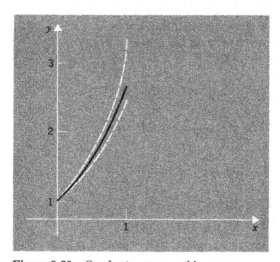

Figure 6-60 Quadratic upper and lower estimates for e^x.

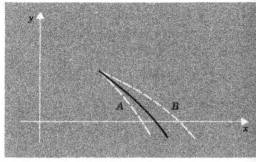

Figure 6-61 Location of zero with aid of quadratic estimates.

Finer Details of the Graphs. We return to (6-110), which locates our graph between the two straight lines at $x = c$. We now ask how the graph is related to the two lines. To that end we take $x \geq c$ and compare f with the lower line, $y = f(c) + A(x - c)$. We set

$$g(x) = f(x) - [f(c) + A(x - c)], \qquad c \leq x \leq b$$

Then

$$g'(x) = f'(x) - A \geq 0$$

since $f'(x) \geq A$. Now $g(c) = f(c) - f(c) = 0$, so that g starts at 0 and either remains there or rises to positive values. If g remained at 0 over an interval, we would have $g'(x) \equiv 0$, $f'(x) \equiv A$, and f would have to be a linear function $Ax + C$ over that interval. Except for this trivial case, g rises to positive values and, in fact, is steadily increasing. In the same way, if we set

$$G(x) = f(c) + B(x - c) - f(x)$$

then $G'(x) \geq 0$ and G is steadily increasing (except for the trivial case when f coincides with a linear function $Bx + C$ over an interval).

Thus, *in general, the graph of f is located between the two lines in such a way that the vertical distance from the graph of f to each line is steadily expanding as x increases.* In particular, at $x = b$, $f(b)$ is definitely greater than $f(c) + A(b - c)$ and less than $f(c) + B(b - c)$; the only exception—an equality in one case or the other—arises when f is a linear function for $c \leq x \leq b$. These statements are illustrated in the graphs of Figures 6-58 and 6-59. The same reasoning applies for $a \leq x \leq c$; in this case, the distance expands as x decreases, as shown in the graphs.

For the second derivative estimates we proceed similarly, as suggested by (6-111). We let

$$g(x) = f(x) - \left[f(c) + f'(c)(x - c) + A \frac{(x - c)^2}{2} \right], \qquad c \leq x \leq b$$

Then

$$g'(x) = f'(x) - [f'(c) + A(x - c)]$$
$$g''(x) = f''(x) - A \geq 0$$

Now $g'(c) = f'(c) - f'(c) = 0$ and, since $g'' \geq 0$, g' is steadily increasing (except for the trivial case in which f' is linear and hence f is quadratic over an interval). Hence $g' \geq 0$ and, if we rule out the trivial case, in fact $g' > 0$ for $c < x \leq b$, so that g is also steadily increasing. Hence *the vertical distance of f from the lower parabola is also expanding and, since $g''(x) \geq 0$, at an ever increasing rate.* (The graph of the difference, $g(x)$, is concave upward.) A similar analysis applies to the upper curve and to the interval $a \leq x \leq c$. Hence in all but the trivial cases, the vertical distances increase, at an ever increasing rate, as one moves to the left or right from $x = c$. This is suggested in the graph of Figure 6-60. In particular, at the end points a and b the graph of f is definitely below the B-parabola and above the A-parabola; an exception—equality in one case—can arise only in the trivial case when

f' is linear: $f' = f(c) + A(x - c)$ or $f' = f(c) + B(x - c)$ in $[a, c]$ or $[c, b]$. In an exceptional case, we can verify that f coincides with an upper or lower parabola over the whole interval (Problem 8 below).

By generalizing the process that led to (6-111) we get a similar rule for a function $f(x)$ whose nth derivative $f^{(n)}(x)$ is bounded between A and B ($A \leq B$) and is continuous in $[a, b]$. We find that $f(x)$ is between

$$f(c) + f'(c)(x - c) + f''(c)\frac{(x - c)^2}{2}$$

$$+ \cdots + f^{(n-1)}(c)\frac{(x - c)^{n-1}}{(n - 1)!} + A\frac{(x - c)^n}{n!} \qquad (6\text{-}112)$$

and

$$f(c) + f'(c)(x - c) + f''(c)\frac{(x - c)^2}{2}$$

$$+ \cdots + f^{(n-1)}(c)\frac{(x - c)^{n-1}}{(n - 1)!} + B\frac{(x - c)^n}{n!} \qquad (6\text{-}113)$$

When n is even, the A-curve is below throughout, the B-curve is above throughout. When n is odd, the A-curve is below for $x \geq c$, above for $x \leq c$, the B-curve is above for $x \geq c$, below for $x \leq c$, as in Figure 6-62. If we rule out the trivial case in which f coincides with a polynomial of degree n over some interval including c, then the graph of f meets the graphs of the A-curve and B-curve only at $x = c$ and the vertical distance of f from each curve increases as x moves away from c. The proof is left as a problem (Problems 7 and 8 below).

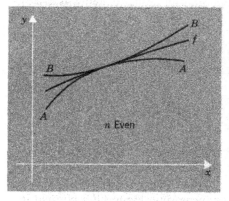
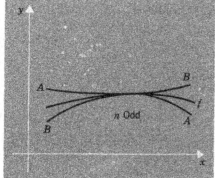

Figure 6-62　Upper and lower estimates of degree n.

PROBLEMS

1. For each of the following functions verify that with the given A, B, $A \leq f'(x) \leq B$ in the interval $[a, b]$ given. Graph the function and verify that the graph lies between the lines of slope A, B through $[a, f(a)]$ and $[b, f(b)]$.

(a) $f(x) = 2x - (x^3/3)$, $-1 \leq x \leq 1$, $A = 1, B = 2$
(b) $f(x) = 1 - 2x + 3x^2 - x^3$, $\frac{1}{2} \leq x \leq \frac{3}{2}$, $A = \frac{1}{4}, B = 1$
(c) $f(x) = \cos x$, $0 \leq x \leq \pi/4$, $A = -0.7, B = 0$
(d) $f(x) = \ln x$, $1 \leq x \leq 3$, $A = \frac{1}{3}, B = 1$

2. A function $f(x), 0 \leq x \leq 1$, is such that $f''(x) \equiv 1/(1 + x^4), f(0) = 1$, and $f'(0) = 0$. Show that $1 + (x^2/4) < f(x) < 1 + (x^2/2)$ for $0 < x \leq 1$ and illustrate by a graph.

3. A function $f(x), 0 \leq x < \infty$ is such that $f''(x) \equiv e^{-x^2}, f(0) = 0$, and $f'(0) = 1$. Show that $x < f(x) < x + (x^2/2)$ for $0 < x < \infty$ and illustrate by a graph.

4. A function $y = f(x), 0 \leq x \leq 2\pi$, is such that $f(0) = 1$ and $y' = 5 + \sin(xy)$, where $y = f(x)$. Show that $1 + 4x \leq f(x) \leq 1 + 6x$ for $0 \leq x \leq 2\pi$.

5. (a) Let $f(x) = x^3 + x - 1$. Show that for $0 \leq x \leq 1$

$$-0.375 + 1.75(x - 0.5) \leq f(x) \leq -0.375 + 1.75(x - 0.5) + 3(x - 0.5)^2$$

and, hence, that f has a zero between $\frac{2}{3} = 0.667$ and $\frac{5}{7} = 0.714$.

(b) Use the method suggested by part (a) to locate approximately a zero of $f(x) = x^2 - \cos x$, in $0 \leq x \leq \pi/2$.

6. Show that (6-111) is valid for $a \leq x \leq c$, under the assumptions stated.

7. Prove that $f(x)$ is between the values (6-112) and (6-113), under the assumptions stated.

8. (a) Show that in (6-111), under the hypotheses stated, one has strict inequality ($<$) in both cases, unless f is a quadratic function of x over the interval from c to x.

(b) Generalize the result of part (a) to the nth degree estimates (6-112), (6-113).

6-12 TAYLOR'S FORMULA WITH REMAINDER

Let f have a continuous derivative f' for $a \leq x \leq b$. We return to the inequalities (6-110) and set $x = b$. We can then write the inequalities as follows:

$$A \leq \frac{f(b) - f(c)}{b - c} \leq B \tag{6-120}$$

We can choose A as the absolute minimum of $f'(x)$ in $[c, b]$, B as the absolute maximum of $f'(x)$ in $[c, b]$, and (6-120) remains valid. (These choices might fail for $[a, c]$, but we now consider only $[c, b]$). We can describe (6-120) as stating that $[f(b) - f(c)]/(b - c)$ is a number between A and B. But $f'(x)$, being continuous, takes on every value between its minimum and maximum in $[c, b]$. Hence we can write, for some ξ,

$$f'(\xi) = \frac{f(b) - f(c)}{b - c}, \qquad c \leq \xi \leq b$$

This is the Mean Value Theorem, with a minor difference: we should have $c < \xi < b$. If, in fact, we were forced to choose $\xi = c$ or b, then $f'(\xi)$ would have to be the maximum or minimum of f', that is

$$\frac{f(b) - f(c)}{b - c} = A \text{ or } B$$

But we saw that the graph of f is definitely between the two lines, except when f is linear; in that case $A = B$ and f' is identically constant, so that even in that case we can take ξ in the open interval (c, b)—in fact anywhere in that interval! Thus we conclude: in all cases there is a ξ such that

$$f'(\xi) = \frac{f(b) - f(c)}{b - c}, \qquad c < \xi < b \qquad (6\text{-}121)$$

A similar reasoning applies to the interval $[a, c]$, where we now choose A, B as the minimum and maximum of f' in that interval. We conclude that

$$A \leq \frac{f(a) - f(c)}{a - c} \leq B$$

and in the same way as before

$$f'(\xi) = \frac{f(a) - f(c)}{a - c}, \qquad a < \xi < c \qquad (6\text{-}121')$$

We can summarize our conclusions as follows. *Let f have a continuous deriva-tive on the interval whose end points are c and x. Then*

$$f(x) = f(c) + f'(\xi)(x - c) \qquad (6\text{-}122)$$

for some ξ between c and x.

The reasoning extends to the higher order estimates. Thus let f have a con-tinuous second derivative on $[a, b]$. Then (6-111), applied at $x = b$, gives

$$A \leq \frac{f(b) - f(c) - f'(c)(b - c)}{(b - c)^2/2} \leq B \qquad (6\text{-}123)$$

and if we take A, B as the absolute minimum and maximum of f'' in $[c, b]$, we conclude that the middle expression in (6-123) must equal $f''(\xi)$, $c \leq \xi \leq b$. As before we can rule out the $=$ signs, and have

$$f(b) = f(c) + f'(c)(b - c) + f''(\xi)\frac{(b - c)^2}{2}, \qquad c < \xi < b$$

A similar relation holds true for $[a, c]$, and we can again make a general rule: *If f has a continuous second derivative on an interval whose end points are c and x, then*

$$f(x) = f(c) + f'(c)(x - c) + f''(\xi)\frac{(x - c)^2}{2} \qquad (6\text{-}124)$$

for some ξ between c and x.

The reasoning extends to the nth derivative estimates (6-113) without change, and (replacing n by $n + 1$) we obtain the following general theorem:

THEOREM 10 *(TAYLOR'S THEOREM WITH REMAINDER).* Let f be *defined and have continuous derivatives through order $n + 1$ on the inter-val from c to x ($c < x$ or $c > x$). Then there is some ξ between c and x for which*

$$f(x) = f(c) + f'(c)(x - c) + f''(c)\frac{(x - c)^2}{2!}$$

$$+ \cdots + f^{(n)}(c)\frac{(x - c)^n}{n!} + f^{(n+1)}(\xi)\frac{(x - c)^{n+1}}{(n + 1)!} \qquad (6\text{-}125)$$

The formula (6-125) is called Taylor's Formula with Remainder in Lagrange's form. We often write (6-125) as follows:

$$f(x) = f(c) + f'(c)(x - c) + \cdots + f^{(n)}(c)\frac{(x - c)^n}{n!} + R_n \quad (6\text{-}125')$$

and call R_n the *remainder term*. For many functions R_n is negligible for n large (more precisely, $\lim R_n = 0$ as $n \to \infty$), as is shown in Chapter 8. Our formula (6-125) gives Lagrange's expression for the remainder. We can write it

$$R_n = \frac{f^{(n+1)}(\xi)(x - c)^{n+1}}{(n + 1)!}, \qquad \xi \text{ between } c \text{ and } x$$

or

$$R_n = f^{(n+1)}[c + \theta(x - c)]\frac{(x - c)^{n+1}}{(n + 1)!}, \qquad 0 < \theta < 1$$

since any number ξ between c and x can be expressed as $c + \theta(x - c)$ with $0 < \theta < 1$.

EXAMPLE 1. Let $f(x) = \sin x$, $0 \le x \le \pi/2$. Then we apply (6-125') with $c = 0$ and $n = 4$:

$$\sin x = 0 + x + 0 - \frac{x^3}{6} + 0 + R_4$$

where

$$R_4 = \frac{f^{(5)}(\xi)x^5}{5!} = \cos \xi \frac{x^5}{5!}, \qquad 0 < \xi < x$$

Since $0 \le \cos \xi \le 1$ for $0 \le \xi \le \pi/2$, R_4 lies between 0 and $x^5/120$. Thus we can say that

$$\sin x \sim x - \frac{x^3}{6}, \qquad 0 \le x \le \frac{\pi}{2}$$

with an error at most $x^5/120$. For example, $\sin 0.1 \sim 0.1 - (0.001/6) = 0.99983$, $\sin 0.5 \sim 0.47917$, $\sin 1 \sim 0.83333$, and $\sin(\pi/2) \sim 0.92483$. From tables we find $\sin 0.1 = 0.99983$, $\sin 0.5 = 0.47943$, $\sin 1 = 0.84147$, and $\sin(\pi/2) = 1$. Thus the errors are clearly within the amount stated. In Figure 6-63, we graph $\sin x$ and the approximating function $x - (x^3/6)$. It is clear that the two curves separate more and more as x increases.

Remark. Let f have continuous derivatives through order $n + 1$ on the interval from c to b. Then, on this interval, we can write

$$f(x) = f(c) + f'(c)(x - c) + \cdots + \frac{f^{(n)}(c)}{n!}(x - c)^n + (x - c)^{n+1}p(x)$$

$$(6\text{-}126)$$

where $p(x)$ is continuous on the interval. For division of Equation (6-126) by $(x - c)^{n+1}$ gives an expression for $p(x)$, for $x \ne c$; as a quotient of two con-

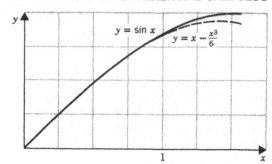

Figure 6-63 Approximation of sin x by Taylor's formula.

tinuous functions, this expression is continuous. By Theorem 10, this same expression can be equated to $f^{(n+1)}(\xi)/(n+1)!$, where ξ is between c and x. As $x \to c$, $f^{(n+1)}(\xi) \to f^{(n+1)}(c)$. Hence, if we set $p(c) = f^{(n+1)}(c)/(n+1)!$, then $p(x)$ is continuous on the given interval.

The expression (6-126) is simpler than the previous remainder formulas and is often just what is needed to study the function near $x = c$. Applications are made in Section 6-14 below. The following are illustrations of the formula:

$$e^x = 1 + x + \frac{x^2}{2} + x^3 p(x), \qquad \cos x = 1 - \frac{x^2}{2} + \frac{x^4}{24} + x^6 p(x)$$

$$\sin x = x - \frac{x^3}{6} + x^5 p(x), \qquad \ln x = (x-1) - \frac{(x-1)^2}{2} + (x-1)^3 p(x)$$

For the last of these, $p(x)$ is continuous for $x > 0$; for the others it is continuous for all x.

PROBLEMS

1. (a) Show that $\sin x = x - (x^3/6) + (x^5/120) + R_6$, $0 < x \le \pi/2$, where $-x^7/7! < R_6 < 0$ and compare the values of $\sin x$ and $p(x) = x - (x^3/6) + (x^5/120)$ for $x = 0.1$, $x = 0.5$, and $x = 1$.
   (b) Graph $\sin x$ and $p(x)$.
2. (a) Show that $\ln(1 + x) = x - (x^2/2) + (x^3/3) + R_3$, $x > -1$, where $-x^4/4 < R_3 < 0$, and compare the values of $\ln(1 + x)$ and $q(x) = x - (x^2/2) + (x^3/3)$ for $x = 0.1, 0.5, 1$, and -0.5.
   (b) Graph $\ln(1 + x)$ and $q(x)$.
3. (a) Apply Taylor's Theorem with Remainder to $f(x) = e^x$ in the interval $[0, b]$ and show that $0 \le R_n \le (e^b b^n)/n!$.
   (b) Show that, for $b > 0$, $\lim\limits_{n \to \infty} (e^b b^n)/n! = 0$. [*Hint.* Let $x_n = (e^b b^n)/n!$ Show that

$$(°) \qquad x_{n+1} = x_n \cdot \frac{b}{n+1}$$

and therefore that, for $n \ge b$, the sequence is strictly decreasing and bounded and hence has a limit x_0. Show by passage to the limit in ($°$) that $x_0 = x_0 \cdot 0 = 0$.]
   (c) Conclude from the results of (a) and (b) that for $0 \le x \le b$,

$$(°°) \qquad e^x = \lim_{n \to \infty} \left(1 + x + \frac{x^2}{2!} + \cdots + \frac{x^n}{n!} \right)$$

[Since b is arbitrary, $(°°)$ is valid for every positive x. The conclusion is usually written in the form

$$e^x = 1 + x + \frac{x^2}{2!} + \cdots + \frac{x^n}{n!} + \cdots$$

The right-hand side is called an infinite series.]

(d) Show that the result of (c) is valid for $a \le x \le 0$ and, hence, that $(°°)$ holds true for all x.

4. Proceed as in Problem 3 to show that

$$\sin x = \lim_{k \to \infty} \left[x - \frac{x^3}{3!} + \frac{x^5}{5!} - \frac{x^7}{7!} + \cdots + (-1)^k \frac{x^{2k+1}}{(2k+1)!} \right]$$

5. Prove the remainder formula (6-124) by applying Rolle's theorem for the interval $[c, b]$ to

$$g(x) = f(b) - f(x) - (b - x)f'(x) - \left(\frac{b - x}{b - c}\right)^2 [f(b) - f(c) - (b - c)f'(c)]$$

6. (a) Use Taylor's Formula (6-124) to prove that, if $f''(x)$ is continuous for $a < x < b$ and $f(c) = 0$, $f'(c) = 0$, and $f''(c) > 0$, where $a < c < b$, then f has a relative minimum at $x = c$ (cf. Theorem 2, Section 6-1).

   (b) Use Taylor's Formula (6-125) to prove that, if $f^{(n+1)}(x)$ is continuous for $a < x < b$ and

$$f'(c) = 0, \ldots, \qquad f^{(n)}(c) = 0, \qquad f^{(n+1)}(c) \ne 0$$

   then f has a relative minimum, a relative maximum, or an inflection point at $x = c$ according as this holds true for

$$y = f^{(n+1)}(c) \frac{(x - c)^{n+1}}{(n + 1)!}$$

6-13 ERROR IN NEWTON'S METHOD

We assume the conditions of Theorem 9 (Section 6-10) are satisfied, and that Case 1 or Case 2 holds true, so that either $x_1 < x_2 < \cdots < x_n < x_0$ or $x_0 < \cdots < x_n < \cdots < x_2 < x_1$. We apply Taylor's Formula (6-124) with $c = x_n$ and $x = x_0$. Thus

$$f(x_0) = f(x_n) + f'(x_n)(x_0 - x_n) + f''(\xi_n) \frac{(x_0 - x_n)^2}{2}$$

where ξ_n is between x_0 and x_n. Since $f(x_0) = 0$, this can be written:

$$f(x_n) = f'(x_n)(x_n - x_0) - f''(\xi_n) \frac{(x_n - x_0)^2}{2}$$

Equation (6-101) now gives

$$x_{n+1} = x_n - \frac{f'(x_n)(x_n - x_0) - f''(\xi_n)(x_n - x_0)^2/2}{f'(x_n)}$$

or after simplification

$$x_{n+1} - x_0 = \left(\frac{(x_n - x_0)^2}{2}\right)\left(\frac{f''(\xi_n)}{f'(x_n)}\right)$$

Hence

$$|x_{n+1} - x_0| = \frac{(x_n - x_0)^2}{2} \cdot \frac{|f''(\xi_n)|}{|f'(x_n)|} \tag{6-130}$$

This equation relates the error $x_n - x_0$ at the nth stage to the error $x_{n+1} - x_0$ at the next stage. Now in Case 1 we are sure that all values x_n and x_0 lie in the interval $[x_1, b]$, in Case 2 we are sure they all lie in the interval $[a, x_1]$. In each interval $|f''|$ has a maximum M_1 and $|f'|$ has a minimum $M_2 > 0$. Consequently

$$|x_{n+1} - x_0| \le |x_n - x_0|^2 \frac{M_1}{2M_2}$$

or

$$|x_{n+1} - x_0| \le M|x_n - x_0|^2 \tag{6-131}$$

where $M = M_1/(2M_2)$, or a number larger than $M_1/(2M_2)$. If $M > 1$ and $|x_n - x_0| > 1$, inequality (6-131) is of no use, since it then says only that the error cannot increase more than a certain amount, whereas we know that the error decreases.

To use (6-131) effectively we proceed as follows. We carry out Newton's method as usual, observing the value of $b - x_n$ in Case 1 and of $x_n - a$ in Case 2. We also observe the maximum M_1 of $|f''|$ and minimum M_2 of $|f'|$ in the interval $[x_n, b]$ or $[a, x_n]$ and calculate M, an upper estimate for $M_1/(2M_2)$. As n increases, $|x_n - x_0|$ is approaching 0 and M is decreasing. Hence, we must eventually reach a value of n for which $M|x_n - x_0| = r < 1$. We then renumber x_n as x_1 and start as before. We now have

$$M|x_1 - x_0| = r < 1 \tag{6-132}$$

and, therefore, by (6-131)

$$|x_2 - x_0| \le M|x_1 - x_0|^2 = \frac{r^2}{M}$$

$$|x_3 - x_0| \le M|x_2 - x_0|^2 \le M \frac{r^4}{M^2} = \frac{r^4}{M}$$

$$|x_4 - x_0| \le M|x_3 - x_0|^2 \le M \frac{r^8}{M^2} = \frac{r^8}{M}$$

and, by induction,

$$|x_n - x_0| \le \frac{r^{2^{n-1}}}{M} \qquad (n = 1, 2, 3, \ldots) \tag{6-133}$$

Since $0 < r < 1$, the right-hand side describes a monotone decreasing sequence approaching 0 as $n \to \infty$ (Problem 4 below). The members of this sequence provide the desired error estimates. If, in particular, $M = 1$ and $r = 0.1$, then we have

$$\begin{aligned} |x_1 - x_0| &\le 0.1 \\ |x_2 - x_0| &\le 0.01 \\ |x_3 - x_0| &\le 0.0001, \ldots \end{aligned} \tag{6-134}$$

that is, we double the number of decimal places of accuracy at each step.

EXAMPLE. $f(x) = x^2 - \cos x$. This is Example 2 of Section 6-10. There we found that there was a zero between 0.8 and 0.9 and that Case 2 applies, so that we could take $x_1 = 0.9$. In the interval $[0.8, 0.9]$, we observe that $|f''(x)| = |2 + \cos x| < 2.7$ and $|f'(x)| = |2x + \sin x| > 2.3$, so that

$$\frac{M_1}{2M_2} = \frac{1}{2} \cdot \frac{\max|f''|}{\min|f'|} < \frac{2.7}{4.6} < 1 = M$$

Furthermore, since $0.8 < x_0 < 0.9$, $|x_1 - x_0| < |0.9 - x_0| < 0.1$, so that (6-132) applies with $r = 0.1$ and the conditions (6-134) hold true. In Section 6-10 we found that $x_2 = 0.83$, $x_3 = 0.82411$. Hence we can be sure that x_2 is in error by at most 0.01, x_3 by at most 0.0001, that is,

$$0.82 \le x_0 \le 0.83$$
$$0.82401 \le x_0 \le 0.82411$$

A further iteration would reduce the error to less than 0.00000001.

In the preceding discussion we have pretended that all values $x_1, x_2, \ldots$ are precisely computed. In general, these are decimals rounded off after so many places. The resulting errors may affect our conclusions significantly. The problem of round-off errors is one of the hazards of all numerical computations and must always be kept in mind. It is especially serious for work with digital computers.

PROBLEMS

1. Show on the basis of (6-130) that the worst conditions for Newton's method prevail when near x_0 the derivative f' is small but the graph of f has large curvature.
2. Show by (6-132) that, for Example 1 in Section 6-10 ($x^2 - 5 = 0$), we have

$$|x_{n+1} - x_0| < \tfrac{1}{4}|x_n - x_0|^2$$

and that the errors in x_1, x_2, x_3, x_4 are less than 1, 1/4, 1/64, 1/16,384 = 0.00006.
3. Apply Newton's method to find the solution of each of the following equations in the interval given. In each case, carry the solution far enough to find a starting value x_1 and constants $r < 1$ and M so that (6-132) and (6-133) hold true.
   (a) $x^5 + x - 1 = 0, 0 \le x \le 1$ (b) $x^{10} + 10^{18}x - 10^{20} = 0, 0 \le x \le 100$
4. Let M be a fixed positive number, let r be a fixed number between 0 and 1. Let

$$y_n = \frac{r^{2^{n-1}}}{M} \qquad (n = 1, 2, 3, \ldots)$$

   (a) Show that $y_1 > y_2 > \cdots > y_n > \cdots > 0$
   (b) Show that $\lim_{n \to \infty} y_n = 0$. [*Hint.* Show that $\lim_{n \to \infty} \ln y_n = -\infty$.]

6-14 INDETERMINATE FORMS, L'HOSPITAL'S RULE

Often in the calculus and its applications we are led to a function F given as a quotient:

$$F(x) = \frac{f(x)}{g(x)}$$

and for a particular value of x of interest, say $x = c$, both $f(x)$ and $g(x)$ reduce to 0, so that $F(c) = 0/0$, a meaningless expression. It has been seen that in many cases we can still find the limit of F, as x approaches c, and this limit usually provides a useful number. By tradition, we say that such a function F is an *indeterminate form*, of type $0/0$, at $x = c$, and we refer to finding the limit as "evaluating the indeterminate form" (See Section 2-11).

EXAMPLE 1. $F(x) = [\sin x - x \cos x]/x^3$. Here F is defined and continuous for $x \neq 0$ and numerator and denominator are 0 for $x = 0$, so that F is an indeterminate form, of type $0/0$, at $x = 0$. To find the limit of F as x approaches 0, we can apply Taylor's formula in the form (6-126), choosing n appropriately large. Here we write

$$\sin x = x - \frac{x^3}{6} + x^5 p(x), \qquad \cos x = 1 - \frac{x^2}{2} + x^4 q(x)$$

where p and q are continuous for all x. Hence, for $x \neq 0$,

$$F(x) = \frac{x - \dfrac{x^3}{6} + x^5 p(x) - x\left[1 - \dfrac{x^2}{2} + x^4 q(x)\right]}{x^3} = \frac{1}{3} + x^2 [p(x) - q(x)].$$

Therefore, $\lim\limits_{x \to 0} F(x) = \frac{1}{3}$.

EXAMPLE 2. $$F(x) = \frac{(\ln x)^2}{(x - 1)(e^x - e)}$$

Again we have the indeterminate form $0/0$, but occurring at $x = 1$. Now by (6-126), with $c = 1$,

$$\ln x = (x - 1) + (x - 1)^2 p(x), \qquad e^x = e + e(x - 1) + (x - 1)^2 q(x)$$

where $p(x)$ is continuous for $x > 0$ and $q(x)$ is continuous for all x. Hence, for $x \neq 1$,

$$F(x) = \frac{[(x - 1) + (x - 1)^2 p(x)]^2}{(x - 1)[e(x - 1) + (x - 1)^2 q(x)]}$$

$$= \frac{1 + 2(x - 1)p(x) + (x - 1)^2 [p(x)]^2}{e + (x - 1)q(x)}$$

Therefore, $\lim\limits_{x \to 1} F(x) = 1/e$.

The method just illustrated will take care of the most common indeterminate forms of this type. An alternate method is provided by the following theorem.

THEOREM 11 (*L'Hospital's Rule*). *Let $f(x)$ and $g(x)$ be continuous for $c \leq x \leq b$, with $f(c) = g(c) = 0$. Let f and g be differentiable for $c < x < b$. Then*

$$\lim_{x \to c+} \frac{f(x)}{g(x)} = \lim_{x \to c+} \frac{f'(x)}{g'(x)} \qquad (6\text{-}140)$$

provided that the limit on the right exists.

If we assume that $f'(x)$ and $g'(x)$ are continuous at $x = c$, and that $g'(c) \neq 0$, then the conclusion follows from the Mean Value Theorem. For

$$f(x) = f'(\xi_1)(x - c), \qquad g(x) = g'(\xi_2)(x - c)$$

where ξ_1 and ξ_2 are between c and x. Then

$$\frac{f(x)}{g(x)} = \frac{f'(\xi_1)}{g'(\xi_2)} \longrightarrow \frac{f'(c)}{g'(c)} = \lim_{x \to c+} \frac{f'(x)}{g'(x)}$$

as $x \to c+$. The proof for the general case is given in the next section.

There is an analogous theorem for limits from the left:

$$\lim_{x \to c-} \frac{f(x)}{g(x)} = \lim_{x \to c-} \frac{f'(x)}{g'(x)} \tag{6-140$'$}$$

When the appropriate hypotheses hold true both to the left and right, we have a similar theorem for two-sided limits:

$$\lim_{x \to c} \frac{f(x)}{g(x)} = \lim_{x \to c} \frac{f'(x)}{g'(x)} \tag{6-140$''$}$$

To apply L'Hospital's Rule for a given indeterminate form, f/g, we form the ratio of derivatives (not the derivative of the quotient) and seek its limit as $x \to c$. This limit may again lead us to the indeterminate form $0/0$. In that case we can apply the rule again. Thus it may be necessary to differentiate the numerator and denominator several times. For our Example 1 above we have

$$\frac{f(x)}{g(x)} = \frac{\sin x - x \cos x}{x^3} \qquad \left(= \frac{0}{0} \text{ at } x = 0\right)$$

$$\frac{f'(x)}{g'(x)} = \frac{x \sin x}{3x^2} = \frac{\sin x}{3x} \qquad \left(= \frac{0}{0} \text{ at } x = 0\right)$$

$$\frac{f''(x)}{g''(x)} = \frac{\cos x}{3} \qquad \left(= \frac{1}{3} \text{ at } x = 0\right)$$

Thus

$$\lim_{x \to 0} \frac{\sin x - x \cos x}{x^3} = \lim_{x \to 0} \frac{x \sin x}{3x^2} = \lim_{x \to 0} \frac{\sin x}{3x} = \lim_{x \to 0} \frac{\cos x}{3} = \frac{1}{3}$$

EXAMPLE 3. $\displaystyle \lim_{x \to 0} \frac{2e^x - 2 - 2x - x^2}{x^3}$. Here

$$\lim_{x \to 0} \frac{2e^x - 2 - 2x - x^2}{x^3} = \lim_{x \to 0} \frac{2e^x - 2 - 2x}{3x^2}$$

$$= \lim_{x \to 0} \frac{2e^x - 2}{6x} = \lim_{x \to 0} \frac{2e^x}{6} = \frac{1}{3}$$

We could have used a Taylor's formula for e^x:

$$\frac{2e^x - 2 - 2x - x^2}{x^3} = \frac{2[1 + x + x^2/2 + x^3/6 + x^4 p(x)] - 2 - 2x - x^2}{x^3}$$

$$= \frac{1}{3} + 2xp(x)$$

Thus again the limit as $x \to 0$ is found to be $\frac{1}{3}$.

Warning. It is very tempting to blindly differentiate numerator and denominator, and stop only when simple expressions are obtained, or patience runs out. However, at each stage *differentiation is permitted only if one has an indeterminate form.* If one ignores this condition, wrong or meaningless results are found. For example,

$$\lim_{x \to 1} \frac{x^3 - 1}{x^2 - 1} = \lim_{x \to 1} \frac{3x^2}{2x} \overset{?}{=} \lim_{x \to 1} \frac{6x}{2} = 3$$

The first quotient is an indeterminate form, but the second is not, and the limit is $\frac{3}{2}$, not 3.

Extensions of L'Hospital's Rule. The rules (6-140), (6-140'), and (6-140'') remain valid if $f(x)$ and $g(x)$ are *discontinuous* at c provided that f has limit $+\infty$ or $-\infty$, g has limit $+\infty$ or $-\infty$, and f', g' satisfy the same conditions as before. Furthermore, the rules remain valid if $c = +\infty$ or $c = -\infty$, provided that either $f \to 0$ and $g \to 0$ or $f \to +\infty$ (or $-\infty$) and $g \to +\infty$ (or $-\infty$) as $x \to c$, and the derivatives f', g' are defined in the appropriate infinite interval. Finally, all rules remain valid if the limit of f'/g' is infinite, so that the limit of f/g is also infinite. The proofs are given in the next section. When f and g have $\pm\infty$ as limit, we say the indeterminate form is of type ∞/∞.

EXAMPLE 4
$$\lim_{x \to 0+} \frac{\sqrt{x} - \sqrt[3]{x}}{\sqrt{x} + \sin x} = \lim_{x \to 0+} \frac{f(x)}{g(x)}$$

The functions f and g are not defined for negative x, so that at 0 only the limit as $x \to 0+$ has meaning. We find by (6-140) that

$$\lim_{x \to 0+} \frac{\sqrt{x} - \sqrt[3]{x}}{\sqrt{x} + \sin x} = \lim_{x \to 0+} \frac{(2x^{1/2})^{-1} - (3x^{2/3})^{-1}}{(2x^{1/2})^{-1} + \cos x}$$

Our new expression is complicated. We simplify it by multiplying numerator and denominator by $6x^{2/3}$ to obtain

$$\lim_{x \to 0+} \frac{3x^{1/6} - 2}{3x^{1/6} + 6x^{2/3} \cos x}$$

Now the numerator has limit -2, the denominator is positive and has limit 0, so that we conclude (cf. Section 2-11) that this limit is $-\infty$, and hence

$$\lim_{x \to 0+} \frac{f(x)}{g(x)} = -\infty$$

EXAMPLE 5. $\lim\limits_{x \to 0+} x \ln x$. This has the limiting form $0 \cdot \infty$ [more precisely, $0 \cdot (-\infty)$]. In such cases we "reason" that

$$0 \cdot \infty = \frac{0}{1/\infty} = \frac{0}{0} \quad \text{or} \quad 0 \cdot \infty = \frac{\infty}{1/0} = \frac{\infty}{\infty}$$

that is, we write

$$x \ln x = \frac{x}{1/\ln x} \quad \left(\text{form } \frac{0}{0}\right) \quad \text{or} \quad x \ln x = \frac{\ln x}{1/x} \quad \left(\text{form } \frac{\infty}{\infty}\right)$$

Immediately we observe that the second form is preferable for differentiation. According to (6-140):

$$\lim_{x \to 0+} \frac{\ln x}{1/x} = \lim_{x \to 0+} \frac{1/x}{-1/x^2} = \lim_{x \to 0+} \frac{-x}{1} = 0$$

This last example is important because it shows that, as $x \to 0+$, $\ln x$ becomes infinite "more slowly" than $1/x$, so that the ratio of $\ln x$ to $1/x$ approaches 0. Figure 6-64 shows in graphical form the rapidity of approach to ∞ of different functions, as $x \to 0+$. Figure 6-65 shows comparable rates for $x \to +\infty$. In each case $f/g \to 0$, if f lies below g in the diagram.

EXAMPLE 6. $\lim\limits_{x \to \infty} x^3/e^x$. This has form ∞/∞, but according to the diagram of Figure 6-65, the limit should be 0. We find that

$$\lim_{x \to \infty} \frac{x^3}{e^x} = \lim_{x \to \infty} \frac{3x^2}{e^x} = \lim_{x \to \infty} \frac{6x}{e^x} = \lim_{x \to \infty} \frac{6}{e^x} = 0$$

$$\left(\frac{\infty}{\infty}\right) \qquad \left(\frac{\infty}{\infty}\right) \qquad \left(\frac{\infty}{\infty}\right)$$

Observe that all the indeterminate forms are (∞/∞). At the last step we can write $6/e^x = 6e^{-x} \to 0$: a proof is given in Section 5-7.

In a similar way we justify the other information in the graphs [Problem 3, parts (i), (j), (p), (q), and (s) below].

EXAMPLE 7. $\lim\limits_{x \to \infty} (x - \sqrt{x^2 + 1})$. This has form $\infty - \infty$, not one of our

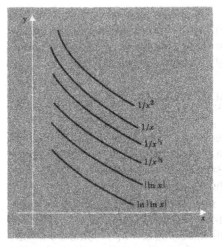

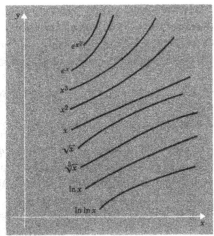

Figure 6-64 Scale of growth rates as $x \to 0+$.

Figure 6-65 Scale of growth rates as $x \to \infty$.

allowed forms $0/0$ or ∞/∞. However, we can write

$$x - \sqrt{x^2 + 1} = x\left(1 - \frac{\sqrt{x^2 + 1}}{x}\right) = \frac{1 - x^{-1}\sqrt{x^2 + 1}}{1/x}$$

Since $x^{-1}\sqrt{x^2 + 1} = \sqrt{1 + x^{-2}} \to 1$ as $x \to \infty$, our form is now $0/0$, and we can apply L'Hospital's Rule:

$$\lim_{x \to \infty} (x - \sqrt{x^2 + 1}) = \lim_{x \to \infty} \frac{1 - x^{-1}\sqrt{x^2 + 1}}{1/x} = \lim_{x \to \infty} \frac{1/(x^2\sqrt{x^2 + 1})}{-1/x^2}$$

$$= \lim_{x \to \infty} \frac{-1}{\sqrt{x^2 + 1}} = 0$$

We can describe our procedure symbolically:

$$\infty - \infty = \infty\left(1 - \frac{\infty}{\infty}\right) = \frac{1 - \infty/\infty}{1/\infty} = \frac{0}{0}$$

However, for the last step, we must check that the ∞/∞ can be replaced by 1, so that the numerator is $1 - 1 = 0$. If the ∞/∞ corresponded to a limit other than 1 (say to 2), our answer would be quite different.

Other forms can be reduced to the standard ones. We indicate two important cases in symbols:

$$1^\infty = e^{\infty \ln 1} = e^{\infty \cdot 0} = e^{0/0} = e^{\infty/\infty}$$

$$0^0 = e^{0 \ln 0} = e^{0 \cdot \infty} = e^{0/0} = e^{\infty/\infty}$$

For example, letting $\exp(u) = e^u$, we have

$$\lim_{x \to 0} (\cos x)^{1/x} = \lim_{x \to 0} \exp\left(\frac{\ln \cos x}{x}\right) = \exp\left(\lim_{x \to 0} \frac{\ln \cos x}{x}\right)$$

provided the limit exists (Section 2-7). We easily find the last limit to be 0, so that our answer is $e^0 = 1$.

Remark. Evaluating limits is one of the central problems of the calculus (consider the derivative and the integral!), and there is no general method for all problems. L'Hospital's Rule is often strong enough for the case at hand, but it may equally well be inadequate (Problem 4 below).

†6-15 PROOFS OF L'HOSPITAL'S RULES

We first consider two special cases:

CASE 1. The form $\varphi(x)/x$ where $\varphi(x)$ is defined for $0 < x \le \alpha$, is continuous and has a derivative in that interval, and $\varphi(x) \to 0$ as $x \to 0+$. Here we have an indeterminate form $0/0$ and want to show that

$$\lim_{x \to 0+} \frac{\varphi(x)}{x} = \lim_{x \to 0+} \varphi'(x) \tag{6-150}$$

provided the limit on the right exists or is infinite. We define $\varphi(0)$ to be 0, so that φ is now continuous in the interval $[0, \alpha]$, and we apply the Mean Value

Theorem (Section 3-21) to the interval $[0, x]$ for $x > 0$:

$$\varphi(x) - \varphi(0) = \varphi(x) = \varphi'(\xi)x, \qquad 0 < \xi < x$$

Hence, for $x > 0$,

$$\frac{\varphi(x)}{x} = \varphi'(\xi), \qquad 0 < \xi < x$$

As $x \to 0+$, $\xi \to 0+$, and therefore (6-150) follows, provided that φ' has a finite or infinite limit as $x \to 0+$. Thus the rule has been proved in this special case.

A similar proof is valid for the interval $\alpha \le x < 0$ with $x \to 0-$ and for the interval $\alpha \le x \le \beta$, where $\alpha < 0 < \beta$ and $x \to 0$.

CASE 2. The form $\varphi(x)/x$, where φ is defined, is continuous, and has a derivative, for large x, say $x \ge \alpha$, and $\varphi(x) \to \infty$ as $x \to \infty$. Thus we have an indeterminate form ∞/∞. We wish to show that

$$\lim_{x \to \infty} \frac{\varphi(x)}{x} = \lim_{x \to \infty} \varphi'(x) \qquad (6\text{-}151)$$

provided that the limit on the right is finite or infinite. Let us assume first that the limit of φ' is finite, equal to A. Then given $\epsilon > 0$, we can choose x_0 so large, and positive, that

$$A - \epsilon < \varphi'(x) < A + \epsilon, \qquad x \ge x_0 \qquad (6\text{-}152)$$

Now by the Mean Value Theorem, $\varphi(x) - \varphi(x_0) = \varphi'(\xi)(x - x_0)$, so that by (6-152)

$$(A - \epsilon)(x - x_0) < \varphi(x) - \varphi(x_0) < (A + \epsilon)(x - x_0)$$

Therefore, if we divide by x,

$$A - \epsilon - (A - \epsilon)\frac{x_0}{x} < \frac{\varphi(x)}{x} - \frac{\varphi(x_0)}{x} < A + \epsilon - (A + \epsilon)\frac{x_0}{x}$$

or

$$A - \epsilon + \frac{\varphi(x_0) - (A - \epsilon)x_0}{x} < \frac{\varphi(x)}{x} < A + \epsilon + \frac{\varphi(x_0) - (A + \epsilon)x_0}{x}$$

As $x \to \infty$ the left and right members approach $A - \epsilon$ and $A + \epsilon$, respectively. Accordingly, we can choose x_1 so large that for $x \ge x_1$ the left member exceeds $A - 2\epsilon$ and the right member is less than $A + 2\epsilon$. Thus

$$A - 2\epsilon < \frac{\varphi(x)}{x} < A + 2\epsilon, \qquad x \ge x_1$$

Accordingly, $\varphi(x)/x$ has limit A, as $x \to \infty$, as was to be proved.

If $\varphi'(x) \to \infty$ as $x \to \infty$, we proceed similarly. For a given K, we can choose a number x_0 so large and positive that

$$K < \varphi'(x), \qquad x \ge x_0$$

Hence

$$K(x - x_0) < \varphi(x) - \varphi(x_0), \qquad x \geq x_0$$

$$K + \frac{\varphi(x_0) - Kx_0}{x} < \frac{\varphi(x)}{x}, \qquad x \geq x_0$$

The lefthand member surely exceeds $K - 1$ for $x \geq x_1$ and x_1 sufficiently large, so that

$$K - 1 < \frac{\varphi(x)}{x}, \qquad x \geq x_1$$

Since K was arbitrary, this implies that $\varphi(x)/x \to \infty$ also, as $x \to \infty$.

Thus the rule is justified for Case 2. We remark that there is a similar proof for the interval $-\infty < x \leq \alpha$, with $x \to -\infty$.

We now turn to the form f/g, as in Theorem 11, but replace x by t, so that we are considering two functions $f(t)$, $g(t)$ satisfying the stated conditions in the interval $[c, b]$. Since f'/g' is assumed to have a limit as $t \to c+$, f'/g' must be defined for $c < t < c + h$, for h sufficiently small. Without loss of generality, we can assume that f'/g' is defined for $c < t < b$. Accordingly, $g'(t) \neq 0$ on this open interval. Since g is continuous on the interval $[c, b]$, g must be monotone strictly increasing or decreasing; for otherwise g would take the same value twice and, by Rolle's theorem, g' would have a zero inside the interval. Let us assume, for example, that g is strictly increasing. Then $x = g(t)$ has a continuous inverse function $t = g^{-1}(x)$, defined on the interval $0 \leq x \leq g(b)$ (see Figure 6-66).

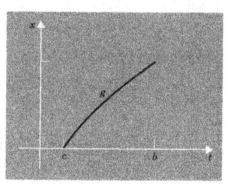

Figure 6-66 Proof of Theorem 12.

We now interpret the equations $x = g(t)$, $y = f(t)$ as parametric equations of a curve, as in Section 3-9. Since g has a continuous inverse g^{-1}, y can be expressed in terms of x, along the path, say $y = \varphi(x)$. Here $\varphi(x) = f[g^{-1}(x)]$. Accordingly,

$$\frac{f(t)}{g(t)} = \frac{y}{x} = \frac{\varphi(x)}{x}$$

Furthermore, as in Section 3-9,

$$\frac{dy}{dx} = \varphi'(x) = \frac{f'(t)}{g'(t)}$$

Accordingly, the rule to be proved:

$$\lim_{t\to c+} \frac{f(t)}{g(t)} = \lim_{t\to c+} \frac{f'(t)}{g'(t)}$$

is equivalent to the rule

$$\lim_{x\to 0+} \frac{\varphi(x)}{x} = \lim_{x\to 0+} \varphi'(x) \qquad (6\text{-}153)$$

for the corresponding interval. In each case as t approaches c, x approaches $0 = g(c)$ (see Problem 7 below). But (6-153) is the same as (6-150), so we are done.

The other cases are treated in exactly the same fashion. In each case, g is monotone strictly increasing or decreasing, and $y = f(t)$ can be expressed in terms of x as $\varphi(x)$, so that $f(t)/g(t) = \varphi(x)/x$ at corresponding values t, x; also $f'(t)/g'(t) = \varphi'(x)$. The proof that $\lim(f/g) = \lim(f'/g')$ is reduced to the proof that $\lim \varphi(x)/x = \lim \varphi'(x)$ for the corresponding interval. One is always led to Cases 1 and 2, so that the conclusion holds true in general.

For example, let us consider the case of $f(t)/g(t)$ on the interval $a \le t < \infty$, with $f(t) \to 0$ and $g(t) \to 0$ as $t \to \infty$. Let us assume that $g(t)$ is monotone decreasing for $a \le t < \infty$; its inverse is therefore defined for $0 < x \le g(a)$ (see Figure 6-67). Accordingly, $y = \phi(x)$ is defined on this interval, and $\phi(x)$ approaches 0 as $x \to 0+$. We are again in Case 1, and the conclusion follows as above.

For the form ∞/∞ we are led to Case 2. If, for example, the interval is $[c, b]$ and $t \to c+$, let us assume g is monotone decreasing. Then Figure 6-68 shows the relation between t and x. As $t \to c+$, $x \to \infty$ and we must show that

$$\lim_{x\to\infty} \frac{\varphi(x)}{x} = \lim_{x\to\infty} \varphi'(x)$$

with $\varphi(x) \to \infty$ as $x \to \infty$. This is precisely Case 2, and we are done.

Thus the rules are justified in all cases.

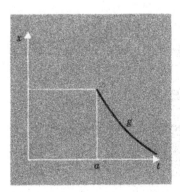

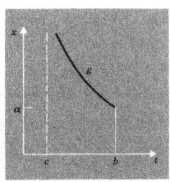

Figure 6-67 Proof for case of $0/0$ as $t \to \infty$.

Figure 6-68 Proof for case of ∞/∞ as $t \to c+$.

PROBLEMS

1. Evaluate with the aid of Taylor's formula:

(a) $\lim\limits_{x \to 0} \dfrac{x^3}{\sin x - x}$ (b) $\lim\limits_{x \to 0} \dfrac{1 - \cos x}{x \sin x}$

(c) $\lim\limits_{x \to 0} \dfrac{e^{x^2} - 1}{x^2}$ (d) $\lim\limits_{x \to 1} \dfrac{\ln x}{x^2 - 1}$

2. Evaluate by definition of the derivative:

(a) $\lim\limits_{x \to 0} \dfrac{\sin x}{x}$ (b) $\lim\limits_{x \to 0} \dfrac{e^x - 1}{x}$ (c) $\lim\limits_{x \to 0} \dfrac{\tan x}{x}$ (d) $\lim\limits_{x \to 1} \dfrac{\ln x}{x - 1}$

3. Evaluate by L'Hospital's Rule:

(a) $\lim\limits_{x \to 1} \dfrac{e^x - e}{\ln x}$ (b) $\lim\limits_{x \to 0+} \dfrac{\sin x - x^2}{\cos x - 1}$ (c) $\lim\limits_{x \to 0+} \dfrac{\sec x - 1}{x^3}$

(d) $\lim\limits_{x \to 0} \dfrac{\ln \sec x}{\sin x}$ (e) $\lim\limits_{x \to 0+} \dfrac{\sqrt{x}}{\sin \sqrt{x} + x}$ (f) $\lim\limits_{x \to 1+} \dfrac{\sqrt{x^2 - 1}}{\ln x}$

(g) $\lim\limits_{x \to 0+} \dfrac{e^{-1/x}}{x}$ (h) $\lim\limits_{x \to \pi-} \dfrac{\sqrt{\pi^2 - x^2}}{\sin x}$ (i) $\lim\limits_{x \to 0+} x^\alpha \ln x$ $(\alpha = \text{const})$

(j) $\lim\limits_{x \to 0+} \dfrac{e^{1/x}}{x^\alpha}$ $(\alpha = \text{const.} < 0)$ (k) $\lim\limits_{x \to 0} \dfrac{\ln x}{\ln \ln(1 + x)}$

(l) $\lim\limits_{x \to 0+} \dfrac{\ln \ln(1 + x)}{\ln \ln \ln(e + x)}$ (m) $\lim\limits_{x \to \infty} \dfrac{x^3 + 5x^2 + 1}{2x^3 - x + 3}$

(n) $\lim\limits_{x \to \infty} \dfrac{x^4 - 1}{x^5 + 2x^2 + 2}$ (o) $\lim\limits_{x \to \infty} \dfrac{x^2}{e^x}$

(p) $\lim\limits_{x \to \infty} \dfrac{x^\alpha}{e^x}$ $(\alpha = \text{const.})$ (q) $\lim\limits_{x \to \infty} \dfrac{x^x}{e^{x^2}}$

(r) $\lim\limits_{x \to \infty} \dfrac{x^\alpha}{\ln x}$ $(\alpha = \text{const.} > 0)$ (s) $\lim\limits_{x \to \infty} \dfrac{\ln \ln x}{\ln x}$

(t) $\lim\limits_{x \to \infty} \dfrac{\ln (x^2 + 1)}{\ln(x^3 + 1)}$ (u) $\lim\limits_{x \to 0+} x^x$

(v) $\lim\limits_{x \to 0+} x^{\sin x}$ (w) $\lim\limits_{x \to \infty} \left(1 + \dfrac{1}{x}\right)^x$

(x) $\lim\limits_{x \to \pi/2} (\sec x - \tan x)$ ‡ (y) $\lim\limits_{x \to \infty} (e^x - e^{\sqrt{x^2+1}})$

4. (a) Show that L'Hospital's Rule is not applicable to $\lim\limits_{x \to \infty} \sin x / x$, but evaluate the limit by showing that $|\sin x/x| \leq 1/x$.

(b) Evaluate $\lim\limits_{x \to \infty} \dfrac{\sin x}{e^x}$ (c) Evaluate $\lim\limits_{x \to \infty} \dfrac{\cos (\ln x)}{x}$

5. Let f have first and second derivatives in the interval $[a, b]$ and let $a < c < b$. Prove

(a) $f'(c) = \lim\limits_{x \to 0} \dfrac{f(c + x) - f(c - x)}{2x}$

(b) $f''(c) = \lim\limits_{x \to 0} \dfrac{f(c + x) - 2f(c) + f(c - x)}{x^2}$

6. *Cauchy's Mean Value Theorem* states the following: If f and g are continuous in the interval $[a, b]$ and differentiable in (a, b), $g(a) \neq g(b)$, and $f'(x)$ and $g'(x)$ are never 0 at the same x, then

$$\frac{f(b) - f(a)}{g(b) - g(a)} = \frac{f'(\xi)}{g'(\xi)}$$

for some ξ in (a, b).

(a) Use the theorem to prove L'Hospital's Rule for the case $0/0$ as $x \to c+$.

(b) Prove Cauchy's Theorem with the aid of the geometric meaning of the equations $x = g(t), y = f(t)$. (Cf. the proof of Rolle's theorem and the Mean Value Theorem in Sections 3-20 and 3-21.)

(c) Prove Cauchy's theorem by applying Rolle's theorem to

$$F(t) = g(t)[f(b) - f(a)] - f(t)[g(b) - g(a)]$$

7. Prove that, if $x = g(t)$ is defined for $c < t \leq b$ and has limit a as $t \to c+$, if $F(x)$ is defined for $a < x \leq d$ and F has a finite or infinite limit as $x \to a+$, and if $F[g(t)]$ is defined for $c < t \leq b$, then

$$\lim_{t \to c+} F[g(t)] = \lim_{x \to a+} F(x)$$

7

APPLICATIONS OF THE INTEGRAL CALCULUS

7-1 AREA BETWEEN TWO CURVES

We observed in Section 4-19 that, if f is continuous in the interval $[a, b]$ and $f(x) > 0$ in this interval, except perhaps at a and b, then

$$\int_a^b f(x)\, dx$$

can be interpreted as the "area under the curve," in the units chosen. If $f(x) < 0$ in the interval, then $-f(x) > 0$ and

$$\int_a^b f(x)\, dx = -\int_a^b [-f(x)]\, dx$$

so that the integral can be interpreted as minus the area under the reflected curve $y = -f(x)$ or, by symmetry, as minus the area above the curve $y = f(x)$ and below the x-axis, for $a \leq x \leq b$ (see Figure 7-1).

If f changes sign several times in the interval, as in Figure 7-2, the integral can be interpreted as the total area above the x-axis minus the area

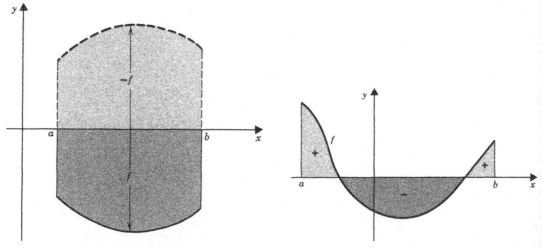

Figure 7-1 Integral as negative of area below x-axis. **Figure 7-2** Integral as sum of signed areas.

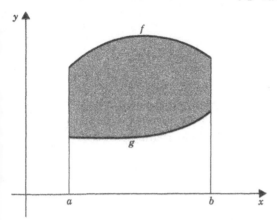

Figure 7-3   Area between two curves.

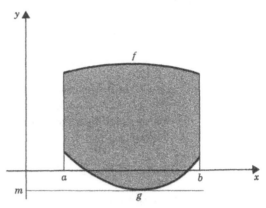

Figure 7-4   Area between two curves, general case.

below the x-axis, as suggested in the figure. If we want all the area counted positively, that is, all the area "between the curve $y = f(x)$ and the x-axis," we must change the sign of f wherever it becomes negative; that is, we must compute

$$\int_a^b |f(x)|\, dx$$

Now let two continuous functions, f and g, be given in the interval $[a, b]$, and let $f(x) > g(x)$ in the interval, except perhaps at a and b. Then it is natural to seek the area between the curves, as suggested in Figure 7-3. Let us assume first that $g(x) > 0$ in the interval. Then the area sought exists, by the rules of Section 4-18 and

area under graph of g + area between curves = area under graph of f

or

$$\int_a^b g(x)\, dx + \text{area between curves} = \int_a^b f(x)\, dx$$

so that

area between curves $= \int_a^b f(x)\, dx - \int_a^b g(x)\, dx = \int_a^b [f(x) - g(x)]\, dx$ (7-10)

If $g(x)$ is not > 0, then g has a minimum value, say m, in the closed interval $[a, b]$, as in Figure 7-4. The same reasoning as that leading us to the original interpretation of integral as area makes us interpret

$$\int_a^b [g(x) - m]\, dx$$

as the area below the graph of g and above the line $y = m$. This area plus the area between the two curves should give the area below the curve $y = f(x)$ and above the line $y = m$:

$$\int_a^b [g(x) - m]\, dx + \text{area between curves} = \int_a^b [f(x) - m]\, dx$$

so that

$$\text{area between curves} = \int_a^b [f(x) - m]\, dx - \int_a^b [g(x) - m]\, dx$$

$$= \int_a^b [f(x) - g(x)]\, dx$$

in agreement with (7-10). Thus (7-10) applies in general, as long as $f(x) > g(x)$, except perhaps at a and b. If $f(x) - g(x)$ changes sign in the interval, we again get area counted positively and negatively as above, and can always use

$$\int_a^b |f(x) - g(x)|\, dx$$

as the area between the curves.

PROBLEMS

1. Find the area of the region bounded by the curves or lines given:
   (a) $y = \cos x$, $\pi/2 \leq x \leq 3\pi/2$; the line $y = 0$
   (b) $y = x^2 - 10$, $0 \leq x \leq 1$; $y = 0$; $x = 0$; $x = 1$
   (c) $y = e^x$; $x = 1$; $y = 1$
   (d) $y = \sin x$, $x = 0$, $x = 2\pi$, $y = -1$
   (e) $y = x^2 - 4$, $y = 12 - x^2$
   (f) $y = x^4 - 3x^2 - 8$, $y = -4$
2. Find the area between the two curves given:
   (a) $y = x$ and $y = \sin x$, $0 \leq x \leq \pi/2$
   (b) $y = \ln x$ and $y = \ln(1/x)$, $1 \leq x \leq e$
   (c) $y = x/(x^2 - 4)$ and $y = 1/(x^2 + 2)$, $0 \leq x \leq 1$
   (d) $y = \sin^2 x$ and $y = \sec x$, $3\pi/4 \leq x \leq 5\pi/4$
   (e) $y = x^3 - 6x^2 + 13x - 5$, $y = 2x + 1$
   (f) $y = \sin x$, $y = \cos x$, $\pi/4 \leq x \leq 5\pi/4$
3. (a) By integration, find a formula for the area of a segment of a circle of radius a, if the chord determining the segment is at distance b from the center.
   (b) From the result of part (a) show that the area of a circular sector of angle γ and radius a is $a^2\gamma/2$.
4. Find a formula for the area of a segment of a parabola $y^2 = 4ax$, $a > 0$, if the chord determining the segment passes through the focus and has slope $m > 0$.
5. Let f and g be continuous for $a \leq x \leq b$ and let $f(x) > g(x)$ except perhaps at a and b. Let a subdivision of $[a, b]$ be chosen: $a \leq x_0 < x_1 < \cdots < x_n = b$. Let f take its maximum in $[x_{i-1}, x_i]$ at ξ_i, its minimum at η_i; let g take its maximum at ξ_i', its minimum at η_i'. Justify the double inequality

$$[f(\eta_i) - g(\xi_i')]\, \Delta_i x \leq \Delta_i A \leq [f(\xi_i) - g(\eta_i')]\, \Delta_i x$$

for $i = 1, \ldots, n$, where $\Delta_i x = x_i - x_{i-1}$, and $\Delta_i A$ is the area between the curves $y = f(x)$, $y = g(x)$, for $x_{i-1} \leq x \leq x_i$. Add these inequalities for $i = 1, \ldots, n$ and conclude from the properties of the Riemann integral that

$$A = \lim_{\text{mesh} \to 0} \sum_{i=1}^n [f(\eta_i) - g(\xi_i')]\, \Delta_i x = \lim_{\text{mesh} \to 0} \sum_{i=1}^n [f(\xi_i) - g(\eta_i')]\, \Delta_i x$$

$$= \int_a^b [f(x) - g(x)]\, dx$$

7-2 AREA IN POLAR COORDINATES

Let a curve be given by a polar coordinate equation:

$$r = f(\theta), \qquad \alpha \leq \theta \leq \beta \tag{7-20}$$

where f is continuous, $f(\theta) > 0$ for $\alpha < \theta < \beta$, and $\alpha < \beta \leq \alpha + 2\pi$.

We seek the area of the region bounded by the given curve (7-20) and the lines $\theta = \alpha$, $\theta = \beta$, as suggested in Figure 7-5.

We imitate the reasoning used in rectangular coordinates (Section 4-19). We subdivide the interval $[\alpha, \beta]$: $\alpha = \theta_0 < \theta_1 < \cdots < \theta_n = \beta$, and let $\Delta_i A$ denote the area of the region bounded by the lines $\theta = \theta_{i-1}$, $\theta = \theta_i$, and the given curve (Figure 7-6). Let m_i be the minimum of f in the interval $[\theta_{i-1}, \theta_i]$, M_i the maximum of f in this interval, so that the region of area $\Delta_i A$ includes the sector $0 \leq r \leq m_i$, $\theta_{i-1} \leq \theta \leq \theta_i$, and is included in the sector $0 \leq r \leq M_i$, $\theta_{i-1} \leq \theta \leq \theta_i$. The area of a circular sector we know from geometry or calculus (Problem 3 following Section 7-1) to be $a^2\gamma/2$, where a is the radius and γ the angle in radians. Hence

$$\frac{m_i^2\,\Delta_i\theta}{2} \leq \Delta_i A \leq \frac{M_i^2\,\Delta_i\theta}{2}$$

where $\Delta_i\theta = \theta_i - \theta_{i-1}$. Accordingly, the area sought (if it exists) is such that

$$\sum_{i=1}^{n} \frac{m_i^2\,\Delta_i\theta}{2} \leq A \leq \sum_{i=1}^{n} \frac{M_i^2\,\Delta_i\theta}{2} \tag{7-21}$$

Now $m_i^2/2$ and $M_i^2/2$ can be interpreted as the minimum m_i' and maximum M_i', respectively, of the function

$$F(\theta) = \frac{[f(\theta)]^2}{2}$$

in the interval $[\theta_{i-1}, \theta_i]$. Thus (7-21) is of the form:

$$\sum m_i'\,\Delta_i\theta \leq A \leq \sum M_i'\,\Delta_i\theta$$

Since this holds true for all subdivisions, we conclude, as in Section 4-19, that the area A does exist and that

$$A = \int_\alpha^\beta F(\theta)\,d\theta = \frac{1}{2} \int_\alpha^\beta [f(\theta)]^2\,d\theta$$

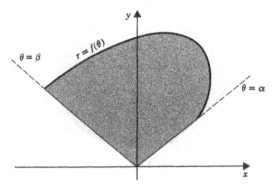

Figure 7-5   Area in polar coordinates.

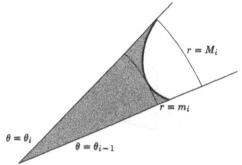

Figure 7-6 $\Delta_i A$ in polar coordinates.

We thus obtain the expression for *area in polar coordinates:*

$$\text{Area} = \frac{1}{2} \int_\alpha^\beta r^2 \, d\theta = \frac{1}{2} \int_\alpha^\beta [f(\theta)]^2 \, d\theta \qquad (7\text{-}22)$$

the area being that of the region bounded by the curve (7-20) and the lines $\theta = \alpha$, $\theta = \beta$.

The reasoning of Section 7-1 leads us to an expression for the area between two such curves:

$$r = f(\theta), \qquad r = g(\theta), \qquad \alpha \leq \theta \leq \beta$$

where $0 < g(\theta) < f(\theta)$, except perhaps at $\theta = \alpha$ and $\theta = \beta$:

$$\text{Area} = \frac{1}{2} \int_\alpha^\beta \{[f(\theta)]^2 - [g(\theta)]^2\} \, d\theta \qquad (7\text{-}23)$$

In the case when $\beta = \alpha + 2\pi$ and $f(\alpha) = f(\beta)$, the curve (7-20) is a *closed curve* and the area given by (7-22) is that of *the region enclosed,* as in Figure 7-7.

Similarly, when $\beta = \alpha + 2\pi$ and $f(\alpha) = f(\beta)$, $g(\alpha) = g(\beta)$, $f(\theta) > g(\theta)$ for $\alpha < \theta < \beta$, Equation (7-23) gives the area of the annuluslike region between two closed curves, as in Figure 7-8.

EXAMPLE 1. The area enclosed by the *cardioid*

$$r = a(1 - \cos \theta), \qquad 0 \leq \theta \leq 2\pi$$

(see Figure 7-9) is

$$\frac{1}{2} \int_0^{2\pi} a^2 (1 - \cos \theta)^2 \, d\theta = \frac{a^2}{2} \int_0^{2\pi} (1 - 2\cos \theta + \cos^2 \theta) \, d\theta = \frac{3\pi a^2}{2}$$

Remark. A mechanical application of (7-22) can lead to meaningless results. For example, the area of the region enclosed by the curve

$$r^2 = \sin \theta, \qquad 0 \leq \theta \leq 2\pi$$

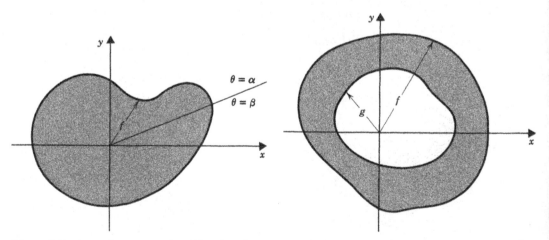

Figure 7-7 Area of region enclosed by closed curve.

Figure 7-8 Area of region between two closed curves.

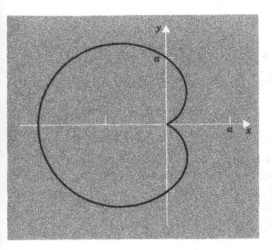

Figure 7-9   Cardioid.

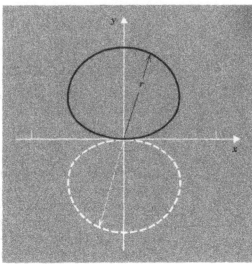

Figure 7-10   Curve $r^2 = \sin \theta$.

might be evaluated as

$$\frac{1}{2} \int_0^{2\pi} r^2 \, d\theta = \frac{1}{2} \int_0^{2\pi} \sin \theta \, d\theta = 0$$

However, for $\pi < \theta < 2\pi$, $\sin \theta$ is negative, so that $r = \pm \sqrt{\sin \theta}$ is imaginary! For $0 < \theta < \pi$, we get two values for r, one positive, one negative. If we follow the usual convention of ignoring the negative values of r, we obtain the upper oval of Figure 7-10, the area enclosed being

$$\frac{1}{2} \int_0^{\pi} \sin \theta \, d\theta = 1$$

The negative values of r can be plotted to give the symmetric oval shown dotted in the figure. By symmetry, both ovals enclose the same area.

PROBLEMS

In all cases, r, θ are polar coordinates.

1. Find the area of the region bounded by the given curves. Check by a rough graph.
   (a) $r = 2 \sec \theta$, $-\pi/4 \le \theta \le \pi/4$; $\theta = -\pi/4$; $\theta = \pi/4$
   (b) $r = \theta$, $0 \le \theta \le \pi$; $\theta = \pi$
   (c) $r = e^\theta$, $0 \le \theta \le 3\pi/2$; $\theta = 0$; $\theta = 3\pi/2$
   (d) $r = 2 \cos \theta$, $0 \le \theta \le \pi/2$; $\theta = 0$
   (e) $r = \sin 2\theta$, $0 \le \theta \le \pi/2$
   (f) $r = 1 - \sin \theta$, $\pi/2 \le \theta \le 5\pi/2$
2. Find the area of the region enclosed by the curve.
   (a) $r = a + b \cos \theta$, $0 < b < a$ (limaçon)
   (b) $r^2 = \sin 2\theta$ (lemniscate)
   (c) $r = \sin 3\theta$
   (d) $r = 2\pi\theta - \theta^2$, $0 \le \theta \le 2\pi$

3. Find the area of the region between the curves:
   (a) $r = 1$ and $r = 1 + \sin\theta, 0 \le \theta \le \pi$
   (b) $r = 1 + \sin(\theta/2), 0 \le \theta \le 2\pi$ and $r = 1 + \sin(\theta/2), 2\pi \le \theta \le 4\pi$

7-3 A GENERAL AREA FORMULA

In Section 1-12 it was pointed out that for two vectors **u**, **v** in the plane, the determinant

$$\begin{vmatrix} u_x & u_y \\ v_x & v_y \end{vmatrix}$$

equals plus or minus the area of the parallelogram whose sides, properly directed, are directed segments representing **u** and **v** (Figure 7-11). The plus sign holds true when the directed angle φ from **u** to **v** can be chosen so that $0 < \varphi < \pi$; the minus sign holds for $\pi < \varphi < 2\pi$. For $\varphi = 0, \pi, 2\pi, \ldots$, **u**, **v** are linearly dependent and the area is zero.

Let us now consider a path

$$x = f(t), \qquad y = g(t), \qquad a \le t \le b \tag{7-30}$$

where f and g have continuous first derivatives, and let us suppose first that the polar angle θ can be defined along the path: $\theta = \theta(t)$, as in Section 6-6, so that θ is continuous and monotone strictly increasing; in particular, we suppose that the path does not pass through the origin; see Figure 7-12. We are now interested in the area bounded by our path and the lines $\theta = \alpha = \theta(a)$, $\theta = \beta = \theta(b)$, as in polar coordinates. To find this we now use a different procedure. We subdivide our t-interval: $a = t_0 < t_1 < t_2 < \cdots < t_n = b$, and consider the area corresponding to the interval $[t_{i-1}, t_i]$, as in Figure 7-13. Let P_{i-1}, P_i be the positions at $t = t_{i-1}$, t_i, respectively, so that P_{i-1} has coordinates $f(t_{i-1})$, $g(t_{i-1})$. Then we take the position vectors $\overline{OP}_{i-1}$, $\overline{OP}_i$ as the vectors **u**, **v** in our determinant formula. Hence *the area of the triangle $OP_{i-1}P_i$ is*

$$\frac{1}{2}\begin{vmatrix} f(t_{i-1}) & g(t_{i-1}) \\ f(t_i) & g(t_i) \end{vmatrix}$$

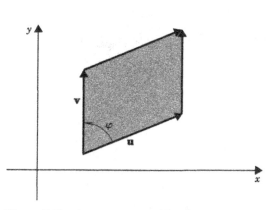

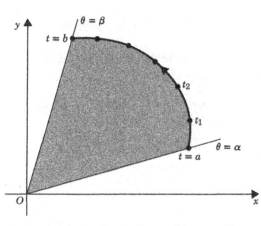

Figure 7-11   Area represented by determinant.

Figure 7-12   Area for path $x = f(t), y = g(t)$.

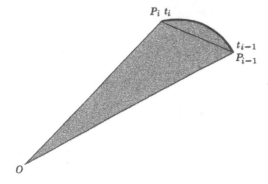

Figure 7-13   Element of area.

Now by the Mean Value Theorem

$$f(t_i) = f(t_{i-1}) + f'(\xi_i)\, \Delta_i t, \qquad t_{i-1} < \xi_i < t_i$$
$$g(t_i) = g(t_{i-1}) + g'(\eta_i)\, \Delta_i t, \qquad t_{i-1} < \eta_i < t_i$$

If we substitute these expressions in our determinant and expand, we obtain

$$\frac{1}{2}\left[f(t_{i-1})g'(\eta_i) - g(t_{i-1})f'(\xi_i) \right] \Delta_i t$$

as an expression for the area of the triangle. It is plausible that the area sought is the limit of the sum of the areas of these triangles:

$$\text{Area} = \lim_{\text{mesh} \to 0} \sum_{i=1}^{n} \frac{1}{2}\left[f(t_{i-1})g'(\eta_i) - g(t_{i-1})f'(\xi_i) \right] \Delta_i t$$

It can be verified (as in Section 4-25, see Problem 5 following Section 4-26) that the limit in question is an integral:

$$\text{Area} = \frac{1}{2}\int_a^b \left[f(t)g'(t) - g(t)f'(t) \right] dt \tag{7-31}$$

Instead of filling in the missing steps, we can justify (7-31) at once by making the substitution $\theta = \theta(t)$ in our polar coordinate area formula:

$$\text{Area} = \frac{1}{2}\int_\alpha^\beta r^2\, d\theta = \frac{1}{2}\int_a^b r^2 \frac{d\theta}{dt}\, dt \tag{7-32}$$

where $r^2 = [r(t)]^2 = [x(t)]^2 + [y(t)]^2 = [f(t)]^2 + [g(t)]^2$. As in Section 6-6,

$$\frac{d\theta}{dt} = \frac{x(dy/dt) - y(dx/dt)}{x^2 + y^2} = \frac{f(t)g'(t) - g(t)f'(t)}{[f(t)]^2 + [g(t)]^2}$$

so that (7-32) becomes

$$\text{Area} = \frac{1}{2}\int_a^b \left\{ [f(t)]^2 + [g(t)]^2 \right\} \cdot \frac{f(t)g'(t) - g(t)f'(t)}{[f(t)]^2 + [g(t)]^2}\, dt$$

and cancellation gives (7-31). *Hence (7-31) is in agreement with the polar coordinate formula for area.*

We are here interested mainly in the case when our path (7-30) is *closed.* If precisely one circuit of the origin is made as t goes from a to b, so that θ

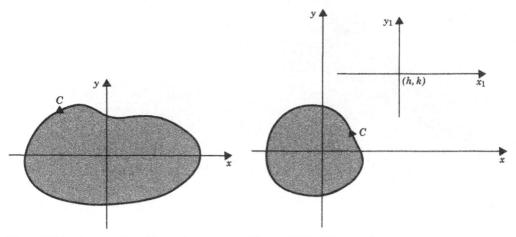

Figure 7-14   Area enclosed by a closed path.

Figure 7-15   Change of origin.

goes from α to $\beta = \alpha + 2\pi$, then (7-31) represents the total area enclosed by the path (Figure 7-14). In such a case, it is clear that we can also find a parametrization of the path starting and returning to any chosen point on the path—and that every parametrization that takes us around once in the counterclockwise direction gives the same value to the integral: namely, the area enclosed. For this reason, we can label the integral more concisely as

$$\frac{1}{2} \oint_C \left(x \frac{dy}{dt} - y \frac{dx}{dt} \right) dt$$

where C denotes the closed path in question and the circular arrow indicates the counterclockwise direction. We can go one step further and write the integral as

$$\frac{1}{2} \oint_C x \, dy - y \, dx \qquad (7\text{-}33)$$

where we understand: $dx = (dx/dt) \, dt$, $dy = (dy/dt) \, dt$, in some allowed parametrization. Such an integral (7-33) is called a *line integral*. Its value depends on the path C chosen and the direction in which C is to be traversed, but not on the particular parametrization chosen. Thus our area formula can be written:

$$\text{Area enclosed by } C = \frac{1}{2} \oint_C x \, dy - y \, dx \qquad (7\text{-}34)$$

We shall see that this formula has a remarkable range of validity.

Shift of Origin. We first verify that the form of (7-34) is unaffected by a translation of axes to a new origin (h, k) as in Figure 7-15. As in Section 6-5, this amounts to a substitution:

$$x = x_1 + h, \qquad y = y_1 + k$$

in the new coordinate system, x_1, y_1 become functions of the parameter:

$$x_1 = x - h = f(t) - h, \qquad y_1 = y - k = g(t) - k$$

Accordingly

$$\int_a^b \left(x\frac{dy}{dt} - y\frac{dx}{dt} \right) dt = \int_a^b \left[(x_1 + h)\frac{dy_1}{dt} - (y_1 + k)\frac{dx_1}{dt} \right] dt$$

$$= \int_a^b \left[x_1\frac{dy_1}{dt} - y_1\frac{dx_1}{dt} \right] dt + \int_a^b \left[h\frac{dy_1}{dt} - k\frac{dx_1}{dt} \right] dt$$

$$= \int_a^b \left[x_1\frac{dy_1}{dt} - y_1\frac{dx_1}{dt} \right] dt + \left. [hy_1(t) - kx_1(t)] \right|_{t=a}^{t=b}$$

But the last term is 0, since $y_1(b) = y_1(a)$, $x_1(b) = x_1(a)$, because the path is closed. Hence we conclude: for a closed path,

$$\frac{1}{2} \int_a^b \left(x\frac{dy}{dt} - y\frac{dx}{dt} \right) dt = \frac{1}{2} \int_a^b \left(x_1\frac{dy_1}{dt} - y_1\frac{dx_1}{dt} \right) dt$$

Now by shifting the origin as in Figure 7-15, we can obtain a configuration in which the path in question no longer goes around the origin. Accordingly, we conclude: *Equation (7-34) is correct whether C encloses the origin or not.* The direction in which C is traversed is, however, usually given with reference to a point inside C. Thus we can say: *The area enclosed by C is given by (7-34) as long as there is some point inside C relative to which, as origin, the polar angle θ is steadily increasing in the direction chosen and increases by 2π in one circuit.*

EXAMPLE 1. Let C be the path: $x = 3 + \cos t$, $y = \sin t$, $0 \le t \le 2\pi$. The path is a circle of center $(3, 0)$ and radius 1. Relative to the center as origin, t is the polar angle, so that C has the properties required. Hence, the area enclosed is

$$\frac{1}{2} \oint_C x\, dy - y\, dx = \frac{1}{2} \int_0^{2\pi} [(3 + \cos t) \cos t + \sin^2 t]\, dt = \pi$$

θ not steadily increasing. Let us return to the original origin, inside our closed path C, but let us not require that θ be steadily increasing; we allow $\theta(t)$ to go up and down as t increases from a to b, but do still require that $\theta(b) = \theta(a) + 2\pi$ (Figure 7-16). However, we do not permit C to cross itself, as in Figure 7-17. A path that does not cross itself is called *simple*. The complete definition is as follows:

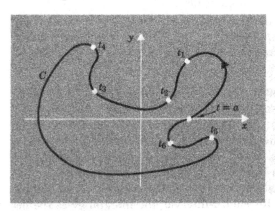

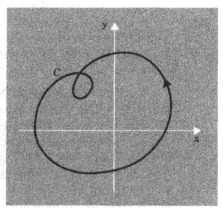

Figure 7-16 Simple closed path.

Figure 7-17 Nonsimple closed path.

Definition: *A simple closed path is a path:* $x = f(t)$, $y = g(t)$, $a \leq t \leq b$ *for which the conditions*

$$f(t_1) = f(t_2), \qquad g(t_1) = g(t_2), \qquad t_1 < t_2$$

are satisfied for $t_1 = a$, $t_2 = b$ *and only for these values.*

The area formula (7-34) is correct for a simple closed path as described. A complete proof would take us far afield. However, we can easily verify the statement for the case of a path as in Figure 7-16, for which $\theta(t)$ is alternately increasing and decreasing, with only a finite number of changes from increase to decrease or vice-versa. Let θ increase from $t = a$ to $t = t_1$, then decrease from $t = t_1$ to $t = t_2$, and so on. Let $\theta(a) = \alpha$, $\theta(t_1) = \theta_1$, $\theta(t_2) = \theta_2$, ... Then

$$\frac{1}{2} \int_a^{t_1} \left(x \frac{dy}{dt} - y \frac{dx}{dt} \right) dt = \frac{1}{2} \int_0^{\theta_1} r^2 \, d\theta$$

and this is the area of the "curved sector" shaded in Figure 7-18. Next

$$\frac{1}{2} \int_{t_1}^{t_2} \left(x \frac{dy}{dt} - y \frac{dx}{dt} \right) dt = \frac{1}{2} \int_{\theta_1}^{\theta_2} r^2 \, d\theta$$

Since $\theta_2 < \theta_1$, this is minus the integral from θ_2 to θ_1 and hence minus the area shaded heavily in Figure 7-18.

Thus

$$\frac{1}{2} \int_a^{t_2} \left(x \frac{dy}{dt} - y \frac{dx}{dt} \right) dt$$

gives the area of the first curved sector minus that of the second. It is clear that, if we proceed in this way, we find that the whole integral is obtained by successively adding and subtracting such areas of curved sectors and that the net result is simply the total area inside C.

We can also think of the area in question as being swept out by the variable line segment OP, P being the point $[x(t), y(t)]$; as θ increases, OP sweeps out areas positively; as θ decreases, OP sweeps out areas negatively. The net area swept out is the total area inside C.

We can again shift origin as above and now conclude: *The formula (7-34) is valid for every simple closed curve C, as long as for some point inside C as origin the polar angle increases by 2π as C is traversed once in the given direction.*

Piecewise Smooth Paths. We have thus far assumed that f and g have continuous derivatives, so that the path is *smooth* (see Section 4-30). The results remain valid when C is merely piecewise smooth: that is, when f and g are continuous and the interval $[a, b]$ can be divided into a finite number of intervals in each of which f and g have continuous first derivatives. Such a path is shown in Figure 7-19. To justify the extension to this case we need only reason as above in the case of θ alternately increasing and decreasing, to show that, for each t-interval, the integral gives the net area swept out. Then the sum of the integrals

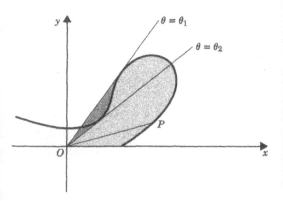

Figure 7-18   Case of θ alternately increasing and decreasing.

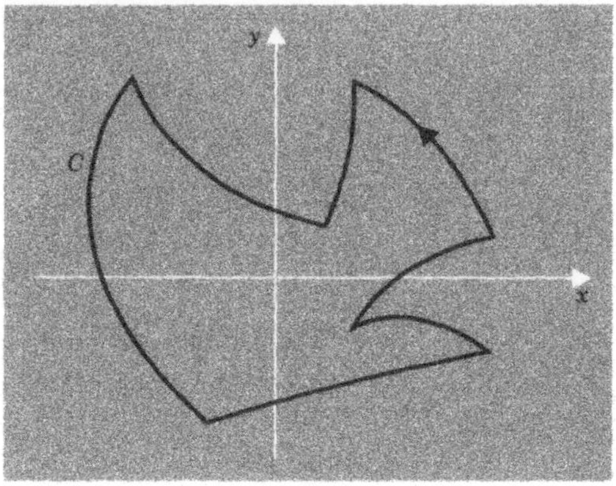

Figure 7-19   Piecewise smooth path.

$$\frac{1}{2} \int_a^{t_1} \left(x\frac{dy}{dt} - y\frac{dx}{dt} \right) dt + \frac{1}{2} \int_{t_1}^{t_2} \left(x\frac{dy}{dt} - y\frac{dx}{dt} \right) dt + \cdots + \frac{1}{2} \int_{t_{m-1}}^b (\ ) \, dt$$

is, as in Section 4-31, precisely the integral

$$\frac{1}{2} \int_a^b \left(x\frac{dy}{dt} - y\frac{dx}{dt} \right) dt$$

and, therefore, also equals the total net area swept out—that is, the area inside C.

Vector Form of the Area Formula. If we write $\overrightarrow{OP} = \mathbf{r}(t) = x\mathbf{i} + y\mathbf{j} = f(t)\mathbf{i} + g(t)\mathbf{j}$, then

$$\frac{d\mathbf{r}}{dt} = \mathbf{v} = f'(t)\mathbf{i} + g'(t)\mathbf{j}$$

and $\mathbf{v}$ can be interpreted as the velocity vector of the moving point P. The integrand in our area formula can be written:

$$f(t)g'(t) - g(t)f'(t) = \begin{vmatrix} f(t) & g(t) \\ f'(t) & g'(t) \end{vmatrix}$$

It is thus the area of the parallelogram with sides $\mathbf{r}(t)$, $\mathbf{v}(t)$. If we introduce the "left-turn" vector $\mathbf{r}^{\dashv} = -y\mathbf{i} + x\mathbf{j}$, it can also be written as

$$\mathbf{r}^{\dashv} \cdot \mathbf{v} = [-g(t)\mathbf{i} + f(t)\mathbf{j}] \cdot [f'(t)\mathbf{i} + g'(t)\mathbf{j}]$$

Thus, for a closed path as above

$$\text{Area} = \frac{1}{2} \oint_C \mathbf{r}^{\dashv}(t) \cdot \mathbf{v}(t) \, dt \tag{7-35}$$

We can interpret the integral as a limit of a sum of terms of form

$$\frac{1}{2} \mathbf{r}^{\dashv}(\xi_i) \cdot \mathbf{v}(\eta_i) \, \Delta t, \qquad t_{i-1} < \xi_i < t_i, \qquad t_{i-1} < \eta_i < t_i$$

Since $\mathbf{v}$ is velocity, $\mathbf{v}\Delta t$ (or $\Delta t\, \mathbf{v}$) represents approximately a displacement along the path, as in Figure 7-20. The expression $\frac{1}{2}\mathbf{r}^{\dashv} \cdot \mathbf{v}\Delta t$ thus represents approximately the area of the triangle swept out as in Figure 7-20. By summing these areas, we obtain, in the limit, the exact formula (7-35).

We saw above that our area formula does not depend on choice of origin. The vector formula (7-35) shows that it does not depend on the choice of coordinate axes; it does depend on the *orientation* of the axes, which fixes the positive direction for angles.

We summarize all we have found:

THEOREM 1. *Let the equations* $x = f(t)$, $y = g(t)$, $a \le t \le b$, *describe a piecewise smooth simple closed path C, traced in the counterclockwise direction as t increases from a to b. Then the area A enclosed by C is given by*

$$A = \frac{1}{2} \oint_C x \, dy - y \, dx = \frac{1}{2} \int_a^b \left(x \frac{dy}{dt} - y \frac{dx}{dt} \right) dt$$

or by the vector formula

$$A = \frac{1}{2} \int_a^b \mathbf{r}^{\dashv} \cdot \mathbf{v} \, dt$$

where $\mathbf{r} = \mathbf{r}(t) = f(t)\mathbf{i} + g(t)\mathbf{j}$ *and* $\mathbf{v} = \mathbf{r}'(t)$.

EXAMPLE 2. Find the area enclosed by the curve $y = \sin x$, $-\pi \le x \le \pi$ and the semicircle $x^2 + y^2 = \pi^2$, $y \ge 0$ (Figure 7-21). Here we can use a parametrization: $x = t + \pi$, $y = \sin(t + \pi)$, $-2\pi \le t \le 0$; $x = \pi \cos t$, $y = \pi \sin t$, $0 \le t \le \pi$. Hence the area is

$$\frac{1}{2} \oint_C x \, dy - y \, dx = \frac{1}{2} \int_{-2\pi}^0 [(t + \pi) \cos(t + \pi) - \sin(t + \pi)] \, dt$$

$$+ \frac{1}{2} \int_0^\pi (\pi^2 \cos^2 t + \pi^2 \sin^2 t) \, dt = \pi^3/2$$

(Can the answer be found without calculus?)

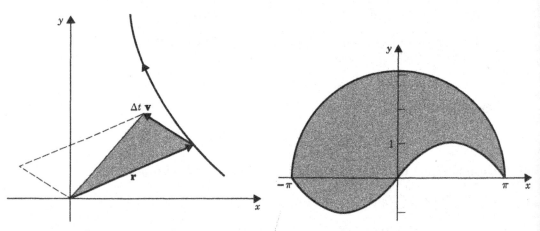

Figure 7-20   Element of area in terms of **r** and **v**.   **Figure 7-21**   Example 2.

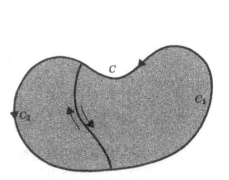

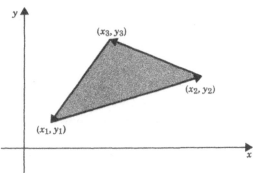

Figure 7-22   Area of whole equals sum   **Figure 7-23**   Area of triangle.
of areas of parts.

†7-4 A NEW APPROACH TO AREA

In the preceding section we developed a formula for area inside a piecewise smooth simple closed path, and justified its interpretation as area by reference to polar coordinates. However, it is possible to regard the new formula as the starting point and define the area by the equation:

$$\text{Area inside } C = \frac{1}{2} \oint_C x \, dy - y \, dx = \frac{1}{2} \int_a^b \mathbf{r}^\perp \cdot \mathbf{v} \, dt \qquad (7\text{-}40)$$

We can justify the definition by showing that it leads to an assignment of area to a large class of figures in agreement with the principles of Section 4-18. We do not carry this out here, but merely demonstrate two basic consequences of the definition:

A. Equation (7-40) gives the correct area for a triangle.

B. If the region inside C is divided into two parts by an auxiliary curve, as in Figure 7-22, so that the region enclosed by C is divided into two regions bounded by piecewise smooth simple closed paths, then the area of the whole is the sum of the areas of the parts.

For A. we consider a triangle with vertexes P_1: (x_1, y_1), P_2: (x_2, y_2), P_3: (x_3, y_3), as in Figure 7-23. Our integral is then the sum of three integrals, corresponding to the three sides of the triangle. For the integral along the side from P_1 to P_2, we parametrize by the equations

$$x = x_1 + (x_2 - x_1)t, \qquad y = y_1 + (y_2 - y_1)t, \qquad 0 \le t \le 1$$

(see Section 1-15). We obtain

$$\frac{1}{2} \int_0^1 \left(x \frac{dy}{dt} - y \frac{dx}{dt} \right) dt$$

$$= \frac{1}{2} \int_0^1 \{ [x_1 + (x_2 - x_1)t][y_2 - y_1] - [x_2 - x_1][y_1 + (y_2 - y_1)t] \} \, dt$$

$$= \frac{1}{2} \int_0^1 (x_1 y_2 - x_2 y_1) \, dt = \frac{1}{2} (x_1 y_2 - x_2 y_1)$$

Similarly, the integral on the second side gives $\frac{1}{2}(x_2 y_3 - x_3 y_2)$, on the third gives $\frac{1}{2}(x_3 y_1 - x_1 y_3)$. Hence the sum is

$$\frac{1}{2}(x_1y_2 - x_2y_1 + x_2y_3 - x_3y_2 + x_3y_1 - x_1y_3)$$

$$= \frac{1}{2}[(x_2 - x_1)(y_3 - y_1) - (x_3 - x_1)(y_2 - y_1)]$$

$$= \frac{1}{2}\begin{vmatrix} x_2 - x_1 & y_2 - y_1 \\ x_3 - x_1 & y_3 - y_1 \end{vmatrix} = \frac{1}{2}\overrightarrow{P_1P_2}\dashv\cdot\overrightarrow{P_1P_3}$$

and this is the area because of our choice of the counterclockwise direction (Section 1-12).

Property B. follows from the fact that the counterclockwise direction for the two regions (Figure 7-22) determines opposite directions for the integration along the common part of the boundary curves C_1, C_2. Hence

(Area of region inside C_1) + (Area of region inside C_2)

$$= \frac{1}{2}\oint_{C_1} x\,dy - y\,dx + \frac{1}{2}\oint_{C_2} x\,dy - y\,dx = \frac{1}{2}\oint_{C} x\,dy - y\,dx$$

For when we add the integrals, the portion of each coming from the common part of the boundary cancels out, and we are left with the integral along the outer boundary C.

From A. and B. it follows (by induction) that, for a polygon, cut into triangles, the total area is the sum of the areas of the parts. Typical cases are suggested in Figure 7-24.

The area formula (7-40) can be extended to more general regions by subtraction. Thus the area of the region of Figure 7-25 can be evaluated as

$$\frac{1}{2}\oint_{C_1} x\,dy - y\,dx - \frac{1}{2}\oint_{C_2} x\,dy - y\,dx$$

and that of Figure 7-26 as

$$\frac{1}{2}\oint_{C_1} x\,dy - y\,dx - \frac{1}{2}\oint_{C_2} x\,dy - y\,dx - \frac{1}{2}\oint_{C_3} x\,dy - y\,dx$$

The region of Figure 7-25 is said to be *doubly connected* (it has one "hole"), that of Figure 7-26 is called *triply connected* (two "holes"); a region enclosed by one simple closed curve is called *simply connected* (no "holes").

The general class suggested by these examples consists of all regions bounded by a finite number of nonintersecting simple closed curves (piecewise smooth), of which one encloses the others.

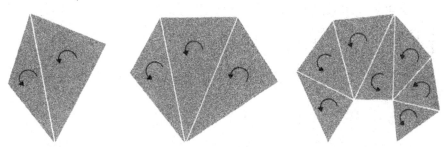

Figure 7-24   Area of polygon.

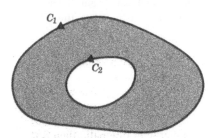

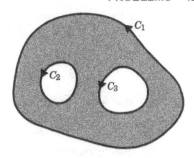

Figure 7-25 Doubly connected region.

Figure 7-26 Triply connected region.

PROBLEMS

1. Graph and find the area enclosed. Throughout a and b are positive constants.
   (a) $x = h + a \cos t$, $y = k + a \sin t$, $0 \leq t \leq 2\pi$ (circle)
   (b) $x = h + a \cos t$, $y = k + b \sin t$, $0 \leq t \leq 2\pi$ (ellipse)
   (c) $x = a \cos^3 t$, $y = a \sin^3 t$, $0 \leq t \leq 2\pi$ (hypocycloid)
   (d) The x-axis and the curve $x = a(t - \sin t)$, $y = a(1 - \cos t)$, $0 \leq t \leq 2\pi$ (cycloid)
   (e) $x = a(\cos t + t \sin t)$, $y = a(\sin t - t \cos t)$, $0 \leq t \leq 2\pi$, (evolute of circle) and the line segment $x = a$, $-2\pi a \leq y \leq 0$
   (f) $x = 2 \cos t + \cos^3 t$, $y = 2 \sin t + \sin t \cos^2 t$, $0 \leq t \leq 2\pi$
   (g) $x = 1 - t$ for $0 \leq t \leq 2$, $x = t - 3$ for $2 \leq t \leq 4$, $y = t$ for $0 \leq t \leq 1$, $y = 2 - t$ for $1 \leq t \leq 3$, $y = t - 4$ for $3 \leq t \leq 4$
   (h) The curve $y = x^3$, $-1 \leq x \leq 1$, and the portion of the circle $x^2 + y^2 = 2$ below the line $x = y$.
2. Find the area enclosed in each case:
   (a) The triangle of vertexes $(3, 2)$, $(5, 4)$, $(1, 7)$.
   (b) The quadrilateral of vertexes $(2, 1)$, $(4, 3)$, $(5, 5)$, $(3, 7)$.
   (c) The quadrilateral of successive vertexes (x_1, y_1), (x_2, y_2), (x_3, y_3), (x_4, y_4).
   (d) The pentagon of successive vertexes (x_1, y_1), $\ldots (x_5, y_5)$.
   (e) The polygon of successive vertexes (x_1, y_1), (x_2, y_2), $\ldots$, (x_n, y_n).
3. Show that, if $\mathbf{r} = \mathbf{F}(t)$, $a \leq t \leq b$, describes a smooth simple closed path on which the polar angle θ increases steadily from 0 to 2π and if $k(t) > 0$, then the path $\mathbf{r} = k(t)\mathbf{F}(t)$ encloses the area:

$$A = \frac{1}{2} \int_a^b [k(t)]^2 \mathbf{F}(t) \cdot \mathbf{F}'(t) \, dt$$

4. Graph and find the area enclosed without integration:
   (a) $x = (\cos t)|\cos t|$, $y = (\sin t)|\sin t|$, $0 \leq t \leq 2\pi$
   (b) $x = \cos(2\pi \sin t)$, $y = \sin(2\pi \sin t)$, $0 \leq t \leq \pi/2$
5. Let C be a simple closed path as in Theorem 1.
   (a) Show that

$$\int_a^b \left(x \frac{dy}{dt} + y \frac{dx}{dt} \right) dt = 0$$

   (b) With the aid of the result of part (a) show that the area A enclosed by C is given by

$$A = \oint_C x \, dy = \int_a^b x \frac{dy}{dt} \, dt \text{ and } A = -\oint_C y \, dx = -\int_a^b y \frac{dx}{dt} \, dt$$

6. Let a point P move on a smooth path not passing through the origin O, let $\overrightarrow{OP} = \mathbf{r} = \mathbf{F}(t)$, $a \leq t \leq b$. Let $A(t)$ denote the area swept out by $\overrightarrow{OP}$ in the interval $[a, t]$, so that

$$A(t_1) = \frac{1}{2} \int_a^{t_1} \mathbf{r}^{\perp} \cdot \mathbf{v} \, dt$$

(a) Show that $A'(t) = \frac{1}{2}\mathbf{r}^{\perp}(t) \cdot \mathbf{v}(t)$

(b) Show that $A'(t) = \frac{1}{2}\left(x\dfrac{dy}{dt} - y\dfrac{dx}{dt}\right)$

(c) Show that, if the polar angle θ varies continuously on the path, then $A'(t) = \frac{1}{2}r^2 \, d\theta/dt$, where $r = |\overrightarrow{OP}|$.

(d) Show that $A'(t) = \frac{1}{2}rv_\theta$, where v_θ is the transverse component of velocity $\mathbf{v}$ (Section 6-8).

7. *Kepler's second and third laws.* In Section 6-9 it is shown that each planet P moves in a conic section having the polar coordinate equation

$$r = \frac{h^2/k}{1 + e \cos \theta}$$

where h and k are positive constants, e is the eccentricity, and one focus is at the sun, taken as origin. Let $0 < e < 1$, so that the orbit is an ellipse. It is shown in Section 6-9 that r and θ are functions of t for which

$$r^2 \frac{d\theta}{dt} = h$$

The choice of h varies with the planet, but k is the same for all planets.

(a) Show with the aid of the results of Problem 6 that the radius vector $\overrightarrow{OP}$ sweeps out area at the constant rate $h/2$ (Kepler's second law).

(b) Let T be the length of a year (time for one circuit of the ellipse). Conclude from (a) that the area of the ellipse is $hT/2$.

(c) Show that the ellipse has semi-axis $a = h^2/[k(1 - e^2)]$ and semi-axis $b = h^2\sqrt{1 - e^2}/[k(1 - e^2)]$ (see Section 0-14), and deduce that the area is $\pi ab = \pi h^4/[k^2(1 - e^2)^{3/2}]$.

(d) From (b) and (c) conclude that

$$T = \frac{2\pi}{\sqrt{k}} a^{3/2}$$

Hence, *the squares of the periods of the planets are proportional to the cubes of their major axes* (Kepler's third law).

8. (a) Prove that if $t = \varphi(\tau)$, $\alpha \leq \tau \leq \beta$, has a continuous derivative $\varphi'(\tau) > 0$ in $[\alpha, \beta]$ and $\varphi(\alpha) = a$, $\varphi(\beta) = b$, then for a smooth path $x = f(t)$, $y = g(t)$, $a \leq t \leq b$,

$$\frac{1}{2} \int_a^b \left(x\frac{dy}{dt} - y\frac{dx}{dt}\right) dt = \frac{1}{2} \int_\alpha^\beta \left(x\frac{dy}{d\tau} - y\frac{dx}{d\tau}\right) d\tau$$

where on the right

$$\frac{dx}{d\tau} = \frac{d}{d\tau} f\,[\varphi(\tau)], \qquad \frac{dy}{d\tau} = \frac{d}{d\tau} g\,[\varphi(\tau)]$$

(This shows that the value of the integral is not dependent on the parametrization; equivalent paths give the same value.)

(b) Show that the result of part (a) remains valid for any integral of form:

$$\int_a^b \left[p_1(x)q_1(y)\frac{dx}{dt} + p_2(x)q_2(y)\frac{dy}{dt}\right] dt$$

9. Figure 7-27 shows 5 piecewise smooth curves in the xy plane. Let

$$\frac{1}{2} \oint_{C_k} x \, dy - y \, dx = a_k$$

Find the area of the region bounded by the curves named:

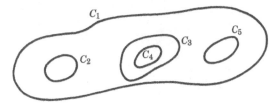

Figure 7-27

(a) C_1 (b) C_1 and C_3 (c) C_1, C_2, and C_5 (d) C_3 and C_4

10. Show that the region between two curves: $y = f(x)$, $y = g(x)$, $a \le x \le b$, as in Figure 7-3, can be considered as a region bounded by a simple closed path C and that

$$\frac{1}{2} \oint_C x \, dy - y \, dx$$

equals

$$\frac{1}{2} \left\{ \int_a^b (xg' - g) \, dx + b \int_{g(b)}^{f(b)} dy + \int_b^a (xf' - f) \, dx + a \int_{f(a)}^{g(a)} dy \right\}$$

Simplify the expression to conclude that the area equals

$$\int_a^b [f(x) - g(x)] \, dx$$

11. The path C of Theorem 1 can be represented by means of a *complex function* of t: $z(t) = x(t) + iy(t) = f(t) + ig(t)$ (See Section 5-8).

(a) Show that $A = \frac{1}{2} \int_a^b \overline{iz(t)} \, z'(t) \, dt$, where the bar denotes the complex conjugate.

(b) Find the area enclosed by the path $z = 3e^{it}$, $0 \le t \le 2\pi$.

(c) Find the area enclosed by the path $z = 1 + 2e^{it} + ie^{2it}$, $0 \le t \le 2\pi$.

(d) Show that, if C has equation

$$z = \sum_{k=1}^n a_k e^{ikt}, \quad \text{then} \quad A = \pi \sum_{k=1}^n k|a_k|^2$$

[*Hint.* Show first that, if k and l are integers, then $\int_0^{2\pi} e^{(k-l)it} \, dt$ equals 0 if $k \ne l$ and equals 2π if $k = l$.]

12. *A project.* Find out how a *planimeter* works and relate it to the methods of the text.

7-5 VOLUME OF SOLID OF REVOLUTION

The concept of volume of a solid object is analogous to that of area of a plane figure and can be developed by axioms as in Section 4-18. Here we assume that volume can be defined for a class of solid regions, including all those to be considered in this section, and then deduce that the volume is a definite integral. Thus our procedure parallels that of Section 4-19 in representing area as a definite integral.

By a *solid of revolution* we mean a region in space obtained by rotating a planar region about a line in the plane, as in Figure 7-28. Thus each point of the planar region traces out a circle whose center is on the axis of revolution; the points on these circles form the solid of revolution. One normally

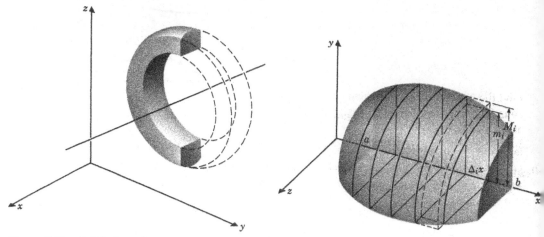

Figure 7-28   Solid of revolution.   **Figure 7-29**   Slicing solid of revolution.

assumes that the planar region lies on one side of the axis, so that each circle meets the region in one point (and not in two).

Let us first assume axes chosen in space so that our planar region lies in the xy-plane and that the x-axis is the axis of revolution. We further assume that the planar region is precisely that "beneath a curve"; that is, that it is bounded by the interval $[a, b]$ of the x-axis, the lines $x = a$, $x = b$, and a curve $y = f(x)$, $a \leq x \leq b$, where f is continuous and positive, except perhaps at a and b. The solid of revolution is then as in Figure 7-29.

To obtain a formula for the volume of the solid, we subdivide our interval as usual: $a = x_0 < x_1 < \cdots < x_n = b$, and thereby slice our solid into pieces, the piece between the planes $x = x_{i-1}$ and $x = x_i$ having volume $\Delta_i V$. The piece resembles a cylinder. In fact, if f has minimum m_i and maximum M_i in $[x_{i-1}, x_i]$, the piece contains the cylinder of altitude $\Delta_i x$ and base πm_i^2, and is contained in the cylinder of altitude $\Delta_i x$ and base πM_i^2, as suggested in Figure 7-29. Hence

$$\pi m_i^2 \, \Delta_i x \leq \Delta_i V \leq \pi M_i^2 \, \Delta_i x$$

so that

$$\sum_{i=1}^{n} \pi m_i^2 \, \Delta_i x \leq V \leq \sum_{i=1}^{n} \pi M_i^2 \, \Delta_i x \tag{7-50}$$

where $V = \Sigma \, \Delta_i V$ is the total volume of our solid. Now if $F(x) = \pi [f(x)]^2$, then (7-50) can be written

$$\sum_{i=1}^{n} m_i' \, \Delta_i x \leq V \leq \sum_{i=1}^{n} M_i' \, \Delta_i x$$

where m_i', M_i' are, respectively, the minimum and maximum of $F(x)$ in $[x_{i-1}, x_i]$. Since F is continuous, we conclude that

$$V = \int_a^b F(x) \, dx = \pi \int_a^b [f(x)]^2 \, dx \tag{7-51}$$

This is the desired integral representation of the volume.

EXAMPLE 1. $y = kx$, $0 \le x \le h$, where $k > 0$. The planar region is a right triangle, the solid of revolution is a *right circular cone*, as in Figure 7-30. By (7-51)

$$V = \pi \int_0^h (kx)^2 \, dx = \pi k^2 \frac{h^3}{3}$$

Now the radius of the base is $a = kh$. Hence we can write our result in the familiar form

$$V = \frac{\pi a^2 h}{3} = \frac{1}{3} \text{ base} \times \text{altitude}$$

EXAMPLE 2. $y = \sqrt{a^2 - x^2}$, $-a \le x \le a$. Here our planar region is a semicircle, the solid of revolution is a sphere, as in Figure 7-31. The volume is

$$V = \pi \int_{-a}^a (a^2 - x^2) \, dx = \frac{4}{3} \pi a^3$$

The power of the calculus is shown in the ease with which these familiar results are obtained—results obtained only with great difficulty in geometry.

Method of Cylindrical Shells. Let us consider a solid of revolution obtained by rotating about the x-axis a region in the xy-plane bounded by a curve $x = g(y)$, $c \le y \le d$, the lines $y = c$, $y = d$, and the interval $[c, d]$ of the y-axis, as in Figure 7-32. We assume g is continuous and positive, except possibly at c and d, and that $c \ge 0$. We now subdivide the interval $[c, d]$: $c = y_0 < y_1 < \cdots < y_n = d$, and let $\Delta_i V$ be the portion of the volume obtained from the part of the region for which $y_{i-1} \le y \le y_i$. Thus $\Delta_i V$ is the volume of a solid resembling a cylindrical shell—a solid region between two coaxial cylinders of equal altitude. The volume of such a shell is given by the difference of the volumes of the two cylinders, hence by $\pi r_2^2 h - \pi r_1^2 h$, where $r_1 < r_2$ and r_1, r_2 are the two radii and h is the altitude. Hence the volume also equals $(\pi r_2^2 - \pi r_1^2)h = $ (base of shell) $\times$ (altitude of shell).

For $\Delta_i V$ we have no precise altitude, but the base is known: $\pi y_i^2 - \pi y_{i-1}^2$.

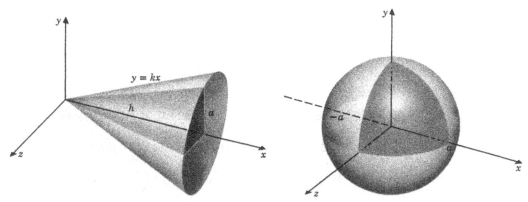

Figure 7-30 Circular cone as solid of revolution.   **Figure 7-31** Sphere as solid of revolution.

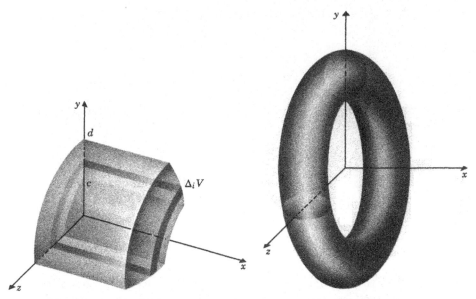

Figure 7-32   Solid of revolution decomposed into cylindrical shells.

Figure 7-33   Solid torus.

We can reason that $\Delta_i V$ is less than or equal to the volume of the shell with this base and altitude M_i, where M_i is the maximum of g in $[y_{i-1}, y_i]$; similarly $\Delta_i V$ is greater than or equal to the volume of the shell with this base and altitude m_i, where m_i is the minimum of g in $[y_{i-1}, y_i]$. Hence

$$m_i(\pi y_i^2 - \pi y_{i-1}^2) \leq \Delta_i V \leq M_i(\pi y_i^2 - \pi y_{i-1}^2)$$

or, with $\Delta_i y = y_i - y_{i-1}$,

$$\pi m_i(y_i + y_{i-1})\,\Delta_i y \leq \Delta_i V \leq \pi M_i(y_i + y_{i-1})\,\Delta_i y$$

Hence, by the Intermediate Value Theorem,

$$\Delta_i V = \pi g(\xi_i)(y_i + y_{i-1})\,\Delta_i y$$

for some ξ_i, $y_{i-1} < \xi_i < y_i$. We can also write

$$y_i + y_{i-1} = 2\eta_i$$

where $y_{i-1} < \eta_i < y_i$; η_i is simply the midpoint of $[y_{i-1}, y_i]$. Hence we can write

$$\Delta_i V = f(\eta_i)g(\xi_i)\,\Delta_i y, \qquad y_{i-1} < \eta_i < y_i, \qquad y_{i-1} < \xi_i < y_i$$

where $f(y) = 2\pi y$ and $g(y)$ is as given. Therefore, as in Section 4-26,

$$\lim_{\text{mesh}\to 0} \sum_{i=1}^{n} \Delta_i V = \lim_{\text{mesh}\to 0} \sum_{i=1}^{n} f(\eta_i)g(\xi_i)\,\Delta_i y = \int_c^d f(y)g(y)\,dy$$

Since $\sum_{i=1}^{n} \Delta_i V = V$, the total volume, we are led to the formula:

$$V = \int_c^d f(y)g(y)\,dy = 2\pi \int_c^d y g(y)\,dy \qquad (7\text{-}52)$$

EXAMPLE 3. *Solid torus.* The circular region bounded by the circle $x^2 + (y - b)^2 = a^2$, $0 < a < b$, is rotated about the x-axis (Figure 7-33). The solid obtained is called a solid torus of cross-sectional radius a and meridianal radius b. By symmetry the volume is twice that obtained by rotating the semicircular region for which $x \geq 0$. Thus we can take $g(y) = \sqrt{a^2 - (y - b)^2}$, $b - a \leq y \leq b + a$, and (7-52) gives for the whole torus

$$V = 4\pi \int_{b-a}^{b+a} y \sqrt{a^2 - (y - b)^2} \, dy$$

The substitution $y = b + a \sin \varphi$, $-\pi/2 \leq \varphi \leq \pi/2$ reduces this to

$$4\pi a^2 \int_{-\pi/2}^{\pi/2} (b + a \sin \varphi) \cos^2 \varphi \, d\varphi = 2\pi^2 a^2 b$$

Thus *the volume of the torus equals $2\pi^2 a^2 b$.*

Remark 1. Interchanging x and y, we conclude that the volume generated by rotating the area beneath a curve $y = f(x)$, $a \leq x \leq b$, about the y-axis is given by

$$V = 2\pi \int_a^b x f(x) \, dx \tag{7-53}$$

Remark 2. Let f be continuous and monotone strictly increasing for $a \leq x \leq b$, so that f has a continuous inverse $x = g(y)$, $c \leq y \leq d$ [Figure 7-34(a)]. Let f and g both be positive, except possibly for $f(a) = 0$, and $g(c) = 0$. Let V_1 be the volume of the solid obtained by rotating the region beneath the graph of f about the x-axis; let V_2 be the volume of the solid obtained by rotating the area between the graph of g and the y-axis about the x-axis. Then by geometry

$$V_1 + V_2 = \pi(d^2 b - c^2 a) \tag{7-54}$$

If f is monotone decreasing, as in Figure 7-34(b), the formula becomes

$$V_1 - V_2 = \pi(bc^2 - ad^2) \tag{7-55}$$

These relations can be used to reduce one integration problem to another or to check results.

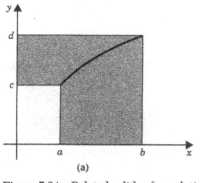
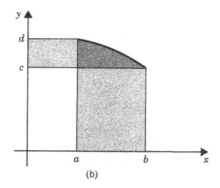

(a)   (b)

Figure 7-34 Related solids of revolution.

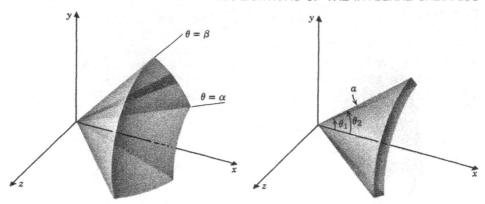

Figure 7-35 Solid of revolution defined by polar coordinate equations.   **Figure 7-36** Conical shell.

†7-6 SOLIDS OF REVOLUTION: POLAR COORDINATES AND PARAMETRIC FORMULA

Now let a region in the xy-plane be given as that bounded by a curve $r = f(\theta)$, $\alpha \le \theta \le \beta$, and the rays $\theta = \alpha$, $\theta = \beta$, as in Section 7-2. We assume f is continuous and that f is positive, except perhaps at α and β, and that $0 \le \alpha < \beta \le \pi$. We can thus consider the solid of revolution obtained by rotating our region about the x-axis, as in Figure 7-35. Here it is natural to partition our θ interval: $\alpha = \theta_0 < \theta_1 < \theta_2 < \cdots < \theta_n = \beta$. Our solid is thereby partitioned into pieces, each of which is roughly a "conical shell," a solid of revolution obtained by rotating a circular sector about the x-axis, as in Figure 7-36. For the case pictured the volume of the shell is found (Problem 33 below) to be $2\pi a^3(\cos \theta_1 - \cos \theta_2)/3$. We can now reason as for the cylindrical shells in the preceding section. Let f have minimum and maximum m_i, M_i in $[\theta_{i-1}, \theta_i]$. Then

$$\frac{2\pi m_i^3(\cos \theta_{i-1} - \cos \theta_i)}{3} \le \Delta_i V \le \frac{2\pi M_i^3(\cos \theta_{i-1} - \cos \theta_i)}{3}$$

$$\Delta_i V = 2\pi [\, f(\xi_i)]^3 \frac{(\cos \theta_{i-1} - \cos \theta_i)}{3} = \frac{2\pi}{3} [\, f(\xi_i)]^3 \sin \eta_i \, \Delta_i\theta$$

where $\theta_{i-1} \le \xi_i \le \theta_i$, $\theta_{i-1} \le \eta_i \le \theta_i$. For the last step the Mean Value Theorem was applied to the cosine function. Finally, we obtain

$$V = \frac{2\pi}{3} \int_\alpha^\beta [\, f(\theta)]^3 \sin \theta \, d\theta \qquad (7\text{-}60)$$

This is the desired formula.

EXAMPLE 1. The region enclosed by the cardioid $r = a(1 + \cos \theta)$ is rotated about the x-axis. By symmetry the volume of the solid obtained (Figure 7-37) is equal to the volume of the solid obtained by rotating the upper half ($y \ge 0$) of the region about the x-axis. Thus we can take $\alpha = 0$, $\beta = \pi$ and (7-60) gives

$$V = \frac{2\pi}{3} \int_0^\pi a^3 (1 + \cos\theta)^3 \sin\theta \; d\theta = \frac{2\pi a^3}{3} \cdot \frac{(1 + \cos\theta)^4}{-4} \Big|_0^\pi = \frac{8}{3}\pi a^3$$

Case of Parametric Equations. Our goal is to find a way of expressing the volume of a solid of revolution when the region being rotated is bounded by a simple closed piecewise smooth curve C. The method we shall use is to start with the polar coordinate formula (7-60) and proceed as in Section 7-3.

Let a piecewise smooth path $x = x(t)$, $y = y(t)$, $a \le t \le b$ be given, not passing through the origin, and let us assume the angle function $\theta(t)$ can be defined along the path so that $\theta'(t) > 0$, $\theta(a) = \alpha$, $\theta(b) = \beta$, $0 \le \alpha < \beta \le \pi$. Consequently, θ can be used as parameter along the path, so that the path can be expressed in terms of polar coordinates in the form $r = f(\theta)$, $\alpha \le \theta \le \beta$:

$$r = \sqrt{x^2 + y^2} = \sqrt{\{x[t(\theta)]\}^2 + \{y[t(\theta)]\}^2} = f(\theta)$$

The solid in question is that of Figure 7-35. Hence its volume is

$$V = \frac{2\pi}{3} \int_\alpha^\beta [f(\theta)]^3 \sin\theta \; d\theta$$

$$= \frac{2\pi}{3} \int_a^b r^3 \frac{y}{r} \frac{xy' - yx'}{r^2} \; dt = \frac{2\pi}{3} \int_a^b (xyy' - y^2x') \; dt$$

This last expression is now an alternative to the polar coordinate formula for the volume. We can write it thus

$$V = \frac{2\pi}{3} \int_C xy \; dy - y^2 \; dx \tag{7-61}$$

since its value must be the same for each parametrization, as we can verify directly (Problem 8 following Section 7-4). Now for a simple closed curve as in Figure 7-38, for which the two portions C_1, C_2 are of the type of C, but θ decreases along C_2, the volume generated is the difference of the two volumes and, therefore, equal to

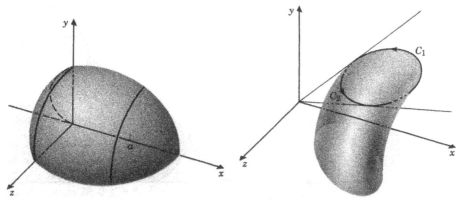

Figure 7-37   Solid of revolution obtained from cardioid.

Figure 7-38   Solid of revolution generated by region bounded by simple closed curve.

$$V = \frac{2\pi}{3} \int_{C_1} xy\,dy - y^2\,dx + \frac{2\pi}{3} \int_{C_2} xy\,dy - y^2\,dx$$

the plus sign in the second term being implied by the direction of C_2. Accordingly, we can write

$$V = \frac{2\pi}{3} \oint_C xy\,dy - y^2\,dx \tag{7-62}$$

where C is the combined path. If we have a parametrization for C:

$$x = x(t), \qquad y = y(t), \qquad a \le t \le b$$

with t increasing in the direction chosen for C (counterclockwise), then (7-62) becomes

$$V = \frac{2\pi}{3} \int_a^b [xy\,y'(t) - y^2 x'(t)]\,dt \tag{7-63}$$

However, just as in computing areas in Section 7-3, we can use any equivalent parametrization for each portion of C. Also, as in Section 7-3, the polar angle need not be steadily increasing relative to a chosen origin C.

EXAMPLE 2. *The solid torus.* We test our formula on this solid, already treated as Example 3 in Section 7-5. This time we represent the circular boundary C parametrically:

$$x = h + a \cos t, \qquad y = b + a \sin t, \qquad 0 \le t \le 2\pi$$

as in Figure 7-39 so that $0 < a < b$. By Equation (7-63),

$$V = \frac{2\pi}{3} \int_0^{2\pi} [(h + a \cos t)(b + a \sin t)a \cos t + (b + a \sin t)^2 a \sin t]\,dt$$

With the aid of entry No. 107 in Appendix Table I, we find easily that

$$V = 2\pi^2 a^2 b$$

in agreement with our previous result.

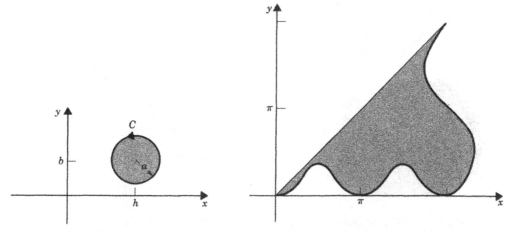

Figure 7-39   Region generating solid torus.   **Figure 7-40**   Curve of Example 3.

EXAMPLE 3. Let C be formed of the three curves:

$$y = \sin^2 x, \qquad 0 \le x \le 2\pi$$
$$x = 2\pi + \sin y, \qquad 0 \le y \le 2\pi$$
$$y = x, \qquad 0 \le x \le 2$$

as in Figure 7-40. We can use x as parameter t for the first curve, y as parameter t for the second, and $-x$ as parameter t for the third:

$$x = t, y = \sin^2 t, \qquad 0 \le t \le 2\pi$$
$$x = 2\pi + \sin t, y = t, \qquad 0 \le t \le 2\pi$$
$$x = -t, y = -t, \qquad -2\pi \le t \le 0$$

If we rotate about the x axis the region enclosed by C, we obtain a solid whose volume is

$$V = \frac{2\pi}{3} \int_0^{2\pi} [t \sin^2 t \cdot 2 \sin t \cos t - \sin^4 t]\, dt$$

$$+ \frac{2\pi}{3} \int_0^{2\pi} [t(2\pi + \sin t) - t^2 \cos t]\, dt$$

$$+ \frac{2\pi}{3} \int_{-2\pi}^0 (-t^2 + t^2)\, dt$$

The third integral is 0; the first is found to equal $-9\pi/8$, the second $4\pi^2 - 6\pi$, so that $V = \pi^2(32\pi - 57)/24$.

PROBLEMS

In Problems 1-14 find the volume of the solid generated by rotation about the x-axis of the region beneath the curve given. Throughout, a and b are positive constants.

1. $y = ax^2, 0 \le x \le 1$
2. $y = \sqrt{4ax}, 0 \le x \le b$ (segment of paraboloid of revolution)
3. $y = \sin x, 0 \le x \le \pi$
4. $y = \text{Sin}^{-1} x, 0 \le x \le 1$
5. $y = \tan x, 0 \le x \le \pi/4$
6. $y = e^x, 0 \le x \le 1$
7. $y = \cosh x, -1 \le x \le 1$
8. $y = \ln x, 1 \le x \le e$
9. $y = \text{Cosh}^{-1} x, 1 \le x \le 5$
10. $y = 1/(x^2 + 1), -1 \le x \le 1$
11. $y = mx + b, 0 \le x \le h, mh + b > 0$ (frustum of cone)
12. $y = \sqrt{a^2 - x^2}, b \le x \le a$ (segment of sphere)
13. $y = (b/a)\sqrt{a^2 - x^2}, -a \le x \le a$ (ellipsoid of revolution)
14. $y = (b/a)\sqrt{x^2 - a^2}, a \le x \le h$ (segment of hyperboloid of revolution)

In Problems 15 to 26 find the volume obtained by rotating about the x-axis the region bounded by the given curves; r and θ are polar coordinates, $a, b, h,$ and k are positive constants.

15. $r = a \cos \theta, 0 \le \theta \le \pi/2; \theta = 0$
16. $r = a \sin \theta, 0 \le \theta \le \pi$
17. $r = a + b \cos \theta, 0 \le \theta \le \pi, \theta = 0, \theta = \pi$ $(b < a)$
18. $r^2 = a \cos 2\theta, 0 \le \theta \le \pi/4, \theta = 0$
19. $r = e^{a\theta}, 0 \le \theta \le \pi; \theta = 0$ and $\theta = \pi$
20. $r = a\theta, 0 \le \theta \le \pi; \theta = 0$ and $\theta = \pi$

21. $x = k + a \cos t, \ y = h + b \sin t, \ 0 \le t \le 2\pi, \ b < h$

22. $x = \cos^3 t, \ y = \sin^3 t, \ 0 \le t \le 2\pi$

23. $x = 4 \cos t + \cos 3t, \ y = 6 + 4 \sin t, \ 0 \le t \le 2\pi$

24. $x = -4t - \sin 3t, \ y = 4t - \sin 3t, \ 0 \le t \le 2\pi; \ x = -4, \ y = 4 + 2\pi - t,$
$2\pi \le t \le 2\pi + 4; \ x = t - 2\pi - 8, \ y = 0, \ 2\pi + 4 \le t \le 2\pi + 8$

25. The square of vertexes $(1, 1)$, $(1, 2)$, $(2, 2)$, $(2, 1)$.

26. The triangle of vertexes $(2, 1)$, $(3, 5)$, $(4, 3)$.

In Problems 27 to 30, find the volume obtained by rotating the region specified about the y-axis.

27. The region below the curve $y = 4 - x^2, \ -2 \le x \le 2$

28. The region below the curve $y = e^x, \ 0 \le x \le 1$.

29. The region enclosed by the curve $x^{2/5} + y^{2/5} = 1$. (*Hint.* Find a parametrization.)

30. The region enclosed by the curve $y(y - 1)^2 - x^2 = 0, \ 0 \le y \le 1$.

31. (a) Let $f(x)$ be continuous for $a \le x \le b$ and let $f(x) \ge k = $ const in the interval. Show that, if the region below the graph of f and above the line $y = k$ is rotated about the line $y = k$, then the volume obtained is $\pi \int_a^b [f(x) - k]^2 \, dx$.

(b) Let $x = g(y)$ be continuous and positive for $c \le y \le d$, and let $k < c$. Let the region between the graph of g and the y-axis be rotated about the line $y = k$. Show that the volume obtained is $2\pi \int_a^b (y - k)g(y) \, dy$.

32. Let C be a smooth simple closed path as in Section 7-6. Show that the volume V of the corresponding solid of revolution is given not only by (7-61) but also by the formulas

$$V = -\pi \oint_C y^2 \, dx, \qquad V = 2\pi \oint_C xy \, dy$$

[*Hint.* Show first that $\int_a^b [y^2(dx/dt) + 2xy(dy/dt)] \, dt = 0$, since $x(a) = x(b)$, $y(a) = y(b)$.]

33. The volume of a segment of a sphere of radius a bounded by a plane at distance b from the center is known to be $(\pi/3)(a - b)^2(2a + b)$ (Problem 12 above). Use this result and formulas of geometry to show that the volume of a conical shell (Figure 7-36) is $(2\pi a^3/3)(\cos \theta_1 - \cos \theta_2)$.

7-7 VOLUME OF OTHER SOLIDS

For the volume of a solid other than one of revolution we can adapt the method of slices first used in Section 7-5.

EXAMPLE 1. Find the volume of the tetrahedron whose vertexes have (x, y, z) coordinates $(0, 0, 0)$, $(a, 0, 0)$, $(a, b, 0)$, $(a, 0, c)$, respectively. The solid is shown in Figure 7-41. The edges are along the line $y = (b/a)x$ in the xy-plane, the line $z = (c/a)x$ in the xz-plane, the x-axis, and a right triangle in the plane $x = a$. To find the volume, we subdivide the interval $[0, a]$: $0 = x_0 < x_1 < \cdots < x_n = a$. The solid is thus cut into slices, the ith slice, with volume $\Delta_i V$, corresponding to the interval $[x_{i-1}, x_i]$. Let $A(x)$ denote the area of the cross section of our solid by a plane perpendicular to the x-axis at x. Then, for our example, $A(x)$ is the area of a right triangle

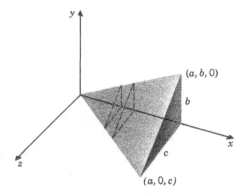

Figure 7-41 Tetrahedron.

of legs $y = (b/a)x$ and $z = (c/a)x$, so that $A(x) = \frac{1}{2}yz = bcx^2/(2a^2)$. Thus $A(x)$ is monotone strictly increasing. It is also clear that the slice corresponding to $\Delta_i V$ contains a right prism of base $A(x_{i-1})$ and altitude $\Delta_i x$, and is contained in a right prism of base $A(x_i)$ and altitude $\Delta_i x$. Thus

$$A(x_{i-1})\,\Delta_i x \le \Delta_i V \le A(x_i)\,\Delta_i x \qquad (7\text{-}70)$$

Since $A(x)$ has its minimum m_i in the interval $[x_{i-1}, x_i]$ at x_{i-1} and its maximum M_i at x_i, the inequality (7-70) is of the standard form for a definite integral, and we conclude that

$$V = \int_0^a A(x)\,dx$$

so that

$$V = \int_0^a \frac{bcx^2}{2a^2}\,dx = \frac{abc}{6}$$

The result is in agreement with the formula of geometry for a tetrahedron: *volume equals one third of base times altitude.*

For other solids we can proceed as suggested in the example: We consider slices formed by planes $x = \text{const} = x_i$, $i = 0, 1, \ldots, n$ (or by planes perpendicular to some other axis). The volume of the ith slice, $\Delta_i V$, should be given by $A(\xi_i)\,\Delta_i x$, where $x_{i-1} \le \xi_i \le x_i$ and $A(x)$ is the area of cross section at x. The total volume is thus expected to be

$$V = \int_a^b A(x)\,dx \qquad (7\text{-}71)$$

For many solids this formula can be justified as we did for the tetrahedron, but for many others the geometric relationships are more complicated and the simple reasoning is insufficient. A complete proof, depending on *multiple integrals,* is given in Chapter 13. For the present we shall accept the formula as valid and use it.

We can check the correctness of the formula for certain solids of revolution. Let the solid be obtained by rotation about the x-axis of the region "below a curve" $y = f(x)$, $a \le x \le b$, as in Section 7-5. Then the area of cross section, $A(x)$, is the area of a circle of radius $y = f(x)$, so that $A(x) = \pi y^2 = \pi[f(x)]^2$ and (7-71) gives

$$V = \pi \int_a^b [f(x)]^2 \, dx \tag{7-72}$$

as in Section 7-5.

Remark. "Cavalieri's Principle," formulated in 1635, states (in our language) that two solids having the same cross-sectional area $A(x)$ for each x have equal volumes. From this principle the correctness of (7-71) can be demonstrated, whenever $A(x)$ is continuous (or piecewise continuous). For we can always construct a solid of revolution generated as in the preceding paragraph, with $f(x)$ chosen so that $A(x) = \pi[f(x)]^2$. Hence, the volume of the solid of revolution is given by (7-72), so that by the Cavalieri principle the volume of the given solid is as in (7-71). This argument *assumes* the correctness of Cavalieri's principle, a proof of which requires multiple integrals.

EXAMPLE 2. Let a solid be such that the cross section at each x is a square, one edge of which is the segment from $(x, 0)$ to $[x, f(x)]$, as in Figure 7-42. Then by (7-71) the volume is

$$V = \int_a^b [f(x)]^2 \, dx$$

EXAMPLE 3. Find the volume of a solid lying between the planes $z = 0$ and $z = 1$ if the cross section at each z is the region beneath the curve $y = z^2 - x^2$ for $-z \le x \le z$, as in Figure 7-43. (We remark that the solid can be described as the region in space bounded by the surfaces $y = z^2 - x^2$, $y = 0$ and $z = 1$.)

To find the volume we first find the area of cross section, which we denote by $A(z)$:

$$A(z) = \int_{-z}^{z} (z^2 - x^2) \, dx = \left(z^2 x - \frac{x^3}{3} \right) \Big|_{x=-z}^{x=z} = \frac{4}{3} z^3$$

Then

$$V = \int_0^1 A(z) \, dz = \frac{4}{3} \int_0^1 z^3 \, dz = \frac{1}{3}$$

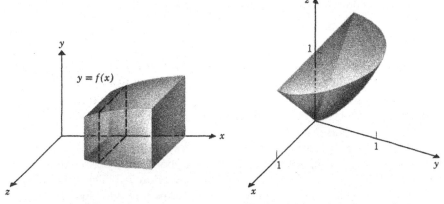

Figure 7-42 Solid with square cross section. **Figure 7-43** Solid of Example 3.

PROBLEMS

In Problems 1 to 12 find the volume of the solid described. Make a sketch in each case.

1. The cross section by a plane perpendicular to the x-axis is a triangle whose vertexes are $(x, 0, 0)$, $[x, f(x), 0]$, $[x, 0, g(x)]$, where $f(x) = 2x$, $g(x) = x^2$, $0 \le x \le 2$.

2. The cross section by a plane perpendicular to the z-axis is a triangle whose vertexes are $(0, 0, z)$, $[f(z), 0, z]$, $[0, g(z), z]$, where $f(z) = z^2$, $g(z) = 2z^2$, and $0 \le z \le 3$.

3. The cross section by a plane perpendicular to the x-axis is a square three of whose vertexes are $(x, 0, 0)$, $(x, \sin x, 0)$, $(x, 0, \sin x)$, $0 \le x \le \pi$.

4. The solid is formed of the points common to two intersecting right circular cylinders, both of radius a, whose axes meet at right angles. (*Hint.* Take the axes to be the x- and y-axes. Consider first the portion in the first octant: all coordinates positive.)

5. The cross section by a plane perpendicular to the x-axis is a circle whose center is at the point $[x, e^x, 0]$ and whose radius is e^x, $0 \le x \le 1$.

6. The cross-section perpendicular to the z-axis is a circle tangent to the z-axis with radius z^2 and passing through the point $(2z^2 \cos z^2, 2z^2 \sin z^2, z)$, $0 \le z \le \pi/2$.

7. The solid enclosed by the ellipsoid $(x^2/a^2) + (y^2/b^2) + (z^2/c^2) = 1$.

8. The solid bounded by the surface $x^2 + 16y^2 - 16z = 0$ and the plane $z = 1$.

9. The solid whose cross-section perpendicular to the x-axis is the region beneath the curve $z = xy^2$, $0 \le y \le x$, for $0 \le x \le 1$.

10. The solid whose cross-section perpendicular to the z-axis is the interior of the curve $x = z \cos^3 t$, $y = z \sin^3 t$, $0 \le t \le 2\pi$, for $1 \le z \le 2$.

11. The solid bounded by the surfaces $z = x^2 + 2y^2$, $y = x^2$, $x = 1$, $y = 0$, $z = 0$.

12. The solid below the plane $z = 4 - x - y$ and above the region enclosed by the circle $(x - 1)^2 + (y - 1)^2 = 1$ in the xy-plane.

13. Let a solid consist of the points below a surface $z = f(x, y)$ and above the xy-plane, for (x, y) in the region below a curve $y = g(x)$, $a \le x \le b$. Assume that f is continuous and positive for (x, y) in this region, and that g is continuous and positive for $a \le x \le b$. Justify the formula for the volume V of the solid:

$$V = \int_a^b \left(\int_0^{g(x)} f(x, y) \, dy \right) dx$$

(Here we are dealing with a double integral. See Chapter 13.)

7-8 AREA OF SURFACE OF REVOLUTION

The area of a surface in space is a familiar concept—we use it whenever we discuss the areas of geographical regions. The mathematical theory of surface area is surprisingly difficult, far more subtle than that of plane area or of the volume of a solid. The theory can be regarded as an extension of arc length to higher dimension. Accordingly, it is natural to attempt to define surface area by inscribing polyhedral surfaces (like the polygons used for arc length), and then passing to the limit. A straightforward application of this procedure turns out to be unsuccessful—the limit may fail to exist, even for simple surfaces. However, when certain modifications are made, a reasonable definition is obtained that provides a measure of surface area for all "reasonable" surfaces. For details, see Chapter 13.

Here we confine attention to *surfaces of revolution*, each generated by revolving a curve in a plane about an axis in the plane. We give a special definition of surface area for this case. It can be shown to agree with the general definition.

We start with a special case, when the curve being rotated is a line segment: $y = f(x) = ax + b$, $c \le x \le d$, and the axis of revolution is the x-axis (Figure 7-44). We assume that $y \ge 0$ on the segment, $y \not\equiv 0$. Accordingly, our surface is either

> A cone ($y = 0$ at one end point)
> A cylinder ($a = 0$, $b \neq 0$)

or

> A frustum of a cone ($y > 0$ for $c \le x \le d$ and $a \neq 0$)

In all three cases elementary geometry gives us a formula for the surface area; in fact, the one formula:

$$\text{Surface area} = 2\pi s \frac{f(c) + f(d)}{2}, \qquad s = \text{length of segment} \quad (7\text{-}80)$$

(average circumference of the two bases times slant height) covers all three cases. We shall assume this formula to be correct. It is made very plausible by the fact that a cone or cylinder, when physically constructed by a sheet of paper, can be laid flat to give a plane region of area equal to that stated.

Now let a general curve $y = f(x)$, $a \le x \le b$, be given, with $f(x) > 0$ except perhaps at a and b. We assume that f has a continuous derivative, so that the curve is *smooth* (Section 4-27). The theory is easily extended to *piecewise smooth curves*, as was done for arc length.

Under these assumptions on f, arc length s can be defined along the curve, regarded as a path with x as parameter. As in Section 4-28, we can take

$$s = \int_a^x \sqrt{1 + [f'(u)]^2}\, du \qquad (7\text{-}81)$$

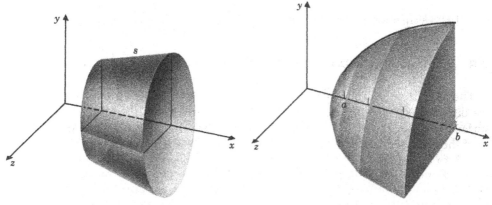

Figure 7-44   Surface area of frustum of a cone.

Figure 7-45   Definition of area of surface of revolution.

The length of the whole curve or any section of it is obtained as the limit of the lengths of inscribed polygons; each polygon joins the successive points (x_i, y_i), where $y_i = f(x_i)$ and

$$a = x_0 < x_1 < \cdots < x_n = b$$

is a subdivision of $[a, b]$; the limit is taken as the mesh $\to 0$. We now define the surface area to be the analogous limit, as the mesh $\to 0$, of the surface area of the surfaces of revolution generated by the inscribed polygons. For each polygon, the corresponding surface is formed of portions (frustum of cone, cylinder, or cone) generated by the separate line segments, as in Figure 7-45. Hence the surface area is defined in each case as the sum of the separate areas computed by the rule (7-80). Hence for the ith piece, the area is

$$\Delta_i S = 2\pi \sqrt{(x_i - x_{i-1})^2 + (y_i - y_{i-1})^2} \, \frac{y_{i-1} + y_i}{2}$$

By the Mean Value Theorem

$$y_i - y_{i-1} = f(x_i) - f(x_{i-1}) = f'(\xi_i)(x_i - x_{i-1})$$

where $x_{i-1} < \xi_i < x_i$. Also $(y_{i-1} + y_i)/2$ is a number between y_{i-1} and y_i; therefore, by the Intermediate Value Theorem,

$$\frac{y_{i-1} + y_i}{2} = f(\eta_i), \qquad x_{i-1} < \eta_i < x_i$$

Hence we obtain

$$\Delta_i S = 2\pi \sqrt{1 + [f'(\xi_i)]^2} \, f(\eta_i) \, \Delta_i x$$

By the same reasoning as in Section 4-27

$$\lim_{\text{mesh} \to 0} \sum_{i=1}^{n} \Delta_i S = 2\pi \int_a^b \sqrt{1 + [f'(x)]^2} \, f(x) \, dx$$

Accordingly, by our definition, every surface of revolution as given above has a surface area

$$S = 2\pi \int_a^b \sqrt{1 + [f'(x)]^2} \, f(x) \, dx \tag{7-82}$$

From (7-81) $s'(x) = \sqrt{1 + [f'(x)]^2}$, so that (7-82) can also be written

$$S = 2\pi \int_a^b s'(x) f(x) \, dx \tag{7-82'}$$

which resembles (7-80).

EXAMPLE 1. *Area of surface of a sphere.* Here we take the generating curve to be the semicircle

$$y = \sqrt{a^2 - x^2}, \qquad -a \le x \le a$$

so that $y' = -x/\sqrt{a^2 - x^2}$ and hence

$$S = 2\pi \int_{-a}^{a} \sqrt{1 + \frac{x^2}{a^2 - x^2}} \, \sqrt{a^2 - x^2} \, dx = 2\pi \int_{-a}^{a} \sqrt{a^2} \, dx = 4\pi a^2$$

in agreement with the familiar formula (Section 0-5).

Remark. In this example y' is discontinuous at $x = \pm a$, as is $(1 + y'^2)^{1/2}$. However, the integrand $(1 + y'^2)^{1/2} y$ is continuous (more precisely, it becomes continuous at $x = \pm a$ if we define it properly at these points), and the steps in the limit process leading to (7-82) remain valid under these conditions, just as they do for integration of a piecewise continuous function (Section 4-30).

The Use of s or t as Parameter. We can use arc length s as parameter along our curve, as defined by (7-81). Since $ds = (1 + y'^2)^{1/2} \, dx$, our formula (7-82) becomes

$$S = 2\pi \int_0^L y \, ds \tag{7-83}$$

where y is expressed in terms of s: $y = f[x(s)]$ and L is the length of the curve. If now another equivalent parametrization is chosen so that $s = s(t)$, $\alpha \leq t \leq \beta$, with say $ds/dt > 0$, and $s(\alpha) = 0$, $s(\beta) = L$, our formula becomes

$$S = 2\pi \int_\alpha^\beta y \frac{ds}{dt} \, dt \tag{7-84}$$

and, if the path is given originally as $x = x(t)$, $y = y(t)$, $\alpha \leq t \leq \beta$, then (7-84) can be written:

$$S = 2\pi \int_\alpha^\beta y(t) \sqrt{\left(\frac{dx}{dt}\right)^2 + \left(\frac{dy}{dt}\right)^2} \, dt \tag{7-84'}$$

Furthermore, (7-84') can be used to evaluate the surface area of a surface of revolution obtained from a general smooth (or piecewise smooth) path when the path is given parametrically: $x = x(t)$, $y = y(t)$, $\alpha \leq t \leq \beta$, with $y(t) \geq 0$. The area is now obtained from inscribing polygons exactly as in obtaining the arc length (Figure 7-46) and then applying the same reasoning as that leading to (7-82). We usually assume that the path is simple (does

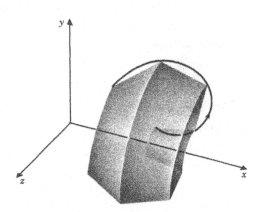

Figure 7-46 Surface obtained from a parametrized path.

not cross itself), although some meaning can be given to (7-84') as "surface area swept out" (like distance traversed) when there are crossings.

EXAMPLE 2. *Surface of a torus.* Here our generating curve can be taken as before (Example 2 of Section 7-6):

$$x = b + a \cos t, \qquad y = b + a \sin t, \qquad 0 \le t \le 2\pi$$

with $0 < a < b$. Equation (7-84') gives

$$S = 2\pi \int_0^{2\pi} (b + a \sin t) \sqrt{a^2 \sin^2 t + a^2 \cos^2 t} \, dt$$

$$= 2\pi a \int_0^{2\pi} (b + a \sin t) \, dt = 4\pi^2 \, ab$$

If the surface is obtained by rotating about the y-axis rather than the x-axis, then x and y are interchanged in our formulas. Thus (7-82) and (7-84') become

$$S = 2\pi \int_c^d \sqrt{1 + [g'(y)]^2} \, g(y) \, dy \qquad (7\text{-}85)$$

$$S = 2\pi \int_\alpha^\beta x(t) \sqrt{[x'(t)]^2 + [y'(t)]^2} \, dt \qquad (7\text{-}86)$$

We can also use $y = f(x)$ to obtain the area for rotation about the y-axis. We can regard x as the parameter in (7-86) [that is, $x = t$, $y = f(t)$], and obtain

$$S = 2\pi \int_a^b x \sqrt{1 + [f'(x)]^2} \, dx \qquad (7\text{-}87)$$

PROBLEMS

In Problems 1 to 7, find the area of the surface of revolution obtained by rotating the curve described about the axis stated. Throughout, a, b, and h are positive constants.

1. $y = \sqrt{a^2 - x^2}$, $b \le x \le b + h$, where $-a \le b \le b + h \le a$ (zone of sphere), about the x-axis.
2. $y = \sqrt{4ax}$, $0 \le x \le b$.   (a) About the x-axis.   (b) About the y-axis.
3. $y = (b/a)\sqrt{a^2 - x^2}$, $0 \le x \le a$, $b < a$.
   (a) About the x-axis.
   (b) About the y-axis.
   (Surface of ellipsoid).
4. $y = (b/a)\sqrt{x^2 - a^2}$, $a \le x \le a + ah$.
   (a) About the x-axis.
   (b) About the y-axis.
   (Surface of hyperboloid)
5. $x = 2\sqrt{2} \, a \cos t$, $y = a \sin t \cos t$, $0 \le t \le \pi/2$.
   (a) About the x-axis.
   (b) About the y-axis.
6. $x = a \cos^3 t$, $y = a \sin^3 t$, $0 \le t \le \pi/2$ about the x-axis.
7. $y = \cosh x$, $0 \le x \le a$.   (a) About the x-axis.   (b) About the y-axis.
8. Let a curve be given by an equation in polar coordinates: $r = f(\theta)$, $\alpha \le \theta \le \beta$.

Under appropriate assumptions on f, α and β, obtain the formulas for the area of the following surfaces of revolution obtained from the curve as described:

(a) $S = 2\pi \int_\alpha^\beta r \sin\theta \sqrt{r^2 + r'^2}\, d\theta$, when rotated about x-axis.

(b) $S = 2\pi \int_\alpha^\beta r \cos\theta \sqrt{r^2 + r'^2}\, d\theta$, when rotated about y-axis.

In Problems 9 to 12 use the results of Problem 8 to find the area of the surface of revolution obtained by rotating the curve described about the axis stated. Throughout, a and b are positive constants.

9. $r = 2a \cos\theta, 0 \le \theta \le \pi$ about the x-axis.
10. $r = a(1 + \cos\theta), 0 \le \theta \le \pi$ (cardioid).
   (a) About the x-axis.
   (b) About the y-axis. (Is this meaningful?)
11. $r^2 = a^2 \cos^2\theta + b^2 \sin^2\theta, b < a, 0 \le \theta \le \pi/2$.
   (a) About the x-axis.
   (b) About the y-axis.
12. $r^2 = a^2 \cos^2\theta - b^2 \sin^2\theta, b < a, 0 \le \theta \le \alpha = \mathrm{Tan}^{-1}(a/b)$.
   (a) About the x-axis.
   (b) About the y-axis.
‡13. Show that, in deriving (7-82), it is plausible to take $\Delta S = 2\pi f(\xi_i)\, \Delta_i s$, and show that the limit process leads to (7-83) and, hence, obtain (7-82), and (7-82').

14. Let a smooth path C: $x = f(t)$, $y = g(t)$, $a \le t \le b$ be given in an xy-plane. Let a positive function z be defined along the path, so that z can be considered as a function of t, $z = \psi(t)$, or of arc length s, $z = \varphi(s)$. Justify the following formulas for the cylindrical surface (Figure 7-47), given by $0 \le z \le \varphi(s)$, $x = x(s)$, and $y = y(s)$:

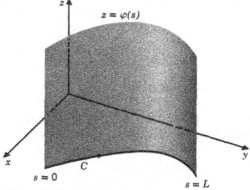

Figure 7-47   Cylindrical surface.

(a) $S = \int_0^L \varphi(s)\, ds$

(b) $S = \int_a^b \psi(t) \sqrt{[f'(t)]^2 + [g'(t)]^2}\, dt$

15. Apply the formulas of Problem 14 to find the areas of the following cylindrical surfaces.
   (a) $x = a \cos t, y = a \sin t, 0 \le z \le c, 0 \le t \le 2\pi$ (Circular cylinder)
   (b) $x = t, y = t^2, 0 \le z \le 3t, 0 \le t \le 1$
   (c) $x = a \cos t, y = b \sin t, 0 \le z \le c \sin t, 0 \le t \le \pi/2$. ($a$, b, c are positive constants)

7-9 DISTRIBUTION OF MASS AND OTHER DISTRIBUTIONS

In physics we assign mass to solid objects, including the limiting case of a particle: a solid concentrated at a single point. At times, we conveniently idealize a very thin wirelike object as a curve, having a certain mass, or idealize a thin membranelike object as a surface having a definite mass. In each case, the mass is given as a positive number, in terms of a chosen unit (usually the gram). We then view the various objects of the physical universe, at a given time, as representing a certain *distribution of mass* in three-dimensional space. In particular, each region of space will have a certain total mass, determined by the objects or parts of objects within the region.

Let us consider the special case of mass distributed along a straight line, the x-axis. The simplest case is that of *a discrete distribution:* there are n particles, with coordinates $x_1, \ldots, x_n$ and positive masses $m_1, \ldots, m_n$, as in Figure 7-48. By the *kth moment* of this mass distribution, we mean the quantity

$$M_k = \sum_{i=1}^{n} m_i x_i^{k}, \; k = 0, 1, 2, \ldots \tag{7-90}$$

Thus the 0th moment is the *total mass.* The first moment is related to the turning effect of forces perpendicular to the x-axis at each x_i and proportional to the mass in each case. In particular, we can interpret the x-axis as a horizontal lever with fulcrum at 0, as in Figure 7-49. The force of gravity (on the surface of the earth) at each particle is a downward vector of magnitude $m_i g$ and moment $m_i g x_i$ about the fulcrum. Hence

$$\sum_{i=1}^{m} m_i g x_i = g \sum_{i=1}^{n} m_i x_i = g M_1$$

is the total moment of the forces. The lever is in equilibrium precisely when the total moment is 0.

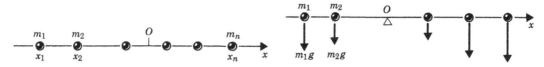

Figure 7-48 *n* particles on a line.   Figure 7-49   Moment of forces.

The *center of mass* of the mass distribution of Figure 7-48 is defined as the coordinate $\bar{x}$ at which we could place *all* the mass: $m_1 + \cdots + m_n$ and have the same first moment. Thus $\bar{x}$ is to be such that

$$(m_1 + \cdots + m_n)\bar{x} = \sum_{i=1}^{n} m_i x_i$$

Hence, $\bar{x}$ is uniquely determined and

$$\bar{x} = \frac{m_1 x_1 + \cdots + m_n x_n}{m_1 + \cdots + m_n} \tag{7-91}$$

We describe $\bar{x}$ also as a *weighted mean* of $x_1, \ldots, x_n$, with weights $m_1, \ldots, m_n$. From (7-91) it follows (Problem 6 below) that $\bar{x}$ lies between the minimum and maximum of $x_1 \ldots x_n$. Also, if we concentrate the masses in bunches at the corresponding centers of mass (masses m_1, m_2, m_3 at the center of mass of particles 1, 2, 3, masses m_4, m_5 at the center of mass of particles 4, 5, and so on) we obtain a new set of particles with the *same* center of mass as the original set (Problem 7 below).

The second moment $\sum_{i=1}^{n} m_i x_i^2$ is known as the *moment of inertia* of the particles about the origin (or about a line perpendicular to the x-axis at the origin). This has application in mechanics as a measurement of how easily the set of particles can be accelerated in angular motion in circles about the origin. The *radius of gyration* ρ of the set of particles about the origin is the distance from the origin at which a single particle of mass $m_1 + \cdots + m_n$ could be placed and yield the same moment of inertia about the origin. Thus

$$(m_1 + \cdots + m_n)\rho^2 = \sum_{i=1}^{n} m_i x_i^2$$

or

$$\rho^2 = \frac{\sum_{i=1}^{n} m_i x_i^2}{m_1 + \cdots + m_n}$$

EXAMPLE 1. Let a particle of mass 2 grams have x-coordinate -1, let another of mass 5 grams have x-coordinate 1. Then the center of mass $\bar{x}$ and radius of gyration ρ^2 about the origin are given by

$$(2 + 5)\bar{x} = 2(-1) + 5(1), \quad (2 + 5)\rho^2 = 2(1) + 5(1)$$

Thus $\bar{x} = 3/7$, $\rho^2 = 1$.

It should be remarked that the location of the center of mass is independent of the coordinates chosen, whereas the radius of gyration depends on the choice of origin (Problem 8 below).

We now wish to generalize these concepts to a *continuous mass distribution* along the x-axis. We assume all the mass is contained in the interval $[a, b]$, but that the mass is spread out—each interval contains a certain mass, but at each point there is 0 mass. Let $F(x)$ be the amount of mass in the interval $[a, x]$. Then $F(x)$ is a monotone strictly increasing function of x; $F(a) = 0$ and $F(b)$ is the total mass. We assume F is continuous (in fact, a discontinuity would indicate the presence of a particle—see Problem 10 below), and we assume further that $F'(x)$ exists and is continuous. Now

$$F'(x) = \lim_{h \to 0} \frac{F(x + h) - F(x)}{h}$$

and, for $h > 0$, $F(x + h) - F(x)$ is the mass in the interval $[x, x + h]$, h is the length of the interval. Hence the derivative is the limit of mass divided by length, or *limit of average density*. Accordingly, we interpret $F'(x)$ as *the density at the point x* and denote it by $\delta(x)$:

$$\delta(x) = F'(x) = \text{density at } x$$

The total mass in an interval $[c, d]$ is then

$$F(d) - F(c) = \int_c^d F'(x)\, dx = \int_c^d \delta(x)\, dx \qquad (7\text{-}92)$$

or roughly in words "mass is the integral of density." The density $\delta(x)$ is here measured in units of mass per unit of length (for example, gm/cm).

The *kth moment* of our continuous mass distribution along the x-axis is now defined as follows:

$$M_k = \int_a^b x^k\, \delta(x)\, dx, \qquad k = 0, 1, 2, \ldots \qquad (7\text{-}93)$$

This can be justified by considering the integral as limit of a sum

$$\sum_{i=1}^n \xi_i{}^k\, \delta(\xi_i)\, \Delta_i x \qquad (7\text{-}94)$$

If ξ_i is chosen so that $F(x_i) - F(x_{i-1}) = \delta(\xi_i)(x_i - x_{i-1})$, as is possible since $\delta(x) = F'(x)$, then $\delta(\xi_i)\, \Delta_i x$ is just the mass m_i between x_{i-1} and x_i, and the sum (7-94) is just the kth moment of a set of particles of masses m_i concentrated at the positions ξ_i. Hence, we reasonably define the kth moment by (7-93). The 0th moment is then the total mass:

$$M_0 = \int_a^b \delta(x)\, dx$$

The *center of mass* is defined as before in terms of the first moment: it has x-coordinate $\bar{x}$ so chosen that concentration of all the mass, M_0, at $\bar{x}$ yields the same first moment as the given distribution:

$$M_0\bar{x} = \int_a^b x\, \delta(x)\, dx$$

$$\bar{x} = \frac{\int_a^b x\, \delta(x)\, dx}{M_0} = \frac{\int_a^b x\, \delta(x)\, dx}{\int_a^b \delta(x)\, dx} \qquad (7\text{-}95)$$

Its location is independent of the coordinate system (Problem 9 below).

More generally, we can consider a mass distribution over several distinct intervals. The kth moment of such a distribution is the sum of the kth moments of the parts. Equivalently, we can allow $\delta(x)$ to be piecewise continuous, and set $\delta = 0$ in any interval containing no mass. Then (7-93) still defines the kth moment M_k. For M_1, we then have the rule that its value is unchanged if we replace the given distribution by one obtained by concentrating the mass of each part at the center of mass of that part. For the case pictured in Figure 7-50 this means that

$$m_1\bar{x}_1 + m_2\bar{x}_2 + m_3\bar{x}_3 = (m_1 + m_2 + m_3)\bar{x}$$

Figure 7-50 Combination of first moments.

where $\bar{x}_1$, $\bar{x}_2$, $\bar{x}_3$ locate the three centers of mass, and m_1, m_2, m_3 are the masses in the three intervals. If we let $\delta(x)$ denote the density throughout the interval $[a_1, b_3]$, so that $\delta = 0$ between b_1 and a_2 and between b_2 and a_3, then our condition is equivalent to the clearly correct equation

$$\int_{a_1}^{b_3} x\, \delta(x)\, dx = \int_a^{b_1} x\, \delta(x)\, dx + \int_{a_2}^{b_2} x\, \delta(x)\, dx + \int_{a_1}^{b_3} x\, \delta(x)\, dx$$

and a similar result holds true for an arbitrary finite number of intervals.

For the second moment M_2, we define the *radius of gyration* about $x = 0$ (or about an axis perpendicular to the x-axis at $x = 0$) as a value $\rho \geq 0$ for which

$$M_0 \rho^2 = M_2$$

so that

$$\rho^2 = \frac{M_2}{M_0} = \frac{\displaystyle\int_a^b x^2\, \delta(x)\, dx}{\displaystyle\int_a^b \delta(x)\, dx}$$

Again, concentration of all the mass at position ρ yields the same second moment. We have a rule about concentrating parts of the mass at the corresponding ρ-values. In fact, for arbitrary k

$$\int_{a_1}^{b_3} x^k\, \delta(x)\, dx = \int_a^{b_1} x^k\, \delta(x)\, dx + \int_{a_2}^{b_2} x^k\, \delta(x)\, dx + \int_{a_3}^{b_3} x^k\, \delta(x)\, dx$$

for the case of Figure 7-50, and in general

$$M_0 \rho_k^k = m_1 \rho_{k1}^k + m_2 \rho_{k2}^k + \cdots + m_n \rho_{kn}^k \tag{7-96}$$

where ρ_k, ρ_{k1}, ..., ρ_{kn} are chosen so that

$$M_0 \rho_k^k = \int_{a_1}^{b_n} x^k\, \delta(x)\, dx, \qquad m_1 \rho_{k1}^k = \int_{a_1}^{b_1} x^k\, \delta(x)\, dx, \qquad \cdots$$

For M_2, we have the valuable rule

$$M_2 = \int_a^b (x - \bar{x})^2\, \delta(x)\, dx + M_0 \bar{x}^2 \tag{7-97}$$

or in words: *the second moment about O equals the second moment about the center of mass plus the second moment about O of the total mass concentrated at the center of mass.* The proof follows from expanding the $(x - \bar{x})^2$:

$$\int_a^b (x - \bar{x})^2\, \delta(x)\, dx = \int_a^b x^2\, dx - 2\bar{x} \int_a^b x\, \delta(x)\, dx + \bar{x}^2 \int_a^b \delta(x)\, dx$$

$$= M_2 - 2\bar{x} \cdot M_0 \bar{x} + M_0 \bar{x}^2 = M_2 - M_0 \bar{x}^2$$

From (7-97) we conclude that, if the origin O is placed at the center of mass, then the second moment about O has its smallest possible value. There is a rule analogous to (7-97) for discrete distributions:

$$M_2 = \sum_{i=1}^{n} m_i(x_i - \bar{x})^2 + M_0\bar{x}^2 \tag{7-97'}$$

The proof is left as an exercise (Problem 11 below).

EXAMPLE 2. Let mass be spread over the interval $[0, 2]$ with density $\delta(x) = kx(2 - x)$, where k is a positive constant. Then

$$M_0 = \int_0^2 kx(2 - x)\, dx = 4k/3 = \text{total mass}$$

$$M_1 = \int_0^2 kx^2(2 - x)\, dx = 4k/3 = M_0\bar{x}$$

$$M_2 = \int_0^2 kx^3(2 - x)\, dx = 8k/5 = M_0\rho^2$$

Thus $\bar{x} = 1$, $\rho^2 = 6/5$. By (7-97) the second moment about the center of mass is $M_2 - M_0\bar{x}^2 = 4k/15$. Dividing by M_0, we obtain the squared radius of gyration about the center of mass as $1/5$.

The ideas discussed here have a much broader application than to masses in physics. One can consider *distribution of electric charge*, either at points or continuously, with associated charge density; in this case the density and charge can be both positive and negative. One can consider *distribution of energy* in the same way. One can also analyze concentration of a chemical substance, concentration of biological units, population densities, and many other concepts in the sciences as examples of the same type. Thus far we have considered the concepts only in one dimension, but all can be extended to two and three dimensions (Sections 7–10 to 7–12).

PROBLEMS

1. Find the center of mass for each of the following mass distributions on the x-axis (in given units of mass and distance).
   (a) Mass 3 at $x = 0$, 5 at $x = -1$, 7 at $x = 3$.
   (b) Mass 1 at $x = j$, $j = 1, \ldots, n$.
   (c) Mass $1/j$ at $x = j$, $j = 1, \ldots, n$.
   (d) Mass 2^j at $x = 2^j$, $j = 1, \ldots, n$.
   (e) Density 1 for $0 \le x \le 1$ and for $2 \le x \le 3$, density 0 otherwise.
   (f) Density 1 for $0 \le x \le 1$, 2 for $2 \le x \le 3$, 4 for $4 \le x \le 5$, density 0 otherwise.
   (g) Density e^x for $0 \le x \le 1$, 0 otherwise.
   (h) Density $\sin x$ for $0 \le x \le \pi/2$, 0 otherwise.
   (i) Density e^x for $0 \le x \le 1$, 1 for $-1 \le x \le 0$, density 0 otherwise.
   (j) Mass in interval $[0, x]$ is x^2 for $0 \le x \le 1$, is 1 for $x > 1$, no mass in the interval $(-\infty, 0)$.
   (k) Mass in interval $[0, x]$ is $1 - e^{-x}$ for $0 \le x \le 2$, is $1 - e^{-2}$ for $x \ge 2$, no mass in $(-\infty, 0)$.
   (l) Mass in interval $[0, x]$ is 1 for $0 \le x < 1$, 2 for $x \ge 1$, no mass in $(-\infty, 0)$.
2. (a) $\ldots$ (l). Find the second moment and radius of gyration about $x = 0$ for each of the mass distributions of Problem 1.

3. Find a general expression for the kth moment M_k for each of the following mass distributions:
   (a) As in Problem 1(f)   (b) As in Problem 1(g).

4. (a) Mass is distributed at $x = 2$ and $x = 5$ so that $M_0 = 3$, $M_1 = 9$. Find the mass distribution.
   (b) Show generally: a mass distribution at distinct points x_1, x_2 ($x_1 < x_2$) is determined, and determined uniquely, by giving the moments M_0 and M_1. When are the masses m_1, m_2 positive?
   (c) Show generally: for three distinct points x_1, x_2, x_3 there is one and only one mass distribution at these points with given moments M_0, M_1, M_2.
   (d) Is a mass distribution at distinct points x_1, x_2 determined by giving the moments M_1 and M_2?

5. Charge is distributed continuously on the interval $[0, \pi]$ with density $\delta(x)$.
   (a) Find $\delta(x)$ if it is known that δ has form $c_1 \cos x + c_2 \sin x$, and that $M_0 = 3$, $M_1 = 5$.
   (b) Find δ, if it is known that δ has form $c_1 + c_2 x + c_3 x^2$ for certain constants c_1, c_2, c_3, and that $M_0 = 7$, $M_1 = 9$, $M_3 = 2$.

6. Prove that, for a positive mass distribution at $x_1, \ldots, x_n$, the center of mass $\bar{x}$ is between the minimum and maximum of $x_1, \ldots, x_n$. When is $\bar{x}$ at the minimum or maximum?

7. Prove that, for a discrete mass distribution at $x_1, \ldots, x_n$, the center of mass is unaffected if the masses at $x_1, \ldots, x_h$ ($h < n$) are replaced by the single mass $m_1 + \cdots + m_h$ placed at the center of mass of the distribution on $x_1, \ldots, x_h$.

8. Let a new origin be chosen on the x-axis, so that each point has both an old coordinate x and a new coordinate x', related by $x = x' + h$ ($h = $ const). Show that, for a discrete mass distribution at $x_1, \ldots, x_n$, the center of mass when computed from the new coordinates has new coordinate $\bar{x}' = \bar{x} - h$, where $\bar{x}$ is the old coordinate. Thus the position is unchanged.

9. Extend the result of Problem 8 to the case of a continuous distribution of mass on $[a, b]$.

10. For a positive mass distribution in the interval $[a, b]$, let $F(x)$ be the total mass in $[a, x]$.
    (a) Show that F is monotone nondecreasing. What is the significance of an interval of constancy of F?
    (b) Show that, if $a < c < b$ and $\lim\limits_{x \to c-} F(x) < \lim\limits_{x \to c+} F(x)$, then there must be a particle at $x = c$.

11. Prove (11-97').

7-10 MASS DISTRIBUTIONS IN THE PLANE

For the case of particles in the xy-plane, we have n points $(x_1, y_1), \ldots, (x_n, y_n)$, with mass m_i at (x_i, y_i). We can locate the points also by position vectors $\overrightarrow{OP_i}$, where P_i is (x_i, y_i) and O is a chosen origin (Figure 7-51). The kth x-moment of the mass distribution is then defined as

$$M_{kx} = \sum_{l=1}^{n} m_l x_l^k \qquad (k = 0, 1, 2, \ldots)$$

and M_{ky} is defined similarly. The definition can be given in a more

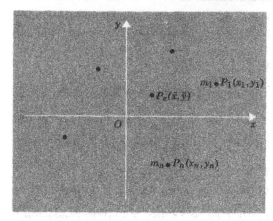

Figure 7-51 Particles in a plane.

geometric form thus: The *kth* moment with respect to origin O and unit vector $\mathbf{u}$ is

$$M_{k\mathbf{u}} = \sum_{l=1}^{n} m_l(\overrightarrow{OP_l} \cdot \mathbf{u})^k, \qquad k = 0, 1, 2, \dots$$

When $\mathbf{u} = \mathbf{i}$, $\overrightarrow{OP_l} \cdot \mathbf{u} = \overrightarrow{OP_l} \cdot \mathbf{i} = x_l$, and we obtain the previous expression. Notice that $M_{0\mathbf{u}} = M_0 = $ total mass.

The *center of mass* P_c of the distribution is now defined as a point for which

$$M_0\overrightarrow{OP_c} \cdot \mathbf{u} = \sum_{l=1}^{n} m_l(\overrightarrow{OP_l} \cdot \mathbf{u}) \tag{7-100}$$

for every unit vector $\mathbf{u}$. By taking $\mathbf{u} = \mathbf{i}$ and $\mathbf{j}$ in turn, we conclude that P_c must have coordinates $(\bar{x}, \bar{y})$ so that

$$M_0\bar{x} = \sum_{l=1}^{n} m_l x_l, \qquad M_0\bar{y} = \sum_{l=1}^{n} m_l y_l$$

Thus $\bar{x}$, $\bar{y}$ are uniquely determined. We are led to the following equation for $\overrightarrow{OP_c}$:

$$M_0\overrightarrow{OP_c} = M_0(\bar{x}\mathbf{i} + \bar{y}\mathbf{j}) = \sum_{l=1}^{n} m_l(x_l\mathbf{i} + y_l\mathbf{j}) = \sum_{l=1}^{n} m_l\overrightarrow{OP_l}$$

and hence

$$\overrightarrow{OP_c} = \frac{\displaystyle\sum_{l=1}^{n} m_l\overrightarrow{OP_l}}{M_0} \tag{7-101}$$

Thus far, we are sure only that $\overrightarrow{OP_c}$ satisfies (7-100) for $\mathbf{u} = \mathbf{i}$ or $\mathbf{j}$, but our result (7-101) is independent of the choice of direction of the axes and must hold true in general; in fact P_c has a location independent of the choice of origin O. (See Problem 10 below.)

The second moment M_{2x} is called *the moment of inertia* of the mass distribution about the y-axis; similarly the second moment M_{2y} is the moment of inertia about the x-axis. The *radius of gyration* of the mass distribution about the y-axis is a number $\rho_y > 0$ so that

$$M_0 \rho_y{}^2 = \sum_{l=1}^{n} m_l x_l{}^2$$

and ρ_x is defined similarly. In general, the moment of inertia about a line L in the plane is $\sum_{l=1}^{n} m_l d_l{}^2$, where d_l is the distance from P_l to L, and the radius of gyration about L is a number ρ so that $M_0 \rho^2$ equals the moment of inertia about L.

For continuous mass distribution in the plane we are in general led to double integrals. We shall here consider mainly one special case, in which double integrals are not needed, and refer the reader to Chapter 13 for the general case. The special case is that of *homogeneous mass distributions*, in which the density is constant over each of certain regions $R_1, R_2, \ldots$, and 0 outside these regions. The assumptions are suggested in Figure 7-52. The physical counterpart is a set of metal plates of different metals and different shapes. For each region in R_l having area A the corresponding mass is $\delta_l A$. Thus δ_l is mass per unit area.

Let us consider a single region R, of constant density δ, and of the type "area under a curve $y = f(x)$," as in Figure 7-53. Let f be continuous in $[a, b]$. In order to define the kth x-moment of our mass distribution over R, we subdivide $[a, b]$ as usual: $a = x_0 < x_1 < \cdots < x_n = b$. Then the mass of the ith piece is δ times the area or

$$\delta \int_{x_{i-1}}^{x_i} f(x)\, dx = \delta f(\xi_i)\, \Delta_i x$$

by the Mean Value Theorem for integrals (Section 4-24). Within this piece, the x-coordinate lies between x_{i-1} and x_i, so that it is reasonable to take the kth moment to be

$$\eta_i{}^k \delta f(\xi_i)\, \Delta_i x$$

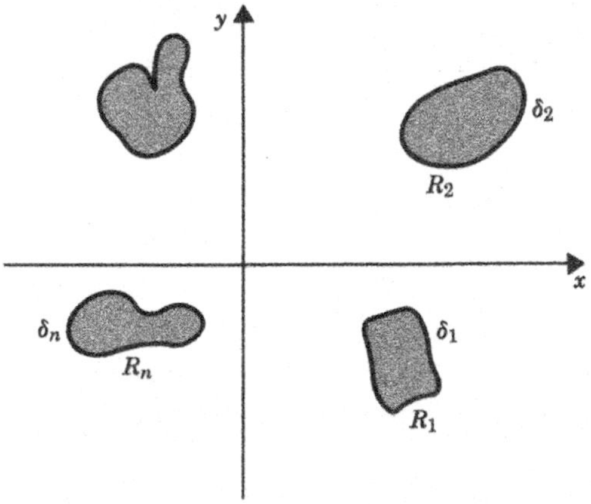

Figure 7-52   Mass distribution in the plane. 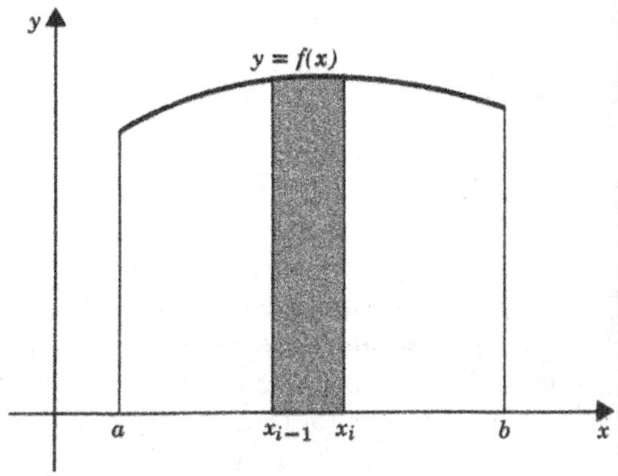   Figure 7-53   Moments for planar region.

where η_i is some value between x_{i-1} and x_i. Hence, we are led to the integral expression:

$$M_{kx} = \delta \int_a^b x^k f(x) \, dx \qquad (7\text{-}102)$$

We here regard this as a definition; it can be justified as consistent with a general theory (not related to the shape or location of the region) through double integrals (Chapter 13).

EXAMPLE 1. Let f be constant, equal to h, and let a be 0. Then

$$M_{kx} = \delta \int_0^b h x^k \, dx = \frac{h \delta b^{k+1}}{k+1} = \frac{M_0 b^k}{k+1}$$

since $M_0 = \delta b h$. Our result can be stated in words: *The kth moment of a homogeneous rectangular plate about an edge is equal to the mass times $b^k/(k+1)$, where b is the length of the other edge.*

We now seek a definition of M_{ky} for a homogeneous mass distribution as in Figure 7-53. We subdivide as before. Our ith piece includes a rectangle of base $\Delta_i x$ and altitude equal to the minimum of f in $[x_{i-1}, x_i]$, and is included in a similar rectangle of altitude equal to the maximum of f. Hence it is reasonable to take the y-moment of the piece to be that of a rectangle of base $\Delta_i x$ and altitude $f(\xi_i)$ about the x-axis. But Example 1 gives us that moment: mass times $[f(\xi_i)]^k/(k+1)$, or

$$\delta f(\eta_i) \cdot \frac{[f(\xi_i)]^k}{k+1} \, \Delta_i x, \qquad x_{i-1} < \xi_i < x_i, \ x_{i-1} < \eta_i < x_i$$

Hence we are led to take the moment to be

$$M_{ky} = \delta \int_a^b \frac{[f(x)]^{k+1}}{k+1} \, dx \qquad (7\text{-}103)$$

Remark 1. We can easily generalize the analysis leading to (7-102) and (7-103) to the case in which δ depends on x. We find that

$$M_{kx} = \int_a^b x^k f(x) \delta(x) \, dx, \qquad M_{ky} = \int_a^b \frac{[f(x)]^{k+1}}{k+1} \delta(x) \, dx$$

Remark 2. By the methods of Section 7-3 above, we can obtain line integral expressions for the moments:

$$M_{kx} = \frac{\delta}{k+2} \oint_C x^{k+1} \, dy - x^k y \, dx \qquad (7\text{-}104)$$

$$M_{ky} = \frac{\delta}{k+2} \oint_C x y^k \, dy - y^{k+1} \, dx \qquad (7\text{-}105)$$

Here C is a piecewise smooth simple closed path enclosing the region in question. Instead of concerning ourselves with the process leading to these formulas, we can verify that they are correct for rectangles and more generally for regions as in Figure 7-53 (Problem 9 below). By the reasoning of Section 7-4, we then show that the formulas give a proper definition of moments for an extended class of regions.

For a combination of regions, as in Figure 7-52, we now compute each moment by summing the moments for the several parts.

We can now define the center of mass for such a mass distribution as a point $(\bar{x}, \bar{y})$ for which

$$M_0\bar{x} = M_{1x}, \qquad M_0\bar{y} = M_{1y} \tag{7-106}$$

Since the moment for a combination of several distributions in different regions is the sum of the moments for the individual distributions, it follows that the center of mass can always be found by concentrating the mass of each part at its center of mass and then finding the center of mass of the resulting set of particles. Furthermore, *the location of the center of mass does not depend on the coordinate system chosen* (see Problem 11 below).

The radius of gyration about each axis is defined as for particles:

$$M_0\rho_y{}^2 = M_{2x}, \qquad M_0\rho_x{}^2 = M_{2y} \qquad (\rho_x \geq 0, \qquad \rho_y \geq 0)$$

In computing each of them, we can again concentrate mass at a point. However, there is no single point at which the mass can be located to give the second moment, or moment of inertia, about *all* axes.

THEOREM 2 (*PARALLEL AXIS THEOREM*). *The moment of inertia of a mass distribution in the plane with respect to a given axis in the plane equals that with respect to a parallel axis through the center of mass plus the moment of inertia about the given axis of a particle obtained by concentrating all mass at the center of mass.*

Thus for a system of particles with the y-axis as chosen axis

$$M_{2x} = \sum_{l=1}^{n} m_l(x_l - \bar{x})^2 + M_0\bar{x}^2 \tag{7-107}$$

and for mass spread over a region so that (7-104) holds true:

$$M_{2x} = \frac{\delta}{4} \oint_C (x - \bar{x})^3 \, dy - (x - \bar{x})^2 y \, dx + M_0\bar{x}^2 \tag{7-107'}$$

The proofs are left as a problem (Problem 12 below).

7-11 CENTROID

For a plane region on which there is a homogeneous mass distribution with constant density δ, the location of the center of mass does not depend on δ, since M_0, M_{1x}, M_{1y} are all proportional to δ and $\bar{x} = M_{1x}/M_0$, $\bar{y} = M_{1y}/M_0$. Thus the center of mass is associated with the region and not with the particular density δ. For that reason the center of mass is given a special name: *the centroid of the region*. We can also use this term for the center of mass of a combination of several regions, on which mass is spread with the same density throughout, or more generally for the center of mass of any homogeneous mass distribution. In all these cases the radius of gyration of the distribution about a line is also independent of the particular density, but no special terminology is used.

THEOREM 3 (*A THEOREM OF PAPPUS*). *The volume of a solid of revolution generated by a plane region is equal to the area of the region times the length of the path through which the centroid of the region is rotated; that is,*

$$V = A \cdot 2\pi d \qquad (7\text{-}110)$$

where A is the area of the region, d is the distance of the centroid from the axis of revolution.

PROOF. Let us assume our region is as in Figure 7-53, so that by (7-102)

$$M_{1y} = M_0 \bar{y} = \delta \int_a^b \frac{[f(x)]^2}{2} \, dx$$

But by (7-51) the volume generated is

$$V = \pi \int_a^b [f(x)]^2 \, dx$$

Hence

$$V = \frac{2\pi M_0 \bar{y}}{\delta} = 2\pi A \bar{y}$$

since $M_0 = A\delta$. But $2\pi \bar{y}$ is precisely the path length of the center of mass (Figure 7-54). Hence the theorem is proved for this case.

For a more general region, bounded by a piecewise smooth simple closed curve, the proof is left as an exercise (Problem 13 below).

EXAMPLE. Let an ellipse in the xy-plane have center at (h, k), major axis $2a$, minor axis $2b$, and let the ellipse not meet the x-axis. Let the region enclosed be rotated about the x-axis. Then, regardless of the directions of the axes of the ellipse, the solid of revolution has volume $(\pi ab)(2\pi k) = 2\pi^2 abk$.

†7-12 MASS DISTRIBUTION ON CURVES

Let a piecewise smooth path $x = x(t)$, $y = y(t)$, $a \leq t \leq b$ be given in the xy-plane, and let arc length s be defined along the path. We assume

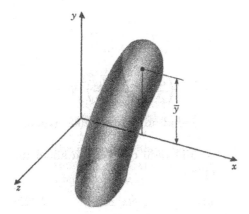

Figure 7-54 Pappus theorem on volumes.

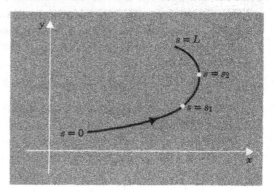

Figure 7-55   Mass distribution
on a curve.

that the path is simple and that $ds/dt > 0$ (except at corners), so that s can
be used as parameter, with $s = 0$ at $t = a$ and $s = L$ at $t = b$ (Figure 7-55).
We then consider a continuous mass distribution along the "wire": that is,
to each interval $[0, s]$, $0 \leq x \leq L$, a mass $F(s)$ is to be assigned, so that F is
continuous and monotone strictly increasing with $F(0) = 0$. Thus, for
$s_2 > s_1$, $F(s_2) - F(s_1)$ can be interpreted as the mass on the portion of the
curve for which $s_1 \leq s \leq s_2$. Finally, we assume that $\delta(s) = F'(s)$ exists
and is continuous, so that

$$\int_{s_1}^{s_2} \delta(u)\, du = \int_{s_1}^{s_2} F'(u)\, du = F(s_2) - F(s_1)$$

is the mass on the portion of the curve. Then, just as for a continuous
distribution on a line, we can interpret $\delta(s)$ as the density (in units of mass
per unit of length).

By a reasoning just like that for the case of the line (Section 7-9) we then
define moments

$$M_{kx} = \int_0^L x^k \delta(s)\, ds$$
$$\qquad\qquad k = 0, 1, 2, \ldots \qquad\qquad (7\text{-}120)$$
$$M_{ky} = \int_0^L y^k \delta(s)\, ds$$

where $x = x(s)$, $y = y(s)$. If s is expressed in terms of t, then x, y, and δ can
also be expressed in terms of t, and (7-120) becomes

$$M_{kx} = \int_a^b x^k \cdot \delta \cdot \frac{ds}{dt}\, dt = \int_a^b x^k \cdot \delta \cdot \sqrt{\left(\frac{dx}{dt}\right)^2 + \left(\frac{dy}{dt}\right)^2}\, dt$$
$$(7\text{-}121)$$
$$M_{ky} = \int_a^b y^k \cdot \delta \cdot \frac{ds}{dt}\, dt = \int_a^b y^k \cdot \delta \cdot \sqrt{\left(\frac{dx}{dt}\right)^2 + \left(\frac{dy}{dt}\right)^2}\, dt$$

But the value of these moments does not depend on the particular parametrization chosen (as long as s increases with t).

We can now define the center of mass and radii of gyration about the
axes, exactly as in Section 7-11. For example

$$M_0 \bar{x} = \int_0^L x\, \delta(s)\, ds, \qquad M_0 \bar{y} = \int_0^L y\, \delta(s)\, ds \qquad (7\text{-}122)$$

where, as usual, M_0 is the total mass:

$$M_0 = \int_0^L \delta(s)\, ds$$

We can then verify that $(\bar{x}, \bar{y})$ has a location independent of the choice of axes (Problem 14 below). For the second moments the parallel axis theorem continues to hold true (Problem 15 below).

We can also consider several distributions of mass on different curves. The kth moment of the combined distribution is defined to be the sum of the kth moments of the separate distributions. We can then verify that the center of mass of the whole distribution is the same as that of the set of particles obtained by concentrating each mass at the center of mass.

The *centroid of a curve*, or system of curves, is defined to be the center of mass of a distribution of constant density δ (homogeneous distribution).

THEOREM 4 (*SECOND THEOREM OF PAPPUS*). *Let C be a piecewise smooth simple path in the xy-plane, on which $y \geq 0$. Let S be the surface area of the surface of revolution obtained by rotating C about the x-axis. Then S equals the length of C times the length of the circular path through which the centroid of C is rotated:*

$$S = L \cdot 2\pi\bar{y} \qquad (7\text{-}123)$$

The proof is left as a problem (Problem 16 below).

All the theory of moments can be extended to three-dimensional space. Illustrations are given in Problems 17 to 22 below.

PROBLEMS

1. Find the center of mass for the following mass distributions in the xy-plane:
   (a) Mass 1 at $(3, 0)$, mass 2 at $(2, 4)$, mass 7 at $(-1, -3)$.
   (b) Mass 5 at $(3, 2)$, mass 2 at $(1, 1)$, mass 1 spread with constant density over the interior of the circle $(x - 5)^2 + (y - 3)^2 = 1$.
   (c) Mass fills the region below the curve $y = 4 - x^2$, $0 \leq x \leq 2$ with density $\delta = 3x$.
   (d) Mass fills the region below the curve $y = e^x$, $0 \leq x \leq 1$, with density $\delta = x^2$.
2. Find the centroid of each of the following plane regions:
   (a) A rectangle. (b) A triangle. [Do right triangle first. See hint for (i).]
   (c) A semicircle. (d) A quarter-circle.
   (e) The region below $y = (b/a)\sqrt{a^2 - x^2}$, $-a \leq x \leq a$, (semi-ellipse).
   (f) The region below $y = 4 - x^2$, $-2 \leq x \leq 2$.
   (g) The region enclosed by the curve $x = 2\cos t - \sin t$, $y = 2\sin t$, $0 \leq t \leq 2\pi$.
   (h) The region enclosed by the curve: $x = t^2$, $y = t^3$, $-1 \leq t \leq 1$; $x = \sqrt{2}\cos[t - 1 + (\pi/4)]$, $y = \sqrt{2}\sin[t - 1 + (\pi/4)]$, $1 \leq t \leq 1 + (3\pi/2)$.
   (i) The rectangle with vertexes $(\pm 1, 0)$, $(\pm 1, 2)$ combined with a triangle with vertexes $(\pm 1, 2)$ and $(0, 3)$. [*Hint.* Centroid of triangle is where medians meet.]
3. Let a homogeneous mass distribution be given in a region enclosed by the curves

$$r = f(\theta), \qquad \alpha \leq \theta \leq \beta; \qquad \theta = \alpha; \qquad \theta = \beta.$$

Let f be continuous and positive, except possibly for $f(\alpha) = 0$ and $f(\beta) = 0$; let $\alpha < \beta \le \alpha + 2\pi$. Show from (7-105) that the corresponding kth moments are

$$M_{kx} = \frac{\delta}{k+2} \int_\alpha^\beta r^{k+2} \cos^k \theta \, d\theta, \qquad M_{ky} = \frac{\delta}{k+2} \int_\alpha^\beta r^{k+2} \sin^k \theta \, d\theta$$

4. Apply the result of Problem 3 to find the centroid of the region specified:
   (a) A circular sector of radius a and angle γ.
   (b) The region enclosed by the cardioid $r = 1 + \sin \theta$.
   (c) The region enclosed by the curves $r = e^\theta$, $0 \le \theta \le \pi$, $\theta = 0$, $\theta = \pi$.

5. Find the moment of inertia about the line specified of a homogeneous mass distribution in the region specified:
   (a) A right triangle about a side.
   (b) A semicircle about its bounding diameter.
   (c) A circular sector of radius a and angle γ about its bisector.
   (d) A rectangle about a diagonal.

6. Apply the Parallel Axis Theorem to obtain the moment of inertia about the axis specified of a homogeneous mass distribution in the given region. In each case use the moment of inertia given.
   (a) A rectangle of sides a, b about a line parallel to side a through the centroid. Use moment of inertia about side a: $M_0 b^2/3$.
   (b) A circle of radius a about a tangent line. Use moment of inertia about a diameter: $M_0 a^2/4$.
   (c) The circle of part (b) about a line that is the perpendicular bisector of a radius.

7. Evaluate the following with the aid of theorems of Pappus:
   (a) The volume of a torus.
   (b) The volume of a solid of revolution obtained by rotating a square region of side a about a line b units from the center, $b > a/\sqrt{2}$.
   (c) The centroid of a semicircular region [see Problem 2(c)].
   (d) The centroid of a triangular region.
   (e) The surface area of a torus.
   (f) The surface area of the solid of part (b).

8. Find the center of mass for each of the following mass distributions on curves.
   (a) The semicircle $y = \sqrt{a^2 - x^2}$, $-a \le x \le a$, density constant.
   (b) The quarter-circle $y = \sqrt{a^2 - x^2}$, $0 \le x \le a$, density constant.
   (c) The circle $x = 2 \cos t$, $y = 2 \sin t$, $0 \le t \le 2\pi$ with $\delta = 2t$.
   (d) The straight line segment $x = 5 - 4t$, $y = 4 + 3t$, $0 \le t \le 1$, with $\delta = t^3$.

9. Show that Equations (7-104) and (7-105) give the correct moments for a region as in Figure 7-53. (*Hint.* See Problem 10 following Section 7-4.)

10. (a) Show that, if P_c satisfies (7-101), it also satisfies (7-100) for every unit vector $\mathbf{u}$.
    (b) Show that, if P_c satisfies (7-101), then for a new origin O' we have

$$M_0 \overrightarrow{O'P_c} = \sum_{l=1}^n m_l \overrightarrow{O'P_l}$$

    Thus the location of P_c is independent of the choice of origin.

11. (a) Show that the formulas (7-104), (7-105), for $k = 1$, and (7-106) lead to the following equation for the center of mass $P_c : (\bar{x}, \bar{y})$:

$$M_0 \overrightarrow{OP_c} = \frac{\delta}{3} \int_a^b (\mathbf{r}^\perp \cdot \mathbf{v}) \mathbf{r} \, dt$$

and, hence, conclude that the location of P_c is not affected by the choice of directions for the coordinate axes.

(b) Let a new origin be chosen at (h, k), so that $x_1 = x - h$, $y_1 = y - k$ are coordinates relative to the new origin as in Section 6-5 (translation of axes), and on a given path $x_1 = x_1(t) = x(t) - h$, $y_1 = y_1(t) = y(t) - k$. In the new coordinates (7-104) and (7-106) give

$$M_0 \bar{x}_1 = \frac{\delta}{3} \oint_C x_1^2 \, dy_1 - x_1 y_1 \, dx_1$$

Show that the right side equals

$$\frac{\delta}{3} \oint_C x^2 \, dy - xy \, dx - M_0 h = M_0(\bar{x} - h)$$

and, hence, that $\bar{x}_1 = \bar{x} - h$. This result and the analogous one for $\bar{y}$ show that the location of the center of mass is unaffected by a translation of axes.

12. (a) Prove the parallel axis theorem (7-107).

‡(b) Prove the parallel axis theorem (7-107′). [Hint. Use the relations:

$$M_0 = (\delta/2) \oint_C x \, dy - y \, dx, \qquad M_0 \bar{x} = (\delta/3) \oint_C x^2 \, dy - xy \, dx]$$

13. Prove the Pappus theorem (7-110) for the case of a region enclosed by a piecewise smooth simple closed curve C. [Hint. The volume can be obtained from (7-62), the location of the centroid from (7-106).]

14. From (7-122) obtain a single equation for $M_0 \overrightarrow{OP}_c$ and, thus, show that P_c has a location independent of direction of axes. Also show from the vector formula that the location of P_c is unaffected by a change of origin (as in Problem 10).

15. Formulate the parallel axis theorem for the case of a mass distribution on a curve. Choose the given axis as the y-axis and prove the theorem.

16. Prove the Pappus theorem (7-123). [Hint. The surface area is given by (7-83), $\bar{y}$ by (7-122).]

17. *Centroid of solid of revolution.* Let the region below the curve $y = f(x)$, $a \leq x \leq b$, where f is continuous and nonnegative, be rotated about the x-axis. Justify the assertion that the centroid of the solid obtained must lie on the x-axis and have x-coordinate $\bar{x}$, where

$$V\bar{x} = \pi \int_a^b x[f(x)]^2 \, dx$$

18. Find the centroid of the solid of revolution specified (see Problem 17): (a) a hemisphere, (b) a right circular cone, (c) the solid obtained by rotating the region beneath the curve $y = a^2 - x^2$, $0 \leq x \leq a$, about the x-axis, (d) the solid obtained by rotating the region of part (c) about the y-axis.

19. Set up an appropriate integral and deduce that the centroid of a right square pyramid is on the altitude, one fourth of the way from the base to the apex.

20. *Moment of inertia of solid of revolution.* Let a region as in Problem 17 be rotated about the y-axis. Show by the method of cylindrical shells (Section 7-5) that the homogeneous solid obtained has moment of inertia about the y-axis given by

$$2\pi\delta \int_a^b x^3 f(x) \, dx$$

21. (a) Use the result of Problem 20 to show that the moment of inertia of a homogeneous right cylinder of mass M_0 and radius a about its axis is $M_0 a^2/2$.

(b) Use the result of part (a) to show that, for the homogeneous solid of revolution of Problem 17, the moment of inertia about the x-axis is given by

$$\frac{1}{2}\pi\delta\int_a^b [f(x)]^4\, dx$$

22. (a) to (d) Use the results of Problems 20 and 21 to find the moment of inertia about the axis of revolution of the homogeneous solids of revolution in Problem 18.

†7-13 OTHER APPLICATIONS OF INTEGRATION

The concept of integration finds wide application in the sciences. The typical reasoning leading to an integral is as follows: A simple model of a physical process leads us to calculate a sum $u_1 + \cdots + u_n$. A more precise model leads us to represent each u_i as $f(\xi_i)\, \Delta_i x$, where $f(x)$ is a function defined on an interval $[a, b]$ and $f(\xi_i)\, \Delta_i x$ is as in the definition of the integral. We then "pass to the limit," so that

$$u_1 + \cdots + u_n = f(\xi_1)\, \Delta_1 x + \cdots + f(\xi_n)\, \Delta_n x \longrightarrow \int_a^b f(x)\, dx$$

In some cases, we can say that the final result, a definite integral, is a more precise value for the quantity sought. In others, we can only say it is more convenient to work with, even though greatest precision is obtained with a sum (with many terms). For example, the mass of a solid is computed as an integral of density. However, the solid is more accurately described as a finite collection of molecules. The density concept thus appears as an idealization. For the mass per unit of volume would be zero for all small volumes except for those containing molecules!

We here discuss briefly some typical applications. Their purpose is to illustrate the great variety of situations in which integrals arise.

(a) **Work.** In Section 4-20 we showed that, for a particle of mass m moving along the x-axis from $x = a$ to $x = b$, subject to a force $F = F(x)$, the work done by the force F is

$$W = \int_a^b F(x)\, dx \qquad\qquad (7\text{-}130)$$

We now consider a particle of mass m moving on a path in the xy-plane, as in Figure 7-56. Let the path be given by parametric equations

$$x = f(t), \qquad y = g(t), \qquad a \le t \le b$$

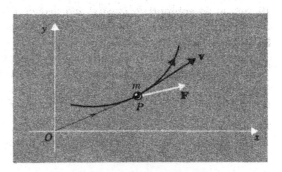

Figure 7-56 Work for motion on a path.

where we interpret t as time. We assume that f and g have continuous derivatives, so that arc length s can be defined on the path, increasing with t, with $s = 0$ for $t = a$ and $s = L$ for $t = b$. Let the particle be acted on by a force $\mathbf{F}$ (a vector), varying with position on the path, so that $\mathbf{F}$ can be expressed as a continuous function of t and, hence, also of s. We introduce the position vector $OP = \mathbf{r} = x\mathbf{i} + y\mathbf{j}$, so that $\mathbf{r} = f(t)\mathbf{i} + g(t)\mathbf{j}$. The velocity vector $\mathbf{v} = f'(t)\mathbf{i} + g'(t)\mathbf{j}$ is then tangent to the path at the point $\mathbf{r}(t)$. We assume that $\mathbf{v} \neq 0$ throughout. At each point $\mathbf{F}$ has then a tangential component

$$F_{\text{tang}} = \mathbf{F} \cdot \frac{\mathbf{v}}{|\mathbf{v}|}$$

Generalizing the case of the motion on a straight line, we are led to approximate the work done in moving a short distance Δs along the path by $F_{\text{tang}}(\xi) \, \Delta s$, where ξ is a value of s in the little piece considered. By summing and passing to the limit, we are led to define the work to be

$$W = \int_0^L F_{\text{tang}}(s) \, ds \tag{7-131}$$

When $\mathbf{F}$ is expressed in terms of t, we can change to t as variable of integration and write

$$W = \int_a^b F_{\text{tang}}(t) \frac{ds}{dt} \, dt \tag{7-131'}$$

Since $ds/dt = |\mathbf{v}|$ and $F_{\text{tang}} = (\mathbf{F} \cdot \mathbf{v})/|\mathbf{v}|$, we can write simply

$$W = \int_a^b [\mathbf{F}(t) \cdot \mathbf{v}(t)] \, dt \tag{7-131''}$$

From (7-131'') we can prove the important rule of mechanics: *Work equals gain in kinetic energy.* For by Newton's second law, $\mathbf{F} = m\mathbf{a} = m \, d\mathbf{v}/dt$, if $\mathbf{F}$ is the total force applied, so that

$$W = \int_a^b m \frac{d\mathbf{v}}{dt} \cdot \mathbf{v} \, dt = \frac{1}{2} m \int_a^b \frac{d}{dt} (\mathbf{v} \cdot \mathbf{v}) \, dt$$

$$= \frac{1}{2} m \int_a^b \frac{d}{dt} |\mathbf{v}|^2 \, dt = \frac{1}{2} m|\mathbf{v}(t)|^2 \Big|_a^b$$

But $m|\mathbf{v}|^2/2$ is just the kinetic energy of the particle. Thus the rule is proved.

(b) Impulse and Momentum. For the particle moving in the xy-plane, as above, the impulse of the force $\mathbf{F}$ is defined to be the vector:

$$\int_a^b \mathbf{F}(t) \, dt$$

This can again be obtained by a limiting process, starting with the case of a constant force. By applying Newton's second law as above, we can write:

$$\text{Impulse} = \int_a^b \mathbf{F}(t) \, dt = \int_a^b m \frac{d\mathbf{v}}{dt} \, dt = m\mathbf{v}(t) \Big|_a^b$$

Thus we obtain the rule of mechanics: *Impulse is gain in linear momentum mv.* This relation is valuable in analyzing collisions of particles.

(c) **Potential Energy.** We consider only the case of a particle moving on a line as above, with force $F = F(x)$. The potential energy associated with F is then defined as minus the indefinite integral of F with respect to x. However, the arbitrary constant is usually fixed in some manner, so that the potential energy becomes a function $U(x)$ for which

$$\frac{dU}{dx} = -F(x)$$

From (7-130) we obtain

$$W = \int_a^b F\,dx = -\int_a^b \frac{dU}{dx}\,dx = -U(x)\Big|_a^b$$

Thus the work done is the loss in potential energy. Hence

gain in kinetic energy = loss in potential energy

or

change in (kinetic energy + potential energy) = 0

Thus the "total energy" remains constant for every motion. This is *the law of conservation of energy.* We stress that the conclusion is valid only when F depends only on position x, and not when F depends on velocity (as in the case of friction). Also for motion in a plane a potential energy need not exist (see Chapter 12).

(d) **Force and Pressure.** The total force exerted by the weight of the earth's atmosphere on a certain portion of the earth's surface can be regarded as the total of forces acting on tiny portions of the region. We thus are led naturally to the concept of pressure as force per unit area, just as density is mass per unit volume. When the pressure is constant, we can compute the total force by simply multiplying the pressure by the area. When the pressure varies over the region, we are forced to subdivide the region into pieces over which the pressure is approximately constant and add the forces for the pieces.

An interesting example is provided by the force of a fluid in a container against a vertical wall of the container (Figure 7-57). By hydrostatics, the pressure at a given depth below the surface depends only on the depth and, for an incompressible fluid, is simply proportional to the depth, being equal to the weight per unit area of a column of fluid of that depth. Thus, if x and y axes are introduced as in Figure 7-57, so that x is the depth, then

Pressure $= wx$

where w is the weight per unit volume. Let the vertical wall be precisely the region under a curve $y = f(x)$, $0 \le x \le h$. Then we cut this region into pieces by subdiving the x-interval, exactly as in finding the area under the curve. Between x_{i-1} and x_i the pressure is approximately constant, hence, the force equals $\Delta_i A \cdot w\xi_i$, where $\Delta_i A$ is the area and $x_{i-1} < \xi_i < x_i$. Now

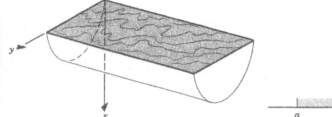

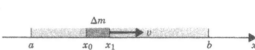

Figure 7-57 Force due to fluid pressure.   **Figure 7-58** Motion of a rubber band.

$\Delta_i A = f(\eta_i)\, \Delta_i x$, $x_{i-1} \le \eta_i \le x_i$ and consequently, we are led to the sum

$$\sum_{i=1}^{n} w \xi_i f(\eta_i)\, \Delta_i x$$

Passage to the limit gives the desired expression:

$$\text{Total force} = w \int_a^b x f(x)\, dx$$

We observe that the integral is the same as one encountered earlier in finding the centroid (Section 7-11); in fact, it equals $A\bar{x}$, where A is the area of the region and $(\bar{x}, \bar{y})$ is the centroid. Hence the total force is the same as if the wall were horizontal, at a depth equal to that of the centroid.

(e) **Kinetic Energy of a Continuous Medium.** We consider a rubber band being stretched back and forth along the x-axis. At any instant the portion of the band between two nearby points x_0 and x_1 can be idealized as a particle of mass Δm moving with velocity v (measured positive when the motion is in the direction of increasing x, as in Figure 7-58). Therefore, we are led to assign a kinetic energy $(1/2)v^2\, \Delta m$ to this piece of the band. At the instant the band will also have a density $\delta(x)$ so that $\Delta m = \delta(\xi)\, \Delta x$, where $\Delta x = x_1 - x_0$ and $x_0 < \xi < x_1$. Accordingly, the total kinetic energy is a sum of form

$$\frac{1}{2} \sum_{i=1}^{n} \delta(\xi_i)[v(\eta_i)]^2\, \Delta_i x$$

where we have subdivided the whole interval $[a, b]$ as usual and $x_{i-1} < \eta_i < x_i$, $x_{i-1} < \xi_i < x_i$. Accordingly, in the limit we obtain an integral

$$\frac{1}{2} \int_a^b \delta(x)[v(x)]^2\, dx$$

This is the total kinetic energy at the instant considered. As the motion continues, $\delta(x)$ will change, $v(x)$ will change, and even the limits of integration a and b will change.

(f) **Conduction of Heat.** We consider a rod along the x-axis and assume temperature T varies along the rod, so that $T = T(x)$. We assume the surrounding medium is at constant temperature T_0. One model of heat conduction states that heat flows across the boundary surface between two media at a rate proportional to the temperature difference of the media. Hence, each little piece of the rod, between x_0 and x_1, will lose heat at a rate proportional to $T(\xi) - T_0$, $x_0 < \xi < x_1$. The loss of heat for the piece must also be proportional to the length. We are therefore led to an expression $c[T(\xi) - T_0]\,\Delta x$, and finally to an integral for the rate of heat loss:

$$\int_a^b c[T(x) - T_0]\,dx$$

Again, this is an instantaneous value. In general, $T(x)$ and T_0 will vary as time increases.

(g) **Cost of Building a Road.** We are familiar with quotations of so many millions of dollars per mile for building a highway. The cost for a given piece in general will depend on the landscape, geological formations, accessibility of supplies and labor, and many other factors. However, for a given highway on which distance s varies from 0 to L, the cost of the short piece from s_0 to s_1 can be expressed as $f(\xi)\,\Delta s$, with $s_0 < \xi < s_1$. Here $f(s)$ is a local cost density: the cost per mile near the particular value s. The total cost then appears as an integral

$$\int_0^L f(s)\,ds$$

(h) **Cost of Heating a House.** The cost of heating a given house depends principally on the outdoor temperature T; there are other factors such as wind velocity (the "chill factor"), which we shall ignore. For a given value of T, we then have a cost rate f, equal to the number of cents per hour needed to pay for maintaining the indoor temperature at a given value. Hence, $f = f(T)$. As time increases, T will vary, so that we must write: $T = T(t)$. The cost for a small time interval, from t_0 to t_1, is given by an expression $f[T(\xi)]\,\Delta t$, with $t_0 < \xi < t_1$ and $\Delta t = t_1 - t_0$. Therefore, the total cost from $t = a$ to $t = b$ is

$$\int_a^b f[T(t)]\,dt$$

(i) **Electric Energy Consumed.** The rate at which any unit using electricity (a motor, a house, a factory, a city) draws energy in that form is the power P, usually measured in watts or kilowatts. One watt equals ten million ergs (energy) per second. For a given unit, we have $P = P(t)$, where t is time. Typically $P(t)$ is only piecewise continuous, since $P(t)$ jumps each time a switch is opened or closed. However, in an interval of continuity, we obtain an integral

$$\int_a^b P(t)\,dt$$

for the total energy consumed. If there are jumps within the interval, the integral still has meaning and is simply the sum of the integrals over the intervals between jumps; this clearly corresponds to the total energy consumption.

The detailed reasoning leading to the integral is the same as above.

(j) **Probability.** We give one example to illustrate the occurrence of integrals in probability theory. Imagine a person throwing darts at a circular target of radius a, with an aim so poor that we are inclined to say that he is as likely to strike any one portion of the target as any other. To be more precise, we should then say that the probability of hitting a portion of the target should be proportional to its area.

Now let coordinates (x, y) be introduced so that the edge of the target is the circle $x^2 + y^2 = a^2$. The probability of striking the target at a point to the left of a particular line $x = x_1$ should be proportional to the corresponding area (shaded in Figure 7-59). Hence it should be

$$F(x_1) = 2k \int_{-a}^{x_1} \sqrt{a^2 - x^2} \, dx$$

where k is a constant. When $x_1 = a$, the probability should be 1; that is, we assume that we are sure the person will at least hit the target. Therefore

$$1 = 2k \int_{-a}^{a} \sqrt{a^2 - x^2} \, dx = \pi k a^2$$

Thus, $k = 1/(\pi a^2)$, and we can write

$$F(x_1) = \int_{-a}^{x_1} \frac{2\sqrt{a^2 - x^2}}{\pi a^2} \, dx = \int_{-a}^{x_1} f(x) \, dx$$

where $f(x) = F'(x) = 2\sqrt{a^2 - x^2}/(\pi a^2)$. Similarly, the probability of striking at a point (x, y) with $x_1 \leq x \leq x_2$ is

$$\text{Prob}(x_1 \leq x \leq x_2) = F(x_2) - F(x_1) = \int_{x_1}^{x_2} f(x) \, dx \qquad (7\text{-}132)$$

The function $f(x)$ is called the *probability density function* of our problem, $F(x)$ the *distribution function*. They are graphed in Figures 7-60 and 7-61. If the person's aim is better, the probability density for values close to 0 will be increased. In actual experience we find that the probability density function is close to a *normal* density (a bell-shaped curve), as in Figure 7-62. Here

$$f(x) = \frac{e^{-x^2/2\sigma^2}}{\sqrt{2\pi}\,\sigma}$$

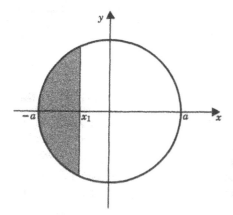

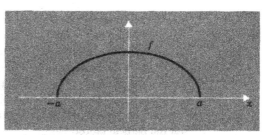

Figure 7-59 Probability of hitting part of a target.

Figure 7-60 Probability density function for target problem.

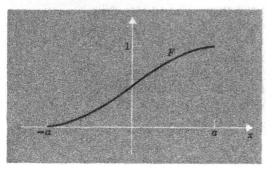

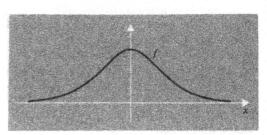

Figure 7-61 Distribution function for target problem.

Figure 7-62 Normal probability density function.

and σ (the standard deviation) is a positive constant. The smaller the σ, the more accurate the aim, the larger f is for small $|x|$. For all choices of the probability density function, Equation (7-132) remains valid; however, in general x_1 and x_2 need not be restricted to a finite interval. The rule can be stated: the probability that x is between x_1 and x_2 equals the area under the graph of the probability density function between x_1 and x_2.

Remarks. The examples show the very great variety of situations in which the integral appears, and we could easily give many more illustrations of this variety. In all cases, we are concerned basically with a summation process. A finite sum will usually represent the quantity sought with the accuracy needed. However, when a very large number of terms must be added, it is simpler to pass to the limit and work with the integral. This process leads us to construct various models for physical processes in which the integral is treated as the "true" expression for a physical quantity. The justification is only partly our belief that the laws of nature refer to the limiting case in which we integrate rather than sum. Very often, it is simply more convenient to use integrals, conceptually and for computations. For the latter, the main advantage lies in the techniques of the calculus, as in Part II of Chapter 4. When these techniques prove inadequate, we are usually forced to use a numerical method (see Sections 4-25 and 7-21). Thus, in effect, we are going back to the sums from which we started. Section 7-21 describes these numerical methods.

PROBLEMS

1. A particle moves on the x-axis subject to a force of $-4x$ lb, where x is in feet. Find the work done by the force in each of the following cases:
   (a) The particle moves from $x = 3$ to $x = 10$.
   (b) The particle moves from $x = 5$ to $x = -5$.
   (c) The particle moves from $x = 0$ to $x = 3$ and back to $x = 0$.
2. A particle of weight 2 lb slides on a horizontal surface subject only to a force of friction of 1.5 lb. If initially the particle has a speed of 10 ft per sec, how far will the particle move?
3. (a) Show that, for a particle moving in the xy-plane subject to a constant force $\mathbf{F}$, the work done by $\mathbf{F}$ when the particle moves from (x_1, y_1) to (x_2, y_2) is

the same for all paths from (x_1, y_1) to (x_2, y_2). [*Hint.* Use (7-131″) and notice that, when $\mathbf{F}$ is constant, $\mathbf{F} \cdot \mathbf{v}(t) = (d/dt)(\mathbf{F} \cdot \mathbf{r})$.]

(b) Show that, for a particle moving in a vertical plane near the earth's surface, the gravity force $\mathbf{F}$ is approximately constant and that, if the particle moves from a position of altitude h_1 above the surface to a position of altitude h_2, then the work done by gravity is approximately $mg(h_1 - h_2)$, where m is the mass of the particle and g is the acceleration of gravity.

4. Let two particles, of masses m_1 and m_2, respectively, move in the xy-plane subject only to equal but opposite forces (action and reaction) $\mathbf{F}$ and $-\mathbf{F}$, varying with time.

(a) Show that the center of mass of the two particles moves on a straight line at constant velocity. (*Hint.* Apply Newton's law—force equals mass times acceleration—to each particle and add the two equations.)

(b) At time t_0 the particles are observed to have velocities v_1, v_2, respectively. The particles are then observed to collide, so that at time t_1 the first particle has velocity $-v_1$. What is the velocity of the second particle at time t_1? (*Hint.* Assume the total momentum is unaffected by the collision.)

5. A particle of mass 1 gram moves on the x-axis subject to a force $10 \sin^3 t$ dynes, where t is in seconds. For $t = 0$ the particle has velocity 5 cm per sec. What is the velocity for $t = 10$?

6. A particle of mass m moves on the x-axis subject to a spring force $-kx$.

(a) Show that

$$\frac{mv^2}{2} + \frac{kx^2}{2} = \text{const} = E \text{ (total energy)}$$

(b) If initially $x = 0$ and $v = v_0$, show that x can never exceed $|v_0| \sqrt{m/k}$.

7. A spring is tested and found to be "nonlinear"; that is, the spring does not obey Hooke's law: force is proportional to displacement from equilibrium. Instead the force is found to be proportional to the square of the displacement. Find the results analogous to (*a*) and (*b*) of Problem 6 for the nonlinear spring.

8. The potential energy of a molecule influenced by a second molecule is found to be

$$\frac{a}{r^{12}} - \frac{b}{r^6}$$

where a and b are positive constants and r is the distance between the molecules. If the first molecule moves on the positive x-axis and the second molecule is fixed at the origin, show that there is a unique value x_0 for which the first molecule is attracted to the second for $x > x_0$ and repelled from the origin for $x < x_0$.

9. A horizontal beam, along the x-axis from $x = 0$ to $x = L$, is unevenly loaded (Figure 7-63). The force on the piece between x_0 and x_1 is found to be representable as $f(\xi) \, \Delta x$, where $\Delta x = x_1 - x_0$, $x_0 < \xi < x_1$, and $f(x) = k/(1 + e^x)$ where k is a constant. Find the total load on the beam.

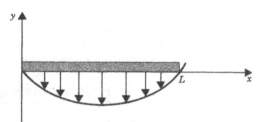

Figure 7-63 Load on a beam.

10. The wall of a dam is vertical and representable as the region below a curve $y = f(x)$ as in (d) above. Let the fluid density (weight of unit volume) be w. Find the total force on the wall for each of the following cases:

    (a) $f(x) = a^2 - x^2$, $0 \le x \le a$ (b) $f(x) = b \cos cx$, $0 \le x \le \frac{1}{2}\pi/c$

    (c) $f(x) = c_1 = $ const for $0 \le x \le h_1$, $f(x) = c_1(x - h_2)/(h_1 - h_2)$

       for $h_1 \le x \le h_2$

‡11. A homogeneous elastic band is stretched initially from $x = 0$ to $x = 2$ (in cm) and has total mass of 3 gr. It then is stretched and relaxed so that the particle of the band at position x_0 for $t = 0$ occupies the position $x = x_0(1 + 0.1 \sin t)$ at time t (in sec). Find the total kinetic energy at time t. (*Hint.* Notice that the density also varies with t.)

12. Give an example of an integral occurring naturally in each of the following contexts:

    (a) Consumption of fuel by a jet plane.

    (b) Total earnings of a company over a year.

    (c) Total precipitation in a given small geographical area in a year.

    (d) Total number of phone calls handled by a telephone exchange over a 24-hour period.

    (e) Total amount of salt in a river near the ocean.

    (f) The number of cars on the New York Thruway at a given time.

13. The grades on a typical examination fall in the interval $0 \le x \le 100$ and have a probability density function $f(x)$ and distribution function $F(x)$ so that Equation (7-132) holds true. Choose a reasonable function f and use it to compute the probabilities for the following cases:

    (a) $90 \le x \le 100$ (b) $80 \le x \le 90$ (c) $60 \le x \le 61$ (d) $0 \le x \le 100$

7-14 IMPROPER INTEGRALS

Up to this point we have given a meaning to the definite integral $\int_a^b f(x)\,dx$ only when a and b are finite and f is continuous or, more generally, piecewise continuous. We shall now show that a meaning can be given in certain other cases, in particular, when f has an infinite discontinuity (Figure 7-64) or when $b = \infty$ (Figure 7-65). In these cases and in the others to be considered here, the integral is called an *improper integral.*

We first consider an integral $\int_a^\infty f(x)\,dx$, where f is continuous for $x \ge a$. For the case of Figure 7-65, f is positive and, hence, we can try to interpret the integral as area. Since f appears to approach zero rapidly as $x \to \infty$, we should obtain most of the area by stopping at a large value c. This suggests our definition for integrals of this type:

$$\int_a^\infty f(x)\,dx = \lim_{c \to \infty} \int_a^c f(x)\,dx \qquad (7\text{-}140)$$

provided that the limit exists.

When the limit does exist (as a finite number) the improper integral

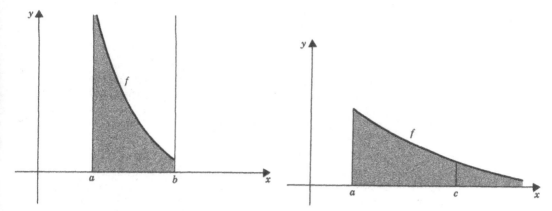

Figure 7-64 Function with infinite discontinuity. **Figure 7-65** Integration over infinite interval.

is said to be *convergent;* otherwise it is called *divergent.* This terminology will be used for all the improper integrals of this section.

EXAMPLE 1. $\int_a^\infty \frac{1}{x^p}\,dx$, $a > 0$. Here we consider

$$\lim_{c \to \infty} \int_a^c \frac{1}{x^p}\,dx = \lim_{c \to \infty} \frac{-1}{(p-1)x^{p-1}}\Big|_a^c$$

for $p \neq 1$. The limit exists for $p > 1$:

$$\lim_{c \to \infty} \frac{-1}{(p-1)x^{p-1}}\Big|_a^c = \lim_{c \to \infty} \frac{1}{p-1}\left(\frac{1}{a^{p-1}} - \frac{1}{c^{p-1}}\right) = \frac{1}{(p-1)a^{p-1}}$$

and there is no limit for $p < 1$. For $p = 1$, we have

$$\int_a^\infty \frac{dx}{x} = \lim_{c \to \infty} \int_a^c \frac{dx}{x} = \lim_{c \to \infty} \ln x \Big|_a^c = \lim_{c \to \infty} (\ln c - \ln a) = \infty$$

Accordingly, there is no limit in this case. We conclude that

$$\int_a^\infty \frac{dx}{x^p} = \begin{cases} \dfrac{1}{(p-1)a^{p-1}} & \text{for} \quad p > 1 \quad \text{(convergent)} \\ \text{no value} & \text{for} \quad p \leq 1 \quad \text{(divergent)} \end{cases}$$

We next consider an integral $\int_{-\infty}^b f(x)\,dx$, where f is continuous for

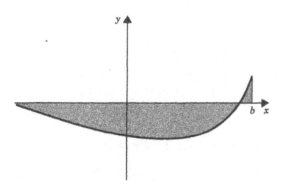

Figure 7-66 $\int_{-\infty}^b f(x)\,dx$

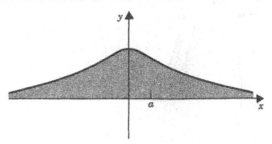

Figure 7-67 $\int_{-\infty}^{\infty} f(x)\, dx$

$x \leq b$, as suggested in Figure 7-66. By analogy with the previous case, we define

$$\int_{-\infty}^{b} f(x)\, dx = \lim_{k \to -\infty} \int_{k}^{b} f(x)\, dx \qquad (7\text{-}141)$$

Remark. For the two types of integrals thus far considered, we have the rules

$$\int_{a}^{\infty} f(x)\, dx = \int_{a}^{k} f(x)\, dx + \int_{k}^{\infty} f(x)\, dx \qquad (7\text{-}142)$$

$$\int_{-\infty}^{b} f(x)\, dx = \int_{-\infty}^{k} f(x)\, dx + \int_{k}^{b} f(x)\, dx \qquad (7\text{-}143)$$

provided that f is continuous on the intervals concerned. More precisely, (7-142) is understood to mean that the integral from a to ∞ converges if, and only if, the integral from k to ∞ converges, and when both converge they are related by (7-142). There is a similar statement for (7-143). To prove (7-142), for example, we write

$$\int_{a}^{c} f(x)\, dx = \int_{a}^{k} f(x)\, dx + \int_{k}^{c} f(x)\, dx$$

and let $c \to \infty$. The integral from a to k is a constant. Hence, convergence of one improper integral implies convergence of the other, and validity of (7-142).

EXAMPLE 2

$$\int_{-\infty}^{0} e^{x}\, dx = \lim_{c \to -\infty} \int_{c}^{0} e^{x}\, dx = \lim_{c \to -\infty} e^{x} \Big|_{c}^{0} = \lim_{c \to -\infty} (1 - e^{c}) = 1$$

The integral is convergent.

We next consider an integral $\int_{-\infty}^{\infty} f(x)\, dx$, where f is continuous for $-\infty < x < \infty$, as in Figure 7-67. We are here concerned with limit processes to the left and right. Accordingly, we break up the integral into two integrals and set

$$\int_{-\infty}^{\infty} f(x)\, dx = \int_{-\infty}^{a} f(x)\, dx + \int_{a}^{\infty} f(x)\, dx \qquad (7\text{-}144)$$

If *both* integrals on the right are convergent, the one on the left is called convergent and has the indicated value; if *either* of the integrals on the right is divergent, the one on the left is said to be divergent. The convergence or divergence, and the value for $\int_{-\infty}^{\infty} f(x)\, dx$ obtained in the former case, are unaffected by the choice of a. For let us suppose that we have convergence for one value a. Then, for $a' \neq a$,

$$\int_{a'}^{\infty} f(x)\, dx = \int_{a'}^{a} f(x)\, dx + \int_{a}^{\infty} f(x)\, dx$$

$$\int_{-\infty}^{a'} f(x)\, dx = \int_{-\infty}^{a} f(x)\, dx + \int_{a}^{a'} f(x)\, dx$$

by (7-142) and (7-143), where all four improper integrals converge. If we add the two equations, we obtain

$$\int_{-\infty}^{a'} f(x)\, dx + \int_{a'}^{\infty} f(x)\, dx = \int_{-\infty}^{a} f(x)\, dx + \int_{a}^{\infty} f(x)\, dx$$

Thus the convergence and value of the integral are unaffected by where we divide the infinite interval.

EXAMPLE 3

$$\int_{-\infty}^{\infty} \frac{dx}{x^2 + 1} = \int_{-\infty}^{0} \frac{dx}{x^2 + 1} + \int_{0}^{\infty} \frac{dx}{x^2 + 1}$$

$$= \lim_{k \to -\infty} \int_{k}^{0} \frac{dx}{x^2 + 1} + \lim_{c \to \infty} \int_{0}^{c} \frac{dx}{x^2 + 1} = \lim_{k \to -\infty} \operatorname{Tan}^{-1} x \Big|_{k}^{0} + \lim_{c \to \infty} \operatorname{Tan}^{-1} x \Big|_{0}^{c}$$

$$= \lim_{k \to -\infty} - \operatorname{Tan}^{-1} k + \lim_{c \to \infty} \operatorname{Tan}^{-1} c = (\pi/2) + (\pi/2) = \pi$$

Now we turn to the integral $\int_{a}^{b} f(x)\, dx$, where f is continuous for $a < x \leq b$ *but discontinuous at* $x = a$, as in Figure 7-64. In this case we define

$$\int_{a}^{b} f(x)\, dx = \lim_{c \to a+} \int_{c}^{b} f(x)\, dx \tag{7-145}$$

if the limit exists.

EXAMPLE 4. $\int_{0}^{b} 1/x^p\, dx$, where $b > 0$. For $p \neq 1$, we find (cf. Example 1 above) that

$$\lim_{c \to 0+} \int_{c}^{b} \frac{1}{x^p}\, dx = \lim_{c \to 0+} \frac{1}{p-1} \left(\frac{1}{c^{p-1}} - \frac{1}{b^{p-1}} \right) = \lim_{c \to 0+} \frac{1}{p-1} (c^{1-p} - b^{1-p})$$

and there is a limit for $p < 1$, equal to $b^{1-p}/(1 - p)$; for $p > 1$ there is no limit. For $p = 1$ there is no limit:

$$\lim_{c \to 0+} \int_{c}^{b} \frac{1}{x}\, dx = \lim_{c \to 0+} (\ln b - \ln c) = \infty$$

Hence

$$\int_{0}^{b} \frac{1}{x^p}\, dx = \begin{cases} \dfrac{b^{1-p}}{1 - p}, & p < 1 \quad \text{(convergent)} \\[2mm] \text{no value}, & p \geq 1 \quad \text{(divergent)} \end{cases}$$

Remark. If f has a limit as $x \to a+$, then f is piecewise continuous for $a \leq x \leq b$ and becomes continuous if we define $f(a)$ to be the limiting value; in this case the integral $\int_a^b f(x)\, dx$ exists even as a Riemann integral, as pointed out in Section 4-30. More generally, the Riemann integral exists if f is continuous for $a < x \leq b$ and bounded, $|f(x)| < K$, even though f may be discontinuous at $x = a$. Furthermore, in these cases (7-145) gives the same value as that given by the Riemann integral. These assertions are proved by the reasoning of Section 4-30.

EXAMPLE 5. Let f be the continuous piecewise linear function of Figure 7-68. Thus $f(x) = 0$ for $x = 1, 1/2, 1/4, \ldots, 1/2^n, \ldots, f(x) = 1$ for $x = 3/4, 3/8, \ldots, 3/2^n, \ldots$ and f is linear between the values previously given. Accordingly, f is bounded: $|f(x)| \leq 1$, f is continuous for $0 < x \leq 1$, f is discontinuous at $x = 0$, with no limit as $x \to 0$. Thus $\int_0^1 f(x)\, dx$ can be obtained either as a limit:

$$\int_0^1 f(x)\, dx = \lim_{c \to 0+} \int_c^1 f(x)\, dx \tag{7-146}$$

or as a Riemann integral:

$$\int_0^1 f(x)\, dx = \lim_{\text{mesh} \to 0} \sum_{i=1}^n f(\xi_i)\, \Delta_i x \tag{7-146'}$$

$f(0)$ being given some fixed value (say 0). For (7-146), since we know the limit exists, it is sufficient to take $c = 1/2^n$ and let $n \to \infty$ through integer values. Then

$$\int_c^1 f(x)\, dx = \int_{2^{-n}}^1 f(x)\, dx$$
$$= \int_{2^{-n}}^{2^{-n+1}} f(x)\, dx + \int_{2^{-n+1}}^{2^{-n+2}} f(x)\, dx + \cdots + \int_{2^{-1}}^1 f(x)\, dx$$
$$= \left(\frac{1}{4}\right) + \left(\frac{1}{8}\right) + \cdots + \frac{1}{2^{n+1}}$$

by addition of the areas of the triangles. Hence, by the formula for sum of a geometric progression (Problem 2 following Section 0-21),

$$\lim_{c \to 0+} \int_c^1 f(x)\, dx = \lim_{n \to \infty} \left(\frac{1}{4} + \frac{1}{8} + \cdots + \frac{1}{2^{n+1}}\right) = \lim_{n \to \infty} \left[\left(\frac{1}{2}\right) - \frac{1}{2^{n+1}}\right] = \frac{1}{2}$$

Therefore, the integral in question equals $\frac{1}{2}$.

Remark. The term "improper integral" would normally not be used in this case, since the Riemann integral exists.

For an integral $\int_a^b f(x)\, dx$, where f is continuous for $a \leq x < b$ but discontinuous at $x = b$, we define

$$\int_a^b f(x)\, dx = \lim_{c \to b-} \int_a^c f(x)\, dx \tag{7-147}$$

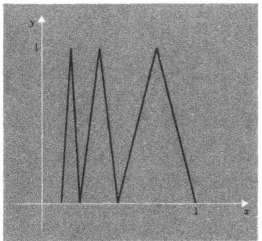

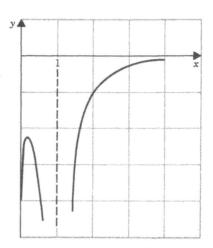

Figure 7-68 Bounded function f having a Riemann integral.

Figure 7-69 Function of Example 6.

The discussion is analogous to that for (7-145).

For an integral $\int_a^b f(x) \, dx$, where f is continuous for $a < x < b$ but discontinuous at both a and b, we proceed as in (7-144) and define

$$\int_a^b f(x) \, dx = \int_a^k f(x) \, dx + \int_k^b f(x) \, dx, \qquad a < k < b$$

If both integrals on the right converge, then so does the one on the left, and the value does not depend on the choice of k. Otherwise the given integral diverges.

For integrals of the form $\int_a^\infty f(x) \, dx$ or $\int_{-\infty}^b f(x) \, dx$, where f is continuous except at a or b, respectively, we again introduce an auxiliary point to reduce the integral to a sum of two integrals of the previous types. For an integral with several discontinuities, we introduce several auxiliary points for the same purpose. In all cases, we reduce the problem to several integrals, each involving a limit process. If all limits exist, the given integral converges; if one fails to exist, the given integral diverges.

EXAMPLE 6. $\int_0^\infty f(x) \, dx$ where

$$f(x) = \frac{d}{dx}\left(\frac{x^{1/3}}{x-1}\right) = \frac{-1-2x}{3x^{2/3}(x-1)^2}$$

The graph of f is sketched in Figure 7-69. There are discontinuities at $x = 0$ and $x = 1$, and we hence write

$$\int_0^\infty f(x) \, dx = \int_0^{\frac{1}{2}} f(x) \, dx + \int_{\frac{1}{2}}^1 f(x) \, dx + \int_1^2 f(x) \, dx + \int_2^\infty f(x) \, dx$$

The integrals on the right are of the types considered above and, hence,

$$\int_0^\infty f(x)\,dx = \lim_{a\to 0+}\int_a^{\frac{1}{2}} f(x)\,dx + \lim_{b\to 1-}\int_{\frac{1}{2}}^b f(x)\,dx + \lim_{c\to 1+}\int_c^2 f(x)\,dx$$

$$+ \lim_{k\to\infty}\int_2^k f(x)\,dx$$

$$= F(x)\Big|_{0+}^{\frac{1}{2}} + F(x)\Big|_{\frac{1}{2}}^{1-} + F(x)\Big|_{1+}^2 + F(x)\Big|_2^\infty$$

where the value $0+$ means $\lim_{x\to 0+} F(x)$ and so on, and $F(x) = x^{1/3}/(x-1)$.
Accordingly

$$F(x)\Big|_{0+}^{\frac{1}{2}} = F\left(\frac{1}{2}\right) - \lim_{x\to 0+} F(x) = -2^{2/3} - 0 = -2^{2/3}$$

$$F(x)\Big|_{\frac{1}{2}}^{1-} = \lim_{x\to 1-} F(x) - F\left(\frac{1}{2}\right) = -\infty + 2^{2/3} = -\infty$$

$$F(x)\Big|_{1+}^2 = F(2) - \lim_{x\to 1+} F(x) = 2^{1/3} - (\infty) = -\infty$$

$$F(x)\Big|_2^\infty = \lim_{x\to\infty} F(x) - F(2) = 0 - 2^{1/3} = -2^{1/3}$$

(see Table 2-3 in Section 2-11). Consequently, the second and third integrals are infinite, and the given integral diverges.

Properties of Improper Integrals. Many of the properties of the definite integral carry over to improper integrals. We formulate several of them for the integrals from a to ∞. Similar results hold true for the other types:

THEOREM 5. *Let $f(x)$ and $g(x)$ be continuous for $a < x < \infty$ and let the integrals $\int_a^\infty f(x)\,dx$ and $\int_a^\infty g(x)\,dx$ converge. If k_1 and k_2 are constants then*

$$\int_a^\infty [k_1 f(x) + k_2 g(x)]\,dx = k_1 \int_a^\infty f(x)\,dx + k_2 \int_a^\infty g(x)\,dx \quad (7\text{-}148)$$

In particular, the integral on the left converges. If $f(x) \le g(x)$ for $a \le x < \infty$, then

$$\int_a^\infty f(x)\,dx \le \int_a^\infty g(x)\,dx \quad (7\text{-}149)$$

PROOF. By the theorem on limits of a sum and product (Sections 2-7 and 2-10),

$$\int_a^\infty [k_1 f(x) + k_2 g(x)]\,dx = \lim_{b\to\infty}\int_a^b [k_1 f(x) + k_2 g(x)]\,dx$$

$$= \lim_{b\to\infty}\left\{ k_1 \int_a^b f(x)\,dx + k_2 \int_a^b g(x)\,dx \right\}$$

$$= k_1 \lim_{b\to\infty}\int_a^b f(x)\,dx + k_2 \lim_{b\to\infty}\int_a^b g(x)\,dx = k_1\int_a^\infty f(x)\,dx + k_2\int_a^\infty g(x)\,dx$$

Thus (7-148) is proved. Next, if $f(x) \leq g(x)$ for all x, then for all b

$$\int_a^b f(x) \, dx \leq \int_a^b g(x) \, dx$$

so that (7-149) follows by letting $b \to \infty$ (see Remarks at the end of Section 2-5).

The systematic study of improper integrals is closely related to the study of infinite series. The two topics are studied together in Chapter 8.

PROBLEMS

In Problems 1 to 24 determine convergence or divergence and give the value when the integral converges. The results of Sections 2–10, 2–11, and 6–14 will be found helpful.

1. $\int_1^\infty \dfrac{dx}{x(x+1)}$

2. $\int_0^\infty \dfrac{x \, dx}{3x^2 + 1}$

3. $\int_2^\infty \dfrac{x+2}{x^3 - x} \, dx$

4. $\int_1^\infty \dfrac{x-1}{x(x^2+1)} \, dx$

5. $\int_{-\infty}^{-1} \dfrac{15(x+2)}{x(x-3)(x-5)} \, dx$

6. $\int_{-\infty}^0 \dfrac{x}{(x^2+1)^{4/3}} \, dx$

7. $\int_0^\infty x e^{-x} \, dx$

8. $\int_0^\infty x^2 e^{-3x} \, dx$

9. $\int_0^\infty e^{-x} \sin x \, dx$

10. $\int_{-\infty}^0 e^{2x} \cos x \, dx$

11. $\int_0^\infty \sin 2x \, dx$

12. $\int_{-\infty}^0 \cos^3 x \, dx$

13. $\int_1^\infty \dfrac{\ln x}{x^2} \, dx$

14. $\int_1^\infty \dfrac{\mathrm{Tan}^{-1} x}{x^2} \, dx$

15. $\int_0^\infty \dfrac{1 + x + x^2}{x^3} \, dx$

16. $\int_1^2 \dfrac{1}{\sqrt{x-1}} \, dx$

17. $\int_1^2 \dfrac{dx}{\sqrt[3]{2-x}}$

18. $\int_0^1 x \ln x \, dx$

19. $\int_{-1}^1 \dfrac{dx}{x^2}$

20. $\int_{-1}^1 \dfrac{dx}{x^3}$

21. $\int_0^2 \dfrac{1}{\sqrt[3]{x-1}} \, dx$

22. $\int_0^2 \dfrac{1}{(x-1)^{4/3}} \, dx$

23. $\int_0^\infty \dfrac{dx}{x^2 - 4}$

24. $\int_0^\infty \dfrac{dx}{x(x^2+1)}$

25. (a) Let f be continuous and positive for $a \leq x \leq \infty$ and let a constant K exist so that the definite integral of f from a to b is less than or equal to K for all b, $a \leq b < \infty$. Show that $\int_a^\infty f(x) \, dx$ converges and has value at most K.

(b) With the aid of the result of part (a) prove *the comparison rule*: Let f and g be continuous and positive for $a \leq x < \infty$ and let $f(x) \leq g(x)$ for all x. If $\int_a^\infty g(x) \, dx$ converges, then $\int_a^\infty f(x) \, dx$ converges. If $\int_a^\infty f(x) \, dx$ diverges, then $\int_a^\infty g(x) \, dx$ diverges.

26. Apply the result of Problem 25 to determine convergence or divergence of the following integrals, but do not evaluate:

(a) $\int_1^\infty \dfrac{dx}{x^2(1 + x^3)}$

(b) $\int_1^\infty \dfrac{e^{-x}}{x} \, dx$

(c) $\int_1^\infty \dfrac{2 + \sin x}{x} \, dx$

(d) $\int_1^\infty \dfrac{x^3 + 2}{2x^4(x^3 + 1)} \, dx$

(e) $\int_1^\infty \dfrac{\ln x}{x^2} \, dx$

(f) $\int_1^\infty \dfrac{x \sin(1/x)}{x + 1} \, dx$

(g) $\int_0^\infty e^{-x^2} \, dx$

(h) $\int_0^\infty x^2 e^{-x^2} \, dx$

27. Formulate and prove the comparison rule, as in Problem 25, for each of the following types of improper integral:
    (a) Type of (7-141)   (b) Type of (7-145)

28. With the aid of appropriate comparison rules (Problems 25 and 27), determine convergence or divergence but do not evaluate:

    (a) $\int_0^1 \frac{\cos x}{\sqrt{x}}\, dx$ (b) $\int_0^1 \frac{\cos x}{x}\, dx$ (c) $\int_{-\infty}^{-1} \frac{\sin^2 x}{x^2}\, dx$

    (d) $\int_0^\infty \frac{e^{-x^2}}{x}\, dx$ (e) $\int_{-\infty}^\infty \frac{dx}{(x-1)^{1/3}(x-2)^{1/5}(x^2+1)}$ (f) $\int_{-\infty}^\infty \frac{e^{-x^2}\, dx}{(x^2-1)^{1/3}}$

29. If $f(x) \geq 0$ for $a \leq x < \infty$, we can interpret $\int_a^\infty f(x)\, dx$ as the area beneath the curve $y = f(x)$, $a \leq x < \infty$. Similarly, other improper integrals can be interpreted as volumes and moments. Evaluate each of the following that is meaningful.
    (a) The area under the curve $y = x^2 e^{-2x}$ for $0 \leq x < \infty$.
    (b) The volume of the solid of revolution obtained by rotating the region beneath the curve $y = 1/x$, $1 \leq x < \infty$, about the x-axis. What is the area of the region?
    (c) The volume of the solid of revolution obtained by rotating the region beneath the curve $y = 1/\sqrt{x}$, $0 < x \leq 1$ about the x-axis. What is the area of the region?
    (d) The mass spread along the x-axis if the density is $\delta = 1/(x^2 + 1)$. Also find the moments M_1 and M_2.

30. For certain divergent improper integrals, a value can sometimes be assigned by using a modified limit process, to be described here. When the limit exists, we say that the value obtained is the *principal value* of the integral and prefaces the integral symbol by (P). If f is continuous for all finite x, we write

$$(P)\int_{-\infty}^\infty f(x)\, dx = \lim_{c \to \infty} \int_{-c}^c f(x)\, dx$$

If f is continuous from a to b except at $x = x_1$, where $a < x_1 < b$, we write

$$(P) \int_a^b f(x)\, dx = \lim_{h \to 0+} \left[\int_a^{x_1-h} f(x)\, dx + \int_{x_1+h}^b f(x)\, dx \right]$$

Other improper integrals can be assigned principal values by splitting the interval of integration into appropriate pieces.
Show that the following principal values exist and find them:

    (a) $(P)\int_{-\infty}^\infty x\, dx$ (b) $(P)\int_{-\infty}^\infty \sin x\, dx$

    (c) $(P)\int_{-\infty}^\infty \frac{x^3 + x + 1}{x^2 + 1}\, dx$ (d) $(P)\int_{-\infty}^\infty \frac{x^3 + 4x}{x^4 + 8x^2 + 1}\, dx$

    (e) $(P)\int_{-1}^1 \frac{1}{x}\, dx$ (f) $(P)\int_{-1}^1 \frac{1}{x^3}\, dx$

    (g) $(P)\int_{-\pi/2}^{\pi/2} \csc x\, dx$ (h) $(P)\int_{-1}^1 \frac{dx}{x(x-2)}$

7-15 DIFFERENTIAL EQUATIONS

An *ordinary differential equation* is an equation relating an unspecified function and its derivatives up to a certain order. Thus each of the following is an ordinary differential equation for the unspecified function $y(x)$:

1. $y' = xe^x$
2. $y'' + y = 0$
3. $y''' - 2y'' + y' - y = \sin x$
4. $y'^2 + 2yy' = x^2$

The *order* of the differential equation is the order of the highest derivative appearing. Thus equation No. 1 is of order 1, as is No. 4; No. 2 has order 2, No. 3 has order 3. When the equation is an algebraic equation for the highest order derivative, the degree of that equation is called the *degree* of the differential equation. Thus No. 4 has degree 2 (quadratic in y'), the others have degree 1.

The word "ordinary" will generally be omitted here; it is needed to distinguish the equations described above from partial differential equations, which contain partial derivatives (Chapter 12).

The theory of differential equations is an extended one. In this chapter a few simple types of equations will be considered. A more complete discussion is given in Chapter 14.

Differential equations have many applications in physics and other sciences. In particular, many basic laws of physics are expressible as differential equations. An example is Newton's Second Law for motion of a particle on a line. The statement: "force equals mass times acceleration" is equivalent, in this case, to a differential equation of form:

$$\frac{d^2x}{dt^2} = F\left(t, x, \frac{dx}{dt}\right) \tag{7-150}$$

Here F is a function of the three variables t, x, dx/dt.

Solution, General solution. By a *solution* (or *particular solution*) of a differential equation is meant a function, defined in an interval, which satisfies the equation. Thus $y = \cos x$, $-\infty < x < \infty$, is a solution of the differential equation $y'' + y = 0$, since

$$y' = -\sin x, \qquad y'' = -\cos x, \qquad y'' + y \equiv -\cos x + \cos x \equiv 0$$

By the *general solution* is meant the collection of all solutions of the equation. For No. 1 above, the solutions are simply the indefinite integrals of the right-hand side:

$$y' = xe^x \text{ is equivalent to } y = \int xe^x \, dx + C$$

where C is an arbitrary constant. By evaluating the indefinite integral, we find that

$$y = e^x(x - 1) + C$$

and this equation gives the general solution of the given equation. Similar expressions will be obtained for many other differential equations. For equations of order n, n arbitrary constants normally appear. Thus the general solution of No. 2 will be seen to be

$$y = c_1 \cos x + c_2 \sin x$$

where c_1 and c_2 are arbitrary constants. In each case, to show that we have the general solution, we must show that the asserted formula does provide functions that are solutions and that all solutions are contained in the formula. It is not always easy to find the general solution.

The Existence Theorem. For an equation of order n, we are often led to seek a solution $y = \varphi(x)$ for which $\varphi, \varphi', \ldots, \varphi^{(n-1)}$ have prescribed values at a chosen value of x; that is, for which

$$\varphi(x_0) = y_0, \; \varphi'(x_0) = y_0', \ldots, \qquad \varphi^{(n-1)}(x_0) = y_0^{(n-1)} \qquad (7\text{-}151)$$

where $y_0, \ldots, y_0^{(n-1)}$ are n given numbers. In fact, it can be shown that, under appropriate hypotheses, there is one and only one solution $\varphi(x)$ satisfying the conditions (7-151). This assertion is the basic *existence theorem* for ordinary differential equations. It will be considered in detail in Chapter 14. In physical applications, x is often replaced by time t and t_0 is a chosen "initial time." For this reason conditions (7-151) are usually termed *initial conditions*, and the numbers $y_0, \ldots, y_0^{(n-1)}$ are called *initial values*.

For the case of Equation (7-150), we are asserting that the motion of the particle is completely determined by its position and velocity at a chosen instant of time.

The Equation $y^{(n)} = f(x)$. Let f be defined and continuous in an interval. Then the equation can be solved by integration:

$$y^{(n-1)} = \int f(x) \, dx + c_1 = f_1(x) + c_1$$

where $f_1(x)$ is a choice of the indefinite integral and c_1 is an arbitrary constant. Similarly

$$y^{(n-2)} = \int [f_1(x) + c_1)] \, dx = f_2(x) + c_1 x + c_2$$

and so on, until finally y is obtained.

EXAMPLE. $y''' = x^3 - x$. We find

$$y'' = \frac{x^4}{4} - \frac{x^2}{2} + c_1$$

$$y' = \frac{x^5}{20} - \frac{x^3}{6} + c_1 x + c_2$$

$$y = \frac{x^6}{120} - \frac{x^4}{24} + c_1 \frac{x^2}{2} + c_2 x + c_3$$

The result provides the general solution for all x. We observe that $y(0) = c_3$, $y'(0) = c_2$, $y''(0) = c_2$; thus, in accordance with the existence theorem, there is one and only one solution with given initial values for $x = 0$.

7-16 FIRST ORDER DIFFERENTIAL EQUATIONS

A differential equation of first order relates x, y, and y'. The existence theorem (when applicable) states that there is a unique solution $\varphi(x)$ with

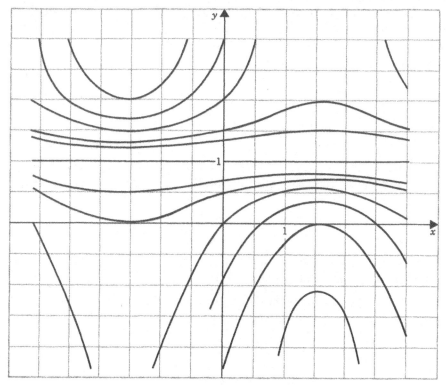

Figure 7-70 Solutions of first order differential equation.

given value y_0 for $x = x_0$. If we consider the solutions as curves in the xy-plane, the theorem thus asserts that there is a unique solution through each point (x_0, y_0) (Figure 7-70). We here consider only three of the commonly occurring types of first order equations.

(a) **Separation of Variables: the Equation** $y' = f(x)g(y)$. This equation has already been considered in Problems 68 to 72 following Section 4-8. The formal procedure is to separate the x and y with the aid of differentials:

$$\frac{dy}{dx} = f(x)g(y)$$

$$\frac{dy}{g(y)} = f(x)\,dx$$

$$\int \frac{dy}{g(y)} = \int f(x)\,dx + C$$

$$G(y) = F(x) + C \qquad (7\text{-}160)$$

where G and F are particular indefinite integrals. As shown in Chapter 4, the formal procedure does give all solutions, provided that f and g are continuous and that $g \neq 0$ for the values considered. If g has zeros at a_1, $a_2, \ldots$, then the functions

$$y = a_1 = \text{const}, \qquad y = a_2, \ldots \qquad (7\text{-}161)$$

are also solutions. Equations (7-160) and (7-161) together give all solutions [(7-160) defining solutions implicitly]. Thus the general solution is described in a rather complicated manner.

EXAMPLE 1. $y' = (y - 1)^2 \cos x$. We write

$$\frac{dy}{(y - 1)^2} = \cos x \, dx, \qquad \int \frac{dy}{(y - 1)^2} = \int \cos x \, dx$$

$$-\frac{1}{y - 1} = \sin x + C, \qquad y = 1 - \frac{1}{\sin x + C}$$

The general solution is given by the last equation and the constant function $y = 1$. For the solution through the point $(\pi/2, 5)$, we obtain a condition on C:

$$5 = 1 - \frac{1}{1 + C}$$

Hence $C = -\frac{5}{4}$, and the solution sought is $y = 1 - [4/(4 \sin x - 5)]$. All the solutions are graphed in Figure 7-70.

(b) The Homogeneous Equation $y' = f(y/x)$. The equation $y' = (x^2 + xy + y^2)/x^2$ can be written as

$$y' = 1 + \frac{y}{x} + \frac{y^2}{x^2} = 1 + v + v^2$$

where $v = y/x$. If we set $v(x) = y(x)/x$, then

$$y(x) = xv(x), \qquad y' = xv' + v$$

and v satisfies the equation

$$xv' + v = 1 + v + v^2$$

or

$$xv' = 1 + v^2$$

We can now separate variables:

$$x \frac{dv}{dx} = 1 + v^2, \qquad \frac{dv}{1 + v^2} = \frac{dx}{x}$$

and obtain (for $x \neq 0$)

$$\mathrm{Tan}^{-1} v = \ln |x| + C$$
$$v = \tan(\ln |x| + C)$$
$$y = xv = x \tan(\ln |x| + C)$$

The method applies to every equation of form

$$y' = f(v), \qquad v = \frac{y}{x}$$

provided that f is continuous. The substitution $y = xv$ always leads to a separation of variables. The equation is said to be *homogeneous* (see Problem 6 below).

(c) **The Linear Equation of First Order.** An equation of form

$$y' + p(x)y = q(x) \tag{7-162}$$

is said to be a linear equation of first order. We assume that p and q are continuous in some interval. We first consider an example: $y' + (y/x) = 4x^2$. We multiply through by x and obtain the equation:

$$xy' + y = 4x^3$$

Now, $xy' + y = (xy)'$. Hence we can write the equation as

$$\frac{d}{dx}(xy) = 4x^3$$

Accordingly, we obtain $xy = x^4 + C$ or $y = x^3 + (C/x)$ as expression for the general solution (throughout, the value $x = 0$ is excluded).

The example suggests that we try to multiply (7-162) by a factor $v(x)$ (called *integrating factor*) so that the left side becomes the derivative of a product: (function of x) times y. We obtain the equation:

$$v(x)y' + p(x)v(x)y = q(x)v(x) \tag{7-163}$$

The left side is the derivative of $v(x)y$, provided that $v(x)$ is such that $v'(x) = p(x)v(x)$—that is, so that

$$\frac{dv}{dx} = p(x)v$$

This is a differential equation for v, which we solve by separating variables. We want only one solution, so that we write:

$$\int \frac{dv}{v} = \int p(x)\,dx = P(x) + C, \qquad \ln v = P(x), \qquad v = e^{P(x)}$$

where $P(x)$ is an indefinite integral of $p(x)$. Accordingly, $v(x) = e^{P(x)}$ is the integrating factor sought. With $v(x)$ so chosen, Equation (7-163) becomes

$$e^{P(x)}y' + p(x)e^{P(x)}y = q(x)e^{P(x)}$$

The left side is the derivative of $v(x)y$—that is, of $e^{P(x)}y$. Hence the equation can be written

$$[e^{P(x)}y]' = q(x)e^{P(x)}$$

and all solutions are given by

$$e^{P(x)}y = \int q(x)e^{P(x)}\,dx + C \qquad \text{or} \qquad y = e^{-P(x)}\int q(x)e^{P(x)}\,dx + Ce^{-P(x)}$$

EXAMPLE 2. $y' + xy = x$. Here $p(x) = x$, we can take $P(x) = x^2/2$, and our integrating factor is $e^{x^2/2}$. After multiplying by this factor, we obtain

$$e^{x^2/2}y' + e^{x^2/2}xy = xe^{x^2/2}, \qquad (e^{x^2/2}y)' = xe^{x^2/2}$$

$$e^{x^2/2}y = \int xe^{x^2/2}\,dx = e^{x^2/2} + C, \qquad y = 1 + Ce^{-x^2/2}$$

Check. $y' = -Cxe^{x^2/2}$, $y' + xy = -Cxe^{x^2/2} + x(1 + Ce^{-x^2/2}) = x$.

PROBLEMS

1. For each of the following differential equations state the order and degree, and determine whether the function f given is a solution:

   (a) $y' = 2y$, $f(x) = 5e^{2x}$, $-\infty < x < \infty$

   (b) $y'' + y' = 0$, $f(x) = 1$, $-\infty < x < \infty$

   (c) $y'^2 + y^3 - y = 0$, $f(x) = 1$, $-\infty < x < \infty$

   (d) $y'' - y = 0$, $f(x) = c_1 e^x + c_2 e^{-x}$, $-\infty < x < \infty$, c_1, c_2 constant

   (e) $xy' + y = 0$, $f(x) = c/x$, $x \neq 0$, $c = $ constant

   (f) $2x^2 yy' + xy^2 - 1 = 0$, $f(x)$ defined implicitly by $e^{xy^2} = x$ for $x > 1$, $y > 0$

   (g) $y''' - 3y'' + 3y' - y = 0$, $f(x) = e^x(c_1 + c_2 x + c_3 x^2)$, $-\infty < x < \infty$, c_1, c_2, c_3 constant

2. Find the general solution:

   (a) $y' = 3x^2 - 2x + 5$ (b) $xy' = 3x + 5$ (c) $y'' = 1$

   (d) $y'' = \sin 3x$ (e) $y''' = e^{2x} + 5$ (f) $xy^{(iv)} = 1$

3. Find the general solution:

   (a) $xy' + 3y = 0$ (b) $y' + y = 0$ (c) $y' + e^x y = 0$

   (d) $y' + y^2 = 0$ (e) $y' + y^2 - 1 = 0$ (f) $y' = y \cos x$

   (g) $y' = xy^2$ (h) $y' = x^2 \tan y$

4. Find the general solution:

   (a) $y' = (x + y)/x$ (b) $y' = (x^2 + y^2)/xy$ (c) $xy' = y + xe^{y/x}$

   (d) $xy' = 2y - x$ (e) $x^2 y' = y^2 + xy - x^2$ (f) $(x + 2y)y' = 2x - y$

5. Find the general solution:

   (a) $y' + 3y = x$ (b) $y' + xy = x$ (c) $y' + y \sin x = \sin x$

   (d) $xy' + y = x^3$ (e) $x(x - 1)y' + y = x^2$ (f) $y' \sin x + y \cos x = \sin x$

6. Show that the substitution $y = xv$ always leads to a separation of variables in a homogeneous first order equation $y' = f(y/x)$.

7. Find the solution satisfying the given initial conditions:

   (a) $y'' = e^x$, $y = 2$ and $y' = 0$ for $x = 0$

   (b) $y''' = \sin x$, $y = 1$, $y' = 0$ and $y'' = -1$ for $x = \pi$

   (c) $y' = xy^3$, $y = 0$ for $x = 2$

   (d) $xy' + 2y = 3x$, $y = 0$ for $x = 0$

   (e) $xy' = y$, $y = 1$ for $x = 1$

   (f) $y' = (y/x) + (y^3/x^3)$, $y = 3$ for $x = -2$

8. For a descending parachute, Newton's Second Law (7-150) can be applied. Here x can be taken as the distance (in feet) that the parachute has descended from a reference altitude, t as the time in seconds. The forces acting are mg ($m = $ mass of parachute and load, $g = 32$ ft. per sec^2) and air resistance, which we assume to be proportional to the speed. Hence we obtain a differential equation

$$m \frac{dv}{dt} = mg - kv$$

where k is a positive constant and $v = dx/dt$. (a) The acceleration is observed to be $-0.2g$ when $v = 16$ ft per sec. Show that $k = 2.4\, m$. (b) For $t = 0$, $v = 0$. Use the result of part (a) to find v as a function of t and to find the limiting velocity as $t \to \infty$.

9. The rate of formation of a certain chemical in a reaction is known to be governed by the equation

$$\frac{dx}{dt} = (a - x)(x - b)$$

where x is the amount (mass) of the chemical present at time t and a, b are the amounts of certain other chemicals present when $t = 0$, with $0 < b < a$. For $t = 0$, $x = \frac{1}{2}(a + b)$. Find x as a function of t and determine $\lim_{t \to \infty} x(t)$.

10. The rate of growth of a certain population can be described by the equation

$$\frac{dx}{dt} = kx + be^{-t}$$

Here the term be^{-t} represents the effect of immigration, diminishing with time t, in years, and kx, where k is a positive constant, describes the effect of natural growth of families. The number x can be considered to be an estimate of population size. (a) If $b = 0$ and $k = 0.03$, how many years does it take for the population to double in size? (b) If $k = 0.03$ and $b = 10,000$, and $x = 500,000$ for $t = 0$, find x for $t = 10$ years.

7-17 LINEAR DIFFERENTIAL EQUATIONS OF SECOND ORDER

By a *linear differential equation of order n* we mean an equation of form

$$p_0(x)y^{(n)} + p_1(x)y^{(n-1)} + \cdots + p_{n-1}(x)y' + p_n(x)y = q(x)$$

where $p_0(x), \ldots, p_n(x)$, $q(x)$ are defined on a certain interval and $p_0(x) \not\equiv 0$. For $n = 1$, the equation can be written in the form (7-162) and can be solved as in Section 7-16. We here concentrate on the equation of second order

$$p_0(x)y'' + p_1(x)y' + p_2(x)y = q(x) \tag{7-170}$$

When $q(x) \equiv 0$, the equation is said to be *homogeneous* (not to be confused with the homogeneous first order equation of Section 7-16). In particular, the equation

$$p_0(x)y'' + p_1(x)y' + p_2(x)y = 0 \tag{7-171}$$

is said to be the *homogeneous equation related to* (7-170). We shall assume that our interval is of form $a < x < b$, where a may be $-\infty$ and b may be $+\infty$, that $p_0(x)$, $p_1(x)$, $p_2(x)$, $q(x)$ are continuous on the interval, and that $p_0(x)$ has no zeros on the interval. The conditions on the p_i are satisfied, in particular, when these functions are constants, with $p_0 \neq 0$, so that Equation (7-170) can be written

$$a_0 y'' + a_1 y' + a_2 y = q(x) \tag{7-172}$$

where a_0, a_1, a_2 are constants, $a_0 \neq 0$, and $q(x)$ is continuous for $a < x < b$. We call such an equation (7-172) an equation with *constant coefficients*.

The existence theorem of Section 7-15 is applicable to Equation (7-170). Because of the linearity, a stronger theorem can be stated: under the hypotheses stated above, for each x_0 in the interval (a, b) and each choice of y_0, y_0' there is a unique solution $y = y(x)$ of (7-170) for $a < x < b$, with the given initial values y_0, y_0' at x_0. This theorem is considered in

Chapter 14. For the special cases to be considered here, the truth of the theorem can be directly verified (see, for example, Problem 6 below). We notice that, because of the existence theorem, we need only consider solutions defined on the whole interval (a, b).

THEOREM 6. *Under the hypotheses stated, the solutions of the homogeneous linear differential equation* (7-171) *on* (a, b) *form a vector space V of functions. Furthermore, there exist two linearly independent solutions* $y_1(x)$, $y_2(x)$ *on* (a, b), *and every solution* $y(x)$ *on* (a, b) *is a linear combination of* $y_1(x)$, $y_2(x)$:

$$y(x) = c_1 y_1(x) + c_2 y_2(x), \qquad a < x < b$$

Thus the solutions of (7-171) *form a two-dimensional vector space V.*

PROOF. Let $y_1(x)$, $y_2(x)$ be solutions of (7-171) on (a, b). Then

$$p_0(x)y_1''(x) + p_1(x)y_1'(x) + p_2(x)y_1(x) = 0,$$
$$p_0(x)y_2''(x) + p_1(x)y_2'(x) + p_2(x)y_2(x) = 0$$

in the interval. From these equations we deduce at once that

$$p_0(x)[c_1 y_1(x) + c_2 y_2(x)]'' + p_1(x)[c_1 y_1(x) + c_2 y_2(x)]'$$
$$+ p_2(x)[c_1 y_1(x) + c_2 y_2(x)] = 0$$

where c_1 and c_2 are scalars. Therefore, the solutions of (7-171) on (a, b) satisfy the conditions for a vector space of functions (see Section 2-9).

To show that V is two-dimensional, we apply the existence theorem. We select x_0, $a < x_0 < b$, and let $y_1(x)$, $y_2(x)$ be the two solutions on (a, b) such that

$$y_1(x_0) = 1, \; y_1'(x_0) = 0 \qquad \text{and} \qquad y_2(x_0) = 0, \; y_2'(x_0) = 1 \quad (7\text{-}173)$$

Then, $y_1(x)$, $y_2(x)$ are linearly independent. For if $c_1 y_1(x) + c_2 y_2(x) \equiv 0$, then also $c_1 y_1'(x) + c_2 y_2'(x) \equiv 0$. If we set $x = x_0$ in these equations and use (7-173), we obtain

$$c_1 + c_2 \cdot 0 = 0, \qquad c_1 \cdot 0 + c_2 = 0$$

so that $c_1 = 0$, $c_2 = 0$. Therefore $y_1(x)$, $y_2(x)$ are linearly independent. To show that they are a basis for V, we let $y(x)$ be a solution of (7-171) on (a, b), and try to determine c_1, c_2 so that

$$y(x) = c_1 y_1(x) + c_2 y_2(x), \qquad a < x < b \qquad (7\text{-}174)$$

We must then also have

$$y'(x) = c_1 y_1'(x) + c_2 y_2'(x), \qquad a < x < b$$

If we set $x = x_0$ in the last two equations and apply (7-173), we obtain

$$y(x_0) = c_1, \qquad y'(x_0) = c_2$$

Hence, c_1, c_2 are uniquely determined. We claim that, with these choices of c_1 and c_2, Equation (7-174) is valid. For both sides of the equation are solutions of (7-171) and, by the choices of c_1, c_2, they have the same value and derivative at x_0. By the existence theorem, they must be the same solution! Thus, $y_1(x)$, $y_2(x)$, form a basis for V, and V is two-dimensional.

As a consequence of this theorem, to find the general solution of (7-171), we need only find two linearly independent solutions.

EXAMPLE 1. $y'' + y = 0$. Here the interval is $-\infty < x < \infty$. By "inspection" we observe that $y = y_1(x) = \cos x$, $y = y_2(x) = \sin x$ are solutions. These functions are linearly independent, since neither is a scalar times the other. Hence, the general solution is $y = c_1 \cos x + c_2 \sin x$.

EXAMPLE 2. $y'' - y = 0$. The interval is again $(-\infty, \infty)$. By "inspection," we find the linearly independent solutions $y_1(x) = e^x$, $y_2(x) = e^{-x}$. Consequently, the general solution is $y = c_1 e^x + c_2 e^{-x}$.

THEOREM 7. *Under the hypotheses stated above, the nonhomogeneous equation (7-170) has solutions on (a, b). If $y = y_p(x)$ is one particular solution of (7-170) on (a, b), then the general solution is given by*

$$y = y_p(x) + c_1 y_1(x) + c_2 y_2(x) \tag{7-175}$$

where $y = c_1 y_1(x) + c_2 y_2(x)$ is the general solution of the related homogeneous equation (7-171).

PROOF. By the existence theorem, the nonhomogeneous equation (7-170) has solutions on (a, b). Let $y = y_p(x)$ be one solution and let $y = y(x)$ be an arbitrary solution. Then

$$p_0 y'' + p_1 y' + p_2 y = q(x), \qquad p_0 y_p'' + p_1 y_p' + p_2 y_p = q(x),$$
$$a < x < b \tag{7-176}$$

If we subtract, we obtain

$$p_0 (y - y_p)'' + p_1 (y - y_p)' + p_2 (y - y_p) = 0, \qquad a < x < b \tag{7-177}$$

Hence $y - y_p$ is a solution of the related homogeneous equation (7-171). Therefore, by Theorem 6, $y - y_p = c_1 y_1(x) + c_2 y_2(x)$, so that (7-175) holds true, for the appropriate choices of c_1, c_2. Conversely, if a function $y(x)$ is defined by Equation (7-175) for some choice of c_1, c_2, then $y - y_p$ is a solution of the homogeneous equation, so that (7-177) holds true. Since the second equation in (7-176) also holds true, the first must hold true, that is, $y(x)$ is a solution of the nonhomogeneous differential equation. Therefore, (7-175) does give the general solution of (7-170).

EXAMPLE 3. $y'' + y = x$, on the interval $(-\infty, \infty)$. Here we can guess a solution: $y = x = y_p(x)$. For $y_p'' + y_p = 0 + x = x$. The related homogeneous equation is solved as Example 1 above. Therefore, the general solution is $y = x + c_1 \cos x + c_2 \sin x$.

7-18 THE HOMOGENEOUS LINEAR DIFFERENTIAL EQUATION OF SECOND ORDER WITH CONSTANT COEFFICIENTS

We consider the equation

$$a_0 y'' + a_1 y' + a_2 y = 0, \qquad -\infty < x < \infty \qquad (7\text{-}180)$$

Here a_0, a_1, and a_2 are constants and $a_0 \neq 0$. Example 2 of Section 7-17 suggests that we seek solutions of form $e^{\lambda x}$, where λ is a constant. Substitution of $y = e^{\lambda x}$ in (7-180) gives

$$e^{\lambda x}(a_0\lambda^2 + a_1\lambda + a_2) = 0$$

Consequently, $y = e^{\lambda x}$ is a solution provided that λ satisfies the quadratic equation

$$a_0\lambda^2 + a_1\lambda + a_2 = 0 \qquad (7\text{-}181)$$

This equation is called the *characteristic equation*, or *auxiliary equation*. Its roots (called *characteristic roots*) may be real and distinct, real and equal, or complex and distinct. We consider these three cases in turn.

I. Characteristic roots real and distinct. Let the roots be λ_1, λ_2. Then $y_1(x) = e^{\lambda_1 x}$ and $y_2(x) = e^{\lambda_2 x}$ are solutions and are linearly independent (Problem 7 below). Hence by Theorem 6 the general solution is

$$y = c_1 e^{\lambda_1 x} + c_2 e^{\lambda_2 x}$$

EXAMPLE 1. $y'' - 3y' + 2y = 0$. The characteristic equation is $\lambda^2 - 3\lambda + 2 = 0$, the characteristic roots are $\lambda_1 = 1$, $\lambda_2 = 2$, and the general solution is

$$y = c_1 e^x + c_2 e^{2x}$$

II. Characteristic roots real and equal. Let the roots be $\lambda_1 = \lambda_2$. Then we appear to have only one solution, $e^{\lambda_1 x}$, not enough for a basis. We must somehow find a second linearly independent solution. To this end, we observe that the characteristic equation is equivalent to the equation $(\lambda - \lambda_1)^2 = 0$ or

$$\lambda^2 - 2\lambda_1\lambda + \lambda_1{}^2 = 0$$

Thus the differential equation, after division by a_0, becomes

$$y'' - 2\lambda_1 y' + \lambda_1{}^2 y = 0$$

or

$$y'' - \lambda_1 y' = \lambda_1(y' - \lambda_1 y)$$

If we set $u = y' - \lambda_1 y$, the equation becomes $u' = \lambda_1 u$. Hence, $u = c_1 e^{\lambda_1 x}$ and

$$y' - \lambda_1 y = c_1 e^{\lambda_1 x}$$

This is a first order linear equation, with integrating factor $e^{-\lambda_1 x}$. We solve as in Section 7-16 above:

$$e^{-\lambda_1 x}(y' - \lambda_1 y) = c_1, \qquad (e^{-\lambda_1 x}y)' = c_1, \qquad e^{-\lambda_1 x}y = c_1 x + c_2$$

Hence

$$y = c_1 x e^{\lambda_1 x} + c_2 e^{\lambda_1 x}$$

Our method has revealed that the missing solution is $xe^{\lambda_1 x}$. We can now go back and directly verify that $e^{\lambda_1 x}$ and $xe^{\lambda_1 x}$ are linearly independent solutions, so that we have the general solution as in Theorem 6. (Notice that we have in fact found the general solution here without use of the existence theorem or of Theorem 6.)

EXAMPLE 2. $y'' - 6y' + 9y = 0$. The characteristic equation is $\lambda^2 - 6\lambda + 9 = 0$ or $(\lambda - 3)^2 = 0$. The roots are 3, 3 and the general solution is $y = c_1 e^{3x} + c_2 x e^{3x}$.

III. Characteristic roots complex and unequal. Since the coefficients a_0, a_1, a_2 are real, the two roots must be conjugate complex numbers $\alpha \pm \beta i (\beta \neq 0)$. The characteristic equation is equivalent to the equation $\lambda^2 - 2\alpha\lambda + \alpha^2 + \beta^2 = 0$, so that, after division by a_0, the differential equation can be written

$$y'' - 2\alpha y' + (\alpha^2 + \beta^2)y = 0 \qquad (7\text{-}182)$$

Formally we obtain two solutions:

$$y_1(x) = e^{(\alpha - \beta i)x} \qquad \text{and} \qquad y_2(x) = e^{(\alpha - \beta i)x}$$

[In fact, they can be interpreted as *complex* solutions of (7-180), as in Section 5-8.] It is natural to try to combine these two solutions to obtain real solutions. Now by Euler's identity (Section 5-8)

$$y_1(x) = e^{(\alpha + \beta i)x} = e^{\alpha x}e^{\beta i x} = e^{\alpha x}(\cos \beta x + i \sin \beta x)$$
$$y_2(x) = e^{(\alpha - \beta i)x} = e^{\alpha x}e^{-\beta i x} = e^{\alpha x}(\cos \beta x - i \sin \beta x)$$

Now if we add and divide by 2 or subtract and divide by $2i$, we obtain

$$\tfrac{1}{2}y_1(x) + \tfrac{1}{2}y_2(x) = e^{\alpha x} \cos \beta x, \qquad \frac{1}{2i}y_1(x) - \frac{1}{2i}y_2(x) = e^{\alpha x} \sin \beta x$$

Even though we are dealing with complex coefficients, we expect these linear combinations of $y_1(x)$, $y_2(x)$ to be solutions of (7-182). Indeed, direct substitution in (7-182) shows that the two functions $e^{\alpha x} \cos \beta x$, $e^{\alpha x} \sin \beta x$ are solutions; the functions are also linearly independent (Problem 7 below). Therefore, the general solution of (7-182) is

$$y = c_1 e^{\alpha x} \cos \beta x + c_2 e^{\alpha x} \sin \beta x$$

EXAMPLE 3. $y'' + 4y' + 13y = 0$. The characteristic equation is $\lambda^2 + 4\lambda + 13 = 0$, the characteristic roots are $-2 \pm 3i$, so that $\alpha = -2$, $\beta = 3$, and the general solution is $y = c_1 e^{-2x} \cos 3x + c_2 e^{-2x} \sin 3x$.

7-19 THE NONHOMOGENEOUS LINEAR EQUATION OF SECOND ORDER WITH CONSTANT COEFFICIENTS

We consider the equation

$$a_0 y'' + a_1 y' + a_2 y = q(x) \qquad (7\text{-}190)$$

where a_0, a_1, a_2 are constants, $a_0 \neq 0$, and $q(x)$ is continuous on a given interval. By Theorem 7, the general solution of (7-190) can be written

$$y = y_p(x) + c_1 y_1(x) + c_2 y_2(x) = y_p(x) + y_c(x)$$

where y_p is one solution of (7-190) and $y_c(x) = c_1 y_1(x) + c_2 y_2(x)$ is the general solution of the related homogeneous equation

$$a_0 y'' + a_1 y' + a_2 y = 0 \qquad (7\text{-}191)$$

The expression $c_1 y_1(x) + c_2 y_2(x)$ is often called the *complementary function*. The previous section tells us how to find the complementary function. We now consider how to find $y_p(x)$, a particular solution of (7-190).

We here describe one general method for finding y_p; the method is called *variation of parameters*. We assume that we know the complementary function $c_1 y_1(x) + c_2 y_2(x)$. We now introduce two new functions $v_1(x)$, $v_2(x)$, and consider the function

$$y = v_1(x) y_1(x) + v_2(x) y_2(x)$$

(Thus the constants c_1, c_2 are replaced by "variables," and thus we speak of "variation of constants" or "variation of parameters.") We now try to choose $v_1(x)$, $v_2(x)$ so that $y = v_1 y_1 + v_2 y_2$ is a solution of (7-190). Since we seek two functions v_1, v_2, we must impose one other condition.

Now, by the product rule,

$$y' = v_1 y_1' + v_2 y_2' + v_1' y_1 + v_2' y_2$$

The extra condition is that the last two terms have sum zero:

$$v_1' y_1 + v_2' y_2 = 0 \qquad (7\text{-}192)$$

This leads to the equation

$$y' = v_1 y_1' + v_2 y_2'$$

and insures that v_1'', v_2'' do not appear at the next stage. We find that

$$y'' = v_1 y_1'' + v_2 y_2'' + v_1' y_1' + v_2' y_2'$$

We replace y by $v_1 y_1 + v_2 y_2$ and y', y'' by the expressions just obtained in Equation (7-190):

$$a_0(v_1 y_1'' + v_2 y_2'' + v_1' y_1' + v_2' y_2')$$
$$+ a_1(v_1 y_1' + v_2 y_2') + a_2(v_1 y_1 + v_2 y_2) = q(x)$$

We collect terms to obtain

$$v_1(a_0 y_1'' + a_1 y_1' + a_2 y_1) + v_2(a_0 y_2'' + a_1 y_2' + a_2 y_2)$$
$$+ a_0(v_1' y_1' + v_2' y_2') = q(x)$$

But $y_1(x)$ and $y_2(x)$ are solutions of the homogeneous Equation (7-191). Hence, the first two parentheses reduce to zero, and we obtain

$$a_0(v_1' y_1' + v_2' y_2') = q(x) \qquad (7\text{-}193)$$

The equations (7-192) and (7-193) provide two simultaneous equations for v_1' and v_2' (Problem 9 below). We solve them for v_1', v_2', and then integrate to obtain v_1, v_2:

$$v_1 = g_1(x), \qquad v_2 = g_2(x)$$

No arbitrary constants are needed, since we seek only one solution. That solution is now given by

$$y = v_1y_1 + v_2y_2 = g_1(x)y_1(x) + g_2(x)y_2(x) = g(x)$$

Finally, our general solution is

$$y = c_1y_1(x) + c_2y_2(x) + g_1(x)y_1(x) + g_2(x)y_2(x)$$
$$= y_c(x) + g(x) = y_c(x) + y_p(x)$$

EXAMPLE 9. $y'' - 3y' + 2y = 4x$. The related homogeneous equation was solved in Example 1 of Section 7-18. We found its solutions to be

$$y = c_1e^x + c_2e^{2x}$$

Therefore, we set

$$y = v_1e^x + v_2e^{2x}$$

where $v_1 = v_1(x)$, $v_2 = v_2(x)$. Our Equations (7-192) and (7-193) become

$$v_1'e^x + v_2'e^{2x} = 0$$

$$v_1'e^x + 2v_2'e^{2x} = \frac{4x}{1} = 4x$$

By elimination, we find that

$$v_1' = -4xe^{-x}, \qquad v_2' = 4xe^{-2x}$$

and, hence, by integrating (without arbitrary constants)

$$v_1 = -\int 4xe^{-x}\, dx = 4e^{-x}(x + 1)$$

$$v_2 = -\int 4xe^{-2x}\, dx = -e^{-2x}(2x + 1)$$

Accordingly, our particular solution is

$$y = v_1e^x + v_2e^{2x} = 4x + 4 - (2x + 1) = 2x + 3$$

Thus, finally, the general solution is $y = y_p(x) +$ complementary function or $y = 2x + 3 + c_1e^x + c_2e^{2x}$.

Remark. The method of variation of parameters can be applied to equations whose coefficients are not constant, provided that one has found the complementary function.

PROBLEMS

1. Find the general solution:
   (a) $y'' - 9y = 0$
   (b) $y'' + 9y = 0$
   (c) $y'' + 2y' - 3y = 0$
   (d) $y'' - 2y' + 2y = 0$
   (e) $y'' + 2y' + 2y = 0$
   (f) $y'' + y' - y = 0$
   (g) $9y'' + 6y' + y = 0$
   (h) $2y'' + 3y' + 5y = 0$

2. Find the general solution:

(a) $y'' - 9y = e^x$ (b) $y'' - 9y = e^{3x}$ (c) $y'' - 9y = x$

(d) $y'' + 9y = e^x$ (e) $y'' + 9y = \cos 2x$ (f) $y'' + 9y = \cos 3x$

(g) $y'' + 2y' + y = x$ (h) $y'' + 2y' + y = e^{-x}$

3. Show that the functions given are linearly independent solutions of the differential equation on the stated interval and find the general solution:

(a) $x^2 y'' - 2xy' + 2y = 0, \, x > 0; \, y = x, \, y = x^2$

(b) $(x^2 + 1)^2 y'' - 4(x^3 + x)y' + (6x^2 - 2)y = 0, \, -\infty < x < \infty; \, y = x^2 + 1,$
$y = x^3 + x$

(c) $(x - x^2 \ln x)y'' + (x^2 \ln x + 1)y' - (x + 1)y = 0, \, x > 0; \, y = e^x, \, y = \ln x$

(d) $2x^2 y'' - xy' + y = 0, \, x > 0; \, y = x, \, y = \sqrt{x}$

4. Find the general solution, with the aid of the results of Problem 3:

(a) $x^2 y'' - 2xy' + 2y = x^3$ (b) $2x^2 y'' - xy' + y = x^2$

5. Find the general solution:

(a) $y''' + y' = 0$ [Hint: Set $u = y'$.]   (b) $y''' - y' = 0$

(c) $y''' + 3y'' + 2y' = e^x$ (d) $y^{(iv)} + 2y''' + 2y'' = 1$

6. Let the characteristic equation (7-181) have distinct real roots λ_1, λ_2, so that we can assume the characteristic equation is $\lambda^2 - (\lambda_1 + \lambda_2)\lambda + \lambda_1\lambda_2 = 0$ and the differential equation (7-180) is $y'' - (\lambda_1 + \lambda_2)y' + \lambda_1\lambda_2 y = 0$. Let $y(x)$ be a solution and let $u(x) = y(x)e^{-\lambda_2 x}$. Show that $u(x)$ satisfies the equation $u'' + (\lambda_2 - \lambda_1)u' = 0$ and that this implies that $u = c_1 e^{(\lambda_1 - \lambda_2)x} + c_2$, so that $y(x) = c_1 e^{\lambda_1 x} + c_2 e^{\lambda_2 x}$. (This is a proof, without use of the existence theorem, that for the equation considered the general solution has the form asserted in the text.)

7. Prove linear independence of the given functions on the interval (a, b):

(a) $e^{\lambda_1 x}, \, e^{\lambda_2 x}$ for $\lambda_1 \neq \lambda_2$ (b) $e^{\lambda x}, \, xe^{\lambda x}$

(c) $e^{ax} \cos \beta x, \, e^{ax} \sin \beta x$ for $\beta \neq 0$

8. Verify that the functions of Problem 7(c) are solutions of Equation (7-182).

9. Let $y_1(x), \, y_2(x)$ be a basis for the vector space V of Theorem 6, so that $y = c_1 y_1(x) + c_2 y_2(x)$ is the general solution of Equation (7-171) on the given interval (a, b).

(a) Show that there is no x_0 for which both $y_1(x_0) = 0$ and $y_2(x_0) = 0$. [Hint. Apply the existence theorem.]

(b) Show that there is no x_0 in (a, b) for which $y_1'(x_0) = 0$ and $y_2'(x_0) = 0$.

(c) Show that there is no x_0 in (a, b) for which $W(x_0) = 0$, where

$$W(x) = \begin{vmatrix} y_1(x) & y_2(x) \\ y_1'(x) & y_2'(x) \end{vmatrix} = y_1(x)y_2'(x) - y_2(x)y_1'(x)$$

[$W(x)$ is called the *Wronskian determinant* of the two solutions.]

(d) Show that $p_0(x)W'(x) + p_1(x)W(x) = 0$ and, hence, that $W(x) = ce^{-\int [p_1(x)/p_2(x)] \, dx}$.

Remark. The result in (c) shows that the method of variation of parameters leads to simultaneous equations for $v_1', \, v_2'$ which always have a unique solution.

7-20 VIBRATIONS

The second order linear differential equation with constant coefficients has important applications to physical problems. We here consider one basic class of those problems—the motion of a mass-spring system (Figure 7-71). A particle of mass m moves on the x-axis, subject to a spring force in accordance with Hooke's law: force is proportional to displacement. The

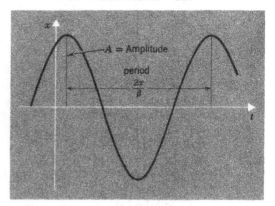

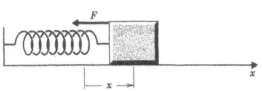

Figure 7-71 Mass-spring system.   Figure 7-72 Simple harmonic motion.

displacement is x, measured from the position of the particle when the spring is in equilibrium (neither stretched nor compressed). Therefore the force can be represented as $-k^2x$, where k is a positive constant. We also allow for friction proportional to velocity, hence, a force of form $-h\,dx/dt$, where h is a positive constant or 0. Finally, we allow for an external driving force $F(t)$. By Newton's Second Law, the motion of the particle is governed by the equation

$$m \frac{d^2x}{dt^2} = -k^2x - h \frac{dx}{dt} + F(t)$$

or

$$m \frac{d^2x}{dt^2} + h \frac{dx}{dt} + k^2x = F(t) \qquad (7\text{-}200)$$

This is a linear differential equation of second order with constant coefficients. Therefore all solutions can be found by the methods of Sections 7-18 and 7-19.

When $F(t) = 0$, the motion is *unforced* and the differential equation is the homogeneous equation:

$$m \frac{d^2x}{dt^2} + h \frac{dx}{dt} + k^2x = 0 \qquad (7\text{-}201)$$

Let first $h = 0$ (no friction) so that we have the equation

$$m \frac{d^2x}{dt^2} + k^2x = 0 \qquad (7\text{-}202)$$

The characteristic equation is $m\lambda^2 + k^2 = 0$, the characteristic roots are $\pm\beta i$, where $\beta = k/\sqrt{m}$, and the solutions are

$$x = c_1 \cos \beta t + c_2 \sin \beta t \qquad (7\text{-}203)$$

Here we can write $c_1 = A \sin \gamma$, $c_2 = A \cos \gamma$ [so that A, γ are polar coordinates of the point (c_2, c_1)]. Thus

$$x = A(\sin \gamma \cos \beta t + \cos \gamma \sin \beta t) = A \sin(\beta t + \gamma) \qquad (7\text{-}203')$$

This equation shows that the motion follows a sine curve, changed in scale and position, as in Figure 7-72. We say that the particle moves in *simple*

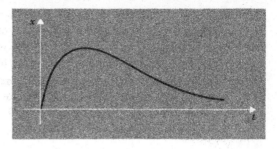

Figure 7-73 Exponential decay.

harmonic motion. The time for a complete cycle (from one maximum to the next) is $2\pi/\beta$; this is called the *period* of the oscillation. The number β is called the *frequency* (measured in radians per unit time). The maximum value of x is A, which we call the *amplitude* of the oscillation.

For $h > 0$, the characteristic equation becomes $m\lambda^2 + h\lambda + k^2 = 0$ and the roots are

$$\lambda = \frac{-h \pm \sqrt{h^2 - 4k^2m}}{2m}$$

Here we have three cases, as in Section 7-18.

I. $h^2 - 4k^2m > 0$. The roots λ_1, λ_2 are real and distinct. We observe that, since m, h, and k^2 are positive, no characteristic root can be positive. Hence, $\lambda_1 < 0$, $\lambda_2 < 0$. The solutions are given by $x = c_1e^{\lambda_1 t} + c_2e^{\lambda_2 t}$, and x approaches 0 as $t \to \infty$. A typical solution is graphed in Figure 7-73. We refer to the motion as *exponential decay.* The friction force (proportional to h) is so large that vibrations have disappeared. In general, friction has a tendency to at least slow down or *dampen* vibrations. Here we say the motion is *overdamped.*

II. $h^2 - 4k^2m = 0$. The roots λ_1, λ_2 are real, equal, and negative. The solutions have form $x = c_1e^{\lambda_1 t} + c_2te^{\lambda_1 t}$ and, since $\lambda_1 < 0$, they again approach 0 as $t \to \infty$. The motion is much like that of Figure 7-73. We here say that the vibrations are *critically damped.* This is a borderline case, as the next paragraph shows.

III. $h^2 - 4k^2m < 0$. Now the characteristic roots are complex: $\alpha \pm \beta i$, where

$$\alpha = -\frac{h}{2m} < 0, \qquad \beta = \frac{\sqrt{4k^2m - h^2}}{2m} > 0$$

The solutions have form $x = c_1e^{\alpha t} \cos \beta t + c_2e^{\alpha t} \sin \beta t = e^{\alpha t}A \sin(\beta t + \gamma)$, as for simple harmonic motion. The motion is like that of Figure 7-71 except that, because $\alpha < 0$, the factor $e^{\alpha t}$ forces the vibrations to gradually die down, approaching 0 as $t \to \infty$, as in Figure 7-74. We call the motion a *damped vibration.*

The Effect of Driving Force. When we have an external driving force, the motion of the particle is described by adding to the complementary function a particular solution. We consider only the case of a sinusoidal driving force of frequency $\omega > 0$; that is, the equation

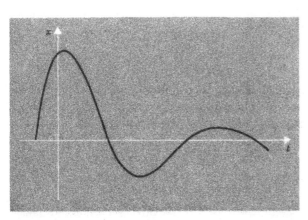

Figure 7-74 Damped vibrations.   **Figure 7-75** Resonance.

$$m \frac{d^2x}{dt^2} + h \frac{dx}{dt} + k^2x = B \sin \omega t \qquad (7\text{-}204)$$

We give two examples and leave further analysis to problems.

EXAMPLE 1.

$$\frac{d^2x}{dt^2} + 3 \frac{dx}{dt} + 2x = 2 \sin 2t$$

By the methods of Sections 7-18 and 7-19, the general solution is found to be

$$x = c_1 e^{-t} + c_2 e^{-2t} + \frac{2}{5} (-\sin 2t + 2 \cos 2t)$$

At t increases, the first two terms approach 0 (consequently, they are called *transients*), and, hence, for large positive t, only the last two terms are important. They represent a simple harmonic motion of the same frequency as the driving force.

EXAMPLE 2.

$$\frac{d^2x}{dt} + x = 2 \sin t$$

We determine the general solution to be $x = c_1 \cos t + c_2 \sin t - t \cos t$. The first two terms represent a simple harmonic motion, not damped. However, the last term represents an oscillation steadily growing in size, and hence dominating the motion for large positive t; this term is graphed in Figure 7-75. We have here an example of *resonance* or *sympathetic vibrations*. The driving force has the same frequency, 1, as the "natural" vibrations $x = c_1 \cos t + c_2 \sin t$. The forcing term and the natural vibrations reinforce each other and lead to the ever-growing oscillations.

PROBLEMS

1. In each of the following equations determine whether the motion is simple harmonic, overdamped, critically damped, or a damped vibration, and plot a typical nonzero solution.

(a) $d^2x/dt^2 + 2(dx/dt) + 2x = 0$ (b) $d^2x/dt^2 + 9x = 0$
(c) $d^2x/dt^2 + 4(dx/dt) + 3x = 0$ (d) $d^2x/dt^2 + 8(dx/dt) + 16x = 0$
(e) $d^2x/dt^2 + 16x = 0$ (f) $d^2x/dt^2 + 4(dx/dt) + 5x = 0$

2. (a) Example 1 suggests that, in general, Equation (7-204) has a solution of the form $a \cos \omega t + b \sin \omega t$, where a and b are constants. By substituting this expression in Equation (7-204), show that such a solution does exist with

$$a = \frac{-Bh\omega}{(k^2 - m\omega^2)^2 + h^2\omega^2}, \qquad b = \frac{B(k^2 - m\omega^2)}{(k^2 - m\omega^2)^2 + h^2\omega^2}$$

provided that the denominators are not 0. Show also that the latter case arises only when $h = 0$, $\omega^2 = k^2/m$.

(b) Show that, in the exceptional case of part (a), the driving function and natural vibrations both have frequency ω. Show, by substituting in the equation, that the equation has a particular solution of form $x = at \cos \omega t$, with $a = -B/(2m\omega)$. This is the case of resonance.

3. Determine the solution of Equation (7-202) so that $x = x_0$ and $dx/dt = v_0$ when $t = 0$.

4. Show that in Case I (overdamped motion) each nonzero solution has at most one critical point (where $dx/dt = 0$).

5. In mechanics it is shown that the motion of a *pendulum* (Figure 7-76) is governed by the equation

$$mL \frac{d^2\theta}{dt^2} = -mg \sin \theta$$

For small oscillations, we can replace $\sin \theta$ by θ to get an approximate equation for the motion. Use this approximation to show that the *period* of the pendulum is $T = 2\pi \sqrt{L/g}$.

6. *Torsional vibrations.* A body of mass m when suspended by a vertical wire can move by rotating about the wire as axis, so that the wire is twisted or untwisted. It is shown in mechanics that the motion is governed by the equation

$$I \frac{d^2\theta}{dt^2} + C\theta = 0$$

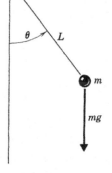

Figure 7-76
Pendulum.

where I and C are positive constants and θ is the angle through which the wire has been twisted. Describe the motion for the case in which $\theta = \pi/2$ and $d\theta/dt = 0$ for $t = 0$.

7-21 NUMERICAL EVALUATION OF INTEGRALS, TRAPEZOIDAL RULE

As we pointed out in Chapter 4 (see especially Sections 4-2, 4-14, and 4-25), a particular definite or indefinite integral may be difficult or impossible to evaluate by the techniques of indefinite integrals, and we are then forced to use some approximate method. In particular, the fact that

$$\int_a^b f(x)\,dx = \lim_{\text{mesh}\to 0} \sum_{i=1}^n f(\xi_i)\,\Delta_i x$$

allows us to approximate the integral sought by a sum

$$f(\xi_1)\,\Delta_1 x + \cdots + f(\xi_n)\,\Delta_n x$$

We call such a sum a *rectangular sum,* since it represents the sum of the areas of rectangles. As shown in Theorem 24 in Section 4-25, if we know that $|f'(x)| \leq K$ over the interval, then we can be sure that the rectangular sum differs from the integral by less than a given $\epsilon > 0$, if only the mesh of the subdivision is less than δ where $\delta = \epsilon/[K(b-a)]$. An example is worked out in that section.

The Trapezoidal Rule. Let $f(x)$ now be a given continuous function in the interval $[a, b]$. To evaluate the integral of f over this interval, we can subdivide the interval $[a, b]$ as usual but use, instead of rectangles, trape-

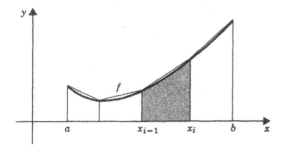

Figure 7-77   Trapezoidal rule.

zoids, as suggested in Figure 7-77. The shaded trapezoid has bases $f(x_{i-1})$ and $f(x_i)$ and altitude $\Delta_i x = x_i - x_{i-1}$. Hence its area is

$$\frac{f(x_{i-1}) + f(x_i)}{2} \cdot \Delta_i x$$

From the figure, it appears that the area of the trapezoid should very closely approximate the area under the curve. Hence we are led to approximate the integral sought by the sum

$$\frac{f(x_0) + f(x_1)}{2} \cdot \Delta_1 x + \frac{f(x_1) + f(x_2)}{2} \cdot \Delta_2 x + \cdots + \frac{f(x_{n-1}) + f(x_n)}{2} \cdot \Delta_n x$$

If we use equal subdivision intervals of length Δx, our expression becomes

$$\frac{\Delta x}{2}[f(x_0) + 2f(x_1) + 2f(x_2) + \cdots + 2f(x_{n-1}) + f(x_n)]$$

In either case, we describe the expression as a trapezoidal rule for computing the integral.

We discuss the accuracy of the rule below, but first we consider an example.

EXAMPLE. We first consider

$$\int_0^1 \frac{4}{1 + x^2}\, dx = 4\, \text{Tan}^{-1} x \Big|_0^1 = \pi = 3.1415927\ldots$$

To test the trapezoidal rule, we use it to compute this integral. We first use just one subinterval, taking $\Delta x = 1$, so that we compute

$$\frac{f(0) + f(1)}{2} \times 1 = \frac{4 + 2}{2} \times 1 = 3$$

This is too small, by about 5 percent. We next take $\Delta x = 0.5$, and compute

$$\frac{f(0) + 2f(0.5) + f(1)}{2} \cdot \frac{1}{2} = \frac{4 + 6.4 + 2}{4} = 3.1$$

This is still too small, but the error is only about 1 percent. We take $\Delta x = 0.25$ and obtain the results shown in Table 7-1. The parentheses in

Table 7-1

x	$f(x)$	$2f(x)$	
0	4	(4.0000)	
0.25	3.7647	7.5294	$\dfrac{25.0494}{2} \times 0.25 = 3.1312.$
0.50	3.2000	6.4000	
0.75	2.5600	5.1200	
1	2.0000	(2.0000)	
		Sum 25.0494	

the last column indicate that $f(x)$, not $2f(x)$, is used. The value obtained is now only about 0.3 of 1 percent too small.

We now examine the accuracy of the method. We first remark that each trapezoidal sum can be interpreted as a rectangular sum for the same subdivision. Indeed, $[f(x_{i-1}) + f(x_i)]/2$ is the average of the values $f(x_{i-1})$, $f(x_i)$, and hence must lie between these values. By the Intermediate Value Theorem we can thus write

$$\frac{1}{2} [f(x_{i-1}) + f(x_i)] = f(\xi_i), \qquad x_{i-1} < \xi_i < x_i$$

If this replacement is made in each term, the trapezoidal sum becomes a rectangular sum. We can therefore conclude that, as the mesh tends to zero, the trapezoidal sums also have the integral sought as limit. Furthermore, as in Section 4-25, the error is at most ϵ for a subdivision of mesh less than δ, where $\delta = \epsilon/[K(b - a)]$ and $|f'(x)| \le K$ on the interval.

We next notice that if $f''(x)$ exists and is positive throughout the interval $[a, b]$, then the graph of $y = f(x)$ will lie below the chord in each sub-interval, as in Figure 7-77 (see Section 6-3). Hence, in this case, the trapezoidal rule will give too large a value (error positive). Similarly, if $f''(x) < 0$ in the interval, the rule gives too small a value (error negative). When $f''(x)$ is alternately positive and negative, we thus have some cancellation of errors, which will help. This is the case for our example, for which we find that

$$f''(x) = 4 \frac{6x^2 - 2}{(1 + x^2)^3}, f'''(x) = 96x(1 + x^2)^{-4}(1 - x^2)$$

From this expression we observe that $f''(x)$ increases as x increases, with $f''(0) = -8, f''(1) = 2$, and $f''(x) = 0$ only for $x = \sqrt{3}/3 = 0.58$ in the interval $[0, 1]$.

Further information concerning the error is given by the following result:

THEOREM 8. *Let $f(x)$ be defined and have a continuous second derivative for $a \leq x \leq b$. Let a subdivision: $x_0 = a < x_1 < \cdots < x_n = b$ of the interval $[a, b]$ be chosen and let T_n denote the value of the trapezoidal sum for this subdivision.*

Let I denote the value of $\int_a^b f(x) \, dx$. Then

$$T_n - I = -\frac{1}{2} \int_a^b f''(x)\varphi_n(x) \, dx \tag{7-210}$$

where $\varphi_n(x)$ is the continuous function equal to $(x - x_{i-1})(x - x_i)$ for $x_{i-1} \leq x \leq x_i$, $i = 1, 2, \ldots, n$. If the subdivision intervals are equal, then one can write, for some ξ,

$$T_n - I = f''(\xi)\frac{(b - a)^3}{12n^2}, \qquad a < \xi < b \tag{7-211}$$

Hence if $|f''(x)| \leq L$ throughout the interval

$$|T_n - I| \leq \frac{L(b - a)^3}{12n^2} \tag{7-212}$$

The proof is given in Section 7-23. The relation (7-211) shows the influence of the sign of f'' on the error; this can also be seen in (7-210), since $\varphi_n(x)$ is always negative or zero. The rule (7-212) tells us how n can be chosen to insure that the error is less than a prescribed error ϵ. We find that the error is less than ϵ in absolute value if

$$n > \left[\frac{L(b - a)^3}{12\epsilon}\right]^{1/2} \tag{7-213}$$

For our example, $|f''(x)| \leq 8 = L$, $a = 0$, $b = 1$, so that (7-212) gives

$$|T_n - I| < \frac{2}{3n^2}$$

Thus, for $n = 1$, the error is less than $2/3 = 0.67$; for $n = 2$, the error is less than $2/12 = 0.17$; for $n = 4$, the error is less than $2/48 = 0.042$. We observed that the errors were actually much smaller than these values (for example 0.0103 for $n = 4$). Clearly, the change in sign of $f''(x)$ helped.

7-22 SIMPSON'S RULE

The trapezoidal rule was based on replacement of $f(x)$ by a linear function whose graph joins two successive points of the graph of f. If we take three successive points, we cannot use a linear function, but we can use a quadratic function: $y = Ax^2 + Bx + C$. This leads to Simpson's rule.

To formulate the rule in detail, we consider a subdivision of our interval $[a, b]$ into an even number of subintervals and require that each successive pair of subintervals be formed of two equal intervals: $\Delta_1 x = \Delta_2 x$, $\Delta_3 x = \Delta_4 x, \ldots$. We consider a typical pair of subintervals, which we can denote by $[x_{2k-2}, x_{2k-1}]$, $[x_{2k-1}, x_{2k}]$ (Figure 7-78). We then wish to find a quadratic

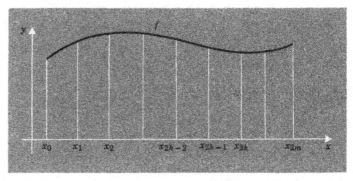

Figure 7-78   Simpson's rule.

function passing through the three points $[x_i\, f(x_i)]$ for $i = 2k$, $2k - 1$, $2k - 2$.

To simplify our work we can translate parallel to the x-axis and take the case when our three x-values are $-h$, 0, h, as in Figure 7-79. We let $f(-h) = p$, $f(0) = q$, $f(h) = r$. We then seek a quadratic function

$$g(x) = Ax^2 + Bx + C$$

such that

$$g(-h) = p, \qquad g(0) = q, \qquad g(h) = r$$

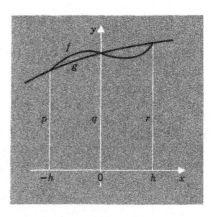

Figure 7-79   Special case for Simpson's rule.

as in Figure 7-79. We obtain the equations

$$Ah^2 - Bh + C = p, \qquad C = q, \qquad Ah^2 + Bh + C = r$$

and, hence,

$$A = \frac{p + r - 2q}{2h^2}, \qquad B = \frac{r - p}{h}, \qquad C = q$$

Thus $g(x)$ is well determined. Of course, A may be 0, so that g is actually a linear function. (When will this happen?)

We now compute the integral of $g(x)$ from $-h$ to h:

$$\int_{-h}^{h} g(x)\, dx = \int_{-h}^{h} (Ax^2 + Bx + C)\, dx = \frac{2Ah^3}{3} + 2Ch$$

With the values for A and C from above, we find that

$$\int_{-h}^{h} g(x)\, dx = \frac{2h^3}{3} \cdot \frac{p + r - 2q}{2h^2} + 2qh = \frac{h}{3} \cdot (p + 4q + r)$$

$$= \frac{h}{3}[f(-h) + 4f(0) + f(h)]$$

If we now translate back to the original x-values, we obtain the expression

$$\frac{\Delta_{2k}x}{3}[f(x_{2k-2}) + 4f(x_{2k-1}) + f(x_{2k})]$$

as an approximation to the integral of f from x_{2k-1} to x_{2k}. If we add these expressions for $k = 1, 2, \ldots, m$, where $n = 2m$, we obtain *Simpson's rule*:

$$\int_{a}^{b} f(x)\, dx \sim \frac{1}{3} \sum_{k=1}^{m} \Delta_{2k}x[f(x_{2k-2}) + 4f(x_{2k-1}) + f(x_{2k})] \qquad (7\text{-}220)$$

When all subdivision intervals are equal, the rule becomes

$$\int_{a}^{b} f(x)\, dx \sim \frac{\Delta x}{3} \sum_{k=1}^{m} [f(x_{2k-2}) + 4f(x_{2k-1}) + f(x_{2k})]$$

$$\qquad\qquad\qquad\qquad\qquad\qquad\qquad (7\text{-}221)$$

$$\sim \frac{\Delta x}{3}[f(x_0) + 4f(x_1) + 2f(x_2) + 4f(x_3) + \cdots + 2f(x_{n-2})$$

$$+ 4f(x_{n-1}) + f(x_n)]$$

EXAMPLE. We again consider the integral

$$\int_{0}^{1} \frac{4}{1 + x^2}\, dx = \pi = 3.1415927\ldots$$

With $n = 2$, (7-221) gives the value

$$\frac{0.5}{3}\left[4 + 4 \cdot \frac{16}{5} + 2\right] = 3.1333$$

The accuracy is even better than the accuracy obtained by the trapezoidal rule with $n = 4$. If we take $n = 4$, (7-221) gives (with the aid of Table 7-1 above)

$$\frac{0.25}{3}[4 + 4 \times 3.764706 + 2 \times 3.2 + 4 \times 2.56 + 2] = 3.141569$$

The error is in the fifth decimal place.

Thus we have a method for obtaining a very accurate value of π with ease.

We now consider the accuracy of Simpson's rule. We first notice that, as for the trapezoidal rule, Simpson's rule can be considered as providing a rectangular sum. We consider the subdivision by the points $x_0, x_2, x_4, \ldots, x_n$, where $n = 2m$. Simpson's rule provides a term for each pair of successive subintervals, which we can write as

$$\frac{f(x_{2k-2}) + 4f(x_{2k-1}) + f(x_{2k})}{6} \cdot (x_{2k} - x_{2k-2})$$

the fraction is a weighted average of the values of f at the three points $x_{2k-2}, x_{2k-1}, x_{2k}$ and, hence, lies between the maximum and minimum of f in the interval $[x_{2k-2}, x_{2k}]$ (see Problem 5 below). Thus we can write the whole term as

$$f(\xi_k)(x_{2k} - x_{2k-2}), \qquad x_{2k-2} < \xi_k < x_{2k}$$

and Simpson's rule approximates the integral by the rectangular sum

$$\sum_{k=1}^{m} f(\xi_k)(x_{2k} - x_{2k-2})$$

It now follows, as for the trapezoidal rule that, as the mesh approaches 0, the values obtained by Simpson's rule approach the exact value of the integral.

From the way we obtained Simpson's rule, it must give an exact value when f itself is a quadratic function. Surprisingly, the rule continues to give an exact value when f is a cubic polynomial (Problems 9, 10 below). We have a general theorem on the error like that for the trapezoidal rule.

THEOREM 9. *Let $f(x)$ be defined and have a continuous fourth derivative for $a \leq x \leq b$. Let a subdivision of the interval $[a, b]$ be chosen, with equal pairs of subintervals: $x_0 < x_1 = x_0 + \Delta_1 x < x_2 = x_0 + 2\Delta_1 x < x_3 < \cdots < x_n = b$, and let S_n denote the value of the sum given by Simpson's rule for this subdivision. Let I denote the value of $\int_a^b f(x)\, dx$.*

Let $n = 2m$. Then

$$S_n - I = -\int_a^b f^{(iv)}(x)\psi_n(x)\, dx \tag{7-222}$$

where $\psi_n(x)$ denotes the continuous function defined as follows:

$$\psi_n(x) = \begin{cases} \dfrac{1}{72}(x - x_{2k-2})^3(3x - 3x_{2k-1} - \Delta_{2k}x), & x_{2k-2} \leq x \leq x_{2k-1} \\[2mm] \dfrac{1}{72}(x - x_{2k})^3(3x - 3x_{2k-1} + \Delta_{2k}x), & x_{2k-1} \leq x \leq x_{2k} \end{cases} \tag{7-223}$$

for $k = 1, 2, \ldots, m$. Furthermore, if all subdivision intervals are equal, for some ζ

$$S_n - I = f^{(iv)}(\xi) \frac{(b-a)^5}{180n^4}, \qquad a < \xi < b \qquad (7\text{-}224)$$

Hence, if $|f^{(iv)}(x)| \leq M$ throughout the interval,

$$|S_n - I| \leq \frac{M(b-a)^5}{180n^4} \qquad (7\text{-}225)$$

The proof is given in the next section. The function $\psi_n(x)$ is zero at the points x_0, x_2, and x_4, . . . and is negative between these values (see Figure 7-80). Hence the sign of the error is now determined by the fourth derivative of f. If that derivative is always positive, the error must be positive (value too large); if that derivative is always negative, the error must be negative (value too small).

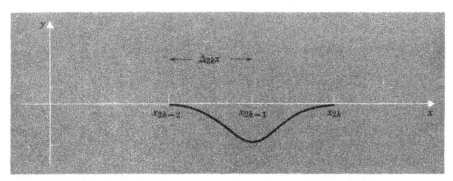

Figure 7-80   The function $\psi_n(x)$ in error expression for Simpson's rule.

‡7-23   PROOFS OF EXPRESSIONS FOR ERROR IN TRAPEZOIDAL AND SIMPSON'S RULES

We first proceed to prove the Theorem 8 of Section 7-21. We are given a function f in the interval $[a, b]$, having a continuous second derivative f''. We subdivide the interval $[a, b]$ by points $a < x_0 < x_1 \cdots < x_n = b$ and let $g(x)$ denote the continuous function that agrees with f at the subdivision points and is linear between these values. The trapezoidal rule then provides $\int_a^b g(x)\,dx$ as an approximation to $\int_a^b f(x)\,dx$. The error made is

$$T_n - I = \int_a^b g(x)\,dx - \int_a^b f(x)\,dx = \int_a^b [g(x) - f(x)]\,dx = \int_a^b F(x)\,dx$$

where $F(x) = g(x) - f(x)$. Accordingly, F has zeros at all subdivision points, and in each interval $[x_{i-1}, x_i]$, F has a continuous second derivative. We integrate by parts, taking advantage of the zeros at x_{i-1} and x_i:

$$\int_{x_{i-1}}^{x_i} F(x)\,dx = F(x)(x-k)\Big|_{x_{i-1}}^{x_i} - \int_{x_{i-1}}^{x_i} F'(x)(x-k)\,dx = -\int_{x_{i-1}}^{x_i} F'(x)(x-k)\,dx$$

Here k is a constant to be adjusted. We integrate by parts again:

$$\int_{x_{i-1}}^{x_i} F(x)\,dx = -F'(x)[(\tfrac{1}{2})x^2 - kx + l]\Big|_{x_{i-1}}^{x_i} + \int_{x_{i-1}}^{x_i} F''(x)[(\tfrac{1}{2})x^2 - kx + l]\,dx$$

Here l is another constant to be adjusted. We now choose k and l so that the first term on the right drops out; that is, so that the quadratic function has zeros at x_{i-1} and x_i. This leads to the relation

$$\frac{1}{2} x^2 - kx + l = \frac{1}{2}(x - x_{i-1})(x - x_i)$$

from which k and l can be found. We do not need the values explicitly. We can now write

$$\int_{x_{i-1}}^{x_i} F(x)\, dx = \frac{1}{2} \int_{x_{i-1}}^{x_i} F''(x)(x - x_{i-1})(x - x_i)\, dx$$

But $F(x) = g(x) - f(x)$, so that $F''(x) = g''(x) - f''(x) = -f''(x)$, since g is linear and has second derivative 0. Thus, finally,

$$\int_{x_{i-1}}^{x_i} F(x)\, dx = -\frac{1}{2} \int_{x_{i-1}}^{x_i} f''(x)(x - x_{i-1})(x - x_i)\, dx$$

If we add these relations for $i = 1, 2, \ldots, n$ we obtain

$$T_n - I = \int_a^b F(x)\, dx = -\frac{1}{2} \int_a^b f''(x)\varphi_n(x)\, dx$$

so that (7-210) is proved, with $\varphi_n(x)$ as defined in the theorem. Let f'' have absolute minimum A and absolute maximum B in $[a, b]$. Then, since $\varphi_n(x)$ is always negative or 0 in $[a, b]$, we can write $B\varphi_n(x) \leq f''(x)\varphi_n(x) \leq A\varphi_n(x)$ in $[a, b]$, so that, as in Section 4-23,

$$B \int_a^b \varphi_n(x)\, dx \leq \int_a^b f''(x)\varphi_n(x)\, dx \leq A \int_a^b \varphi_n(x)\, dx$$

Hence, by the Intermediate Value Theorem,

$$\int_a^b f''(x)\varphi_n(x)\, dx = f''(\xi) \int_a^b \varphi_n(x)\, dx, \qquad a < \xi < b$$

so that

$$T_n - I = -\frac{1}{2} f''(\xi) \int_a^b \varphi_n(x)\, dx$$

The integral of $\varphi_n(x)$ is easily evaluated (Problem 6 below), and we obtain (7-211), from which (7-212) follows.

Next we prove Theorem 9 of Section 7-22. Now f has a continuous fourth derivative $f^{(iv)}(x)$ in $[a, b]$ and the number n of subdivision intervals is even: $n = 2m$. We let $g(x)$ be the function that agrees with f at each subdivision point and that is a quadratic function of x in each interval $[x_0, x_2], [x_2, x_4], \ldots, [x_{n-2}, x_n]$. Then Simpson's rule tells us to compute

$$S_n = \int_a^b g(x)\, dx$$

as an approximation to

$$\int_a^b f(x)\, dx$$

The error is

$$S_n - I = \int_a^b g(x)\, dx - \int_a^b f(x)\, dx = \int_a^b F(x)\, dx$$

where $F(x) = g(x) - f(x)$. Thus, in particular, F has zeros at all the subdivision points. Also within each interval $[x_{2k-2}, x_{2k}]$, F is a function with a continuous fourth derivative. We now assert that

$$\int_{x_{2k-2}}^{x_{2k}} F(x)\, dx = \int_{x_{2k-2}}^{x_{2k}} F^{(iv)}(x)\psi_n(x)\, dx \qquad (7\text{-}230)$$

for $k = 1, 2, \ldots, m$, where $\psi_n(x)$ is defined by (7-223). It is somewhat simpler to prove this formula starting with the right-hand side (rather than with the left, as above, for the trapezoidal rule). To simplify writing, we let $c = x_{2k-1}$, and $x_{2k-2} = c - h$, $x_{2k} = c + h$. Then by (7-223)

$$\psi_n(x) = \begin{cases} \dfrac{1}{72}(x - c + h)^3(3x - 3c - h), & c - h \leq x \leq c \\[2mm] \dfrac{1}{72}(x - c - h)^3(3x - 3c + h), & c \leq x \leq c + h \end{cases} \qquad (7\text{-}231)$$

We verify easily that the function defined by (7-231) is continuous, along with its first and second derivatives, in the interval $[c - h, c + h]$ (Problem 7 below). The right side of (7-230) can be written as a sum of two terms:

$$\int_{c-h}^c F^{(iv)}(x)\psi_n(x)\, dx + \int_c^{c+h} F^{(iv)}(x)\psi_n(x)\, dx$$

We evaluate each separately by integration by parts:

$$\int_{c-h}^c F^{(iv)}(x)\psi_n(x)\, dx = F'''(x)\psi_n(x)\Big|_{c-h}^c - \int_{c-h}^c F'''(x)\psi_n'(x)\, dx$$

$$= F'''(c)\psi_n(c) - \int_{c-h}^c F'''(x)\psi_n'(x)\, dx$$

Here we used the fact that $\psi_n(c - h) = 0$, as follows from (7-231). Since also $\psi_n'(c - h) = 0$, $\psi_n''(c - h) = 0$ (Problem 7 below), we find similarly that

$$\int_{c-h}^c F^{(iv)}(x)\psi_n(x)\, dx = F'''(c)\psi_n(c) - F''(c)\psi_n'(c) + \int_{c-h}^c F''(x)\psi_n''(x)\, dx$$

$$= F'''(c)\psi_n(c) - F''(c)\psi_n'(c) + F'(c)\psi_n''(c) - \int_{c-h}^c F'(x)\psi_n'''(x)\, dx$$

$$= F'''(c)\psi_n(c) - F''(c)\psi_n'(c) + F'(c)\psi_n''(c)$$

$$- [F(x)\psi_n'''(x)]\Big|_{c-h}^c + \int_{c-h}^c F(x)\psi_n^{(iv)}(x)\, dx$$

Now $\psi_n(x)$ is a polynomial in x in the interval $[c - h, c]$. Hence its third and fourth derivatives exist (these are right-hand derivatives at $c - h$, left-hand derivatives at c). Since F has zeros at c and at $c - h$ the bracket drops out. Also $\psi_n^{(iv)}(x) \equiv 1$. Hence, finally,

$$\int_{c-h}^{c} F^{(iv)}(x)\psi_n(x) \, dx = F'''(c)\psi_n(c) - F''(c)\psi_n'(c) + F'(c)\psi_n''(c) + \int_{c-h}^{c} F(x) \, dx$$

Similarly, we find that

$$\int_{c}^{c+h} F^{(iv)}(x)\psi_n(x) \, dx = -F'''(c)\psi_n(c) + F''(c)\psi_n'(c) - F'(c)\psi_n''(c) + \int_{c}^{c+h} F(x) \, dx$$

If we add (observing the continuity of ψ_n, ψ_n', ψ_n''), all terms except the integrals cancel, and we obtain (7-230).

Now $F^{(iv)}(x) = g^{(iv)}(x) - f^{(iv)}(x) = -f^{(iv)}(x)$ in $[x_{2k-2}, x_{2k}]$. Hence (7-230) can be written

$$\int_{x_{2k-2}}^{x_{2k}} F(x) \, dx = -\int_{x_{2-2}}^{x_{2k}} f^{(iv)}(x)\psi_n(x) \, dx$$

If we add these relations for $k = 1, 2, \ldots, m$, we find that

$$S_n - I = \int_{a}^{b} F(x) \, dx = -\int_{a}^{b} f^{(iv)}(x)\psi_n(x) \, dx$$

so that (7-222) is proved. As for the trapezoidal rule, we now reason that

$$S_n - I = -f^{(iv)}(\xi) \int_{a}^{b} \psi_n(x) \, dx$$

For equal subdivision intervals, the integral of ψ_n is found to be $(b - a)^5 n^{-4}/180$ (Problem 8 below), so that (7-224) and also (7-225) follow.

PROBLEMS

1. Let the integral $\int_0^{\pi} (x/\pi) \sin x \, dx$ be given.

   (a) Find the exact value.
   (b) Evaluate by the trapezoidal rule, with equal subdivisions, for $n = 1$, then for $n = 2$, and for $n = 4$.
   (c) Evaluate by Simpson's rule, with equal subdivisions, for $n = 2$, then for $n = 4$ and for $n = 6$.

2. Let the integral $\int_1^{10} (1/x) \, dx = \ln 10 = 2.3025850930$ be given.

   (a) Evaluate by the trapezoidal rule with the subdivision 1, 5, 10, and then with the subdivision 1, 3, 5, 7, 9, and 10.
   (b) Evaluate by Simpson's rule with the subdivision 1, 5.5, 10, then with the subdivision 1, 3, 5, 7.5, and 10.

3. Each of the following integrals is to be evaluated by the trapezoidal rule with n equal subdivisions. Give a value of n which insures that the error is in absolute value less than the value ϵ given:

   (a) $\int_0^1 \dfrac{1}{1 + x^3} \, dx$, $\epsilon = 0.5$ (b) $\int_0^1 \sin x^2 \, dx$, $\epsilon = 0.1$

   (c) $\int_0^1 \sqrt{1 + x^2} \, dx$, $\epsilon = 0.001$ (d) $\int_0^{\pi} e^{\sin x} \, dx$, $\epsilon = 0.02$

4. Each of the following integrals is to be evaluated by Simpson's rule with $n = 2m$

equal subdivisions. Give a value of n which insures that the error is in absolute value less than the value ϵ given:

(a) $\int_0^{\pi/2} \ln(1 + \sin x)\, dx,$ $\epsilon = 0.01$

(b) $\int_1^2 \text{Tan}^{-1} x^2\, dx,$ $\epsilon = 0.05$

5. (a) Show that the number $(y_0 + 4y_1 + y_2)/6$ lies between the smallest and the largest of the numbers y_0, y_1, y_2.

(b) Show that if $w_1, \ldots, w_k$ are positive numbers and $A \leq y_1 \leq B$, $A \leq y_2 \leq B$, $\ldots, A \leq y_k \leq B$, then

$$A \leq \frac{w_1 y_1 + w_2 y_2 + \cdots + w_k y_k}{w_1 + w_2 + \cdots + w_k} \leq B$$

(c) Interpret the result of (b) in terms of the center of mass of k particles.

6. (a) Show that

$$\int_{x_{i-1}}^{x_i} (x - x_{i-1})(x - x_i)\, dx = -\frac{(x_i - x_{i-1})^3}{6}$$

(b) Let $\varphi_n(x)$ be defined as in Section 7-21 for equal subdivisions. Show, with the aid of the result of part (a), that

$$\int_a^b \varphi_n(x)\, dx = -\frac{(b - a)^3}{6n^2}$$

7. Let $\psi_n(x)$ be defined for $c - h \leq x \leq c + h$ by (7-231).

(a) Show that $\psi_n, \psi_n', \psi_n''$ are continuous, that ψ_n''' and $\psi_n^{(iv)}$ have a discontinuity at $x = c$, but $\psi_n^{(iv)}$ has limit 1 as x approaches c.

(b) Show that $\psi_n(c \pm h) = 0$, $\psi_n'(c \pm h) = 0$, $\psi_n''(c \pm h) = 0$.

(c) Show that $\int_{c-h}^{c+h} \psi_n(x)\, dx = -h^5/90$.

8. Let $\psi_n(x)$ be defined as in Section 7-22, with n equal subdivisions. Show with the aid of the result of part (c) of Problem 7 that

$$\int_a^b \psi_n(x)\, dx = \frac{(b - a)^5}{180n^4}$$

9. Show that Simpson's rule gives the exact value for $\int_a^b f(x)\, dx$ when f is a cubic polynomial. [Hint. Take $a = -h$, $b = h$ and let $g(x)$ be the quadratic function that agrees with f at $-h$, 0, and h. Then $f(x) - g(x)$ is a cubic polynomial with zeros at $-h$, 0, and h. Show by symmetry that $\int_{-h}^h [f(x) - g(x)]\, dx = 0.$]

10. Prove the result of Problem 9 by applying (7-222).

11. (a) Show from (7-210) that in applying the trapezoidal rule with unequal subdivisions, it is best to place the subdivision points where $|f''(x)|$ has its largest values.

(b) Show from (7 222) and (7-223) that in applying Simpson's rule with unequal subdivisions (but with equal pairs of subintervals), it is best to place the subdivision points $x_2, x_4, \ldots$ where $|f^{(iv)}(x)|$ has its largest values.

8

INFINITE SERIES

8-1 INTRODUCTION

An infinite series is a limit process given in terms of successive steps of addition. An example is the infinite series

$$\frac{1}{2} + \frac{1}{4} + \cdots + \frac{1}{2^n} + \cdots \tag{8-10}$$

Here we can think of a person walking one-half mile, then one-quarter mile, and so on; his first step is $\frac{1}{2}$, his second is $\frac{1}{4}$, ..., his nth step is $1/2^n$ (all in miles). When we write down the series as in (8-10), we are describing the process and at the same time saying: find the limit, if there is one. The limit is the total approached as the number of steps approaches ∞. After one step the total is $\frac{1}{2}$, after two it is $\frac{3}{4}$, after three it is $\frac{7}{8}$, after four it is $\frac{15}{16}$, and so on. Thus we seem to be approaching 1, and we would then write

$$1 = \frac{1}{2} + \frac{1}{4} + \cdots + \frac{1}{2^n} + \cdots \tag{8-11}$$

We shall see below that the limit is indeed 1, so that we are justified in writing (8-11).

We can give another interpretation of (8-10) that makes the conclusion very plausible: A person at a certain distance from a wall walks $\frac{1}{2}$ the distance to the wall, then $\frac{1}{2}$ of the remaining distance, then $\frac{1}{2}$ of the remaining distance, and so on. If we take the original distance from the wall as unit, then the distances moved are $\frac{1}{2}, \frac{1}{4}, \frac{1}{8}$, and so on, so that we obtain the infinite series (8-10). The person is clearly approaching the wall as limit—that is, his total distance is approaching 1, the given unit, and (8-11) follows. In ancient times this process and similar ones gave rise to disputes among philosophers, who found it strange that a finite number, 1, could be obtained by making an "infinite number of additions."

A very common example of an infinite series is an infinite decimal: for example, 0.33333 . . . . We all know that this decimal equals $\frac{1}{3}$. When we write $\frac{1}{3} = 0.33333$ . . . , we are saying that

$$\frac{1}{3} = \frac{3}{10} + \frac{3}{100} + \cdots + \frac{3}{10^n} + \cdots \tag{8-12}$$

or that the sequence of numbers

$$\frac{3}{10}, \quad \frac{3}{10} + \frac{3}{100}, \ldots, \quad \frac{3}{10} + \frac{3}{100} + \cdots + \frac{3}{10^n}, \ldots$$

has $\frac{1}{3}$ as limit. The way in which the decimal is defined assures us that the limit is $\frac{1}{3}$. For example, the first term $\frac{3}{10}$ is selected because $\frac{1}{3}$ lies between $\frac{3}{10}$ and $\frac{4}{10}$; the second term $\frac{3}{100}$ is chosen because $\frac{1}{3}$ lies between

$$\frac{3}{10} + \frac{3}{100} \quad \text{and} \quad \frac{3}{10} + \frac{4}{100}$$

In each case we are dividing the number axis more and more finely—into tenths, then hundredths, and so on, as suggested in Figure 8-1. At each step

Figure 8-1 Meaning of $1/3 = 0.33333\ldots$

we move as close to $\frac{1}{3}$ as we can get, by using the subdivisions of that stage, stopping to the left of $\frac{1}{3}$. Thus, after the first step, we know our position is within $\frac{1}{10}$ of $\frac{1}{3}$; after the second step, we are within $\frac{1}{100}$, and so on. Thus the limiting position is $\frac{1}{3}$, and (8-12) is correct.

The same interpretation is valid for every infinite decimal: for example,

$$\frac{1}{7} = 0.142857\ 142857\ldots, \qquad \sqrt{2} = 1.414214\ldots$$
$$e = 2.71828\ 18285\ldots, \qquad \pi = 3.1415927\ldots$$

In each case, we have an expression of the form

$$c = A + \frac{d_1}{10} + \frac{d_2}{10^2} + \cdots + \frac{d_n}{10^n} + \cdots \tag{8-13}$$

where A is an integer and the numbers $d_1, d_2, \ldots, d_n, \ldots$ are digits—that is are chosen from the integers $0, 1, \ldots, 9$. The expression (8-13) expresses the limit relation:

$$c = \lim_{n \to \infty} \left(A + \frac{d_1}{10} + \cdots + \frac{d_n}{10^n} \right)$$

All these examples indicate that the limit process involved is that of an *infinite sequence*. The infinite series is simply a way of obtaining an infinite sequence: by forming the sums of 1 term, 2 terms, $\ldots$, n terms, $\ldots$; the "sum" of the infinite series is the limit of this sequence:

$$\text{sum of series} = \lim_{n \to \infty} (\text{sum of first } n \text{ terms})$$

We know very well that an infinite sequence may or may not converge; that is, have a limit. For infinite series it is essential to know when the limit does exist; we then call the series *convergent*. In this chapter we shall present a number of "convergence tests" for this purpose.

In the practical applications of mathematics one is very frequently led to infinite series. For example, the motion of the planets (or of artificial satellites) is governed by such complicated equations that the positions at any given

time are best calculated as infinite series, which must then be "summed"—
that is, the corresponding limit is the desired numerical value in each case.
When we know that the series converges, we know that we can obtain a
numerical value as close as desired to the true value by simply adding enough
terms. The question of how many terms are needed for given accuracy can
also be studied mathematically; in practice, one often develops a good instinct for this, but instinct alone can lead to serious errors.

8-2 INFINITE SEQUENCES

We have seen that infinite series are closely related to infinite sequences.
Hence we begin our study of infinite series by considering sequences. The
basic properties of infinite sequences are developed in Sections 2-12 and 2-13.
A review of these sections at this point will be found helpful. We shall assume
familiarity with the basic concepts and properties of infinite sequences. We
shall often write simply "sequence" for "infinite sequence."

In particular, we assume familiarity with the following concepts:

1. The convergence and divergence of a sequence.
2. The limit of a sequence.
3. The meaning of $\lim x_n = \infty$ and $\lim x_n = -\infty$.
4. Bounded sequences, sequences bounded above or below.
5. Monotone nondecreasing or strictly increasing sequences.
6. Monotone nonincreasing or strictly decreasing sequences.
7. The rules for limit of a sum, difference, product, or quotient.

We now introduce some new terminology:

A sequence x_n such that $x_n \to \infty$ or $x_n \to -\infty$ is termed *properly divergent*, or *divergent to* ∞ or *to* $-\infty$. A sequence that is divergent but not properly divergent is said to be *oscillatory-divergent*, or simply *oscillating*. For
example, the sequence $\{2^n\}$ is properly divergent, whereas the sequence
$\{(-1)^n\}$ or $\{-1, 1, -1, 1, \ldots\}$ is oscillatory-divergent.

A sequence that is not bounded is termed *unbounded*. Every unbounded
sequence diverges (Problem 12 below), but not every bounded sequence converges; for example, $\{(-1)^n\}$ is bounded but divergent.

A sequence with limit 0 is said to be a *null sequence*. Thus $x_n \to c$ if, and
only if, $\{x_n - c\}$ is a null sequence. Also two convergent sequences $\{x_n\}$,
$\{y_n\}$ have the same limit if, and only if, $\{x_n - y_n\}$ is a null sequence (Problem
2 below).

THEOREM 1. *Every bounded monotone sequence* $\{x_n\}$ *is convergent.*

This is Theorem H of Section 2-12. As shown in Section 2-13, the limit c of
the monotone sequence $\{x_n\}$ is the least upper bound of the set of values x_n,
when the sequence is nondecreasing; it is the greatest lower bound of these
values, when the sequence is nonincreasing.

COROLLARY OF THEOREM 1. *A monotone nondecreasing sequence
either converges or diverges to* ∞; *a monotone nonincreasing sequence
either converges or diverges to* $-\infty$.

PROOF. Let $\{x_n\}$ be nondecreasing, so that $x_1 \leq x_2 \leq \cdots x_n \leq \cdots$. If $\{x_n\}$ is bounded, then $\{x_n\}$ is convergent by Theorem 1. If $\{x_n\}$ is unbounded, then it must be unbounded above, since it is bounded below by x_1. Hence, for every K, we can choose N so that $x_n > K$ for $n > N$; that is, $x_n \to \infty$. There is a similar proof for $\{x_n\}$ nonincreasing.

THEOREM 2. *Let the real function f be such that f is continuous at x_0 and*

$$\lim_{x \to x_0} f(x) = c$$

Let $\{x_n\}$ be a sequence such that $f(x_n)$ is defined for $n > n_0$ and $x_n \to x_0$. Then

$$\lim_{n \to \infty} f(x_n) = c$$

PROOF. This theorem is related to Theorem C of Chapter 2 (see Section 2-12), but we give a separate proof. Given $\epsilon > 0$, we can choose $\delta > 0$ so that $|f(x) - c| < \epsilon$ for $|x - x_0| < \delta$. Since $\{x_n\}$ has limit x_0, we can choose N so large (and larger than n_0) that $|x_n - x_0| < \delta$ for $n > N$; hence also $|f(x_n) - c| < \epsilon$ for $n > N$, and the conclusion follows.

Remark. The condition that f be continuous at x_0 can be omitted if we know that $x_n \neq x_0$ for $n > n_0$. There are analogous results for the cases

$$\lim_{x \to x_0+} f(x) = c, \quad \lim_{x \to x_0-} f(x) = c, \quad \lim_{x \to \infty} f(x) = c, \quad \lim_{x \to -\infty} f(x) = c$$

and for the cases in which c is replaced by ∞ or $-\infty$.

EXAMPLE 1. The sequence $\{3^{1/n}\}$. Here we take $f(x) = 3^x$ and $x_n = 1/n$. The function f is continuous for all x, and f has limit $3^0 = 1$ as $x \to 0$. Since $x_n \to 0$, we conclude that

$$\lim_{n \to \infty} 3^{1/n} = 3^0 = 1$$

EXAMPLE 2. The sequence $\{n^2/(3n^2+1)\}$. Here we take $f(x) = x^2/(3x^2+1)$ and $x_n = n$. Thus $x_n \to \infty$ and

$$\lim_{x \to \infty} f(x) = \lim_{x \to \infty} \frac{x^2}{3x^2 + 1} = \lim_{x \to \infty} \frac{1}{3 + (1/x^2)} = \frac{1}{3}$$

Accordingly,

$$\lim_{n \to \infty} \frac{n^2}{3n^2 + 1} = \frac{1}{3}$$

We could also have taken $f(x)$ to be $1/(3 + x^2)$ and x_n to be $1/n$. Now f is continuous at $x = 0$, $x_n \to 0$ and the same conclusion follows.

Subsequences. If we omit some terms in a sequence, but leave infinitely many terms, we can renumber the remaining terms to form a new sequence. For example, in the sequence $\{n^2\}$, or $\{1^2, 2^2, 3^2, 4^2, \ldots \}$, we can omit the squares of the even integers and obtain the new sequence $\{1^2, 3^2, 5^2, \ldots \}$. A new sequence obtained in this way is termed a *subsequence* of the original sequence. In general, to obtain a subsequence of a sequence $\{x_n\}$, we must choose a sequence of integers $\{t_n\}$ to indicate which terms are to be kept. For the example just given we can take $t_n = 2n - 1$ for $n = 1, 2, \ldots$ so that $\{t_n\}$ is the sequence $\{1, 3, 5, \ldots \}$. The subsequence is then simply the sequence $\{x_{t_n}\}$. In general, $\{t_n\}$ is a monotone strictly increasing sequence of positive integers.

THEOREM 3. *If $\{x_n\}$ is a sequence with limit c (or $\pm \infty$) as n tends to ∞, then each subsequence of $\{x_n\}$ has c (or $\pm \infty$) as its limit as n tends to ∞.*

PROOF. We shall give a proof for the case where the limit is a finite number c. The other cases are proved similarly and are left for the reader. We shall also assume that the values of n are $1, 2, \ldots$, as this case is typical.

Let $\{x_{t_n}\}$ be a subsequence of $\{x_n\}$. If $\lim_{n \to \infty} x_n = c$, then for each $\epsilon > 0$ there is an integer N so that $|x_n - c| < \epsilon$ for all $n > N$. But $\{t_n\}$ is a monotone strictly increasing sequence of positive integers and hence $t_n \geq n$, for all n (Problem 4 below). Consequently $|x_{t_n} - c| < \epsilon$ for all $n > N$ and, therefore, $\lim_{n \to \infty} x_{t_n} = c$.

COROLLARY TO THEOREM 3. *If a sequence has two subsequences with different limits, then the sequence is oscillatory-divergent.*

PROOF. If $x_n \to c$ or $\pm \infty$, then all subsequences of $\{x_n\}$ would have the same property. Hence, if two subsequences have different limits, the sequence can have no limit and must be oscillatory-divergent.

EXAMPLE 3. The sequence $\{1, -1, 1, -1, \ldots, (-1)^{n-1}, \ldots \}$ is oscillatory divergent, since it contains the subsequences $\{1, 1, \ldots, 1, \ldots \} \to 1$ and $\{-1, -1, \ldots, -1, \ldots \} \to -1$.

THEOREM 4. *If $\{x_n\}$ is a monotone sequence, and a subsequence of $\{x_n\}$ has limit c_1 (or $\pm \infty$), then $x_n \to c_1$ (or $\pm \infty$).*

PROOF. By the Corollary to Theorem 1, $x_n \to c$ or $\pm \infty$. If $x_n \to c$, then by Theorem 3 each subsequence has limit $c_1 = c$. There is a similar reasoning for the cases $x_n \to \pm \infty$. Hence the conclusion follows.

Thus far we have considered only sequences whose terms are real numbers. However, the theory is easily extended to complex sequences, such as $\{(1 + i)^n\}$. If $\{z_n\}$ is a sequence of complex numbers, then we write

$$\lim_{n \to \infty} z_n = c = a + bi$$

if, for each $\epsilon > 0$, we can choose N so large that $|z_n - c| < \epsilon$ for $n > N$. Thus, $z_n \to c$ precisely when $|z_n - c| \to 0$. [As in Section 0-17, the absolute value of $x + iy$ is $(x^2 + y^2)^{1/2}$, the distance from the origin to $x + iy$.] As in Section 5-8, one verifies that $z_n = x_n + iy_n \to c = a + bi$ if, and only if, $x_n \to a$ and $y_n \to b$. Hence, the study of convergence of complex sequences can be reduced to the study of real sequences. As in Section 5-8, we verify that the rules on limit of a sum, product, and quotient apply to complex sequences. A complex sequence $\{z_n\}$ is termed bounded if $|z_n| \le K = \text{const}$ for all n.

EXAMPLE 3. The sequence $\{n^2 e^{-n} + i[n^2/(n^2 + 1)]\}$ has limit i, since

$$\lim_{n \to \infty} \frac{n^2}{e^n} = 0, \qquad \lim_{n \to \infty} \frac{n^2}{n^2 + 1} = 1$$

The first of these limits is obtained by first using L'Hospital's rule, as in Section 6-14, to evaluate $\lim x^2 e^{-x}$ and then replacing x by n as in Theorem 2.

We can also consider other types of sequences: for example, sequences of vectors, and sequences of functions. For vectors in the plane, the theory is essentially the same as for complex numbers; in particular, $\mathbf{v}_n = x_n \mathbf{i} + y_n \mathbf{j} \to \mathbf{v}_0 = x_0 \mathbf{i} + y_0 \mathbf{j}$ precisely when $|\mathbf{v}_n - \mathbf{v}_0| \to 0$ and precisely when $x_n \to x_0$ and $y_n \to y_0$. The theory of sequences of functions will be considered later in this chapter.

In the first part of the chapter, the emphasis will be on sequences and series of real numbers. However, many of the theorems (and, in most cases, their proofs) carry over to the complex case without change. We shall point this out in the most important cases. For the present, we remark that Theorem 3 holds true for complex sequences.

‡8-3 THE CAUCHY CONDITION FOR SEQUENCES

We now consider an important criterion for convergence. We say that a sequence $\{x_n\}$ satisfies the *Cauchy condition* provided that, for each $\epsilon > 0$, there is an N for which

$$|x_n - x_m| < \epsilon \qquad \text{whenever } m, n > N$$

The condition can be interpreted as saying that, far out in the sequence, all terms are close together, as suggested in Figure 8-2.

$x_1 \qquad x_2 \qquad x_3 \qquad x_4 \qquad x_n\ x_m \qquad\qquad x$

Figure 8-2 Cauchy condition.

The sequence $\{1/n\}$ satisfies the Cauchy condition. For

$$\left| \frac{1}{n} - \frac{1}{m} \right| \le \frac{1}{n} + \frac{1}{m} < \frac{1}{N} + \frac{1}{N} = \frac{2}{N}$$

for $m > N$, $n > N$. Hence for $N > 2/\epsilon$, we are sure that

$$\left| \frac{1}{n} - \frac{1}{m} \right| < \epsilon \qquad \text{for } m > N, n > N$$

THEOREM 5. (*Cauchy criterion*). (*a*) *If the sequence* $\{x_n\}$ *converges, then* $\{x_n\}$ *satisfies the Cauchy condition.* (*b*) *If* $\{x_n\}$ *satisfies the Cauchy condition, then* $\{x_n\}$ *is convergent.*

PROOF. (*a*) Let $x_n \to c$. Then given $\epsilon > 0$ we can choose N so that $|x_n - c| < \epsilon/2$ for $n > N$. Hence, if also $m > N$, we have

$$|x_n - c| < \frac{\epsilon}{2}, \qquad |x_m - c| < \frac{\epsilon}{2}$$

so that for $n > N$, $m > N$,

$$|x_n - x_m| = |x_n - c + c - x_m| \le |x_n - c| + |c - x_m| < \frac{\epsilon}{2} + \frac{\epsilon}{2} = \epsilon$$

(*b*) Let $\{x_n\}$ satisfy the Cauchy condition. Then $\{x_n\}$ must be bounded. For we can choose $\epsilon = 1$ and then choose N so large that $|x_n - x_m| < 1$ for $n > N$ and $m > N$. Hence, in particular, for this N, $|x_{N+1} - x_m| < 1$ for $m > N$ and, hence, $|x_m| < |x_{N+1}| + 1$ for $m > N$. If now K is chosen larger than all of $|x_1|, \ldots, |x_N|, |x_{N+1}| + 1$, then we have $|x_m| < K$ for all m. Therefore, the sequence is bounded.

Now let the set E_n be the set of all numbers $x_n, x_{n+1}, \ldots$ Then E_n is a bounded set, and E_{n+1} is contained in E_n for $n = 1, 2, \ldots$. Let a_n be glb E_n, and let $b_n =$ lub E_n. Then $a_n \le b_n$ and, as n increases, a_n cannot decrease and b_n cannot increase, so that in general $a_n \le a_{n+1} \le b_{n+1} \le b_n$. By Theorem 1, $\{a_n\}$ has a limit a, $\{b_n\}$ has a limit b, and $a_n \le a \le b \le b_n$ for all n. Now, by the Cauchy condition, for each $\epsilon > 0$ we can find N so that $|x_n - x_m| < \epsilon$ for $n > N$ and $m > N$. This implies that, for each $n > N$, the set E_n lies in the interval $[x_n - \epsilon, x_n + \epsilon]$. It follows that $|b_n - a_n| \le 2\epsilon$ for $n > N$, so that $b_n - a_n \to 0$. Therefore, $a = b$ and also $|x_n - a| \le 2\epsilon$ for $n > N$. Therefore, the sequence $\{x_n\}$ converges to a.

Remark. The theorem remains valid for complex sequences. For (*a*) the same proof can be used. For (*b*) the proof is left as an exercise (Problem 13 below).

PROBLEMS

Throughout assume $n = 1, 2, \ldots$ in all sequences.

1. Show that the following sequences converge and find their limits:

(a) $\left\{ \dfrac{n^2 + 1}{n^3 + 1} \right\}$ (b) $\left\{ \dfrac{5n^3 - 7}{7n^3 + 2n - 1} \right\}$ (c) $\left\{ \dfrac{\ln n}{n} \right\}$

(d) $\left\{ \dfrac{n}{2^n} \right\}$ (e) $\left\{ \dfrac{11}{12}, \dfrac{13}{14}, \dfrac{15}{16}, \ldots \right\}$ (f) $\left\{ \dfrac{1}{\sqrt{n}} \right\}$

(g) $\left\{ \sqrt{\dfrac{4n + 1}{n}} \right\}$ (h) $\{1, 1.1, 1.01, 1.001, \ldots\}$ (i) $\{\sqrt{n + 1} - \sqrt{n}\}$

(j) $\left\{ \dfrac{(-1)^n 2^n}{3^n - 5} \right\}$ (k) $\left\{ \dfrac{1 + 2 + \cdots + n}{n + 2} - \dfrac{n}{2} \right\}$ (l) $\left\{ 0, \dfrac{1}{2}, 0, \dfrac{1}{3}, 0, \dfrac{1}{4}, \ldots \right\}$

2. Prove the following:
   (a) If $x_n \to c$ and $y_n \to c$, then $\{x_n - y_n\}$ is a null sequence.
   (b) If $x_n \to c$ and $\{x_n - y_n\}$ is a null sequence, then $y_n \to c$.
   (c) If $\{x_n\}$ and $\{y_n\}$ are null sequences, then so are $\{x_n + y_n\}$, $\{x_n y_n\}$, $\{k x_n\}$.
   (d) If $\{x_n\}$ is a null sequence and $\{y_n\}$ is a bounded sequence, then $\{x_n y_n\}$ is a null sequence. [*Hint.* $|x_n y_n| = |x_n||y_n| \le k|x_n|$ for some constant k.]
   (e) If $x_n \to c$, $\{y_n\}$ is bounded and $\{z_n\}$ is a null sequence, then $\{x_n + y_n z_n\} \to c$.

3. Show, with the aid of the results of Problem 2, that the following sequences converge:

   (a) $\left\{ \dfrac{\sin n}{n} \right\}$

   (b) $\{ e^{-n} \cos 3n \}$

   (c) $\left\{ \dfrac{1}{n + \sin n} \right\} = \left\{ \dfrac{1/n}{1 + [(\sin n)/n]} \right\}$

   (d) $\left\{ \dfrac{1}{n^2 + n \cos n} \right\}$

   (e) $\left\{ \dfrac{1}{n} \displaystyle\int_1^n \dfrac{1}{x^3 + 3x + 1}\, dx \right\}$. (*Hint.* Let y_n denote the integral; show that $|y_n|$ is bounded by first showing that $|y_n| < \displaystyle\int_1^n x^{-3}\, dx$.)

4. Prove that, if $\{t_n\}$ is a monotone strictly increasing sequence of integers and $t_1 \ge 1$, then $t_n \ge n$ for all n. [*Hint.* Use induction.]

5. Which of the following sequences converge, which diverge to ∞ or $-\infty$, which diverge with no infinite limit?

   (a) $\left\{ \dfrac{(-1)^n 2^n}{2^n + 1} \right\}$

   (b) $\left\{ \cos \dfrac{n\pi}{3} \right\}$

   (c) $\left\{ \dfrac{n! - 1}{n! + 1} \right\}$

   (d) $\left\{ \dfrac{\sqrt{n} - 1}{\sqrt{n} + 1} \right\}$

   (e) $\left\{ \dfrac{2^n}{\pi^{n-1}} \right\}$

   (f) $\left\{ \dfrac{5^n - 4^n}{3^n} \right\}$

   (g) $\left\{ \left(\dfrac{2n - 1}{3n} \right)^n \right\}$

   (h) $\left\{ \dfrac{\cos \pi n}{n^2} \right\}$

   (i) $\left\{ \dfrac{2^n}{n^2 \ln 2n} \right\}$

   (j) $\{ \sqrt{n} \cos \pi n \}$

   (k) $\left\{ \dfrac{n}{2^n} - \dfrac{\ln n}{n} \right\}$

   (l) $\left\{ \dfrac{n}{n + \sin n} \right\}$

   (m) $\left\{ \dfrac{n}{\ln (n + 1)!} \right\}$

   (n) $\{ (-1)^n \sqrt[n]{n} \}$

   (o) $\left\{ \dfrac{1 \cdot 3 \cdots (2n - 1)}{2 \cdot 4 \cdots (2n)} \right\}$

   (p) $\left\{ \dfrac{n!}{n^n} \right\}$

6. Which of the following complex sequences converge?

   (a) $\left\{ \dfrac{1}{2^n} - \dfrac{in}{\ln n} \right\}$

   (b) $\left\{ \dfrac{n - i \sin n}{n^2 + 3} \right\}$

   (c) $\left\{ 1 - \dfrac{i \cos \pi n}{2^n} \right\}$

   (d) $\{ (1 + i) \cos n \}$

   (e) $\{ (1 + i)^n \}$

   (f) $\left\{ \left(\dfrac{1 + i}{\sqrt{2}} \right)^n \right\}$

7. Give examples where $\{x_n\}$ and $\{y_n\}$ diverge to infinity and
   (a) $\lim(x_n - y_n) = 0$
   (b) $\{x_n - y_n\}$ converges to a finite nonzero value
   (c) $\{x_n - y_n\}$ diverges to $+\infty$
   (d) $\{x_n - y_n\}$ diverges with no infinite limit

8. If $\lim\limits_{n \to \infty} \mathbf{u}_n = \mathbf{u}$, $\lim\limits_{n \to \infty} \mathbf{v}_n = \mathbf{v}$, where $\mathbf{u}_n, \mathbf{v}_n, \mathbf{u}, \mathbf{v}$ are vectors in the plane, prove:

   (a) $\lim\limits_{n \to \infty} (\mathbf{u}_n + \mathbf{v}_n) = \mathbf{u} + \mathbf{v}$

   (b) $\lim\limits_{n \to \infty} (c\mathbf{u}_n) = c\mathbf{u}$

   (c) $\lim\limits_{n \to \infty} (\mathbf{u}_n \cdot \mathbf{v}_n) = \mathbf{u} \cdot \mathbf{v}$

9. Prove:

(a) If $a \geq 1$, then $\lim_{n \to \infty} \sqrt[n]{a} = 1$.

(b) If $0 < a < 1$, then $\lim_{n \to \infty} \sqrt[n]{a} = 1$.

(c) If $a > 1$ and k is a positive integer, then $\lim_{n \to \infty} \dfrac{a^n}{n^k} = +\infty$.

10. Show that, as $n \to \infty$,

(a) $\left(1 + \dfrac{1}{2n}\right)^{2n} \to e$ (b) $\left(1 - \dfrac{1}{n^2}\right)^{n} \to 1$

(c) $\left(1 + \dfrac{3}{n}\right)^{n} \to e^3$ (d) $\left(1 - \dfrac{1}{n-3}\right)^{n+7} \to e^{-1}$

‡11. (a) Show that $\lim_{n \to \infty} s_n = 1$, where $s_{n+1} = \sqrt{2 - s_n}$ and $s_1 = 0$.

(b) Show that $\lim_{n \to \infty} s_n = 1$, where $s_{n+1} = (1/2) + (1/2)\, s_n^2$ and $s_1 = 1/4$.

(c) If $0 < a < 1/4$, $s_1 = 0$ and $s_{n+1} = a + s_n^2$, find $\lim_{n \to \infty} s_n$.

12. Prove: Every unbounded sequence diverges.

‡13. Prove: If a complex sequence $\{z_n\}$ satisfies the Cauchy condition, then $\{z_n\}$ converges. [*Hint.* Let $z_n = x_n + iy_n$ and show that $\{x_n\}$ and $\{y_n\}$ satisfy the Cauchy condition.]

8-4 INFINITE SERIES

An *infinite series* is an expression of the form:

$$a_m + a_{m+1} + a_{m+2} + \cdots + a_n + \cdots$$

where the a_n are real (or complex) numbers and m is some integer. Thus an infinite series is an infinite sequence of numbers connected by the $+$ sign. We usually abbreviate a series by use of the Σ symbol:

$$\sum_{n=m}^{\infty} a_n = a_m + a_{m+1} + \cdots + a_n + \cdots$$

If we put $b_1 = a_m$, $b_2 = a_{m+1}, \ldots$, then

$$\sum_{n=m}^{\infty} a_n = \sum_{n=1}^{\infty} b_n$$

and, therefore, there is no loss in generality in assuming that an infinite series begins with the index $m = 1$. Similarly, we can renumber to make the series begin with index $m = 0$.

The following are examples of infinite series:

(a) $1 + \dfrac{1}{3} + \dfrac{1}{9} + \dfrac{1}{27} + \cdots + \dfrac{1}{3^{n-1}} + \cdots$

(b) $1 + \dfrac{1}{2} + \dfrac{1}{3} + \dfrac{1}{4} + \cdots + \dfrac{1}{n} + \cdots$

(c) $1 - 1 + 1 - 1 + 1 - 1 + \cdots + (-1)^{n-1} + \cdots$

(d) $1 - \dfrac{1}{2} + \dfrac{1}{3} - \dfrac{1}{4} + \dfrac{1}{5} - \dfrac{1}{6} + \cdots + (-1)^{n-1}\dfrac{1}{n} + \cdots$

(e) $1 + i + \dfrac{i^2}{2} + \cdots + \dfrac{i^n}{n!} + \cdots$

Associated with each infinite series $\displaystyle\sum_{n=1}^{\infty} a_n$ is its *sequence of partial sums* $\{s_n\}$:

$$s_n = \sum_{m=1}^{n} a_m = a_1 + \cdots + a_n, \qquad (n = 1, 2, \ldots)$$

Thus, s_n is the sum of the first n terms of the series.

For the series above, we have the following partial sums:

(a) $s_1 = 1,\ s_2 = \dfrac{4}{3},\ s_3 = \dfrac{13}{9},\ \ldots$

(b) $s_1 = 1,\ s_2 = \dfrac{3}{2},\ s_3 = \dfrac{11}{6},\ s_4 = \dfrac{25}{12},\ \ldots$

(c) $s_1 = 1,\ s_2 = 0,\ s_3 = 1,\ s_4 = 0,\ \ldots$

(d) $s_1 = 1,\ s_2 = \dfrac{1}{2},\ s_3 = \dfrac{5}{6},\ s_4 = \dfrac{7}{12},\ s_5 = \dfrac{47}{60},\ \ldots$

(e) $s_1 = 1,\ s_2 = 1 + i,\ s_3 = \dfrac{1}{2} + i,\ s_4 = \dfrac{1}{2} + \dfrac{5}{6}i,\ \ldots$

An infinite series $\displaystyle\sum_{n=1}^{\infty} a_n$ is said to be *convergent* if its sequence of partial sums $\{s_n\}$ is convergent. If the sequence $\{s_n\}$ is convergent and $\lim\limits_{n\to\infty} s_n = S$, we say that S is *the sum of the infinite series*, and we write

$$S = \sum_{n=1}^{\infty} a_n$$

Hence, by definition,

$$\sum_{n=1}^{\infty} a_n = \lim_{n\to\infty}\left(\sum_{m=1}^{n} a_m\right) = \lim_{n\to\infty} s_n = S$$

If $s_n \to +\infty$ or $s_n \to -\infty$, the infinite series $\displaystyle\sum_{n=1}^{\infty} a_n$ is said to be (*properly*) *divergent*, and we write $\displaystyle\sum_{n=1}^{\infty} a_n = +\infty$ (or $-\infty$, as the case may be). If the sequence $\{s_n\}$ is oscillating and has no limit (finite or infinite) we say that the infinite series $\displaystyle\sum_{n=1}^{\infty} a_n$ is (*oscillatory*) *divergent*.

In this chapter all series considered will be infinite series, and we often abbreviate and write "series" for "infinite series." We also occasionally write simply Σa_n for $\displaystyle\sum_{n=1}^{\infty} a_n$. The index n can be replaced by m or any other symbol (as can the variable in a definite integral).

THEOREM 6. *If $a_n \geq 0$ for $n = 1, 2, \ldots$, then the series $\displaystyle\sum_{n=1}^{\infty} a_n$ is con-*

vergent or properly divergent according as the sequence of partial sums is bounded or unbounded.

PROOF. Since each $a_n \geq 0$, we have

$$s_{n+1} = a_1 + \cdots + a_n + a_{n+1} = s_n + a_{n+1} \geq s_n$$

Hence, the sequence of partial sums for $\sum_{n=1}^{\infty} a_n$ is a monotone nondecreasing sequence. If that sequence is bounded, then by Theorem 1, $\{s_n\}$ is convergent, so that Σa_n is convergent. If the sequence $\{s_n\}$ is unbounded, then by the Corollary of Theorem 1, $s_n \to +\infty$ and the series is properly divergent.

EXAMPLE 1. The infinite series $\sum_{n=1}^{\infty} (\tfrac{1}{2})^{n-1}$ is convergent and has 2 as its sum. Each partial sum of this series is of the form: $s_n = 1 + (\tfrac{1}{2}) + \cdots + (\tfrac{1}{2})^{n-1}$, and so is the sum of the terms of a geometric progression. By Problem 2 following Section 0-21, we have the general rule

$$a + ar + \cdots + ar^{n-1} = a\frac{1 - r^n}{1 - r}, \qquad r \neq 1$$

Here $a = 1$ and $r = 1/2$, so that

$$s_n = \frac{1 - (1/2)^n}{1 - (1/2)} = 2\left(1 - \frac{1}{2^n}\right)$$

Therefore, the sequence of partial sums is bounded: $0 \leq s_n \leq 2$, and by Theorem 6 the series converges. From the expression for s_n, we see that $s_n \to 2$, so that the sum of the series is 2.

EXAMPLE 2. The infinite series $1 + (1/2) + (1/3) + \cdots = \sum_{n=1}^{\infty} (1/n)$ is properly divergent. We observe that

$$s_1 = 1,\ s_2 = 1 + \frac{1}{2} = \frac{3}{2},\ s_4 = 1 + \frac{1}{2} + \frac{1}{3} + \frac{1}{4} > 1 + \frac{1}{2} + \frac{1}{4} + \frac{1}{4} = 2$$

$$s_8 = 1 + \frac{1}{2} + \frac{1}{3} + \frac{1}{4} + \frac{1}{5} + \frac{1}{6} + \frac{1}{7} + \frac{1}{8}$$

$$> 1 + \frac{1}{2} + \frac{1}{4} + \frac{1}{4} + \frac{1}{8} + \frac{1}{8} + \frac{1}{8} + \frac{1}{8} = \frac{5}{2}$$

More generally for $n = 2^k$

$$s_n = 1 + \cdots + \frac{1}{2^k} > 1 + \frac{1}{2} + \frac{1}{4} + \frac{1}{4} + \frac{1}{8}$$

$$+ \cdots + \underbrace{\frac{1}{2^k} + \cdots + \frac{1}{2^k}}_{2^{k-1}\ \text{times}} = 1 + \frac{k}{2}$$

Hence the sequence $\{s_n\}$ is unbounded and, by Theorem 6, the series is properly divergent.

The series $\sum_{n=1}^{\infty} 1/n$ is called the *harmonic series*. Accordingly, *the harmonic series is divergent.*

EXAMPLE 3. The infinite series $\sum\limits_{n=1}^{\infty} (-1)^{n-1}$ is oscillatory divergent, since its sequence of partial sums oscillates between 0 and 1: $s_1 = 1$, $s_2 = 0$, $s_3 = 1, \ldots$.

EXAMPLE 4. Test for convergence: $\sum\limits_{n=1}^{\infty} \dfrac{1}{n(n+1)}$. We can write

$$\frac{1}{n(n+1)} = \frac{1}{n} - \frac{1}{n+1}$$

Hence

$$s_n = \left(1 - \frac{1}{2}\right) + \left(\frac{1}{2} - \frac{1}{3}\right) + \cdots + \left(\frac{1}{n} - \frac{1}{n+1}\right) = 1 - \frac{1}{n+1}$$

and $s_n \to 1$. Thus the series converges and has 1 as its sum.

THEOREM 7. *If an infinite series $\sum\limits_{n=1}^{\infty} a_n$ converges, then the sequence $\{a_n\}$ is a null sequence; that is, the nth term of a convergent series has limit 0.*

PROOF. We observe that $\lim\limits_{n\to\infty} s_{n-1} = \lim\limits_{n\to\infty} s_n = \Sigma a_n$ and therefore the sequence $\{s_n - s_{n-1}\} = \{a_n\}$ must have limit 0. (Theorem and proof are valid in the complex case.)

Remark 1. The converse of Theorem 7 is false as can be seen from Example 2. Thus, convergence of Σa_n implies $a_n \to 0$, but $a_n \to 0$ **does not imply convergence.** This result is at first surprising. To convince one's self of its truth, one should examine Example 2 and its analysis very carefully, and should compute a number of partial sums of the harmonic series (see also Problem 5 below).

Remark 2. One must be careful to distinguish the notions of a *series: Σa_i,* the *sequence of terms* of the series: $\{a_i\}$, and the *sequence of partial sums* of the series: $\{s_n\}$, where $s_n = \sum\limits_{i=1}^{n} a_i$. The series is essentially just another way of describing the sequence of partial sums, whereas the terms of the series specify the number to be added to each partial sum to obtain the next partial sum: $s_{n+1} = s_n + a_{n+1}$. This is suggested in Figure 8-3.

Figure 8-3 Sequence of terms and sequence of partial sums of an infinite series.

A sequence $\{x_n\}$ can *always be realized as the sequence of partial sums of a series;* for let $a_1 = x_1$, $a_2 = x_2 - x_1, \ldots$, and in general $a_n = x_n - x_{n-1}$. Then the series $\sum\limits_{i=1}^{\infty} a_i$ has $s_n = a_1 + \cdots + a_n = x_1 + (x_2 - x_1) + \cdots + (x_n - x_{n-1}) = x_n$.

Geometric Series. The infinite series $\sum\limits_{n=1}^{\infty} ar^{n-1}$, where $a \neq 0$, is called a geometric series with ratio r.

THEOREM 8. *If $|r| < 1$, the geometric series $\sum\limits_{n=1}^{\infty} ar^{n-1}$ converges to the sum $a/(1-r)$. If $|r| \geq 1$, the geometric series diverges.*

PROOF. As in Example 1 above, the partial sum of the series is

$$s_n = a + ar + \cdots + ar^{n-1} = a\frac{1-r^n}{1-r} = \frac{a}{1-r} - a\frac{r^n}{1-r}$$

If $|r| < 1$, then $r^n \to 0$, so that $s_n \to a/(1-r)$. If $|r| \geq 1$, then $|ar^n| \geq |a| > 0$ and, hence, ar^n cannot $\to 0$, so that by Theorem 7 the series diverges.

Remark. Theorem 8 and its proof remain valid when a and r are allowed to be complex numbers.

EXAMPLE 5

$$\sum_{n=1}^{\infty} nr^{n-1} = 1 + 2r + \cdots + nr^{n-1} + \cdots$$

Here

$$s_n = 1 + 2r + \cdots + nr^{n-1} = \frac{d}{dr}(1 + r + \cdots + r^n) = \frac{d}{dr}\left(\frac{1-r^{n+1}}{1-r}\right)$$

$$= \frac{1}{(1-r)^2} - \frac{r^{n+1}}{(1-r)^2} - \frac{(n+1)r^n}{1-r}$$

Now for $|r| < 1$, $r^{n+1} \to 0$ and $(n+1)r^n \to 0$ as $n \to \infty$ [in the latter case, since $(x+1)r^x \to 0$ as $x \to \infty$]. Hence for $|r| < 1$, $s_n \to (1-r)^{-2}$, and this is the sum of the series. If $|r| \geq 1$, then $|nr^{n-1}| \geq 1$ for all n and by Theorem 7 the series diverges.

PROBLEMS

1. Find the first 5 partial sums of the following series:

(a) $1 - \dfrac{1}{2} + \dfrac{1}{3} - \dfrac{1}{4} - \dfrac{1}{5} + \dfrac{1}{6} + \cdots$

(b) $1 + \dfrac{1}{2} + \dfrac{2}{3} + \dfrac{3}{4} + \dfrac{4}{5} + \dfrac{5}{6} + \cdots$

(c) $1 - \dfrac{1}{2} + \dfrac{2}{3} - \dfrac{3}{4} + \dfrac{4}{5} - \dfrac{5}{6} + \cdots$

(d) $1 + \dfrac{1}{2} + \dfrac{1}{2 \cdot 3} + \dfrac{1}{2 \cdot 3 \cdot 4} + \dfrac{1}{2 \cdot 3 \cdot 4 \cdot 5} + \cdots$

(e) $\dfrac{1}{2} + \dfrac{1}{2 \cdot 3} + \dfrac{1}{3 \cdot 4} + \dfrac{1}{4 \cdot 5} + \dfrac{1}{5 \cdot 6} + \cdots$

(f) $\dfrac{1}{2} - \dfrac{1}{2 \cdot 3} - \dfrac{1}{3 \cdot 4} + \dfrac{1}{4 \cdot 5} - \dfrac{1}{5 \cdot 6} - \dfrac{1}{6 \cdot 7} + \cdots$

2. Show that the following infinite series converge:

(a) $3 + \dfrac{3}{2} + \dfrac{3}{4} + \cdots + \dfrac{3}{2^{n-1}} + \cdots$

(b) $1 + \dfrac{1}{3} + \dfrac{1}{9} + \cdots + \dfrac{1}{3^{n-1}} + \cdots$

(c) $\dfrac{1}{16} + \dfrac{1}{32} + \dfrac{1}{64} + \cdots + \dfrac{1}{2^{n+3}} + \cdots$

(d) $5 + 4 + 3 + 2 + 1 + \dfrac{2}{3} + \dfrac{2}{9} + \dfrac{2}{27} + \cdots + \dfrac{2}{3^{n-5}} + \cdots$

3. The following series diverge. Explain why, and for (a)-(d) determine if they are properly divergent or oscillatory-divergent:

(a) $1 + \dfrac{1}{2} + \dfrac{2}{3} + \dfrac{3}{4} + \cdots + \dfrac{n}{n+1} + \cdots$

(b) $1 - \dfrac{1}{2} + \dfrac{2}{3} - \dfrac{3}{4} + \cdots + (-1)^{n-1}\dfrac{n-1}{n} + \cdots$

(c) $1 + \dfrac{5}{2} + \dfrac{5}{3} + \dfrac{5}{4} + \cdots + \dfrac{5}{n} + \cdots$

(d) $1 - \dfrac{1}{2} - \dfrac{2}{3} + \dfrac{3}{4} - \dfrac{4}{5} - \dfrac{5}{6} + \dfrac{6}{7} + \cdots + \epsilon_n\dfrac{n-1}{n} + \cdots,$

$\epsilon_n = 1$ for $n = 1, 4, 7, \ldots,$ $\epsilon_n = -1$ otherwise

(e) $\displaystyle\sum_{n=1}^{\infty} i^n$ (f) $\displaystyle\sum_{n=1}^{\infty} \left(\cos n\dfrac{\pi}{7} + i \sin n\dfrac{\pi}{7}\right)$

4. Test for convergence as in Example 4:

(a) $\displaystyle\sum_{n=1}^{\infty} \dfrac{1}{n^2 + 3n + 2}$

(b) $\displaystyle\sum_{n=1}^{\infty} \dfrac{1}{n(n+3)}$

(c) $\displaystyle\sum_{n=1}^{\infty} \dfrac{1}{n(n+1)(n+2)}$ $\left\{Hint: \dfrac{1}{n(n+1)(n+2)}\right.$

$$= \dfrac{1}{2}\left[\left(\dfrac{1}{n} - \dfrac{1}{n+1}\right) + \left(\dfrac{1}{n+2} - \dfrac{1}{n+1}\right)\right]\Bigg\}$$

(d) $\displaystyle\sum_{n=1}^{\infty} \dfrac{1}{n(n+3)(n+5)}$

5. (a) Show that the series $\displaystyle\sum_{n=1}^{\infty} [f(n+1) - f(n)]$ converges if, and only if, $f(n)$ has a limit c as $n \to \infty$ and that, in the case of convergence, the sum is $c - f(1)$.

(b) Show that for $f(n) = \ln n$, the series of part (a) diverges, but the nth term has limit 0. Interpret this result in terms of the graph of $y = \ln x$.

(c) Proceed as in part (b) for $f(x) = \sqrt{x}$.

6. Determine which of the following series converge. Compute the sum of the ones that converge:

(a) $\displaystyle\sum_{n=1}^{\infty} \dfrac{3n+1}{5n-3}$ (b) $\displaystyle\sum_{n=1}^{\infty} \dfrac{1}{5n}$ (c) $\displaystyle\sum_{n=1}^{\infty} \cos\dfrac{n\pi}{2}$

(d) $\sum_{n=1}^{\infty} \sin \dfrac{n\pi}{6}$ (e) $\sum_{n=1}^{\infty} \dbinom{512}{n}$, $\dbinom{m}{n} = \dfrac{m(m-1)\dots(m-n+1)}{1 \cdot 2 \dots n}$

(f) $\sum_{n=1}^{\infty} \dfrac{n}{\cos n}$ (g) $\sum_{n=1}^{\infty} \dfrac{1}{\sqrt{n+1} + \sqrt{n}}$ (h) $\sum_{n=2}^{\infty} \left(1 - \dfrac{2}{n(n-1)}\right)$

(i) $1 - \dfrac{2}{3} + \dfrac{4}{9} - \dfrac{16}{22} + \cdots + \left(\dfrac{-2}{3}\right)^{n-1} + \cdots$

(j) $\sum_{n=1}^{\infty} \left(\dfrac{i}{3}\right)^n$ (k) $\sum_{n=1}^{\infty} \left(\dfrac{1}{2^n} + \dfrac{i}{n}\right)$

(l) $1 - \dfrac{1}{3} - \dfrac{1}{9} + \dfrac{1}{27} - \dfrac{1}{81} - \dfrac{1}{243} + \cdots + \dfrac{\epsilon_n}{3^{n-1}} + \cdots$, ϵ_n as in Problem 3(d)

(m) $\dfrac{3}{2} + \dfrac{7}{12} + \dfrac{11}{30} + \cdots + \left(\dfrac{1}{2n-1} + \dfrac{1}{2n}\right) + \cdots$

(n) $\dfrac{1}{6} + \dfrac{1}{10} + \dfrac{1}{14} + \cdots + \dfrac{1}{4n+2} + \cdots$

8-5 PROPERTIES OF INFINITE SERIES

We first point out two simple ways in which infinite series can be changed without affecting convergence.

I. Introducing or Removing Zero Terms. If we replace the series $a_1 + a_2 + \cdots + a_n + \cdots$ by the series $a_1 + 0 + a_2 + 0 + a_3 + \cdots$, then we clearly have not affected either convergence of the series or the sum, if it does converge. Similarly, zero terms can be deleted without any effect.

EXAMPLE 1. $\sum_{n=1}^{\infty} 2^{-n}[1 - (-1)^n]$. For n even, $(-1)^n = 1$, so that the term is 0. For n odd, the nth term is $2/2^n$. Hence, if we drop the zero terms, we have the series

$$\frac{2}{2} + \frac{2}{2^3} + \frac{2}{2^5} + \cdots + \frac{2}{2^{2n-1}} + \cdots$$

which we recognize as a geometric series with ratio $\frac{1}{4}$. Hence the series converges.

II. Changing the First k Terms. If in a series Σa_n we replace a_1 by b_1, a_2, by b_2, $\dots$, a_k by b_k, but leave all other terms unchanged, then the nth partial sum (for $n > k$) becomes

$$
\begin{aligned}
b_1 + \cdots + b_k &+ a_{k+1} + \cdots + a_n \\
&= a_1 + \cdots + a_n + [(b_1 + \cdots + b_k) - (a_1 + \cdots + a_k)] \\
&= s_n + c
\end{aligned}
$$

where s_n is the nth partial sum of the original series and c is a constant. Hence the modified series diverges if the given series diverges, and converges to $S + c$ if the original series converges to S. If some of the b's are 0, we have in effect dropped some of the a's. Thus, omission of a finite number of terms has no effect on convergence. Similarly, the introduction of a finite number of new terms has no effect on convergence. In general, convergence is determined by what happens far out in the series.

EXAMPLE 2. The two series

$$1 + 2 + 3 + \frac{1}{4} + \frac{1}{5} + \cdots + \frac{1}{n} + \cdots$$

$$1 + \frac{1}{2} + \frac{1}{3} + \frac{1}{4} + \frac{1}{5} + \cdots + \frac{1}{n} + \cdots$$

diverge, since they agree after the third term, and the second series is the harmonic series, which diverges as shown in Section 8-4 above. We notice that for $n \geq 3$ the two partial sums are

$$s_n = 1 + 2 + 3 + \frac{1}{4} + \cdots + \frac{1}{n}, \qquad s'_n = 1 + \frac{1}{2} + \frac{1}{3} + \frac{1}{4} + \cdots + \frac{1}{n}$$

and hence that

$$s'_n - s_n = \left(1 + \frac{1}{2} + \frac{1}{3}\right) - (1 + 2 + 3) = -4\frac{1}{6} = c, \qquad n \geq 3$$

THEOREM 9. *Let an infinite series Σa_n converge and have sum S. Then for each n the series $a_{n+1} + a_{n+2} + \cdots$ converges. If we write*

$$s_n = a_1 + \cdots + a_n, \qquad r_n = a_{n+1} + a_{n+2} + \cdots = \sum_{k=1}^{\infty} a_{n+k},$$

then

$$S = s_n + r_n$$

and $r_n \to 0$ as $n \to \infty$

PROOF. The series for r_n is simply the given series minus the first n terms. Hence this series converges and its sum r_n is related to S by the equation $S = s_n + r_n$, by the reasoning above. Since $s_n \to S$, we conclude that $r_n = S - s_n \to 0$.

Given a series $\sum_{n=1}^{\infty} a_n$, we can *group the terms* of this series to form a new series. For example, we might put $b_1 = a_1 + a_2$, $b_2 = a_3 + a_4, \ldots, b_n = a_{2n-1} + a_{2n}, \ldots$, then $\sum_{n=1}^{\infty} b_n = \sum_{n=1}^{\infty} (a_{2n-1} + a_{2n})$. Or we might put $c_1 = a_1$, $c_2 = a_2 + a_3$, $c_3 = a_4 + a_5 + a_6 + a_7, \ldots, c_n = a_{2n-1} + \cdots + a_{2^n - 1}$, $\ldots$, whence

$$\sum_{n=1}^{\infty} c_n = \sum_{n=1}^{\infty} (a_{2n-1} + \cdots + a_{2^n - 1})$$

When we considered finite sums, we saw (by the Associative Law) that the sum did not depend on the manner of grouping the terms. Thus $(a_1 + a_2) + (a_3 + a_4 + a_5) = a_1 + (a_2 + a_3) + (a_4 + a_5)$. We shall now show that this is also true for convergent infinite series.

THEOREM 10. *Let* $\sum_{n=1}^{\infty} a_n$ *be a convergent series with sum k and let $\{t_n\}$ be a monotone strictly increasing sequence of positive integers. Put*

$$b_1 = a_1 + \cdots + a_{t_1}, \qquad b_2 = a_{t_1+1} + \cdots + a_{t_2}$$

and in general

$$b_n = a_{t_{n-1}+1} + \cdots + a_{t_n}$$

Then the series $\sum_{n=1}^{\infty} b_n$ is convergent and has sum k.

PROOF. Let $\{s_n\}$ and $\{S_n\}$ be the sequences of partial sums for Σa_n and Σb_n, respectively. Then

$$S_n = s_{t_n}, \qquad \text{for all } n$$

and therefore $\{S_n\}$ is a subsequence of $\{s_n\}$. Since $s_n \to k$, it follows from Theorem 3 that $S_n \to k$. Thus Σb_n is convergent and has k as its sum.

EXAMPLE 3. Since the series $\Sigma 1/2^{n-1}$ converges, we can conclude that the series

$$\left(1 + \frac{1}{2}\right) + \left(\frac{1}{4}\right) + \left(\frac{1}{8} + \frac{1}{16}\right) + \left(\frac{1}{32}\right) + \left(\frac{1}{64} + \frac{1}{128}\right) + \cdots$$

$$= \frac{3}{2} + \frac{1}{4} + \frac{3}{16} + \frac{1}{32} + \frac{3}{128} + \frac{1}{256} + \frac{3}{1024} + \cdots$$

is also convergent.

Remark 1. The converse of Theorem 10 is not true, for the series $\sum_{n=1}^{\infty} (-1)^n$ diverges while the series $\sum_{n=1}^{\infty} [(-1)^{2n-1} + (-1)^{2n}] = \Sigma 0$ converges.

Remark 2. As a consequence of Theorem 10, we have: *If, after parentheses have been inserted in a series, the new series thus formed diverges, then the original series diverges.*

THEOREM 11. *Let $\sum_{n=1}^{\infty} a_n$ be a convergent series with sum A, and let $\sum_{n=1}^{\infty} b_n$ be a convergent series with sum B.*

(1) *If for each n, $d_n = ca_n$, then $\sum_{n=1}^{\infty} d_n$ converges and has cA as its sum; that is, $\sum_{n=1}^{\infty} ca_n = c \sum_{n=1}^{\infty} a_n$.*

(2) *If for each n, $e_n = a_n + b_n$, then $\sum_{n=1}^{\infty} e_n$ converges and has $A + B$ as its sum; that is, $\sum_{n=1}^{\infty} (a_n + b_n) = \sum_{n=1}^{\infty} a_n + \sum_{n=1}^{\infty} b_n$.*

PROOF. Let $s_m = \sum_{n=1}^{m} a_n$, $t_m = \sum_{n=1}^{m} b_n$, $u_m = \sum_{n=1}^{m} d_n$, $w_m = \sum_{n=1}^{m} e_n$. Then $u_m = cs_m$, $w_m = s_m + t_m$. Consequently

$$\sum_{m=1}^{\infty} d_m = \lim_{m \to \infty} u_m = c \lim_{m \to \infty} s_m = cA$$

$$\sum_{m=1}^{\infty} e_m = \lim_{m \to \infty} w_m = \lim_{m \to \infty} (s_m + t_m) = \lim_{m \to \infty} s_m + \lim_{m \to \infty} t_m = A + B$$

EXAMPLE 4. Discuss the convergence of the series $\sum_{n=1}^{\infty} 1/(2n)$.

Solution. If the series $\Sigma 1/(2n)$ converged, then by Theorem 11, the series $2 \cdot \Sigma 1/(2n) = \Sigma 1/n$ would be convergent. But we observed in Example 2 that the harmonic series $\Sigma 1/n$ diverges and, hence, we have a contradiction. Consequently, $\Sigma 1/(2n)$ must diverge.

The argument in Example 4 can be used to show: *If $c \neq 0$ and Σa_n diverges, then $\Sigma(ca_n)$ diverges.*

‡8-6 CAUCHY CRITERION FOR INFINITE SERIES

THEOREM 12 (*Cauchy condition for convergence of a series*). *The series $\sum_{n=1}^{\infty} a_n$ is convergent if, and only if, for each $\epsilon > 0$ there is an integer N for which*

$$|a_{n+1} + \cdots + a_m| < \epsilon \qquad \textit{whenever} \qquad m > n > N$$

PROOF. This is simply Theorem 5 restated for infinite series, in view of the relation $|s_m - s_n| = |a_{n+1} + \cdots + a_m|$. (Theorem and proof are valid for complex series also.)

Note. Theorem 7 is a Corollary of Theorem 12 ($n = m - 1$ in Theorem 12).

PROBLEMS

1. Prove convergence:

   (a) $1 + 0 + \dfrac{1}{3} + 0 + \cdots + \dfrac{1}{3^n} + 0 + \cdots$ (b) $\sum_{n=1}^{\infty} \dfrac{\sin(n\pi/2)}{4^n}$

   (c) $\sum_{n=1}^{\infty} \ln \tan^2[(2n+1)\pi/4]$ (d) $\sum_{n=1}^{\infty} [(n-1)(n-2) \dots (n-n^2)]$

2. Test for convergence:

   (a) $\dfrac{1}{3} + \dfrac{1}{4} + \dfrac{1}{5} + \cdots + \dfrac{1}{n+2} + \cdots$

   (b) $3 + 2 + 1 + \dfrac{1}{2} + \dfrac{1}{4} + \cdots + \dfrac{1}{2^{n-3}} + \cdots$

3. Evaluate where meaningful:

   (a) $\lim_{n \to \infty} \left(\dfrac{1}{2^n} + \dfrac{1}{2^{n+1}} + \cdots \right)$ (b) $\lim_{n \to \infty} \left(\dfrac{1}{n} + \dfrac{1}{n+1} + \dfrac{1}{n+2} + \cdots \right)$

4. Prove convergence:

   (a) $\dfrac{1}{2} + \left(\dfrac{1}{4} + \dfrac{1}{8} \right) + \left(\dfrac{1}{16} + \dfrac{1}{32} + \dfrac{1}{64} \right) + \cdots$

   (b) $\sum_{n=1}^{\infty} \left(\dfrac{1}{3^{n^2}} + \dfrac{1}{3^{n^2+1}} + \cdots + \dfrac{1}{3^{n^2+2n}} \right)$

(c) $\displaystyle\sum_{n=1}^{\infty} \frac{(\frac{2}{3})^n + (\frac{1}{3})^n}{2^n}$

(d) $\displaystyle\sum_{n=0}^{\infty} \left(2^{-n} + \frac{1}{(n+1)(n+2)}\right)$

(e) $\displaystyle\sum_{n=0}^{\infty} a^n(1+a^n), \qquad |a| < 1$

(f) $\displaystyle\sum_{n=0}^{\infty} a^n(1+a^n)(1+a^{2n}), \qquad |a| < 1$

(g) $\displaystyle\sum_{n=1}^{\infty} na^n, \qquad |a| < 1$

‡(h) $\displaystyle\sum_{n=2}^{\infty} n(n+1)a^n, \qquad |a| < 1$

(i) $\displaystyle\sum_{n=1}^{\infty} \left[\frac{(1+i)^n}{3^n} + \frac{(1+2i)^n}{5^n}\right]$

(j) $\displaystyle\sum_{n=1}^{\infty} \left(\frac{1}{n+1+i} - \frac{1}{n+i}\right)$

6. If $1 > |a| > |b| > 0$, show that the series

$$a + b + a^2 + b^2 + a^3 + b^3 + \cdots + a^n + b^n + \cdots$$

converges.

7. Assuming the following relations

$$\sum_{n=1}^{\infty} \frac{1}{n^2} = \frac{\pi^2}{6}, \qquad \sum_{n=1}^{\infty} \frac{1}{n^4} = \frac{\pi^4}{90}, \qquad \sum_{n=1}^{\infty} \frac{1}{n^6} = \frac{\pi^6}{945}$$

evaluate the following series:

(a) $\displaystyle\sum_{n=1}^{\infty} \frac{12}{n^2}$

(b) $\displaystyle\sum_{n=1}^{\infty} \frac{n^2 + 3}{n^4}$

(c) $\displaystyle\sum_{n=1}^{\infty} \frac{5n^2 - 6}{n^6}$

8. Prove that, if $\displaystyle\sum_{n=1}^{\infty} a_n$ is convergent and $\displaystyle\sum_{n=1}^{\infty} b_n$ is divergent, then $\displaystyle\sum_{n=1}^{\infty} (a_n + b_n)$ is divergent.

9. Suppose $a_n > 0$. Show that the two series $\displaystyle\sum_{n=1}^{\infty} a_n$ and $\displaystyle\sum_{n=1}^{\infty} (a_n + a_{n+1})$ either both converge or both diverge.

10. Give examples of properly divergent series Σa_n, Σb_n so that
    (a) $\Sigma(a_n + b_n)$ is properly divergent   (b) $\Sigma(a_n + b_n)$ is convergent
    (c) $\Sigma(a_n + b_n)$ is oscillatory-divergent

11. The number 1.000 . . . has two decimal representations. Find the other one and prove that you are correct.

8-7 COMPARISON TESTS FOR SERIES WITH NONNEGATIVE TERMS

In this and the next four sections we shall discuss a number of tests for determining whether a series is convergent or divergent. The tests are not always conclusive and there are series for which all the tests here given fail to determine whether or not the series converges. However, these tests do cover a wide class of series, and they have the virtue of being easily checked. For more elaborate and sophisticated tests, see Knopp: *Infinite Series* (Blackie and Sons, Glasgow, 1928).

Definition. A series $\displaystyle\sum_{n=k}^{\infty} a_n$ with real or complex terms is said to be *dominated* by the series $\displaystyle\sum_{n=k}^{\infty} b_n$ with nonnegative real terms provided that

$|a_n| \leq b_n$ for all $n \geq k$. Thus the series $\sum\limits_{n=1}^{\infty} 1/(n^2 + n)$ is dominated by the series $\sum\limits_{n=1}^{\infty} 1/n^2$ [since $1/(n^2 + n) < 1/n^2$], and in turn $\Sigma 1/n^2$ is dominated by the series $\Sigma 1/n$.

THEOREM 13. *If* $\sum\limits_{n=1}^{\infty} a_n$ *is a series with nonnegative terms and* Σa_n *is dominated by the convergent series* Σb_n *with the sum B, then* Σa_n *is convergent and has sum A at most equal to B. Furthermore*

$$a_1 + \cdots + a_n \leq A \leq a_1 + \cdots + a_n + \sum_{k=n+1}^{\infty} b_k \qquad (8\text{-}70)$$

PROOF. Let $s_n = a_1 + \cdots + a_n$, $t_n = b_1 + \cdots + b_n$. Since $0 \leq a_n \leq b_n$ for all n, we have $0 \leq s_n \leq t_n$, and both sequences $\{s_n\}$, $\{t_n\}$ are monotone nondecreasing. Now Σb_n is convergent; therefore, by Theorem 6, $\{t_n\}$ is bounded and $t_n \to B$. Since $s_n \leq t_n$, $\{s_n\}$ is also bounded, so that $s_n \to A \leq B$. Clearly $s_n \leq A$ for all n. Also the series Σa_n is dominated by the series $a_1 + \cdots + a_n + b_{n+1} + b_{n+2} + \cdots$, so that A is at most equal to the sum of this series. The last two statements give (8-70).

THEOREM 14. *If* $a_n \geq 0$ *for all n,* Σa_n *diverges, and* Σa_n *is dominated by* Σb_n*, then* Σb_n *diverges.*

PROOF. If Σb_n were to converge then, by Theorem 13, Σa_n would converge, contrary to assumption.

EXAMPLE 1. Show that the series $\sum\limits_{n=1}^{\infty} (\cos^2 n)/2^n$ is convergent.

Solution: Since $|\cos n| \leq 1$, $0 \leq (\cos^2 n)/2^n \leq 1/2^n$ and, therefore, $\Sigma(1/2^n)$ dominates $\Sigma(\cos^2 n)/2^n$. Since $\sum\limits_{n=1}^{\infty} (1/2^n)$ converges to 1, we can conclude from the Theorem 13 that $\sum\limits_{n=1}^{\infty} (\cos^2 n)/2^n$ converges to a sum A that is less than or equal to 1. By (8-70)

$$\frac{\cos^2 1}{2} + \cdots + \frac{\cos^2 n}{2^n} \leq A \leq \frac{\cos^2 1}{2} + \cdots + \frac{\cos^2 n}{2^n} + \sum_{k=n+1}^{\infty} \frac{1}{2^k}$$

Since the last term is a geometric series with sum $1/2^n$, A can be found with an error of less than $1/2^n$ by computing the nth partial sum of the given series.

EXAMPLE 2. Show that the series $\sum\limits_{n=1}^{\infty} 1/\sqrt{n}$ is divergent.

Solution. Clearly $1/\sqrt{n} > 1/n > 0$. We saw in Example 2 of Section 8-4 that $\Sigma 1/n$ is divergent and, hence, by Theorem 14, the series $\Sigma 1/\sqrt{n}$ is properly divergent.

EXAMPLE 3. Determine whether $\sum\limits_{n=1}^{\infty} [1 \cdot 3 \cdot 5 \cdots (2n - 1)]/(2 \cdot 4 \cdot 6 \cdots 2n)$ is convergent or divergent.

Solution. We observe that $[1 \cdot 3 \cdots (2n - 1)]/(2 \cdot 4 \cdots 2n) = (3/2) \cdot$

$(5/4) \cdots [(2n - 1)/(2n - 2)] \cdot [1/(2n)] \geq 1/(2n)$. Since $\Sigma 1/(2n)$ is divergent (see Example 4, Section 8-5) it follows that the given series is divergent.

Our next tests are slight variations on those of Theorems 13 and 14.

THEOREM 15. *If Σb_n is a convergent series with positive terms and if the a_n are nonnegative real numbers so that*

$$\lim_{n \to \infty} \frac{a_n}{b_n} = L < \infty$$

then Σa_n converges.

PROOF. Since $\lim_{n \to \infty} a_n/b_n = L < \infty$, the terms a_n/b_n are arbitrarily close to L for n sufficiently large. In particular, we can find an integer N for which

$$L - 1 \leq \frac{a_n}{b_n} \leq L + 1, \qquad \text{for} \qquad n \geq N$$

Hence

$$0 < a_n \leq (L + 1)b_n, \qquad \text{for} \qquad n \geq N$$

But then the series $\displaystyle\sum_{n=N}^{\infty} a_n$ is dominated by the series $\displaystyle\sum_{n=N}^{\infty} (L + 1)b_n$. Since $\displaystyle\sum_{n=1}^{\infty} b_n$ converges, the series $\displaystyle\sum_{n=N}^{\infty} b_n$ converges and, hence, $\displaystyle\sum_{n=N}^{\infty} (L + 1)b_n$ converges (Theorem 11). But then, by Theorem 13, $\displaystyle\sum_{n=N}^{\infty} a_n$ is convergent and, hence, $\displaystyle\sum_{n=1}^{\infty} a_n$ is convergent.

THEOREM 16. *If $\displaystyle\sum_{n=1}^{\infty} a_n$ is a properly divergent series with positive real terms and if the b_n are positive numbers so that $\displaystyle\lim_{n \to \infty} \frac{b_n}{a_n} = K > 0$, then*

$$\sum_{n=1}^{\infty} b_n \text{ is properly divergent.}$$

PROOF. This follows from Theorem 14 (Problem 40 below).

In practice, Theorems 15 and 16 are usually easier to apply than are Theorems 13 and 14. All four theorems are referred to as *comparison tests.*

EXAMPLE 4. Test the series $\displaystyle\sum_{n=2}^{\infty} (n^3 + 1)/(n^4 + n^2 + n - 3)$ for convergence.

Solution. We compare $(n^3 + 1)/(n^4 + n^2 + n - 3)$ with $1/n$ and find that

$$\lim_{n \to \infty} \left(\frac{n^3 + 1}{n^4 + n^2 + n - 3} \right) \bigg/ \left(\frac{1}{n} \right) = \lim_{n \to \infty} \frac{n^4 + n}{n^4 + n^2 + n - 3} = 1$$

Since the series $\Sigma 1/n$ is divergent, it follows from Theorem 16 that the given series diverges.

8-8 THE INTEGRAL TEST

In Section 7-14 we discussed the convergence and divergence of the improper integral $\int_a^\infty f(t)dt$. Our next theorem relates such integrals to series of the form $\Sigma f(n)$, for certain functions $f(x)$.

THEOREM 17 (*Integral Test*). *Let the function f be defined on the interval* $1 \leq x < \infty$ *and let f be continuous, monotone nonincreasing, and nonnegative. Then* $\int_1^\infty f(x)dx$ *converges or diverges according as the infinite series* $\sum_{n=1}^\infty f(n)$ *converges or diverges. In the case of convergence, the sum S of the series satisfies*

$$f(1) + \cdots + f(n) \leq S \leq f(1) + \cdots + f(n) + \int_n^\infty f(x)\, dx \quad (8\text{-}80)$$

and the value L of the integral satisfies

$$\int_1^n f(x)dx \leq L \leq \int_1^n f(x)\, dx + \sum_{m=n}^\infty f(m) \quad (8\text{-}81)$$

PROOF. The improper integral converges and has value L when

$$\lim_{b \to \infty} \int_1^b f(x)\, dx$$

exists and equals L. Since f is nonnegative, the integral from 1 to b is a nondecreasing function of b and, hence, is either unbounded (and thus approaches ∞ as $b \to \infty$) or bounded with limit L as $b \to \infty$ (Theorem H' in Section 2-12). When it is unbounded, the sequence

$$\int_1^n f(x)\, dx, \qquad n = 1, 2, \ldots \quad (8\text{-}82)$$

must also be unbounded and tend to ∞ as $n \to \infty$; when the integral from 1 to b is a bounded function of b, the sequence (8-82) is a bounded monotone sequence with limit L (cf. Theorem 2 in Section 8-2). Hence, we can write

$$\int_1^\infty f(x)dx = \lim_{n \to \infty} \int_1^n f(x)\, dx \quad (8\text{-}83)$$

with the understanding that either both sides have the same finite value L or both sides have no value (one could assign the value ∞ in this case). Now

$$\int_1^n f(x)\, dx = \int_1^2 f(x)\, dx + \int_2^3 f(x)\, dx + \cdots + \int_{n-1}^n f(x)\, dx$$

Hence, $\int_1^n f(x)\, dx$ is the $(n-1)$-st partial sum of the series

$$\sum_{n=1}^\infty A_n, \qquad A_n = \int_n^{n+1} f(x)\, dx$$

Therefore, (8-83) is equivalent to the statement:

$$\int_1^\infty f(x)\, dx = \sum_{n=1}^\infty A_n \tag{8-83'}$$

This is illustrated in Figure 8-4.

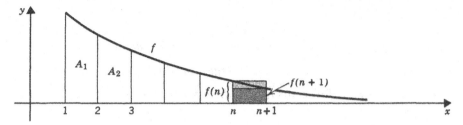

Figure 8-4   Improper integral and infinite series.

Now f is assumed to be nonincreasing. Hence, $f(n + 1) \le f(x) \le f(n)$ for $n \le x \le n + 1$. Therefore

$$\int_n^{n+1} f(n + 1)\, dx \le \int_n^{n+1} f(x)\, dx \le \int_n^{n+1} f(n)\, dx$$

or

$$f(n + 1) \le A_n \le f(n) \tag{8-84}$$

that is, the area A_n is greater than or equal to that of the smaller shaded rectangle in Figure 8-4, and the area A_n is at most equal to the area of the larger shaded rectangle.

From (8-84) we conclude by Theorems 13 and 14 that the series $\Sigma f(n)$ converges if, and only if, the series ΣA_n converges and, hence, by (8-83'), if and, only if, the improper integral converges. In the case of convergence, $f(n) \le A_{n-1}$ for $n > 1$, and (8-70) gives

$$f(1) + \cdots + f(n) \le S \le f(1) + \cdots + f(n) + A_n + A_{n+1} + \cdots$$

which is equivalent to (8-80). Similarly

$$A_1 + \cdots + A_{n-1} \le L \le A_1 + \cdots + A_{n-1} + f(n) + f(n + 1) + \cdots$$

which is equivalent to (8-81).

Remarks. The equation (8-83') is itself of interest. It shows how an improper integral can be rewritten as an infinite series; the only assumptions required are that f be continuous and nonnegative. For the integral test, however, it is essential that f be nonincreasing.

The function f is sometimes given in a different interval, say for $k \le x < \infty$, where k is an integer. Theorem 17 remains valid in this case with the modifications that n must be at least equal to k in (8-80) and (8-81) and that the lower limit 1 of all integrals must be replaced by k.

As a special case of Theorem 17, we have the following useful result.

THEOREM 18. *The series* $\sum\limits_{n=1}^{\infty} n^{-\alpha} = 1 + (1/2^{\alpha}) + (1/3^{\alpha}) + \cdots$ *is convergent for $\alpha > 1$ and divergent for $0 < \alpha \leq 1$.*

PROOF. We saw earlier that Σn^{-1} is divergent. So suppose $\alpha \neq 1$. Theorem 18 now follows from Theorem 17 on observing that $x^{-\alpha}$ is a monotone non-increasing function and that the integral of $x^{-\alpha}$ from 1 to ∞ converges for $\alpha > 1$, diverges for $\alpha \leq 1$ (Section 7-14).

It follows from the theorem that the function $\zeta(x) = \sum\limits_{n=1}^{\infty} n^{-x}$ is defined for all $x > 1$. This function is actually continuous and differentiable on the interval $1 < x < \infty$.

EXAMPLE 1. Evaluate $\sum\limits_{n=1}^{\infty} n^{-5}$ to 3 decimal places.

Solution. By Theorem 18, with $\alpha = 5$, the series converges. By Theorem 17, with $f(x) = x^{-5}$,

$$1 + \frac{1}{2^5} + \cdots + \frac{1}{n^5} \leq S \leq 1 + \frac{1}{2^5} + \cdots + \frac{1}{n^5} + \int_n^{\infty} \frac{dx}{x^5}$$

Now

$$\int_n^{\infty} \frac{dx}{x^5} = -\frac{1}{4x^4} \Big|_n^{\infty} = \frac{1}{4n^4}$$

and $1/(4n^4) < 0.0005$ for $n = 5$. Hence we write

$$S = 1 + \frac{1}{2^5} + \frac{1}{3^5} + \frac{1}{4^5} + \frac{1}{5^5} + \epsilon$$

where ϵ, the "error," is between 0 and $1/(4 \cdot 5^4) = 0.0004$. Thus we find that

$$1 + 2^{-5} + \cdots + 5^{-5} = 1.03666 \leq S \leq 1.03666 + \epsilon \leq 1.03706$$

and, hence, $S = 1.037$ to 3 decimal places. (Here we have verified that the addition of ϵ would not affect the third decimal place; sometimes even a very small error, say less than 0.000 001, can affect even the first decimal place, because of rounding off.)

PROBLEMS

In Problems 1 to 8 prove convergence or divergence by a comparison test:

1. $\sum\limits_{n=1}^{\infty} \dfrac{1}{n2^n}$ 2. $\sum\limits_{n=1}^{\infty} \dfrac{\ln n}{n}$ 3. $\sum\limits_{n=1}^{\infty} \dfrac{n+2}{3^n(n+1)}$

4. $\sum\limits_{n=1}^{\infty} \dfrac{1+e^{-n}}{1+e^n}$ 5. $\sum\limits_{n=1}^{\infty} \dfrac{1}{2n+3}$ 6. $\sum\limits_{n=1}^{\infty} \dfrac{n^2}{n^3+7n^2+1}$

7. $\sum\limits_{n=1}^{\infty} \dfrac{n}{5^n}$ 8. $\sum\limits_{n=1}^{\infty} \dfrac{\sqrt{n}}{n+1}$

In Problems 9 to 16 prove convergence or divergence by the integral test:

9. $\displaystyle\sum_{n=2}^{\infty} \frac{1}{n \ln^\alpha n}, \ \alpha > 0$ 10. $\displaystyle\sum_{n=1}^{\infty} \frac{\ln^\alpha n}{n}, \ \alpha > 0$

11. $\displaystyle\sum_{n=1}^{\infty} \frac{1}{n^2 + 1}$ 12. $\displaystyle\sum_{n=1}^{\infty} \frac{n}{e^{n^2}}$ 13. $\displaystyle\sum_{n=1}^{\infty} \frac{n^2}{e^{n^3}}$

14. $\displaystyle\sum_{n=10}^{\infty} \frac{1}{n \ln n \ln \ln n}$ 15. $\displaystyle\sum_{n=1}^{\infty} \frac{2n + 1}{n^2 + n}$

16. $\displaystyle\sum_{n=1}^{\infty} \frac{1}{(n + 1)(n^2 + 1)}$

In Problems 17 to 32 determine the convergence or divergence of the given series:

17. $\displaystyle\sum_{n=1}^{\infty} \frac{1}{\sqrt{n}\, 2^n}$ 18. $\displaystyle\sum_{n=0}^{\infty} \frac{n}{2^n}$ 19. $\displaystyle\sum_{n=2}^{\infty} \frac{1}{\sqrt{n} \ln n}$

20. $\displaystyle\sum_{n=1}^{\infty} \frac{2^n}{4^n - 3^n}$ 21. $\displaystyle\sum_{n=3}^{\infty} \frac{3 + \sin n}{n - 2}$ 22. $\displaystyle\sum_{n=2}^{\infty} \frac{\ln n}{2^{n-3}}$

23. $\displaystyle\sum_{n=0}^{\infty} \frac{1}{\sqrt[3]{4n^3 - n + 2}}$ 24. $\displaystyle\sum_{n=2}^{\infty} \frac{1}{4^n + \ln n}$ 25. $\displaystyle\sum_{n=3}^{\infty} \frac{1}{\sqrt[3]{5n^4 - n^3 - 75n + 1}}$

26. $\displaystyle\sum_{n=1}^{\infty} \frac{1}{n^2 - 7\pi}$ 27. $\displaystyle\sum_{n=1}^{\infty} \frac{\sqrt[n]{e}}{n^2}$ 28. $\displaystyle\sum_{n=1}^{\infty} \frac{1}{n \sqrt[n]{\pi}}$

29. $\displaystyle\sum_{n=2}^{\infty} \frac{2\sin n}{n^2 - 1}$ 30. $\displaystyle\sum_{n=1}^{\infty} \frac{1}{\sqrt[n]{5}}$ 31. $\displaystyle\sum_{n=1}^{\infty} \frac{1}{n + \sqrt[n]{n}}$

32. $\displaystyle\sum_{n=1}^{\infty} \frac{\sqrt[n]{4}}{n^{3/2}}$

In Problems 33 to 38 find the sum of the series correct to 3 decimal places:

33. $\displaystyle\sum_{n=1}^{\infty} \frac{1}{n + 10^n}$ 34. $\displaystyle\sum_{n=1}^{\infty} \frac{1}{n8^n}$ 35. $\displaystyle\sum_{n=2}^{\infty} \frac{1}{n^5 \ln n}$

36. $\displaystyle\sum_{n=1}^{\infty} \frac{n + 1}{n10^n}$ 37. $\displaystyle\sum_{n=1}^{\infty} \frac{1}{2^{n^2}}$ 38. $\displaystyle\sum_{n=2}^{\infty} \frac{\ln n}{n^{11}}$

39. (a) Let $f(x) = \sin^2 \pi x, \ 1 \le x \le \infty$. Show that $\displaystyle\sum_{n=1}^{\infty} f(n) = 0$ but $\displaystyle\int_1^\infty f(x)\, dx$

   diverges to $+\infty$.

   ‡(b) Let $f(x) = n^4(x - n) + n, \ n - n^{-3} \le x \le n, \ f(x) = -n^4(x - n) + n, \ n \le x \le n + n^{-3}$ for $n = 2, 3, 4, \ldots, f(x) = 0$ otherwise for $1 \le x \le \infty$. Show

   that $\displaystyle\sum_{n=1}^{\infty} f(n)$ diverges but $\displaystyle\int_1^\infty f(x)\, dx$ converges.

40. Prove Theorem 16.

‡41. Apply the ideas involved in the integral test to $\displaystyle\int_1^n \ln x\, dx$ and deduce that

$$n^n e^{-n+1} \le n! \le n^{n+1} e^{-n+1}$$

8-9 ABSOLUTE CONVERGENCE

Thus far, most of our tests for convergence and divergence have been for series with positive terms. We now prove a theorem that extends these tests to series with arbitrary real and complex terms.

A series Σa_n such that $\Sigma |a_n|$ converges is called *absolutely convergent*.

THEOREM 19. *Let Σa_n be a series with real (or complex) terms. If Σa_n is absolutely convergent, then it is convergent; that is, if $\Sigma |a_n|$ converges, then Σa_n converges. Moreover, when $\Sigma |a_n|$ converges, $|\Sigma a_n| \leq \Sigma |a_n|$.*

PROOF. Let Σa_n be a real series and let $\Sigma |a_n|$ converge. If we replace the negative terms of the series Σa_n by zeros, we obtain a new series of non-negative terms, whose nth term is at most equal to $|a_n|$. Hence, by the comparison test (Theorem 13), the new series converges. Similarly, if we replace the positive terms by zeros, we obtain a second new series of nonpositive terms. After reversing signs, this can also be compared with the series $\Sigma |a_n|$; hence, the second new series also converges. But addition of the first and second new series gives Σa_n again. Consequently, by Theorem 11, Σa_n converges.

Next let $\Sigma a_n = \Sigma(\alpha_n + i\beta_n)$ be a complex series. Then

$$|\alpha_n| \leq \sqrt{\alpha_n{}^2 + \beta_n{}^2} = |a_n|$$

Therefore, by the rule just proved for real series, $\Sigma \alpha_n$ converges. Similarly, $\Sigma \beta_n$ converges, so that $\Sigma(\alpha_n + i\beta_n)$ converges.

For absolutely convergent real or complex series we can write

$$|\Sigma a_n| = |S| = |s_n + r_n| \leq |s_n| + |r_n| = |a_1 + \cdots + a_n| + |r_n|$$
$$\leq |a_1| + \cdots + |a_n| + |r_n| \leq \left(\sum_{n=1}^{\infty} |a_n| \right) + |r_n|$$

As $n \to \infty$, $|r_n| \to 0$. Hence we conclude that $|\Sigma a_n| \leq \Sigma |a_n|$.

COROLLARY TO THEOREM 19. *A real or complex series which is dominated by a convergent series with nonnegative terms is convergent— in fact, absolutely convergent.*

This Corollary follows from Theorem 13 and Theorem 19.

If Σa_n is convergent, but not absolutely convergent, we say that Σa_n is *conditionally convergent*. It is shown in Section 8-11 that the series $\sum_{n=1}^{\infty} (-1)^n n^{-1}$ is convergent. We have observed earlier that $\sum_{n=1}^{\infty} n^{-1}$ is divergent, and therefore $\sum_{n=1}^{\infty} (-1)^n n^{-1}$ is a conditionally convergent series.

If Σa_n is an absolutely convergent series and $\{e_n\}$ is a sequence of real (or complex) numbers for which $|e_n| \leq 1$, then $\Sigma e_n a_n$ is an (absolutely) convergent series. For the series $\Sigma e_n a_n$ is dominated by the series $\Sigma |a_n|$, and the Corollary to Theorem 19 applies. Thus the series $\sum_{n=0}^{\infty} (\sin n + i \cos n)/2^n$ is an absolutely convergent series, with $e_n = \sin n + i \cos n$, $|e_n| = 1$.

We remark that, as for convergence, absolute convergence is unaffected by modifying the first k terms of a series.

8-10 RATIO AND ROOT TESTS

THEOREM 20 (*Ratio test*). *If a series Σa_n with nonzero real (or complex) terms is such that*

$$\lim_{n \to \infty} \left| \frac{a_{n+1}}{a_n} \right|$$

exists and equals L, then

(i)   *If $L < 1$, the series converges absolutely.*
(ii) *If $L > 1$, the series diverges.*
(iii) *If $L = 1$, the test is inconclusive.*

PROOF. *Part(i).* If $L < 1$, there is a number r between L and $1: L < r < 1$. Then, by the definition of a limit of a sequence, there is an integer N such that

$$\left| \frac{a_{n+1}}{a_n} \right| < r, \qquad \text{for } n \geq N$$

Hence $|a_{N+1}| < r|a_N|, \qquad |a_{N+2}| < r|a_{N+1}| < r^2|a_N|$

and, in general,

$$|a_{N+k}| < r^k|a_N|, \qquad k = 1, 2, \dots$$

Thus the series $\displaystyle\sum_{n=N}^{\infty} a_n$ is dominated by the geometric series $\displaystyle\sum_{n=N}^{\infty} |a_N| r^n$, which converges since $|r| < 1$ (Theorem 8). Accordingly, by the Corollary to Theorem 19, $\displaystyle\sum_{n=N}^{\infty} a_n$ is absolutely convergent, so that $\displaystyle\sum_{n=0}^{\infty} a_n$ is also absolutely convergent.

Part (ii). Since $L > 1$, there is a real number r between 1 and $L: 1 < r < L$. Then, by the definition of limit of a sequence, there is an integer N such that

$$r < \left| \frac{a_{n+1}}{a_n} \right|, \qquad \text{for } n > N$$

so that, as above,

$$|a_N| \, r^k < |a_{N+k}|, \qquad k = 1, 2, \dots$$

But $r > 1$, and hence $|a_{N+k}| > |a_N| > 0$ for $k \geq 1$, so that a_n cannot $\to 0$, whence (by Theorem 7) the series Σa_n diverges.

Part (iii). By Theorem 18, the series $\displaystyle\sum_{n=1}^{\infty} 1/n$ is divergent and the series $\displaystyle\sum_{n=1}^{\infty} 1/n^2$ is convergent, yet in both cases $\lim |(a_{n+1})/a_n| = 1$. Thus, when $L = 1$, the ratio test does not allow us to conclude anything regarding convergence or divergence of the series.

EXAMPLE 1. Test for convergence: $\displaystyle\sum_{n=0}^{\infty} \frac{n^2 \cos n}{4^n}$

Solution.

$$\lim_{n \to \infty} \left(\frac{(n+1)^2}{4^{n+1}}\right)\Big/\left(\frac{n^2}{4^n}\right) = \lim_{n \to \infty} \frac{1}{4}\left(\frac{n+1}{n}\right)^2 = \frac{1}{4}$$

Hence, by the ratio test, the series $\Sigma n^2/4^n$ converges. By the comparison test, it follows that $\Sigma(n^2|\cos n|)/4^n$ converges and, hence, that the given series is absolutely convergent.

We notice that $n^2/4^n < 1/2^n$ for $n \geq 5$ and, therefore, if $S = \displaystyle\sum_{n=0}^{\infty} (n^2 \cos n)/4^n$, then $|S - s_{10}| < \displaystyle\sum_{n=11}^{\infty} (1/2^n) = 1/2^{10} < .001$. Thus s_{10} is a reasonable approximation to S.

THEOREM 21 (*Root test*). *If a series Σa_n with real (or complex) terms is such that*

$$\lim_{n \to \infty} \sqrt[n]{|a_n|}$$

exists and equals R, then
   (i)   *If $R < 1$, the series converges absolutely.*
   (ii) *If $R > 1$, the series diverges.*
   (iii) *If $R = 1$, the test is inconclusive.*

PROOF. *Part (i).* If $R < 1$, then as in the proof of Theorem 20 there is a real number r between R and 1 and there is an integer N such that

$$\sqrt[n]{|a_n|} < r \qquad \text{for} \qquad n \geq N$$

But then $|a_n| < r^n$ for $n \geq N$, and the series $\displaystyle\sum_{n=N}^{\infty} a_n$ is dominated by the geometric series $\displaystyle\sum_{n=N}^{\infty} r^n$ which is convergent since $r < 1$. It follows that the series $\displaystyle\sum_{n=0}^{\infty} a_n$ converges absolutely.

Part (ii). If $R > 1$, there is an integer N so that for $n \geq N$, $\sqrt[n]{|a_n|} \geq 1$ and, hence, $|a_n| \geq 1$. Therefore, by Theorem 7, the series diverges.

Part (iii). The series $\Sigma 1/n$ is divergent, while the series $\Sigma 1/n^2$ is convergent, yet in both cases $\lim \sqrt[n]{|a_n|} = 1$, since $\lim_{n \to \infty} \sqrt[n]{n} = 1$ by De L'Hospital's Rules. Thus the root test is inconclusive when $R = 1$.

EXAMPLE 2. Test for convergence: $\displaystyle\sum_{n=2}^{\infty} \frac{1}{(\ln n)^n}$

Solution. We apply the root test: $\sqrt[n]{1/(\ln n)^n} = 1/\ln n$, which tends to 0 as n tends to $+\infty$, and so by the root test the series converges. Let S be the sum. Then, since $1/\ln n > 1/\ln(n+1)$, we observe that

$$\sum_{n=2}^{7} \frac{1}{(\ln n)^n} < S < \sum_{n=2}^{7} \frac{1}{(\ln n)^n} + \sum_{n=8}^{\infty} \frac{1}{(\ln 8)^n}$$

The last series is a geometric series Σar^{n-1} with $a = (\ln 8)^{-8}$, $r = (\ln 8)^{-1}$. Its sum is $a/(1 - r)$, found to be less than 0.002. Hence, S is given by the sum on the left with error less than 0.002.

PROBLEMS

In problems 1 to 17 determine the convergence or divergence of the series:

1. $\displaystyle\sum_{n=1}^{\infty} \frac{1}{n!}$

2. $\displaystyle\sum_{n=1}^{\infty} \frac{n+3}{2^n}$

3. $\displaystyle\sum_{n=1}^{\infty} \frac{(-1)^n}{n^n}$

4. $\displaystyle\sum_{n=1}^{\infty} \frac{n+1}{n!}$

5. $\displaystyle\sum_{n=1}^{\infty} \frac{(-1)^n \cos n}{n!}$

6. $\displaystyle\sum_{n=1}^{\infty} \frac{n^n}{n!}$

7. $\displaystyle\sum_{n=1}^{\infty} \frac{(-1)^n 3^n}{1 \cdot 4 \cdot 7 \cdots (3n+1)}$

8. $\displaystyle\sum_{n=0}^{\infty} \frac{1 \cdot 3 \cdot 5 \cdots (2n+1)}{1 \cdot 4 \cdot 7 \cdots (3n+1)}$

9. $\displaystyle\sum_{n=1}^{\infty} \frac{2^n}{n^2}$

10. $\displaystyle\sum_{n=0}^{\infty} \frac{n!}{5^n}$

11. $\displaystyle\sum_{n=1}^{\infty} \left(\frac{n}{n+1}\right)^{n^2}$

12. $\displaystyle\sum_{n=1}^{\infty} \frac{2^n + n^2}{2^n n^2}$

13. $\displaystyle\sum_{n=2}^{\infty} \frac{1 + \ln^2 n}{n \ln^2 n}$

14. $\displaystyle\sum_{n=2}^{\infty} \frac{1 + \ln^2 n}{n \ln^3 n}$

15. $\displaystyle\sum_{n=1}^{\infty} \frac{(1 + 3i)^n}{n^2 5^n}$

16. $\displaystyle\sum_{n=1}^{\infty} \frac{(2 + 3i)^n}{n!}$

17. $\displaystyle\sum_{n=1}^{\infty} \frac{(5 - i)^n}{n^n}$

18. *Extended ratio test.* Prove: (a) If there are an $r < 1$ and an integer N such that $|(a_{n+1})/a_n| < r$, for $n \geq N$, then the series Σa_n converges absolutely. Notice: In this case $\lim |(a_{n+1})/a_n|$ need not exist.
(b) If $|(a_{n+1})/a_n| \geq 1$ for $n \geq N$, then the series Σa_n diverges.

19. *Extended root test.* Prove: (a) If there are an $r < 1$ and an integer N such that $\sqrt[n]{|a_n|} < r$, for $n \geq N$, then the series Σa_n converges absolutely. Notice: In this case $\lim \sqrt[n]{|a_n|}$ need not exist.
(b) If $\sqrt[n]{|a_n|} \geq 1$ for all $n \geq N$, then the series Σa_n diverges.

20. Test for convergence:

(a) $\displaystyle\sum_{n=0}^{\infty} \frac{7 + 12(-1)^n}{2^n}$

(b) $\dfrac{1}{2^2} + \dfrac{1}{2 \ln 2} + \dfrac{1}{3^2} + \dfrac{1}{3 \ln 3} + \cdots + \dfrac{1}{n^2} + \dfrac{1}{n \ln n} + \cdots$

(c) $\displaystyle\sum_{n=1}^{\infty} \frac{1}{n \sqrt[n]{n}}$

(d) $\displaystyle\sum_{n=1}^{\infty} n^2 r^n, \; |r| < 1$

21. Show that the series $\displaystyle\sum_{n=1}^{\infty} r^{2n}/[(1 + r^2)^{n-1}]$ converges absolutely for all r.

22. Show that the series $\displaystyle\sum_{n=1}^{\infty} r^n/[(1 + r^n)^2]$ converges for all $r \geq 0$ except 1.

23. Let Σa_n, Σb_n be convergent series with positive real terms. Prove:
(a) A series whose terms are a subsequence of $\{a_n\}$ converges.
(b) If $\{t_n\}$ is a monotone strictly increasing sequence of integers and

$$c_m = \begin{cases} a_{t_n} & \text{if } m = t_n, \\ 0 & \text{otherwise,} \end{cases} \qquad \text{then } \Sigma c_n \text{ converges.}$$

(c) The series $\Sigma \sqrt{a_n b_n}$ converges.

(d) The series $\Sigma \sqrt{a_n a_{n+1}}$ converges.

24. Find the sum of the following series correct to 3 decimal places:

(a) $\displaystyle\sum_{n=0}^{\infty} \frac{(-1)^n}{n!}$ 　(b) $\displaystyle\sum_{n=1}^{\infty} \frac{1}{n^n}$ 　(c) $\displaystyle\sum_{n=1}^{\infty} \frac{n^2}{n!}$

(d) $\displaystyle\sum_{n=1}^{\infty} \frac{2^n}{1 \cdot 3 \cdot 5 \cdots (2n-1)}$

8-11 ALTERNATING SERIES

An alternating series is a real series whose successive terms have opposite signs. Thus the series $1 - 1/2 + 1/3 - 1/4 + \cdots + [(-1)^{n-1}]/n + \cdots$ is an alternating series. There is no simple test that will always determine the convergence or divergence of an alternating series, but the following theorem is a very useful rule for convergence.

THEOREM 22. *If $\{a_n\}$ is a monotone nonincreasing sequence with limit 0, then the alternating series*

$$a_1 - a_2 + a_3 - a_4 + \cdots = \sum_{n=1}^{\infty} (-1)^{n-1} a_n$$

converges. The sum S and the nth partial sum s_n differ by at most a_{n+1}; more precisely,

$$s_n \leq S \leq s_n + a_{n+1} \quad \text{for n even,} \qquad s_n - a_{n+1} \leq S \leq s_n \quad \text{for n odd}$$

PROOF. Since $a_1 \geq a_2 \geq a_3 \geq \cdots \geq 0$, we have

$$s_1 = a_1 \geq s_3 = a_1 - (a_2 - a_3) \geq s_5 = a_1 - (a_2 - a_3) - (a_4 - a_5)$$

and so on; that is, the partial sums of odd index form a monotone nonincreasing sequence, as in Figure 8-5. Similarly, $s_2 \leq s_4 = s_2 + (a_3 - a_4)$ and so

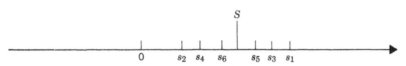

Figure 8-5　Partial sums of alternating series.

on; that is, the partial sums of even index form a monotone nondecreasing sequence, as in Figure 8-5. Furthermore, for $k = 1, 2, \ldots, s_{2k+1} = s_{2k} + a_{2k+1}$, so that $s_{2k+1} \geq s_{2k}$. Hence $s_3 \geq s_2$, $s_5 \geq s_4$, $s_7 \geq s_6$, $\ldots$ and, since $s_2 \leq s_4 \leq s_6 \leq \ldots$ and $s_1 \geq s_3 \geq s_5 \geq \ldots$, we must have

$$s_2 \leq s_4 \leq s_6 \leq \cdots \leq s_5 \leq s_3 \leq s_1$$

that is, all s_n of even index are less than or equal to all s_n of odd index. Thus, those of even index form a bounded nondecreasing sequence and, hence, have limit S; similarly, those of odd index have limit S'. But $s_{2k+1} = s_{2k} + a_{2k+1}$ and $a_{2k+1} \to 0$; consequently

$$\lim_{k\to\infty} s_{2k+1} = S' = \lim_{k\to\infty} s_{2k} = S$$

It follows that s_n converges to S and that S is the sum of the series. Since $s_{2k+1} = s_{2k} + a_{2k+1}$ and S must lie between s_{2k} and s_{2k+1}, we conclude that $s_{2k} \le S \le s_{2k} + a_{2k+1}$. Similarly, $s_{2k+1} - a_{2k+2} \le S \le s_{2k+1}$ and the theorem is proved.

EXAMPLE. Since the sequence $1/n$ is monotone strictly decreasing and $1/n \to 0$, the series

$$1 - \frac{1}{2} + \frac{1}{3} - \frac{1}{4} + \cdots = \sum_{n=1}^{\infty} \frac{(-1)^{n-1}}{n}$$

converges and, for example,

$$1 - \frac{1}{2} + \frac{1}{3} - \frac{1}{4} \le S \le 1 - \frac{1}{2} + \frac{1}{3} - \frac{1}{4} + \frac{1}{5}$$

or $\frac{7}{12} \le S \le (\frac{7}{12}) + (\frac{1}{5})$, or $0.583 \le S \le 0.783$.

It is shown in advanced texts that $S = \ln 2 = 0.693$. Hence the fourth or fifth partial sum has an error of about 0.1. The theorem indicates that to reduce the error to less than 0.0005, we would need 2000 terms! The fact that the partial sums swing back and forth, above and below S, suggests averaging two successive partial sums to estimate S. Thus $\frac{1}{2}(s_4 + s_5) = \frac{1}{2}(0.583 + 0.783) = 0.683$, a value much closer to the limit.

Remark. Since $\Sigma(1/n)$ diverges, the series of the example is conditionally convergent.

PROBLEMS

In Problems 1 to 6, discuss the convergence or divergence of the series:

1. $\displaystyle\sum_{n=1}^{\infty} \frac{n(-1)^{n-1}}{n^2 + 1}$ 2. $\displaystyle\sum_{n=0}^{\infty} \frac{(-1)^n}{n!}$ 3. $\displaystyle\sum_{n=1}^{\infty} \frac{(-1)^{n-1}}{n^2}$

4. $\displaystyle\sum_{n=1}^{\infty} \frac{(-1)^{n-1}n}{\sqrt{n^2 + 1}}$ 5. $\displaystyle\sum_{n=1}^{\infty} \frac{(-1)^n n^3}{e^n}$

6. Σa_n where $a_n = \dfrac{(-1)^{(n-1)/2}}{n}$ if n is odd and $a_n = \dfrac{(-1)^{n/2}}{3n + 1}$ if n is even.

7. Give a value for N so that for $n \ge N$, s_n differs from the sum of the series by less than .005.

(a) $\displaystyle\sum_{n=1}^{\infty} \frac{(-1)^{n+1}}{n!}$ (b) $\displaystyle\sum_{n=0}^{\infty} \frac{(-1)^n}{(n + 1)3^n}$ (c) $\displaystyle\sum_{n=0}^{\infty} \frac{(-1)^n}{5^n}$

(d) $\displaystyle\sum_{n=1}^{\infty} \frac{(-1)^{n+1}}{10n^3}$ (e) $\displaystyle\sum_{n=1}^{\infty} \frac{(-1)^n n}{(2n)!}$

‡8-12 REARRANGEMENT OF SERIES

In this section we show that every absolutely convergent series can be rearranged freely without affecting convergence or the sum.

A series $\sum_{m=1}^{\infty} b_m$ is said to be a rearrangement of the series $\sum_{n=1}^{\infty} a_n$ if there is a one-to-one correspondence between the indexes m and n so that $b_m = a_n$ for corresponding indexes. For example, a rearrangement of

$$\sum_{n=1}^{\infty} a_n = 1 + \frac{1}{2} + \frac{1}{3} + \frac{1}{4} + \frac{1}{5} + \frac{1}{6} + \cdots + \frac{1}{n} + \cdots$$

is the series

$$\sum_{m=1}^{\infty} b_m = \frac{1}{2} + 1 + \frac{1}{4} + \frac{1}{3} + \frac{1}{6} + \frac{1}{5} + \cdots$$

Here $b_1 = a_2$, $b_2 = a_1$, $b_3 = a_4$, $b_4 = a_3$, ... and, in general, $b_m = a_{m-1}$ if m is even, $b_m = a_{m+1}$ if m is odd.

THEOREM 23. *Let Σa_n be a convergent series. If Σa_n is absolutely convergent, then every series obtained from Σa_n by rearranging terms converges absolutely and has the same sum as the series Σa_n.*

PROOF. Let Σa_n be absolutely convergent, with sum S and let Σb_m be a rearrangement of Σa_n. Then for every k,

$$|b_1| + \cdots + |b_k| \leq \sum_{n=1}^{\infty} |a_n|$$

For the k terms on the left appear as differently numbered terms on the right. Hence the series $\Sigma |b_m|$ converges, so that Σb_m is an absolutely convergent series. Also we can choose n so large that

$$|a_1 + \cdots a_n - S| < \frac{\epsilon}{2}, \qquad |a_{n+1}| + \cdots + |a_{n+p}| < \frac{\epsilon}{2}, \text{ for } p = 1, 2, \ldots$$

This is possible since $\Sigma a_n = S$ and since $\Sigma |a_n|$ satisfies the Cauchy condition (Theorem 12, Section 8-6). For m sufficiently large, say $m > M$, $b_1 + \cdots + b_m$ is a sum of n terms equal to $a_1 + \cdots + a_n$ plus $m - n$ terms selected from $a_{n+1} + a_{n+2} + \cdots$. By the Cauchy condition the sum of these $m - n$ terms has absolute value less than $\epsilon/2$. Hence

$$|(b_1 + \cdots + b_m) - (a_1 + \cdots + a_n)| < \frac{\epsilon}{2}$$

and, since $|(a + \cdots + a_n) - S| < \epsilon/2$, we conclude that

$$|(b_1 + \cdots + b_m) - S| < \epsilon, \qquad m > M$$

Hence $\Sigma b_m = S$.

Remark. A conditionally convergent real series can be rearranged to converge to any desired sum, to become oscillatory or to diverge to $+\infty$ or to $-\infty$ (see Problems 2 to 6 below).

PROBLEMS

1. Prove:
   (a) If Σa_n is a convergent real series and all nonzero terms are of the same sign, then Σa_n converges absolutely.
   (b) If Σa_n is a convergent real series and all but finitely many of the nonzero terms have the same sign, then Σa_n converges absolutely.
   (c) If Σa_n is a conditionally convergent real series then infinitely many terms are positive and infinitely many terms are negative.
   (d) If Σa_n, Σb_n are absolutely convergent series and $c \neq 0$, then $\Sigma c a_n$ and $\Sigma (a_n + b_n)$ are absolutely convergent series.
   (e) Let Σa_n be a conditionally convergent series with real terms, let α_n be the larger of 0 and a_n, and let β_n be the smaller of 0 and a_n. Then $a_n = \alpha_n + \beta_n$ and $\Sigma \alpha_n$, $\Sigma \beta_n$ are series with terms of same sign that diverge to infinity. [*Hint.* If they both converge, then by parts (*a*) and (*d*), Σa_n is absolutely convergent. Next examine the case where one converges and the other diverges.]

   Remark. Part (*e*) shows that in a conditionally convergent real series the positive terms form a series properly divergent to ∞, the negative terms form a series properly divergent to $-\infty$.

   In Problems 2 to 6 let the series Σa_n be conditionally convergent and real with no zero terms; let b_1 be the first positive term, b_2 the second positive term, and so on, so that Σb_n is the series formed of the positive terms. Similarly, let Σc_n be the series formed of the negative terms. By the remark above, these series are properly divergent.

2. Prove that the series Σa_n can be rearranged to diverge to ∞. (*Hint.* Choose n_1 so that $b_1 + \cdots + b_{n_1} > |c_1| + 1$, n_2 so that $b_{n_1+1} + \cdots + b_{n_2} > |c_2| + 1$, and so on. Then use the arrangement $b_1 + \cdots + b_{n_1} + c_1 + b_{n_1+1} + \cdots + b_{n_2} + c_2 + b_{n_2+1} + \cdots$.)

3. Prove that the series Σa_n can be rearranged to diverge to $-\infty$.

4. Prove that the series Σa_n can be rearranged to converge to 0. (*Hint.* First use b_1, then add just enough negative terms to have a negative partial sum, then add just enough positive terms to have a positive partial sum, and so on; now use the fact that $a_n \to 0$.)

5. Prove that the series Σa_n can be rearranged to converge to a given number k.

6. Prove that the series Σa_n can be rearranged to become oscillatory divergent.

‡8-13 PRODUCTS OF SERIES

We have observed that infinite series are generalizations of finite sums. Now two finite sums $(a_0 + \cdots + a_n)$ and $(b_0 + \cdots + b_m)$ can be multiplied and, by the distributive law, the product is the finite sum $\displaystyle\sum_{k=0}^{n} \sum_{j=1}^{m} a_k b_j$ Also, by the distributive, associative, and commutative laws, this sum is independent of the order in which we write the terms. The question arises whether, given two convergent series, $\displaystyle\sum_{n=0}^{\infty} a_n$, $\displaystyle\sum_{m=0}^{\infty} b_m$, we can arrange the terms $a_n b_m$ so as to obtain a convergent series and, if so, whether the sum of that series is the product of the sums of the two given series. (Throughout this section our series will be indexed beginning with 0.)

Among the possible arrangements of the terms $a_n b_m$ there are two rather natural arrangements: one suggested by the multiplication of the partial sums for Σa_n and Σb_m, the other suggested by the product of two polynomials or the formal product of two power series.

For the first of these arrangements, the terms are chosen so that the first term is $a_0 b_0$, the first four terms are $(a_0 + a_1)(b_0 + b_1) = a_0 b_0 + a_0 b_1 + a_1 b_1 + a_1 b_0$, and the first nine terms are $(a_0 + a_1 + a_2)(b_0 + b_1 + b_2) = a_0 b_0 + a_0 b_1 + \cdots + a_2 b_0$. Thus, the arrangement is that suggested in Figure 8-6, and we can write the series as

$$a_0 b_0 + a_0 b_1 + a_1 b_1 + a_1 b_0 + \cdots$$
$$+ a_0 b_n + a_1 b_n + \cdots + a_n b_n + a_n b_{n-1} + \cdots + a_n b_0 + \cdots \quad (8\text{-}130)$$

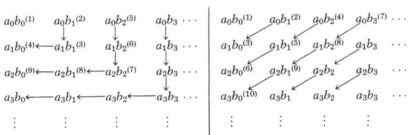

Figure 8-6 First arrangement of terms of product of two series.

Figure 8-7 Second arrangement of terms of product of two series.

We notice that the sum of the first n^2 terms of this series is the product of the partial sums:

$$\sum_{k=0}^{n-1} a_k \quad \text{and} \quad \sum_{j=0}^{n-1} b_j$$

In the second arrangement, the terms are chosen as suggested in Figure 8-7. Thus the first term is $a_0 b_0$, the next two terms are $a_0 b_1 + a_1 b_0$, the next three terms are $a_0 b_2 + a_1 b_1 + a_2 b_0$. We can write the series obtained by this arrangement as

$$a_0 b_0 + a_0 b_1 + a_1 b_0 + a_0 b_2 + a_1 b_1 + a_2 b_0 + \cdots$$
$$+ a_0 b_n + a_1 b_{n-1} + \cdots + a_n b_0 + \cdots \quad (8\text{-}131)$$

We shall prove (Theorem 24) that, if Σa_n and Σb_n are absolutely convergent, then every arrangement of the terms $a_n b_m$ as an infinite series is absolutely convergent. Hence, by Theorem 10, we can group the terms by introducing parentheses without affecting convergence or the sum. For example, the series (8-130) is usually grouped to give the series

$$\sum_{n=0}^{\infty} (a_0 b_n + a_1 b_n + \cdots + a_n b_n + \cdots + a_n b_0) \quad (8\text{-}130')$$

Here the term of index n corresponds to the terms on two edges of the $(n + 1)$-st square of the diagram of Figure 8-6. The series (8-131) is usually grouped to give the series

$$\sum_{n=0}^{\infty} (a_0 b_n + a_1 b_{n-1} + \cdots + a_n b_0) \qquad (8\text{-}131')$$

Here the term of index n corresponds to the $(n + 1)$-st diagonal in the diagram of Figure 8-7. The series (8-131') is called the *Cauchy product* of the series Σa_n, Σb_n. It occurs naturally in multiplying two polynomials or two power series

$$\alpha_0 + \alpha_1 x + \cdots + \alpha_n x^n + \cdots, \qquad \beta_0 + \beta_1 x + \cdots + \beta_n x^n + \cdots$$

If we form all products of terms and collect terms of the same degree, we obtain

$$\alpha_0 \beta_0 + x(\alpha_0 \beta_1 + \alpha_1 \beta_0) + \cdots + x^n(\alpha_0 \beta_n + \cdots + \alpha_n \beta_0) + \cdots$$

This is the Cauchy product of the two series $\Sigma \alpha_n x^n$ and $\Sigma \beta_n x^n$.

THEOREM 24. *If* $\displaystyle\sum_{n=0}^{\infty} a_n$ *and* $\displaystyle\sum_{m=0}^{\infty} b_m$ *are absolutely convergent series with sum A and B, respectively, then each arrangement of the terms* $a_n b_m$ *is an absolutely convergent series and all arrangements have the same sum: AB.*

PROOF. We shall show that the series given by the first arrangement discussed above is an absolutely convergent series with sum AB. Then Theorem 23 implies the conclusion of Theorem 24.

We are given that $\Sigma|a_n|$, $\Sigma|b_m|$ are convergent series. Let s_{n-1}, t_{n-1} be the sum of the first n terms of $\Sigma|a_n|$ and $\Sigma|b_m|$, respectively. Let $\displaystyle\sum_{n=0}^{\infty} c_n$ be the series given by the first arrangement. Then

$$c_0 = a_0 b_0, \qquad c_1 = a_0 b_1, \qquad c_2 = a_1 b_1, \qquad c_3 = a_1 b_0, \qquad \ldots$$

Clearly

$$\sum_{k=0}^{n^2-1} |c_k| = \sum_{k=0}^{n-1} |a_k| \sum_{k=0}^{n-1} |b_k|$$

Let S_{n-1} be the sum of the first n terms of the series $\Sigma|c_k|$. Then

$$S_{n^2-1} = s_{n-1} t_{n-1}$$

Since $\lim_{n \to \infty} s_n$ exists and is $\Sigma|a_n|$ and $\lim_{n \to \infty} t_n$ exists and is $\Sigma|b_m|$, we have

$$\lim_{n \to \infty} S_{n^2-1} = \lim_{n \to \infty} s_{n-1} t_{n-1} = (\lim s_n)(\lim t_n) = \left(\sum_{m=0}^{\infty} |a_m| \right) \left(\sum_{m=0}^{\infty} |b_m| \right)$$

But $|c_k| \geq 0$ and, therefore, S_n is a monotone increasing sequence, and hence (by Theorem 4)

$$\lim_{n \to \infty} S_n = \lim_{n \to \infty} S_{n^2-1} = (\Sigma|a_n|)(\Sigma|b_m|)$$

Thus

$$\sum_{k=0}^{\infty} |c_k| = (\Sigma|a_n|)(\Sigma|b_m|)$$

Hence $\Sigma|c_k|$ is convergent and, therefore, Σc_k is absolutely convergent.

Let T_n be the sum of the first n terms of the series $\Sigma c_n = a_0b_0 + a_0b_1 + \cdots$. Then $\lim_{n\to\infty} T_n$ exists and

$$T_{n^2-1} = (a_0 + \cdots + a_{n-1})(b_0 + \cdots + b_{n-1})$$

whence

$$\lim_{n\to\infty} T_{n^2-1} = \lim_{n\to\infty} (a_0 + \cdots + a_{n-1}) \lim_{n\to\infty} (b_0 + \cdots + b_{n-1}) = AB$$

But a convergent sequence and all its subsequences have the same limit (Theorem 3) and, therefore, $\lim_{n\to\infty} T_n = AB$; or

$$AB = a_0b_0 + a_0b_1 + a_1b_1 + a_2b_0 + a_0b_2 + \cdots$$

This completes the proof of Theorem 24.

Remark. The Cauchy product converges, and has the expected value, under more general conditions than absolute convergence of $\Sigma a_n, \Sigma b_n$. (See K. Knopp: *Infinite Series*, p. 321; London, Blackie, 1928.)

PROBLEMS

1. Show that the Cauchy product of $\sum_{n=0}^{\infty} a_n$ and $\sum_{n=0}^{\infty} b_n$ is $\sum_{n=0}^{\infty} c_n$ where

$$c_n = \sum_{k=0}^{n} a_k b_{n-k}$$

2. (a) Evaluate $\sum_{n=0}^{\infty} (n + 1)r^n$ for $|r| < 1$ by showing it equals $\left(\sum_{n=0}^{\infty} r^n\right)^2$.

   (b) Evaluate $\sum_{n=0}^{\infty} \frac{1}{2}(n + 1)(n + 2)r^n$ for $|r| < 1$ by showing it equals $\left(\sum_{n=0}^{\infty} r^n\right)^3$.

3. Prove with the aid of the binomial theorem:

   (a) $\sum_{k=0}^{n} \frac{1}{k!(n - k)!} = \frac{2^n}{n!}$ (b) $\sum_{k=0}^{n} \frac{(-1)^k}{k!(n - k)!} = 0$

   (c) $\sum_{k=0}^{m} \frac{1}{(2k)!(2m + 1 - 2k)!} = \frac{2^{2m}}{(2m + 1)!}$

   [*Hint.* Apply (a) and (b) for $n = 2m + 1$.]

4. The series

$$e(x) = \sum_{n=0}^{\infty} \frac{x^n}{n!} \qquad c(x) = \sum_{n=0}^{\infty} \frac{(-1)^n x^{2n}}{(2n)!} \qquad s(x) = \sum_{n=0}^{\infty} \frac{(-1)^n x^{2n+1}}{(2n + 1)!}$$

converge absolutely for all real (or complex) x. Prove, with the aid of the results of problem 3, that for all x and y the following relations hold true:
(a) $e^2(x) = e(2x)$ (b) $e(x)e(y) = e(x + y)$ (c) $c^2(x) + s^2(x) = 1$
(d) $2s(x)c(x) = s(2x)$

8-14 SEQUENCES AND SERIES OF FUNCTIONS

We now turn to sequences and series formed not of numbers but of functions. Several such series and sequences have occurred in the examples of earlier sections. For example, the series of Theorem 18:

$$1 + \frac{1}{2^\alpha} + \frac{1}{3^\alpha} + \cdots + \frac{1}{n^\alpha} + \cdots$$

is a series whose terms are functions of α; for each fixed α, the series is a series of numbers, convergent for $\alpha > 1$, divergent for $\alpha \le 1$. The terms of this series form a sequence $1/n^\alpha$, a sequence of functions of α; for each fixed α, the sequence is a sequence of numbers, converging to 0 for $\alpha > 0$, to 1 for $\alpha = 0$, diverging to ∞ for $\alpha < 0$. We can portray this sequence of functions graphically, as in Figure 8-8. This figure also shows a typical sequence for fixed α (chosen as 0.3), converging to 0.

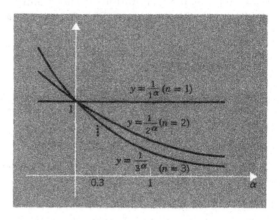

Figure 8-8 The sequence $1/n^\alpha$.

Similarly, the geometric series (with $a = 1$):

$$1 + r + \cdots + r^n + \cdots$$

is a series of functions of r, converging to $1/(1 - r)$ for $|r| < 1$, diverging otherwise. The nth partial sum of this series is also a function of r:

$$s_n = 1 + r + \cdots + r^{n-1}$$

and these functions form a sequence of functions converging to $1/(1 - r)$ for $|r| < 1$, diverging otherwise.

Let a sequence of functions $\{f_n(x)\}$ be given: we assume all functions to be defined on some interval of x. We define the *region of convergence* of the sequence to be the set of x-values for which the sequence converges. For any particular x in the region of convergence the limit of the sequence is a number that we can denote by $f(x)$; we can then write

$$\lim_{n \to \infty} f_n(x) = f(x), \qquad \text{or} \qquad f_n(x) \to f(x)$$

Thus, f is a function defined on the region of convergence of the sequence. We can write simply: $f_n \to f$, with the understanding that all functions are considered only in the region of convergence. We also say that the sequence

f_n converges pointwise to f (in the region of convergence).

These definitions are illustrated by the sequence of partial sums of the geometric series (with r replaced by x and s_n replaced by f_n):

$$f_n(x) = 1 + \cdots + x^{n-1}, \qquad f(x) = \frac{1}{1-x}$$

The region of convergence is the interval $-1 < x < 1$; in this interval

$$\lim_{n \to \infty} (1 + \cdots + x^{n-1}) = \frac{1}{1-x} \qquad \text{or} \qquad \lim_{n \to \infty} f_n(x) = f(x)$$

A similar discussion applies to infinite series of functions

$$\sum_{n=1}^{\infty} u_n(x)$$

Here we assume all functions $u_n(x)$ to be defined on some interval. The set of values of x for which the series converges is the *region of convergence of the series*. The sum of the series is a function f defined in the region of convergence:

$$f(x) = \sum_{n=1}^{\infty} u_n(x)$$

Also f is (by definition) the limit of the sequence of partial sums:

$$f(x) = \lim_{n \to \infty} f_n(x), \qquad f_n(x) = u_1(x) + \cdots + u_n(x)$$

These definitions are again illustrated by the geometric series (numbered starting with $n = 0$):

$$1 + x + \cdots + x^n + \cdots \equiv \sum_{n=0}^{\infty} u_n(x), \qquad u_n(x) = x^n$$

Here

$$\sum_{n=0}^{\infty} x^n = \frac{1}{1-x}, \qquad |x| < 1$$

THEOREM 25. *Let* $\sum_{n=0}^{\infty} M_n$ *be a convergent series with nonnegative real terms and let* $\sum_{n=0}^{\infty} u_n(x)$ *be an infinite series of functions. If*

$$|u_n(x)| \leq M_n, \qquad \text{for all } x \text{ in an interval } I$$

then $\Sigma u_n(x)$ converges absolutely for all x in I and $\left| \sum_{n=0}^{\infty} u_n(x) \right| \leq \sum_{n=0}^{\infty} M_n$

PROOF. By hypothesis, for each x in I, the series $\Sigma u_n(x)$ is dominated by ΣM_n and hence, by Theorem 19 and its corollary, the series $\Sigma u_n(x)$ converges absolutely and $|\Sigma u_n(x)| \leq \Sigma M_n$.

EXAMPLE. Show that the series $\sum_{n=1}^{\infty} \dfrac{\cos nx}{n^2}$ converges absolutely for all real x.

Solution. Clearly $\left| \dfrac{\cos nx}{n^2} \right| \leq \dfrac{1}{n^2}$ for all x. Since the series $\Sigma 1/n^2$ converges, it follows from Theorem 25 that the series $\sum\limits_{n=1}^{\infty} \dfrac{\cos nx}{n^2}$ converges

absolutely for all x. Hence the series $\sum\limits_{n=1}^{\infty} \dfrac{\cos nx}{n^2}$ defines a function $g(x)$

for all real x. Since $\cos nx = \cos(-nx)$, and $\cos n(x + 2\pi) = \cos nx$ for all x, it follows that $g(x) = g(-x)$ (g is an even function) and $g(x + 2\pi) = g(x)$ (g is a periodic function with period 2π).

Remark. The test contained in Theorem 25 is known as the *Weierstrass M-test* (developed by the German mathematician Karl Weierstrass in the latter part of the 19th century). A series $\Sigma u_n(x)$, which satisfies the test, is not only absolutely convergent but is also "uniformly convergent"—that is, we can make the remainder $u_{n+1}(x) + u_{n+2}(x) + \cdots$ uniformly small (i.e., less in absolute value than a prescribed positive ϵ for *all* x in the given interval) by choosing n sufficiently large (say n larger than N).

We shall concentrate on two types of infinite series of functions: *Power series,* where the terms are monomials: $a(x - c)^n$ and *Trigonometric series,* where the terms are the trigonometric function: $a \cos nx$, $b \sin nx$. The power series are natural generalizations of polynomials—in fact, their partial sums are polynomials. Both types of series are very useful in the physical sciences.

8-15 POWER SERIES

A power series is an infinite series of functions of the form:

$$a_0 + a_1(x - c) + \cdots + a_n(x - c)^n + \cdots \qquad (8\text{-}150)$$

Here $c, a_0, a_1, \ldots, a_n, \ldots$ are fixed numbers, whereas x is a variable number, so that the $(n + 1)$-*st* term is a polynomial in x: $a_n(x - c)^n$. More specifically, (8-150) is said to be a *power series in* $(x - c)$, the number c is said to be the *center* of the power series, and the numbers $a_0, a_1, \ldots, a_n, \ldots$ are said to be the *coefficients* of the power series. Each polynomial can be written as a power series with any given number as center (Section 3-15). For example

$$1 + 2x - x^2 = 2 - (x - 1)^2 = -2 - 4(x - 3) - (x - 3)^2$$

For any particular x, the power series (8-150) may or may not converge. The series always converges for $x = c$, since each term after the a_0 term is 0. Hence, in testing for convergence, we need consider only values of x other than c.

EXAMPLE 1. Show that the power series

$$\sum\limits_{n=0}^{\infty} \frac{x^n}{n!}$$

converges absolutely for all x.

Solution. Here $c = 0$. For $x \neq 0$, we can apply the ratio test. Then

$$\left| \frac{x^{n+1}}{(n + 1)!} \bigg/ \frac{x^n}{n!} \right| = \frac{|x|}{n + 1} \qquad \text{for} \qquad n = 0, 1, 2, \ldots$$

and for each $x \neq 0$,

$$\lim_{n \to \infty} \frac{|x|}{n+1} = |x| \cdot \lim_{n \to \infty} \frac{1}{n+1} = |x| \cdot 0 = 0$$

Consequently, by Theorem 20, the series $\Sigma x^n/n!$ converges absolutely for all x. It will be seen that the sum is e^x.

EXAMPLE 2. The power series $\displaystyle\sum_{n=0}^{\infty} x^n$ converges absolutely for $|x| < 1$ and diverges for $|x| \geq 1$. This is the geometric series (with $a = 1$) and the sum is $1/(1 - x)$ for $|x| < 1$.

EXAMPLE 3. Show that the power series $\displaystyle\sum_{n=0}^{\infty} n!\, x^n$ converges only for $x = 0$.

  Solution. For each $x \neq 0$

$$\lim_{n \to \infty} \left| \frac{(n+1)!\, x^{n+1}}{n!\, x^n} \right| = \lim_{n \to \infty} (n+1)|x| = |x| \cdot \lim_{n \to \infty} (n+1) = \infty$$

and hence, by the ratio test, this series diverges for each $x \neq 0$.

EXAMPLE 4. Show that the series $\displaystyle\sum_{n=1}^{\infty} x^n/n$ converges for $-1 \leq x < 1$ and diverges for all other x.

  Solution. If $x \neq 0$, then

$$\lim_{n \to \infty} \left| \frac{x^{n+1}}{n+1} \middle/ \frac{x^n}{n} \right| = \lim_{n \to \infty} \frac{n}{n+1} |x| = |x|$$

and, by the ratio test, the series converges for $|x| < 1$ and diverges for $|x| > 1$. The series $\Sigma x^n/n$ diverges at $x = 1$, since $\Sigma (1)^n/n$ is the harmonic series; it converges for $x = -1$, since the series $\Sigma(-1)^n/n$ converges by the test for alternating series. It will be seen that the sum is $\ln(1 - x)^{-1}$ for $|x| < 1$.

THEOREM 26. *If a power series $\displaystyle\sum_{n=0}^{\infty} a_n(x - c)^n$ converges at a point other than c, then either the series converges absolutely for all x or there is a positive real number R so that the series converges absolutely for $|x - c| < R$ and diverges for $|x - c| > R$. At each of the points $c + R$, $c - R$ the series may converge absolutely, converge conditionally, or diverge (see Figure 8-9).*

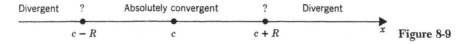

| Divergent | ? | Absolutely convergent | ? | Divergent |

$c - R$ c $c + R$ x **Figure 8-9**

The proof is given in the next section.

The number R is called the *radius of convergence* of the power series. If the series converges for all x, we say the radius of convergence is infinite, and we write $R = +\infty$. If the series diverges for all $x \neq c$, we say the radius of convergence is 0, and we write $R = 0$. Thus the radius of convergence R can be 0, ∞, or any positive real number.

The radius of convergence of Example 1 is $R = \infty$; for Examples 2 and 4, $R = 1$; for Example 3, $R = 0$. In Example 4 the series converges for $x = -1$, but it does not converge absolutely for $x = -1$.

THEOREM 27. *Let the power series* $\sum\limits_{n=0}^{\infty} a_n(x - c)^n$ *be given.*

(a) *If* $\lim\limits_{n \to \infty} \left| \dfrac{a_n}{a_{n+1}} \right| = R$ *(where* $0 \leq R \leq \infty$*), then* R *is the radius of convergence for the power series.*

(b) *If* $\lim\limits_{n \to \infty} \dfrac{1}{\sqrt[n]{|a_n|}} = R$ *(where* $0 \leq R \leq \infty$ *), then* R *is the radius of convergence for the power series.*

PROOF. *Part* (a). If $x = c$, the series converges absolutely. If $x \neq c$, then

$$\lim_{n \to \infty} \left| \frac{a_{n+1}(x - c)^{n+1}}{a_n(x - c)^n} \right| = |x - c| \lim_{n \to \infty} \left| \frac{a_{n+1}}{a_n} \right| = \begin{cases} |x - c|/R, & \text{if } R \neq 0, \infty \\ 0, & \text{if } R = \infty \\ \infty, & \text{if } R = 0 \end{cases}$$

Hence, by the ratio test, when $R = \infty$, the series converges absolutely for all x; when $R = 0$, the series diverges for all $x \neq c$; when $0 < R < \infty$, the series converges absolutely for $|x - c| < R$ and diverges for $|x - c| > R$.

The proof of Part (b) is left as an exercise (Problem 3 below).

Remark 1. The tests are applicable only if $a_n \neq 0$ for all large n. If we are given the series $\Sigma(3x)^{2n}$, then to test its region of convergence we use the ratio test of two successive terms.

$$\lim_{n \to \infty} \left| \frac{(3x)^{2n+2}}{(3x)^{2n}} \right| = \lim_{n \to \infty} 9|x|^2 = 9|x|^2 \begin{cases} < 1 \text{ if } |x| < \frac{1}{3} \\ > 1 \text{ if } |x| > \frac{1}{3} \end{cases}$$

Hence $\Sigma(3x)^{2n}$ converges for $|x| < \frac{1}{3}$, and diverges for $|x| > \frac{1}{3}$. We can easily verify that it also diverges at $x = \pm\frac{1}{3}$. (The series is a geometric series with $r = 9x^2$.)

Remark 2. It may very well occur that a power series is such that neither of the limits in Theorem 27 exists. Other tests must then be used. The tests given in Theorem 27 are the most easily applied and cover a large percentage of the series arising in the physical sciences.

‡8-16 PROOF OF THEOREM ON RADIUS OF CONVERGENCE

In this section we give a proof of Theorem 26 on real power series. For simplicity we consider only the case of the series

$$\sum_{n=0}^{\infty} a_n x^n \tag{8-160}$$

The general case is treated in the same way, with x replaced by $x - c$.

LEMMA 1. *If* $\sum\limits_{n=0}^{\infty} a_n$ *converges, then the sequence* $\{a_n\}$ *is bounded:* $|a_n| \leq B$, $n = 0, 1, 2, \ldots$, *for some constant* B.

PROOF. By Theorem 7 (Section 8-4), $a_n \to 0$, so that we can find N for which $|a_n| < 1$ for $n > N$. We choose B larger than 1 and larger than $|a_0|$, $|a_1|, \ldots, |a_N|$, and conclude that $|a_n| \le B$ for all n.

LEMMA 2. *If the series* (8-160) *converges for* $x = x_0 \ne 0$, *then the series converges absolutely for* $|x| < |x_0|$.

PROOF. The series $\Sigma a_n x_0^n$ converges. Hence by Lemma 1 there is a constant B such that $|a_n x_0^n| \le B$ for all n. Let $|x| < |x_0|$, so that $|x|/|x_0| = r < 1$. Then

$$|a_n x^n| = \left| a_n x_0^n \frac{x^n}{x_0^n} \right| = |a_n x_0^n| \left| \frac{x}{x_0} \right|^n \le B r^n$$

Hence the series $\Sigma |a_n| x^n$ converges by comparison with the geometric series $\Sigma B r^n$.

PROOF OF THEOREM 26. The theorem asserts that, if the series (8-160) converges for one value x_0 other than 0, then the series has a positive or infinite radius of convergence R for which the series converges absolutely for $|x| < R$ and diverges for $|x| > R$. Let E be the set of all r such that it converges absolutely for $|x| < r$. By Lemma 2, E includes $|x_0|$. If E is unbounded, we take $R = \infty$; if E is bounded, we take R to be the least upper bound of E.

For $R = \infty$, the series must converge absolutely for all x. For, since E is unbounded, given x, we can choose r_1 in E so that $|x| < r_1$. Hence, by the definition of E, $\Sigma |a_n x^n|$ converges.

For R finite, the series converges absolutely for $|x| < R$. For, since $R = $ lub E, for each such x, we can find r in E so that $|x| < r < R$, so that again $\Sigma |a_n x^n|$ converges. Also, for R finite, the series diverges for $|x| > R$. For, if the series were to converge for $x = x_1$ with $|x_1| > R$, then by Lemma 2 the series would converge absolutely for $|x| < |x_1|$ and R could not be lub E.

Hence Theorem 26 is proved.

PROBLEMS

1. Determine the values of x for which each of the following series converges.

(a) $\sum_{n=0}^{\infty} \frac{n x^n}{2^n}$

(b) $\sum_{n=1}^{\infty} \frac{(x-2)^n}{n^3}$

(c) $\sum_{n=0}^{\infty} n x^n$

(d) $\sum_{n=1}^{\infty} \frac{x^{2n}}{(1-x)^n}$

(e) $\sum_{n=0}^{\infty} \frac{1}{3^{nx}}$

(f) $\sum_{n=1}^{\infty} \frac{\sin nx}{n^2 + 5}$

(g) $\sum_{n=1}^{\infty} \frac{1}{n} \left(\frac{x+1}{x-1} \right)^n$

(h) $\sum_{n=1}^{\infty} \frac{n}{x^n}$

2. Determine the radius of convergence of each of the following power series:

(a) $\sum_{n=0}^{\infty} n(n+1) x^n$

(b) $\sum_{n=0}^{\infty} 2^n (x-1)^n$

(c) $\sum_{n=0}^{\infty} \frac{2^n x^n}{n!}$

(d) $\sum_{n=0}^{\infty} \frac{n! x^n}{3^n}$

(e) $\displaystyle\sum_{n=0}^{\infty} \frac{n^{2n}}{(3n)!}(x-2)^n$ (f) $\displaystyle\sum_{n=0}^{\infty} \frac{1\cdot3\cdot5\cdots(2n+1)}{1\cdot4\cdot7\cdots(3n+1)}(x+2)^n$

(g) $1 + 2x + 3x^2 + x^3 + 2x^4 + 3x^5 + x^6 + 2x^7 + 3x^8 + \cdots$

3. Prove that if

$$\lim_{n\to\infty} \frac{1}{\sqrt[n]{|a_n|}} = R$$

then the series (8-150) has radius of convergence R.

‡4. Show that every series of the form

$$\sum_{n=0}^{\infty} \frac{a_n}{(x-c)^n}$$

has a "radius of convergence" ρ, so that the series converges absolutely for $|x-c|>\rho$ and diverges for $|x-c|<\rho$.

8-17 PROPERTIES OF POWER SERIES

If a power series $\displaystyle\sum_{n=0}^{\infty} a_n(x-c)^n$ has sum $f(x)$, so that

$$\sum_{n=0}^{\infty} a_n(x-c)^n = f(x)$$

for x in an interval I, we say that the series $\displaystyle\sum_{n=0}^{\infty} a_n(x-c)^n$ *represents the function f on the interval I.*

A power series $\displaystyle\sum_{n=0}^{\infty} a_n(x-c)^n$ with radius of convergence $R \neq 0$ represents one, and only one, function on the open interval $|x-c| < R$, that function being the function that assigns to a point x_0 in the interval the number $\displaystyle\sum_{n=0}^{\infty} a_n(x_0-c)^n$. Thus a power series can be used to define a function on its interval of convergence. This method of defining a function is a very powerful way to obtain new functions. For example, the following functions play a significant role in both pure mathematics and in the application of mathematics to the physical sciences:

$$J_0(x) = 1 + \sum_{n=1}^{\infty} \frac{(-1)^n x^{2n}}{2^{2n}(n!)^2}$$

$$H(x;\,\alpha,\,\beta,\,\gamma) = 1 + \frac{\alpha\beta}{\gamma\cdot1}x + \frac{\alpha(\alpha+1)\beta(\beta+1)}{\gamma(\gamma+1)2!}x^2 + \cdots$$
$$+ \frac{\alpha(\alpha+1)\cdots(\alpha+n)\beta(\beta+1)\cdots(\beta+n)}{\gamma(\gamma+1)\cdots(\gamma+n)n!}x^n + \cdots$$

The function $J_0(x)$ is called the *Bessel function* of the first kind of order 0. The function H is the *hypergeometric series* and is a solution of the differential equation

$$x(1-x)y'' + [\gamma - (1+\alpha+\beta)x]y' - \alpha\beta y = 0$$

which plays an important role in the study of the hydrogen atom.

Many of the standard functions, for example $\ln x$, e^x, $\sin x$, $\cos x$, $\tan x$ can be defined by power series and all their properties can be derived from their series expression.

We shall now determine some of the properties of a function represented (or defined) by a power series with a positive radius of convergence.

THEOREM 28. *If the power series* $\displaystyle\sum_{n=0}^{\infty} a_n(x - c)^n$ *represents the function* $f(x)$ *on the open interval* $|x - c| < r$ *(here r is positive and is less than or equal to the radius of convergence R for the given series), then*

(a) $f(x)$ *is continuous on* $|x - c| < r$.

(b) $f(x)$ *is differentiable on* $|x - c| < r$ *and* $f'(x)$ *is represented by the power series* $\displaystyle\sum_{n=1}^{\infty} n a_n(x - c)^{n-1}$ *on the interval* $|x - c| < r$.

(c) *The indefinite integral* $\displaystyle\int_c^x f(t)\, dt$ *is represented by the power series*

$$\sum_{n=0}^{\infty} \frac{a_n}{n + 1}(x - c)^{n+1} \text{ on the interval } |x - c| < r.$$

The proof will be given in the next section.

This theorem shows that a power series, like a polynomial, can be differentiated and integrated term by term.

COROLLARY 1. *If a function $f(x)$ can be represented by a power series on an open interval* $|x - c| < r$, *then that function possesses derivatives of every order on that interval.*

PROOF. By part (b) of Theorem 28 the derivative of a function represented by a power series is again a power series with the same interval of convergence and hence is differentiable. We can now continue inductively to show that f'', f''', ... exist. In particular

$$f(x) = f^{(0)}(x) = a_0 + a_1(x - c) + \cdots + a_n(x - c)^n + \cdots$$
$$f'(x) = a_1 + 2a_2(x - c) + \cdots + n a_n(x - c)^{n-1} + \cdots$$
$$f''(x) = 2a_2 + 3 \cdot 2a_3(x - c) + \cdots + n(n - 1) a_n(x - c)^{n-2} + \cdots$$
$$\vdots$$
$$f^{(m)}(x) = m! a_m + (m + 1)! a_{m+1}(x - c) + \cdots + n(n - 1) \cdots$$
$$(n - m + 1) a_n(x - c)^{n-m} + \cdots$$

$$= \sum_{n=m}^{\infty} n(n - 1) \cdots (n - m + 1) a_n(x - c)^{n-m} \qquad (8\text{-}170)$$

$$= \sum_{n=0}^{\infty} (n + m)(n + m - 1) \cdots (n + 1) a_{n+m}(x - c)^n$$

The last equation is obtained from the preceding one, first by letting $n - m = k$, then $n = m + k$ and, as n goes from m to ∞, the index k goes from 0 to ∞, so that the series becomes

$$\sum_{k=0}^{\infty} (k + m)(k + m - 1) \cdots (k + 1) a_{k+m}(x - c)^k$$

Now set $k = n$ to obtain the last expression in (8-170).

EXAMPLE 1. The geometric series $\sum\limits_{n=0}^{\infty} x^n = 1 + x + x^2 + \cdots$ converges for $|x| < 1$ and has $1/(1 - x)$ as its sum. Thus

$$\frac{1}{1 - x} = 1 + x + x^2 + \cdots + x^n + \cdots = \sum_{n=0}^{\infty} x^n \qquad \text{for} \qquad |x| < 1$$

Since $d/dx[1/(1 - x)] = 1/(1 - x)^2$, we have

$$\frac{1}{(1 - x)^2} = 1 + 2x + 3x^2 + \cdots = \sum_{n=1}^{\infty} nx^{n-1}, \qquad \text{for} \qquad |x| < 1$$

Differentiating again, we obtain

$$\frac{2}{(1 - x)^3} = 2 + 6x + 12x^2 + \cdots = \sum_{n=2}^{\infty} n(n - 1)x^{n-2}$$

or

$$\frac{1}{(1 - x)^3} = 1 + 3x + 6x^2 + \cdots = \sum_{n=2}^{\infty} \frac{n(n - 1)}{2} x^{n-2}$$

$$= \sum_{n=0}^{\infty} \frac{(n + 2)(n + 1)}{2} x^n, \qquad \text{for } |x| < 1$$

Continuing inductively, we find that

$$\frac{1}{(1 - x)^k} = 1 + kx + \cdots + \frac{(n + k - 1) \cdots (n + 1)}{(k - 1)!} x^n + \cdots$$

$$= \sum_{n=0}^{\infty} \frac{(n + k - 1) \cdots (n + 1)}{(k - 1)!} x^n$$

$$= \sum_{n=0}^{\infty} \binom{n + k - 1}{k - 1} x^n, \qquad \text{for} \qquad |x| < 1$$

(See Problem 4.) Here the binomial coefficient $\binom{m}{k}$ is defined for $k = 0, 1, 2, \ldots$ and all m by the equation:

$$\binom{m}{k} = \frac{m(m - 1) \cdots (m - k + 1)}{k!}$$

If we integrate the function $1/(1 - x)$ we find that $\int_0^x dt/(1 - t) = -\ln(1 - x)$. Hence

$$\ln(1 - x) = -x - \frac{x^2}{2} - \frac{x^3}{3} - \cdots = \sum_{n=1}^{\infty} (-1) \frac{x^n}{n}, \qquad \text{for} \qquad |x| < 1$$

EXAMPLE 2. The series $\sum\limits_{n=0}^{\infty} x^n/n!$ converges for all x. If $f(x)$ is the function represented by the series $\sum\limits_{n=0}^{\infty} x^n/n!$, then

$$f'(x) = \sum_{n=1}^{\infty} \frac{nx^{n-1}}{n!} = \sum_{n=1}^{\infty} \frac{x^{n-1}}{(n - 1)!} = \sum_{n=0}^{\infty} \frac{x^n}{n!} = f(x)$$

Thus, $f(x)$ is a solution of the differential equation $y' = y$. The solutions of this equation are $y = ce^x$, where c is a constant (see Section 7-16). Since $f(0) = 1$, it follows that $f(x) = e^x$. Thus

$$e^x = 1 + \frac{x}{1!} + \frac{x^2}{2!} + \cdots = \sum_{n=0}^{\infty} \frac{x^n}{n!}, \qquad -\infty < x < \infty$$

COROLLARY 2. *If the function $f(x)$ is represented by the power series $\Sigma a_n(x - c)^n$ on an interval $|x - c| < r$, then $f^{(n)}(c) = n! a_n (n = 0, 1, 2, \ldots)$. Hence*

$$f(x) = \sum_{n=0}^{\infty} \frac{f^{(n)}(c)}{n!} (x - c)^n \qquad (8\text{-}171)$$

PROOF. We evaluate (8-170) at $x = c$, to obtain $f^{(m)}(c) = m! a_m$.

COROLLARY 3. *If a function $f(x)$ is represented by both the power series $\Sigma a_n(x - c)^n$ and $\Sigma b_n(x - c)^n$ on some common open interval about c, then $a_n = b_n$ for $n = 0, 1, 2, 3, \ldots$.*

PROOF. By Corollary 2, $f^{(m)}(c) = m! a_m = m! b_m$, hence, $a_m = b_m$.

It follows from Corollary 3 that a function can be represented by at most one power series with center c. Of course, a function can be represented by series with different centers. Thus

$$f(x) = 1 + x + x^2 = 3 + 3(x - 1) + (x - 1)^2$$

are different power series representations of the same polynomial (all terms are 0 after those shown). Also it will be seen that

$$e^x = \sum_{n=0}^{\infty} \frac{x^n}{n!} = \sum_{n=0}^{\infty} \frac{e(x - 1)^n}{n!}$$

are different power series representations for the function e^x. Both of these series converge for all x.

COROLLARY 4. *If $\displaystyle\sum_{n=0}^{\infty} a_n(x - c)^n$ converges to the zero function on an open interval about c, then each $a_n = 0$.*

PROOF. The series $\Sigma 0(x - c)^n$ and $\Sigma a_n(x - c)^n$ represent the same function on an open interval about c. Hence, by Corollary 3, the coefficients of the two series are equal; that is, $a_n = 0$.

THEOREM 29. *Let f be a function represented by the power series $\Sigma a_n(x - c)^n$ on the open interval $|x - c| < r_1$, and g be a function represented by the power series $\Sigma b_n(x - c)^n$ on the open interval $|x - c| < r_2$. Let r denote the smaller of r_1 and r_2 (or $r = \infty$ if $r_1 = r_2 = \infty$). Then*

(a) $\displaystyle\sum_{n=0}^{\infty} (ka_n)(x - c)^n$ *represents the function kf on $|x - c| < r_1$;*

(b) $\displaystyle\sum_{n=0}^{\infty} (a_n + b_n)(x - c)^n$ *represents the function $f + g$ on $|x - c| < r$;*

(c) $a_0 b_0 + (a_1 b_0 + a_0 b_1)(x - c) + (a_2 b_0 + a_1 b_1 + a_0 b_2)(x - c)^2 + \cdots$

$$= \sum_{n=0}^{\infty} \left(\sum_{m=0}^{n} a_{n-m} b_m \right)(x - c)^n \ \textit{represents the function fg on } |x - c| < r$$

PROOF. Parts (a), and (b) are immediate consequences of Theorem 11. For example, $\Sigma a_n (x - c)^n$, $\Sigma b_n (x - c)^n$ converge to $f(x)$, $g(x)$, respectively, on $|x - c| < r$ and, hence, by Theorem 11

$$\Sigma[a_n(x - c)^n + b_n(x - c)^n] = \Sigma(a_n + b_n)(x - c)^n$$

converges to $f(x) + g(x)$ on $|x - c| < r$.

We have observed in Theorem 26 that the given series are absolutely convergent for $|x - c| < r$. Hence, by Theorem 24, the Cauchy product of these two series:

$$\sum_{n=0}^{\infty} \left[\sum_{m=0}^{n} a_{n-m} b_m (x - c)^n \right]$$

converges to $\left[\displaystyle\sum_{n=0}^{\infty} a_n(x - c)^n \right] \left[\displaystyle\sum_{n=0}^{\infty} b_n(x - c)^n \right] = f(x)g(x)$ on $|x - c| < r$.

EXAMPLE 3. We have seen that $e^x = \displaystyle\sum_{n=0}^{\infty} x^n/n!$ for $-\infty < x < \infty$, and $(1 - x)^{-1} = \displaystyle\sum_{n=0}^{\infty} x^n$ for $|x| < 1$. Hence

$$e^x + \frac{2}{1 - x} = \sum_{n=0}^{\infty} \left(2 + \frac{1}{n!} \right) x^n, \qquad |x| < 1$$

$$xe^x = \sum_{n=0}^{\infty} \frac{x^{n+1}}{n!}, \qquad -\infty < x < \infty$$

$$(1 - x)e^x = 1 + (1 - 1)x + \left(\frac{1}{2} - 1 \right)x^2$$

$$+ \cdots + \left[\frac{1}{n!} - \frac{1}{(n - 1)!} \right] x^n + \cdots$$

$$= 1 - \frac{1}{2} x^2 - \frac{1}{3} x^3 - \cdots - \frac{x^n}{n(n - 2)!} - \cdots, \qquad -\infty < x < \infty$$

$$\frac{e^x}{1 - x} = \sum_{n=0}^{\infty} \left(1 + \frac{1}{2!} + \cdots + \frac{1}{n!} \right) x^n, \qquad |x| < 1$$

‡8-18 PROOF OF THEOREM ON PROPERTIES OF POWER SERIES

We now prove Theorem 28 of the preceding section. As in Section 8-16, for simplicity we consider only a power series with center at 0:

$$\sum_{n=0}^{\infty} a_n x^n \qquad\qquad (8\text{-}180)$$

We are given that the series has positive or infinite radius of convergence R and sum $f(x)$, $|x| < R$. We must show that f is continuous, that f' and $\displaystyle\int_0^x f(t)\,dt$

can be obtained by differentiating and integrating the series term by term, and that the integrated and differentiated series

$$\sum_{n=0}^{\infty} \frac{a_n}{n+1} x^{n+1} \tag{8-181}$$

$$\sum_{n=0}^{\infty} n a_n x^{n-1} \tag{8-182}$$

also have radius of convergence R.

We first consider the question of radius of convergence of the series (8-182) or, equivalently, of the series $\Sigma n a_n x^n$ (multiplication by x cannot affect the radius of convergence). Let x be chosen so that $|x| < R$ and choose x_0 so that $|x| < |x_0| < R$. Then, as in the proof of Lemma 2 (Section 8-16), there is a constant B so that $|a_n||x_0|^n \le B$ for all n. We let $r = |x|/|x_0|$, so that $0 < r < 1$. Then for all n

$$n|a_n||x|^n = |a_n||x_0|^n\, n \left(\frac{|x|}{|x_0|}\right)^n \le B n r^n$$

The series $\Sigma n r^n$ converges by the ratio test. Hence the series $\Sigma n a_n x^n$ converges absolutely for $|x| < R$. Thus, the multiplication of the nth coefficient by n cannot decrease the radius of convergence of a power series. It also cannot increase the radius of convergence since for $n \ge 1$

$$|a_n||x|^n \le n|a_n||x^n|$$

Hence the multiplication of the nth coefficient by n does not change the radius of convergence. For the same reason, division of the nth coefficient by n does not change the radius of convergence. It follows that both series (8-181), (8-182) have radius of convergence R.

Now we study the derivative of f. We first note that by Taylor's formula with remainder (Section 6-12, also Section 8-19 below)

$$(x + h)^n = x^n + n x^{n-1} h + \frac{n(n-1)}{2} \xi_n^{n-2} h^2$$

where ξ_n is between x and $x + h$. Throughout the following discussion we choose x and $x + h$ so that $|x| < r$, $|x + h| < r$, where $r < R$. Then

$$f(x + h) = \sum_{n=0}^{\infty} a_n(x + h)^n = \sum_{n=0}^{\infty} a_n \left[x^n + n x^{n-1} h + \frac{n(n-1)}{2} \xi_n^{n-2} h^2\right]$$

$$= \sum_{n=0}^{\infty} a_n x^n + h \sum_{n=0}^{\infty} a_n n x^{n-1} + \frac{h^2}{2} \sum_{n=0}^{\infty} a_n n(n-1) \xi_n^{n-2} \tag{8-183}$$

The first two series on the right converge by what we already know, so that (since the series $\Sigma a_n(x + h)^n$ converges), the last series on the right also converges. In fact

$$|a_n n(n-1) \xi_n^{n-2}| \le n(n-1)|a_n| r^{n-2}$$

and the series

$$\sum_{n=0}^{\infty} n(n-1)|a_n|r^{n-2} \tag{8-184}$$

converges by the reasoning given above (multiplying the nth term by n and then by $n-1$, as well as lowering the power of r, having no effect on the radius of convergence).

By using (8-183), we can now write, for $h \neq 0$,

$$\frac{f(x+h) - f(x)}{h} = \sum_{n=0}^{\infty} n a_n x^{n-1} + \frac{h}{2} \sum_{n=0}^{\infty} n(n-1)a_n \xi_n^{n-2}$$

The last term is in absolute value at most equal to $|h|/2$ times the sum (8-184), and hence this term approaches 0 as $h \to 0$. Therefore

$$\lim_{h \to 0} \frac{f(x+h) - f(x)}{h} = \sum_{n=0}^{\infty} n a_n x^{n-1}$$

Hence $f'(x)$ exists and is obtainable by differentiating the series term by term. Therefore, f is continuous for $|x| < R$. Finally, by the same reasoning, if

$$F(x) = \sum_{n=0}^{\infty} \frac{a_n}{n+1} x^{n+1} \tag{8-185}$$

then

$$F'(x) = \sum_{n=0}^{\infty} a_n(n+1)\frac{x^n}{n+1} = \sum_{n=0}^{\infty} a_n x^n = f(x)$$

But $F(0) = 0$. Therefore

$$F(x) = \int_0^x f(t)\, dt$$

and, by (8-185), we have shown that the integral of f can be obtained by integrating the series term by term.

PROBLEMS

In problems 1, 2, and 3, let

$$s(x) = \sum_{n=0}^{\infty} \frac{(-1)^n x^{2n+1}}{(2n+1)!}, \qquad c(x) = \sum_{n=0}^{\infty} \frac{(-1)^n x^{2n}}{(2n)!}$$

1. Determine the region of convergence of each of the above series.
2. Prove:
   ‡ (a) $c(2x) = c^2(x) - s^2(x)$ (b) $s(-x) = -s(x)$ (c) $c(-x) = c(x)$
   (d) $s'(x) = c(x)$ (e) $c'(x) = -s(x)$

3. (a) Show that $s''(x) = -s(x)$, $c''(x) = -c(x)$.
   (b) Since $s(0) = 0$ and $s'(0) = 1$, and $s(x)$ satisfies the differential equation $y'' + y = 0$, explain why we can conclude that $s(x) = \sin x$ [see Section 7-18].
   (c) Show that $c(x) = \cos x$.
4. Verify in detail the correctness of the various steps made in Example 1, Section 8-17.
5. Using power series representations already found, determine a power series $\Sigma a_n x^n$ representing each of the functions given and give an interval in which the representation is valid.

   (a) e^{x^2}
   (b) $\ln(1 - x^2)$ [*Hint.* See Example 1 for series for $\ln(1 - x)$.],

   (c) $\ln(1 + x)$
   (d) $\dfrac{1}{(1 - 3x)^2}$
   (e) $\dfrac{1}{3x - 5}$

   (f) $\dfrac{1}{5 - x^2}$
   (g) $\dfrac{1}{x^2 - 5x + 6}$ $\left[Hint. \dfrac{1}{x^2 - 5x + 6} = \dfrac{-1}{x - 2} + \dfrac{1}{x - 3} \right]$

   (h) $\dfrac{1}{2x^2 - 3x + 1}$
   (i) $\dfrac{2x + 3}{1 - 2x}$
   (j) $\dfrac{1}{(1 - 2x^2)^4}$

   (k) $\operatorname{Tan}^{-1}x$ $\left[Hint. \int \dfrac{1}{1 + x^2}\, dx = \operatorname{Tan}^{-1}x \right]$

6. From the results of Example 1 in Section 8-17, show that for $|x| < 1$

   (a) $\displaystyle\sum_{n=0}^{\infty} x^n = \dfrac{1}{1 - x}$,
   (b) $\displaystyle\sum_{n=0}^{\infty} nx^n = \dfrac{x}{(1 - x)^2}$,
   (c) $\displaystyle\sum_{n=0}^{\infty} (n^2 - n)x^n = \dfrac{2x^2}{(1 - x)^3}$

   (d) $\displaystyle\sum_{n=0}^{\infty} n^2x^n = \dfrac{x + x^2}{(1 - x)^3}$,
   (e) $\displaystyle\sum_{n=0}^{\infty} (-1)^n x^n = \dfrac{1}{1 + x}$

   (f) $\displaystyle\sum_{n=0}^{\infty} (n + 1)^2x^n = \dfrac{x + 1}{(1 - x)^3}$

Problems 7 to 9 suggest new ways of obtaining power series representations for functions. They can be justified by general theorems in the book of Knopp cited at the end of Section 8-13.

7. Let $f(x) = \displaystyle\sum_{n=0}^{\infty} (n + 1)^2x^n$, $-1 < x < 1$. We seek a power series for $g(x) = 1/f(x)$. We first notice that $f(0) = 1$. Since f is continuous, g is continuous in some interval about $x = 0$. To find a series for g, we write $g(x) = b_0 + b_1x + \cdots$ and try to determine $b_0, b_1, b_2, \ldots$ so that $g(x)f(x) = 1$. Show that this leads to the equations

$$b_0 = 1, \qquad b_1 + 4b_0 = 0, \qquad b_2 + 4b_1 + 9b_0 = 0, \ldots$$

and, hence, to $g(x) = 1 - 4x + 7x^2 - 8x^3 + \cdots$. Use the result of Problem 6(f) to show that the equation is valid for $|x| < 1$ and find the general term in the series for $g(x)$.

8. To find a power series for $f(x) = e^{x/(1-x)}$, we write

$$e^{x/(1-x)} = 1 + \frac{x}{1 - x} + \frac{1}{2!}\left(\frac{x}{1 - x}\right)^2 + \cdots + \frac{1}{n!}\left(\frac{x}{1 - x}\right)^n + \cdots$$

$$= 1 + (x + x^2 + \cdots) + \frac{1}{2!}(x + x^2 + \cdots)^2 + \cdots$$

Show that this leads to the equation

$$f(x) = 1 + x + \frac{3}{2}x^2 + \frac{13}{6}x^3 + \frac{73}{24}x^4 + \cdots$$

This can be shown to be valid for $|x| < 1$.

9. The function $f(x) = x/(1 - x) = x + x^2 + \cdots$ is monotone strictly increasing for $|x| < 1$ and has an inverse $g = f^{-1}$. To find a power series for g, we write $g(x) = f^{-1}(x) = b_1 x + b_2 x^2 + b_3 x^3 + \cdots$ [Notice that $f(0) = 0$ so that $f^{-1}(0) = 0$]. Now

$$x \equiv f[f^{-1}(x)] = (b_1 x + b_2 x^2 + \cdots) + (b_1 x + b_2 x^2 + \cdots)^2 + \cdots$$

$$\equiv b_1 x + (b_2 + b_1^2) x^2 + \cdots$$

(a) Show by Corollary 3 to Theorem 28 that $b_1 = 1$, $b_2 + b_1^2 = 0$, $b_3 + 2b_1 b_2 + b_1^3 = 0$, and hence, ultimately $g(x) = x - x^2 + x^3 - x^4 + \cdots$.

(b) Show that the result of part (a) is valid for $|x| < 1$ by showing that $g(x) = x/(x + 1)$.

‡10. Let $\Sigma a_n x^n$ have radius of convergence 1. With the aid of results proved in the text find the radius of convergence of each of the series

(a) $\displaystyle\sum_{n=0}^{\infty} a_n x^{n+2}$ (b) $\displaystyle\sum_{n=1}^{\infty} a_n x^{n-1}$ (c) $\displaystyle\sum_{n=0}^{\infty} n a_n x^{n+2}$

(d) $\displaystyle\sum_{n=0}^{\infty} n^2 a_n x^{n-1}$ (e) $\displaystyle\sum_{n=0}^{\infty} n a_n 2^n x^{n-1}$ (f) $\displaystyle\sum_{n=0}^{\infty} (n^2 + n) a_n e^n x^n$

8-19 TAYLOR'S FORMULA WITH REMAINDER

At this point we digress to review and to elaborate on Taylor's formula with remainder. This formula will be seen to provide a powerful tool for justifying the representation of functions by power series.

In Section 6-12 it was shown that if the function f is continuous and has continuous derivatives through order $n + 1$ on an interval $|x - c| < r$, then for each x of that interval

$$f(x) = f(c) + f'(c)(x - c) + f''(c)\frac{(x - c)^2}{2!} + \cdots + \frac{f^{(n)}(c)}{n!}(x - c)^n + R_n$$

$$(8\text{-}190)$$

with

$$R_n = f^{(n+1)}(\xi)\frac{(x - c)^{n+1}}{(n + 1)!}, \qquad \text{for some } \xi \text{ between } c \text{ and } x \quad (8\text{-}191)$$

These two equations together constitute *Taylor's formula with remainder in Lagrange's form*. The remainder term, R_n, can be expressed in other forms, as illustrated by the next theorem.

THEOREM 30. (*Taylor's formula with remainder in integral form*). Let f be continuous and have continuous derivatives through order $n + 1$ on the interval $|x - c| < r$. Then, for each x in the interval, (8-190) is valid with

$$R_n = \int_c^x \frac{(x - t)^n}{n!} f^{(n+1)}(t)\, dt \qquad (8\text{-}192)$$

PROOF. For $n = 0$, our theorem is just the formula:

$$f(x) - f(c) = \int_c^x f'(t)\, dt$$

which is just a fundamental rule of calculus [Theorem 11, Section 4-17]. Next we integrate the right-hand side by parts, taking $u = f'(t)$, $v' = 1$, $u' = f''$, $v = (t - x)$ in the rule $\int u(t)v'(t)\, dt = u(t)v(t) - \int u'(t)v(t)\, dt$. We obtain

$$\int_c^x f'(t)\, dt = \int_c^x f'(t) \frac{d}{dt}(t - x)\, dt = (t - x)f'(t)\Big|_c^x - \int_c^x (t - x)f''(t)\, dt$$

$$= (x - c)f'(c) + \int_c^x (x - t)f''(t)\, dt$$

and thereby obtain the rule for the case $n = 1$:

$$f(x) = f(c) + \frac{f'(c)}{1!}(x - c) + \int_c^x (x - t)f''(t)\, dt$$

We integrate the last term by parts, taking $u = f''(t)$, $v' = (x - t)$, $u' = f'''$, and $v = -[(x - t)^2/2]$. We obtain

$$\int_c^x (x - t)f''(t)\, dt = -\frac{(x - t)^2}{2!}f''(t)\Big|_c^x + \int_c^x \frac{(x - t)^2}{2!}f'''(t)\, dt$$

$$= \frac{(x - c)^2}{2!}f''(c) + \int_c^x \frac{(x - t)^2}{2!}f'''(t)\, dt$$

and, hence, obtain the case $n = 2$ of the theorem:

$$f(x) = f(c) + \frac{f'(c)}{1!}(x - c) + \frac{f''(c)}{2!}(x - c)^2 + \int_c^x \frac{(x - t)^2}{2!}f'''(t)\, dt$$

We now obtain the general rule by induction. Suppose that

$$f(x) = f(c) + \frac{f'(c)}{1!}(x - c) + \cdots + \frac{f^{(n-1)}(c)}{(n-1)!}(x - c)^{n-1}$$

$$+ \int_c^x \frac{(x - t)^{n-1}}{(n-1)!}f^{(n)}(t)\, dt$$

We then integrate the last term by parts, letting $u = f^{(n)}(t)$, $v' = (x - t)^{n-1}/(n - 1)!$, $u' = f^{(n+1)}(t)$, and $v = -[(x - t)^n/n!]$. We obtain

$$\int_c^x \frac{(x - t)^{n-1}}{(n-1)!}f^{(n)}(t)\, dt = -\frac{(x - t)^n}{n!}f^{(n)}(t)\Big|_c^x + \int_c^x \frac{(x - t)^n}{n!}f^{(n+1)}(t)\, dt$$

$$= \frac{(x - c)^n}{n!}f^{(n)}(c) + \int_c^x \frac{(x - t)^n}{n!}f^{(n+1)}(t)\, dt$$

and, hence, that

$$f(x) = f(c) + \cdots + \frac{f^n(c)}{n!}(x - c)^n + \int_c^x \frac{(x - t)^n}{n!}f^{(n+1)}(t)\, dt$$

All steps are valid, provided that f, f', $\ldots$, $f^{(n+1)}$ are continuous in the interval concerned.

In the remainder formula (8-192) let us assume $c < x$, so that $x - t \geq 0$ throughout the interval of integration. If we know that $A \leq f^{(n+1)}(t) \leq B$ throughout this interval, then also

$$A\frac{(x-t)^n}{n!} \leq \frac{(x-t)^n}{n!} f^{(n+1)}(t) \leq B\frac{(x-t)^n}{n!}$$

and, hence, if we integrate from $t = c$ to $t = x$

$$A \int_c^x \frac{(x-t)^n}{n!}\,dt \leq R_n \leq B \int_c^x \frac{(x-t)^n}{n!}\,dt$$

The integrals both equal $(x-c)^{n+1}/(n+1)!$. Hence

$$R_n = (\text{const}) \cdot \frac{(x-c)^{n+1}}{(n+1)!}$$

where "const" is a number between A and B. Since $f^{(n+1)}$ is continuous, A can be chosen as the minimum, B as the maximum of $f^{(n+1)}$ in the interval $[c, x]$ and, hence, "const" can be written as $f^{(n+1)}(\xi)$, with ξ between c and x. This gives the Lagrange rule (8-191) again, for $c < x$. The proof for $c > x$ is similar. For the Lagrange rule, it is actually sufficient that $f^{(n+1)}$ exist between c and x, continuity being unnecessary. (A proof of this assertion, for the case $n = 1$, is suggested in Problem 5 following Section 6-12; for $n = 0$, the assertion is just the Mean Value Theorem.)

8-20 TAYLOR'S SERIES

Let the function f be defined on an interval I and let c be an interior point of I. The function f is said to be *analytic at* c if, in some open interval containing c, f is the sum of a convergent power series with center at c. Thus, f is analytic at c exactly when for some $r > 0$

$$f(x) = \sum_{n=0}^{\infty} a_n(x-c)^n, \qquad |x-c| < r$$

The function f is said to be *analytic on an interval* if f is analytic at each point of that interval.

If f is analytic at c, then, by Corollary 1 of Theorem 28, f has derivatives of all orders in an open interval containing c. However, the converse does not hold true: there are functions f for which $f^{(n)}(x)$ exists for all x for $n = 1$, $2, 3, \ldots$, and yet f is not the sum of a power series with center at c.

If f has derivatives of all orders at c, we can form the power series

$$\sum_{n=0}^{\infty} \frac{f^{(n)}(c)}{n!} (x-c)^n \qquad (8\text{-}200)$$

This series is called the *Taylor series of* f *at* c. When $c = 0$, the series is also known as the *Maclaurin series of* f.

From Corollary 2 of Theorem 28 it follows that, if f is analytic at c, then on an open interval containing c, f is the sum of its Taylor series at c. However, in general, the series (8-200) need not converge at any point other than c and, even if it does have a positive radius of convergence, the sum need not be the given function f. (For further discussion of these difficulties, see Remarks at the end of this section.)

The following theorem shows that analyticity of f in an interval follows from analyticity at a single point.

THEOREM 31. *Let*

$$f(x) = \sum_{n=0}^{\infty} \frac{f^{(n)}(c)}{n!} (x - c)^n, \qquad |x - c| < r$$

Then for each point b of the interval $|x - c| < r$

$$f(x) = \sum_{n=0}^{\infty} \frac{f^{(n)}(b)}{n!} (x - b)^n, \qquad |x - b| < r_1$$

where r_1 is the largest number such that the interval $|x - b| < r_1$ is contained in the interval $|x - c| < r$.

For a proof see Knopp: *Infinite Series*, page 173; London, Blackie, 1928.

THEOREM 32. *Let f be a function having derivatives of all orders in an interval $|x - c| < r$. Let $\{M_n\}$ be a convergent sequence with limit 0. If, for $n = 1, 2, \ldots$,*

$$\left| \frac{f^{(n)}(\xi)}{n!} (x - c)^n \right| \leq M_n$$

for all x, ξ in the interval $|x - c| < r$, then f is analytic at c and f is the sum of its Taylor's series (8-200) in the interval $|x - c| < r$.

PROOF. Let x be a point of the given interval. Then by Taylor's remainder formula (Section 8-19),

$$f(x) = s_n(x) + R_n$$

where $\qquad s_n(x) = f(c) + f'(c)(x - c) + \cdots + \frac{f^{(n)}(c)}{n!}(x - c)^n$

and $\qquad R_n = \frac{f^{(n+1)}(\xi)}{(n + 1)!}(x - c)^{n+1}, \qquad \xi$ between c and x

By hypothesis, $|R_n| \leq M_{n+1}$ and $M_n \to 0$. Hence, $R_n \to 0$ and, therefore, $s_n(x) \to f(x)$. Thus, f is the sum of its Taylor series on the interval $|x - c| < r$ and, as sum of this power series on an interval, f is analytic at c.

THEOREM 33. *The functions e^x, $\sin x$, $\cos x$, $\ln(1 - x)$ are analytic at $x = 0$ and have the Maclaurin series representations:*

$$e^x = 1 + \frac{x}{1!} + \frac{x^2}{2!} + \cdots = \sum_{n=0}^{\infty} \frac{x^n}{n!}, \qquad -\infty < x < \infty \qquad (8\text{-}201)$$

$$\sin x = x - \frac{x^3}{3!} + \frac{x^5}{5!} - \cdots = \sum_{n=0}^{\infty} \frac{(-1)^n x^{2n+1}}{(2n + 1)!}, \qquad -\infty < x < \infty \qquad (8\text{-}202)$$

$$\cos x = 1 - \frac{x^2}{2!} + \frac{x^4}{4!} - \cdots = \sum_{n=0}^{\infty} \frac{(-1)^n x^{2n}}{(2n)!}, \qquad -\infty < x < \infty \qquad (8\text{-}203)$$

$$\ln(1 - x) = -x - \frac{x^2}{2} - \cdots = -\sum_{n=1}^{\infty} \frac{x^n}{n}, \qquad -1 < x < 1 \qquad (8\text{-}204)$$

PROOF. In each case we verify that the right side is the Maclaurin series of the function. To show that the series converges to the function we apply Theorem 32.

For e^x we choose a fixed positive r. Since the nth derivative of e^x is e^x, we must choose M_n so that

$$\left| \frac{e^\xi}{n!} x^n \right| \leq M_n, \qquad |x| < r, \qquad |\xi| < r \qquad (8\text{-}205)$$

Now e^x is monotone strictly increasing, so that $e^\xi < e^r$ in the interval considered. Also $|x^n| < r^n$ for $|x| < r$. Hence the conditions (8-205) are satisfied if we choose

$$M_n = \frac{e^r r^n}{n!}$$

Since M_n is the term of index n of the convergent series $\Sigma(e^r \cdot r^n/n!)$, $M_n \to 0$ by Theorem 7 (Section 8-4). Hence, Theorem 32 applies and the series converges to e^x for $|x| < r$. Since this is true for every positive r, the series converges to e^x for all x. Thus (8-201) is proved.

For $\sin x$ or $\cos x$, the nth derivative is $\pm \sin x$ or $\pm \cos x$ and, hence, $|f^{(n)}(\xi)| \leq 1$. Hence

$$\left| \frac{f^{(n)}(\xi) x^n}{n!} \right| \leq M_n = \frac{r^n}{n!}, \qquad |x| < r, \qquad |\xi| < r$$

The rest of the reasoning is the same as for e^x.

The proof of (8-204) is given in Example 1, Section 8-17.

From Theorem 31 it now follows that the functions e^x, $\sin x$, $\cos x$ are analytic for all x, and $\ln(1 - x)$ is analytic for $|x| < 1$. Thus, for example, e^x can be expanded in a power series with center at $x = 1$: for all x

$$e^x = \sum_{n=0}^{\infty} \frac{f^{(n)}(1)}{n!} (x - 1)^n = e \sum_{n=0}^{\infty} \frac{(x - 1)^n}{n!}$$

THEOREM 34. (*Binomial series*). *Let α be a fixed real number and let*

$$\binom{\alpha}{0} = 1, \qquad \binom{\alpha}{n} = \frac{\alpha(\alpha - 1) \cdots (\alpha - n + 1)}{n!}, \qquad n = 1, 2, \ldots$$

Then the function $f(x) = (1 + x)^\alpha$ has the Maclaurin series representation:

$$(1 + x)^\alpha = \sum_{n=0}^{\infty} \binom{\alpha}{n} x^n, \qquad |x| < 1 \qquad (8\text{-}206)$$

PROOF. If α is a positive integer, then $\binom{\alpha}{n} = 0$ for $n > \alpha$ and (8-206) follows from the binomial theorem; in this case the expansion is valid for all x. For $\alpha = 0$ both sides reduce to 1.

For general α we verify that $f^{(n)}(x) = \alpha(\alpha - 1) \cdots (\alpha - n + 1)(1 + x)^{\alpha-n}$ so that the right-hand side of (8-206) is the Maclaurin series of f. To show convergence to f for $|x| < 1$, we choose a fixed x of this interval and apply the integral form of the remainder formula (Theorem 30) to write

$$f(x) = s_n(x) + R_n$$

$$R_n = \int_0^x \frac{(x-t)^n}{n!} f^{(n+1)}(t)\, dt = \binom{\alpha}{n}(\alpha - n) \int_0^x (x-t)^n (1+t)^{\alpha-n-1}\, dt$$

We set $t = xv$ and find that

$$R_n = \binom{\alpha}{n}(\alpha - n)x^{n+1} \int_0^1 (1-v)^n (1+xv)^{\alpha-n-1}\, dv$$

$$= \binom{\alpha}{n}(\alpha - n)x^{n+1} \int_0^1 \left(\frac{1-v}{1+xv}\right)^n (1+xv)^{\alpha-1}\, dv$$

Now for $0 \le v \le 1$ and $-1 < x < 1$, we have

$$0 \le \frac{1-v}{1+xv} \le 1, \qquad 0 < 1 + xv$$

and hence (for $x \ne 0$)

$$|R_n| \le \binom{\alpha}{n}(\alpha - n)|x|^{n+1} \int_0^1 (1+xv)^{\alpha-1}\, dv$$

$$= \binom{\alpha}{n}(\alpha - n)|x|^{n+1} \frac{(1+x)^\alpha - 1}{\alpha x}$$

Now the series $\Sigma \binom{\alpha}{n}(\alpha - n)|x|^n$ converges for $|x| < 1$ (by the ratio test) and hence $R_n \to 0$. Therefore $s_n(x) \to f(x)$, or f is the sum of its Maclaurin series for $|x| < 1$.

A SECOND PROOF. If α is a positive integer, $(1+x)^\alpha$ is a polynomial, and the conclusion is easily established. Therefore, suppose that α is not a positive integer or 0. Then

$$\lim_{n \to \infty} \left| \binom{\alpha}{n} \middle/ \binom{\alpha}{n+1} \right| = \lim_{n \to \infty} \left| \frac{n+1}{\alpha - n} \right| = 1$$

and, by Theorem 27, the series $\Sigma \binom{\alpha}{n} x^n$ converges to a function $g(x)$ on the interval $|x| < 1$. By Theorem 28, the function $g(x)$ is differentiable for $|x| < 1$. Now for $|x| < 1$

$$g'(x) = \sum_{n=1}^{\infty} n\binom{\alpha}{n} x^{n-1} = \sum_{n=0}^{\infty} (n+1)\binom{\alpha}{n+1} x^n$$

$$xg'(x) = \sum_{n=1}^{\infty} n\binom{\alpha}{n} x^n = \sum_{n=0}^{\infty} n\binom{\alpha}{n} x^n$$

$$(1+x)g'(x) = \sum_{n=0}^{\infty} \left[(n+1)\binom{\alpha}{n+1} + n\binom{\alpha}{n} \right] x^n = \alpha \sum_{n=0}^{\infty} \binom{\alpha}{n} x^n = \alpha g(x)$$

Hence

$$(1 + x)^{-\alpha}g'(x) - \alpha(1 + x)^{-\alpha-1}g(x) = 0, \qquad \frac{d}{dx}[(1 + x)^{-\alpha}g(x)] = 0$$

$$(1 + x)^{-\alpha}g(x) = \text{const}$$

But at $x = 0$ both $g(x)$ and $(1 + x)^{\alpha}$ are 1, so the constant is 1, and we have

$$g(x) = (1 + x)^{\alpha}, \qquad |x| < 1$$

Since $g(x)$ is the sum of a power series, that series must be the Maclaurin series for $g(x) = (1 + x)^{\alpha}$.

THEOREM 35. Sin^{-1}x *is the sum of its Maclaurin series:*

$$\text{Sin}^{-1} x = x + \sum_{n=1}^{\infty} \frac{1 \cdot 3 \cdots (2n - 1)}{n!\, 2^n (2n + 1)} x^{2n+1}, \qquad |x| < 1$$

PROOF. We could approach this result as we did for sin x and cos x, instead we recall that

$$\text{Sin}^{-1}x = \int_0^x \frac{dt}{\sqrt{1 - t^2}}$$

Now, by Theorem 34,

$$(1 + x)^{-1/2} = 1 - \frac{1}{2}x + \frac{(-1/2)(-3/2)}{2!} x^2$$

$$+ \cdots + \frac{(-1/2)(-3/2)\cdots[-(2n - 1)/2]}{n!} x^n + \cdots$$

$$= 1 + \sum_{n=1}^{\infty} \frac{(-1)^n 1 \cdot 3 \cdots (2n - 1)}{2^n n!} x^n, \qquad |x| < 1$$

Hence (on setting $x = -t^2$)

$$(1 - t^2)^{-1/2} = 1 + \sum_{n=0}^{\infty} \frac{1 \cdot 3 \cdots (2n - 1)}{2^n n!} t^{2n}, \qquad |t| < 1$$

and, by Theorem 28,

$$\text{Sin}^{-1} x = \int_0^x (1 - t^2)^{-1/2}\, dt$$

$$= x + \sum_{n=1}^{\infty} \frac{1 \cdot 3 \cdots (2n - 1)}{2^n n!\, (2n + 1)} x^{2n+1} \qquad \text{on } |x| < 1$$

Since Sin$^{-1} x$ is the sum of a power series for $|x| < 1$ that series must be the Maclaurin series for Sin$^{-1} x$.

Note. Since Sin$^{-1} 1/2 = \pi/6$, we have

$$\pi = 6\left[\frac{1}{2} + \sum_{n=1}^{\infty} \frac{1 \cdot 3 \cdots (2n - 1)}{n!\, 2^{3n+1}(2n + 1)}\right]$$

Remark. At the beginning of this section we indicated that a function might have a derivative of every order at $x = c$, and so have a Taylor's series with center c, and yet that series may only converge at c, or it may converge on an interval about c, but the given function may not be its sum. The first situation can occur when $f^{(n)}(c)$ increases extremely rapidly as n tends to ∞, say $f^{(n)}(c) > (2n)!$ for sufficiently large n. The second situation occurs when two functions differ in every small interval about c and yet $f^{(n)}(c) = g^{(n)}(c)$ for $n = 0, 1, 2, \ldots$. For then, the two functions have the same Taylor's series with center c and that series cannot possibly have both functions as its sum on any small interval about c. An example of this last situation is provided by the functions

$$f(x) \equiv 0, \qquad g(x) = \begin{cases} 0, & \text{if } x = 0 \\ e^{-1/x^2}, & \text{if } x \neq 0 \end{cases}, \qquad -\infty < x < \infty$$

Clearly $f(x) = g(x)$ only for $x = 0$, and yet $f^{(n)}(0) = g^{(n)}(0)$ for all n (Problem 9 below).

8-21 NUMERICAL EVALUATION OF FUNCTIONS BY POWER SERIES

When a function f is known to be the sum of a power series $\Sigma a_n(x - c)^n$ on an interval $|x - c| < r$, we can use the series to calculate f numerically. We write:

$$f(x) = s_n(x) + R_n(x)$$

where
$$s_n(x) = a_0 + a_1(x - c) + \cdots + a_n(x - c)^n$$

and
$$R_n(x) = \sum_{k=n+1}^{\infty} a_k(x - c)^k$$

as in Theorem 9. For each particular x of the interval we know that $R_n(x) \to 0$; hence n can be chosen so large that $|R_n| < \epsilon$ for a prescribed ϵ. If we evaluate $s_n(x)$ for such an n, then we have found an approximation to the value $f(x)$, with error less than ϵ:

$$f(x) \sim s_n(x), \qquad |f(x) - s_n(x)| < \epsilon$$

To know how large n must be chosen, for given ϵ, we need estimates for R_n. If we know that $|f^{(n)}(x)| \leq K_n = $ const. over the interval considered, then the remainder formula (Lagrange form) gives

$$|R_n(x)| = \left| f^{(n+1)}(\xi) \frac{(x - c)^{n+1}}{(n + 1)!} \right| \leq K_{n+1} \frac{|x - c|^{n+1}}{(n + 1)!} \qquad (8\text{-}210)$$

This estimate often suffices to determine how many terms are needed. The graphical meaning of (8-210) is considered in Section 6-11.

EXAMPLE 1. Find e correct to 5 decimal places.

Solution. We use the series expansion $e^x = \Sigma x^n/n!$. Then (8-210) gives

$$|R_n| = \frac{e^\xi}{(n + 1)!} x^{n+1} \leq \frac{e}{(n + 1)!}, \qquad 0 < \xi < x = 1$$

For $n = 1$ we thus have

$$\left| e - \left(1 + \frac{1}{1!}\right)\right| < \frac{e}{2!} \qquad \text{or} \qquad |e - 2| < \frac{e}{2}$$

Hence $e - 2 < e/2$ or $e < 4$. Thus for general n we have

$$|R_n| < \frac{4}{(n + 1)!}$$

For the desired accuracy we want $|R_n| < 0.000\,005$. The smallest value of n for which $4/(n + 1)!$ is less than $0.000\,005$ is 9:

$$|R_9| < \frac{4}{10!} = 0.000\,001$$

Hence (with the aid of tables) we find that

$$e \sim 1 + \frac{1}{1!} + \frac{1}{2!} + \cdots + \frac{1}{9!} = 2.718282$$

with an error of at most $0.000\,001$, or $e = 2.71828$ to five decimal places.

EXAMPLE 2. Find $\ln(4/3)$ correct to 3 decimal places.

Solution. We use the Maclaurin series for $\ln(1 - x)$ as in Theorem 33, taking $x = -(1/3)$:

$$\ln\left(\frac{4}{3}\right) = \frac{1}{3} - \frac{1}{2 \cdot 3^2} + \frac{1}{3 \cdot 3^3} + \cdots + (-1)^{n-1} \frac{1}{n \cdot 3^n} + \cdots$$

The series is alternating and, hence, by Theorem 22, the error is less than the first term omitted:

$$|R_n| < \frac{1}{(n + 1)3^{n+1}}$$

For $|R_n|$ to be $< .0005$, we must take $n \geq 5$, in particular $|R_5| < 0.00023$. Hence

$$\ln\left(\frac{4}{3}\right) \sim \frac{1}{3} - \frac{1}{2 \cdot 3^2} + \cdots + \frac{1}{5 \cdot 3^5} = 0.28785$$

Thus $\ln(4/3) = 0.288$ to three decimal places.

EXAMPLE 3. Find $\ln 3$ correct to 3 decimal places.

Solution. Here we cannot use the series expansion for $\ln(1 - x) = \Sigma(-x^n/n)$, since it only converges for $-1 < x \leq 1$, and $1 - x = 3$ implies $x = -2$, which is outside the interval of convergence. However, we can reason as follows:

$$\ln\left(\frac{1 + x}{1 - x}\right) = \ln(1 + x) - \ln(1 - x)$$

and therefore

$$\ln\left(\frac{1 + x}{1 - x}\right) = -\sum_{n=1}^{\infty} \frac{(-x)^n}{n} + \sum_{n=1}^{\infty} \frac{x^n}{n} = 2 \sum_{n=0}^{\infty} \frac{x^{2n+1}}{2n + 1}, \qquad \text{for } |x| < 1$$

Now $(1 + x)/(1 - x) = 3$ when $x = \frac{1}{2}$, and this point lies in the interval of convergence for our new series. For this series

$$|R_n(x)| = 2\left|\sum_{m=n+1}^{\infty} \frac{x^{2m+1}}{2m+1}\right| \leq \frac{2|x|}{2n+3} \sum_{m=n+1}^{\infty} x^{2m} = \frac{2|x|^{2n+3}}{(2n+3)(1-|x|^2)}$$

Thus

$$|R_4(\tfrac{1}{2})| < .0002$$

We then evaluate

$$s_4\left(\frac{1}{2}\right) = 2\left[\frac{1}{2} + \frac{1}{3 \cdot 2^3} + \frac{1}{5 \cdot 2^5} + \frac{1}{7 \cdot 2^7} + \frac{1}{9 \cdot 2^9}\right] = 1.0984$$

and, hence, $\ln(3) = 1.0984$ (with error less than 0.0002).

EXAMPLE 4. Evaluate $\int_0^1 e^{-x^2}\,dx$ correct to 4 decimal places.
  Solution.

$$e^{-x^2} = \sum_{n=0}^{\infty} \frac{(-x^2)^n}{n!}$$

and therefore

$$f(x) = \int_0^x e^{-t^2}\,dt = \sum_{n=0}^{\infty} \frac{(-1)^n x^{2n+1}}{(2n+1)n!}$$

and this series converges for all x. The series for $f(1)$ is the alternating series $1 - (1/3) + [1/(5 \cdot 2!)] - [1/(7 \cdot 3!)] + [1/(9 \cdot 4!)] - \cdots$ and, therefore, by Theorem 22, $|f(1) - s_n(1)| < 1/[(2n+3)(n+1)!]$. In particular $|f(1) - s_6(1)| < .00001$. As

$$s_6(1) = 1 - (1/3) + (1/10) - (1/42) + (1/216) - (1/1320) = 0.73773$$

we can conclude that

$$\int_0^1 e^{-x^2}\,dx = 0.7377 \qquad \text{(correct to 4 decimal places)}.$$

PROBLEMS

1. Find the Taylor's series for each of the following functions at the specified center. For those starred, determine the radius of convergence, and state whether the sum is the given function.
   ° (a) $\sqrt{x}$ at $x = 1$ ° (b) $1/x$ at $x = 2$ ° (c) $1/(x+2)$ at $x = 0$
   ° (d) $x \sin x^2$ at $x = 0$ ° (e) $\ln x$ at $x = 1$ ° (f) e^x at $x = 2$
     (g) $\sin x$ at $x = \pi/3$ (h) $\ln x$ at $x = 2$ (i) $\sqrt{2+x}$ at $x = 1$
     (j) $\text{Cos}^{-1} x$ at $x = 0$

2. Find the first 3 nonzero terms of the Maclaurin series for each of the following functions and, for those starred, determine the interval of convergence and state whether the sum is the given function.

° (a) $\ln(1 - x^2)$ (b) $\tan x$ (c) $\ln(\cos x)$
(d) $\ln^2(1 - x)$ ° (e) $(1 + x^2)^{-1}$ ° (f) $\text{Tan}^{-1} x$
(g) $\sec x$ (h) $(1 + x^2)^{-2/3}$ (i) $(2x + 3)^{1/5}$

(j) $\ln(\sec x)$ (k) $\int_0^x t^2 \cos t \, dt$ (l) $\int_0^x \sin t^2 \, dt$

(m) $\int_0^x \frac{\sin t}{t} dt$ (n) $\frac{x + 2}{\sqrt[3]{1 - x}}$ ° (o) $\sinh x$

° (p) $\cosh x$ (q) $[\ln(1 + x)]/\sqrt{1 + x}$ (r) $e^{-x} \sqrt{\frac{1 - x}{1 + x}}$

3. Find the first three nonzero terms for the Maclaurin series for the following functions.

(a) $\sqrt[3]{1 - x^2} \sin x$ (b) $\int_0^x e^t \cos t \, dt$

(c) $\int_0^x \sin t \ln(1 + t) \, dt$ (d) $\int_0^x \frac{dt}{\sqrt{1 - m^2 \sin^2 t}}$, $|m| < 1$

4. Determine the following numbers to within 3 decimal place accuracy.
(a) $\sqrt{84}$ (b) $\sqrt[3]{30}$ (c) $\sqrt{127}$ (d) $\sqrt[5]{33}$
[*Hint.* $\sqrt{84} = \sqrt{81[1 + (1/27)]} = 9\sqrt{1 + (1/27)}$]

5. Evaluate the following numbers correct to within 3-decimal-place accuracy:

(a) $\sin(1/10)$ (b) $\int_0^1 \frac{\sin t}{t} \, dt$ (c) $\ln 2$

(d) $\ln 5/6$ (e) $\ln 7/3$ (f) $\ln 5$

(g) $\int_0^{1/4} \cos t^3 \, dt$ (h) $\int_0^{1/10} e^t \sin t^3 \, dt$

6. (a) Determine the values of x for which $|x - \sin x| < .001$.
(b) Determine how much $\sin x$ differs from $x - x^3/6 + x^5/120$ on the interval $|x| \leq \pi$.

7. (a) If $1/2 \leq \alpha \leq 1$, how accurately must you know α to get $\sqrt{\alpha}$ correct to 3 decimal places?
(b) Find a solution of the equation $10x^2 = \cos x$ correct to 3 decimal places. [*Hint.* Replace $\cos x$ by a polynomial which approximates it quite accurately near $x = 0$.]

8. (a) Show that $\text{Tan}^{-1} x = \sum_{n=0}^{\infty} \frac{(-1)^n x^{2n+1}}{2n + 1}$ for $-1 < x < 1$

(b) Assume (a) valid for $x = 1$ to obtain $\pi = 4 \sum_{n=0}^{\infty} \frac{(-1)^n}{2n + 1}$

(c) If we use part (b) to evaluate π, how many terms are necessary to get π correct to 5 decimal places?
(d) Evaluate $\text{Tan}^{-1}(1/10)$ correct to 5 decimal places.
(e) Prove: If α, β are real numbers for which $\alpha > 0, \beta > 0, \alpha\beta < 1$, then

$$\text{Tan}^{-1}\left(\frac{\alpha + \beta}{1 - \alpha\beta}\right) = \text{Tan}^{-1} \alpha + \text{Tan}^{-1} \beta$$

(f) Justify the following: $\text{Tan}^{-1}(1) = \text{Tan}^{-1}(1/2) + \text{Tan}^{-1}(1/3)$
$\text{Tan}^{-1}(1/2) = \text{Tan}^{-1}(1/3) + \text{Tan}^{-1}(1/7)$

(g) Prove: If p, q, and r are positive integers and $(p - r)(q - r) = r^2 + 1$, then
$\text{Tan}^{-1}(1/r) = \text{Tan}^{-1}(1/p) + \text{Tan}^{-1}(1/q)$.

(h) Prove: $\text{Tan}^{-1}(1) = 3\,\text{Tan}^{-1}(1/7) + 2\,\text{Tan}^{-1}(1/8) + 2\,\text{Tan}^{-1}(1/18)$.

(i) Use part (h) to evaluate π correct to 5 decimal place accuracy.

(j) Evaluate $\text{Tan}^{-1}(1/4)$ correct to 5 decimal places.

(k) Evaluate $\text{Tan}^{-1}(2)$ correct to 5 decimal places.

‡9. Let $F(x) = \begin{cases} 0 & \text{if } x = 0 \\ e^{-1/x^2} & \text{if } x \neq 0 \end{cases}$ $G(x) = \begin{cases} 0 & \text{if } x \leq 0 \\ e^{-1/x^2} & \text{if } x > 0 \end{cases}$

(a) Prove: $\lim\limits_{x \to 0} e^{-1/x^2} = 0$

(b) Prove: $F(x)$ and $G(x)$ are continuous for all x.

(c) Evaluate: $\lim\limits_{h \to 0} e^{-1/h^2}/h$. (*Hint.* Use L'Hospital's Rules.)

(d) Prove: $F'(0) = G'(0) = 0$.

(e) Prove: $F'(x)$ and $G'(x)$ are continuous functions for all x.

(f) Prove: $F''(0) = 0$, $G''(0) = 0$ and G'', F'' are continuous functions for all x.

(g) Prove: For each $n = 0, 1, 2, \ldots$, $F^{(n)}(0) = G^n(0) = 0$.

(h) Graph the equations: $y = F(x)$, $y = G(x)$.

(i) Show that F, G, and the zero function all have the same Taylor series with center at 0. Which of these functions, if any, is the sum of the common Taylor's series?

8-22 POWER SERIES SOLUTION OF DIFFERENTIAL EQUATIONS

We here consider briefly how power series can be used to obtain solutions of differential equations. For a more thorough discussion, see Chapter 14.

We consider an example: the equation

$$y'' - y = 0 \tag{8-220}$$

This is a linear differential equation of second order with constant coefficients, and it can be solved as in Section 7-18. However, we now use power series to find the solutions.

We first observe that if $y = f(x)$ is a solution of (8-220) in an interval, then $f''(x)$ must exist in that interval. Hence $f'(x)$ must be continuous, and f must be continuous in the interval. Since (8-220) states that $f''(x) = f(x)$, we conclude that f'' must also be continuous and, in fact, twice differentiable, with

$$f'''(x) = f'(x), \qquad f^{(iv)}(x) = f''(x)$$

By induction, we conclude that f has continuous derivatives of every order in the interval. Accordingly we can form the Taylor series of f with center at a point c of the interval. To test for convergence, we use the integral form of the remainder:

$$R_n = \int_c^x \frac{(x - t)^n}{n!} f^{(n+1)}(t)\, dt$$

Let us now suppose that n is odd, so that $n + 1$ is even. Then, as above,

$$f^{(n+1)} = f^{(n-1)} = \cdots f'' = f$$

so that

$$R_n = \int_c^x \frac{(x - t)^n}{n!} f(t)\, dt$$

Since f is continuous on the given interval, $|f| \leq K$ for some constant K and, for $x > c$, we can write

$$|R_n| \leq \int_c^x \frac{(x - t)^n}{n!} K\, dt = K \frac{(x - c)^{n+1}}{(n + 1)!}$$

A similar reasoning applies for $x < c$, except that we must replace $x - c$ by $|x - c|$. For n odd we have

$$f^{(n+1)} = f^{(n-1)} = f''' = f'$$

so that

$$R_n = \int_c^x \frac{(x - t)^n}{n!} f'(t)\, dt$$

and, if K is chosen so large that both $|f| < K$ and $|f'| < K$ in the interval from c to x, for all n we have

$$|R_n| \leq \frac{K|x - c|^{n+1}}{(n + 1)!}$$

Since the right side is the general term of a convergent series, $R_n \to 0$. Hence, f equals the sum of its Taylor series and f is analytic. We have proved that *every solution of (8-220) is analytic throughout the interval in which it is defined.*

To find solutions explicitly, we seek them in the form of a power series with center at 0 (Maclaurin series):

$$y = a_0 + a_1 x + a_2 x^2 + \cdots + a_n x^n + \cdots$$

If the series converges in an interval, then we may differentiate successively:

$$y' = a_1 + 2a_2 x + \cdots + na_n x^{n-1} + (n + 1)a_{n+1} x^n + \cdots$$
$$y'' = 2a_2 + \cdots + (n + 1)na_{n+1} x^{n-1} + (n + 2)(n + 1)a_{n+2} x^n + \cdots$$

If we substitute in the equation (8-220) and collect terms of same degree, we obtain

$$2a_2 - a_0 + x(6a_3 - a_1) + \cdots + x^n[(n + 2)(n + 1)a_{n+2} - a_n] + \cdots = 0$$

By Corollary 4 of Theorem 28, we conclude that

$$2a_2 - a_0 = 0, \qquad 6a_3 - a_1 = 0, \ldots, \qquad (n + 2)(n + 1)a_{n+2} - a_n = 0, \ldots$$

Hence

$$a_2 = \frac{a_0}{2}, \qquad a_3 = \frac{a_1}{6}, \qquad a_4 = \frac{a_2}{4 \cdot 3} = \frac{a_0}{4 \cdot 3 \cdot 2}, \qquad a_5 = \frac{a_3}{5 \cdot 4} = \frac{a_1}{5 \cdot 4 \cdot 3 \cdot 2}$$

and in general

$$a_{n+2} = \frac{a_n}{(n+2)(n+1)} = \frac{a_{n-2}}{(n+2)(n+1)n(n-1)} = \cdots$$

Thus

$$a_n = \frac{a_0}{n!} \quad \text{for } n \text{ even}, \qquad a_n = \frac{a_1}{n!} \quad \text{for } n \text{ odd}$$

Accordingly

$$y = a_0 + a_1 x + \frac{a_0}{2!} x^2 + \frac{a_1}{3!} x^3 + \frac{a_0}{4!} x^4 + \cdots$$

$$= a_0 \left(1 + \frac{x^2}{2!} + \frac{x^4}{4!} + \cdots \right) + a_1 \left(x + \frac{x^3}{3!} + \frac{x^5}{5!} + \cdots \right)$$

Here a_0 and a_1 can be chosen arbitrarily, so that

$$y = a_0 y_1(x) + a_1 y_2(x) \tag{8-221}$$

$$y_1(x) = 1 + \frac{x^2}{2!} + \frac{x^4}{4!} + \cdots, \qquad y_2(x) = x + \frac{x^3}{3!} + \frac{x^5}{5!} + \cdots$$

Both series converge for all x and we verify that $y_1'' - y_1 = 0$ for all x, $y_2'' - y_2 = 0$ for all x, so that y is a solution of (8-220) for all x. We verify that, if we had used series with center at $x = c$, we would again have obtained series converging for all x. It follows that every solution is representable by its Maclaurin series, so that (8-221) gives *all solutions* (the general solution).

Remark. The functions $y_1(x)$, $y_2(x)$ are familiar functions; in fact

$$y_1(x) = \frac{1}{2}(e^x + e^{-x}) = \cosh x, \qquad y_2(x) = \frac{1}{2}(e^x - e^{-x}) = \sinh x$$

Hence the general solution is

$$y = a_0 \frac{e^x + e^{-x}}{2} + a_1 \frac{e^x - e^{-x}}{2} = c_1 e^x + c_2 e^{-x}$$

with $c_1 = (a_0 + a_1)/2$, $c_2 = (a_0 - a_1)/2$. The expression $y = c_1 e^x + c_2 e^{-x}$ for the general solution is obtainable by the methods of Section 7-18.

The example is illustrative of methods applicable to a broad class of equations, in particular, to linear differential equations

$$p_0(x) y^{(n)} + p_1(x) y^{(n-1)} + \cdots + p_n(x) y = q(x) \tag{8-222}$$

in which the coefficients $p_0, \cdots, p_n$ and the right-hand member are polynomials. It can be shown that (apart from certain exceptional points) all solutions are analytic and hence, are obtainable as power series; the zeros of $p_0(x)$ are in general *excluded* as centers of the power series. If, for example, $p_0(0) \neq 0$, we can seek a solution of the form:

$$y = a_0 + a_1 x + \cdots + a_n x^n + \cdots$$

Substitution in (8-222) leads to relations between the coefficients, from which we obtain the solutions in the form

$$y = a_0 y_1(x) + a_1 y_2(x) + \cdots + a_{n-1} y_n(x) + y^\circ(x) \qquad (8\text{-}223)$$

where $y_1(x), \cdots, y_n(x), y^\circ(x)$ are power series. Here $a_0, \cdots, a_n$ are "arbitrary constants." The general theory assures that the series all converge in an interval $|x| < r$ (and, in fact, the theory gives a value of r), and that (8-223) represents all solutions in this interval. To obtain solutions in an interval about a point c [not a zero of $p_0(x)$], we proceed similarly with a series

$$y = a_0 + a_1(x - c) + \cdots + a_n(x - c)^n + \cdots$$

PROBLEMS

1. Use power series with center at $x = 0$ to obtain the general solution, and verify convergence for all x.
   (a) $y' - 3y = 0$ (b) $y' - xy = x$ (c) $y'' - 4y = 0$
   (d) $y'' + 4y = 0$ ‡(e) $y'' + 2y' + y = 0$ (f) $y'' + y' + y = 1 + x$
2. Obtain power series solutions with center at $x = 0$ (proof of convergence not required):
   (a) $y'' - xy' + y = 0$ (b) $(x^2 - 1)y'' - 2y = 0$
   (c) $y'' - xy = 0$ (d) $(x^2 + 2)y'' - xy' - 3y = 0$
3. Let α be a fixed real number. Find the general solution of the differential equation $y'' - 2xy' + \alpha y = 0$. Show that when $\alpha = 2m$ ($m = 0, 1, 2, 3, \ldots$) the equation has a polynomial as solution. [These polynomials are constant multiples of Hermite polynomials, which play an important role in quantum mechanics.]
4. Let m be a nonnegative integer. Show that the equation

$$(1 - x^2)y'' - 2xy' + (m^2 + m)y = 0$$

has a polynomial of degree m as one solution. Show that these polynomials do not vanish at $x = 1$ and that there is one $P_m(x)$ for which $P_m(1) = 1$. [The polynomials $P_m(x)$ are the Legendre polynomials. They have the interesting property $\int_{-1}^{1} P_n(x) P_m(x)\, dx = 0$ if $n \neq m$.]

8-23 COMPLEX POWER SERIES

Thus far we have restricted our power series to have real coefficients and the variable x to lie on the real line. There is no need to do so. We can consider complex power series

$$\sum_{n=0}^{\infty} \alpha_n(z - \gamma)^n = \alpha_0 + \alpha_1(z - \gamma) + \alpha_2(z - \gamma)^2 + \cdots \qquad (8\text{-}230)$$

where the coefficients $\alpha_0, \alpha_1, \ldots$ and the center γ are fixed complex numbers, and the variable z ranges over the complex numbers. As with real power series, we find that there is a real number $R \geq 0$ so that the series (8-230) converges for $|z - \gamma| < R$ and diverges for $|z - \gamma| > R$; for each z on the

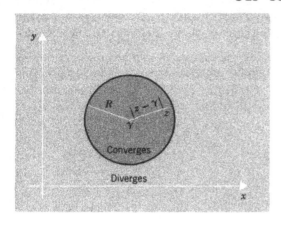

Figure 8-10 Radius of convergence for complex power series.

circle $|z - \gamma| = R$ the series may either converge or diverge (see Figure 8-10). As for real series, R is called the *radius of convergence* for the given power series. More specifically, the series $\Sigma \alpha_n(z - \gamma)^n$ converges absolutely for z in $|z - \gamma| < R$; that is, the series $\Sigma|\alpha_n(z - \gamma)^n|$ converges. As with real power series, absolutely convergent power series can be rearranged without affecting the sum. Also, if

$$\lim_{n \to \infty} \left| \frac{\alpha_n}{\alpha_{n+1}} \right| = R$$

or if

$$\lim_{n \to \infty} \frac{1}{\sqrt[n]{|\alpha_n|}} = R$$

then R is the radius of convergence for the series. The proofs of these statements are very similar to the ones for real power series.

It is of special interest that, when x is replaced by z, the power series for e^x, $\cos x$, $\sin x$ become the complex power series:

$$\sum_{n=0}^{\infty} \frac{z^n}{n!}, \qquad \sum_{n=0}^{\infty} \frac{(-1)^n z^{2n}}{(2n)!}, \qquad \sum_{n=0}^{\infty} \frac{(-1)^n z^{2n+1}}{(2n+1)!}$$

They converge for all complex numbers z. Accordingly, we can define

$$e^z = \sum_{n=0}^{\infty} \frac{z^n}{n!} = 1 + z + \frac{z^2}{2!} + \cdots + \frac{z^n}{n!} + \cdots$$

$$\cos z = \sum_{n=0}^{\infty} \frac{(-1)^n z^{2n}}{(2n)!} = 1 - \frac{z^2}{2!} + \frac{z^4}{4!} - \frac{z^6}{6!} + \cdots \qquad (8\text{-}231)$$

$$\sin z = \sum_{n=0}^{\infty} \frac{(-1)^n z^{2n+1}}{(2n+1)!} = z - \frac{z^3}{3!} + \frac{z^5}{5!} - \frac{z^7}{7!} + \cdots$$

These complex valued functions of a complex variable agree with the real valued functions of a real variable: e^x, cos x, sin x, on the real axis, and so are natural generalizations and extensions of these functions to the complex plane.

THEOREM 36. *If z, w are complex numbers and x, y real numbers, then*

(a) $e^{iz} = \cos z + i \sin z$

(b) $e^{-iz} = \cos z - i \sin z$

(c) $\cos z = \dfrac{e^{iz} + e^{-iz}}{2}$

(d) $\sin z = \dfrac{e^{iz} - e^{-iz}}{2i}$

(e) $e^{z+w} = e^z e^w$

(f) $e^{x+iy} = e^x[\cos y + i \sin y]$

(g) $\cos^2 z + \sin^2 z = 1$

(h) $\cos(z + w) = \cos z \cos w - \sin z \sin w$

(i) $\sin(z + w) = \sin z \cos w + \cos z \sin w$

(j) $e^{z+2\pi i} = e^z$

PROOF. *Part (a)*. We add the series for $\cos z$ and $i \sin z$:

$$\cos z + i \sin z = 1 + iz - \frac{z^2}{2!} - \frac{iz^3}{3!} + \frac{z^4}{4!} + \frac{iz^5}{5!} - \frac{z^6}{6!} - \cdots$$

and observe that it is the series for e^{iz}. *Part (b)* is proved similarly. *Parts (c)*, *and (d)* are obtained from *(a)*, *(b)* by solving for $\cos z$ and $\sin z$.

Part (e). We must show that

$$\left(\sum \frac{z^n}{n!}\right)\left(\sum \frac{w^n}{n!}\right) = \sum \frac{(z+w)^n}{n!}$$

We recall that the product of two absolutely convergent series can be expressed as the Cauchy product. Hence

$$\sum_{n=0}^{\infty} \frac{z^n}{n!} \sum_{n=0}^{\infty} \frac{w^n}{n!} = \sum_{n=0}^{\infty} \left[\sum_{m=0}^{n} \frac{z^m}{m!} \frac{w^{n-m}}{(n-m)!}\right]$$

But, by the Binomial Theorem (Section 0-21),

$$\sum_{m=0}^{n} \binom{n}{m} z^m w^{n-m} = \sum_{m=0}^{n} \frac{n! \, z^m w^{n-m}}{m! \, (n-m)!} = (z+w)^n$$

and therefore we obtain the desired result.

Part (f). By parts *(e)*, and *(a)*,

$$e^{x+iy} = e^x e^{iy} = e^x(\cos y + i \sin y)$$

Part (g). We use parts *(a)*, *(b)*, and *(e)*

$$1 = e^0 = e^{iz-iz} = e^{iz}e^{-iz} = [\cos z + i \sin z][\cos z - i \sin z]$$
$$= \cos^2 z - [i \sin z]^2 = \cos^2 z + \sin^2 z$$

We leave *parts* (*h*), (*i*), (*j*) as exercises.

Remark. We observe that Part (*f*) of Theorem 36 is the definition we gave earlier for *e* to a complex exponent (See Section 5-8). We are free to define e^z either by (*f*) or by the power series (8-231). In either case we are discussing the same function. Actually, we can start with the equations (8-231) as definition and derive all the properties of the functions e^z, cos z, sin z.

PROBLEMS

1. Let $\quad \cosh z = \dfrac{e^z + e^{-z}}{2}, \quad \sinh z = \dfrac{e^z - e^{-z}}{2}$

   Prove the identities:
   (a) $\sin(-z) = -\sin z$ (b) $\cos(-z) = \cos z$ (c) $\cos iz = \cosh z$
   (d) $i \sinh z = \sin iz$ (e) $\cos z = \cos x \cosh y - i \sin x \sinh y$ $(z = x + iy)$
   (f) $\sin z = \sin x \cosh y + i \cos x \sinh y$
   (g) $|\cos z|^2 = \cos^2 x + \sinh^2 y = \cosh^2 y - \sin^2 x$
   (h) $|\sin z|^2 = \sin^2 x + \sinh^2 y = \cosh^2 y - \cos^2 x$
2. Prove the following parts of Theorem 36:
   (a) Part (*h*)   (b) Part (*i*)   (c) Part (*j*)
3. Prove: (a) $|\sinh y| \le |\cos z| \le \cosh y \quad (z = x + iy)$.
   (b) $|\sinh y| \le |\sin z| \le \cosh y$. [*Hint.* Use (g), (h) of Problem 1.]
4. Prove that there is a complex number z such that $|\sin z| > 1$ and $|\cos z| > 1$.
5. Find the radius of convergence of each of the following series.

   (a) $\displaystyle\sum_{n=1}^{\infty} \frac{z^n}{n}$ (b) $\displaystyle\sum_{n=1}^{\infty} \frac{(1 + ni) z^n}{n^2}$ (c) $\displaystyle\sum_{n=0}^{\infty} z^n$

   (d) $\displaystyle\sum_{n=0}^{\infty} (1 + ni) z^n$ (e) $\displaystyle\sum_{n=0}^{\infty} \frac{z^n}{(n + 2^n i)}$ (f) $\displaystyle\sum_{n=1}^{\infty} \frac{z^{2n}}{(in)^3}$

6. (a) Show that Lemma 1 of Section 8-16 remains true for complex series.
   (b) Show that Lemma 2 of Section 8-16 remains true for a complex power series: $\Sigma a_n z^n$; that is, if the series converges for $z = z_0 \ne 0$, then it converges absolutely for $|z| < |z_0|$.
   (c) Prove the counterpart of Theorem 26 (Sections 8-15, and 8-16) for complex power series $\Sigma a_n z^n$.

8-24 FOURIER SERIES

We here present a brief introduction to this topic. By a *trigonometric series* we mean a series of form:

$$\frac{a_0}{2} + a_1 \cos x + b_1 \sin x + \cdots + a_n \cos nx + b_n \sin nx + \cdots \qquad (8\text{-}240)$$

(The first term is written $a_0/2$ to simplify later formulas.) For example

$$1 + \frac{\cos x}{1^2} + \frac{\sin x}{1^2} + \cdots + \frac{\cos nx}{n^2} + \frac{\sin nx}{n^2} + \cdots$$

is a trigonometric series. This series converges absolutely for all x, by comparison with the series:

$$1 + \frac{1}{1^2} + \frac{1}{1^2} + \cdots + \frac{1}{n^2} + \frac{1}{n^2} + \cdots$$

Let a trigonometric series (8-240) converge for all x; let $f(x)$ be the sum. Then f has the important property of periodicity, with period 2π:

$$f(x + 2\pi) = f(x), \qquad \text{for all } x \tag{8-241}$$

For

$$f(x + 2\pi) = \frac{a_0}{2} + a_1 \cos(x + 2\pi) + \cdots + a_n \cos n(x + 2\pi)$$

$$+ b_n \sin n(x + 2\pi) + \cdots$$

$$= \frac{a_0}{2} + a_1 \cos x + \cdots + a_n \cos nx + b_n \sin nx + \cdots$$

$$= f(x)$$

By varying the choice of coefficients in (8-240) we can obtain a large class of series converging to periodic functions. It is a remarkable fact that *essentially all periodic functions (of period 2π) are obtainable in this way.*

This mathematical statement corresponds to a common physical observation. If we think of x as "time," then a term $a_k \cos kx$ in the series (8-240) can be interpreted as a function describing a simple harmonic motion or sinusoidal oscillation, such as that of a spring or a pendulum. This motion has period $2\pi/k$ in the chosen time units. The term $b_k \sin kx$ represents a similar motion of the same period, but shifted in phase (the maxima and minima occur at different times). The two terms together represent a general sinusoidal oscillation of period $2\pi/k$ (see Section 7-20). In a time interval of 2π units, this oscillation completes k cycles. Hence, referred to this time interval, we say that the oscillation has *frequency k.*

By adding together such terms for $k = 1, 2, \ldots$, we are combining sinusoidal oscillations of different frequencies but they all have 2π as period, so that the sum has period 2π. Thus the mathematical statement made above corresponds to the physical statement that every oscillation of period 2π is a combination of sinusoidal oscillations of frequencies $1, 2, 3, \ldots$ (The constant term $a_0/2$ in (8-240) corresponds to a shift in the center about which the oscillation occurs; we can write it as $(a_0/2) \cos 0x$ and assign it the frequency $k = 0$.)

The vibrations of a point on a violin string illustrate the physical situation. When the string is being played, a complicated motion occurs, but this can be shown to be a combination of motions of frequencies 1 (fundamental tone), 2 (octave or first overtone), 3 (second overtone), and so on; here the frequencies refer to the number of cycles in an appropriate time interval (1/440 sec for the open violin A string).

Now let a series (8-240) converge[1] for all x to $f(x)$:

---

[1] In (8-242) we have grouped terms, for convenience. However, the series should be thought of as in (8-240).

$$f(x) = \frac{a_0}{2} + \sum_{n=1}^{\infty} (a_n \cos nx + b_n \sin nx) \qquad (8\text{-}242)$$

We then ask: how are the coefficients $a_0, a_1, b_1, \ldots$ related to f? In the case of power series, the analogous question was answered by differentiating repeatedly. Here our answer is found by integrating.

We multiply both sides of (8-242) by $\cos mx$ $(m = 1, 2, \ldots)$ and integrate term-by-term from $-\pi$ to π. These steps can be justified under quite general conditions. We here assume the steps to be valid and proceed formally:

$$\int_{-\pi}^{\pi} f(x) \cos mx \, dx = \frac{a_0}{2} \int_{-\pi}^{\pi} \cos mx \, dx + a_1 \int_{-\pi}^{\pi} \cos mx \cos x \, dx$$

$$+ b_1 \int_{-\pi}^{\pi} \cos mx \sin x \, dx + \cdots$$

The first term on the right is 0. For the second we use Identity 43 of Appendix V:

$$\int_{-\pi}^{\pi} \cos mx \cos x \, dx = \int_{-\pi}^{\pi} \frac{1}{2} [\cos(m+1)x + \cos(m-1)x] \, dx$$

$$= \frac{1}{2} \left[\frac{\sin(m+1)x}{m+1} + \frac{\sin(m-1)x}{m-1} \right]_{-\pi}^{\pi} = 0$$

provided that $m \neq 1$. For $m = 1$, the value is found in the same way to be π. Similarly, we find that

$$\int_{-\pi}^{\pi} \cos mx \cos nx \, dx = \begin{cases} 0, & m \neq n \\ \pi, & m = n \end{cases}$$

$$\int_{-\pi}^{\pi} \cos mx \sin nx \, dx = 0, \qquad \text{for all } m, n$$

$$\int_{-\pi}^{\pi} \sin mx \sin nx \, dx = \begin{cases} 0, & m \neq n \\ \pi, & m = n \end{cases}$$

Hence, our term-by-term integration of the series gives zeros for all terms on the right except for one term, and we conclude:

$$\int_{-\pi}^{\pi} f(x) \cos mx \, dx = a_m \pi, \qquad m = 1, 2, \ldots$$

The same rule is verified to hold true for $m = 0$, so that

$$a_n = \frac{1}{\pi} \int_{-\pi}^{\pi} f(x) \cos nx \, dx, \qquad n = 0, 1, 2, \ldots \qquad (8\text{-}243)$$

Similarly, we find that

$$b_n = \frac{1}{\pi} \int_{-\pi}^{\pi} f(x) \sin nx \, dx, \qquad n = 1, 2, \ldots \qquad (8\text{-}243')$$

These are the basic formulas for the theory of trigonometric series.

We now make our fundamental definition. Let f be a function for which all the integrals in (8-243), (8-243') have meaning. Then $a_0, a_1, b_1, \ldots$ are called the *Fourier coefficients of f*, and the series (8-240) formed with these coefficients is called the *Fourier series of f*:

The *Fourier series of f* is the series

$$\frac{a_0}{2} + a_1 \cos x + b_1 \sin x + \cdots + a_n \cos nx + b_n \sin nx + \cdots$$

where

$$a_n = \frac{1}{\pi} \int_{-\pi}^{\pi} f(x) \cos nx \, dx, \qquad b_n = \frac{1}{\pi} \int_{-\pi}^{\pi} f(x) \sin nx \, dx$$

If f is continuous for all x and has period 2π, then the Fourier series of f can certainly be formed, and we can ask whether the series converges to f. Mere continuity of f is not enough to insure the conclusion. Among the many sufficient conditions we mention the following, which is adequate for most applications:

THEOREM 37. *Let the function f be defined for all x and have period 2π. In the interval $[-\pi, \pi]$ let there be values*

$$x_0 = -\pi < x_1 < x_2 < \cdots < x_k = \pi$$

such that f is continuous and monotone nondecreasing or nonincreasing in each of the open intervals (x_0, x_1), (x_1, x_2), ..., with limits to the left and right at the end points. Then the Fourier series of f converges to f at each point of continuity of f; at each discontinuity x the series converges to the average of the left and right limits; that is, to the value:

$$\frac{1}{2}[\lim_{h \to 0-} f(x + h) + \lim_{h \to 0+} f(x + h)] \qquad (8\text{-}244)$$

The assumptions are illustrated in Figure 8-11. Here there are discontinuities at $x_0 = -\pi$, x_1, x_2, $x_4 = \pi$ and at the values differing from them by

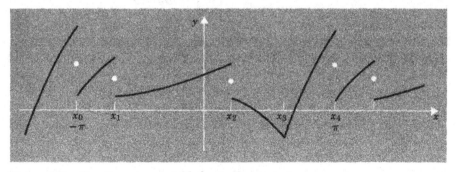

Figure 8-11   Function representable by its Fourier series.

multiples of 2π. At x_3 there is no discontinuity; this value serves merely to separate an interval of increase from one of decrease. At each discontinuity point the series converges to the average value (8-244), shown by a white dot. We can describe the function roughly as a piecewise continuous function of period 2π with a finite number of relative minima and maxima in each finite interval. A proof of the theorem requires advanced theory; see, for example, *Trigonometrical Series* by A. Zygmund (Cambridge University Press, 1959) Vol. I, p. 57.

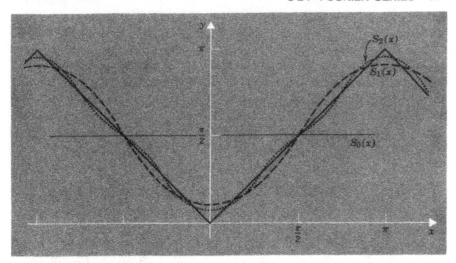

Figure 8-12 Triangular wave.

EXAMPLE 1. f has period 2π and $f(x) = |x|$ for $-\pi \le x \le \pi$. The graph, shown in Figure 8-12, is a "triangular wave". The function clearly satisfies the hypotheses of Theorem 37 (with $x_0 = -\pi, x_1 = 0, x_2 = \pi$). We find that

$$a_0 = \frac{1}{\pi} \int_{-\pi}^{\pi} |x|\, dx = \frac{2}{\pi} \int_0^{\pi} x\, dx = \pi$$

$$a_n = \frac{1}{\pi} \int_{-\pi}^{\pi} |x| \cos nx\, dx = \frac{2}{\pi} \int_0^{\pi} x \cos nx\, dx$$

$$= \frac{2}{\pi} \left[\frac{x}{n} \sin nx \Big|_0^{\pi} - \frac{1}{n} \int_0^{\pi} \sin nx\, dx \right]$$

$$= \frac{2}{\pi n^2} \cos nx \Big|_0^{\pi} = \begin{cases} -\dfrac{4}{\pi n^2}, & \text{for } n \text{ odd} \\ 0, & \text{for } n \text{ even} \end{cases}$$

$$\pi b_n = \int_{-\pi}^{\pi} |x| \sin nx\, dx = \int_{-\pi}^{0} -x \sin nx\, dx + \int_0^{\pi} x \sin nx\, dx$$

$$= -\int_0^{\pi} x \sin nx\, dx + \int_0^{\pi} x \sin nx\, dx = 0, \qquad \text{for } n = 1, 2, \ldots$$

Therefore, for all x,

$$f(x) = \frac{\pi}{2} - \frac{4}{\pi} \sum_{n=1}^{\infty} \frac{\cos(2n-1)x}{(2n-1)^2}$$

The successive partial sums

$$S_0(x) = \frac{\pi}{2}, \quad S_1(x) = \frac{\pi}{2} - \frac{4}{\pi} \cos x, \quad S_2(x) = \frac{\pi}{2} - \frac{4}{\pi} \cos x - \frac{4}{\pi} \frac{\cos 3x}{9}, \ldots$$

must therefore converge to $f(x)$ at each x. The first three partial sums are shown in Figure 8-12.

EXAMPLE 2. Let f have period 2π and let $f(x) = -1$ for $-\pi < x < 0$,

$f(x) = 1$ for $0 \leq x \leq \pi$. The graph (Figure 8-13) is called a "square wave." Here there are jumps at 0, $\pm\pi$, $\pm2\pi$, ... The function is defined to be 1 at each of these points but by Theorem 37 the series converges to the average value (8-244), which is 0 at such points. We notice that the coefficients of the series are computed by the integrals (8-243) and (8-243'), which are unaffected by changes in the value of f at a finite numbers of points. We find that

$$a_0 = \frac{1}{\pi}\int_{-\pi}^{\pi} f(x)\,dx = \frac{1}{\pi}\int_{-\pi}^{0} -1\,dx + \frac{1}{\pi}\int_{0}^{\pi} 1\,dx = 0$$

$$a_n = \frac{1}{\pi}\int_{-\pi}^{\pi} f(x)\cos nx\,dx = \frac{1}{\pi}\int_{-\pi}^{0} -\cos nx\,dx + \frac{1}{\pi}\int_{0}^{\pi} \cos nx\,dx = 0$$

$$b_n = \frac{1}{\pi}\int_{-\pi}^{\pi} f(x)\sin nx\,dx = \frac{1}{\pi}\int_{-\pi}^{0} -\sin nx\,dx + \frac{1}{\pi}\int_{0}^{\pi} \sin nx\,dx$$

$$= \frac{2}{\pi}\int_{0}^{\pi} \sin nx\,dx = \begin{cases} 0, & \text{for } n \text{ even} \\ \dfrac{4}{\pi n}, & \text{for } n \text{ odd} \end{cases}$$

Thus, except at the discontinuity points,

$$f(x) = \frac{4}{\pi}\sum_{n=0}^{\infty} \frac{\sin(2n+1)x}{(2n+1)}$$

At the discontinuity points the series clearly converges to 0. Figure 8-13 shows the first few partial sums of the Fourier series.

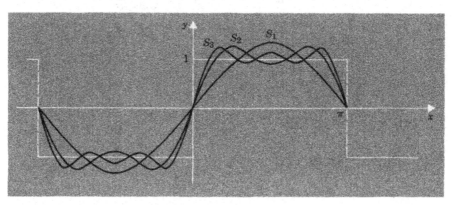

Figure 8-13   Square wave.

Remark. The formulas (8-243) and (8-243') use only the values of f in the interval $[-\pi, \pi]$. For a function f given only in this interval, the Fourier series of f is defined as above. If the series converges (to f) for $-\pi \leq x \leq \pi$, then it also converges outside this interval, since each term of the series has period 2π. The sum of the series is the unique periodic function, of period 2π, coinciding with f for $-\pi \leq x \leq \pi$. We call this function the *periodic extension* of f. This process is involved in Examples 1 and 2; in both cases we are given the values of f for $-\pi \leq x \leq \pi$, are told to repeat periodically outside this interval.

PROBLEMS

1. In each case verify that Theorem 37 is applicable, and represent each function by its Fourier series. Graph the function and the first three partial sums of the series.
   (a) f has period 2π, $f(x) = x^2$ for $-\pi \leq x \leq \pi$.
   (b) f has period 2π, $f(x) = 0$ for $-\pi < x < 0$, $f(x) = 1$ for $0 < x < \pi$, $f(-\pi) = f(0) = f(\pi) = \frac{1}{2}$.
   (c) $f(x) = \sin^2 x$, all x.
   (d) $f(x) = |\sin x|$, all x.
2. Each of the following functions f is defined on the interval $[-\pi, \pi]$ by the expression given. Find the Fourier series of f, state what its sum is, and graph the sum for $-3\pi \leq x \leq 3\pi$.
   (a) $f(x) = x$ (b) $f(x) = 3x - 5$ (c) $f(x) = \cos(x/2)$ (d) $f(x) = x + |x|$
   (e) $f(x) = 0, -\pi \leq x < -\pi/2; f(x) = 1, -\pi/2 \leq x \leq \pi/2; f(x) = 2, \pi/2 < x \leq \pi$
   (f) $f(x) = e^x$
3. Let f be defined and piecewise continuous on the interval $[-\pi, \pi]$.
   (a) Show that, if f is even $[f(x) = f(-x)]$, then $b_n = 0$ for all n and
   $$a_n = \frac{2}{\pi} \int_0^\pi f(x) \cos nx \, dx$$
   (b) Show that, if f is odd $[f(-x) = -f(x)]$, then $a_n = 0$ for all n and
   $$b_n = \frac{2}{\pi} \int_0^\pi f(x) \sin nx \, dx$$
4. (a) Let $y = g(x)$ be given only in the interval $[0, \pi]$ and let g be continuous and monotone in successive intervals $[0, x_1], [x_1, x_2], \ldots, [x_{n-1}, \pi]$. Show that there is a unique even function f in $[-\pi, \pi]$ such that $f(x) = g(x)$ in $[0, \pi]$. Show, with the aid of the results of Problem 3, that
   $$g(x) = \frac{a_0}{2} + \sum_{n=1}^{\infty} a_n \cos nx, \qquad a_n = \frac{2}{\pi} \int_0^\pi g(x) \cos nx \, dx$$
   The series is called the *Fourier cosine series of g.*
   (b) Represent the function $g(x) = 2x - 1$, $0 \leq x \leq \pi$, by its Fourier cosine series and graph the sum of the series for $-3\pi \leq x \leq 3\pi$.
5. (a) Let g be as in Problem 4 (a) with $g(0) = 0$. Show that there is a unique odd function f in $[-\pi, \pi]$ so that $f(x) = g(x)$ for $0 \leq x \leq \pi$. Show with the aid of the results of Problem 3 that
   $$g(x) = \sum_{n=1}^{\infty} b_n \sin nx, \qquad b_n = \frac{2}{\pi} \int_0^\pi g(x) \sin nx \, dx$$
   The series is called the *Fourier sine series of g.*
   (b) Represent the function $g(x) = x^2, 0 \leq x \leq \pi$, by its Fourier sine series and graph the sum of the series for $-3\pi \leq x \leq 3\pi$.
6. (a) Let the function F be defined for all x and have period $\tau > 0$; let F be continuous and monotone in each of the intervals $(0, x_1), (x_1, x_2), \ldots, (x_{n-1}, \tau)$. Let $\omega = 2\pi/\tau$.

Let F have limits to the left and right at each discontinuity. Show that the function f such that

$$f(x) = F\left(\frac{\tau x}{2\pi}\right) = F\left(\frac{x}{\omega}\right)$$

has period 2π and hence, at each point of continuity of F,

$$F(x) = f(\omega x) = \frac{a_0}{2} + \sum_{n=1}^{\infty} (a_n \cos n\omega x + b_n \sin n\omega x)$$

where a_n, b_n are given by (8-243), (8-243'). (We call this series the *Fourier series of F, for period* τ.)

(b) Show that a_n, b_n in part (a) can be evaluated as follows:

$$a_n = \frac{2}{\tau} \int_{-\tau/2}^{\tau/2} F(x) \cos \omega n x\, dx, \qquad n = 0, 1, 2, \ldots$$

$$b_n = \frac{2}{\tau} \int_{-\tau/2}^{\tau/2} F(x) \sin \omega n x\, dx, \qquad n = 1, 2, \ldots$$

(c) Represent F by its Fourier series of period 2 if $F(x) = x^2 - 1$ for $-1 \leq x \leq 1$ and F has period 2.

APPENDIXES

I

TABLE OF INDEFINITE INTEGRALS

(In all formulas, we omit the arbitrary constant; a, b, c, α denote real numbers, and m, n, p, q denote positive integers. When a^2 appears in the integrand, a is to be taken as a positive number. Throughout, $\ln(\ )$ can be replaced by $\ln|\ \ |$.)

1. $\int af(x)dx = a\int f(x)\,dx$

2. $\int \{f(x) + g(x)\}\,dx = \int f(x)\,dx + \int g(x)\,dx$

3. $\int f(x)g'(x)\,dx = f(x)g(x) - \int g(x)f'(x)\,dx$

4. $\int u\,dv = uv - \int v\,du$

5. $\int x^\alpha\,dx = \dfrac{x^{\alpha+1}}{\alpha+1}$ $(\alpha \neq -1)$

6. $\int \dfrac{dx}{x} = \ln|x|$

7. $\int \dfrac{f'(x)}{f(x)}\,dx = \ln|f(x)|$

8. $\int e^{ax}\,dx = \dfrac{e^{ax}}{a}$ $(a \neq 0)$

9. $\int a^x\,dx = \dfrac{a^x}{\ln a}$ $(a \neq 1)$

10. $\int \ln x\,dx = x\ln x - x$

11. $\int \log_a x\,dx = \dfrac{1}{\ln a}[x\ln x - x] = x\log_a x - \dfrac{x}{\ln a}$ $(a \neq 1)$

12. $\int \dfrac{dx}{x^2 + a^2} = \dfrac{1}{a}\,\mathrm{Tan}^{-1}\left(\dfrac{x}{a}\right) = -\dfrac{1}{a}\,\mathrm{Cot}^{-1}\left(\dfrac{x}{a}\right)$

13. $\int \dfrac{dx}{x^2 - a^2} = \dfrac{-1}{a}\tanh^{-1}\left(\dfrac{x}{a}\right) = \dfrac{1}{2a}\ln\left|\dfrac{x-a}{x+a}\right|$

14. $\displaystyle\int \frac{dx}{a + bx^2} = \begin{cases} \dfrac{1}{\sqrt{ab}}\,\mathrm{Tan}^{-1}\dfrac{x\sqrt{ab}}{a}, & ab > 0 \\[3mm] \dfrac{1}{2\sqrt{-ab}}\,\ln\dfrac{\sqrt{-bx} + \sqrt{a}}{\sqrt{-bx} - \sqrt{a}}, & a > 0,\, b < 0 \end{cases}$

15. $\displaystyle\int \frac{x\,dx}{a + bx^2} = \frac{1}{2b}\ln\left(\frac{a + bx^2}{b}\right)$

16. $\displaystyle\int \frac{dx}{(a + bx^2)^m} = \frac{1}{2a(m - 1)}\frac{x}{(a + bx^2)^{m-1}}$

$$+ \frac{2m - 3}{2(m - 1)a}\int \frac{dx}{(a + bx^2)^{m-1}}, \qquad (m > 1)$$

17. $\displaystyle\int \frac{x\,dx}{(a + bx^2)^m} = \frac{-1}{2b(m - 1)(a + bx^2)^{m-1}}, \qquad (m > 1)$

18. $\displaystyle\int \frac{dx}{\sqrt{a^2 - x^2}} = \mathrm{Sin}^{-1}\left(\frac{x}{a}\right) = -\mathrm{Cos}^{-1}\left(\frac{x}{a}\right)$

19. $\displaystyle\int \frac{dx}{\sqrt{x^2 \pm a^2}} = \ln(x + \sqrt{x^2 \pm a^2})$

20. $\displaystyle\int \frac{dx}{x\sqrt{x^2 - a^2}} = \frac{1}{a}\,\mathrm{Cos}^{-1}\left(\frac{a}{x}\right)$

21. $\displaystyle\int \frac{dx}{x\sqrt{a^2 \pm x^2}} = -\frac{1}{a}\ln\left(\frac{a + \sqrt{a^2 + x^2}}{x}\right)$

22. $\displaystyle\int \sqrt{x^2 \pm a^2}\,dx = \frac{1}{2}\left[x\sqrt{x^2 \pm a^2} \pm a^2\ln(x + \sqrt{x^2 \pm a^2})\right]$

23. $\displaystyle\int \frac{\sqrt{x^2 + a^2}}{x}\,dx = \sqrt{x^2 + a^2} - a\ln\frac{a + \sqrt{x^2 + a^2}}{x}$

24. $\displaystyle\int \frac{\sqrt{x^2 - a^2}}{x}\,dx = \sqrt{x^2 - a^2} - a\,\mathrm{Cos}^{-1}\frac{a}{x}$

25. $\displaystyle\int \frac{x\,dx}{\sqrt{x^2 \pm a^2}} = \sqrt{x^2 \pm a^2}$

26. $\displaystyle\int x\sqrt{x^2 \pm a^2} = \frac{1}{3}(x^2 \pm a^2)^{3/2}$

27. $\displaystyle\int (x^2 \pm a^2)^{3/2}\,dx =$

$$\frac{1}{8}\left[2x(x^2 \pm a^2)^{3/2} \pm 3a^2x\sqrt{x^2 \pm a^2} + 3a^4\ln(x + \sqrt{x^2 \pm a^2})\right]$$

28. $\displaystyle\int \frac{dx}{(x^2 \pm a^2)^{3/2}} = \frac{\pm x}{a^2\sqrt{x^2 \pm a^2}}$

29. $\displaystyle\int x^2\sqrt{x^2 \pm a^2}\,dx$

$$= \frac{x}{4}(x^2 \pm a^2)^{3/2} \mp \frac{a^2x}{8}\sqrt{x^2 \pm a^2} - \frac{a^4}{8}\ln(x + \sqrt{x^2 \pm a^2})$$

30. $\displaystyle\int \sqrt{a^2 - x^2}\,dx = \frac{1}{2}\left[x\sqrt{a^2 - x^2} + a^2\,\mathrm{Sin}^{-1}\frac{x}{a}\right]$

31. $\int \dfrac{\sqrt{a^2 - x^2}}{x} \, dx = \sqrt{a^2 - x^2} - a \ln \dfrac{a + \sqrt{a^2 - x^2}}{x}$

32. $\int \dfrac{dx}{x\sqrt{a^2 - x^2}} = -\dfrac{1}{a} \ln \dfrac{a + \sqrt{a^2 - x^2}}{x}$

33. $\int \dfrac{x \, dx}{\sqrt{a^2 - x^2}} = -\sqrt{a^2 - x^2}$

34. $\int f(x, \sqrt{x^2 + a^2}) \, dx = a \int f(a \tan u, a \sec u) \sec^2 u \, du, \qquad (x = a \tan u)$

35. $\int f(x, \sqrt{x^2 - a^2}) \, dx = a \int f(a \sec u, a \tan u) \sec u \tan u \, du, \qquad (x = a \sec u)$

36. $\int f(x, \sqrt{a^2 - x^2}) \, dx = -a \int f(a \cos u, a \sin u) \sin u \, du, \qquad (x = a \cos u)$

37. $\int f(x, X) \, dx = \dfrac{1}{a} \int f\left(\dfrac{y - b}{a}, \dfrac{y^2 + d}{a}\right) dy \qquad \left(\begin{matrix} x = (y - b)/a, & d = ac - b^2 \\ X = ax^2 + 2bx + c \end{matrix} \right)$

38. $\int x^m (a + bx)^n \, dx = \dfrac{x^{m+1}(a + bx)^n}{m + n + 1} + \dfrac{an}{m + n + 1} \int x^m (a + bx)^{n-1} \, dx$

39. $\int \dfrac{dx}{a + bx} = \dfrac{1}{b} \ln(a + bx)$

40. $\int \dfrac{x \, dx}{a + bx} = \dfrac{1}{b^2} [a + bx - a \ln(a + bx)]$

41. $\int \dfrac{x \, dx}{(a + bx)^2} = \dfrac{1}{b^2} \left[\ln(a + bx) + \dfrac{a}{a + bx} \right]$

42. $\int \dfrac{x \, dx}{(a + bx)^m} = \dfrac{1}{b^2} \left[\dfrac{-1}{(m - 2)(a + bx)^{m-2}} + \dfrac{a}{(m - 1)(a + bx)^{m-1}} \right] \qquad (m \geq 3)$

43. $\int \sqrt{a + bx} \, dx = \dfrac{2}{3b} \sqrt{(a + bx)^3}$

44. $\int x^m \sqrt{a + bx} \, dx = \dfrac{2}{(2m + 3)b} \left[x^m \sqrt{(a + bx)^3} - ma \int x^{m-1} \sqrt{a + bx} \, dx \right]$

45. $\int \dfrac{dx}{x^m \sqrt{a + bx}} = \dfrac{-\sqrt{a + bx}}{a(m - 1)x^{m-1}} - \dfrac{(2m - 3)b}{(2m - 2)a} \int \dfrac{dx}{x^{m-1}\sqrt{a + bx}} \qquad (m \neq 1)$

46. $\int f(x, \sqrt{a + bx}) \, dx = \dfrac{2}{b} \int f\left(\dfrac{z^2 - a}{b}, z\right) z \, dz \qquad (z^2 = a + bx)$

47. $\int \dfrac{dx}{a^3 + x^3} = \dfrac{1}{3a^2} \left\{ \dfrac{1}{2} \ln \left[\dfrac{(a + x)^2}{a^2 - ax + x^2} \right] + \sqrt{3} \operatorname{Tan}^{-1} \dfrac{2x - a}{a\sqrt{3}} \right\}$

48. $\int \sin x \, dx = -\cos x$

49. $\int \cos x \, dx = \sin x$

50. $\int \tan x \, dx = -\ln \cos x$

51. $\int \cot x \, dx = \ln \sin x$

52. $\displaystyle\int \sec x \, dx = \int \frac{dx}{\cos x} = \ln(\sec x + \tan x) = \ln\left\{\tan\left(\frac{x}{2} + \frac{\pi}{4}\right)\right\}$

53. $\displaystyle\int \csc x \, dx = \int \frac{dx}{\sin x} = \ln(\csc x - \cot x) = \ln\left(\tan \frac{x}{2}\right)$

54. $\displaystyle\int \sin^2 x \, dx = \frac{1}{2}x - \frac{1}{4}\sin 2x = \frac{1}{2}[x - \cos x \sin x]$

55. $\displaystyle\int \sin^m x \, dx = \frac{-\cos x \sin^{m-1} x}{m} + \frac{m-1}{m}\int \sin^{m-2} x \, dx$

56. $\displaystyle\int \cos^2 x \, dx = \frac{1}{2}x + \frac{1}{4}\sin 2x$

57. $\displaystyle\int \cos^m x \, dx = \frac{\sin x \cos^{m-1} x}{m} + \frac{m-1}{m}\int \cos^{m-2} x \, dx$

58. $\displaystyle\int \frac{dx}{\cos^2 x} = \int \sec^2 x \, dx = \tan x$

59. $\displaystyle\int \frac{dx}{\cos^m x} = \frac{\sin x}{(m-1)\cos^{m-1} x} + \frac{m-2}{m-1}\int \frac{dx}{\cos^{m-2} x} \qquad (m > 1)$

60. $\displaystyle\int \frac{dx}{\sin^2 x} = \int \csc^2 x \, dx = -\cot x$

61. $\displaystyle\int \frac{dx}{\sin^m x} = \frac{-\cos x}{(m-1)\sin^{m-1} x} + \frac{m-2}{m-1}\int \frac{dx}{\sin^{m-2} x} \qquad (m > 1)$

62. $\displaystyle\int \frac{dx}{1 \pm \sin x} = \mp \tan\left(\frac{\pi}{4} \mp \frac{x}{2}\right)$

63. $\displaystyle\int \frac{dx}{1 + \cos x} = \tan\left(\frac{x}{2}\right)$

64. $\displaystyle\int \frac{dx}{1 - \cos x} = -\cot\left(\frac{x}{2}\right)$

65. $\displaystyle\int \frac{dx}{a + b \sin x} = \begin{cases} \dfrac{1}{\sqrt{b^2 - a^2}} \ln\left(\dfrac{a \tan \frac{x}{2} + b - \sqrt{b^2 - a^2}}{a \tan \frac{x}{2} + b + \sqrt{b^2 - a^2}}\right) & (b^2 > a^2) \\[3em] \dfrac{2}{\sqrt{a^2 - b^2}} \, \mathrm{Tan}^{-1} \dfrac{a \tan \frac{x}{2} + b}{\sqrt{a^2 - b^2}} & (a^2 > b^2) \end{cases}$

66. $\displaystyle\int \frac{dx}{a + b \cos x} = \begin{cases} \dfrac{1}{\sqrt{b^2 - a^2}} \ln\left(\dfrac{\sqrt{b^2 - a^2} \tan \frac{x}{2} + a + b}{\sqrt{b^2 - a^2} \tan \frac{x}{2} - a - b}\right) & (b^2 > a^2) \\[3em] \dfrac{2}{\sqrt{a^2 - b^2}} \, \mathrm{Tan}^{-1} \dfrac{\sqrt{a^2 - b^2} \tan \frac{x}{2}}{a + b} & (a^2 > b^2) \end{cases}$

67. $\displaystyle\int \sin nx \sin mx \, dx = \frac{\sin(m - n)x}{2(m - n)} - \frac{\sin(m + n)x}{2(m + n)} \qquad (m^2 \neq n^2)$

68. $\displaystyle\int \sin nx \cos mx \, dx = \frac{\cos(m - n)x}{2(m - n)} - \frac{\cos(m + n)x}{2(m + n)} \qquad (m^2 \neq n^2)$

69. $\displaystyle\int \cos nx \cos mx \, dx = \frac{\sin(m-n)x}{2(m-n)} + \frac{\sin(m+n)x}{2(m+n)}$ $(m^2 \neq n^2)$

70. $\displaystyle\int \tan^n x \, dx = \frac{\tan^{n-1} x}{n-1} - \int \tan^{n-2} x \, dx$

71. $\displaystyle\int \frac{dx}{\sin x \cos x} = \ln \tan x$

72. $\displaystyle\int \frac{dx}{\sin x \cos^m x} = \frac{1}{(m-1)\cos^{m-1} x} + \int \frac{dx}{\sin x \cos^{m-2} x}$ $(m > 1)$

73. $\displaystyle\int x^m \sin x \, dx = -x^m \cos x + m \int x^{m-1} \cos x \, dx$

74. $\displaystyle\int x^m \cos x \, dx = x^m \sin x - m \int x^{m-1} \sin x \, dx$

75. $\displaystyle\int \mathrm{Sin}^{-1} x \, dx = x \, \mathrm{Sin}^{-1} x + \sqrt{1 - x^2}$

76. $\displaystyle\int \mathrm{Cos}^{-1} x \, dx = x \, \mathrm{Cos}^{-1} x - \sqrt{1 - x^2}$

77. $\displaystyle\int \mathrm{Tan}^{-1} x \, dx = x \, \mathrm{Tan}^{-1} x - \frac{1}{2} \ln (1 + x^2)$

78. $\displaystyle\int \mathrm{Cot}^{-1} x \, dx = x \, \mathrm{Cot}^{-1} x + \frac{1}{2} \ln(1 + x^2)$

79. $\displaystyle\int (\mathrm{Sin}^{-1} x)^2 \, dx = x(\mathrm{Sin}^{-1} x)^2 - 2x + 2 \sqrt{1 - x^2} \, \mathrm{Sin}^{-1} x$

80. $\displaystyle\int (\mathrm{Cos}^{-1} x)^2 \, dx = x(\mathrm{Cos}^{-1} x)^2 - 2x - 2 \sqrt{1 - x^2} \, \mathrm{Cos}^{-1} x$

81. $\displaystyle\int x^n \mathrm{Sin}^{-1} x \, dx = \frac{x^{n+1} \mathrm{Sin}^{-1} x}{n+1} - \frac{1}{n+1} \int \frac{x^{n+1} \, dx}{\sqrt{1 - x^2}}$

82. $\displaystyle\int x^n \mathrm{Cos}^{-1} x \, dx = \frac{x^{n+1} \mathrm{Cos}^{-1} x}{n+1} + \frac{1}{n+1} \int \frac{x^{n+1} \, dx}{\sqrt{1 - x^2}}$

83. $\displaystyle\int x \ln x \, dx = \frac{x^2}{2} \ln x - \frac{x^2}{4}$

84. $\displaystyle\int x^m \ln x \, dx = \frac{x^{m+1} \ln x}{m+1} - \frac{x^{m+1}}{(m+1)^2}$ $(m \neq -1)$

85. $\displaystyle\int (\ln x)^q \, dx = x(\ln x)^q - q \int (\ln x)^{q-1} \, dx$

86. $\displaystyle\int \frac{(\ln x)^q}{x} \, dx = \frac{(\ln x)^{q+1}}{q+1}$

87. $\displaystyle\int \frac{dx}{x \ln x} = \ln \ln x$

88. $\displaystyle\int x^m (\ln x)^q \, dx = \frac{x^{m+1}(\ln x)^q}{m+1} - \frac{q}{m+1} \int x^m (\ln x)^{q-1} \, dx$ $(m \neq -1)$

89. $\displaystyle\int \sin \ln x \, dx = \frac{1}{2} x \sin \ln x - \frac{1}{2} x \cos \ln x$

90. $\displaystyle\int \cos \ln x \, dx = \frac{1}{2} x \sin \ln x + \frac{1}{2} x \cos \ln x$

91. $\int xe^{ax}\,dx = \dfrac{e^{ax}}{a^2}(ax - 1)$

92. $\int x^m e^{ax}\,dx = \dfrac{x^m e^{ax}}{a} - \dfrac{m}{a}\int x^{m-1}e^{ax}\,dx \qquad (m > 0)$

93. $\int \dfrac{e^{ax}}{x^m}\,dx = -\dfrac{e^{ax}}{(m-1)x^{m-1}} + \dfrac{a}{m-1}\int \dfrac{e^{ax}}{x^{m-1}}\,dx \qquad (m > 1)$

94. $\int e^{ax}\ln x\,dx = \dfrac{e^{ax}\ln x}{a} - \dfrac{1}{a}\int \dfrac{e^{ax}}{x}\,dx$

95. $\int e^{ax}\sin nx\,dx = \dfrac{e^{ax}(a\sin nx - n\cos nx)}{a^2 + n^2}$

96. $\int e^{ax}\cos nx\,dx = \dfrac{e^{ax}(a\cos nx + n\sin nx)}{x^2 + n^2}$

97. $\int \dfrac{dx}{a + be^{qx}} = \dfrac{x}{a} - \dfrac{1}{aq}\ln(a + be^{qx})$

98. $\int \sinh x\,dx = \cosh x$

99. $\int \cosh x\,dx = \sinh x$

100. $\int \tanh x\,dx = \ln\cosh x$

101. $\int \coth x\,dx = \ln\sinh x$

102. $\int \operatorname{sech} x\,dx = 2\,\mathrm{Tan}^{-1}e^x = \mathrm{Tan}^{-1}(\sinh x)$

103. $\int \operatorname{csch} x\,dx = \ln\tanh\dfrac{x}{2}$

104. $\displaystyle\int f(\sin x)\,dx = 2\int f\!\left(\dfrac{2z}{1+z^2}\right)\dfrac{dz}{1+z^2} \qquad \left(z = \tan\dfrac{x}{2}\right)$

$\displaystyle\qquad\qquad = \int f(u)\,\dfrac{du}{\sqrt{1-u^2}} \qquad (u = \sin x)$

105. $\displaystyle\int f(\cos x)\,dx = 2\int f\!\left(\dfrac{1-z^2}{1+z^2}\right)\dfrac{dz}{1+z^2} \qquad \left(z = \tan\dfrac{x}{2}\right)$

$\displaystyle\qquad\qquad = -\int f(u)\,\dfrac{du}{\sqrt{1-u^2}} \qquad (u = \cos x)$

106. $\displaystyle\int f(\sin x,\ \cos x)\,dx = 2\int f\!\left(\dfrac{2z}{1+z^2},\dfrac{1-z^2}{1+z^2}\right)\dfrac{dz}{1+z^2} \qquad \left(z = \tan\dfrac{x}{2}\right)$

$\displaystyle\qquad\qquad = \int f(u,\ \sqrt{1-u^2})\,\dfrac{du}{\sqrt{1-u^2}} \qquad (u = \sin x)$

107. $\displaystyle\int_0^{\pi/2}\sin^n x\,dx = \int_0^{\pi/2}\cos^n x\,dx = \begin{cases} \dfrac{1}{2}\cdot\dfrac{3}{4}\ \cdots\ \dfrac{n-1}{n}\cdot\dfrac{\pi}{2} & (n = 2, 4, 6, \ldots) \\[2mm] \dfrac{2}{3}\cdot\dfrac{4}{5}\ \cdots\ \dfrac{n-1}{n} & (n = 3, 5, 7, \ldots) \end{cases}$

108. $\displaystyle\int_0^{\pi/2} \sin^m x \cos^n x \, dx =$

$$\left\{\begin{array}{l} \dfrac{1\cdot 3 \, \cdots \, (m-1)\cdot 1\cdot 3 \, \cdots \, (n-1)}{2\cdot 4 \, \cdots \, (m+n)}\cdot\dfrac{\pi}{2} \qquad (m = 2, 4, \ldots, n = 2, 4, \ldots) \\[4mm] \dfrac{2\cdot 4 \, \cdots \, (m-1)}{(n+1)(n+3) \, \cdots \, (n+m)} \qquad (m = 3, 5, 7, \ldots, n = 1, 2, 3, \ldots) \\[4mm] \dfrac{2\cdot 4 \, \cdots \, (n-1)}{(m+1)(m+3) \, \cdots \, (m+n)} \qquad (m = 1, 2, 3, \ldots, m = 3, 5, 7, \ldots) \end{array}\right.$$

II

NATURAL TRIGONOMETRIC FUNCTIONS FOR ANGLES IN RADIANS[a]

[a] From Mathematical Tables, *CRC Handbook of Chemistry and Physics*. Reprinted by permission of the Chemical Rubber Publishing Company.

Rad.	Sin	Tan	Cot	Cos	Rad.	Sin	Tan	Cot	Cos
.00	.00000	.00000	∞	1.0000	.50	.47943	.54630	1.8305	.87758
.01	.01000	.01000	99.997	.99995	.51	.48818	.55936	1.7878	.87274
.02	.02000	.02000	49.993	.99980	.52	.49688	.57256	1.7465	.86782
.03	.03000	.03001	33.323	.99955	.53	.50553	.58592	1.7067	.86281
.04	.03999	.04002	24.987	.99920	.54	.51414	.59943	1.6683	.85771
.05	.04998	.05004	19.983	.99875	.55	.52269	.61311	1.6310	.85252
.06	.05996	.06007	16.647	.99820	.56	.53119	.62695	1.5950	.84726
.07	.06994	.07011	14.262	.99755	.57	.53963	.64097	1.5601	.84190
.08	.07991	.08017	12.473	.99680	.58	.54802	.65517	1.5263	.83646
.09	.08988	.09024	11.081	.99595	.59	.55636	.66956	1.4935	.83094
.10	.09983	.10033	9.9666	.99500	.60	.56464	.68414	1.4617	.82534
.11	.10978	.11045	9.0542	.99396	.61	.57287	.69892	1.4308	.81965
.12	.11971	.12058	8.2933	.99281	.62	.58104	.71391	1.4007	.81388
.13	.12963	.13074	7.6489	.99156	.63	.58914	.72911	1.3715	.80803
.14	.13954	.14092	7.0961	.99022	.64	.59720	.74454	1.3431	.80210
.15	.14944	.15114	6.6166	.98877	.65	.60519	.76020	1.3154	.79608
.16	.15932	.16138	6.1966	.98723	.66	.61312	.77610	1.2885	.78999
.17	.16918	.17166	5.8256	.98558	.67	.62099	.79225	1.2622	.78382
.18	.17903	.18197	5.4954	.98384	.68	.62879	.80866	1.2366	.77757
.19	.18886	.19232	5.1997	.98200	.69	.63654	.82534	1.2116	.77125
.20	.19867	.20271	4.9332	.98007	.70	.64422	.84229	1.1872	.76484
.21	.20846	.21314	4.6917	.97803	.71	.65183	.85953	1.1634	.75836
.22	.21823	.22362	4.4719	.97590	.72	.65938	.87707	1.1402	.75181
.23	.22798	.23414	4.2709	.97367	.73	.66687	.89492	1.1174	.74517
.24	.23770	.24472	4.0864	.97134	.74	.67429	.91309	1.0952	.73847
.25	.24740	.25534	3.9163	.96891	.75	.68164	.93160	1.0734	.73169
.26	.25708	.26602	3.7591	.96639	.76	.68892	.95045	1.0521	.72484
.27	.26673	.27676	3.6133	.96377	.77	.69614	.96967	1.0313	.71791
.28	.27636	.28755	3.4776	.96106	.78	.70328	.98926	1.0109	.71091
.29	.28595	.29841	3.3511	.95824	.79	.71035	1.0092	.99084	.70385
.30	.29552	.30934	3.2327	.95534	.80	.71736	1.0296	.97121	.69671
.31	.30506	.32033	3.1218	.95233	.81	.72429	1.0505	.95197	.68950
.32	.31457	.33139	3.0176	.94924	.82	.73115	1.0717	.93309	.68222
.33	.32404	.34252	2.9195	.94604	.83	.73793	1.0934	.91455	.67488
.34	.33349	.35374	2.8270	.94275	.84	.74464	1.1156	.89635	.66746
.35	.34290	.36503	2.7395	.93937	.85	.75128	1.1383	.87848	.65998
.36	.35227	.37640	2.6567	.93590	.86	.75784	1.1616	.86091	.65244
.37	.36162	.38786	2.5782	.93233	.87	.76433	1.1853	.84365	.64483
.38	.37092	.39941	2.5037	.92866	.88	.77074	1.2097	.82668	.63715
.39	.38019	.41105	2.4328	.92491	.89	.77707	1.2346	.80998	.62941
.40	.38942	.42279	2.3652	.92106	.90	.78333	1.2602	.79355	.62161
.41	.39861	.43463	2.3008	.91712	.91	.78950	1.2864	.77738	.61375
.42	.40776	.44657	2.2393	.91309	.92	.79560	1.3133	.76146	.60582
.43	.41687	.45862	2.1804	.90897	.93	.80162	1.3409	.74578	.59783
.44	.42594	.47078	2.1241	.90475	.94	.80756	1.3692	.73034	.58979
.45	.43497	.48306	2.0702	.90045	.95	.81342	1.3984	.71511	.58168
.46	.44395	.49545	2.0184	.89605	.96	.81919	1.4284	.70010	.57352
.47	.45289	.50797	1.9686	.89157	.97	.82489	1.4592	.68531	.56530
.48	.46178	.52061	1.9208	.88699	.98	.83050	1.4910	.67071	.55702
.49	.47063	.53339	1.8748	.88233	.99	.83603	1.5237	.65631	.54869
.50	.47943	.54630	1.8305	.87758	1.00	.84147	1.5574	.64209	.54030

Rad.	Sin	Tan	Cot	Cos	Rad.	Sin	Tan	Cot	Cos
1.00	.84147	1.5574	.64209	.54030	1.50	.99749	14.101	.07091	.07074
1.01	.84683	1.5922	.62806	.53186	1.51	.99815	16.428	.06087	.06076
1.02	.85211	1.6281	.61420	.52337	1.52	.99871	19.670	.05084	.05077
1.03	.85730	1.6652	.60051	.51482	1.53	.99917	24.498	.04082	.04079
1.04	.86240	1.7036	.58699	.50622	1.54	.99953	32.461	.03081	.03079
1.05	.86742	1.7433	.57362	.49757	1.55	.99978	48.078	.02080	.02079
1.06	.87236	1.7844	.56040	.48887	1.56	.99994	92.621	.01080	.01080
1.07	.87720	1.8270	.54734	.48012	1.57	1.0000	1255.8	.00080	.00080
1.08	.88196	1.8712	.53441	.47133	1.58	.99996	−108.65	−.00920	−.00920
1.09	.88663	1.9171	.52162	.46249	1.59	.99982	−52.067	−.01921	−.01920
1.10	.89121	1.9648	.50897	.45360	1.60	.99957	−34.233	−.02921	−.02920
1.11	.89570	2.0143	.49644	.44466	1.61	.99923	−25.495	−.03922	−.03919
1.12	.90010	2.0660	.48404	.43568	1.62	.99879	−20.307	−.04924	−.04918
1.13	.90441	2.1198	.47175	.42666	1.63	.99825	−16.871	−.05927	−.05917
1.14	.90863	2.1759	.45959	.41759	1.64	.99761	−14.427	−.06931	−.06915
1.15	.91276	2.2345	.44753	.40849	1.65	.99687	−12.599	−.07937	−.07912
1.16	.91680	2.2958	.43558	.39934	1.66	.99602	−11.181	−.08944	−.08909
1.17	.92075	2.3600	.42373	.39015	1.67	.99508	−10.047	−.09953	−.09904
1.18	.92461	2.4273	.41199	.38092	1.68	.99404	−9.1208	−.10964	−.10899
1.19	.92837	2.4979	.40034	.37166	1.69	.99290	−8.3492	−.11977	−.11892
1.20	.93204	2.5722	.38878	.36236	1.70	.99166	−7.6966	−.12993	−.12884
1.21	.93562	2.6503	.37731	.35302	1.71	.99033	−7.1373	−.14011	−.13875
1.22	.93910	2.7328	.36593	.34360	1.72	.98889	−6.6524	−.15032	−.14865
1.23	.94249	2.8198	.35463	.33424	1.73	.98735	−6.2281	−.16056	−.15853
1.24	.94578	2.9119	.34341	.32480	1.74	.98572	−5.8535	−.17084	−.16840
1.25	.94898	3.0096	.33227	.31532	1.75	.98399	−5.5204	−.18115	−.17825
1.26	.95209	3.1133	.32121	.30582	1.76	.98215	−5.2221	−.19149	−.18808
1.27	.95510	3.2236	.31021	.29628	1.77	.98022	−4.9534	−.20188	−.19789
1.28	.95802	3.3413	.29928	.28672	1.78	.97820	−4.7101	−.21231	−.20768
1.29	.96084	3.4672	.28842	.27712	1.79	.97607	−4.4887	−.22278	−.21745
1.30	.96356	3.6021	.27762	.26750	1.80	.97385	−4.2863	−.23330	−.22720
1.31	.96618	3.7471	.26687	.25785	1.81	.97153	−4.1005	−.24387	−.23693
1.32	.96872	3.9033	.25619	.24818	1.82	.96911	−3.9294	−.25449	−.24663
1.33	.97115	4.0723	.24556	.23848	1.83	.96659	−3.7712	−.26517	−.25631
1.34	.97348	4.2556	.23498	.22875	1.84	.96398	−3.6245	−.27590	−.26596
1.35	.97572	4.4552	.22446	.21901	1.85	.96128	−3.4881	−.28669	−.27559
1.36	.97786	4.6734	.21398	.20924	1.86	.95847	−3.3608	−.29755	−.28519
1.37	.97991	4.9131	.20354	.19945	1.87	.95557	−3.2419	−.30846	−.29476
1.38	.98185	5.1774	.19315	.18964	1.88	.95258	−3.1304	−.31945	−.30430
1.39	.98370	5.4707	.18279	.17981	1.89	.94949	−3.0257	−.33051	−.31381
1.40	.98545	5.7979	.17248	.16997	1.90	.94630	−2.9271	−.34164	−.32329
1.41	.98710	6.1654	.16220	.16010	1.91	.94302	−2.8341	−.35284	−.33274
1.42	.98865	6.5811	.15195	.15023	1.92	.93965	−2.7463	−.36413	−.34215
1.43	.99010	7.0555	.14173	.14033	1.93	.93618	−2.6632	−.37549	−.35153
1.44	.99146	7.6018	.13155	.13042	1.94	.93262	−2.5843	−.38695	−.36087
1.45	.99271	8.2381	.12139	.12050	1.95	.92896	−2.5095	−.39849	−.37018
1.46	.99387	8.9886	.11125	.11057	1.96	.92521	−2.4383	−.41012	−.37945
1.47	.99492	9.8874	.10114	.10063	1.97	.92137	−2.3705	−.42185	−.38868
1.48	.99588	10.983	.09105	.09067	1.98	.91744	−2.3058	−.43368	−.39788
1.49	.99674	12.350	.03097	.08071	1.99	.91341	−2.2441	−.44502	−.40703
1.50	.99749	14.101	.07091	.07074	2.00	.90930	−2.1850	−.45766	−.41615

III
EXPONENTIAL FUNCTIONS[a]

[a] Adapted from Mathematical Tables, *CRC Handbook of Chemistry and Physics*. Reprinted by permission of the Chemical Rubber Publishing Company.

x	e^x	e^{-x}	x	e^x	e^{-x}
0.00	1.0000	1.000000	0.50	1.6487	0.606531
0.01	1.0101	0.990050	0.51	1.6653	.600496
0.02	1.0202	.980199	0.52	1.6820	.594521
0.03	1.0305	.970446	0.53	1.6989	.588605
0.04	1.0408	.960789	0.54	1.7160	.582748
0.05	1.0513	0.951229	0.55	1.7333	0.576950
0.06	1.0618	.941765	0.56	1.7507	.571209
0.07	1.0725	.932394	0.57	1.7683	.565525
0.08	1.0833	.923116	0.58	1.7860	.559898
0.09	1.0942	.913931	0.59	1.8040	.554327
0.10	1.1052	0.904837	0.60	1.8221	0.548812
0.11	1.1163	.895834	0.61	1.8404	.543351
0.12	1.1275	.886920	0.62	1.8589	.537944
0.13	1.1388	.878095	0.63	1.8776	.532592
0.14	1.1503	.869358	0.64	1.8965	.527292
0.15	1.1618	0.860708	0.65	1.9155	0.522046
0.16	1.1735	.852144	0.66	1.9348	.516851
0.17	1.1853	.843665	0.67	1.9542	.511709
0.18	1.1972	.835270	0.68	1.9739	.506617
0.19	1.2092	.826959	0.69	1.9937	.501576
0.20	1.2214	0.818731	0.70	2.0138	0.496585
0.21	1.2337	.810584	0.71	2.0340	.491644
0.22	1.2461	.802519	0.72	2.0544	.486752
0.23	1.2586	.794534	0.73	2.0751	.481909
0.24	1.2712	.786628	0.74	2.0959	.477114
0.25	1.2840	0.778801	0.75	2.1170	0.472367
0.26	1.2969	.771052	0.76	2.1383	.467666
0.27	1.3100	.763379	0.77	2.1598	.463013
0.28	1.3231	.755784	0.78	2.1815	.458406
0.29	1.3364	.748264	0.79	2.2034	.453845
0.30	1.3499	0.740818	0.80	2.2255	0.449329
0.31	1.3634	.733447	0.81	2.2479	.444858
0.32	1.3771	.726149	0.82	2.2705	.440432
0.33	1.3910	.718924	0.83	2.2933	.436049
0.34	1.4049	.711770	0.84	2.3164	.431711
0.35	1.4191	0.704688	0.85	2.3396	0.427415
0.36	1.4333	.697676	0.86	2.3632	.423162
0.37	1.4477	.690734	0.87	2.3869	.418952
0.38	1.4623	.683861	0.88	2.4109	.414783
0.39	1.4770	.677057	0.89	2.4351	.410656
0.40	1.4918	0.670320	0.90	2.4596	0.406570
0.41	1.5068	.663650	0.91	2.4843	.402524
0.42	1.5220	.657047	0.92	2.5093	.398519
0.43	1.5373	.650509	0.93	2.5345	.394554
0.44	1.5527	.644036	0.94	2.5600	.390628
0.45	1.5683	0.637628	0.95	2.5857	0.386741
0.46	1.5841	.631284	0.96	2.6117	.382893
0.47	1.6000	.625002	0.97	2.6379	.379083
0.48	1.6161	.618783	0.98	2.6645	.375311
0.49	1.6323	.612626	0.99	2.6912	.371577

x	e^x	e^{-x}		x	e^x	e^{-x}
1.00	2.7183	0.367879		1.50	4.4817	0.223130
1.01	2.7456	.364219		1.51	4.5267	.220910
1.02	2.7732	.360595		1.52	4.5722	.218712
1.03	2.8011	.357007		1.53	4.6182	.216536
1.04	2.8292	.353455		1.54	4.6646	.214381
1.05	2.8577	0.349938		1.55	4.7115	0.212248
1.06	2.8864	.346456		1.56	4.7588	.210136
1.07	2.9154	.343009		1.57	4.8066	.208045
1.08	2.9447	.339596		1.58	4.8550	.205975
1.09	2.9743	.336216		1.59	4.9037	.203926
1.10	3.0042	0.332871		1.60	4.9530	0.201897
1.11	3.0344	.329559		1.61	5.0028	.199888
1.12	3.0649	.326280		1.62	5.0531	.197899
1.13	3.0957	.323033		1.63	5.1039	.195930
1.14	3.1268	.319819		1.64	5.1552	.193980
1.15	3.1582	0.316637		1.65	5.2070	0.192050
1.16	3.1899	.313486		1.66	5.2593	.190139
1.17	3.2220	.310367		1.67	5.3122	.188247
1.18	3.2544	.307279		1.68	5.3656	.186374
1.19	3.2871	.304221		1.69	5.4195	.184520
1.20	3.3201	0.301194		1.70	5.4739	0.182684
1.21	3.3535	.298197		1.71	5.5290	.180866
1.22	3.3872	.295230		1.72	5.5845	.179066
1.23	3.4212	.292293		1.73	5.6407	.177284
1.24	3.4556	.289384		1.74	5.6973	.175520
1.25	3.4903	0.286505		1.75	5.7546	0.173774
1.26	3.5254	.283654		1.76	5.8124	.172045
1.27	3.5609	.280832		1.77	5.8709	.170333
1.28	3.5966	.278037		1.78	5.9299	.168638
1.29	3.6328	.275271		1.79	5.9895	.166960
1.30	3.6693	0.272532		1.80	6.0496	0.165299
1.31	3.7062	.269820		1.81	6.1104	.163654
1.32	3.7434	.267135		1.82	6.1719	.162026
1.33	3.7810	.264477		1.83	6.2339	.160414
1.34	3.8190	.261846		1.84	6.2965	.158817
1.35	3.8574	0.259240		1.85	6.3598	0.157237
1.36	3.8962	.256661		1.86	6.4237	.155673
1.37	3.9354	.254107		1.87	6.4883	.154124
1.38	3.9749	.251579		1.88	6.5535	.152590
1.39	4.0149	.249075		1.89	6.6194	.151072
1.40	4.0552	0.246597		1.90	6.6859	0.149569
1.41	4.0960	.244143		1.91	6.7531	.148080
1.42	4.1371	.241714		1.92	6.8210	.146607
1.43	4.1787	.239309		1.93	6.8895	.145148
1.44	4.2207	.236928		1.94	6.9588	.143704
1.45	4.2631	0.234570		1.95	7.0287	0.142274
1.46	4.3060	.232236		1.96	7.0993	.140858
1.47	4.3492	.229925		1.97	7.1707	.139457
1.48	4.3929	.227638		1.98	7.2427	.138069
1.49	4.4371	.225373		1.99	7.3155	.136695

x	e^x	e^{-x}	x	e^x	e^{-x}
2.00	7.3891	0.135335	2.50	12.182	0.082085
2.01	7.4633	.133989	2.51	12.305	.081268
2.02	7.5383	.132655	2.52	12.429	.080460
2.03	7.6141	.131336	2.53	12.554	.079659
2.04	7.6906	.130029	2.54	12.680	.078866
2.05	7.7679	0.128735	2.55	12.807	0.078082
2.06	7.8460	.127454	2.56	12.936	.077305
2.07	7.9248	.126186	2.57	13.066	.076536
2.08	8.0045	.124930	2.58	13.197	.075774
2.09	8.0849	.123687	2.59	13.330	.075020
2.10	8.1662	0.122456	2.60	13.464	0.074274
2.11	8.2482	.121238	2.61	13.599	.073535
2.12	8.3311	.120032	2.62	13.736	.072803
2.13	8.4149	.118837	2.63	13.874	.072078
2.14	8.4994	.117655	2.64	14.013	.071361
2.15	8.5849	0.116484	2.65	14.154	0.070651
2.16	8.6711	.115325	2.66	14.296	.069948
2.17	8.7583	.114178	2.67	14.440	.069252
2.18	8.8463	.113042	2.68	14.585	.068563
2.19	8.9352	.111917	2.69	14.732	.067881
2.20	9.0250	0.110803	2.70	14.880	0.067206
2.21	9.1157	.109701	2.71	15.029	.066537
2.22	9.2073	.108609	2.72	15.180	.065875
2.23	9.2999	.107528	2.73	15.333	.065219
2.24	9.3933	.106459	2.74	15.487	.064570
2.25	9.4877	0.105399	2.75	15.643	0.063928
2.26	9.5831	.104350	2.76	15.800	.063292
2.27	9.6794	.103312	2.77	15.959	.062662
2.28	9.7767	.102284	2.78	16.119	.062089
2.29	9.8749	.101266	2.79	16.281	.061421
2.30	9.9742	0.100259	2.80	16.445	0.060810
2.31	10.074	.099261	2.81	16.610	.060205
2.32	10.176	.098274	2.82	16.777	.059606
2.33	10.278	.097296	2.83	16.945	.059013
2.34	10.381	.096328	2.84	17.116	.058426
2.35	10.486	0.095369	2.85	17.288	0.057844
2.36	10.591	.094420	2.86	17.462	.057269
2.37	10.697	.093481	2.87	17.637	.056699
2.38	10.805	.092551	2.88	17.814	.056135
2.39	10.913	.091630	2.89	17.993	.055576
2.40	11.023	0.090718	2.90	18.174	0.055023
2.41	11.134	.089815	2.91	18.357	.054476
2.42	11.246	.088922	2.92	18.541	.053934
2.43	11.359	.088037	2.93	18.728	.053397
2.44	11.473	.087161	2.94	18.916	.052866
2.45	11.588	0.086294	2.95	19.106	0.052340
2.46	11.705	.085435	2.96	19.298	.051819
2.47	11.822	.084585	2.97	19.492	.051303
2.48	11.941	.083743	2.98	19.688	.050793
2.49	12.061	.082910	2.99	19.886	.050287
2.50	12.182	0.082085	3.00	20.086	0.049787

IV
NATURAL LOGARITHMS[a]

This table contains logarithms of numbers from 1 to 10 to the base e. To obtain the natural logarithms of other numbers use the formulas:

$$\log_e (10^r \, N) = \log_e N + \log_e 10^r$$

$$\log_e \left(\frac{N}{10^r}\right) = \log_e N - \log_e 10^r$$

$\log_e 10 = 2.302585$ $\log_e 10^4 = 9.210340$
$\log_e 10^2 = 4.605170$ $\log_e 10^5 = 11.512925$
$\log_e 10^3 = 6.907755$ $\log_e 10^6 = 13.815511$

N	0	1	2	3	4	5	6	7	8	9
1.0	0.0 0000	0995	1980	2956	3922	4879	5827	6766	7696	8618
1.1	0.0 9531	*0436	*1333	*2222	*3103	*3976	*4842	*5700	*6551	*7395
1.2	0.1 8232	9062	9885	*0701	*1511	*2314	*3111	*3902	*4686	*5464
1.3	0.2 6236	7003	7763	8518	9267	*0010	*0748	*1481	*2208	*2930
1.4	0.3 3647	4359	5066	5767	6464	7156	7844	8526	9204	9878
1.5	0.4 0547	1211	1871	2527	3178	3825	4469	5108	5742	6373
1.6	0.4 7000	7623	8243	8858	9470	*0078	*0682	*1282	*1879	*2473
1.7	0.5 3063	3649	4232	4812	5389	5962	6531	7098	7661	8222
1.8	0.5 8779	9333	9884	*0432	*0977	*1519	*2078	*2594	*3127	*3658
1.9	0.6 4185	4710	5233	5752	6269	6783	7294	7803	8310	8813
2.0	0.6 9315	9813	*0310	*0804	*1295	*1784	*2271	*2755	*3237	*3716
2.1	0.7 4194	4669	5142	5612	6081	6547	7011	7473	7932	8390
2.2	0.7 8846	9299	9751	*0200	*0648	*1093	*1536	*1978	*2418	*2855
2.3	0.8 3291	3725	4157	4587	5015	5442	5866	6289	6710	7129
2.4	0.8 7547	7963	8377	8789	9200	9609	*0016	*0422	*0826	*1228
2.5	0.9 1629	2028	2426	2822	3216	3609	4001	4391	4779	5166
2.6	0.9 5551	5935	6317	6698	7078	7456	7833	8208	8582	8954
2.7	0.9 9325	9695	*0063	*0430	*0796	*1160	*1523	*1885	*2245	*2604
2.8	1.0 2962	3318	3674	4028	4380	4732	5082	5431	5779	6126
2.9	1.0 6471	6815	7158	7500	7841	8181	8519	8856	9192	9527
3.0	1.0 9861	*0194	*0526	*0856	*1186	*1514	*1841	*2168	*2493	*2817
3.1	1.1 3140	3462	3783	4103	4422	4740	5057	5373	5688	6002
3.2	1.1 6315	6627	6938	7248	7557	7865	8173	8479	8784	9089
3.3	1.1 9392	9695	9996	*0297	*0597	*0896	*1194	*1491	*1788	*2083
3.4	1.2 2378	2671	2964	3256	3547	3837	4127	4415	4703	4990
3.5	1.2 5276	5562	5846	6130	6413	6695	6976	7257	7536	7815
3.6	1.2 8093	8371	8647	8923	9198	9473	9746	*0019	*0291	*0563
3.7	1.3 0833	1103	1372	1641	1909	2176	2442	2708	2972	3237
3.8	1.3 3500	3763	4025	4286	4547	4807	5067	5325	5584	5841
3.9	1.3 6098	6354	6609	6864	7118	7372	7624	7877	8128	8379
4.0	1.3 8629	8879	9128	9377	9624	9872	*0118	*0364	*0610	*0854
4.1	1.4 1099	1342	1585	1828	2070	2311	2552	2792	3031	3270
4.2	1.4 3508	3746	3984	4220	4456	4692	4927	5161	5395	5629
4.3	1.4 5862	6094	6326	6557	6787	7018	7247	7476	7705	7933
4.4	1.4 8160	8387	8614	8840	9065	9290	9515	9739	9962	*0185
4.5	1.5 0408	0630	0851	1072	1293	1513	1732	1951	2170	2388
4.6	1.5 2606	2823	3039	3256	3471	3687	3902	4116	4330	4543
4.7	1.5 4756	4969	5181	5393	5604	5814	6025	6235	6444	6653
4.8	1.5 6862	7070	7277	7485	7691	7898	8104	8309	8515	8719
4.9	1.5 8924	9127	9331	9534	9737	9939	*0141	*0342	*0543	*0744
5.0	1.6 0944	1144	1343	1542	1741	1939	2137	2334	2531	2728
N	0	1	2	3	4	5	6	7	8	9

N	0	1	2	3	4	5	6	7	8	9
5.0	1.6 0944	1144	1343	1542	1741	1939	2137	2334	2531	2728
5.1	1.6 2924	3120	3315	3511	3705	3900	4094	4287	4481	4673
5.2	1.6 4866	5058	5250	5441	5632	5823	6013	6203	6393	6582
5.3	1.6 6771	6959	7147	7335	7523	7710	7896	8083	8269	8455
5.4	1.6 8640	8825	9010	9194	9378	9562	9745	9928	*0111	*0293
5.5	1.7 0475	0656	0838	1019	1199	1380	1560	1740	1919	2098
5.6	1.7 2277	2455	2633	2811	2988	3166	3342	3519	3695	3871
5.7	1.7 4047	4222	4397	4572	4746	4920	5094	5267	5440	5613
5.8	1.7 5786	5958	6130	6302	6473	6644	6815	6985	7156	7326
5.9	1.7 7495	7665	7843	8002	8171	8339	8507	8675	8842	9009
6.0	1.7 9176	9342	9509	9675	9840	*0006	*0171	*0336	*0500	*0665
6.1	1.8 0829	0993	1156	1319	1482	1645	1808	1970	2132	2294
6.2	1.8 2455	2616	2777	2938	3098	3258	3418	3578	3737	3896
6.3	1.8 4055	4214	4372	4530	4688	4845	5003	5160	5317	5473
6.4	1.8 5630	5786	5942	6097	6253	6408	6563	6718	6872	7026
6.5	1.8 7180	7334	7487	7641	7794	7947	8099	8251	8403	8555
6.6	1.8 8707	8858	9010	9160	9311	9462	9612	9762	9912	*0061
6.7	1.9 0211	0360	0509	0658	0806	0954	1102	1250	1398	1545
6.8	1.9 1692	1839	1986	2132	2279	2425	2571	2716	2862	3007
6.9	1.9 3152	3297	3442	3586	3730	3874	4018	4162	4305	4448
7.0	1.9 4591	4734	4876	5019	5161	5303	5445	5586	5727	5869
7.1	1.9 6009	6150	6291	6431	6571	6711	6851	6991	7130	7269
7.2	1.9 7408	7547	7685	7824	7962	8100	8238	8376	8513	8650
7.3	1.9 8787	8924	9061	9198	9334	9470	9606	9742	9877	*0013
7.4	2.0 0148	0283	0418	0553	0687	0821	0956	1089	1223	1357
7.5	2.0 1490	1624	1757	1890	2022	2155	2287	2419	2551	2683
7.6	2.0 2815	2946	3078	3209	3340	3471	3601	3732	3862	3992
7.7	2.0 4122	4252	4381	4511	4640	4769	4898	5027	5156	5284
7.8	2.0 5412	5540	5668	5796	5924	6051	6179	6306	6433	6560
7.9	2.0 6686	6813	6939	7065	7191	7317	7443	7568	7694	7819
8.0	2.0 7944	8069	8194	8318	8443	8567	8691	8815	8939	9063
8.1	2.0 9186	9310	9433	9556	9679	9802	9924	*0047	*0169	*0291
8.2	2.1 0413	0535	0657	0779	0900	1021	1142	1263	1384	1505
8.3	2.1 1626	1746	1866	1986	2106	2226	2346	2465	2585	2704
8.4	2.1 2823	2942	3061	3180	3298	3417	3535	3653	3771	3889
8.5	2.1 4007	4124	4242	4359	4476	4593	4710	4827	4943	5060
8.6	2.1 5176	5292	5409	5524	5640	5756	5871	5987	6102	6217
8.7	2.1 6332	6447	6562	6677	6791	6905	7020	7134	7248	7361
8.8	2.1 7475	7589	7702	7816	7929	8042	8155	8267	8380	8493
8.9	2.1 8605	8717	8830	8942	9054	9165	9277	9389	9500	9611
9.0	2.1 9722	9834	9944	*0055	*0166	*0276	*0387	*0497	*0607	*0717
9.1	2.2 0827	0937	1047	1157	1266	1375	1485	1594	1703	1812
9.2	2.2 1920	2029	2138	2246	2354	2462	2570	2678	2786	2894
9.3	2.2 3001	3109	3216	3324	3431	3538	3645	3751	3858	3965
9.4	2.2 4071	4177	4284	4390	4496	4601	4707	4813	4918	5024
9.5	2.2 5129	5234	5339	5444	5549	5654	5759	5863	5968	6072
9.6	2.2 6176	6280	6384	6488	6592	6696	6799	6903	7006	7109
9.7	2.2 7213	7316	7419	7521	7624	7727	7829	7932	8034	8136
9.8	2.2 8238	8340	8442	8544	8646	8747	8849	8950	9051	9152
9.9	2.2 9253	9354	9455	9556	9657	9757	9858	9958	*0058	*0158
10.0	2.3 0259	0358	0458	0558	0658	0757	0857	0956	1055	1154
N	0	1	2	3	4	5	6	7	8	9

V

TRIGONOMETRIC FORMULAS

All angles are in radians. The trigonometric functions are defined in Section 0-15 and studied in Section 5-1 to 5-3.

1. $\sin(x + 2\pi) = \sin x$

2. $\cos(x + 2\pi) = \cos x$

3. $\tan(x + \pi) = \tan x$

4. $\cot(x + \pi) = \cot x$

5. $\csc(x + 2\pi) = \csc x$

6. $\sec(x + 2\pi) = \sec x$

7. $\csc x = \dfrac{1}{\sin x}$

8. $\sec x = \dfrac{1}{\cos x}$

9. $\cot x = \dfrac{1}{\tan x}$

10. $\tan x = \dfrac{\sin x}{\cos x}$

11. $\cot x = \dfrac{\cos x}{\sin x}$

12. $\sin^2 x + \cos^2 x = 1$

13. $\tan^2 x + 1 = \sec^2 x$

14. $\cot^2 x + 1 = \csc^2 x$

15. $\cos\left(\dfrac{\pi}{2} - x\right) = \sin x$

16. $\sin\left(\dfrac{\pi}{2} - x\right) = \cos x$

17. $\cos\left(\dfrac{\pi}{2} + x\right) = -\sin x$

18. $\sin\left(\dfrac{\pi}{2} + x\right) = \cos x$

19. $\tan\left(\dfrac{\pi}{2} - x\right) = \cot x$

20. $\cot\left(\dfrac{\pi}{2} - x\right) = \tan x$

21. $\tan\left(\dfrac{\pi}{2} + x\right) = -\cot x$

22. $\cot\left(\dfrac{\pi}{2} + x\right) = -\tan x$

23. $\cos(\pi - x) = -\cos x$

24. $\sin(\pi - x) = \sin x$

25. $\tan(\pi - x) = -\tan x$

26. $\cot(\pi - x) = -\cot x$

27. $\sin(-x) = -\sin x$

28. $\cos(-x) = \cos x$

29. $\tan(-x) = -\tan x$

30. $\cot(-x) = -\cot x$

31. $\sin(x + y) = \sin x \cos y + \cos x \sin y$

32. $\sin(x - y) = \sin x \cos y - \cos x \sin y$

33. $\cos(x + y) = \cos x \cos y - \sin x \sin y$

34. $\cos(x - y) = \cos x \cos y + \sin x \sin y$

35. $\tan(x + y) = \dfrac{\tan x + \tan y}{1 - \tan x \tan y}$

36. $\tan(x - y) = \dfrac{\tan x - \tan y}{1 + \tan x \tan y}$

37. $\sin 2x = 2 \sin x \cos x$

38. $\cos 2x = \cos^2 x - \sin^2 x = 2\cos^2 x - 1 = 1 - 2\sin^2 x$

39. $\sin \dfrac{1}{2}x = \pm\sqrt{\dfrac{1 - \cos x}{2}}$

40. $\cos \dfrac{1}{2}x = \pm\sqrt{\dfrac{1 + \cos x}{2}}$

41. $\tan \dfrac{1}{2}x = \pm\sqrt{\dfrac{1 - \cos x}{1 + \cos x}} = \dfrac{1 - \cos x}{\sin x} = \dfrac{\sin x}{1 + \cos x}$

42. $\sin x \sin y = -\dfrac{1}{2}[\cos(x + y) - \cos(x - y)]$

43. $\cos x \cos y = \dfrac{1}{2}[\cos(x + y) + \cos(x - y)]$

44. $\sin x \cos y = \dfrac{1}{2}[\sin(x + y) + \sin(x - y)]$

45. $\sin x + \sin y = 2\sin \dfrac{x + y}{2} \cos \dfrac{x - y}{2}$

46. $\sin x - \sin y = 2\sin \dfrac{x - y}{2} \cos \dfrac{x + y}{2}$

47. $\cos x + \cos y = 2\cos \dfrac{x + y}{2} \cos \dfrac{x - y}{2}$

48. $\cos x - \cos y = -2\sin \dfrac{x + y}{2} \sin \dfrac{x - y}{2}$

49. $\sin 3x = 3\sin x - 4\sin^3 x$ 50. $\cos 3x = 4\cos^3 x - 3\cos x$
51. $\sin 4x = 8\cos^3 x \sin x - 4\cos x \sin x$
52. $\cos 4x = 8\cos^4 x - 8\cos^2 x + 1$

Formulas 53 to 55 refer to a triangle with angles A, B, C and opposite sides a, b, c. Also s denotes $\frac{1}{2}(a + b + c)$.

53. Law of Sines: $\dfrac{\sin A}{a} = \dfrac{\sin B}{b} = \dfrac{\sin C}{c}$

54. Law of Cosines: $c^2 = a^2 + b^2 - 2ab \cos C$

55. Area $= \frac{1}{2}bc \sin A = \sqrt{s(s - a)(s - b)(s - c)}$

Formulas 56 to 57 refer to a circular sector of radius r and angle α.

56. Area $= \frac{1}{2}r^2\alpha$ 57. Arc length $= r\alpha$

ANSWERS TO SELECTED PROBLEMS

Section 0-3, page 5

1. (a) 15, 16 or 17   (c) 1.42 or 1.5, for example
2. (a) $x < y$ (c) $x < y$
3. (a) 3.5   (c) $|x|$ (e) 0
5. (a) 0   (c) -1 or 3
8. (a) No   (c) Yes   (e) Yes

Section 0-4, page 7

1. (a) 1, 2, 3, 4.   (c) 1
2. (b) Does not belong
3. (a) The empty set; 0; 1; 0 and 1
   (c) The empty set, $\{(0, 0)\}$, $\{(0, 1)\}$, $\{(1, 0)\}$, $\{(1, 1)\}$, $\{(0, 0), (0, 1)\}$,
   $\{(0, 0), (1, 0)\}$, $\{(0, 0), (1, 1)\}$, $\{(0, 1), (1, 0)\}$, $\{(0, 1), (1, 1)\}$, $\{(1, 0), (1, 1)\}$,
   $\{(0, 1), (1, 0), (1, 1)\}$, $\{(0, 0), (1, 0), (1, 1)\}$, $\{(0, 0), (0, 1), (1, 1)\}$,
   $\{(0, 0), (0, 1), (1, 0)\}$, $\{(0, 0), (0, 1), (1, 0), (1, 1)\}$
4. (a) $-2 < x < 2$, open interval   (c) $0 \leq x \leq 1$, closed interval   (e) R
   (g) The union of infinite intervals $x < -1$ and $x \geq 2$
5. (a) closed   (c) open   (e) infinite
6. (a) $[0, 1]$, closed interval   (c) $(0, 1)$, open interval   (e) empty set
7. (c) May be empty or an interval which is open, closed or half-open
8. (a) All real numbers except 1
   (c) Union of open intervals $1 < x < 2$, $2 < x < 3$ (e) 0, 3

Section 0-7, page 13

6. (a) slope $\frac{2}{3}$, intercepts 3, -2 (c) slope 0, y-intercept -3
   (e) slope -3, intercepts -1, -3
7. (a) $y - 2 = 5(x - 4)$ (c) $y - 1 = \frac{1}{2}(x - 5)$ (e) $x - 5y = -14$
   (g) $y = x$ (i) $x = 4$
9. (c) $2x - 7y + 5 = 0$

Section 0-9, page 19

1. (a) $x = \frac{13}{5}$, $y = -\frac{11}{5}$ (c) All pairs (x, y) with $y = x$ (e) no solution
2. (a) $x = 1$, $y = 1$, $z = 1$ (c) All triples (x, y, z) with $y = -3x$, $z = -5x$
3. (a) 6   (c) 0   (e) 0
4. (a) unique   (c) not unique   (e) no solution   (g) no solution
   (i) unique

Section 0-12, page 23

1. (a) domain: 0, 1, 2, 3, 4, range: 2, 3, 4, 5, 7　　　(c) domain: 5, 6, 7, 8, 9, range: 2
2. (a) 3　　　(c) $25\sqrt{3}/2$
3. (a) domain: $(-\infty, \infty)$, range: $(-\infty, -2]$
   (c) domain: $(-\infty, -1] \cup [1, \infty)$, range: $[0, \infty)$
   (e) domain: $(-\infty, 0) \cup (0, \infty)$, range: $(-\infty, 0) \cup (0, \infty)$
4. (a) 2　　　(c) 8
5. (a) 0　　　(c) $16a^2 + 30a$
6. (a) $\frac{1}{2}$　　　(c) 1
7. (a) one-to-one　　　(c) not one-to-one
8. (a) one-to-one, $x = y/2$　　　(c) one-to-one, $x = y^{1/3}$　　　(e) one-to-one, $x = 1/y$
9. (a) domain: $(-\infty, \infty)$, range: $[0, \infty)$
10. (a) $A = s^2$　　　(c) $S = 4\pi r^2$　　　(e) $c = a\sqrt{2}$
11. (a) 2, multiplicity 1　　　(c) 1, multiplicity 2
    (e) 1, multiplicity 3, and -2, multiplicity 5
12. (a) $z = xy$, a polynomial　　　(c) $h = \sqrt{4a^2 - c^2}/2$, not a polynomial
    (e) $z = (c - 2b)/a$, not a polynomial

Section 0-14, page 31

1. (a) vertex $(-1, 4)$, focus $(-1, \frac{17}{4})$, directrix $y = \frac{15}{4}$
   (c) vertex $(1, 3)$, focus $(1, \frac{25}{8})$, directrix $y = \frac{23}{8}$
   (e) degenerate parabola, two parallel lines $y = 3$, $y = -1$
2. (a) $(0, 0)$, $\sqrt{7}$　　　(c) $(\frac{5}{2}, -3)$, $\sqrt{61}/2$
4. $2x - y = \pm 30$
5. $x^2 + y^2 - 13x + y + 10 = 0$
6. $(0, 0)$, $(\frac{48}{13}, \frac{12}{13})$
8. (a) vertexes $(\pm 2, 0)$, $(0, \pm\sqrt{3})$, center $(0, 0)$, foci $(\pm 1, 0)$, $e = \frac{1}{2}$,
   directrices $x = \pm 4$
   (c) vertexes $(\pm 2, 0)$, center $(0, 0)$, foci $(\pm 3, 0)$, $e = \frac{3}{2}$, directrices $x = \pm\frac{4}{3}$
9. (a) $(x^2/9) + (y^2/5) = 1$　　　(c) $(x^2/4) + (y^2/3) = 1$
10. (a) $[x - (\frac{3}{4})]^2/(\frac{33}{16}) + [y - (\frac{1}{2})]^2/(\frac{33}{12}) = 1$
    (c) $[(x - 2)^2/1] - [(y - 1)^2/4] = 0$

Section 0-17, page 37

1. $\pi/2, 2\pi, -\pi, 11\pi/180 = 0.192, 1$
3. $r^2\alpha/2$
4. (a) $nr^2 \sin(\pi/n) \cos(\pi/n)$
5. (a) $\sqrt{2}/2$　　　(c) 0　　　(e) -1　　　(g) 1
7. (a) $(\pi/6) + 2k\pi$, $(5\pi/6) + 2k\pi$, $(3\pi/2) + 2k\pi$, $k = 0, \pm 1, \ldots$
9. (a) $(2\sqrt{2}, (\pi/4) + 2k\pi)$　　　(c) $(2, (3\pi/2) + 2k\pi)$
10. (a) $(\sqrt{3}/2, \frac{1}{2})$　　　(c) $(5, 0)$
11. (a) $5 + 12i$　　　(c) 2　　　(e) $(\frac{1}{2}) + (\frac{3}{2})i$　　　(g) 1

Section 0-18, page 38

1. (a) $\pm i$　　　(c) $1 - i, -i, -5$
2. (a) $k(z^5 - 3z^4 + 4z^3 - 2z^2) = 0$

Section 0-19, page 41

1. (a) $x^8 y^{14}$ (c) $x^2 y^9$
2. (a) 1.72   (c) x

Section 0-21, page 45

8. $\dbinom{52}{13}$

Section 1-4, page 51

2. *ABPQ, APBQ, APQB, PAQB, PQAB, PABQ*
3. (a) $\overrightarrow{AB} = \overrightarrow{OC}$, $\overrightarrow{DE} = \overrightarrow{OF}$, $\overrightarrow{AE}$, $\overrightarrow{OE} = \overrightarrow{AF} = \overrightarrow{BO} = \overrightarrow{CD}$, $\overrightarrow{AC} = \overrightarrow{FD}$, $\overrightarrow{BC}$, $\overrightarrow{AD}$
   (c) The first three are equal

Section 1-5, page 54

1. (a) $\mathbf{v} - \mathbf{u}$ (c) $2\mathbf{v} - 3\mathbf{u}$ (e) $2\mathbf{v} - \mathbf{u}$ (g) $2\mathbf{v} - \mathbf{u}$
2. (a) $-\mathbf{u}$ (c) $\frac{1}{2}\mathbf{u}$ (e) $(\frac{2}{3})\mathbf{u} + (\frac{3}{8})\mathbf{v}$ (g) $\frac{1}{2}\mathbf{u} - \mathbf{v}$ (i) $(\frac{1}{3})\mathbf{u} - (\frac{1}{6})\mathbf{v}$
4. (a) $[m/(m + n)]\mathbf{u}$

Section 1-8, page 63

3. (a) linearly independent   (c) linearly dependent
4. (a) $k = 0$ (c) $k = 1$ or -3
5. (b) $a = 2$, $b = -1$
6. (b) $\mathbf{r} = \frac{3}{2}\mathbf{z} + \frac{1}{2}\mathbf{w}$, $\mathbf{s} = \mathbf{z} + 2\mathbf{w}$

Section 1-9, page 67

2. (a) $5\mathbf{i}$ (c) $\sqrt{10}$, $\sqrt{5}$
3. (a) $\sqrt{2}$ (c) $7/\sqrt{65}$
4. $\cos\left(-\dfrac{\pi}{4} \pm \dfrac{2\pi}{3}\right)\mathbf{i} + \sin\left(-\dfrac{\pi}{4} \pm \dfrac{2\pi}{3}\right)\mathbf{j}$

Section 1-11, page 70

1. (a) 1, 1, 2, 1   (c) 3   (e) $3/\sqrt{5}$, $3/\sqrt{2}$
5. $30\sqrt{3}/2$ ft.-lbs.
6. (a) $3aw$

Section 1-12, page 74

1. (a) $\varphi = \dfrac{\pi}{2}$, $\psi = \dfrac{\pi}{2} + 2k\pi$ $(k = 0, \pm 1, \ldots)$

   (c) $\varphi = \dfrac{3\pi}{4}$, $\psi = \dfrac{5\pi}{4} + 2k\pi$ $(9k = 0, \pm 1, \ldots)$
2. (a) 1   (c) $\frac{11}{2}$
4. (a) and (c) similarly
5. (a) $(\frac{1}{13})(12\mathbf{i} + 5\mathbf{j})$ (c) $(\frac{1}{2})(\sqrt{3}\mathbf{i} - \mathbf{j})$

Section 1-13, page 76

1. (a) 44i, where 40i is the velocity vector of the train
2. (a) N.6°46′W, about 1.2 hours
4. At an angle of $\pi/4$ with each of the first two forces
7. $\mathbf{F}_2 = 148\mathbf{j}$, $\mathbf{F}_4 = -63\mathbf{i}$
9. (c) $m_1 = 1$, $m_2 = 1$, $m_3 = 5$

Section 1-16, page 81

2. (a) $x = 3 + 5t$, $y = 2 - t$ (c) $x = 3 - t$, $y = 2t$
3. (a) velocity vector $2\mathbf{i} - \mathbf{j}$, speed $\sqrt{5}$
5. $x + 3y - 1 = 0$
7. (a) $2x - y = 0$
8. (a) $t = 0$, at $(3, 3)$
9. $(11, -11)$
12. (a) $2(x - 1) - (y - 3) = 0$ (c) $x = 0$
13. (a) $m = 2$, $\mathbf{v} = 2\mathbf{i} - \mathbf{j}$ (c) $m = 0$, $\mathbf{v} = \mathbf{j}$
14. (e) $12/\sqrt{10}$

Section 2-1, page 86

1. (a) $y = \sin u = g(u)$, $u = x^2 + 1 = f(x)$, g, f, and $g \circ f$ have all real numbers as domain.   (c) $y = u^{10} = g(u)$, $u = x + 2 = f(x)$; g, f, and $g \circ f$ have all real numbers as domain.
2. (a) $y = \sin^3 x$ (c) $y = 2^{3x}$ (e) $y = \sin^3 2^x$
4. (a) $x > 1$ (c) $1 \leq x \leq 2$ (e) $\dfrac{-3 - \sqrt{5}}{2} \leq x \leq \dfrac{-3 + \sqrt{5}}{2}$
5. (c) and (d) are correct.

Section 2-4, page 93

1. (a) (i) [3, 5], (ii) [1, 3], (iii) local minimum for $x = 3$ ($y = -1.3$), local maxima at $x = 1$ ($y = 2$) and $x = 5$ ($y = 1.2$), (iv) 2, -1.3 (c) (i) [1, 2.5], [4, 5], (ii) [2.5, 4], (iii) local minima for $x = 1$ ($y = 1$), and for $x = 4$ ($y = -2$), local maxima for $x = 2.5$ ($y = 3$) and $x = 5$ ($y = 0.6$), (iv) 3, -2.
8. (a) Function is $f - g - g$.
   (c) Function is $(g + g + g + g + g)(f \cdot f \cdot f) - f \cdot f - f \cdot f$.
10. No.
12. (a) $x = \frac{1}{2}y$ (c) $x = -1 + \sqrt{y + 2}$
15. (a) $\pi/2$ (c) $\pi/2$
17. (a) Has period 2π (c) Has period 2π (e) Has period 6π.
   (g) Has period 2π.
18. (a) unbounded   (c) bounded, $K = 1$

Section 2-6, page 103

2. empirical limits: (a) 0   (c) 0.17
3. (a) continuous   (c) continuous

Section 2-8, page 112

1. (a) 1   (c) 10   (e) $\frac{2}{3}$
2. (a) $2c + 4k$ (c) $c/(1 + k^2)$ (e) 2^c
4. (a) continuous   (c) discontinuous at $x = 0$ (e) continuous
    (g) continuous   (i) continuous
8. (a) no, no, yes.
14. (a) 1, 4   (c) $\frac{1}{2}$, 1

Section 2-9, page 117

4. (a) linearly dependent   (c) linearly independent   (e) linearly dependent
5. no

Section 2-11, page 123

2. (a) 0   (c) 2   (e) 0   (g) 2   (i) $\frac{1}{2}$ (k) 1   (m) 1
5. (a) $\frac{1}{3}$ (c) $\frac{2}{3}$
6. (a) 0   (c) 0   (e) no limit
7. (a) ∞ (c) $-\infty$ (e) ∞ (g) ∞ (i) ∞ (k) 1
15. (a) A vector space.   (c) A vector space.   (e) Not a vector space.

Section 2-13, page 131

1. (a) 6, 13, 20, 27, 34   (c) $a_0, a_1, a_2, \ldots, a_n$ (e) $\sqrt{1}, \sqrt{2}, \ldots, \sqrt{n}$
2. (a) Divergent   (c)   Convergent to 0   (e) Convergent to 1
    (g) Divergent
3. (a) Those of (d), (e) and (g).   (c) All but (d) and (g).
5. Least upper bounds: (a) 1   (c) $\sqrt{3}$ (e) 1   (g) 1

Section 3-2, page 149

2. At B, 0.06; at C, 0.09.
3. (a) $f'(1) = 0.67$, $f'(1.05) = 0.70$ (approximately).
4. (a) $f'(1) = -0.062$, $f'(1.05) = -0.061$ (approximately).
    (c) $f(1 + h) - f(1) = -h/(16 + 4h)$, $f'(1) = -\frac{1}{16}$
5. (a) 13   (c) $8x + 5$ (e) $3x^2 - 4x$ (g) $2(x + 2)^{-2}$ (i) $x^{-2/3}/3$
6. (a) g_3
10. Vertical gradients in degrees per mile: -3 at A, -3.5 at B, -3.5 at C.

Section 3-4, page 160

1. (a) $21x^6$ (c) 0   (e) $6x(x^2 + 1)^2$ (g) $(3x^2 + 1)/x^2$
    (i) $x(x^4 - 1)^2(15x^5 + 14x^4 + 12x^2 - 3x - 2)$
3. (a) $2x - 4x^{-3} + 12x^{-5}$ (c) $-15x^{-4} + 28x^{-5} + 3x^{-2}$
5. (a) $-3(2x + 3)(x^2 + 3x + 2)^{-4}$ (c) $4x(1 - x^2)(x^2 + 2)^{-4}$
10. (a) $3 \cos x - 5 \sin x$ (c) $4 \tan^3 x \sec^2 x$ (e) 0
11. (a) $2t \ln t + t + 3t^2$ (c) $(\ln t)^{-2}(\ln t - 1)$ (e) $4(x \ln 10)^{-1}$
12. (a) $3e^{3x}$ (c) $\cosh x$ (e) $1/\cosh^2 x$

Section 3-5, page 167

1. (a) $6x(x^2 - 5)^2$ (c) $(2t^2 - 2t - 1)(2t - 1)$

2. (a) $x(x^2 - 1)^{-1/2}$ (c) $(14x/3)(x^2 + 5)^{4/3}$ (e) $\frac{1}{2}x[1 + \sqrt{x^2 - 1}]^{-1/2}(x^2 - 1)^{-1/2}$

   (g) $10[(x^2 - 1)/(x^2 + 2x)]^4(x^2 + x + 1)(x^2 + 2x)^{-2}$

   (i) $5[(x + 1)^{-1/2} - 2x][x^2 - 2\sqrt{x + 1}]^4[1 + (x^2 - 2\sqrt{x + 1})^5]^{-2}$

3. (a) $-3\cos^2 x \sin x$ (c) $(1 + \sin \theta)\cos(\theta - \cos \theta)$

   (e) $12\tan^3 t \sec^2 t$ (g) $6\sin x \cos x[\cos^2(\cos^2 x)][\sin(\cos^2 x)]$

4. (a) $2xe^{x^2}$ (c) $2x/(x^2 - 1)$ (e) $-2x/(x^2 - 1)$ (g) $2\cot x$

   (i) $[x \ln x \ln \ln x]^{-1}$ (k) $x^x(1 + \ln x)$

9. (a) 0,   (c) 0.

10. (a) $e^{x^2} + c$ (c) $(\frac{1}{2})e^{x^2} + c$ (e) $(\frac{1}{3})(x^2 - 1)^{3/2} + c, c = $ const

Section 3-6, page 174

1. (a) $(x + 1)^2$ (c) $-2\sqrt{1 - x}$ (e) $1 + e^{-x}$

2. (a) $[f^{-1}(y)]' = (6x^2 - 6x + 6)^{-1}$ (c) $2\sqrt{x + 1}/(2\sqrt{x + 1} + 1)$

6. (a) $1/\sqrt{4 - x^2}, -2 < x < 2$ (c) $[-x\sqrt{-8x^2 - 6x - 1}]^{-1}, -\frac{1}{2} < x < -\frac{1}{4}$

   (e) $(1 + x^2)^{-1}(\pi + \text{Tan}^{-1} x)^{-1}, -\infty < x < \infty$

Section 3-7, page 176

1. (a) $2f'(x) + 3g'(x) = 0$ (c) $f(x)g'(x) + f'(x)g(x) = 0$

   (e) $x^2f'(x) + 2xf(x) - xg'(x) - g(x) + e^x[f(x)g(x) + f(x)g'(x) + f'(x)g(x)] = 0$

2. $5(196\pi)^{-1}$ ft per sec

4. 10 degrees per min

6. The converse does not hold true.

7. It is not one-to-one.

Section 3-8, page 182

1. (a) $-\frac{2}{3}$ (c) $-v/u$

2. (a) $-(3x^2 + 2xy - 2y^2)/(x^2 - 4xy + 3y^2)$

   (c) $-[6x(x^2 + 1)^2y^3 - 5y^2]/[4y^3 + 3y^2(x^2 + 1)^3 - 10xy]$

   (e) $[(1 - y)\sin(xy) + xy\cos(xy)]/[x\sin(xy) - x^2\cos(xy)]$

6. (a) $x\tan^2 x\, e^{3x}(2\tan x + 3x\sec^2 x + 3x\tan x)$

   (c) $x\cos 5x\,(x^2 + 1)^{-3}(x - 1)^{-2}[2 - 5x\tan 5x - 6x^2(x^2 + 1)^{-1} - 2x(x - 1)^{-1}]$

   (e) $x^x(1 + \ln x)$ (g) $(\cos x)^{\sin x}(\cos x \ln \cos x - \sin x \tan x)$

   (i) $f_1'f_2 \cdots f_n + f_1f_2'f_3 \cdots f_n + \cdots + f_1f_2 \cdots f_{n-1}f_n'$

   (k) $\dfrac{f_1^{k_1} \cdots f_n^{k_n}}{g_1^{h_1} \cdots g_m^{h_m}} \left(k_1\dfrac{f_1'}{f_1} + \cdots + k_n\dfrac{f_n'}{f_n} - h_1\dfrac{g_1'}{g_1} - \cdots - h_m\dfrac{g_m'}{g_m}\right)$

7. (a) Same as 6(e)   (c) Same as 6(g)

Section 3-9, page 186

1. (a) t or $(x + 3)/2$ (c) $[(t - 1)/(t + 1)]^{\frac{1}{2}}$ or $x/(x^2 + 2)^{\frac{1}{2}}$

2. (a) slope 3   (c) slope -1

3. (b) $\sin \theta/(1 - \cos \theta)$

Section 3-12, page 194

8. (a) $-\mathbf{i} + 2\mathbf{j}$ (c) $2\mathbf{i} + \mathbf{j}$
9. (a) $\mathbf{i} - \mathbf{j}$ (c) $5(t-2)^4\mathbf{i} + 3(1 + 2t^3)(1 - t^3)^{-2}\mathbf{j}$
10. For $t = 1$ tangent line is $x = -2 + 2t$, $y = 5 + 5t$.
12. (a) $\mathbf{v} = -a\sin t\,\mathbf{i} + b\cos t\,\mathbf{j}$, $|\mathbf{v}| = (a^2\sin^2 t + b^2\cos^2 t)^{\frac{1}{2}}$
13. (c) At $(1, 0)$ velocity vector is $\mathbf{i} - 2\mathbf{j}$, slope is -2.

Section 3-13, page 198

1. (a) tangent $7x - y = 10$, normal $x + 7y = 30$
   (c) tangent $3x + 4y = 5$, normal $4x - 3y = 0$
2. (a) $\text{Tan}^{-1}(\frac{-26}{7})$

4. (a) $x_1x + y_1y = a^2$ (c) $\dfrac{x_1x}{a^2} - \dfrac{y_1y}{b^2} = 1$

   (e) $Ax_1x + \frac{1}{2}B(x_1y + y_1x) + Cy_1y + \frac{1}{2}D(x + x_1) + \frac{1}{2}E(y + y_1) + F = 0$
6. $\text{Cos}^{-1}(16/\sqrt{740})$.

Section 3-16, page 206

1. (a) $y' = \frac{1}{2}x^{-1/2} - 21x^{-4}$, $y'' = -\frac{1}{4}x^{-3/2} + 84x^{-5}$, $y''' = (\frac{3}{8})x^{-5/2} - 420x^{-6}$
   (c) $y' = (x + 1)^{1/2}\,3x^2 + \frac{1}{2}(x + 1)^{-1/2}(x^3 - 1)$
   $y'' = (x + 1)^{1/2}\,6x + (x + 1)^{-1/2}\,3x^2 - (\frac{1}{4})(x + 1)^{-3/2}(x^3 - 1)$
   $y''' = (x + 1)^{1/2}\,6 + (x + 1)^{-1/2}\,9x - (\frac{3}{2})(x + 1)^{-3/2}x^2 + (\frac{3}{8})(x + 1)^{-5/2}(x^3 - 1)$
   (e) $y' = (x + \sin x)/(1 + \cos x)$ $y'' = (2 + 2\cos x + x\sin x)/(1 + \cos x)^2$
   $y''' = (x\sin^2 x + x + x\cos x + 3\sin x + 3\sin x\cos x)(1 + \cos x)^{-3}$
9. (a) through x^2 (c) through x^3
10. (a) $1 + 3x + 5x^2 + x^3$ (c) $2 + 5x$ (e) $1 + 2x - (x^3/4)$
11. 1,111.1 lb
12. (a) $v = 3$ cm per sec $a = -2$ cm per sec^2 (c) 1000 dynes

Section 3-18, page 213

1. (a) $(60u^3 + 6u)(14x^6 + 3x^2)^2 + (15u^4 + 3u^2)(84x^5 + 6x)$,
   where $u = 2x^7 + x^3 + 1$
   (c) $-[t^2\sin y/(t^2 - 1)] - \cos y(t^2 - 1)^{-3/2}$
2. (a) $f''(u)(2x + 2 - x^{-2})^2 + f'(u)(2 + 2x^{-3})$ (c) $-g'^2\sin u + g''\cos u$
6. (a) velocity vector $2t\mathbf{i} + 3t^2\mathbf{j}$, acceleration vector $2\mathbf{i} + 6t\mathbf{j}$
   (c) velocity vector $-2\pi t\sin \pi t^2\mathbf{i} + 2\pi t\cos \pi t^2\mathbf{j}$,
   acceleration vector $2\pi[(-\sin \pi t^2 - 2\pi t^2\cos \pi t^2)\mathbf{i} + (\cos \pi t^2 - 2\pi t^2\sin \pi t^2)\mathbf{j}]$

Section 3-19, page 218

1. (a) $x = \frac{3}{2}$, $y = \frac{13}{4}$ (c) $x = 2$, $y = 1 - 2\sqrt{2}$
2. (a) Absolute minimum at $x = 0$, maximum at $x = 2$, local minimum at $x = 4$
   (c) Absolute minimum at $x = 0$, absolute maximum at $x = 4$
   (e) Absolute minimum at $x = 3$, maximum at $x = 2$; local minimum at $x = 0$,
   maximum at $x = 4$
   (g) Absolute minimum at $x = 0$, maximum at $x = 4$

6. (a) none   (c) $x = \frac{1}{2}$, local maximum
7. 5 and 5
9. altitude = radius of base

Section 3-21, page 223

1. (a) 3   (c) 3
6. (a) strictly increasing   (c) strictly increasing
17. (a) and (c) are not vector spaces

Section 3-25, page 233

1. (a) $\Delta y = 2\,\Delta x = dy$ (c) $\Delta y = 3x^2\,\Delta x + 3x\,\overline{\Delta x^2} + \overline{\Delta x^3}$, $dy = 3x^2\,\Delta x$
2. (a) $\Delta y = 0.005001, 0.0501, 0.51, 2.75, 6$, $dy = 0.005, 0.05, 0.5, 2.5, 5$
3. (a) $(3x^2 - 6x)\,dx$ (c) $(2x - 3)^{-1/2}\,dx$ (e) $2e^{2x}\,dx$
   (g) $-(x^2 + y^2)(2xy + y^2)^{-1}\,dx$
5. (a) $du/dx = -(2x + u)/(x + 2u)$, $dy/du = -y/(u + 3y^2)$, $dv/dy = -y/v$
   (c) $-v(x + 2u)(u + 3y^2)y^{-2}(2x + u)^{-1}$
6. (a) 9.95   (c) 1.2   (e) 1
9. (a) $x + 2y = 3$ (c) $Ax_1x + \frac{1}{2}B(x_1y + y_1x) + Cy_1y = 1$
   (e) $x = 0$

Miscellaneous Problems, page 234

2. (a) e^2 (c) 0   (e) $2\sin x\cos x$
3. (a) $21x^6 - 6x^2$ (c) $3 + 8x^{-3}$ (e) $(1 + \cos x)^{-1}$ (g) $5e^x(\sin x + \cos x)$
   (i) $2\cos(2u + 1)$ (k) $2t\sin(2 + 2t^2)$ (m) $(x^2 - 1)(x^4 + 3x^2 + 1)^{-1}$
   (o) $-\tan x\cos(\ln\cos x)$ (q) $(\frac{3}{2})e^{3x}(1 + e^{3x})^{-1/2}\cot[(1 + e^{3x})^{1/2}]$
   (s) $e^{x^2}[2\ln(1 - 2x)\{x\cos 2x - \sin 2x\} - 2(1 - 2x)^{-1}\cos 2x)]$
   (u) $-12x[3x^2 + 2(x^2 - 1)^{-3}]^{-3}[1 - 2(x^2 - 1)^{-4}]$ (w) $(2x^2 + 36)(x^2 + 9)^{-2}$
   (y) $x^2(1 + x)^{x^2 - 1} + 2x(1 + x)^{x^2}\ln(1 + x)$
5. (a) $dx/dy = (3x^2 + 1)^{-1}$ (c) $dx/dy = -e^{x^2}(2x)^{-1}$
6. (a) $(y^2 - 2xy - 1)/(x^2 - 2xy - 3)$
   (c) $(1 + y)[\ln(1 + y) - y\cos x]\{(6y^2 + \sin x)(1 + y) - x\}^{-1}$
7. (a) $y' = (3t^2 - 1)/(3t^2 + 1)$, $y'' = 12t/(3t^2 + 1)^3$
   (c) $y' = (t^2 - 1)/(2t)$, $y'' = (1 + t^2)^3(-2t)^{-3}$
8. (a) $-2\sin 2t\mathbf{i} + 2\cos 2t\mathbf{j}$,   (c) $3\cos t(1 + \sin t)^2(3\mathbf{i} - 5\mathbf{j})$
10. (a) $y = -3$ (c) $5x - 27y + 18 = 0$
11. (a) $27\cos 3x(3\sin^2 3x - 1)$ (c) $-x^{-2}(\sin\ln x + \cos\ln x)$

Section 4-2, page 239

2. $x^3 - x + 2$
3. (a) $(x^2 + 5)/2$ (c) $x + \cos x + 1$ (e) $x^3 + x + 5$
4. (a) $x = (t^3/3) + \cos t - 1$ (c) $\ln 2$ (e) $\frac{4}{3}$ (g) $\frac{1}{2}t^2 - \frac{1}{2}t + 1$
   (i) $\frac{1}{2}t^2 - \cos t - \pi t + 1$
5. 0.123

Section 4-3, page 244

1. (a) $(1^2 + 2^2 + \cdots + n^2)/n^3$ (c) $(1^2 + 3^2 + \cdots + (2n - 1)^2)/(4n^3)$
2. (a) $\frac{1}{3}$ (c) $c^3/3$

3. (a) $(3c^2)/2 + 2c$
4. (a) $\frac{397}{10}$ (c) $\ln 2$
5. 45 mi. approximately
6. (a) $e^c - 1$

Section 4-4, page 249

1. (a) $\frac{7}{3}$ (c) $\ln 2 - (\frac{1}{2})$
2. $\frac{1}{6}$
3. (a) 1,   (c) $(\frac{2}{3})(2\sqrt{2} - 1) - \ln 2$
4. $(\pi^2/4) + 1$

Section 4-6, page 253

1. (a) $\frac{1}{4}x^4 - \frac{1}{2}x^2 + 7x + C$ (c) $(x^3/3) - x^2 + x + C$
   (e) $-x^{-1} - x^{-2} + C, x \neq 0$ (g) $2 \ln |x| - x^{-1} - \frac{1}{2}x^{-2} + C, x \neq 0$
   (i) $\ln |x| + 2\sqrt{x} + C, x > 0$ (k) $(\frac{3}{8})x^{8/3} - (\frac{3}{2})x^{2/3} + C$
   (m) $\sin x - \frac{1}{2}x^2 + C$ (o) $-\cos x + C$ (q) C
   (s) $-\cos[x + (\pi/3)] + C$ (u) $(2^x/\ln 2) + C$ (w) $\text{Sin}^{-1} x + C$
6. (a) $C_1 x + C_2$ (c) $-\cos x + C_1 x + C_2$
7. (a) 1

Section 4-8, page 261

1. $(x + 3)^4/4 + C$
3. $(\frac{2}{3})(x^2 + 1)^{3/2} + C$
5. $(\frac{1}{5}) \ln |5x - 1| + C$
7. $-e^{-3x}/3 + C$
9. $-(x^5 - x - 7)^{-8}/8 + C$
11. $\ln |1 + \sin x| + C$
13. $\ln |x^2 + x - 1| + C$
15. $\ln |4 + x| + C$
17. $(x^3/3) - x^2 + 4x - 8 \ln |x + 2| + C$
19. $[486(3x + 2)^2]^{-1}[(3x + 2)^4 - 16(3x + 2)^3 + 118(3x + 2) - 151$
    $+ 48(3x + 2)^2 \ln |3x + 2|] + C$
21. $\ln(e^x + 1) + C$
23. $\sin x - [(\sin^3 x)/3] + C$
25. $\sin x - (\frac{2}{3}) \sin^3 x + (\frac{1}{5}) \sin^5 x + C$
27. $\ln |\sin x| + C$
29. $(\frac{1}{3}) \tan^3 x - \tan x + x + C$
31. $\ln |\ln x| + C$
33. $-\cos(\ln x) + C$
35. $-(\ln x)^{-1} + C$
37. $\text{Sin}^{-1}(x/2) + C$
39. $\text{Sin}^{-1}(x - 2) + C$
41. $(\frac{1}{2}) \text{Sin}^{-1}(2x - 3) + C$
43. $x - 2 \text{Tan}^{-1}(x/2) + C$
45. $\text{Tan}^{-1}(x + 2) + C$
47. $\ln(x + \sqrt{x^2 + 1}) + C$

49. $(\frac{1}{2}) \ln|2x + 3 + \sqrt{(2x + 3)^2 + 9}| + C$

51. $2\sqrt{x^2 + 4} + 3 \ln|x + \sqrt{x^2 + 4}| + C$

53. $(\frac{3}{40})[2(1 + 2x)^{5/3} - 5(1 + 2x)^{2/3}] + C$

55. $3 \sin(1 + x)^{1/3} + C$

57. $(2x^3/3) + x + [2(1 + x^2)^{3/2}/3] + C$

59. $-(2/x) - x + (2\sqrt{1 - x^2}/x) + 2 \operatorname{Sin}^{-1} x + C$

68. (a) $-\ln |C - \sin x|$

69. (a) $y = \pm\sqrt{x^2 + C}$ (c) $y = \operatorname{Sin}^{-1}[C + \sin x]$
   (e) $y = \pm(-\ln |C - 2e^x|)^{1/2}$

71. (a) $y = e^{(x^2/2)+C}, y = 0$ (c) $y = e^{3 \ln |x|+C}$ or $Cx^3, y = 0$
   (e) $y = \operatorname{Sin}^{-1}(e^{\sin x+C}), y = k\pi, (k = 0, \pm1, \dots)$

Section 4-9, page 266

1. $x \sin x + \cos x + C$

3. $2x \cos x + (x^2 - 2) \sin x + C$

5. $-e^{-x}(x + 1) + C$

7. $-(\frac{1}{2})e^{-x^2}(x^2 + 1) + C$

9. $x (\ln x)^2 - 2x \ln x + 2x + C$

11. $\ln x \ln \ln x - \ln x + C$

13. $-2(1 + \ln x)x^{-1} + C$

15. $(x/5)(\sin \ln x^2 - 2 \cos \ln x^2) + C$

17. (a) $(e^{2x}/8)(4x^5 - 10x^4 + 20x^3 - 30x^2 + 30x - 15) + C$
   (c) $-(\frac{1}{15})(3 \sin^4 x \cos x + 8 \cos x + 4 \sin^2 x \cos x) + C$

21. (a) $\frac{1}{2}e^x(\sin x - \cos x) + C$ (c) $e^{ax}(a^2 + b^2)^{-1}(a \sin bx - b \cos bx) + C$

Section 4-11, page 272

(1) $\ln |x - 2| - \ln |x - 1| + C$

(3) $\frac{1}{4} \ln |(x - 2)^3(x + 2)^5| + C$

(5) $-(x - 2)^{-1} + \ln |(x - 1)(x - 2)^{-1}| + C$

(7) $(\frac{1}{4}) \ln |x^2(x + 1)^{-1}(x + 5)^{-1}| + C$

(9) $\ln |(x - 2)(x - 1)^{-1}| + (x - 3)(x - 2)^{-2} + C$

(11) $(\frac{2}{3}) \ln |x - 1| + (\frac{19}{21}) \ln |x + 5| - (\frac{1}{7}) \ln |x - 2| + C$

(13) $(\frac{3}{32}) \ln |(2x + 1)(2x - 1)^{-1}| - (\frac{5}{8})x(4x^2 - 1)^{-1} + C$

(15) $(\frac{1}{9}) \ln|(x - 1)(x + 2)^8| + (\frac{7}{3})(x + 2)^{-1} + C$

(17) $(x^2/2) - 2x + \ln |x^2 + x| + C$

(19) $(\frac{1}{16})(4x^4 - 8x^3 + 14x^2 - 30x + 15 \ln |2x + 1|) + 16 \ln |x + 1| + C$

(21) $(x^5/5) + (4x^3/3) + 17x + (\frac{1}{2}) \ln |x| + (\frac{67}{4}) \ln |x - 2| - (\frac{69}{4}) \ln |x + 2| + C$

26. (a) $(\frac{1}{2}) \ln|(1 + \sin x)(1 - \sin x)^{-1}| + C$
   (c) $(\frac{1}{4}) \ln|(1 + \sin x)(1 - \sin x)^{-1}| + (\frac{1}{2}) \sin x(\cos x)^{-2} + C$
   (e) $\sqrt{x^2 + 1} - \ln |(1 + \sqrt{x^2 + 1})x^{-1}| + C$

Section 4-12, page 277

1. (a) $(\frac{1}{5})[(x - 1)^{-1} + (4x + 9)(x^2 + 4)^{-1}]$
   (c) $-2x^{-1} + 2x(x^2 + 1)^{-1} + 3x(x^2 + 1)^{-2}$
   (e) $x^{-1} + (-x - 1)(x^2 + x + 1)^{-1} + (-x - 1)(x^2 + x + 1)^{-2}$

2. (a) $\operatorname{Tan}^{-1}(x + 1) + C$ (c) $(\frac{1}{54})[3x(x^2 + 9)^{-1} + \operatorname{Tan}^{-1}(x/3)] + C$

(e) $(5x - 1)[2(x^2 + 1)]^{-1} + (\frac{5}{2})\,\mathrm{Tan}^{-1}x + C$

3. (a) $(\frac{1}{10})[2\ln(x - 2) - \ln(x^2 + 1) - 4\,\mathrm{Tan}^{-1}x] + C$

(c) $x^{-1} + \ln|x| - (\frac{1}{2})\ln(x^2 + 1) + \mathrm{Tan}^{-1}x + C$　　(e) $\ln|x^5 + 2x^3 + x| + C$

Section 4-14, page 284

1. (a) $F(x) = x^2/2,\ 0 \le x \le 1;\ F(x) = 2x - 1 - (x^2/2),\ 1 < x \le 2$

(c) $F(x) = -x^2/2,\ x \le 0;\ F(x) = x^2/2,\ x > 0$

3. (a) $F(x) = 0,\ 0 \le x \le 1;\ F(x) = x - 1,\ 1 < x \le 2$

(c) $F(x) = x^2/2,\ 0 \le x \le 1;\ F(x) = \frac{1}{2},\ 1 < x \le 2$

Section 4-15, page 289

1. (a) 11　　　(c) 3

2. (a) $(n/2)[2a + (n - 1)d]$　　　(c) $\ln(n + 1)$

4. (a) $n^2 - (k - 1)^2$　　　(c) $n! - 1$

5. (a) 1　　　(c) π

Section 4-17, page 298

1. (a) $\frac{1}{3}$　　(c) $(e^3 - 1)/3$　　(e) $(\frac{1}{2})\,\mathrm{Tan}^{-1}(\frac{1}{2})$

(g) $\ln 2 - \ln 3 = -0.40546$　　(i) $\pi/6$　　(k) 0

(m) $\sin\sqrt{2} - \sin 1 = 0.146$　　(o) $2\sec^2 1\tan 1$

Section 4-19, page 304

1. (a) $\ln 2$　　(c) 4π

2. (a) $\frac{4}{3}$　　(c) $\pi - 2$

3. (a) $\frac{4}{3}$　　(c) $\frac{16}{3}$

6. (a) $e + e^{-1} - 2$

7. (a) $\frac{5}{12}$

8. (a) $47\pi/4$　　(c) $3\pi - 4$

Section 4-20, page 310

1. 237,600 cu ft

2. (b) $v = (\frac{3}{2})(1 - \cos 2t),\ y(5) = 7.9$ ft　　(d) $v_1 = -v_0$

3. (b) $b = 0.04$ (approximately)

4. $(0.001 + 75k\pi^{-1})^{1/3}$ cm

6. (a) 40 mph

7. (b) part (i) 153 in

Section 4-22, page 315

1. (a) $\pi^2 - 4$　　(c) $(e^2 - 1)/4$

3. (a) $\pi/4$　　(c) $(\frac{3}{2}) + 3\ln(\frac{3}{2})$

4. (a) $(\frac{1}{3})\displaystyle\int_1^9 (1/u)\,du$

8. (a) $b\,\mathrm{Sin}^{-1}b + \sqrt{1 - b^2} - 1,\ -1 \le b \le 1$　　(c) $\frac{5}{4}$

11. (a) 0

Section 4-24, page 320

4. (a) Average is $\frac{15}{4}$, R.M.S. is $(\frac{127}{7})^{1/2} = 4.20$ (c) Average is 0, R.M.S. is $\sqrt{2}/2$

Section 4-26, page 327

1. (a) $\delta = 1$ (c) $\delta = 0.178$
2. (a) $\delta = (2e)^{1/2}\epsilon/2$ (c) $\delta = \epsilon(2\pi^2)^{-1}$
8. (a) $\frac{1}{3}$
9. (a) and (c) are bounded

Section 4-29, page 336

1. (a) $(13\sqrt{13} - 8)/27$ (c) same as (a)   (e) 6   (g) 4   (i) $\frac{1}{2}(e^b - e^{-b})$
3. (a) $\int_0^t \sqrt{4u^2 + 25u^8}\,du$ (c) $\int_1^t \sqrt{u^2 - 1}\,u^{-2}\,du$
4. (a) $s = [(4 + 9t^2)^{3/2} - 8]/27, 0 \le t \le 1; t = [(8 + 27s)^{2/3} - 4]^{1/2}/3,$
   $x = [(8 + 27s)^{2/3} - 4]/9, y = [(8 + 27s)^{2/3} - 4]^{3/2}/27,$
   $0 \le s \le (13\sqrt{13} - 8)/27$
8. (a) $t = \sqrt{\tau}, 1 \le \tau \le 4$ (c) $t = 1 - \tau, 0 \le \tau \le 1$

Section 4-30, page 343

1. (a) 2   (c) $\pi - 3$
2. (a) $F(x) = x, 0 \le x \le 1, F(x) = 1, 1 \le x \le 2$
   (c) $F(x) = e^x - 1, 0 \le x \le 1, F(x) = e - 1 + e^{-1} - e^{-x}, 1 \le x \le 2$
3. $\int_0^t f(t)\,dt = 190$, where $f(t) = 50, 0 \le t \le 2, f(t) = 60, 2 < t \le 3,$
   $f(t) = 30, 3 < t \le 4$
10. (a) $2\sqrt{2}$
11. (a) $s = \sqrt{2}t^2, 0 \le t \le 1, s = \sqrt{2}t, 1 \le t \le 2; x = (s/\sqrt{2}) - 1, 0 \le s \le 2\sqrt{2};$
    $y = s/\sqrt{2}, 0 \le s \le \sqrt{2}, y = 2 - (s/\sqrt{2}), \sqrt{2} \le s \le 2\sqrt{2}. s(t), x(s), y(s)$ and
    dx/ds are continuous, dy/ds is not continuous at $s = \sqrt{2}$ with left-hand limit
    equal to $1/\sqrt{2}$ and right-hand limit equal to $-1/\sqrt{2}$.

Section 4-31, page 348

1. (a) $(\frac{4}{3})i + (\frac{15}{2})j$ (c) $(\frac{1}{2})a + (\frac{1}{3})b$
2. (a) $[(e - 1)/2]i + [3(2^{4/3} - 1)/4]j$
3. Increases by $5(e - 1)^2 + 15$
4. (a) $(t - \text{Tan}^{-1} t)i + \ln(1 + t^2)j + C$
5. (a) Infinite interval on y-axis: if $a > 0$, $-v_0^2/(2a) \le y < \infty$; if $a < 0$,
   $-\infty < y \le v_0^2/(2a)$.

Section 5-5, page 364

4. $\text{Tan}^{-1} t$
5. value 2π
6. (a) 4π

Section 5-6, page 367

1. (a) $(\frac{1}{5}) \ln |(2 \tan \frac{1}{2}x + 1)(2 \tan \frac{1}{2}x - 4)^{-1}| + C$
   (c) $\ln |\tan x/2 - 1| + C$ (e) $\ln |\tan x/2| + C$ (g) $(\frac{1}{3})\tan^3 \frac{1}{2}x + C$
2. (a) $(-\frac{1}{40}) \ln |(\tan^2 x - 4) \sec^8 x \tan^{-10} x| + C$

Section 5-7, page 374

5. (a) $x = e^y - 1, \, -\infty < y < \infty$ (c) $x = y^{1/3}, \, y > 0$

Section 5-10, page 381

1. (a) -1 (c) 1 (e) $0.073 - 0.11i$
5. (a) $(\pi/2) + n\pi, \, n = 0, \pm 1, \pm 2, \ldots$
16. (a) $(-x\sqrt{x^2 - 1}) + C$
   (c) $(\frac{1}{8})[3 \sinh^{-1} x + x\sqrt{1 + x^2} \, (5 + 2x^2)] + C$
19. (a) $0.83 - 0.99i$ (c) -1

Section 6-1, page 392

1. (a) $x_0 = 1$, minimum, $m = -1, M = 0$
   (c) $x_0 = 1$, local maximum, $m = -3, M = 1$,
   (e) $x_0 = 2^{-\frac{1}{3}}$, local minimum, $m = 3 \cdot 2^{-\frac{2}{3}}$, no M
   (g) $x_0 = (\ln 2)/3$, local minimum, $m = 3 \cdot 2^{-\frac{2}{3}}$, no M
   (i) $x = \pi/4$, local maximum, $m = -1, M = 2^{\frac{1}{2}}$
2. (a) $x = 0$, local maximum; $x = 1$, local minimum
   (c) $x = -1$, local minimum; $x = 1$, local minimum; $x = 0$, local maximum
   (e) $x = -2\pi/3$, local maximum; $x = 0$, inflection point; $x = 2\pi/3$, local minimum
   (g) $x = 2$, local maximum
3. (a) a square   (c) none
4. $r_1 = a/[1 + (m_2/m_1)^{\frac{1}{2}}], \, r_2 = a - r_1$
7. $50\sqrt{2}$ mph
9. midpoint
11. $(\frac{1}{6})[a + b - (a^2 + b^2)^{\frac{1}{2}}]$
13. $2\sqrt{2}a/3$
20. (a) $y = 2.04x$
22. 2

Section 6-2, page 400

1. (a) minimum $\frac{1}{2}$ for $x = \frac{1}{2}, y = -\frac{1}{2}$ (c) minimum 1 for $x = 1, y = 0$
   (e) maximum $\frac{7}{16}$ for $x = \pm\sqrt{6}/4, y = -1/4$, no absolute maximum
   (g) minimum 2 for $x = 1, y = 1$
   (i) minimum $-\sqrt{2}$ for $x = -\sqrt{2}/2, y = \sqrt{2}/2$; maximum 1 for $x = 1, y = 0$
   (k) maximum $17^{\frac{3}{4}}$ for $x = 17^{-\frac{1}{4}}, y = 2 \cdot 17^{-\frac{1}{4}}$, minimum -1 for $x = -1, y = 0$
2. (a) major axis: $(\pm 1, \mp 1)$, minor axis: $(\pm 1/\sqrt{3}, \pm 1/\sqrt{3})$
   (c) major axis: $(\pm 3/\sqrt{13}, \mp 2/\sqrt{13})$, minor axis: $(\pm 2/\sqrt{182}, \pm 3/\sqrt{182})$
5. area $2ab$, sides $\sqrt{2}a, \sqrt{2}b$

Section 6-3, page 403

1. (a) up: $x > 0$, down: $x < 0$, inflection: $x = 0$ (c) same as (a)
   (e) up: $x < x_1$ and $x > x_2$, down: $x_1 < x < x_2$, inflection: $x = x_1$, $x = x_2$, where $x_1 = (3 - \sqrt{3})/6$, $x_2 = (3 + \sqrt{3})/6$
   (g) up: $-1 < x < 0$ and $x > 1$, down: $x < -1$ and $0 < x < 1$, inflection: $x = 0$, $x = \pm 1$
   (i) up: $n\pi < x < n\pi + (\pi/2)$, down: $n\pi - (\pi/2) < x < n\pi$, inflection: $x = n\pi$, $n = 0, \pm 1, \pm 2, \ldots$
   (k) up: $x > e^{-\frac{3}{2}}$, down; $0 < x < e^{-\frac{3}{2}}$, inflection: $x = e^{-\frac{3}{2}}$

Section 6-5, page 419

1. (a) $x = x' + 3$, $y = y' - 2$
   (c) $x' - y' + 7 = 0$, $x'^2 + y'^2 + 6x' - 4y' + 12 = 0$, $x'y' - 2x' + 3y' - 8 = 0$
2. (a) $x = (x' - \sqrt{3}y')/2$, $y = (\sqrt{3}x' + y')/2$; $x' = (x + \sqrt{3}y)/2$, $y' = (-\sqrt{3}x + y)/2$
   (c) $y' = 0$, $x'^2 + y'^2 = 1$, $\sqrt{3}x'^2 - 2x'y' - \sqrt{3}y'^2 = 4$
3. (a) $x'^2 + y'^2 = \frac{31}{3}$, new origin $(1, -2)$
   (c) $x'^2 + (y'^2/4) = 1$, new origin $(-1, 3)$
7. (a) hyperbola, new equation $-5x'^2 + 5y'^2 = 4$, $\cos 2\alpha = -3/5$, $\sin 2\alpha = 4/5$,
   (c) imaginary ellipse
11. (a) period 2π with respect to x (c) symmetric with respect to y-axis
    (e) symmetric with respect to origin
    (g) symmetric with respect to line $y = (\sqrt{2} - 1)x$

Section 6-7, page 430

1. (a) $12|t|(9t^4 - 18t^2 + 13)^{-\frac{3}{2}}$ (c) $2[1 + 3\cos^2 2t]^{-\frac{3}{2}}$
   (e) $|\sin x|(1 + \cos^2 x)^{-\frac{3}{2}}$ (g) 1 (path is a circle)
6. $km(r_0 v_0)^{-2}$

Section 6-9, page 441

1. (b) $r' = 5t(5 + 5t^2)^{-\frac{1}{2}}$, $\theta' = -(1 + t^2)^{-1}$, $v_{rad} = r'$, $v_{trans} = -[5/(1 + t^2)]^{\frac{1}{2}}$
   (c) as $t \to +\infty$, $\theta \to -.46$, as $t \to -\infty$, $\theta \to 2.68$
2. (b) $r' = -2\sin t(5 + 4\cos t)^{-\frac{1}{2}}$, $\theta' = 6(1 + \cos t)(5 + 4\cos t)^{-1}$
5. $\rho m \omega_0 / k$

Section 6-15, page 468

1. (a) -6 (c) 1
2. (a) 1   (c) 1
3. (a) e (c) ∞ (e) 1   (g) 0   (i) $-\infty$ if $\alpha \le 0$ and 0 if $\alpha > 0$ (k) 1
   (m) $\frac{1}{2}$ (o) 0   (q) 0   (s) 0   (u) 1   (w) e (y) $-\infty$
4. (a) 0   (c) 0

Section 7-1, page 472

1. (a) 2   (c) $e - 2$ (e) $128\sqrt{2}/3$
2. (a) $(\pi/2) - 1$ (c) $(1/\sqrt{2})\text{Tan}^{-1}(1/\sqrt{2}) - \frac{1}{2}\ln(\frac{3}{4})$ (e) $\frac{1}{2}$

3. (a) $a^2 \operatorname{Sin}^{-1}(c/a) - bc$, $c = \sqrt{a^2 - b^2}$
4. $[(8a^2)/3|m|^3](m^2 + 1)^{\frac{3}{2}}$

Section 7-2, page 475

1. (a) 4   (c) $(e^{3\pi} - 1)/4$ (e) $\pi/8$
2. (a) $\pi(a^2 + \frac{1}{2}b^2)$ (c) $\pi/4$
3. (a) $2 + (\pi/4)$

Section 7-4, page 485

1. (a) πa^2 (c) $3\pi a^2/8$ (e) $\pi a^2 + [4\pi^3 a^2/3]$ (g) 2
2. (a) 7   (c) $\frac{1}{2}|D|$, where $D = \begin{vmatrix} x_3 - x_1 & y_3 - y_1 \\ x_4 - x_2 & y_4 - y_2 \end{vmatrix}$
   (e) $\frac{1}{2}|\overrightarrow{OP_1} \cdot \overrightarrow{P_1P_2}{}^{\dashv} + \overrightarrow{OP_2} \cdot \overrightarrow{P_2P_3}{}^{\dashv} + \cdots + \overrightarrow{OP_n} \cdot \overrightarrow{P_nP_1}{}^{\dashv}|$
4. (a) 2 (figure is a square)
9. (a) a_1 (c) $a_1 - a_2 - a_5$
11. (c) 6π

Section 7-6, page 495

1. $\pi a^2/5$
3. $\pi^2/2$
5. $\pi - (\pi^2/4)$
7. $\pi(1 + \frac{1}{2}\sinh 2)$
9. $\pi(5d^2 - 4\sqrt{3}d + 8)$, $d = \operatorname{Cosh}^{-1} 5$
11. $(\pi h/3)(m^2 h^2 + 3mbh + 3b^2)$
13. $4\pi ab^2/3$
15. $\pi a^3/6$
17. $(4\pi a/3)(a^2 + b^2)$
19. $2\pi(1 + e^{3a\pi})[3(1 + 9a^2)]^{-1}$
21. $2\pi^2 abh$
23. $192\pi^2$
25. 3π
27. 8π
29. $512\pi/9009$

Section 7-7, page 499

1. 4
3. $\pi/2$
5. $\pi(e^2 - 1)/2$
7. $(\frac{4}{3})\pi abc$
9. $\frac{1}{15}$
11. $\frac{31}{105}$

Section 7-8, page 503

1. $2\pi ah$
2. (a) $(8\pi\sqrt{a}/3)[(b + a)^{\frac{3}{2}} - a^{\frac{3}{2}}]$

3. (a) $\pi b[b + (a/\epsilon) \text{Sin}^{-1} \epsilon], \epsilon = \sqrt{a^2 - b^2}/a$

4. (a) $(\pi ab/\epsilon)[(1 + h)\delta - \gamma + \epsilon^2 \ln\{(1 + \delta)(1 + h + \delta)^{-1}\}], \epsilon = a(a^2 + b^2)^{-\frac{1}{2}},$
   $\delta = [(1 + h)^2 - \epsilon^2]^{\frac{1}{2}}, \gamma = b\epsilon/a$

5. (a) $3\pi a^2/2$

7. (a) $\pi(a + \frac{1}{2} \sinh 2a)$

10. (a) $32\pi a^2/5$

11. (a) $\pi a^2 + \pi b^4(a^4 - b^4)^{-\frac{1}{2}} \ln\{b^{-2}[a^2 + (a^4 - b^4)^{\frac{1}{2}}]\}$

12. (a) $\pi\{a^2 - ab^2\delta^{-1} + b^4\epsilon^{-1} \ln[(a^2 + \epsilon)(a\delta - \epsilon)b^{-3}\delta^{-1}]\}, \delta = (a^2 + b^2)^{\frac{1}{2}},$
    $\epsilon = (a^4 - b^4)^{\frac{1}{2}}$

15. (b) $(5^{\frac{3}{2}} - 1)/4$

Section 7-9, page 509

1. (a) $\frac{16}{15}$ (c) $n\left(1 + \frac{1}{2} + \cdots + \frac{1}{n}\right)^{-1}$ (e) $\frac{3}{2}$ (g) $(e - 1)^{-1}$ (i) $(2e)^{-1}$
   (k) $(e^2 - 3)/(e^2 - 1)$

2. (a) $M_2 = 68, \rho = (68/15)^{\frac{1}{2}}$ (c) $M_2 = n(n + 1)/2$
   $\rho = \left\{[n(n + 1)/2]\left(1 + \frac{1}{2} + \cdots + \frac{1}{n}\right)^{-1}\right\}^{\frac{1}{2}}$ (e) $M_2 = 20/3, \rho = (\frac{10}{3})^{\frac{1}{2}}$
   (g) $M_2 = e - 2, \rho = [(e - 2)/(e - 1)]^{\frac{1}{2}}$
   (i) $M_2 = (3e - 2)/3, \rho = [(3e - 2)/(3e)]^{\frac{1}{2}}$
   (k) $M_2 = 2 - 10e^{-2}, \rho = [(2e^2 - 10)(e^2 - 1)^{-1}]^{\frac{1}{2}}$

3. (a) $(1 + 6 \cdot 3^k - 4 \cdot 2^k + 20 \cdot 5^k - 16 \cdot 4^k)/(k + 1)$

4. (a) mass 2 at $x = 2$, 1 at $x = 5$
   (b) Masses positive for $M_0 > 0$ and $x_1 < M_1/M_0 < x_2$.

Section 7-12, page 517

1. (a) $(0, -\frac{13}{10})$ (c) $(\frac{16}{15}, \frac{4}{3})$

2. (a) Where diagonals intersect.   (c) On radius perpendicular to diameter,
   distance $4a/(3\pi)$ from diameter.   (e) $[0, 4b/(3\pi)]$ (g) $(0, 0)$
   (i) $(0, \frac{13}{15})$

4. (a) On bisector, distance $4a \sin(\gamma/2)/(3\gamma)$ from center.
   (c) $\bar{x} = -2k/5, \bar{y} = 2k/15, k = (e^{3\pi} + 1)/(e^{2\pi} - 1)$

5. (a) $M_0h^2/6$, where h is the altitude on the side.   (c) $M_0a^2(2\beta - \sin 2\beta)/(8\beta)$

7. (e) $4\pi^2ab$

8. (a) $(0, 2a/\pi)$ (c) $(0, -2/\pi)$

18. (a) $OP_c = 3a/8$, where O is center, a is radius.   (c) $\bar{x} = 5a/16$

22. (a) $2M_0a^2/5$ (c) $8M_0a^4/21$

Section 7-13, page 526

1. (a) -182 ft-lb   (c) 0 ft-lb

2. 2.1 ft

4. (b) $(m_2v_2 + 2m_1v_1)/m_2$

5. 18.1 cm/sec

6. $|v_0| \sqrt{m/k}$

7. (a) $[(mv^2)/2] + (1/3)k|x|^3 = E$

9. $k \ln[2e^L/(1 + e^L)]$

10. (a) $wa^4/4$ (c) $wc_1(h_1^2 + h_1h_2 + h_2^2)/6$

Section 7-14, page 535

1. $\ln 2$
3. $\ln 4 - \frac{1}{2}\ln 3$
5. $10.5 \ln 1.5 - 4 \ln 2$
7. 1
9. $\frac{1}{2}$
11. div.
13. 1
15. div.
17. $\frac{3}{2}$
19. div.
21. 0
23. div.
26. (a) conv.   (c) div.   (e) conv.   (g) conv.
28. (a) conv.   (c) conv.   (e) conv.
29. (a) $\frac{1}{4}$ (c) volume undefined, area 2
30. (a) 0   (c) π (e) 0   (g) 0

Section 7-16, page 542

1. (a) Order 1, degree 1, f is a solution   (c) Order 1, degree 2, f is a solution
   (e) Order 1, degree 1, f is a solution   (g) Order 3, degree 1, f is a solution
2. (a) $y = x^3 - x^2 + 5x + C$ (c) $y = (x^2/2) + c_1 x + c_2$
   (e) $y = (e^{2x}/8) + (5x^3/6) + c_1 x^2 + c_2 x + c_3$
3. (a) $y = cx^{-3}$ (c) $y = ce^{-e^x}$ (e) $y = (e^{2x} + c)/(e^{2x} - c)$ and $y = -1$
   (g) $y = -2/(c + x^2)$ and $y = 0$
4. (a) $y = x\ln|x| + cx,\ x \neq 0$ (c) $y = -x\ln(\ln|x|^{-1} + c),\ x \neq 0$
   (e) $y = (cx^3 + x)/(1 - cx^2)$
5. (a) $y = [(3x - 1)/9] + ce^{-3x}$ (c) $y = 1 + ce^{\cos x}$
   (e) $y = (x^2 + cx)/(x - 1),\ x \neq 1$
7. (a) $y = e^x + 1 - x$ (c) $y = 0$ (e) $y = x$
9. $x = (ae^{at} + be^{bt})/(e^{at} + e^{bt}),\ x \to a$ as $t \to \infty$
10. (a) 23.1 years

Section 7-19, page 549

1. (a) $y = c_1 e^{3x} + c_2 e^{-3x}$ (c) $y = c_1 e^x + c_2 e^{-3x}$
   (e) $y = e^{-x}(c_1 \cos x + c_2 \sin x)$ (g) $y = e^{-x/3}(c_1 + c_2 x)$
2. (a) $y = c_1 e^{3x} + c_2 e^{-3x} - (e^x/8)$ (c) $y = c_1 e^{3x} + c_2 e^{-3x} - (x/9)$
   (e) $y = c_1 \cos 3x + c_2 \sin 3x + (1/5)\cos 2x$ (g) $y = x - 2 + e^{-x}(c_1 + c_2 x)$
3. (a) $y = c_1 x + c_2 x^2$ (c) $y = c_1 e^x + c_2 \ln x$
4. (a) $y = (x^3/2) + c_1 x + c_2 x^2$
5. (a) $y = c_1 + c_2 \cos x + c_3 \sin x$ (c) $y = (e^x/6) + c_1 e^{-x} + c_2 e^{-2x} + c_3$

Section 7-20, page 554

1. (a) damped vibration   (c) overdamped   (e) simple harmonic
3. $x = x_0 \cos \beta t + (v_0/\beta)\sin \beta t$
6. Simple harmonic, amplitude $\pi/2$, period $2\pi(I/C)^{\frac{1}{2}}$

Section 8-3, page 572

1. (a) 0   (c) 0   (e) 1   (g) 2   (i) 0   (k) $-\frac{1}{2}$
5. (a) oscillatory divergent   (c) converges to 1   (e) converges to 0
    (g) converges to 0   (i) diverges to $+\infty$ (k) converges to 0
    (m) converges to 0   (o) converges to 0
6. (a) diverges   (c) converges to 1   (e) diverges

Section 8-4, page 578

1. (a) $1, \frac{1}{2}, \frac{5}{6}, \frac{7}{12}, \frac{23}{60}$ (c) $1, \frac{1}{2}, \frac{7}{6}, \frac{7}{12}, \frac{73}{60}$ (e) $\frac{1}{2}, \frac{2}{3}, \frac{3}{4}, \frac{4}{5}, \frac{5}{6}$
2. (a) The series is a geometric series and converges to 6.
    (d) The terms after the 5th belong to a geometric series, sum $= 16$.
3. (a) Diverges to $+\infty$, since terms are positive and do not tend to 0.
    (c) Diverges to $+\infty$; except for the first term the terms are 5 times terms of the
    divergent harmonic series.
    (e) The nth term does not tend to zero; nth partial sum has absolute value 0, 1 or
    $\sqrt{2}$.
4. (a) $\dfrac{1}{n^2 + 3n + 2} = \dfrac{1}{n + 1} - \dfrac{1}{n + 2}$, series converges to $\frac{1}{2}$.

    (c) converges to $\frac{1}{4}$
6. (a) diverges to $+\infty$,   (c) oscillatory divergent,
    (e) converges to 2^{512}, [finite sum of 513 terms],   (g) diverges to $+\infty$,
    (i) converges to $\frac{3}{5}$,   (k) diverges,   (m) diverges.

Section 8-6, page 583

2. (a) diverges.
3. (a) 0.
7. (a) $2\pi^2$,   (c) $(\pi^4/18) - (2\pi^6/315)$.

Section 8-8, page 589

1. converges
3. converges
5. diverges
7. converges
9. converges for $\alpha > 1$, diverges for $\alpha \leq 1$
11. converges
13. converges
15. diverges
17. converges
19. diverges

21. diverges
23. diverges
25. converges
27. converges
29. converges
31. diverges
33. 0.102
35. 0.050
37. 0.564

Section 8-10, page 594

1. converges
3. converges
5. converges
7. converges

9. diverges
11. converges
13. diverges
15. converges

17. converges
20. (a) converges   (c) diverges
24. (a) 0.368   (c) 5.436

Section 8-11, page 596

1. converges   **3.** converges   **5.** converges
7. (a) $N = 5$ **(c)** $N = 4$ **(e)** $N = 2$

Section 8-16, page 607

1. (a) $|x| < 2$ **(c)** $|x| < 1$ **(e)** $x > 0$ **(g)** $x \leq 0$
2. (a) 1   **(c)** ∞ **(e)** ∞ **(g)** 1

Section 8-18, page 614

1. All x

5. (a) $\sum_{n=0}^{\infty} (x^{2n}/n!)$, all x **(c)** $\sum_{n=1}^{\infty} (-1)^{n-1} x^n/n$, $|x| < 1$

(e) $-\sum_{n=0}^{\infty} 3^n x^n/5^{n+1}$, $|x| < \frac{5}{3}$ **(g)** $\sum_{n=0}^{\infty} x^n(2^{-n-1} - 3^{-n-1})$, $|x| < 2$

(i) $3 + \sum_{n=1}^{\infty} 2^{n+2}x^n$, $|x| < \frac{1}{2}$ **(k)** $\sum_{n=0}^{\infty} (-1)^n x^{2n+1}/(2n + 1)$, $|x| < 1$

10. (a) 1   **(c)** 1   **(e)** $\frac{1}{2}$

Section 8-21, page 625

In all sums n goes from 0 to ∞.

1. (a) $\sqrt{x} = \sum \binom{1/2}{n}(x - 1)^n$, $|x - 1| < 1$

(c) $(x + 2)^{-1} = \sum (-1)^n x^n/2^{n+1}$, $|x| < 2$
(e) $\ln x = \sum (-1)^{n-1}(x - 1)^n/n$, $|x - 1| < 1$
(g) $\sin x = \sum \sin\left(\frac{\pi}{3} + n\frac{\pi}{2}\right)\left(x - \frac{\pi}{3}\right)^n/n!$, all x

(i) $\sqrt{2 + x} = \sqrt{3}\sum \binom{1/2}{n}\frac{(x - 1)^n}{3^n}$, $|x - 1| < 3$

2. (a) $\ln (1 - x^2) = \sum (-1)x^{2n}/n$, $|x| < 1$
(c) $-(x^2/2) - (x^4/12) - (x^6/45) - (17x^8/2520) - (31x^{10}/14{,}175) + \cdots$
(e) $(1 + x^2)^{-1} = \sum (-1)^n x^{2n}$, $|x| < 1$
(g) $1 + (x^2/2) + (5x^4/24) + (61x^6/720) + \cdots$

(i) $(2x + 3)^{1/5} = \sum 3^{1/5}\binom{1/5}{n}\left(\frac{2x}{3}\right)^n$, $|x| < \frac{3}{2}$

(k) $(x^3/3) - (x^5/10) + (x^7/168) - (x^9/6480) + (x^{11}/443{,}520) + \cdots$
(m) $x - (x^3/18) + (x^5/600) - (x^7/35{,}280) + (x^9/3{,}265{,}920)$
(o) $\sinh x = \sum x^{2n+1}/(2n + 1)!$, all x
(q) $x - x^2 + (23x^3/24) - (11x^4/12) + (563x^5/640) + \cdots$

3. (a) $x - (x^3/2) - (17x^5/360) + \cdots$ **(c)** $(x^3/8) - (x^4/8) + (x^5/30) + \cdots$
4. (a) 9.165   **(c)** 11.269
5. (a) 0.100 (0.0998)   **(c)** 0.693   **(e)** 0.847   **(g)** 0.250
6. (a) $|x| \leq (0.006)^{1/3} = 0.182$ (approximately)
7. (a) Error of at most 0.0007
8. (d) 0.0997   **(i)** 3.14159   **(k)** 1.10712

Section 8-22, page 630

1. (a) $y = a_0 \sum_{n=0}^{\infty} 3^n x^n / n!$

   (c) $y = a_0 \sum_{n=0}^{\infty} \frac{4^n x^{2n}}{(2n)!} + a_1 \sum_{n=0}^{\infty} \frac{2^{2n+1} x^{2n+1}}{(2n+1)!} = a_0 \cosh 2x + a_1 \sinh 2x$

   (e) $y = a_0 \sum_{n=0}^{\infty} \frac{(-1)^n (1-n) x^n}{n!} + a_1 \sum_{n=1}^{\infty} \frac{(-1)^{n-1} x^n}{(n-1)!} = a_0(1+x)e^{-x} + a_1 x e^{-x}$

2. (a) $y = a_0 \left(1 + \sum_{n=1}^{\infty} \frac{1 \cdot (-1) \cdots (3-2n)}{(2n)!} (-1)^n x^{2n} \right) + a_1 x$

   (c) $y = a_0 + a_0 \sum_{m=1}^{\infty} \frac{x^{3m}}{2 \cdot 3 \cdot 5 \cdot 6 \cdots (3m-1)(3m)}$

   $\qquad + a_1 x + a_1 \sum_{m=1}^{\infty} \frac{x^{3m+1}}{3 \cdot 4 \cdot 6 \cdot 7 \cdots (3m)(3m+1)}$

3. General solution $y = a_0 \left[1 + \sum_{n=1}^{\infty} (-1)^n x^{2n} \frac{\alpha(\alpha-4) \cdots (\alpha+4-4n)}{(2n)!} \right]$

   $\qquad + a_1 \left[x + \sum_{n=1}^{\infty} (-1)^n x^{2n+1} \frac{(\alpha-2)(\alpha-6) \cdots (\alpha+2-4n)}{(2n+1)!} \right]$

Section 8-23, page 633

5. (a) $R = 1$ **(c)** $R = 1$ **(e)** $R = 2$

Section 8-24, page 639

1. (a) $\dfrac{\pi^2}{3} + 4 \sum_{n=1}^{\infty} \dfrac{(-1)^n \cos nx}{n^2}$ **(c)** $(1 - \cos 2x)/2$

2. (a) $2 \sum_{n=1}^{\infty} (-1)^{n-1} (\sin nx)/n$ **(c)** $\dfrac{2}{\pi} + \dfrac{4}{\pi} \sum_{n=1}^{\infty} \dfrac{(-1)^n}{1 - 4n^2} \cos nx$

   (e) $\dfrac{1}{2} + \dfrac{2}{\pi} \sum_{n=1}^{\infty} \dfrac{\sin(2n-1)x}{2n-1} - \dfrac{4}{\pi} \sum_{n=1}^{\infty} \dfrac{\sin(4n-2)x}{4n-2}$

4. (b) $\pi - 1 - \dfrac{8}{\pi} \sum_{n=1}^{\infty} \dfrac{\cos(2n-1)x}{(2n-1)^2}$

5. (b) $\dfrac{2}{\pi} \sum_{n=1}^{\infty} \dfrac{(-1)^{n+1}(n^2 \pi^2 - 2) - 2}{n^3} \sin nx$

INDEX

CPSIA information can be obtained at www.ICGtesting.com
Printed in the USA
BVOW06s1118250615

405981BV00017B/305/P